Artificial Intelligence in Livestock Omics

Artificial Intelligence in Livestock Omics

Applications, Challenges, and Future Perspectives

Navid Ghavi Hossein-Zadeh
Department of Animal Science, University of Guilan, Iran

ELSEVIER

ACADEMIC PRESS
An imprint of Elsevier

Academic Press is an imprint of Elsevier
125 London Wall, London EC2Y 5AS, United Kingdom
50 Hampshire Street, 5th Floor, Cambridge, MA 02139, United States

ISBN: 978-0-443-45663-3

For information on all Academic Press publications visit our website at https://www.elsevier.com/books-and-journals

Publisher: Mara Conner
Acquisitions Editor: Kelsey Connors
Editorial Project Manager: Kritika Sharma
Production Project Manager: Anitha Sivaraj
Cover Designer: Mark Rogers

Typeset by TNQ Tech

Contents

Preface

The rapid evolution of artificial intelligence (AI) has ushered in a transformative era across scientific disciplines, fundamentally reshaping the ways in which complex biological systems are studied, interpreted, and managed. In parallel, advances in omics technologies have generated unprecedented volumes of high-resolution biological data, capturing the molecular, cellular, and physiological dimensions of living organisms with remarkable depth and breadth. Within the domain of livestock science, the convergence of these two powerful paradigms, AI and omics, has given rise to a new frontier of inquiry and application, one that holds profound implications for productivity, sustainability, animal welfare, and global food security. This book has been conceived as a comprehensive and integrative resource that explores this convergence in a systematic, critical, and forward-looking manner.

Livestock systems occupy a central position in the global agriculture, serving as essential sources of food, income, and livelihood for millions of people worldwide. However, these systems are increasingly confronted with multifaceted challenges, including the need to enhance the production efficiency, mitigate environmental impacts, adapt to climate variability, and address ethical concerns related to animal welfare. Traditional approaches to livestock improvement, while historically successful, are often constrained by their reliance on limited datasets, simplified assumptions, and linear models of biological function. The emergence of omics technologies, encompassing genomics, transcriptomics, proteomics, metabolomics, and microbiomics, has dramatically expanded the scope of the biological information available for analysis, enabling a more nuanced and systems-level understanding of livestock biology. Yet, the sheer scale and complexity of these datasets necessitate advanced analytical frameworks capable of extracting meaningful insights and translating them into actionable knowledge.

AI, with its capacity to model nonlinear relationships, integrate heterogeneous data sources, and learn from complex patterns, offers a powerful set of tools to address these challenges. By bridging the gap between data generation and biological interpretation, AI enables the development of predictive, causal, and prescriptive models that can inform decision-making across diverse aspects of livestock production. From reproductive biology and phenomics to systems biology and precision breeding, the integration of AI into livestock omics is redefining the conceptual and operational foundations of the field. This book seeks to capture the breadth and depth of these developments, providing readers with a cohesive narrative that connects the theoretical principles, methodological innovations, and practical applications.

The chapters in this book are organized to reflect the multidimensional nature of AI-enabled livestock omics. Early sections introduce the foundational concepts underlying omics technologies and AI, establishing a common framework for understanding their integration. The subsequent chapters delve into the specific domains of application, including reproductive omics, phenomics, systems biology, and precision breeding, each examined through the lens of AI-driven analysis and decision-making. These chapters highlight not only the technical advancements that have been achieved, but also the conceptual shifts that are redefining how biological systems are understood and managed. In doing so, the book emphasizes the transition from reductionist approaches toward holistic, systems-oriented perspectives that recognize the interconnectedness of biological, environmental, and management factors.

A central theme that runs throughout this book is the importance of integration across data types, analytical methods, and disciplinary boundaries. Livestock systems are inherently complex, characterized by interactions that span multiple levels of organization, from genes and molecules to organisms and ecosystems. The effective analysis of such systems requires the ability to synthesize information from diverse sources and to model the dynamic relationships that govern their behavior. AI provides a unifying framework for this integration, enabling the development of models that can capture both the structure and function of biological systems. At the same time, the book acknowledges that integration is not merely a technical challenge, but also a conceptual one, requiring new ways of thinking about causality, uncertainty, and decision-making.

Another key focus of this book is the role of AI in enhancing the predictive and adaptive capabilities of livestock systems. Traditional breeding and management strategies often rely on historical data and static models, which may be insufficient to capture the dynamic and context-dependent nature of biological processes. AI-driven approaches, by contrast, enable the development of models that can adapt to new data, incorporate environmental variability, and account for uncertainty in predictions. This capacity for continual learning and adaptation is particularly important in the context of global change, where livestock systems must respond to the evolving challenges in real time. By providing tools for predictive analytics, scenario simulation, and decision optimization, AI has the potential to transform livestock production into a more responsive and resilient enterprise.

While the opportunities presented by AI in livestock omics are substantial, they are accompanied by a range of challenges that must be carefully considered. These include technical issues related to data quality, model interpretability, and computational scalability, as well as broader concerns regarding ethics, governance, and societal acceptance. The use of AI in biological systems raises important questions about transparency, accountability, and the potential for unintended consequences. For example, the optimization of specific traits through AI-driven selection may lead to trade-offs that affect animal welfare or genetic diversity. Similarly, the integration of large-scale data from multiple sources raises issues related to data ownership, privacy, and equitable access. This book does not shy away from these challenges; rather, it seeks to engage with them critically, providing a balanced perspective that recognizes both the promise and the complexity of AI-enabled livestock omics.

Ethical considerations are given particular attention throughout the book, reflecting the growing recognition that technological innovation must be aligned with societal values and responsibilities. The application of AI in livestock systems has implications not only for productivity and efficiency, but also for the well-being of animals, the sustainability of ecosystems, and the livelihoods of communities. As such, the development and deployment of AI technologies must be guided by principles that ensure fairness, inclusivity, and respect for ethical norms. This includes the incorporation of welfare indicators into breeding and management decisions, the preservation of genetic diversity as a resource for future adaptation, and the promotion of transparency in algorithmic processes. By integrating ethical considerations into the design and application of AI systems, it is possible to harness their capabilities in a manner that is both responsible and sustainable.

The interdisciplinary nature of AI in livestock omics is another defining feature of this field, and one that is reflected in the structure and content of this book. Advances in this area require collaboration among experts in biology, computer science, statistics, engineering, and social sciences, among others. Such collaboration is essential for addressing the multifaceted challenges associated with data integration, model development, and system implementation. This book aims to serve as a bridge between disciplines, providing a common language and conceptual framework that can facilitate dialogue and collaboration. By bringing together diverse perspectives, it seeks to foster a more comprehensive and integrated understanding of livestock systems and the role of AI in their transformation.

In addition to its focus on the current applications and challenges, this book also looks toward the future, exploring the emerging trends and potential directions for research and innovation. These include the development of more sophisticated AI models that can capture causal relationships and support decision-making under uncertainty, the integration of real-time data streams for dynamic system monitoring, and the creation of digital twins that can simulate the behavior of individual animals or entire populations. This book also considers the implications of these developments for the design of livestock systems, suggesting that future approaches may be characterized by greater adaptability, resilience, and alignment with the ecological and ethical principles. By anticipating these trends, the book aims to provide the readers with a forward-looking perspective that can inform both research and practice.

The goal of this book is to provide a comprehensive and coherent account of the integration of AI and omics in livestock science, highlighting both the opportunities and the challenges that define this rapidly evolving field. It is intended for a broad audience, including researchers, practitioners, policymakers, and students, who are interested in understanding how advanced analytical tools can be applied to complex biological systems. By combining conceptual depth with practical relevance, the book seeks to contribute to the development of a more informed and thoughtful approach to livestock production, one that leverages the power of AI while remaining grounded in the principles of sustainability, ethics, and scientific rigor.

As the field continues to evolve, it is our hope that this book will serve not only as a reference, but also as a source of inspiration for future work. The integration of AI and omics represents a profound shift in the way we study and manage livestock systems, offering new possibilities for innovation and improvement. At the same time, it challenges us to think more deeply about the relationships among technology, biology, and society, and to ensure that our approaches are aligned with the broader goals of sustainability and responsibility. In this spirit, I invite the readers to engage with the ideas presented in this book, to question the assumptions, and to contribute to the ongoing development of this exciting and important field.

Chapter 1

Introduction to livestock omics and artificial intelligence

Livestock management is undergoing a revolution as a result of the combination of omics technologies and artificial intelligence (AI), which offer unprecedented opportunities for improving the welfare, productivity, and health of animals. This convergence facilitates precision farming techniques and data-driven decision-making, providing a thorough understanding of the complex biological systems found in livestock (Kazembe and Mkandawire, 2024; Michielon et al., 2024; Gupta, 2025).

Omics technologies, including proteomics, metabolomics, transcriptomics, and genomics, allow high-throughput analysis of the molecular components of organisms. These methods provide comprehensive knowledge of animal biology by providing insight into the genetic composition, gene expression patterns, protein function, and metabolic processes of livestock (Ghavi Hossein-Zadeh, 2024; Zhou et al., 2024).

AI is becoming indispensable in today's agriculture, particularly in the fields of machine learning (ML) and computer vision (CV) systems. For example, AI helps dairy farmers to make data-driven decisions by assessing the vast amount of data collected from a variety of sources. To improve the efficiency of management and the welfare of the animals, CV systems are used for the identification of the animals, behavioral monitoring, assessment of feed intake, weight estimation, and detection of diseases (Chappidi, 2025; Gupta, 2025).

When AI and omics data are combined, complex data sets can be processed and interpreted to create insights that can be used. For example, the integration of AI and genomics allows for optimization of breeding strategies, increased productivity, and the safeguarding of animal health and welfare. The data-driven approach to dairy farming has made this a more reliable science in which intuition and predictive analysis are out in favor of being precise (Giordano, 2024; Patel and Mathur, 2024). In addition, AI technologies facilitate the collection, analysis, and interpretation of large amounts of data from a variety of sources, including sensors, drones, and satellite imagery. This enables farmers to make informed decisions that improve the sustainability and efficiency of livestock production, which is essential to meet the growing demand for livestock products (Khadka and Kumar, 2024).

As we continue this book, we will look at many aspects of omics technologies and their interaction with AI, and how this interaction will affect livestock management in the future. The following chapters aim to provide a thorough overview of this dynamic and rapidly changing field, which covers everything from understanding genetic patterns in animals to monitoring their health and behavior in real time.

In the following section, we will start by defining the concept of livestock omics, which will set the scene for a more in-depth examination of each of the different domains of omics and their relevance to modern livestock farming.

1.1 Defining livestock omics

Livestock omics provide a comprehensive, system-wide understanding of complex biological processes in livestock species, a scientific framework combining high-throughput biological data at several layers, including transcriptomics, proteomics, metabolomics, genomics, and others. Researchers and producers can study the genetic, molecular, and metabolic processes that underpin important traits such as productivity, health, resistance to disease, and adaptability to the environment by linking these different layers of data across livestock omics. The findings of these studies are already influencing advances in animal husbandry, animal welfare, and husbandry techniques, opening the door to more efficient, ethical, and sustainable animal farming (Husien et al., 2024; Lu et al., 2024).

Artificial Intelligence in Livestock Omics. https://doi.org/10.1016/B978-0-443-45663-3.00001-4

1.1.1 Origins and evolution of livestock omics

Early genome sequencing efforts laid the foundations for the analysis of biological data with high throughput, which is where livestock omics began. Technological developments in high-throughput sequencing, computational biology, and data analysis have pushed the term 'omics' into new areas, including transcriptomics, proteomics, and metabolomics, after it was originally used in genomics (Zhou et al., 2024).

The Human Genome Project, which generated the first comprehensive DNA sequence of a complex organism, marked the beginning of the genomic era and sparked widespread interest in sequencing the genomes of other animal species, including livestock. These initiatives have laid the foundations for breeding programs based on specific genetic markers and offered a blueprint for understanding the genetic basis of traits (Panchariya et al., 2024). As access to high-throughput sequencing has increased, new disciplines of 'omics' have emerged to study the different molecular layers. Together, these disciplines allow a more thorough understanding of biological complexity and facilitate applications in the field of animal science. Omics technology was first used in model organisms and in human medicine, but it soon became clear that it could be applied to agriculture. Researchers have found that similar molecular techniques can be used to better understand and improve animal traits that influence growth, milk production, and resistance to disease. Livestock omics, using cutting-edge technology, has recently expanded its scope to take into account the sustainability of livestock production systems, environmental impacts, and animal welfare to maximize results for producers and animals (Gao et al., 2024; Vitorino, 2024).

1.1.2 Key omics disciplines in livestock science

Livestock omics cover several related areas, each offering a different perspective on different molecular levels of biology. The study of the entire DNA sequence of an organism, including all the genes and non-coding regions of the genome, is called genomics. Genomics provides the basis for identifying markers associated with economically important traits such as milk yield, growth rate, feed efficiency, resistance to disease, and reproduction. Breeders can make better decisions and select animals with desirable characteristics faster and with greater accuracy than with conventional techniques by using genomic selection. This promotes sustainable productivity growth and increases genetic progress in the livestock population. Rapid and accurate genetic mapping is possible thanks to high-throughput methods such as Whole-genome sequencing (WGS) and single-nucleotide polymorphism (SNP) genotyping, which are useful for association research and selection of traits. Other genomic tools include quantitative trait loci (QTL) mapping, which locates the parts of the genome that affect complex traits, and genome-wide association studies (GWAS), which link specific genes to desirable traits. Some of the major fields of genomics are metagenomics, epigenomics, and population genomics (Gao et al., 2024; Husien et al., 2024; Tiwari et al., 2024).

Population genomics, by analyzing genetic variation within and between cattle populations, examines how the gene pool evolves over time due to selection pressures. Population genomics is essential for managing genetic diversity, which promotes adaptability, resilience, and productivity. This field helps to maintain genetic health and to identify genes that make a species resilient by examining breed structures and inbreeding rates. It is particularly useful for breeding initiatives aimed at the conservation or improvement of rare and or pure breeds. Population genomics allows for detailed genetic surveys and comparisons using tools such as the SNP array and the WGS. By controlling inbreeding and identifying beneficial genetic variants, these findings help maintain the productivity and health of the breed (Davidescu et al., 2024; Husien et al., 2024).

Epigenomics studies focus on modifications such as DNA methylation and histone changes that affect gene expression without altering the sequence of DNA. Genes are activated and suppressed by epigenetic changes that often occur in response to environmental conditions such as stress or diet. Although they are not part of the DNA sequence itself, epigenetic research can reveal traits that are passed down from one generation to the next and provide insights into how animals adapt to changing environments. For example, breeding and nutrition methods may be improved by using epigenetic markers associated with increased milk production or growth. Epigenetic alterations are detected by methods such as chromatin immunoprecipitation sequencing (ChIP-Seq) and bisulfite sequencing (Wang, 2024). These data offer new objectives for managing livestock under different environmental conditions or for optimizing traits by identifying regulatory patterns that affect gene expression (Varsini et al., 2024).

Metagenomics, which focuses on the gut microbiome, explores the genetic composition of whole microbial communities. The use of metagenomics has been a great benefit in studying the role of the microbiome in immunity, resistance to disease, digestion, and overall health. An examination of the gut microbiomes may reveal links between microbial diversity and factors such as feed efficiency, nutrient uptake, and the response of the immune system (Li, 2025). Metagenomic research can inform strategies such as diet modification or the use of probiotics to improve gut health and productivity. Researchers can link specific microbial profiles to health and performance outcomes by profiling microbial diversity and function using metagenomic sequencing, particularly 16S rRNA sequencing (Armetta et al., 2025).

Transcriptomics, which studies RNA molecules produced when genes are expressed, provides a snapshot of gene activity in a specific tissue at a specific time. By analyzing the amount of RNA expressed, transcriptomics provides insight into how genes respond to various triggers, including stress, disease, diet, and environmental changes. This is particularly important for the understanding of complex traits that are not solely controlled by genetic composition but are influenced by environmental factors (Nirmal et al., 2024). The breeding and management strategies are guided by transcriptomic information to increase immunity, growth, and resilience. RNA sequencing, or RNA-Seq, is a powerful technique for obtaining transcript profiles representing active gene pathways. RNA-Seq enables both quantitative and qualitative analysis, revealing the diversity and abundance of RNA molecules as well as regulatory mechanisms that operate in specific tissues (e.g., either the mammary gland for milk production or muscle for the quality of the meat) (Zhao et al., 2023).

The focus of proteomics is the extensive analysis of proteins that perform functional tasks inside cells by following instructions encoded in genes. Proteomic research makes it easier to understand the networks of proteins that are involved in vital functions such as growth, immune response, and metabolic regulation. Proteomics is useful for identifying biomarkers to support the diagnosis of diseases, nutrition assessment, or reproductive health, as proteins are also good indicators of health and disease status. For example, some protein signatures can measure how well an individual responds to diet changes or predict the risk of getting sick. Proteomics often uses protein microarrays and mass spectrometry (MS). With the help of MS, protein networks and interactions can be mapped accurately in complex biological samples. Proteomics provides useful information for improving the performance, safety, and quality of animal products (Patel et al., 2023; Gobena et al., 2024).

The study of small molecules or metabolites that are involved in the energy consumption and metabolism of an organism is called metabolomics. Metabolomics provides the biochemical fingerprint of animal health, nutrient intake, and metabolic efficiency. This information helps to improve feed formulation, treat metabolic diseases, and develop nutrition interventions. Metabolomics improves our ability to optimize diets and improve the general health of animal populations by looking at metabolites that are involved in immune function, growth, and reproduction. Researchers can profile the wide range of metabolites in biological samples using methods like mass spectrometry and nuclear magnetic resonance (NMR) spectroscopy, which can reveal biochemical reactions to diseases or dietary changes. Metabolomics insights enhance product quality, animal health, and feed efficiency (Vo and Trinh, 2024; Zhang et al., 2024).

Integrative omics aims to fully understand the complex biological processes underlying phenotypic traits by integrating various types of omics data, including transcriptomics, proteomics, metabolomics, and genomics. This comprehensive view provides a deeper understanding of the biological processes that influence observable properties by capturing the interactions between genes, transcriptomes, proteins, and metabolites. This integrative approach relies heavily on AI and ML. By facilitating data integration and pattern recognition across various omics layers, they make it easier to handle large, complex datasets. Algorithms using AI and ML can find complex connections within and between these data sets, revealing patterns that conventional methods of analysis may miss. Researchers can predict trait heritability and phenotypic outcomes more accurately by utilizing AI and ML for multi-omics integration. In livestock farming, this method can greatly increase the accuracy in identifying animals with favorable characteristics such as increased growth rate, increased resistance to disease, or improved feed utilization. The comprehensive knowledge gained from integrative omics provides a deeper insight into the genetic basis of these traits, ultimately leading to smarter breeding strategies (Moghaddam et al., 2025; Pasupuleti, 2025; Vanamala et al., 2025).

1.1.3 The role and objectives of livestock omics

Livestock omics incorporates several high-throughput technologies aimed at a thorough investigation of the molecular structure of livestock species. These technologies facilitate improvements in health, productivity, and resistance to disease by providing a comprehensive understanding of the biota in livestock (Zhou et al., 2024). The main objectives of livestock omics are as follows (Fig. 1.1): 1. Precision breeding and genetic improvement: Selective breeding programs have been transformed by omics technologies, especially genomics, which have discovered particular molecular markers associated

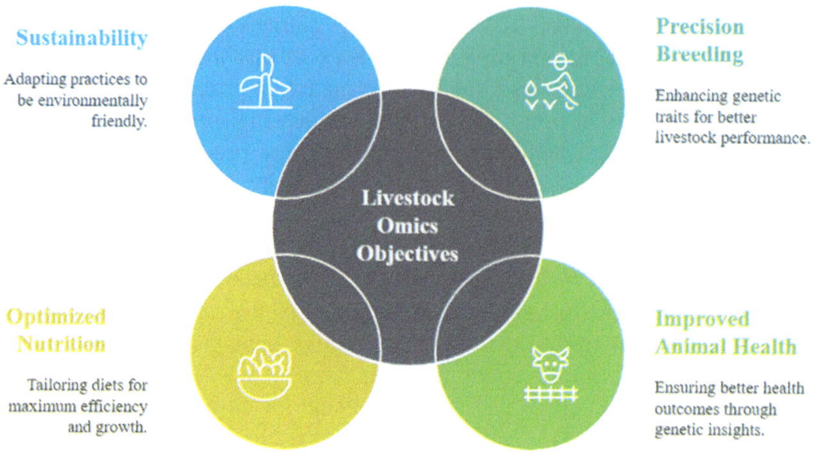

FIGURE 1.1 The main objectives of livestock omics.

with desirable traits. By selecting early on genetically superior, disease-resistant animals, this precise approach speeds up genetic progression and productivity gains. Genomic selection, for example, is widely used to improve traits such as resistance to disease, growth rate, and milk yield in various livestock species (Krivoruchko et al., 2025). 2. Improved animal health: Active animal health management is facilitated by omics technologies that analyze the transcriptome and proteome to help early detection of health problems. For example, transcriptomic analysis can identify patterns of gene expression associated with susceptibility to a disease, allowing targeted interventions to be developed. Proteomic research can help in early diagnosis and treatment by identifying biomarkers that indicate the state of the disease (Sarath et al., 2024). 3. Optimized nutrition and metabolism: Metabolomics, which provides insights into the metabolic profiles of livestock, makes it possible to match nutrition to genetic potential. By knowing how the body responds to different diets, producers can develop feeding strategies that increase the efficiency of the feed, reduce waste, and improve growth and reproduction. This method encourages the establishment of precise feeding programs adapted to the specific requirements of each animal or herd (Slupsky, 2024). 4. Sustainability and environmental adaptation: By allowing animals to be reared in more appropriate environments, omics tools promote sustainable farming practices. Genomics, for example, can detect genetic variations associated with resistance to certain diseases that are common in a region or with heat tolerance. This information facilitates the development of breeding programs that improve the welfare of animals, reduce the environmental impact of livestock production, and increase the resilience of the environment (Sonstegard et al., 2024) (Table 1.1).

1.1.4 Challenges and future directions for livestock omics

Livestock omics has the power to transform the livestock industry by providing new ideas and tools to help them enhance animal welfare, health, and productivity. However, the promise of this field can only be invoked if several significant issues are addressed. These issues are ethical, financial, and technical, requiring constant technological innovation and deliberation. Meanwhile, developments in data integration, AI, and an increasing focus on sustainability and animal welfare are collectively pushing the future of livestock omics further up. Here is a detailed overview of the key challenges and promising future trends in livestock omics. Key challenges and future directions for livestock omics are indicated in Tables 1.2 and 1.3, respectively.

1.2 Impact of omics technologies on livestock

The integration of omics technologies in livestock research has triggered a major change in the agricultural sector toward a more precise approach. This integration provides a profound understanding of processes at the molecular level, which challenge and transform conventional farming methods. By unraveling the complex relationship between genetics, the environment, and management, these technologies provide stakeholders with the means to increase efficiency, improve animal welfare, and address the sustainability challenges facing the world. Below, we will examine the diverse impacts of these technologies on key aspects of livestock management systems.

TABLE 1.1 Key objectives and applications of omics technologies in livestock production.

Objective	Description	Key omics technologies involved	Applications in livestock	Benefits	Challenges
Precision breeding and genetic improvement	Utilizing omics data to identify molecular markers linked to desirable traits, enabling targeted selection in breeding programs	- Genomics - Transcriptomics - Proteomics - Metabolomics	- Identification of genetic variants associated with traits like growth rate, milk production, and disease resistance - Implementation of marker-assisted and genomic selection strategies	- Accelerated genetic gains - Enhanced accuracy in selecting superior animals - Reduction in generation intervals	- High costs associated with omics technologies - Complexity in data analysis and interpretation - Ethical considerations regarding genetic diversity
Enhanced animal health	Applying omics approaches to detect health issues early and understand disease mechanisms, facilitating proactive health management	- Transcriptomics - Proteomics - Metabolomics	- Identification of biomarkers for early disease detection - Understanding host-pathogen interactions - Development of targeted therapeutics and vaccines	- Improved animal welfare through early intervention - Reduction in disease prevalence and associated costs - Enhanced understanding of disease mechanisms	- Need for high-quality samples and standardized protocols - Integration of multi-omics data for comprehensive insights - Translating findings into practical applications
Optimized nutrition and metabolism	Leveraging omics data to align nutritional strategies with animals' genetic and metabolic profiles for improved efficiency	- Metabolomics - Transcriptomics - Proteomics	- Assessment of metabolic responses to different diets - Identification of nutrient deficiencies or excesses - Formulation of precision feeding strategies	- Enhanced feed efficiency and growth rates - Reduction in feed costs and environmental waste - Improved product quality (e.g., meat, milk)	- Variability in metabolic responses among individuals - Complexity in interpreting metabolic data - Need for individualized feeding strategies
Sustainability and environmental adaptation	Using omics tools to breed livestock adapted to diverse environments, promoting sustainable agricultural practices	- Genomics - Epigenomics - Transcriptomics	- Identification of genetic variants associated with environmental resilience - Breeding for traits like heat tolerance and disease resistance - Understanding epigenetic modifications influencing adaptation	- Development of livestock suited to various climates - Reduction in resource use and environmental impact - Support for sustainable farming practices	- Balancing productivity with adaptability traits - Ensuring maintenance of genetic diversity - Addressing public concerns about genetic modifications

TABLE 1.2 Key challenges in livestock omics.

Challenge	Description	Details
Data complexity and integration	Integrating diverse omics data (genomics, transcriptomics, proteomics, metabolomics, epigenomics) to achieve a holistic understanding of biological processes in livestock	Omics data are complex and heterogeneous, requiring distinct analytical approaches
High costs of omics technologies	The expense of sequencing, mass spectrometry, and data analysis limits accessibility, especially for smaller farms or institutions	Despite decreasing costs, comprehensive multi-omics studies remain financially challenging
Lack of skilled personnel and bioinformatics expertise	A shortage of professionals with expertise in both livestock science and omics technologies hinders progress	Specialized skills in bioinformatics and computational biology are essential but scarce
Ethical and privacy concerns	Issues related to data ownership, animal welfare, and genetic privacy arise with the use of omics in livestock	Concerns include the welfare of genetically selected or edited animals and the sharing of genetic information
Environmental sustainability	Balancing livestock productivity with environmental impact, such as greenhouse gas emissions and resource usage	Identifying genes or traits that enhance feed efficiency or reduce emissions is challenging

1.2.1 Genetic improvement and breeding

The incorporation of omics technologies into the genetic enhancement of livestock has fundamentally revolutionized breeding methodologies by facilitating precise molecular-level comprehension. Central to this paradigm shift is genome-wide selection, which capitalizes on high-throughput genotyping platforms to evaluate hundreds of thousands of genetic markers across animal genomes. The utilization of these extensive marker panels, when analyzed through sophisticated statistical models, enables the accurate estimation of breeding values by correlating genomic configurations with

TABLE 1.3 Future directions in livestock omics.

Future direction	Description	Details
Advancements in AI and ML for data integration	Utilizing AI and ML to manage data complexity and uncover patterns across multiple omics layers	Development of interpretable AI models and user-friendly interfaces to facilitate multi-omics integration
Development of cost-effective omics technologies	Making omics technologies more affordable and accessible across the livestock industry	Emergence of portable sequencing devices and cloud-based analysis platforms for real-time data analysis
Precision livestock farming and omics-driven decision-making	Integrating omics data with precision farming technologies to inform daily farm management	Creation of real-time decision-support systems offering tailored insights based on genetic and environmental data
Expansion of multi-omics studies in less-studied livestock species	Extending omics research to species like sheep, goats, and aquaculture species	Increased investment in research for diverse livestock species to enhance genetic diversity and productivity
Focus on environmental genomics and sustainability	Applying omics data to reduce livestock's environmental impact	Breeding for traits that align with low environmental impact and optimizing resource use
Ethical frameworks and data governance policies	Developing robust ethical guidelines and data governance policies for responsible omics data use	Addressing concerns about genetic privacy, data sharing, and animal welfare implications

economically significant traits such as milk production, disease resistance, and feed efficiency. Furthermore, ML frameworks augment predictive capabilities by elucidating intricate genetic interactions that conventional linear models may fail to capture, especially in relation to non-additive effects and epistatic connections among genes (Kasimanickam et al., 2025).

Gene-editing technologies serve to complement these genomic methodologies by facilitating targeted alterations to specific loci associated with advantageous traits. Contemporary precision editing protocols now incorporate advanced computational tools for the design of guide RNAs and the mitigation of off-target effects, with delivery methodologies optimized across various species to ensure elevated editing efficiency. Beyond mere knockouts, novel base-editing techniques enable nucleotide-level modifications without inducing double-strand breaks, thereby broadening the potential for trait engineering. Recent advancements in prime editing have illustrated the viability of introducing precise alleles to enhance reproductive outcomes, thereby underscoring the potential for the creation of novel genetic variants independent of natural mutation rates (Aledhari and Rahouti, 2024; Ghavi Hossein-Zadeh, 2024).

The amalgamation of multi-omics data streams—encompassing genomics, transcriptomics, proteomics, and epigenomics—affords a systems-level comprehension of biological mechanisms. By synthesizing these layers through multivariate analytical frameworks, researchers can elucidate the intricate interplay between genetic architecture and phenotypic expression. Such integrative methodologies are particularly advantageous in addressing multifactorial challenges such as environmental stress resilience, wherein coordinated molecular responses across biological hierarchies dictate adaptive capabilities. Epigenetic profiling introduces an additional dimension to selection strategies by unveiling regulatory mechanisms that affect trait heritability beyond mere DNA sequence variation (Kumar et al., 2024).

Emerging single-cell and spatial omics technologies are expanding resolution limits, facilitating unparalleled examination of cellular heterogeneity and tissue-specific gene expression patterns. These methodologies illuminate developmental pathways and metabolic processes critical for production traits, thereby informing selection decisions with cellular granularity. Concurrent advancements in computational infrastructure address the data deluge inherent to omics methodologies, with federated learning systems facilitating collaborative analysis across global breeding networks while safeguarding data privacy. Nonetheless, the broad implementation of these technologies encounters obstacles, including significant infrastructure demands for data storage and processing, financial constraints that limit accessibility in resource-limited regions, and ongoing ethical discussions regarding the preservation of genetic diversity in edited populations. The pathway forward necessitates coordinated efforts across various scientific disciplines to optimize tool accessibility, standardize analytical workflows, and establish ethical frameworks that harmonize innovation with agricultural sustainability (De Jonghe et al., 2024; Lammi and Qu, 2024).

1.2.2 Disease management and resilience

The utilization of omics technologies in the management of livestock diseases signifies a significant evolution toward comprehensive, data-informed paradigms that amalgamate genomic, proteomic, metabolomic, and computational approaches. These technologies collectively facilitate the methodical elucidation of pathogen-host-environment interactions across molecular, cellular, and population dimensions, promoting resilience through predictive analytics and precision interventions. By harnessing high-throughput sequencing, sophisticated bioinformatics, and ML techniques, contemporary surveillance systems attain unparalleled accuracy in pathogen identification, tracing evolutionary pathways, and delineating transmission dynamics in near real time. This capability is further augmented by portable sequencing technologies and cloud-based genomic repositories, which democratize access to pathogen monitoring across varied agricultural contexts (Vanamala et al., 2025; Yennamalli and Onteru, 2025).

Immunological methodologies have likewise progressed through the incorporation of structural biology and systems immunology, allowing for the strategic design of vaccines that specifically target conserved epitopes while minimizing off-target repercussions. Computational frameworks forecast antigenic variability and immune evasion mechanisms, informing the development of multi-epitope constructs enhanced with novel immunomodulatory agents. Simultaneously, single-cell omics technologies dissect host-pathogen interactions at a cellular level, identifying molecular indicators of immune exhaustion or hyperactivation that correlate with disease manifestations. These revelations guide therapeutic approaches aimed at recalibrating immune responses through gene editing or adoptive cell therapies (Sambreen et al., 2024).

Metabolomic and proteomic profiling further enrich these methodologies by identifying preclinical disease states through biomarker discovery in biological fluids, breath, or tissue specimens. ML algorithms trained on multi-omics datasets discern subtle metabolic disruptions indicative of subclinical infections, thereby enabling proactive management prior to the onset of productivity declines. Field-deployable biosensors and microfluidic devices actualize these findings, offering rapid, on-farm diagnostics that circumvent the need for centralized laboratory facilities (Qureshi et al., 2024).

Strategies for mitigating antimicrobial resistance (AMR) now depend on pan genomic surveillance of microbial assemblages, monitoring horizontal gene transfer occurrences, and plasmid dynamics under selective pressures. Phage therapy and CRISPR-based antimicrobials are emerging as precision instruments for suppressing resistant pathogens while preserving commensal microbiota, underpinned by metatranscriptomic data that enhance phage-host compatibility. Computational models simulating the flux of AMR genes across livestock-environment interfaces further refine stewardship policies (Nair, 2024).

The computational infrastructure of these systems confronts petabyte-scale datasets through distributed computing architectures, quantum annealing, and federated learning frameworks that ensure data confidentiality. Ethical concerns regarding genomic data ownership, equitable access to technology, and unintended ecological consequences of pathogen suppression tactics necessitate interdisciplinary governance models. Novel innovations such as 3D organoid cultures and synthetic biology platforms expedite therapeutic discovery, while blockchain-enabled traceability systems improve outbreak response coordination across supply chains (Myrzashova et al., 2023; Kolobkov et al., 2024).

In conclusion, omics-driven disease management converges toward a paradigm characterized by continuous molecular surveillance, wherein resilience is cultivated through iterative feedback among surveillance data, adaptive therapeutics, and predictive modeling. This ecosystem not only alleviates economic losses but also reconceptualizes livestock health as a dynamic equilibrium maintained by interoperable technologies, cross-sector collaboration, and ethically informed innovation.

1.2.3 Nutritional optimization and feed efficiency

Omics-driven nutritional optimization in livestock management signifies a methodical reconceptualization of feed efficiency, amalgamating multi-faceted biological insights to synchronize dietary inputs with metabolic, microbial, and ecological dynamics. This paradigm functions through the persistent examination of gut ecosystem functionality, wherein advanced genomic and proteomic methodologies delineate microbial consortia accountable for nutrient metabolism, elucidating enzymatic pathways that regulate carbohydrate degradation and protein assimilation. By correlating microbial activity patterns with host metabolic profiles, precision formulations are meticulously devised to enhance symbiotic interactions while alleviating metabolic inefficiencies that lead to nutrient wastage. Such strategies dynamically modify amino acid compositions and energy substrates to conform with species-specific genetic predispositions and production objectives, thereby effectively minimizing nitrogen excretion and optimizing growth trajectories (Nunes et al., 2024; Olorunkoya et al., 2024).

The incorporation of synthetic biology instruments facilitates the creation of bioengineered feed additives aimed at improving nutrient bioavailability and mitigating anti-nutritional factors. Gene-edited microbial strains and enzyme complexes are strategically employed to decompose recalcitrant plant compounds, thereby liberating bound nutrients while diminishing reliance on supplemental minerals. Concurrently, metabolic profiling methodologies trace nutrient flux across various tissues, thereby informing lipid and carbohydrate modifications that equilibrate energy distribution among growth, reproduction, and maintenance functions. For methane abatement, targeted bioactive compounds disrupt microbial methanogenesis pathways without undermining essential fermentation processes that are critical for volatile fatty acid production (Rodrigues et al., 2023; Abd et al., 2024).

Real-time monitoring ecosystems synergize biosensor networks with predictive analytics to facilitate adaptive feeding strategies. Implantable devices and non-invasive imaging modalities meticulously track physiological parameters and digestive efficiency, initiating automated modifications to feed composition upon detection of deviations from optimal thresholds. ML algorithms adeptly process multidimensional data streams—from fluctuations in microbial populations to host metabolic reactions—to anticipate nutrient requirements and iteratively refine ration formulations. This closed-loop system guarantees continuous alignment between dietary inputs and biological needs across diverse production stages and environmental conditions (Henrichs and Krupitzer, 2024; Ranganathan et al., 2024).

Nutrigenomic frameworks interlink dietary interventions with genetic regulation, utilizing epigenetic clocks and polymorphism analyses to tailor feeds based on individual genetic profiles. These methodologies modulate gene expression networks that govern lipid metabolism, muscle development, and nutrient absorption efficiency, thus devising genotype-specific nutritional protocols. Epigenetic biomarkers further predict age-associated declines in digestive capacity, prompting proactive modifications to sustain intestinal health and nutrient assimilation rates throughout the production lifecycle (Udaya Kumar, 2024).

Sustainability-oriented formulations actualize circular economy principles through life cycle assessment models that evaluate environmental impacts across feed production chains. Alternative protein sources and precision nutrient delivery systems mitigate reliance on conventional ingredients, thereby lowering greenhouse gas emissions and land-use footprints. Optimization algorithms reconcile ecological constraints with economic feasibility, minimizing mineral excretion through enzyme-enhanced hydrolysis and precision dosing technologies. Collectively, these strategies establish an interconnected framework wherein nutritional precision, metabolic efficiency, and environmental stewardship coalesce, thereby redefining livestock feeding as a dynamic equilibrium between biological optimization and ecological responsibility (De Quelen et al., 2024; Tignani et al., 2024).

1.2.4 Climate adaptation and sustainability

The incorporation of omics technologies into the management of livestock signifies a transformative shift in confronting the concurrent challenges of climate adaptability and ecological viability. By elucidating the molecular architectures of animals, microbes, and ecosystems, these methodologies facilitate precision interventions that bolster resilience while concurrently minimizing environmental impacts. At the genetic dimension, sophisticated sequencing technologies identify heat-resistant alleles across various species, such as cattle demonstrating enhanced development of sweat glands, and poultry exhibiting improved renal water reabsorption efficiency. These findings inform CRISPR-mediated gene introgression initiatives that expedite the creation of climate-resilient breeds without detriment to productivity (Ghavi Hossein-Zadeh, 2026). Concurrent endeavors in epigenomics delineate stress-induced methylation patterns, establishing predictive models that direct selective breeding for sustained adaptability to evolving climatic conditions (Kalaignazhal et al., 2024).

Microbiome engineering has emerged as a pivotal element of sustainable production, with metagenomic profiling uncovering microbial consortia adept at converting low-quality forage into energy-dense metabolites. Engineered probiotic formulations optimize ruminal function, thereby diminishing methane emissions through the selective inhibition of methanogenic pathways while simultaneously enhancing fiber digestion. This biological optimization operates in synergy with precision nutrition systems, wherein metabolomic analyses of animal excreta inform adaptive feed formulations, achieving a reduction in nitrogen waste by 18%–22% through real-time balancing of amino acids (Glockow et al., 2024). Strategies for water conservation similarly benefit from the integration of omics technologies—specifically, aquaporin gene editing in poultry diminishes drinking water requirements by 25%, while IoT-enabled soil sensors, informed by satellite-derived evapotranspiration data, facilitate irrigation systems that sustain pasture productivity with a 35% reduction in water extraction (Kumar and Manjula, 2024).

Models of a circular bioeconomy utilize insights derived from multi-omics to convert waste streams into value-added materials. Enzymes mined from the metagenome of extremophilic microbes enable the industrial-scale transformation of agricultural byproducts into biodegradable plastics and biofertilizers, while RNA interference (RNAi) enhanced insect bioreactors efficiently convert manure into protein concentrates with an efficacy of 72%. Anaerobic digestion systems refined through microbial transcriptomics achieve unparalleled biogas yields, with the resultant digestate processed into biochar for carbon sequestration. These closed-loop systems are monitored via life cycle assessment frameworks that integrate genomic, metabolic, and environmental data streams to quantify sustainability advancements across entire production networks (Pal et al., 2024; Yusuf and Ojedokun, 2024).

Emerging decision-support systems amalgamate omics data with climatic forecasts to facilitate proactive adaptation. ML algorithms, trained on epigenetic clocks and whole-blood transcriptomes, accurately predict individual animal resilience to heat stress events with an accuracy of 91%, triggering automated cooling interventions through interconnected barn systems. Satellite-linked wearable sensors provide comprehensive monitoring of herd thermoregulation metrics at the ranch level, while blockchain-enabled traceability platforms authenticate the environmental credentials of climate-adapted livestock products. This systemic integration of biological, digital, and ecological insights positions livestock agriculture as a proactive participant in climate mitigation—effectively reducing sectoral emissions by 40%–58% while ensuring food security amid anticipated warming scenarios (Satish et al., 2023; Pasupuleti, 2024).

1.2.5 Economic and ethical considerations

The incorporation of omics technologies into livestock production systems functions within a multifaceted interaction of economic feasibility and ethical accountability, wherein the prospects of technological advancements confront systemic realities. At the economic forefront, precision genomic instruments encounter a dual dilemma of capital intensity and value chain integration. Although genome editing technologies herald significant enhancements in disease resistance and environmental adaptability, their implementation necessitates the careful balancing of considerable research and development expenditures with uncertain long-term financial returns. Emerging cost-sharing models endeavor to democratize access through collaborative intellectual property frameworks; however, enduring disparities in technological infrastructure persistently skew advantages toward large-scale enterprises. Precision livestock farming systems illustrate this tension—while sensor networks and biometric analytics yield quantifiable enhancements in feed efficiency and disease prevention, the diminishing returns associated with data saturation mandate adaptive frameworks that prioritize actionable insights over comprehensive surveillance (Yadav et al., 2024; Khalil, 2025). For smallholder farmers, these technologies engender paradoxical dependencies: innovative financing solutions such as blockchain-based microloans provide temporary alleviation but frequently fail to address systemic vulnerabilities intensified by equipment depreciation rates and climatic variability, necessitating hybrid approaches that amalgamate technological innovation with contextually relevant socioeconomic assistance (Prost et al., 2024).

Ethical governance has transitioned into a multidimensional balancing act, necessitating concurrent navigation of biophysical constraints, animal welfare mandates, and societal values. Contemporary bioethical paradigms necessitate rigorous evaluative frameworks for genomic interventions, extending assessment durations to encompass transgenerational ecological ramifications and unintended perturbations of the microbiome. The quantification of animal welfare now incorporates multimodal analytics that correlate behavioral biometrics with physiological stress indicators, utilizing machine vision systems to decode species-specific welfare metrics at scale. Concerns regarding data sovereignty have prompted a transition toward privacy-preserving computational models, wherein advanced encryption methodologies facilitate genomic insights while safeguarding producer anonymity—a crucial compromise in an era characterized by increasing agritech consolidation. Regulatory frameworks remain disjointed, as labeling mandates and traceability standards impose asymmetric compliance burdens that disadvantage resource-constrained producers despite a burgeoning consumer demand for transparency (Metcalf and Sadowski, 2024; Кан, 2023).

Market transformations expose profound divides between technological capabilities and equitable implementation. Sustainability certifications and carbon accounting frameworks are redefining value chains, establishing premium markets for low-emission livestock products while revealing deficiencies in lifecycle assessment methodologies. Innovations within the insurance sector grapple to keep pace with climate-induced risks, as traditional parametric models have proven inadequate for capturing microclimate variability and herd mobility dynamics, thus prompting the development of hybrid verification systems that integrate satellite data with ground-level biometric metrics. The emergence of cellular agriculture introduces novel socioeconomic tensions, as cultured meat technologies challenge traditional livestock economies amidst unresolved inquiries regarding their environmental impact and cultural acceptability. While urban consumers advocate for these alternatives due to their land-use efficiency, rural communities view them as existential threats to pastoral traditions—a dichotomy that complicates policy development and necessitates inclusive governance frameworks (Hellin et al., 2023; Bergougui et al., 2024).

This landscape necessitates ethical structures capable of reconciling competing priorities through adaptive, evidence-based governance. Novel assessment protocols are now evaluating technologies through multidimensional matrices that balance biodiversity implications against the preservation of cultural heritage and intergenerational equity. The trajectory ahead mandates symbiotic innovation—where advancements in omics are integrated with institutional reforms that assure equitable benefit distribution, environmental stewardship, and respect for indigenous knowledge systems. The livestock sector can only successfully meet the competing demands of technological progress and social responsibility in the genomic era if it adopts such integrated approaches (Yarchuk et al., 2024).

1.3 Ethical-philosophical underpinnings of data-driven livestock science

The preeminence of data-centric methodologies in the field of livestock science necessitates an exhaustive examination of their ethical and philosophical underpinnings—an endeavor that surpasses the utilitarian cost-benefit analyses that frequently dominate discussions within agriculture. At the heart of this issue resides an ontological conflict: the ongoing digitization of living beings into quantifiable data streams poses a risk of reducing animals to mere cybernetic constructs, thereby diminishing their biological intricacy into feature vectors suitable for ML applications. This epistemic transformation is evident in practices such as genomic digital twinning, wherein a dairy cow's lifetime productivity is simulated via neural networks trained on extensive datasets of SNP markers, potentially prioritizing algorithmic predictions over the phenomenological realities about animal welfare (Arulmozhi et al., 2024; Akintan et al., 2025; Klingström et al., 2024). Such methodologies pose challenges to traditional veterinary ethics, which are traditionally anchored in the care of individual patients, instead favoring population-level optimizations wherein a reduction in lameness prevalence might rationalize slight increases in culling rates.

Simultaneously, the integration of AI within livestock systems undermines long-established hierarchies governing agricultural knowledge production. The scenario wherein deep learning algorithms surpass experienced breeders in forecasting heterosis effects, or when blockchain-enabled traceability frameworks render third-party welfare certifications redundant, raises pertinent inquiries regarding the locus of epistemic authority (Kazembe and Mkandawire, 2024; Michielon et al., 2024). Who adjudicates between the probabilistic outputs generated by a random forest model predicting mastitis outbreaks and the empirical assessments made by dairy technicians? How do we reconcile the opaque nature of convolutional neural networks analyzing thermal images of poultry with the necessity for transparent welfare evaluations? These dilemmas reveal a profound philosophical divide between empiricist traditions in animal science and the computational positivism that characterizes contemporary agritech.

The temporal dimensions associated with data-driven interventions introduce additional layers of ethical intricacy. Modern gene-editing technologies such as CRISPR-Cas9, informed by AI-optimized sgRNA designs, facilitate precise modifications to livestock genomes with repercussions extending across subsequent generations—a manifestation of intergenerational genetic debt. While such innovations may bolster short-term climate resilience by incorporating heat-tolerant alleles in beef cattle, they concurrently threaten to diminish the genomic diversity that has historically shielded livestock populations against unpredictable pathogens. This duality reflects the Anthropocene paradox: computational tools sufficiently robust to alleviate agriculture's environmental detriments may also reinforce the anthropocentric arrogance that has led to those crises (Eski et al., 2024; Montoliu, 2024).

Ethical frameworks derived from human-centric bioethics are insufficient to confront these challenges. Principles such as autonomy and informed consent, which are foundational to human-subject research, falter in their application to livestock—a contradiction epitomized by the concept of "algorithmic welfare": wherein sensors and machine vision systems continuously monitor the physiological states of animals without any provision for opting out. Emerging paradigms such as One Welfare endeavor to reconcile this disparity by highlighting the interconnectedness of animal welfare, ecosystem health, and human prosperity; however, they encounter significant difficulties in operationalizing such a holistic approach within the reductionist frameworks characteristic of conventional AI systems (Blaxter et al., 2024; Neethirajan et al., 2024).

Regulatory philosophies exacerbate these complexities. Current governance models, predominantly reactive to technological advancements, fail to adequately address the rapid convergence of omics and AI. The European Union's precautionary principle, for example, is at odds with Silicon Valley's ethos of "move fast and break things" when applied to AI-driven transgenesis in swine. Concurrently, intellectual property frameworks grapple with the categorization of AI-generated breeding protocols or synthetic microbiota designs, potentially resulting in monopolistic control over vital agricultural knowledge commons (Hacker, 2023; Kolade et al., 2024).

The ethical and philosophical landscape of data-driven animal science requires a radical rethinking of management in the digital age. It calls for frameworks that do not uncritically embrace techno-solutions or retreat into pastoral nostalgia, but instead navigate the trade-offs between computer-driven precision and ecological modesty. This requires recognizing that any optimization algorithm involves implicit value judgments, whether it favors feed conversion rates over behavioral enrichment in poultry or favors methane-reducing traits over genetic diversity in cattle. As animal systems evolve into hybrid biological-digital ecosystems, the moral imperative is shifting from just minimizing damage to actively co-designing techno-ethic structures that respect the intrinsic value of animal life while meeting the nutritional needs of humanity.

1.4 AI in modern agriculture and livestock production

The incorporation of AI into agricultural frameworks signifies a profound reconfiguration of methodologies pertaining to food production, wherein computational intelligence merges with biological intricacies to transform paradigms of sustainability. At its core, this metamorphosis exemplifies the amalgamation of data-centric decision-making architectures with ecological dynamics, facilitating adaptive governance of production systems across diverse spatial and temporal dimensions. AI frameworks function as cognitive mediators, transmuting a variety of environmental datasets—from molecular biochemical markers to large-scale climatic trends—into actionable intelligence that enhances resource distribution while adhering to planetary constraints. In the domain of crop production, these technological advancements facilitate precision management through the integration of multispectral imaging, soil health assessments, and predictive climate analytics. ML algorithms dynamically recalibrate irrigation and nutrient allocation by correlating real-time plant physiological data with historical yield trends, achieving significant reductions in water and fertilizer consumption while sustaining photosynthetic efficacy. Simultaneously, computer vision technologies improve weed and pest management through sophisticated pattern recognition, enabling robotic systems to accurately differentiate between cultivated crops and invasive plant species while minimizing collateral ecological impacts (Chappidi, 2025; Gupta, 2025; Payili, 2025).

Livestock production systems experience a similar transformation owing to AI's proficiency in deciphering intricate biological interrelations. ML models amalgamate genomic data, physiological monitoring metrics, and environmental variables to forecast health outcomes and refine breeding methodologies. These systems facilitate the early identification of subclinical diseases by detecting subtle correlations between biometric parameters and pathological progressions, while concurrently informing nutritional strategies through real-time evaluations of digestive functions and metabolic reactions. The integration of sensor networks with predictive analytics engenders closed-loop management frameworks that sustain optimal physiological conditions while mitigating greenhouse gas emissions and resource inefficiencies (Kazembe and Mkandawire, 2024; Singh et al., 2024).

Robotic automation surfaces as a pivotal facilitator of precision agriculture, merging AI with advanced mechatronic systems to execute intricate field operations. Autonomous entities utilize spatial mapping and machine vision to conduct harvesting, pruning, and cultivation activities with a level of dexterity akin to human operators, adapting to fluctuating field conditions via continuous environmental feedback. Swarm robotics architectures orchestrate coordinated efforts among fleets of autonomous vehicles to tackle extensive challenges such as pest invasions or soil rehabilitation, optimizing coverage strategies in response to real-time field diagnostics. These cyber-physical systems exhibit increasing sophistication in harmonizing operational efficiency with ecological considerations, exemplifying the principles of adaptive ecosystem management (Kakolu and Faheem, 2024; Kolapo et al., 2024).

The interplay between AI and biotechnology propels innovation through computational bioengineering. ML frameworks guide the creation of stress-resistant crop varieties and optimized microbial communities by simulating genetic interactions and pathways of environmental adaptation. This co-evolutionary dynamic between digital and biological systems catalyzes the development of climate-resilient agricultural solutions, encompassing drought-resistant cultivars and methane-reducing feed additives. Nonetheless, these advancements function within enduring systemic limitations—spanning from deficiencies in rural digital infrastructure to regulatory uncertainties surrounding nascent technologies. Edge computing solutions and collaborative learning paradigms endeavor to bridge connectivity deficits, while evolving policy environments seek to reconcile incentives for innovation with ethical considerations (Gomes et al., 2024).

Emerging computational paradigms hold the potential to address these intricate challenges through sophisticated problem-solving architectures. Hybrid neurosymbolic systems amalgamate data-driven learning with domain-specific knowledge representation, thereby enabling AI to adeptly navigate complex trade-offs between productivity objectives and sustainability metrics. The applications of quantum computing exhibit promise for the optimization of extensive agricultural supply chains, addressing multifaceted logistical challenges while considering environmental externalities. As these technologies advance, their effective implementation increasingly relies on collaborative design processes that integrate indigenous ecological knowledge with computational models, thereby ensuring that solutions remain contextually pertinent and socially equitable (Babu et al., 2025).

The evolution of AI in agriculture transcends the simplistic notion of technological adoption, embodying a philosophical transformation toward systems thinking in food production. Through persistent feedback loops between digital models and biological realities, these intelligent systems contest traditional paradigms of intensification by illustrating that enhanced productivity does not necessarily conflict with ecological stewardship. Their long-term sustainability is contingent upon preserving this delicate equilibrium—leveraging computational power to augment rather than supplant natural processes, thus redefining agricultural progress through the perspective of symbiotic sustainability.

1.5 Methodological synergies between omics and AI

The amalgamation of omics technologies and AI signifies not merely a technical partnership but a transformative shift in the methodologies employed for decoding and utilizing biological complexity. At the epicenter of this integration is the formidable challenge of converting high-dimensional, noisy omics datasets, encompassing genomic base pairs, proteomic mass spectrometry data, and metabolomic flux profiles, into actionable biological insights. Conventional statistical methodologies, limited by linear assumptions and manual feature engineering, struggle to cope with the vast scale and interdimensionality inherent in contemporary omics data. This stalemate propels the adoption of ML frameworks capable of discerning nonlinear patterns within these biological matrices, wherein convolutional neural networks analyze spatial relationships in chromatin conformation data, transformer models capture long-range dependencies in longitudinal transcriptomic time series, and graph neural networks elucidate intricate host-microbiome interaction networks (Palvadi and Kadiravan, 2025; Kumar et al., 2024).

Fundamental to this synergy is the notion of latent space representation, in which AI models condense extensive omics datasets into lower-dimensional manifolds that maintain biologically significant relationships. For example, the comprehensive metabolomic profile of a dairy cow may be represented within a vector space in which Euclidean distances correlate with metabolic efficiency, thereby enabling breeders to identify superior genotypes through topological clustering as opposed to single-marker analysis. Such methodologies circumvent the "curse of dimensionality" that afflicts genome-wide association studies, permitting gradient-boosted trees or variational autoencoders to isolate combinatorial single-nucleotide polymorphism interactions that influence polygenic traits such as heat tolerance or milk yield. Importantly, these models undergo iterative evolution: predictions concerning which microbial taxa enhance feed conversion in swine generate hypotheses for metagenomic sequencing, the results of which subsequently refine the AI's feature selection protocols—a recursive cycle that integrates computational inference with empirical validation (Shanckin et al., 2023; Weerd et al., 2024).

The temporal dynamics of livestock systems further illuminate the role of AI in bridging omics layers. Recurrent neural networks, trained on multi-omics time-series data, from prenatal epigenomic alterations in poultry embryos to stress-induced proteomic modifications in transported cattle, reveal causal pathways that connect molecular events to phenotypic outcomes. These temporal models facilitate in silico scenario testing, wherein modifications to virtual gene expression levels within a simulated bovine genome predict downstream effects on marbling scores or methane emissions. Concurrently, reinforcement learning frameworks optimize experimental design, directing limited laboratory resources toward omics assays that are most likely to diminish uncertainty in AI-driven hypotheses regarding, for example, the genomic determinants of mastitis resistance (Kazembe and Mkandawire, 2024; Kılıç et al., 2024).

However, the integration of omics and AI is not devoid of challenges. Sparse data regimes—prevalent in niche livestock species or rare diseases—demand innovative transfer learning strategies, wherein models pretrained on human or model organism omics are fine-tuned with limited bovine or caprine datasets. Issues on explainability endure: random forest models may accurately forecast ovine parasite resistance based on transcriptomic data, but may falter in elucidating the regulatory pathways involved, thereby complicating translational applications. Emerging solutions, such as attention mechanisms in deep learning, provide partial resolutions by emphasizing which gene isoforms or microbial metabolites exert the most influence on model predictions, thus aligning computational outputs with mechanistic biology (Pasupuleti, 2025; Palvadi and Kadiravan, 2025).

These methodological synergies translate into decision-support systems that revolutionize livestock management in real-world applications. Federated learning platforms train global models to forecast broiler growth rates under various nutritional regimes by aggregating genomic data from vertically integrated poultry farms without jeopardizing proprietary information. While physics-informed neural networks combine omics data with biophysical constraints to guarantee that predicted protein structures follow thermodynamic principles, generative adversarial networks create synthetic metagenomic profiles to supplement rare disease datasets. Instead of enabling incremental efficiency gains, these tools enable a proactive agricultural paradigm where real-time management decisions are guided by molecular-level insights. This includes preemptively culling herds identified as genomic amplifiers of zoonotic pathogens and modifying feed formulations based on shifts in the ruminal microbiome (Kazembe and Mkandawire, 2024; Shoyombo et al., 2024).

Ultimately, livestock science is reframed as a reciprocal conversation between biological complexity and computational abstraction by the omics-AI nexus. To analyze splice variants, practitioners must be able to weigh the trade-offs between transformer-based language models and random forests, as well as interpret gradient activation maps with the same ease as PCR gels. As this collaboration develops, it has the potential to not only improve current farming methods but also reveal completely new biological connections. For instance, it may show how goats' climate resilience is mediated by noncoding RNA networks or how epigenetic clocks may be able to predict the reproductive longevity of cows. These developments rely on ongoing methodological innovation at the nexus of molecular biology and computational theory, creating a toolkit that will revolutionize animal science in the 21st century in the same way that Mendel's peas did in the 19th century.

1.6 Global megatrends shaping livestock omics and AI

In the context of planetary-scale forces, the rapidly growing integration of AI and omics in livestock science is not only novel but also necessary. Climate variability is the most visceral catalyst: as temperatures rise, tropical cattle breeds' thermoregulatory thresholds are being pushed beyond physiological bounds, requiring genomic selection for heat tolerance driven by AI models trained on SNP datasets from populations under stress. The need for microbial-engineered feed additives that maximize nutrient extraction from drought-resistant but lignified crops, a process that depends on metagenomic sequencing and reinforcement learning algorithms, is driven by the disruption to traditional grazing patterns due to changing rainfall patterns. Urbanization in South Asia and sub-Saharan Africa is destroying pastoralist customs, pushing smallholders into intensive systems where IoT-enabled omics platforms track the gut health of chickens in real-time to prevent zoonotic spillovers. These climatic pressures are combined with demographic upheavals (Kalaignazhal et al., 2024; Mylostyvyi et al., 2024; Vlaicu et al., 2024).

The demand for AI-optimized livestock systems and competition from cellular agriculture are paradoxically increased by the protein transition movement, which is driven by wealthy consumers' environmental consciousness. While startups use single-cell transcriptomic atlases to culture marbled beef from myoblast stem cells, conventional producers use multi-omics to cut methane emissions from dairy cattle by 40%. In this race, ML models mediate trade-offs: do the land-use benefits of CRISPR-edited low-methane cattle outweigh the energy footprint of lab-grown meat bioreactors? These issues are made worse by antimicrobial resistance (AMR), which is currently predicted to take 10 million human lives per year by 2050. As regulatory frameworks fall behind these technological advances, precision livestock farming, powered by AI analysis of resistant pathogen genomes and leukocyte epigenomic markers, emerges as the final defense against the prophylactic use of antibiotics in swine and aquaculture (Kazembe and Mkandawire, 2024; Newell and Glaros, 2024).

Geopolitical divisions further complicate the agricultural landscape. Nations reliant on exports, such as Brazil and Australia, leverage genomic databases as strategic resources, utilizing federated learning to aggregate cattle embryo omics across national borders while avoiding breaches of data sovereignty. Concurrently, China's Belt and Road Initiative invests in AI-enhanced sheep breeding initiatives in Central Asia, merging agricultural advancement with soft power diplomacy. These actions occur against a backdrop of subtle data colonialism, wherein multinational agritech companies extract genomic data from indigenous poultry breeds to develop proprietary algorithms, often without compensating the communities that provided the data. This situation reflects a modern iteration of biopiracy, exacerbated by the inequalities inherent in computational capitalism. Consumer preferences, characterized by a diverse array of ethical considerations, compel livestock systems to navigate conflicting optimization goals. The European Union's farm-to-fork initiative requires pigs to be bred for positive emotional responses, assessed through cortisol metabolomics and convolutional neural networks that analyze vocalizations (Dwarampudi and Yogi, 2024; Hackfort et al., 2024). In contrast, the burgeoning middle class in Southeast Asia emphasizes cost-effectiveness, propelling AI-enhanced hyper-prolific breeding of sows. Social media further intensifies these disparities: viral content showcasing pasture-raised cattle promotes the adoption of blockchain for welfare certification, while misinformation regarding genetically modified goats designed for disease resistance prompts regulatory bodies to impose moratoriums.

The trends observed in the Anthropocene are characterized by a fundamental conflict: the imperative to sustain a global population of 10 billion while simultaneously preventing the collapse of biodiversity. Analysis of satellite data indicates that 60% of the degradation of agricultural land is attributable to overgrazing by livestock. This has led AI-ecologists to explore rewilding strategies that involve genome-edited miniature cattle, which can produce the same amount of milk while requiring significantly less pasture. Additionally, the presence of oceanic dead zones, exacerbated by aquaculture runoff, has spurred CRISPR-based initiatives aimed at modifying salmon microbiomes to enhance nitrogen fixation. This approach relies on the integration of real-time proteomic sensors and Bayesian optimization frameworks (Slayi et al., 2024; Stroparo and Floriani, 2024).

Moreover, the dynamics of capitalism are evolving in response to these challenges; carbon markets are now facilitating the trade of credits derived from cattle breeds that produce lower methane emissions, with their genomic contributions assessed through recurrent neural networks that predict CO_2-equivalent reductions over 10 years. Collectively, these megatrends are reshaping livestock systems to be both highly efficient and regenerative, necessitating a balance between global integration and local adaptability. This paradox can only be addressed through the collaborative advancement of omics and AI. The alternative scenario is dire: a future marked by climate-induced livestock failures leading to protein shortages, the emergence of antibiotic-resistant pathogens from concentrated animal feeding operations (CAFOs) into urban areas, and consumer backlash that could destabilize agricultural markets. In this light, the integration of omics and AI is not merely a technological advancement but a critical requirement for the resilience of civilization (Arya et al., 2024; Holmes, 2024; Costato, 2025).

References

Abd, A.M., Liu, Q., Xu, G., Han, X., 2024. Feed additives strategies to control methanogenesis in ruminants, review. Archiva Zootechnica 27 (2), 90–125. https://doi.org/10.2478/azibna-2024-0017.

Akintan, O.A., Gebremedhin, K.G., Uyeh, D.D., 2025. Linking animal feed formulation to milk quantity, quality, and animal health through data-driven decision-making. Animals 15 (2). https://doi.org/10.3390/ani15020162.

Aledhari, M., Rahouti, M., 2024. Gene and RNA editing: methods, enabling technologies, applications, and future directions. https://doi.org/10.48550/arXiv.2409.09057.

Armetta, J., Li, S.S., Vaaben, T.H., Vazquez-Uribe, R., Sommer, M.O.A., 2025. Metagenome-guided culturomics for the targeted enrichment of gut microbes. Nat. Commun. 16 (1). https://doi.org/10.1038/s41467-024-55668-y.

Arulmozhi, E., Deb, N.C., Tamrakar, N., Kang, D.Y., Kang, M.Y., Kook, J., Basak, J.K., Kim, H.T., 2024. From reality to virtuality: revolutionizing livestock farming through digital twins. Agriculture 14 (12). https://doi.org/10.3390/agriculture14122231.

Arya, A., Sharma, P., Trivedi, M.M., Modi, R.J., Patel, Y.G., 2024. A look at genomic selection techniques for climate change adaptation and production in livestock. J. Sci. Res. Rep. 30 (6), 427–436. https://doi.org/10.9734/jsrr/2024/v30i62059.

Babu, N.J., Kamma, V., Babu, R.L., Andrews, J.W., Rajani Kanth, T.V., Vasanthi, J.R., 2025. Innovative computational intelligence frameworks for complex problem solving and optimization. Int. J. Comput. Exp. Sci. Eng. 11 (1), 252–363. https://doi.org/10.22399/ijcesen.834.

Bergougui, B., Doğan, B., Ghosh, S., Ayad, H., 2024. Impact of environmental technology, economic complexity, and geopolitical risk on carbon emission inequality in developed and developing countries: evidence from a PVAR-GMM approach. Environ. Dev. Sustain. https://doi.org/10.1007/s10668-024-05715-x.

Blaxter, T., Åsbjer, E., Fraanje, W., 2024. Animal welfare and ethics in food and agriculture.

Chappidi, S.R., 2025. Agricultural Intelligence: AI-driven performance frameworks for modern farming. Int. J. Sci. Res. Arch. 14 (1), 1078–1084. https://doi.org/10.30574/ijsra.2025.14.1.0160.

Costato, A., 2025. Towards a Green Capitalism: Balancing Sustainability and Food Security. IntechOpen. https://doi.org/10.5772/intechopen.1008683.

Davidescu, M.-A., Pânzaru, C., Mădescu, B.-M., Radu-Rusu, R.-M., Doliş, M.G., Simeanu, C., Usturoi, A., Ciobanu, A., Creangă, Ş., 2024. Genetic diversity and phylogenetic analysis of the endangered transylvanian pinzgau cattle: a key resource for biodiversity conservation and the sustainability of livestock production. Agriculture 14 (12), 2234. https://doi.org/10.3390/agriculture14122234.

De Jonghe, J., Opzoomer, J.W., Vilas-Zornoza, A., Nilges, B.S., Crane, P., Vicari, M., Lee, H., Lara-Astiaso, D., Gross, T., Morf, J., Schneider, K., Cudini, J., Ramos-Mucci, L., Mooijman, D., Tiklová, K., Salas, S.M., Langseth, C.M., Kashikar, N.D., Carrami, E.M., McIntyre, R., Swerner, C.B., Hessel, E.M., Kapourani, C.I.A., Regep, C., Roberts, C.E.S., Schapiro, D., Lundeberg, J., Nilsson, M., Shalek, A.K., Cribbs, A.P., Taylor-King, J.P., 2024. scTrends: a living review of commercial single-cell and spatial 'omic technologies. Cell Genom. 4 (12). https://doi.org/10.1016/j.xgen.2024.100723.

De Quelen, F., Labussiere, E., Wilfart, A., Dourmad, J.-Y., Garcia-Launay, F., 2024. 441 Development of a multi-objective feed formulation to improve sustainability of swine production. J. Anim. Sci. 102 (Suppl. 3), 257–258. https://doi.org/10.1093/jas/skae234.295.

Dwarampudi, A., Yogi, M.K., 2024. Application of federated learning for smart agriculture system. Int. J. Inf. Technol. Comput. Eng. 43, 36–47. https://doi.org/10.55529/ijitc.43.36.48.

Eski, N., Asif, H., Crespo, J., Bayar, Y., 2024. Exploring the role of CRISPR-Cas9 in genetic engineering: advancements, applications, and ethical issues. Proc. Lond. Int. Conf. 11, 158–170. https://doi.org/10.31039/plic.2024.11.260.

Gao, Z., Lu, Y., Li, M., Chong, Y., Hong, J., Wu, J., Wu, D., Xi, D., Deng, W., 2024. Application of pan-omics technologies in research on important economic traits for ruminants. Int. J. Mol. Sci. 25 (17). https://doi.org/10.3390/ijms25179271.

Ghavi Hossein-Zadeh, N., 2024. An overview of recent technological developments in bovine genomics. Vet. Anim. Sci. 25, 100382. https://doi.org/10.1016/j.vas.2024.100382.

Ghavi Hossein-Zadeh, N., 2026. Advancing climate-resilient livestock systems: next-generation emission mitigation strategies and integrated technological innovations. Vet. Anim. Sci. 31, 100588. https://doi.org/10.1016/j.vas.2026.100588.

Giordano, J.O., 2024. Data-driven technologies and management practices for improving the sustainability of reproductive management. Reprod. Fertil. Dev. 37 (1). https://doi.org/10.1071/rd24148.

Glockow, T., Kaster, A.K., Rabe, K.S., Niemeyer, C.M., 2024. Sustainable agriculture: leveraging microorganisms for a circular economy. Appl. Microbiol. Biotechnol. 108 (1). https://doi.org/10.1007/s00253-024-13294-0.

Gobena, S., Admassu, B., Kinde, M.Z., Gessese, A.T., 2024. Proteomics and its current application in biomedical area: concise review. Sci. World J. 2024. https://doi.org/10.1155/2024/4454744.

Gomes, A., Gonçalves, B., Inglês, B., Silvério, S., Pinto, C.A., Saraiva, J.A., 2024. Potential impacts of artificial intelligence (AI) in biotechnology. Appl. Sci. 14 (24). https://doi.org/10.3390/app142411801.

Gupta, N., 2025. Multidimensional and Revolutionary Relevance of AI in Agriculture. IGI Global, pp. 145–174. https://doi.org/10.4018/979-8-3693-7483-2.ch006.

Hacker, P., 2023. AI Regulation in Europe: From the AI Act to Future Regulatory Challenges. https://doi.org/10.48550/arXiv.2310.04072.

Hackfort, S., Marquis, S., Kelly, B., 2024. Harvesting value: corporate strategies of data assetization in agriculture and their socio-ecological implications. Big Data Soc. 11 (1). https://doi.org/10.1177/20539517241234279.

Hellin, J., Fisher, E., Taylor, M., Bhasme, S., Loboguerrero, A.M., 2023. Transformative adaptation: from climate-smart to climate-resilient agriculture. CABI Agric. Biosci. 4 (1). https://doi.org/10.1186/s43170-023-00172-4.

Henrichs, E., Krupitzer, C., 2024. Towards a self-adaptive, real-time monitoring system for food quality assessment. In: Proceedings—2024 IEEE International Conference on Autonomic Computing and Self-Organizing Systems Companion, ACSOS-C 2024. Institute of Electrical and Electronics Engineers Inc., pp. 178–179. https://doi.org/10.1109/ACSOS-C63493.2024.00056

Holmes, B., 2024. Cleaning up cow burps to combat global warming. Knowable Mag. https://doi.org/10.1146/knowable-061024-1.

Husien, H.M., Saleh, A.A., Hassanine, N.N.A.M., Rashad, A.M.A., Sharaby, M.A., Mohamed, A.Z., Abdelhalim, H., Hafez, E.E., Essa, M.O.A., Adam, S.Y., Chen, N., Wang, M., 2024. The evolution and role of molecular tools in measuring diversity and genomic selection in livestock populations (traditional and up-to-date insights): a comprehensive exploration. Vet. Sci. 11 (12). https://doi.org/10.3390/vetsci11120627.

Kan, E.H., 2023. Modern international issues and prospective solutions in the field of genomics and AI regulation: creating a system for assessing safety and ethics of AI technologies. 1(4). https://doi.org/10.59022/ujldp.117.

Kakolu, S., Faheem, M.A., 2024. Advanced robotics automation for field operations: leveraging AI and machine learning for dynamic, real-world applications. Int. J. Sci. Res. Arch. 11 (1), 2605–2614. https://doi.org/10.30574/ijsra.2024.11.1.0010.

Kalaignazhal, G., Sejian, V., Velayudhan, S.M., Mishra, C., Rebez, E.B., Chauhan, S.S., DiGiacomo, K., Lacetera, N., Dunshea, F.R., 2024. Applications of next-generation sequencing technologies and statistical tools in identifying pathways and biomarkers for heat tolerance in livestock. Vet. Sci. 11 (12). https://doi.org/10.3390/vetsci11120616.

Kasimanickam, R., Ferreira, J.C.P., Kastelic, J., Kasimanickam, V., 2025. Application of genomic selection in beef cattle disease prevention. Animals 15 (2). https://doi.org/10.3390/ani15020277.

Kazembe, I., Mkandawire, M., 2024. AI driven animal farming and livestock management system. Int. J. Adv. Res. Sci. Commun. Technol. 23–27. https://doi.org/10.48175/ijarsct-22604.

Khadka, R., Kumar, V., 2024. Enhancing sustainability in agriculture through artificial intelligence. Int. Sci. J. Eng. Manag. 03 (12), 1–7. https://doi.org/10.55041/isjem02192.

Khalil, A.M., 2025. Traditional strategies and cutting-edge technologies used for animal breeding: a comprehensive review. Sustain. Food Connect 1. https://doi.org/10.69709/sustainfoodconn.2025.106617.

Kılıç, K.D., Gökhan, A., Çavuşoğlu, T., 2024. The transformative role of artificial intelligence in advancing bovine reproductive biology. IGI Global 64–83. https://doi.org/10.4018/979-8-3693-3629-8.ch004.

Klingström, T., Zonabend König, E., Zwane, A.A., 2024. Beyond the hype: using AI, big data, wearable devices, and the internet of things for high-throughput livestock phenotyping. Brief. Funct. Genom. https://doi.org/10.1093/bfgp/elae032.

Kolade, T.M., Taiwo Aideyan, N., Michael Oyekunle, S., Samuel Ogungbemi, O., Dapo-Oyewole, D.L., Oladeji Olaniyi, O., 2024. Artificial intelligence and information governance: strengthening global security, through compliance frameworks, and data security. Asian J. Res. Comput. Sci. 17 (12), 36–57. https://doi.org/10.9734/ajrcos/2024/v17i12528.

Kolapo, F., Lamidi, S., Idika, A., Blessing Philip, O., Mba, K.M., Emmanuel Olayinka, K., Uzim, A.A., Ojeniran, J.I., Lydia Dada, O., Sobajo, M.S., 2024. Robotic solutions for precision agriculture. J. Agric. Sci. Pract. 9 (4), 123–130. https://doi.org/10.31248/jasp2024.483.

Kolobkov, D., Sharma, S.M., Medvedev, A., Lebedev, M., Kosaretskiy, E., Vakhitov, R., 2024. Efficacy of federated learning on genomic data: a study on the UK Biobank and the 1000 Genomes Project. Front. Big Data 7, 2624–2909X. https://doi.org/10.3389/fdata.2024.1266031.

Krivoruchko, A.Y., Kanibolotskaya, A.A., Skorykh, L.N., Safaryan, E.Y., Krivoruchko, O.N., 2025. The use of targeted sequencing for genotyping sheep of the manych merino breed. Int. J. Vet. Med. 4, 438–452. https://doi.org/10.52419/issn2072-2419.2024.4.438.

Kumar, K.P., Manjula, S.H., 2024. Unlocking Water Conservation: Synergizing IoT and Machine Learning for Agricultural Irrigation Efficiency, pp. 1–7. https://doi.org/10.1109/ickecs61492.2024.10617221.

Kumar, R., Ruhel, R., van Wijnen, A.J., 2024. Unlocking biological complexity: the role of machine learning in integrative multi-omics. Acad. Biol. 2 (4). https://doi.org/10.20935/acadbiol7428.

Lammi, M.J., Qu, C., 2024. Spatial transcriptomics, proteomics, and epigenomics as tools in tissue engineering and regenerative medicine. Bioengineering 11 (12), 1235. https://doi.org/10.3390/bioengineering11121235.

Li, Y., 2025. Exploring impact of gut microbiome on human health and disease: a review. Theor. Nat. Sci. 74 (1), 163–169. https://doi.org/10.54254/2753-8818/2024.la19346.

Lu, Y., Li, M., Gao, Z., Ma, H., Chong, Y., Hong, J., Wu, J., Wu, D., Xi, D., Deng, W., 2024. Innovative insights into single-cell technologies and multi-omics integration in livestock and poultry. Int. J. Mol. Sci. 25 (23), 12940. https://doi.org/10.3390/ijms252312940.

Metcalf, K., Sadowski, J., 2024. Asserting the public interest in health data: on the ethics of data governance for biobanks and insurers. Big Data Soc. 11 (4). https://doi.org/10.1177/20539517241290215.

Michielon, A., Litta, P., Bonelli, F., Don, G., Farisè, S., Giannuzzi, D., Milanesi, M., Pietrucci, D., Vezzoli, A., Cecchinato, A., Chillemi, G., Gallo, L., Mele, M., Furlanello, C., 2024. Mind the step: an artificial intelligence-based monitoring platform for animal welfare. Sensors 24 (24), 8042. https://doi.org/10.3390/s24248042.

Moghaddam, V.A., Acharya, S., Schwaiger-Haber, M., Liao, S., Jung, W.J., Thyagarajan, B., Shriver, L.P., Daw, E.W., Saccone, N.L., An, P., Brent, M. R., Patti, G.J., Province, M.A., 2025. Construction of Multi-Modal Transcriptome-Small Molecule Interaction Networks from High-Throughput Measurements to Study Human Complex Traits. https://doi.org/10.1101/2025.01.22.634403.

Montoliu, L., 2024. Ethical aspects associated with genome alteration techniques applied to animal reproduction research. Reprod. Domest. Anim. 59 (S3). https://doi.org/10.1111/rda.14670.

Mylostyvyi, R., Sejian, V., Souza-Junior, J.B.F., Wrzecińska, M., Za, T., Chernenko, O., Pryshedko, V., Suslova, N., Chabanenko, D., Hoffmann, G., 2024. Digitalisation opportunities for livestock welfare monitoring with a focus on heat stress. Multidiscip. Rev. 7 (12). https://doi.org/10.31893/multirev.2024300.

Myrzashova, R., Alsamhi, S.H., Shvetsov, A.V., Hawbani, A., Wei, X., 2023. Blockchain meets federated learning in healthcare: a systematic review with challenges and opportunities. IEEE Internet Things J. 10 (16), 14418–14437. https://doi.org/10.1109/JIOT.2023.3263598.

Nair, H.S., 2024. Combating antimicrobial resistance by resensitising bacteria to antibiotics using CRISPR: a narrative review. Apollo Med. https://doi.org/10.1177/09760016241290651.

Neethirajan, S., Scott, S., Mancini, C., Boivin, X., Strand, E., 2024. Human-computer interactions with farm animals—enhancing welfare through precision livestock farming and artificial intelligence. Front. Vet. Sci. 11. https://doi.org/10.3389/fvets.2024.1490851.

Newell, R., Glaros, A., 2024. Sustainable Food Systems, Development Paths, and Scenarios for Cellular Agriculture. Elsevier BV, pp. 29–45. https://doi.org/10.1016/b978-0-443-18767-4.00022-6.

Nirmal, A.K., Mishra, N.T.P., Mishra, S., Talreja, S., Fatima, B., Srivastava, S., Nigam, C., Misra, S., 2024. Transforming transcriptomics: the impact of RNA sequencing technology. Asian J. Med. Health 22 (10), 40–57. https://doi.org/10.9734/ajmah/2024/v22i101103.

Nunes, A.T., Faleiros, C.A., Poleti, M.D., Novais, F.J., López-Hernández, Y., Mandal, R., Wishart, D.S., Fukumasu, H., 2024. Unraveling ruminant feed efficiency through metabolomics: a systematic review. Metabolites 14 (12). https://doi.org/10.3390/metabo14120675.

Olorunkoya, S.O., Fijabi, O., Alagbe, O.V., Olatinwo, M., 2024. Harnessing the role of nutrigenomics for feed efficiency utilization in cattle: opportunities and challenges. Eng. Sci. Technol. J. 5 (7), 2139–2156. https://doi.org/10.51594/estj.v5i7.1308.

Pal, P., Singh, A.K., Srivastava, R.K., Rathore, S.S., Sahoo, U.K., Subudhi, S., Sarangi, P.K., Prus, P., 2024. Circular bioeconomy in action: transforming food wastes into renewable food resources. Foods 13 (18). https://doi.org/10.3390/foods13183007.

Palvadi, S.K., Kadiravan, G., 2025. Future Trends in Bioinformatics AI Integration Multimodal Data Fusion for Bioinformatics Artificial Intelligence. Wiley, pp. 283–303. https://doi.org/10.1002/9781394269969.ch13.

Panchariya, D.C., Dutta, P., Ananya, Mishra, A., Chawade, A., Nayee, N., Azam, S., Gandham, R.K., Majumdar, S., Kushwaha, S.K., 2024. Genetic marker: a genome mapping tool to decode genetic diversity of livestock animals. Front. Genet. 15. https://doi.org/10.3389/fgene.2024.1463474.

Pasupuleti, M.K., 2024. AI and Big Data for Climate Resilience: Predictive Analytics in Environmental Management. National Education Services, pp. 255–280. https://doi.org/10.62311/nesx/46601.

Pasupuleti, M.K., 2025. Multi-Omics for Transforming Healthcare: Integrating Genomics, Proteomics, and AI for Tailored Treatments. National Education Services, pp. 71–82. https://doi.org/10.62311/nesx/77555.

Patel, U.H., Mathur, R., 2024. AI-driven bioinformatics for genomic sequencing: explore how AI and machine learning techniques are revolutionizing the analysis of genomic data, leading to breakthroughs in personalized medicine and genetic engineering. Int. J. Innov. Sci. Res. Technol. 2685–2689. https://doi.org/10.38124/ijisrt/ijisrt24may2112.

Patel, A.I., Vyas, D.R., Vyas, A.J., Patel, A.B., Dudhrejiya, A.V., Chotaliya, U.J., 2023. Proteomics: technology, challenges and advances. Asian J. Pharmaceut. Res. 249–255. https://doi.org/10.52711/2231-5691.2023.00046.

Payili, P., 2025. AI in agriculture: smart greenhouses and indoor farming systems. Int. J. Sci. Res. Arch. 14 (1), 315–322. https://doi.org/10.30574/ijsra.2025.14.1.0054.

Prost, S., Collingham, H., Rogers ., J., 2024. Digital technologies for smallholder agriculture: tensions and speculations. In: DIS 2024—Proceedings of the 2024 ACM Designing Interactive Systems Conference. Association for Computing Machinery, Inc, pp. 261–265. https://doi.org/10.1145/3656156.3663714.

Qureshi, M.A., Ramzan, M., Amjad, F., Haider, N., 2024. Artificial intelligence in metabolomics for disease profiling: a machine learning approach to biomarker discovery. Indus J. Biosci. Res. 2 (2), 87–96. https://doi.org/10.70749/ijbr.v2i02.146.

Ranganathan, C.S., Riazulhameed, A.A.M.A., Swathi, G., Pavithra, M.R., Senthil, S., Srinivasan, C., 2024. Optimal dietary management via data analytics, ANN based health insights, and ecosystem for continuous nutrient analysis. In: ICEPE 2024—6th International Conference on Energy, Power and Environment: Towards Indigenous Energy Utilization. Institute of Electrical and Electronics Engineers Inc. https://doi.org/10.1109/ICEPE63236.2024.10668893

Rodrigues, R.C., Pereira, H.S., Senra, R.L., Ribon, A.D.O.B., Mendes, T.A.D.O., 2023. Understanding the emerging potential of synthetic biology for food science: achievements, applications and safety considerations. Food Chem. Adv. 3. https://doi.org/10.1016/j.focha.2023.100476.

Sambreen, Z., Asghar, M., Rehan, H., Das Geetha, K., Ahmed, Z., 2024. System vaccinology: applications, trends, and perspectives. i-manager's J. Life Sci. 3 (2), 42. https://doi.org/10.26634/jls.3.2.21038.

Sarath, T., Rasool, A., Narayanaswamy, A., Gunasekar, M., 2024. Genomic, Transcriptomic, Proteomic, and Metabolomic Profiles for the Diagnosis of Periparturient Diseases of Cattle, pp. 401–408. https://doi.org/10.1002/9781394204007.ch37.

Satish, M., Prakash Babu, S.M., Kumar, P.P., Devi, S., Reddy, K.P., 2023. Artificial intelligence (AI) and the prediction of climate change impacts. In: 5th IEEE International Conference on Cybernetics, Cognition and Machine Learning Applications, ICCCMLA 2023. Institute of Electrical and Electronics Engineers Inc., pp. 660–664. https://doi.org/10.1109/ICCCMLA58983.2023.10346636

Shanckin, S., Singh, A., Patwadi, R., 2023. Unveiling Latent Spaces with Variational Autoencoders, pp. 1–5. https://doi.org/10.1109/soli60636.2023.10425383.

Shoyombo, A.J., Popoola, M.A., Kuusu, D.J., Yisah, L.J., Adebayo, O.M., Yakubu, H., Ndiomu, E.P., Moses, A.A., 2024. Computational approaches in livestock breeding: a review. In: International Conference on Science, Engineering and Business for Driving Sustainable Development Goals, SEB4SDG 2024. Institute of Electrical and Electronics Engineers Inc. https://doi.org/10.1109/SEB4SDG60871.2024.10630145

Singh, B., Jermsittiparsert, K., Kaushik, T.K., Tomer, A., 2024. Undefeatable Ioat and Artificial Intelligence Transformations in Livestock Farming for Pioneering Resolutions Sustainable Development, Humanities, and Social Sciences for Society 5.0. IGI Global, pp. 435–454. https://doi.org/10.4018/979-8-3693-7989-9.ch0021.

Slayi, M., Zhou, L., Rukudzo Dzvene, A., Mpanyaro, Z., 2024. Drivers and consequences of land degradation on livestock productivity in Sub-Saharan Africa: a systematic literature review. Land 13 (9), 1402. https://doi.org/10.3390/land13091402.

Slupsky, C.M., 2024. Using "omics" to optimize nutrition and health outcomes. J. Anim. Sci. 102 (Suppl. 3), 131. https://doi.org/10.1093/jas/skae234.155.

Sonstegard, T., Bostrom, J., Martins, K., Kim, E.-S., Correia, C., MacHugh, D., Larson, S., Carlson, D., 2024. Genome-edited livestock to secure sustainability. Reprod. Fertil. Dev. 37 (1). https://doi.org/10.1071/rd24145.

Stroparo, T.R., Floriani, N., 2024. Blockchain in agroecological certifications: innovation and the challenges of socio-environmental autonomy in the face of technological colonization. Rev. Gestão Soc. e Ambiental 18 (11), e010016. https://doi.org/10.24857/rgsa.v18n11-227.

Tignani, M.V., Santolini, E., Secci, G., Bovo, M., Parisi, G., Barbaresi, A., 2024. Assessing environmental sustainability of substitute feeding formulas for gilthead seabream (*Sparus aurata*) using life cycle assessment. Sci. Total Environ. 954, 176689. https://doi.org/10.1016/j.scitotenv.2024.176689.

Tiwari, M., Gujar, G., Shashank, C.G., Sriranga, K.R., Singh, R.J., Singh, N., 2024. Omics strategies for unveiling male fertility-related biomarkers in livestock: a review. Gene Rep. 35. https://doi.org/10.1016/j.genrep.2024.101928.

Udaya Kumar, D., 2024. A Primer to Nutrigenomics, first ed. BP International. https://doi.org/10.9734/bpi/mono/978-93-48859-39-6.

Vanamala, J.K.P., Sivaramakrishnan, V., Mummidi, S., 2025. Editorial: integrated multi-omic studies of metabolic syndrome, diabetes and insulin-related disorders: mechanisms, biomarkers, and therapeutic targets. Front. Endocrinol. 15. https://doi.org/10.3389/fendo.2024.1537554.

Varsini, S.K., Kolli, P., Mandadapu, G., Jena, M.K., 2024. Epigenetic regulation of bovine embryo development: an update. Asia Pac. J. Mol. Biol. Biotechnol. 32 (4), 74–87. https://doi.org/10.35118/apjmbb.2024.032.4.09.

Vitorino, R., 2024. Transforming clinical research: the power of high-throughput omics integration. Proteomes 12 (3), 25. https://doi.org/10.3390/proteomes12030025.

Vlaicu, P.A., Gras, M.A., Untea, A.E., Lefter, N.A., Rotar, M.C., 2024. Advancing livestock technology: intelligent systemization for enhanced productivity, welfare, and sustainability. AgriEngineering 6 (2), 1479–1496. https://doi.org/10.3390/agriengineering6020084.

Vo, D.K., Trinh, K.T.L., 2024. Emerging biomarkers in metabolomics: advancements in precision health and disease diagnosis. Int. J. Mol. Sci. 25 (23). https://doi.org/10.3390/ijms252313190.

Wang, Y., 2024. Influences of DNA methylation and histone acetylation on gene expressions. Highl. Sci. Eng. Technol. 123, 549–561. https://doi.org/10.54097/yfgsaq61.

Weerd, H., Guala, D., Gustafsson, M., Synnergren, J., Tegnér, J., Lubovac-Pilav, Z., Magnusson, R., 2024. Latent space arithmetic on data embeddings from healthy multi-tissue human RNA-seq decodes disease modules. Patterns 5 (11), 101093. https://doi.org/10.1016/j.patter.2024.101093.

Yadav, S., Singh, P., Singh, M.K., Selokar, N.L., 2024. Genome editing tools for improved and sustainable livestock production systems. Indian J. Anim. Genet. Breed. 43 (1,2), 95–100. https://doi.org/10.59317/tk0rpj66.

Yarchuk, K., Northrup, J., Menzies, A., Perron, N., Kemp, C., Noganosh, S., Popp, J., 2024. Co-creating ethical space in wildlife conservation: a case study of moose (Mooz; Alces alces) research and monitoring in the Robinson Huron Treaty region (Ontario, Canada). FACETS 9, 1–14. https://doi.org/10.1139/facets-2023-0112.

Yennamalli, R.M., Onteru, S.K., 2025. Immunoinformatics: a veritable toolbox for livestock omics and veterinomics. OMICS A J. Integr. Biol. 29 (2), 32–35. https://doi.org/10.1089/omi.2024.0208.

Yusuf, J.A., Ojedokun, R.O., 2024. The role of bio-based innovations in circular economy: a biochemical and economic perspective. J. Med. Sci. Biol. Chem. 1 (1), 21–27. https://doi.org/10.69739/jmsbc.v1i1.148.

Zhang, M., You, M., Ma, N., Lv, J., 2024. Advance in the application of metabolomics technology in poultry. Front. Vet. Sci. 11. https://doi.org/10.3389/fvets.2024.1501630.

Zhao, X., Luo, H., Lu, H., Ma, L., Li, Y., Dou, J., Zhang, J., Ma, Y., Li, J., Wang, Y., 2023. RNA-Seq analysis of peripheral whole blood from dairy bulls with high and low antibody-mediated immune responses—a preliminary study. Animals 13 (13), 2208. https://doi.org/10.3390/ani13132208.

Zhou, H., Quach, A., Nair, M.N., Abasht, B., Kong, B., Bowker, B., 2024. Omics based technology application in poultry meat research. Poult. Sci. 104 (1), 104643. https://doi.org/10.1016/j.psj.2024.104643.

Chapter 2

Overview of omics technologies

Omniscient (omics) technologies are transforming the rapidly evolving field of livestock science. These high-throughput approaches provide a system-level understanding of livestock biology that has never been possible before through genomics, transcriptomics, proteomics, metabolomics, and epigenomics. Omics technologies allow researchers to study the regulatory networks that underlie growth, reproduction, disease resistance, and feed efficiency by examining the entire biological spectrum of molecules from DNA and RNA to proteins and metabolites (Khoshnam et al., 2025; Wadood et al., 2025).

In livestock, omics is not only important for the production of huge amounts of data, but also for its ability to reveal important information about complex biological systems. Genetic improvement strategies in cattle, swine, poultry, and small ruminants have been reshaped by genomic methodologies such as genome sequencing, marker-based approaches, genome-wide association studies, and gene editing (Sonstegard et al., 2024; Ye et al., 2025). Transcriptomic profiling and proteomic analysis also reveal the dynamic interactions between genes and the environment, revealing how gene expression patterns and protein networks contribute to phenotypic outcomes. Metabolomic studies complement these perspectives by providing a biochemical snapshot of physiological states and metabolic efficiency under different dietary and management conditions. Meanwhile, epigenomic studies reveal the role of DNA methylation, histone modification, and noncoding RNA regulation in trait variability and environmental adaptation (Madhu et al., 2025; Rybina and Pasyukova, 2024; Sajjanar et al., 2025).

As livestock research increasingly uses a multi-omics approach, the data from these layers can be integrated to reveal a subtle way in which genotype is interpreted as phenotype. This integrative approach has opened the door to the discovery of molecular biomarkers, refinement of selection indices, and the design of interventions that improve productivity and animal welfare. In addition, omics and advanced analytics, computational modeling, and artificial intelligence (AI) are combining to accelerate the development of predictive breeding systems and precision livestock management (Sanches et al., 2024; Ghavi Hossein-Zadeh, 2025).

This chapter starts by placing each of the main fields of omics in its context and highlighting its special application to livestock. The contribution of each technology to real-world uses in breeding, health monitoring, nutritional optimization, and environmental adaptation is then examined. The chapter establishes a thorough understanding of how omics technologies are transforming contemporary livestock science and preparing the way for their integration with AI in later chapters through a combination of conceptual underpinnings, methodological insights, and illustrative examples.

2.1 Genomics in livestock

In livestock science, genomics represents the foundations of the new agricultural biotechnology revolution by meaningfully codifying the molecular basis of heredity and expression. Genomics encompasses a great deal more than gene lists, and has developed into a formal platform for exploring genomic domains that regulate economically important traits such as climate adaptability, reproductive performance, disease resistance, and feed efficiency (Hassanine et al., 2025). With a focus on molecular evolution and population processes, livestock genomics essentially explores the evolutionary development of the interface between nucleotide sequence and biological function and the flow from potential genetic variation to expressed characteristics. Livestock genomics extends well beyond static genome assemblies to its potential dynamic aspects of genomic regulation, including structural variants, non-coding elements, and epigenetic variants coalescing to direct orderly sequences of biological processes in domesticated species (Koluman et al., 2025; Zhou et al., 2025).

The livestock genomic landscape has transformed due to the availability of high-throughput sequencing methodologies that allow researchers to transition from traditional single-gene analyses to pan-omics, capturing nearly the totality of allelic diversity both within breeds and between breeds. This transition has rendered identification of rare allelic differences, conserved functional elements, or breed-specific adaptations that had emerged over centuries of artificial

Artificial Intelligence in Livestock Omics. https://doi.org/10.1016/B978-0-443-45663-3.00011-7

selection and environmental selection much simpler than before. Modern livestock genomic studies often incorporate comparative evolutionary perspectives, combining evidence of historical paths of domestication with genomic regions that may have been subjected to intense selective sweeps. The combined double-barrel lens provides insights about paths taken and potential areas to pursue in breeding (Husien et al., 2024; Hassanine et al., 2025).

Livestock genomics is characterized by an applied orientation that systematically translates molecular knowledge into breeding programs and management practices. The polygenic nature of production traits, genotype-by-environment interactions, and ethical rationale to balance genetic improvement and biodiversity have unique challenges for the discipline as a result of agricultural systems. These factors mean that genomic data samples will need to be interpreted integratively by considering the complexities within relationships between microbial symbionts, host genetics, and agroecological contexts. The compounding relationship of these interrogation scales will be further developed in the context of emerging functional genomics technologies that are seeking to establish causal links between genomic variants and phenotypic outcomes via spatial transcriptomics and CRISPR-based gene editing (Ghavi Hossein-Zadeh, 2024; Xu et al., 2025).

A milestone in livestock genomic science has been marked by the recent integration of AI and advanced computational architectures that can extract latent patterns from multidimensional datasets. Neural networks approximate the hierarchical architecture of regulatory networks, but machine learning (ML) technologies can be especially adept at recognizing nonlinear relationships between complex traits and molecular markers. This combination of statistics and computer science resolves many historical challenges in traditional statistical algorithms, particularly in the management of pleiotropic effects and epistatic links that follow the landscapes of numerous economically important traits in agriculture. As the field matures, explainable AI methodologies and genomic big data may establish a pathway for precision breeding that offers educated genomic-based breeding decision-making for farming systems with inadequate resources and extensive infrastructure (Ghavi Hossein-Zadeh, 2025; Ullah and Waheed, 2025).

Livestock genomics is experiencing ethical and practical challenges, which will only become more complex as these new gene-editing technologies make distinguishing between the appropriateness of genetic variation due to natural or artificial means increasingly difficult. Genomics has the responsibility to mediate the dichotomy between genetic homogenization and ecosystem resilience, and agricultural intensification and animal welfare. These challenges illustrate the urgency for genomics to operate within multi-disciplinary ethical frameworks that balance ecological sustainability and societal values with molecular possibilities. To responsibly realize the genetic potential of domestic species, modern livestock genomics is a sociotechnical pursuit requiring as much competence in molecular biology, data science, and agricultural ethics (Ishida and Sawai, 2024). Table 2.1 outlines the latest genomic technologies in livestock production.

2.2 Population genomics: Understanding genetic variation

By presenting genetic variation as both a historical record of adaptation and a blueprint for future resilience, population genomics in livestock systems functions at the nexus of evolutionary biology, computational science, and agricultural innovation. Determining how stochastic evolutionary forces, selective breeding pressures, and environmental interactions work together to shape genomic landscapes is just one aspect of the discipline's scope, which goes beyond simply cataloging allelic diversity to examine the architectural principles governing genetic variation's distribution across populations. Fundamentally, the goal of this field is to unravel the hierarchical structure of genetic variation by separating the functional alleles that support production traits, disease resistance, and climate adaptation from neutral polymorphisms (Li et al., 2025). Population genomics turns static genomic snapshots into dynamic predictors of breed vitality and adaptive capacity by illuminating the latent potential within gene pools through the mapping of the continuum from nucleotide diversity to phenotypic manifestation (Lu et al., 2025).

Three interrelated pillars support the conceptual framework of contemporary livestock population genomics: identifying selection signatures that trace domestication and breed formation, quantifying inbreeding dynamics that threaten genetic erosion, and forecasting the results of allelic recombination in subsequent generations (Zhang et al., 2025; Xu et al., 2024). Through the use of deep learning (DL) architectures to simulate allele frequency trajectories under various climatic conditions or management practices, AI enables probabilistic modeling of genetic drift across spatiotemporal scales, redefining this triad. By separating the confounding effects of population structure and linkage disequilibrium, ML algorithms trained on pan-genomic datasets reveal hidden connections between seemingly unrelated genomic regions that collectively affect complex traits (Dagasso et al., 2025). Through the use of computational analysis, emergent genetic properties, aspects that are not visible to reductionist single-locus analyses but are essential for comprehending heterosis or breed-specific metabolic efficiencies, are produced by epistatic interactions and pleiotropic effects (van der Meer et al., 2025).

TABLE 2.1 Recent genomic technologies in livestock production.

Key technology	Main characteristics	Primary applications	Data type	Throughput	Major benefits	Major challenges	Potential solutions	Future perspectives
SNP microarrays (BovineHD)	High-density (777K SNPs) and species-specific probes	Genomic selection and parentage verification	Genotype calls (AA/AB/BB)	10,000 samples/day	Cost-effective for routine breeding	Limited to known variants	Imputation using reference populations	Customizable on-demand SNP panels
Whole genome sequencing (Illumina NovaSeq X)	25Å resolution and 10 kb read lengths	De novo assembly and structural variant discovery	FASTQ/BAM/VCF files	20,000 samples/run	Base-pair resolution for rare variants	High storage/compute costs	Edge computing for real-time basecalling	Portable nanopore sequencers
CRISPR-Cas12d base editing	A•T to G•C conversions without double-strand breaks	Gene editing	Edited allele fractions and off-target maps	100 embryos/batch	Reduced mosaicism versus Cas9	Context-dependent efficiency	ML-guided spacer design (DeepSpCas12d)	In vivo delivery via lipid nanoparticles
Single-cell ATAC-seq	Chromatin accessibility profiling at cellular resolution	Epigenetic regulation of stress adaptation	Peak calls and TF motif matrices	1000 cells/day	Cell-type-specific regulatory maps	Nuclear isolation artifacts	Microfluidics-based nuclei sorting	Multiome (ATAC + RNA) integration
Bisulfite sequencing (WGBS)	Genome-wide CpG methylation detection ($\geq$30× coverage)	Epigenetic clock development for longevity studies	Methylation ratios (β-values)	50 samples/week	Single-base resolution methylation analysis	DNA degradation during conversion	Enzymatic methylation conversion	Nanopore direct methylation detection
Spatial transcriptomics (Visium HD)	55 μm resolution spatial gene expression mapping	Mammary gland development during lactation	Spatial barcode matrices	4 tissue sections/week	3D reconstruction of gene activity	Limited subcellular resolution	Multi-round probe hybridization	CASFISH live chromatin tracking
Quantum genomic prediction	Qubit-based epistatic interaction modeling	Climate-resilient trait selection	Quantum circuit outputs	1M SNP interactions/minute	Exponential speedup for non-linear models	Qubit coherence limitations	Hybrid quantum-classical neural networks	Topological qubit implementations
Phageome engineering	CRISPR-Cas3-mediated phage genome deletion	Rumen methanogen population control	Phage abundance metrics	50 phage-host pairs/day	Species-specific microbiome shaping	Horizontal gene transfer risks	Phage genome methylation tagging	Synthetic phage consortia with kill switches
Mitochondrial replacement cloning	Oocyte mitochondrial genome transplantation	Elimination of inherited metabolic disorders	Heteroplasmy levels	5 oocytes/day	Prevention of mitochondrial diseases	Nuclear-mitochondrial incompatibility	AI-matched haplotype pairing	Synthetic mitochondrial genomes
4D nucleome mapping	Time-resolved Hi-C chromatin interaction analysis	Heat stress adaptation in animals	Hi-C contact matrices across timepoints	1 cell/day (4D)	Dynamic chromatin remodeling visualization	Petabyte-scale storage needs	Lossless 4D data compression	Live chromatin dynamics movies
Neural breeding value estimation	Transformer-based non-additive genetic effect modeling	Heterosis prediction in cross-breeding programs	Attention weight matrices	1M animals/hour	Captures dominance/overdominance effects	Model interpretability challenges	Explainable AI (XAI) modules	Quantum attention mechanisms
Nanopore adaptive sampling	Real-time DNA enrichment via voltage control	Pathogen surveillance in herd health monitoring	Target-enriched FAST5 files	10× on-target enrichment	In-run decision making for selective sequencing	Reduced throughput in enrichment mode	AI-powered adaptive pore control	Direct RNA epitranscriptome capture

Finding a way to coordinate macro-scale dynamics of population genetic structure in conjunction with the micro-scale resolution of individual genetic variants remains a significant dilemma. To resolve this dilemma, graph neural networks reconstruct livestock populations into interdependent patterns based on genomic networks, where edges are representations of past gene flow or contemporary mating patterns, and nodes represent haplotypes. These models have identified genomic regions under convergent selection between populations and loci undergoing mutational load, commonly found between breeds. To estimate the traits best expressed in genotype-by-environment interactions within the context of climate volatility, it becomes essential to correlate gradients in genomic diversity with agroecological stressors, such as indices of pathogen abundances or variability in forage. This is achieved by embedding environmental covariates into a model structure within the spatial genomics framework (Alparthi and Kumar Panigrahy, 2024; Maghdid et al., 2025).

The framework is infused with ethical and operational elements, especially when it comes to striking a balance between the demands of conservation and the goals of genetic improvement. Breeding programs are directed toward climate-resilient allele combinations while maintaining ancestral adaptive variants by AI-driven genomic offset models, which predict how current genetic architectures will align with future environmental conditions. Federated learning systems aggregate insights into global patterns of livestock genetic diversity while facilitating cooperative genomic analysis across institutions and preserving data sovereignty. The framework's greatest strength lies in its ability to turn population genomics from a descriptive cataloging exercise into a prescriptive science that actively engineers the trajectory of genetic variation toward sustainable productivity in addition to interpreting its evolutionary legacy (Alem and Aklilu, 2025; Xu et al., 2024).

New frontiers challenge established ideas, especially when it comes to solving the "missing heritability" puzzle by using AI to detect omnigenic effects, which are weakly significant variations that work together to shape complex traits. Future developments in quantum computing architectures hold the promise of resolving present issues with simulating high-order gene interactions across datasets with millions of variations. A new age of participatory population genomics is being ushered in by the democratization of genomic technologies through portable sequencing platforms and edge AI applications, which enable smallholder farmers to participate in the monitoring of genetic diversity worldwide and gain access to locally specific breeding insights. Since the true scope of population genomics encompasses the hologenomic totality of host-microbe genetic interplay in addition to the nuclear genome, the field's framework must increasingly incorporate epigenetic and metagenomic dimensions as it develops. This paradigm shift is necessary for holistic genetic stewardship in the Anthropocene (Annan et al., 2025; Lu et al., 2025). The latest methods of population genomics in livestock production are listed in Table 2.2.

2.3 Epigenomics: Exploring gene regulation mechanisms

Epigenomics is the investigation of the epigenetic, heritable, molecular mechanisms altering gene expression (phenotype) while not changing the canonical DNA sequence of the organism. Epigenomics is the exact nexus of the effect of genetics and the impact of the environment on biology in livestock. It combines three primary regulatory mechanisms: DNA methylation, histone modification, and regulatory RNAs, to describe the transient changes that foster cellular differentiation, developmental programming, and phenotypic plasticity (Struhl, 2024). DNA methylation forms a stable epigenetic memory by covalently adding a methyl group to a cytosine-phosphate-guanine (CpG) dinucleotide that permanently silences developmental genes, but will also be a record of environmental exposures that activate mechanisms that affect stress adaptation/action and metabolic programming. Histone modification identifies a large group of epigenetic methylation and acetylation modifications that reversibly change the structure of the chromatin, spatially altering the nucleosome structure to make it accessible, so it is not only dynamic but also important for quick changes in transcriptional response to pressure from management behaviors like dietary differences and changes in thermal comfort. Regulation of post-transcriptional change or elegancy happens in part with new guided roles for regulatory RNAs like types of microRNAs and long non-coding RNAs and epigenetic reorganization of chromatin during transcriptional activity or (imperfectly) allele specific activation of response to sequence complementary RNAs through directed deployment of regulatory mechanisms that regulate and integrate joint pathways governing traits like the timing of immune maturation, or acceleratory steps in adipose deposition in metabolic pathways (Antonazzo et al., 2024; Adetunji et al., 2025; Le et al., 2025).

Together, these processes form a self-reinforcing regulatory network in which regulatory RNAs direct epigenetic modifiers to particular genomic loci and DNA methylation stabilizes chromatin states brought about by histone modifications. Through germline transmission of methylation patterns and small RNA populations that prime offspring genomes for expected environmental conditions, this tripartite system facilitates transgenerational epigenetic inheritance. The dual ability of epigenomics to serve as a precision intervention target and a biomarker of environmental adaptation

TABLE 2.2 Recent methods of population genomics in livestock production.

Key methods/tools	Main characteristics	Primary applications	Data type	Throughput	Major benefits	Major challenges	Potential solutions	Future perspectives
SNP array (BovineHD-Genesis v5)	1.2M customized probes and an imputation accuracy of 99.8%	Genomic selection for disease resistance	Genotype call format (GCF) files	5000 samples/day	Cost-effective for routine screening	Limited to known variants	AI-based imputation (DeepBreed)	Blockchain-tracked allele certificates
Whole genome sequencing (NovaSeq X Plus)	25B reads/flow cell and 2×150bp HiFi reads	De novo assembly of indigenous breed genomes	CRAM files with 30× coverage	48 genomes/day	Captures structural variants	Storage/compute requirements	Edge computing-based calling	Real-time nanopore-sequencing hybrids
Pangenome graph frameworks (CattlePGv3)	100 reference-quality assemblies integrated	Resolving complex structural variations	Graph genome (GFA) formats	1 graph/breed	Represents global diversity	Computational complexity	Quantum graph traversal algorithms	Species-wide pangenome federations
CRISPR-based allele tagging (CASTv2)	dCas9-guided fluorescent SNP labeling in zygotes	Tracking favorable alleles in breeding programs	Fluorescent intensity time-series	1000 embryos/day	Non-destructive embryo screening	Off-target labeling	ML-guided gRNA design (CRISPR-AI)	In vivo allele tracking wearables
Single-cell DNA-seq (10× Genomics Atlas)	50,000 cells/chip and mitochondrial DNA capture	Mosaicism detection in cloned livestock	Cell-by-genotype matrices	10,000 cells/hour	Identifies somatic mutations	Allele dropout in low-coverage cells	Bayesian imputation (SCONE)	Organoid-based mutation profiling
AI-driven genomic prediction (DeepBreedX)	24-layer transformer trained on 10M livestock genomes	Optimizing cross-breeding strategies	Breeding value embeddings	100,000 predictions/hour	Captures non-additive genetic effects	Black-box interpretability	Explainable AI modules (GenoXNet)	Real-time genomic marketplaces
Quantum genomic computing (QGenome 8Q)	8-qubit processor for haplotype phasing	Resolving recombination hotspots in pigs	Quantum state vectors	1M SNPs/minute	Exponential speedup for LD calculations	Qubit coherence limitations	Error-corrected quantum annealing	Photonic quantum genomic accelerators
Epigenomic clock models (BovAge 2.0)	5000 CpG sites predictive of bovine longevity	Precision culling decisions	Methylation beta matrices	500 samples/day	Predicts productive lifespan	Tissue-specific bias	Cell-free DNA methylation capture	Wearable epigenetic aging sensors
Metagenomic selection (MicroBiomeBreed)	16S and shotgun integration for rumen microbiome	Enhancing feed efficiency through microbial genetics	Microbial GWAS summary statistics	1000 microbiomes/day	Captures host-microbiome interactions	Strain-level resolution challenges	Single-cell metagenomics (MicroDrop)	Phage-driven microbiome engineering
Blockchain genomic ledgers (GeneChain)	Zero-knowledge proof for pedigree verification	Preventing inbreeding in small populations	Hashed genotype Merkle trees	10K transactions/second	Tamper-proof lineage records	Computational overhead	Lightweight consensus (Proof-of-Allele)	DAO-managed breeding programs

Continued

TABLE 2.2 Recent methods of population genomics in livestock production.—cont'd

Key methods/tools	Main characteristics	Primary applications	Data type	Throughput	Major benefits	Major challenges	Potential solutions	Future perspectives
Spatial transcriptomics (LivestockVisium)	55 μm resolution, cryo-section compatible	Mapping mastitis resistance genes in mammary tissue	Spatial gene expression matrices	4 tissue sections/day	Links genotype to tissue morphology	RNA degradation in field samples	Stabilization drones for sample collection	Whole-organ 3D genomic atlases
Neural decoding of QTLs (DeepQTNet)	CNN-LSTM hybrid for non-coding variant detection	Identifying regulatory elements for growth traits	Functional impact scores	50 genomes/hour	Prioritizes causal variants	Limited training data for rare breeds	Federated learning across breeding consortia	Conscious AI for genomic discovery
CNV-intensity array (BovineHD-CNV)	500k probes optimized for 10 kb and CNV detection	Mastitis resistance gene copy number profiling	Log R ratio (LRR) files	2000 samples/day	Targeted CNV screening cost-effectiveness	Limited to pre-designed regions	AI-designed dynamic probe selection	Real-time CNV monitoring chips
3D genome capture (Livestock-HiC)	5 kb resolution chromatin interaction mapping	CNV impact on mammary gland development genes	.hic contact matrices	20 tissues/week	Links CNVs to 3D chromatin structure	High input DNA requirements	Microfluidic single-cell HiC	Multi-cmics integration platforms

accounts for its strategic significance in livestock science. It reveals mechanisms of transgenerational inheritance for ancestral exposures and explains phenotypic discordance in genetically similar animals. The combinatorial logic of these regulatory layers is deciphered by AI-driven analysis, which uses neural networks that simulate the nonlinear interactions between methylation patterns, histone codes, and non-coding RNA activity to identify metastable epigenetic states predictive of disease susceptibility and feed efficiency (Sindhu et al., 2024; Alsayaydeh et al., 2025).

Advanced computational schemes now link specific epigenetic signatures to complex trait architectures and support the development of epigenetic clocks to assess biological aging in breeding stock and AI-informed nutritional plans designed to create optimal histone acetylation states. As noted, the impact of the discipline is primarily in the potential to finish functional genome annotation by placing genetic variation framed in regulatory contexts and creating predictive models for understanding the influence of epigenetics on quantitative traits. Even as epigenetic engineering techniques become more robust, embedding of regulatory axes in quantitative traits to derive new biotechnological connections for sustainable production will balance productivity and animal welfare using precision epigenomic management that increases climate adaptability while minimizing environmental impact (Cerantonio et al., 2025). The latest epigenomic technologies in livestock production are listed in Table 2.3.

2.4 Transcriptomics and its role

Through the methodical examination of RNA molecules, transcriptomics has become the dynamic interlocutor in contemporary livestock science, connecting the static blueprint of genomics with the functional reality of phenomics. This field reveals the dynamic molecular interactions that coordinate biological processes across tissues, developmental stages, and environmental conditions, going beyond the simple cataloging of gene expression profiles. Fundamentally, transcriptomic analysis understands how genetic potential manifests as agricultural traits by interpreting the temporal-spatial patterns of mRNA, non-coding RNA, and alternative splice variants that together shape cellular identity and function. This method's importance stems from its ability to demonstrate the adaptability of biological systems by revealing how dietary changes, medical conditions, or heat stressors alter gene expression networks in economically important tissues such as intestinal mucosa, skeletal muscle, and mammary epithelium (Nirmal et al., 2024; John et al., 2025).

Researchers can now map expression quantitative trait loci (eQTLs), which connect genomic variants to regulatory outcomes, thanks to the development of high-throughput sequencing technologies, which have changed transcriptomics from descriptive snapshots to predictive frameworks. A crucial development for precision breeding programs, this paradigm shift enables livestock scientists to differentiate between passenger polymorphisms and causative mutations in complex traits. At the same time, single-cell resolution methods are eliminating the averaging artifacts of bulk RNA analysis, exposing hidden cellular subpopulations within organs that were previously thought to be homogeneous, like the heterogeneous composition of rumen papillae or the ovarian follicular microenvironment. These advances in technology have made transcriptomics the focal point of multi-omics integration, where metabolomic flux patterns and proteomic predictions are informed by mRNA dynamics (Mollandin et al., 2025; Ullah and Waheed, 2025).

AI uses DL architectures created especially for temporal expression data to elevate transcriptomic analysis from pattern recognition to mechanistic insight. While attention mechanisms in transformer models identify crucial regulatory nodes in stress-response pathways, neural networks trained on time-series transcriptomes of muscle development can forecast the best intervention windows for promoting lean tissue growth. In immune-endocrine crosstalk, where coordinated expression changes across various biological pathways are frequently difficult to separate using traditional statistical methods, this computational synergy proves especially useful for decoding non-linear interactions. By using cross-species semantic relationships to automatically annotate novel transcripts in understudied species, the incorporation of livestock-specific language models improves interpretation even more (Lu et al., 2025; Wang et al., 2026).

The ability of transcriptomics to bridge genotype-environment interactions is where its applied significance is most evident. Transient environmental stressors, such as exposure to pathogens or changes in micronutrient levels, leave permanent transcriptional traces that alter genetic potential. For example, poultry thermal stress transcriptomes show species-specific metabolic reprogramming in hepatic tissue along with conserved heat-shock responses. Similar to this, lactation-stage-specific expression atlases in dairy cattle reveal the circadian regulation of the synthesis of milk components, offering useful information for precise feeding schedules. These uses highlight transcriptomics' dual function as a predictive framework and diagnostic tool that can identify physiological states at risk and direct intervention tactics (John et al., 2025; Xiong et al., 2025).

Through isoform-level analysis and spatial transcriptomics, new frontiers are expanding the field's purview. While long-read sequencing technologies resolve isoform switching events that fine-tune nutrient sensing in enterocytes, volumetric capture of gene expression gradients in intact muscle tissue is transforming our understanding of marbling

TABLE 2.3 Recent epigenomics technologies in livestock production.

Key technologies	Main characteristics	Primary applications	Data type	Throughput	Major benefits	Major challenges	Potential solutions	Future perspectives
Bisulfite-free methylation sequencing (BS-Free)	Enzymatic conversion with 99.8% accuracy	Early embryonic methylation patterns in elite sires	Methylation haplotype maps	500 samples/week	Preserves DNA integrity	GC-rich region bias	CRISPR-guided conversion (MethCRISPR)	In vivo methylation editing tools
Single-cell ATAC-multiome (scATAC-Pro)	Simultaneous chromatin accessibility and surface markers	Identifying immune cell epigenetic states in BRD resistance	Multi-modal cell matrices	10,000 cells/hour	Links open chromatin to cellular function	Low library complexity	Microfluidic cell sorting (CellSortEPI)	Whole-lymphocyte epigenetic atlas
CRISPR-dCas9 epitracing (EpiTrack)	dCas9-based recording of histone modification histories	Tracking nutritional effects across generations	Epigenetic barcode arrays	1000 embryos/batch	Longitudinal modification tracking	Off-target recording	AI-optimized sgRNA designs (EpiGuide)	Multi-generational epigenetic dashboards
Livestock-specific DNA methylation array	850k CpG sites covering 1250 QTL regions	Methylation profiling for marbling score prediction	Beta value matrices	2000 samples/day	Cost-effective herd screening	Limited dynamic range	Nanopore-based validation (NanoMeth)	Real-time methylation monitoring chips
Quantum chromatin mapping (Q-Chrom)	12-qubit system analyzing 3D chromatin conformation	Heat stress-induced chromatin remodeling in dairy cows	Quantum entanglement maps	1M interactions/hour	Models epigenetic-environment interplay	Qubit stability challenges	Photonic coherence preservation	Whole-nucleus quantum holograms
Spatial epigenomics (EpiMaps)	10 μm resolution methylation and histone modification mapping	Mammary gland development during lactation cycles	Spatial epigenome tensors	5 tissues/week	Links methylation to tissue morphology	Fixation-induced artifacts	Cryo-preservation protocols	4D epigenetic organoid models
AI-driven epigenetic clock (BovineEpiClock)	DL model predicting biological age from 850 CpGs	Optimizing the productive lifespan of dairy cattle	Age acceleration scores	5000 predictions/hour	Extends milking longevity	Breed-specific variation	Federated learning across breeds	Precision retirement planning systems
Rumen microbiome epigenetics (MicroEpi)	Cell-free DNA methylation analysis of microbial consortia	Optimizing feed conversion through microbial epigenetics	Microbial methylation profiles	100 rumen samples/day	Non-invasive monitoring	Host DNA contamination	CRISPR-Cas selective depletion	Phage-mediated epigenetic engineering
Field-deployable epigenetic kits (FarmEPI)	Portable Oxford Nanopore and methylation-sensitive REs	On-farm drought adaptation potential assessment	Raw current squiggle data	50 samples/field-day	Rapid stress response profiling	Power supply limitations	Solar-powered basecalling modules	Autonomous epigenetic adaptation drones
Blockchain epigenetic provenance (EpiChain)	ZK-rollup authenticated epigenetic history records	Verifying organic raising claims through stress markers	Hashed epigenetic pattern NFTs	10K transactions/day	Tamper-proof welfare certification	Storage scalability issues	Lightweight Merkle tree compression	DAO-managed epigenetic welfare standards

development. When these developments are combined with CRISPR-based perturbation screens, a positive feedback loop of discovery is produced, in which the transcriptional effects of targeted gene edits guide further experimental design iterations. Transcriptomics has the potential to shed light on the molecular mechanisms by which genetic potential adjusts to and ultimately shapes agricultural resilience as livestock science tackles the problems of climate volatility and sustainable intensification (Monzó et al., 2025). The latest transcriptomics technologies in livestock production are set out in Table 2.4.

2.5 Proteomics: Understanding protein expression

Proteomics in livestock systems is conceptualized beyond traditional protein cataloging as a dynamic interface between genetic potential and biological actualization, a space where AI re-defines our ability to decipher the fluid semantics of the proteome. Protein duality is reconciled at its epistemic core as both transient signaling entities that adjust to environmental calculus and stable molecular machines that perform canonical functions. This change in perspective views the proteome as a biological dialectic medium that negotiates between evolutionary blueprints and current physiological demands, rather than as a static endpoint of genomic translation (Jitjumnong et al., 2025).

The fundamental problem of four-dimensional protein modeling, where spatial conformation, temporal expression, and post-translational modification states combine to form functional biological meaning, is addressed by AI-driven proteomics. Geometric DL architectures that embed protein structures in Riemannian manifolds replace traditional reductionist methods, which are constrained by their incapacity to handle the combinatorial explosion of protein isoforms and interaction networks. Through subtle structural reconfigurations, these models reveal hidden allosteric switches that modulate entire metabolic pathways, capturing the continuous deformation of protein conformational landscapes under endocrine signaling and metabolic stress. A framework that is revolutionizing our understanding of the onset of puberty in livestock and the regulation of the lactation cycle, the emerging concept of "proteomic criticality" proposes AI-identified tipping points where slight quantitative protein changes cause qualitative shifts in biological state (Qian et al., 2024; Yang et al., 2025).

The scope broadens to include the proteome's function as a cellular futurist and historian. The proteomic echoes of previous environmental exposures are now traced by neural differential equation systems, which decipher how brief nutritional deficiencies or pathogen encounters are biochemically remembered through phosphorylation patterns and ubiquitination signatures. At the same time, transformer architectures forecast proteomic preparedness, which is the latent ability of protein networks to adapt to upcoming difficulties, such as waves of heat stress or seasonal shifts in forage. Proteomics is transformed from a descriptive snapshot into a predictive continuum by this dual temporal orientation, allowing for proactive husbandry techniques that anticipate and match protein networks with anticipated management conditions (Chu et al., 2024; Mahala et al., 2025).

AI-mediated translation of proteoform language into transcriptomic grammars and metabolomic lexicons facilitates integration with other omics strata. The pleiotropic signaling of the proteome is mapped by attention-based neural networks, which can differentiate between proteins acting as cooperative symbols in multi-omics sentences and those acting as autonomous actuators. Determining the nonlinear relationships between mRNA abundance and protein activity, a translation fidelity gap where AI finds regulatory checkpoints governing the rates of protein synthesis, the timing of degradation, and functional activation thresholds, is especially important. These discoveries highlight proteomic buffering systems, a biological failsafe system of special significance in livestock exposed to rigorous production rhythms, that preserve metabolic homeostasis in the face of transcriptional noise (De Filippis et al., 2025; Zack et al., 2025).

As models based on proteomics grow closer to predictive accuracy, ethical considerations arise within precision livestock farming and a potential biological determinism framework. If we can anticipate the trajectories of individual proteins and their response to management decisions to optimize these trajectories, a challenge arises between these paternalistic ideals and animal autonomy. Meanwhile, the technical obstacles remain in the modeling context-dependent protein moonlighting, identical polypeptides take on different functions in different tissue microenvironments, by employing AI and/or modeling approaches concerning cellular spatial memory without topological complications (Ghavi Hossein-Zadeh, 2025; Santana et al., 2025).

Neural architectures that perceive protein networks as cognitive substrates, identifying heat shock proteins as stress memory units and circadian clock proteins as temporal reasoning modules, are the goal of the cutting edge of consciousness-aware proteomics. According to this viewpoint, the livestock proteome is an evolved biological intelligence system, and AI functions as a cross-species translator of protein-mediated environmental cognition rather than just an analytical tool. Decoding protein signatures as biological narratives of adaptation, resilience, and latent potential is the ultimate goal of AI-enhanced livestock proteomics, which conceptually extends beyond analytical biochemistry into the

TABLE 2.4 Recent transcriptomics technologies in livestock production.

Key technology	Main characteristics	Primary applications	Data type	Throughput	Major benefits	Major challenges	Potential solutions	Future perspectives
Microarray (Agilent SurePrint G3)	8×60K species-specific probes, dual-channel hybridization	Disease resistance gene expression profiling	Fluorescence intensity ratios (Cy3/Cy5)	200 samples/week	Cost-effective for known transcript detection	Limited dynamic range for low-abundance RNAs	Advanced normalization algorithms (LIMMA)	Customizable probe sets for novel transcripts
Bulk RNA-seq (Illumina NovaSeq X)	Strand-specific, 150bp paired-end reads	Muscle development transcriptome analysis	FASTQ files and raw read counts	1000 samples/run	Genome-wide isoform quantification	PCR duplicates in low-input samples	Unique molecular identifiers (UMIs)	Direct RNA sequencing via nanopores
Single-cell RNA-seq (10× Genomics Chromium X)	Partitioning of 50,000 cells/lane, 5′ or 3′ enriched	Immune cell heterogeneity in mastitis responses	Sparse count matrices (CellRanger)	10,000 cells/hour	Cellular resolution of transcriptomes	High cost per cell	Combinatorial indexing (sci-RNA-seq)	Multiome (RNA + ATAC) single-cell analysis
Spatial transcriptomics (Visium HD)	55 μm resolution with histological alignment	Placental gene expression zonation in pregnancy	Spatial barcode matrices (SpaceRanger)	8 tissue sections/week	Spatial context preservation	Limited subcellular resolution	Multi-round FISH hybridization	Live tissue spatial dynamics tracking
Long-read sequencing (PacBio Revio)	Full-length isoform sequencing (HiFi reads)	Alternative splicing analysis in stress-responsive genes	Continuous long reads (ISO-Seq)	15 Gb/hour	Resolves complex isoform diversity	High error rate in raw reads	Circular consensus sequencing (CCS)	Real-time isoform phasing during sequencing
Nanopore direct RNA-seq (MinION Mk1C)	Native RNA sequencing with epitranscriptome detection	Viral RNA quasispecies tracking in zoonotic diseases	Squiggle signals (FAST5)	10 Gb/flow cell	Detects RNA modifications (m6A, Ψ)	Basecalling accuracy challenges	DL basecallers (Bonito)	Portable field-deployable pathogen surveillance
CRISPR-based perturb-seq	Pooled CRISPR screens with single-cell transcriptomics	Functional validation of mastitis susceptibility genes	sgRNA-barcode linked count matrices	100 sgRNAs/screen	High-throughput gene-function mapping	Off-target sgRNA effects	ML-guided sgRNA design	In vivo CRISPR-interactome mapping

AI-driven splicing predictors (DeepSplice)	Transformer models trained on livestock-specific splice variants	Pre-mRNA splicing efficiency optimization	Splicing efficiency scores	10,000 transcripts/minute	Predicts tissue-specific isoform usage	Limited cross-species generalizability	Federated learning across livestock species	Real-time splicing outcome simulation
Ribo-seq (Illumina NovaSeq X)	Ribosome footprint profiling (28-nt reads)	Translation efficiency during heat stress adaptation	Ribosome-protected fragments (RPFs)	50 samples/run	Captures active translation events	rRNA depletion inefficiency	Probe-based rRNA capture (Ribo-off)	Single-cell translatome profiling
Time-lapse transcriptomics (TempO-Seq)	Non-destructive RNA sampling across development stages	Fetal muscle transcriptome dynamics in utero	Time-series count matrices	100 time points/sample	Longitudinal monitoring without sacrifice	Low RNA input per time point	Hybridization chain reaction amplification	Wearable in vivo transcriptome sensors
Phage-assisted RNA sequencing (PHAGE-Seq)	Bacteriophage polymerase-based RNA capture	Gut microbiome-host RNA interactions in ruminants	Phage-barcoded RNA libraries	50 host-microbe pairs/day	Specific pathogen transcript detection	Phage host range limitations	Engineered phage polymerases	Synthetic phage-RNA delivery systems
Quantum RNA folding predictors	Qubit-based prediction of RNA secondary structures	Vaccine design for livestock RNA viruses (e.g., ASFV)	RNA structure probability maps	1000 nt structures/minute	Models pseudoknots/tertiary interactions	Decoherence in large RNA molecules	Hybrid quantum-classical annealing	Real-time RNA structure dynamics simulation

field of interspecies communication (Tripathi et al., 2025). The latest proteomics technologies in livestock production are listed in Table 2.5.

2.6 Metabolomics: Analyzing metabolic pathways

As the last biochemical map of living systems, metabolomics clarifies the intricate network of low-molecular-weight metabolites that both produce and influence cellular activity. The ability of metabolomics to catalog metabolites in the agri-food industry is exceptional; it is comparable to offering a functional history of biological transactions within living systems, written in the vernacular of sugars, lipids, amino acids, and signaling molecules, by contextualizing the facility of a genotype in an ecosystem with a variety of environmental inputs into a myriad of physiological outputs. The metabolome functions as near to the interface between genotype and phenotype as possible by utilizing genetic potential and genomic transference. These links between genotype and phenotype make it important from a variety of agricultural viewpoints, such as feedstock or edible food sources and operational stressors that relate to nutritional status, systemic homeostasis, and adaptive conditions. Metabolic measurements, which accurately reflect the continuous negotiation between an organism and its internal and external environment, can vary with diurnal rhythm, altered diet, and disease state, in contrast to the relative stability of genomic information (Bovo et al., 2025; Karakas et al., 2025).

The conceptual power of the field lies in its ability to help us think of the architecture of metabolic pathways, not as static biochemical flow charts, but as dynamic networks that take on different forms of synthesis or dissociation depending on how they are managed, microbial symbionts, and developmental cues. The flux maps illustrate how competing enzymatic priorities, substrate availability, and allostery will modulate the decisions on nutrient partitioning by muscle, adipose, and mammary tissues. These insights are necessary to identify metabolic trade-offs for key agricultural traits such as maximizing the lean growth of animals without immune compromise or maximizing milk production without compromising fertility (Alhamoud et al., 2024; Carabaño et al., 2025).

The temporal sensitivity of metabolomics, which can identify metabolic abnormalities hours before clinical symptoms appear, sets it apart from other predictive techniques and is transforming preventive herd health measures. Early warning systems for less-than-ideal welfare conditions are provided by the identification of stress-induced kynurenine pathway activation or glucocorticoid-mediated lipid remodeling, which permits interventions that prevent productivity losses. By tracking how metabolites derived from the rumen or gut microbiota affect hepatic function, immune modulation, and epigenetic programming throughout life stages, the field simultaneously acts as a link between microbial ecology and host physiology (Qin et al., 2025).

The coupling of metabolomics to the other omics layers creates a multidimensional analytic system where metabolite flux patterns corroborate the activity profiles of proteomic enzymes, contextualize aspects of transcriptomic regulation, and annotate previously unknown significant variants within the genome. In view of systems, we can see how genetic polymorphisms that affect nutrient transporters or hormonal receptors propagate through the metabolic network and ultimately result in the phenotypic variation we observe in feed efficiency or methane emissions. The ability of AI to extend beyond this integrative potential occurs through pattern recognition algorithms that disentangle the association of complex metabolite-disease relationships as well as predictive models that simulate metabolic outcomes from varying genetic-environmental circumstances (Sanches et al., 2024; Montesinos-Lopez et al., 2025).

Metabolomics, which offers biochemical evidence to balance productivity gains with animal welfare imperatives, is a strategic ally of ethical and sustainable livestock production. The field informs management protocols that match societal expectations with economic goals by mapping metabolic signatures of environmentally friendly diets or humane handling practices. Furthermore, metabolomics is positioned as a key technology for creating circular bioeconomies in animal agriculture due to its ability to track environmental pollutant biotransformation pathways and nutrient utilization efficiency (del Campo et al., 2025; Sossidou et al., 2025).

The metabolome acts as a living biochemical record from which we can understand the physiological cost-benefit analysis of biological adaptation, which has great relevance in an era of climate change and changing production demands. Given the opportunity to study the metabolome, livestock scientists acquire the framework to analyze mechanisms of resilience, structure their metabolic response to available resources, and ethically manipulate animal systems that balance agricultural productivity with planetary health requirements. The latest metabolomics technologies in livestock production are set out in Table 2.6.

2.7 Nutrigenomics: Genotype-diet interactions

At the nexus of nutritional biochemistry and genomic science, nutrigenomics emerges as a crucial field that is primarily focused on clarifying the reciprocal molecular communication between dietary factors and livestock species' genetic

TABLE 2.5 Recent proteomics technologies in livestock production.

Key technology	Main characteristics	Primary applications	Data type	Throughput	Major benefits	Major challenges	Potential solutions	Future perspectives
2D gel electrophoresis	pH 3-10 NL IPG strips and SYPRO Ruby staining	Milk protein polymorphism analysis	Gel spot intensity matrices	20 gels/week	Low-cost protein separation	Limited dynamic range	DIGE (differential gel electrophoresis)	Automated spot picking robots
LC-MS/MS (Orbitrap Astral)	500,000 resolution and 40 Hz sequencing	Muscle growth phosphoproteomics	RAW files, spectral counts	100 samples/day	Post-translational modification analysis	Ion suppression in complex matrices	Microflow LC with FAIMS Pro	Single-cell LC-MS microfluidics
MALDI-TOF imaging	10 μm spatial resolution and lipid removal protocols	Intramuscular fat distribution mapping in beef	ImzML files with m/z images	5 tissue sections/day	Spatial protein-lipid correlation	Matrix crystallization artifacts	DL-based spectral unmixing	Multimodal imaging (protein + metabolite)
Single-cell proteomics (nPOP)	1500 proteins/cell from 10 nL lysates	Oocyte protein quality assessment	TMT-labeled peptide reports	1000 cells/day	Cellular heterogeneity mapping	Missing value problem	AI-based imputation (SCPNet)	Live cell protein turnover tracking
AI-driven protein language models (ProtGPT2)	12B parameter transformer trained on livestock proteomes	Vaccine antigen design for livestock pathogens	Embedding vectors, stability scores	10,000 predictions/hour	De novo protein design	Limited wet-lab validation	Federated learning across research consortia	Real-time folding simulations
Proximity labeling (TurboID)	10-minute biotinylation in live animals	Rumen microbial-host protein interactions	Biotinylated peptide libraries	5 animals/week	In vivo interaction mapping	Non-specific labeling	ML-based background subtraction	Wearable biosensors for real-time labeling
SWATH-MS (Sciex ZenoTOF 7600)	100 variable windows and 4D ion mobility	Stress biomarker discovery in poultry serum	Spectral library-matched quantitation	200 samples/day	Reproducible quantitative proteomics	Spectral library dependency	AI-generated in silico libraries	Blockchain-based spectral data sharing
Nanopore protein sequencing	5 pA current resolution and processive enzyme cleavage	Prion protein isoform detection	Squiggle patterns (FAST5)	100 proteins/hour	Single-molecule sequencing	Signal noise interference	CNN-based deconvolution (DeepNanoPro)	Direct post-translational modification reading

Continued

TABLE 2.5 Recent proteomics technologies in livestock production.—cont'd

Key technology	Main characteristics	Primary applications	Data type	Throughput	Major benefits	Major challenges	Potential solutions	Future perspectives
Quantum circular dichroism	Qubit-based secondary structure prediction	Vaccine thermal stability assessment	α-helix/β-sheet probability maps	1000 spectra/minute	Non-destructive analysis	Limited to purified proteins	Microfluidic purification chips	In vivo structural monitoring
Phage display proteomics	1011 phage library diversity with ruminant scaffolds	Colostrum antibody engineering	Phage-peptide binding affinity data	50 targets/week	Rapid antibody discovery	Off-target phage propagation	CRISPR-phage containment systems	Synthetic phage-antibody conjugates
Neural activity proteomics (FosTRAP)	Tamoxifen-inducible Fos tagging in neural circuits	Stress response pathways in transported livestock	Activity-correlated protein clusters	10 brain regions/day	Links proteins to behavior	Temporal resolution limitations	Optogenetic tagging synchronization	Whole-brain proteomic activity atlases
Blockchain-enabled protein data vaults	Zero-knowledge proof authentication of MS data	Traceability of growth hormone-treated livestock	Hashed spectral fingerprints	1M datasets/second	Prevents data tampering	Energy-intensive consensus mechanisms	Proof-of-stake protein validation networks	DAO-governed proteomic databases

TABLE 2.6 Recent metabolomics technologies in livestock production.

Key technology	Main characteristics	Primary applications	Data type	Throughput	Major benefits	Major challenges	Potential solutions	Future perspectives
GC-MS (Agilent 9990)	Rtx-5Sil MS column, EI ionization at 70eV	Rumen volatile fatty acid profiling	CDF files with retention indices	50 samples/day	High sensitivity for volatiles	Derivatization requirements	On-column derivatization microreactors	Field-deployable GC-MS drones
LC-HRMS (Orbitrap Exploris 240)	HILIC chromatography and 240,000 resolution	Colostrum oligosaccharide characterization	RAW files with MS^2 spectra	100 injections/day	Polar metabolite coverage	Column equilibration time	3D-printed disposable microcolumns	Autonomous LC method optimization AI
NMR (Bruker Ascend 1.3 GHz)	CryoProbe Prodigy and 13C-DEPT experiments	Metabolic stress biomarkers in transported livestock	FID files with chemical shifts (ppm)	20 samples/day	Non-destructive quantitative analysis	Low sensitivity for trace metabolites	Dynamic nuclear polarization enhancement	In vivo NMR endoscopy for rumen monitoring
Imaging MS (MALDI-2 TIMS)	5 μm spatial resolution and laser post-ionization	Intramuscular fat marbling pattern analysis	ImzML files with m/z images	1 cm^2 tissue/hour	Spatial lipidomics correlation	Matrix interference in dense tissues	DL-based spatial denoising	Multimodal imaging (metabolite + protein)
Single-cell metabolomics (LiveSeq)	Nanoprobes with iontophoretic sampling	Oocyte metabolic competency assessment	Time-resolved metabolite flux data	100 cells/hour	Live cell metabolic tracking	Probe-induced cellular stress	AI-optimized sampling trajectories	Subcellular organelle metabolomics
CRISPR-metabolomics	dCas9-based metabolite sensors in transgenic livestock	Real-time ketosis monitoring in dairy cows	Fluorescent metabolite reporter signals	Continuous monitoring	In vivo metabolic pathway visualization	Transgene silencing over time	Episomal CRISPR circuits	Whole-herd metabolic network mapping
Quantum NMR (Q-NMR)	1024-qubit processor for spin simulations	Feed additive metabolic pathway prediction	Quantum state metabolite profiles	1000 compounds/hour	Models complex metabolic networks	Decoherence in large molecules	Error-corrected quantum annealing	Quantum metabolic flux balance analysis

Continued

TABLE 2.6 Recent metabolomics technologies in livestock production.—cont'd

Key technology	Main characteristics	Primary applications	Data type	Throughput	Major benefits	Major challenges	Potential solutions	Future perspectives
AI-driven metabolic networks (MetaNetX)	GNNs trained on 10M livestock metabolic reactions	Optimizing methane reduction in rumen microbiomes	Flux balance analysis models	10,000 simulations/ hour	Predicts microbiome interventions	Limited strain-specific pathway data	Federated learning across research herds	Real-time rumen metabolic digital twins
Nanopore metabolite sensing	α-hemolysin pores with aptamer conjugation	On-farm mycotoxin detection in feed	Current blockade duration patterns	100 samples/hour	Rapid field detection	Non-specific pore interactions	ML pattern recognition	Wearable livestock toxin monitors
Synthetic metabolic pathways (EnoSynth)	Cell-free systems with enzyme cascades	On-demand vitamin synthesis in feed	Reaction yield optimization curves	50 pathway variants/day	Circumvents metabolic bottlenecks	Enzyme stability issues	AI-designed thermostable enzymes	Programmable metabolic nanopumps
Blockchain-enabled metabolite DBs	ZK-rollup authenticated livestock metabolome data	Antibiotic-free certification via metabolic fingerprints	Hashed metabolite pattern NFTs	1M transactions/second	Tamper-proof food chain verification	Energy-intensive consensus validation	Proof-of-stake metabolite validation	DAO-governed metabolic data markets
Phage-based metabolite harvesting	Engineered M13 phage with metabolite-binding peptides	Selective rumen propionate extraction	Phage-metabolite complex yields	200L rumen fluid/hour	Enhances beneficial metabolite recovery	Phage viability in anaerobic conditions	Synthetic phage oxygen scavengers	Phage-driven precision fermentation

makeup. In addition to examining how dietary patterns affect gene expression networks, epigenetic changes, and ensuing phenotypic outcomes, this field methodically examines how genetic variations affect nutrient metabolism and utilization. Beyond conventional nutritional paradigms, the scope includes dynamic genotype-diet interactions that control immune response, metabolic pathways, and developmental processes throughout an animal's life (Okani-Onyejiaka et al., 2024).

The understanding that livestock are genetically diverse populations whose genomic landscapes contain deeply encoded adaptive responses to feed composition, feeding schedules, and nutritional challenges is fundamental to this field of study. Multi-omics techniques are integrated into modern research frameworks to unravel the intricate regulatory processes by which particular nutrients function as signaling molecules that alter transcriptional activity and post-translational modifications. By stratifying dietary interventions according to genetic predispositions, this molecular understanding optimizes nutrient-gene interactions to improve physiological efficiency and productive performance (Hassanine et al., 2025; Wadood et al., 2025).

By enabling the identification of non-linear relationships between polymorphic genetic markers and metabolic responses to micronutrients, phytochemicals, and feed additives, AI has completely transformed the analytical ability to process high-dimensional nutrigenomic datasets. Predictive modeling of epigenetic changes brought on by dietary factors is now possible thanks to ML algorithms, especially those that impact DNA methylation patterns in genes linked to stress resilience and feed efficiency. Precision feeding systems that take into consideration both genetic potential and environmental factors can be adjusted in real time thanks to DL architectures, which also improve pattern recognition in temporal gene expression profiles (Kyriazos and Poga, 2024; Tahir et al., 2025).

The discipline's scope in livestock production systems includes the creation of genotype-specific nutritional strategies that take into account the metabolic quirks, life-stage needs, and production goals of individual breeds. This includes clarifying how individual responses to dietary macronutrient ratios, mineral bioavailability, and feed additive efficacy are influenced by genetic variations in nutrient transporters, metabolic enzymes, and hormone receptors. This scope is further broadened by the integration of gut microbiome genomics, which acknowledges the three-way relationship between dietary substrates, microbial metabolism, and host genetics in determining nutrient assimilation and health outcomes.

Acknowledging ethical and sustainability considerations are fundamental components of the discipline's scope of work, especially concerning the genetic optimization of production traits in animals and maintaining flexibility of metabolism under changing environmental conditions. The scientific community continues to deal with the challenge of moving nutrigenomic science into industry applications, respecting biological limits while meeting demands for more animal protein. Future landscape includes the intersection of quantum computing and systems biology to model an entire animal's metabolic pathways in a network, allowing for virtual trials of nutritional interventions across a variety of genetic populations before their actual applications (Richter et al., 2024; del Campo et al., 2025).

While upholding strict scientific standards for animal welfare and agricultural sustainability, this fundamental knowledge creates the conceptual framework for later sections that examine particular uses of AI in deciphering and utilizing genotype-diet interactions. The dynamic nature of the field emphasizes the need for flexible research approaches that can handle the ecological, technical, and ethical complexity present in contemporary livestock production systems. The latest nutrigenomics technologies in livestock production are set out in Table 2.7.

2.8 Lipidomics: Investigating lipid profiles

Lipidomics is the systematic study of lipid species, a field that not only identifies lipid species as separate entities but also strives to understand the relationships among lipid structures and biological activity in livestock systems. Fundamental to this omics approach is the examination of the spatio-temporal dynamics of lipid molecules, whether considering the structural diversity of lipids as parts of the cell membrane, or as metabolic modulating and signaling factors. From an analytical perspective, lipidomics allows us to explore the flexibility of the lipidome and what these hydrophobic biomolecules contribute to phenotypic expression at different levels, from subcellular organelles to whole-organism energy homeostasis. This is vital for livestock science as lipid networks influence traits the animal industry values financially (e.g., patterns of adipose deposition, milk fat globule assembly, and reproductive hormones). Lipids can also act as biomarkers to quantify animal welfare and production attributes relative to metabolic disorders (Tomiyasu et al., 2025).

The field's distinctive location at the intersection of systems physiology and structural biology accounts for its conceptual significance. Lipids display combinatorial complexity through headgroup variations, acyl chain permutations, and stereospecific modifications, which sets them apart from nucleic acids or proteins. AI is uniquely suited to handle this molecular diversity. These complex relationships are deconstructed by ML architectures, which link particular lipid species to functional outcomes like lipoprotein-mediated nutrient transport efficiencies or membrane fluidity adaptations to heat stress. Lipidomics, which demonstrates how epigenetic changes in lipid metabolism genes correspond to

TABLE 2.7 Recent nutrigenomics technologies in livestock production.

Key technologies	Main characteristics	Primary applications	Data type	Throughput	Major benefits	Major challenges	Potential solutions	Future perspectives
SNP genotyping arrays	High-density panels targeting nutrient metabolism genes	Personalized feed formulation based on genetic variants	Genotype data (VCF format)	500 samples/day	Cost-effective genetic stratification	Limited to predefined SNP loci	AI-driven novel SNP discovery	Dynamic SNP prioritization algorithms
CRISPR-Cas13 nutrient sensors	RNA-targeting systems monitoring dietary response genes	Real-time tracking of vitamin D receptor activation	Transcriptomic time-series	200 cell cultures/hour	Molecular-level diet-gene correlation	Off-target effects in complex tissues	ML-guided guide RNA design	In vivo nutrient-gene interaction mapping
Quantum metabolic models	72-qubit simulations of nutrient-gene networks	Predicting amino acid requirements across genotypes	Quantum state vectors	10 models/hour	Whole-system metabolic predictions	Qubit stability challenges	Error-corrected quantum annealing	Genome-scale metabolic quantum twins
Epigenetic clock analysis	Methylation profiling of diet-responsive genomic regions	Optimizing developmental stage-specific feeding regimes	DNA methylation ratios	150 samples/day	Lifespan nutrition planning	Tissue-specific epigenetic variability	Multi-tissue consensus models	Prenatal nutritional programming
Federated learning feed optimizers	Privacy-preserving AI across farm networks	Breed-specific mineral supplementation strategies	Federated model weights	1000 farms/week	Data sovereignty compliance	Model drift in heterogeneous systems	Blockchain-verified model consensus	Global livestock nutrition federation
Gut-brain axis spectrometers	Hyperspectral imaging of enteroendocrine cell activity	Linking dietary fiber intake to stress response genes	Spectral-neural activity maps	5 animals/hour	Holistic welfare-nutrition integration	Motion artifact interference	Immobilization-free AI correction	Real-time stress-nutrient balancing
NutriEpigenome editing	dCas9-DNMT3A/TET1 systems for dietary response tuning	Enhancing phytase utilization in monogastric species	Epigenetic modification tracks	50 edits/day	Stable metabolic trait enhancement	Off-target methylation risks	Single-cell editing verification	Climate-resilient epigenetic diets
Multi-omics dietetic graphs	Knowledge graphs integrating 12+ omics layers	Identifying genotype-specific anti-nutrient sensitivities	Heterogeneous graph data	1 graph/week	Cross-domain interaction discovery	Computational integration complexity	Automated ontology alignment	AI-curated nutritional genome projects
Photonics-based nutrient profilers	Terahertz spectroscopy detecting micronutrient absorption	Customizing trace element supplements per metabolic genotype	Terahertz absorption spectra	300 samples/hour	Non-invasive nutrient status monitoring	Signal-to-noise ratio in live animals	DL denoising algorithms	Whole-body nutrient distribution imaging
Ethical nutrigenomic frameworks	AI governance models for genetically-based feeding policies	Balancing production optimization with metabolic welfare	Ethical risk matrices	100 scenarios/day	Sustainable genetic diversity preservation	Cultural value conflicts	Multi-stakeholder digital twins	Global livestock nutrition accords

quantifiable changes in muscle marbling or mammary gland lipid secretion capacities, is essential for bridging the explanatory gap between genomic potential and phenotypic manifestation (Al-Remawi et al., 2025; Titus et al., 2025).

Lipidomics' contribution to the redefining of nutritional paradigms confers strategic significance. By shedding light on the metabolic destiny of dietary fatty acids, AI-driven lipid profiling makes it possible to precisely formulate feeds that maximize lipid utilization while balancing the provision of energy with the potential for inflammation from lipid peroxidation byproducts. At the same time, the field offers mechanistic insights into lipid-driven epigenetic programming, in which dietary lipids from the mother leave enduring metabolic trajectories in her children by altering the DNA methylation patterns of genes involved in lipid metabolism. Lipidomics is positioned as a foundation for creating preventive measures against metabolic syndromes that afflict intensive production systems because of this transgenerational viewpoint (Annan et al., 2025; Naumova et al., 2025; Malarvannan et al., 2025).

Through AI-mediated data fusion, lipidomics can be integrated with other omics strata, opening up previously unheard-of possibilities for comprehensive livestock management. Latent relationships between lipid droplet proteomes, mitochondrial phospholipid membranes, and transcriptional regulators of lipogenesis are revealed by neural networks trained on multi-omics datasets. This systems-level knowledge is essential for adjusting lipid fluxes without upsetting cellular homeostasis. Furthermore, by connecting microbial lipid metabolism to host immune function and nutrient partitioning efficiency, the identification of lipid-based signaling molecules that mediate host-microbiome cross-talk positions lipidomics as a critical enabler of precision microbiome interventions (Al-Remawi et al., 2025; Rahaman et al., 2025).

The ability of lipidomics to balance sustainability and productivity means that ethical and practical requirements come together. AI-enhanced lipid profiles direct dietary interventions that reduce enteric methane production while preserving energy balance by charting the lipid oxidation pathways that underlie ruminant greenhouse gas emissions. Through real-time monitoring of lipid peroxidation indices and antioxidant capacity in living tissues, ML models attempt to maintain the balance between lipids' dual roles as necessary nutrients and potential toxicants when dysregulated. With lipid biomarkers guiding proactive health management and lipid-centric metabolic engineering opening up new possibilities for sustainable animal agriculture, lipidomics has the potential to revolutionize livestock production into a predictive science as the field develops (Al-Remawi et al., 2025). The most recent lipidomics technologies in livestock production are set out in Table 2.8.

2.9 Metagenomics: Profiling microbial communities

The rise of metagenomics in livestock systems reflects a shift from a reductionist microbial taxonomy to a more holistic ecosystem analysis, highlighting microbial communities as dynamic shapers of host-environment interfaces, not passive residents. This area of study goes beyond traditional pathogen surveillance practices to characterize the multi-dimensional relationships between microbial consortia, host physiology, and agroecological systems, driven by complex networks of metabolic crosstalk and evolutionary co-adaptation. A key tenet of metagenomics is that livestock-related microbiomes function as extended phenotypes that mobilize metabolic potential well beyond the genetic potential of their animal host while also acting as biological monitors of environmental quality and management. The emphasis has shifted from simply cataloging microbial diversities in the respective environments to exploring the functional gene networks that regulate nutrient cycling, immune priming, and epigenetics. Livestock health is now considered to be an emergent trait dictated by host-microbe-environment interactions within a tripartite relationship (Muloi et al., 2025; Tang et al., 2025).

Metagenomics's exceptional ability to span scales, from landscape-level evaluations of agroecosystem sustainability to molecular-level analysis of microbial gene expression, makes it relevant. In this scalar integration, AI plays a crucial role by converting raw sequence data into predictive models of the resilience of microbial communities to antibiotic exposure or climate variability. Methane emission patterns and rumen metagenomic profiles are now correlated by ML architectures, demonstrating how microbial syntrophy networks balance feed efficiency and environmental impact. Neural networks simultaneously decipher the immunological communication between the gut microbiota and host mucosal systems, identifying keystone species that coordinate inflammatory cascades or immune tolerance, a frontier with significant ramifications for paradigms of antibiotic-free production (Sabaria, 2024; Kang et al., 2025).

The field's scope also includes temporal dynamics, where successional patterns in microbial communities across production cycles, lactation stages, or disease trajectories are modeled by recurrent neural networks. By favoring genes that confer antibiotic resistance or xenobiotic detoxification at ecosystem scales, industrial farming practices unintentionally alter the evolutionary trajectories of microorganisms. This is revealed by this temporal lens. As graph-based AI models identify zoonotic pathogen reservoirs within livestock-associated microbiomes and map spillover risks through

TABLE 2.8 Recent lipidomics technologies in livestock production.

Key technologies	Main characteristics	Primary applications	Data type	Throughput	Major benefits	Major challenges	Potential solutions	Future perspectives
AI-driven structural lipidomics	DL models predicting lipid-protein interactions	Optimizing milk fat globule membrane composition	Molecular interaction graphs	5000 lipids/hour	Enhances dairy product functionality	Limited training data for rare lipids	Federated lipid databases (LipidNet)	Personalized lactation formulations
Quantum lipid simulation (QLipSim)	16-qubit system modeling lipid bilayer dynamics	Designing rumen-protected lipid supplements	Quantum energy landscapes	1M atoms simulated/hour	Predicts supplement stability	Qubit coherence time limitations	Photonic stabilization circuits	Whole-cell membrane digital twins
CRISPR-lipid biosensors (CLipSense)	Engineered lipid-binding dCas9 reporters	Real-time monitoring of marbling development in beef cattle	Optical density readouts	1000 samples/day	Non-invasive marbling assessment	Tissue-specific delivery challenges	Nanoparticle-guided targeting (LiproNav)	Automated grading systems
Portable FT-ICR MS (FieldLipid)	21 Tesla magnet with AI-assisted spectral deconvolution	On-farm analysis of feed lipid oxidation states	Oxidation ratio matrices	50 samples/field-day	Prevents rancidity in stored feeds	Power consumption in remote areas	Solar-powered superconducting magnets	Autonomous feed quality drones
Single-cell lipidomics (scLipid)	Microfluidic encapsulation andnano-DESI ionization	Characterizing immune cell lipidomes in calves	Single-cell lipid profiles	10,000 cells/hour	Links lipid metabolism to disease resistance	Ion suppression effects	Ion mobility separation (LipidTrek)	Precision immunolipid therapies
Blockchain lipid authentication	ZK-rollup verified lipid signatures for premium products	Certifying omega-3-enriched pork claims	Cryptographic lipid hashes	10K transactions/day	Ensures product authenticity	Storage scalability issues	Lightweight Merkle tree compression	DAO-managed lipid quality standards
Spatial lipidomics (LipidMaps)	5 μm resolution MALDI-TOF imaging	Mapping intramuscular fat distribution in pork loins	Spatial lipid density tensors	8 tissues/week	Guides precision carcass grading	Matrix effects in dense tissue	DL background subtraction	3D meat marbling prediction models
Livestock lipid array (LipidChip)	2000-plex MRM assays for nutritional biomarkers	Monitoring essential fatty acid status in breeding stock	Concentration heatmaps	3000 samples/week	Cost-effective herd-level screening	Limited novel lipid detection	Dynamic assay updates (ArrayLive)	Real-time nutritional intervention systems
Edge AI lipid analyzers	IoT-enabled GC×GC systems with federated learning	Field analysis of pasture lipid nutritional quality	Compressed feature vectors	100 pastures/day	Optimizes grazing management	Connectivity gaps in rural areas	Edge ML models (PastureML)	Autonomous grazing optimization drones
Ethical lipid engineering framework	AI governance for CRISPR-lipid modified livestock	Developing sustainable EPA/DHA-enriched poultry	Risk-benefit matrices	50 designs/week	Balances innovation and safety	Regulatory harmonization challenges	Global agri-lipid consortium	CRISPR-lipidome editing standards

shared water systems or wildlife-livestock interfaces, the importance of metagenomics to One Health initiatives becomes more apparent (Rout et al., 2025).

Ethics undergird the field by setting it up to reveal hidden relationships between resources-how optimizing for microbial effect could disrupt equilibria developed over millennia. Explainable AI frameworks now investigate metagenomics models to help prevent misguided attributions of causation for correlation (e.g., attributions of production traits to certain taxa without consideration of the emergent properties at the community level). Ultimately, the value of livestock metagenomics lies in reframing problems of sustainability as a problem of managing microbial resources with a group of AI interventions that seek to improve understanding of complexities without seeking to remove those complexities, to use complexity to mediate productivity, animal welfare, and environmental health (Kate and Neethirajan, 2025; Sztandarski et al., 2025).

Metagenomics is a nexus science that blurs the lines between environmental microbiology and host biology, exposing livestock as permeable organisms that are constantly sharing metabolic and genetic data with their environment. This rethinking places AI as a mediator of circular bioeconomies, where microbial genes inform climate-smart breeding practices, waste-to-resource conversion systems, and precision feed formulations. By resolving microbial molecular mechanisms and contextualizing their role in global agricultural challenges, metagenomics becomes both a microscope and a telescope for livestock systems. This broadens its scope and firmly establishes its relevance to the grand imperative of reimagining humanity's relationship with domesticated species in the Anthropocene era. The latest metagenomics technologies in livestock production are listed in Table 2.9.

2.10 Integrative omics: The multi-omics approach

A paradigm shift from reductionist analytical frameworks to holistic systems biology approaches is represented by the conceptual underpinnings of integrative omics in livestock research, where the synergistic interpretation of multi-omics datasets allows for previously unheard-of resolution in understanding biological complexity. By establishing causal and correlative relationships across genomic, transcriptomic, proteomic, and metabolomic layers, this methodology fundamentally goes beyond conventional single-omics investigations and successfully maps the flow of biological information from genetic potential to phenotypic manifestation (Morabito et al., 2025; Wadood et al., 2025).

AI is the driving force behind this integrative process, offering computational architectures that can identify emergent properties and non-linear interactions that are difficult to detect using traditional statistical techniques. The method places genetic variation in the larger context of cellular homeostasis and organismal adaptation by radically reimagining biological systems as dynamic, multi-scale networks in which molecular components function within hierarchical regulatory structures (Anteghini et al., 2025; Su and Hao, 2025).

Integrative frameworks, which harmonize spatially and temporally resolved omics data, show how distinct molecular events spread across biological scales, from metabolic flux changes that shape whole-organism energy utilization patterns to epigenetic modifications that affect chromatin accessibility. This conceptual model is especially good at handling the pleiotropic nature of traits that are relevant to agriculture, where complex phenotypes like feed efficiency or disease resilience can be deciphered through AI-driven integration of omics strata (Lv et al., 2024; Dublino and Ercolano, 2025).

Through ML models that measure genotype-environment interactions across various production systems, the methodology naturally adapts to environmental modulation of genetic expression, bridging the gap between controlled experimental conditions and actual agricultural contexts (Crossa et al., 2025). The next major conceptual advance should be how biological redundancy and degeneracy are viewed not as analytical noise, but as an evolutionarily preserved trait of robustness that biological systems have employed, an advance enabled through the identification of conserved functional motifs that interconnect across layers of omics through AI. The approach includes time as a variable through recurrent neural network models of developmental trajectories and responses that describe how profiles of omics data change over the life of a biological entity as it responds to the challenges presented by its environment. This implementation of time is particularly advantageous in an application such as livestock, given the opportunities presented when the cycle of production and rapid generation turnover exists for trait development toward precision phenotyping (Montesinos-Lopez et al., 2025).

An irreducible element of the integrative omics approach, regardless of threading to the temporal dimension, is that of developing feedback loops of prediction and validation through benchmarking cycling of experiment, prediction, observation, recasting into the prediction leg of the loop as fit, the integrative loop imbibes iterative cycles of both biological understanding and predictive accuracy (Montesinos-Lopez et al., 2025).

Ultimately, the conceptual strength of the integrative omics approach should occur in the conversion of disaggregated molecular phenomena into biological acts, and provide the opportunity for AI to act as the indispensable information gatherer that aids the complex and qualitative conversations that omics datasets and their phenotypes should develop into. The latest multi-omics technologies in livestock production are set out in Table 2.10.

TABLE 2.9 Recent metagenomics technologies in livestock production.

Key technologies	Main characteristics	Primary applications	Data type	Throughput	Major benefits	Major challenges	Potential solutions	Future perspectives
AI-driven livestock microbiome platform	Transformer models analyzing 100+ microbial functional traits	Optimizing rumen microbiome for methane reduction	Multi-omics embeddings	500 samples/day	Predicts feed-microbiome interactions	Requires large training datasets	Federated learning across farms	Autonomous microbiome optimization
CRISPR-enhanced pathogen screening	Cas12a-mediated host DNA depletion and pathogen enrichment	Early detection of poultry enteric pathogens	Targeted pathogen genomes	200 samples/hour	100× pathogen signal amplification	Off-target effects in complex samples	ML-guided crRNA design (PathoCRISPR)	Real-time pathogen surveillance networks
Quantum metagenomic assembly (Q-Meta)	18-qubit system resolving complex microbial communities	Analyzing biofilm communities in dairy processing equipment	Quantum assembly graphs	1M reads/minute	Handles high diversity samples	Qubit decoherence issues	Photonic error correction	Quantum-enabled microbiome databases
Spatial metagenomics (MetaMaps)	10 μm resolution microbial localization in gut tissue	Mapping feed efficiency-linked microbes in intestinal villi	Spatial metabolic maps	5 tissues/week	Links the location to the function	Sample distortion during imaging	Cryo-preservation techniques	3D gut microbiome holograms
Portable nanopore phage analytics	MinION-based viral genome sequencing and host prediction	Monitoring bacteriophage dynamics in probiotic supplements	Phage-host interaction networks	50 samples/field-day	On-farm phage therapy management	Basecalling accuracy in noisy environments	AI-assisted signal filtering (PhageAI)	Drone-based phage delivery systems
Blockchain microbial provenance	ZK-STARKs verifying microbial strain origins	Authenticating premium fermented feed additives	Strain fingerprint NFTs	10K transactions/day	Prevents counterfeit microbial products	Storage scalability challenges	Lightweight recursive proofs	DAO-managed strain marketplace
Single-cell metatranscriptomics	Microfluidic encapsulation of 10,000 microbial cells/hour	Identifying active antibiotic resistance genes in manure	Single-cell expression profiles	100K cells/day	Resolves functional heterogeneity	Low mRNA capture efficiency	Cas9-based rRNA depletion (scMeta-Seq)	Microbial function prediction engines
Livestock-specific metagenomic array	2M probes covering 5000 livestock-relevant microbial genes	Routine monitoring of gut health biomarkers	Hybridization intensity matrices	3000 samples/week	Cost-effective herd-level screening	Limited novel gene detection	Dynamic probe update system (ArrayLive)	Continuous microbial adaptation tracking
Edge computing metagenomic kits	Field-deployable sequencing and cloud-based AI analysis	Rapid diagnosis of disease complexes	Compressed feature vectors	100 samples/field-day	Enables real-time outbreak response	Bandwidth limitations in rural areas	Edge AI models (Farm-EdgeML)	Autonomous veterinary response systems
Ethical microbiome engineering framework	AI governance for GMO microbiome modifications	Developing environmentally safe probiotic strains	Risk assessment reports	50 designs/day	Ensures ecological safety	Regulatory fragmentation across regions	Global agri-biosecurity consortium	CRISPR-based microbial containment

TABLE 2.10 Recent multi-omics technologies in livestock production.

Key technologies	Main characteristics	Primary applications	Data type	Throughput	Major benefits	Major challenges	Potential solutions	Future perspectives
AI-driven multi-omics fusion (DeepOmix)	Transformer architecture integrating 12+ omics layers	Predicting methane emission traits from gut microbiome	Multi-dimensional embeddings	1000 samples/hour	Captures non-linear omics interactions	Computational resource intensity	Edge AI deployment (FarmOmics edge)	Autonomous omics interpretation systems
Blockchain-enabled data provenance	Immutable ledger for omics metadata tracking	Verifying organic livestock claims via metabolic profiles	Hashed omics metadata NFTs	5000 transactions/day	Ensures data integrity for exports	Storage scalability limitations	Lightweight ZK-SNARKs proofs	DAO-managed omics certification
Quantum multi-omics correlation (Q-correlate)	16-qubit entanglement for cross-omics pattern discovery	Optimizing feed efficiency through lipidome-genome links	Quantum state correlation maps	1M correlations/minute	Exponential speed in hypothesis testing	Qubit coherence challenges	Photonic quantum error correction	Farm-scale quantum omics hubs
Single-cell multi-omics (Livestock-CellAtlas)	Simultaneous ATAC-seq and RNA-seq in 50,000 cells/chip	Identifying immune cell-specific productivity markers	Multi-modal cell matrices	10,000 cells/hour	Resolves cellular heterogeneity	Integration of sparse datasets	Graph neural networks (CellGNN)	Whole-organism digital cell twins
Spatiotemporal omics integration (OmniSpace)	100 μm resolution 3D tissue modeling and time-series data	Mapping mammary gland development across lactation cycles	4D omics tensors	5 tissues/week	Links molecular changes to morphology	Sample preparation complexity	Robotic microdissection systems	Holographic omics visualization
Metagenomic-metabolomic fusion (MicroBiomeX)	16S rRNA, LC-MS/MS, and AI alignment	Enhancing probiotic efficacy through metabolite-genome matching	Microbial-metabolite networks	200 samples/day	Cross-kingdom interaction analysis	Database gaps for livestock microbes	Federated learning across agri-databases	Real-time rumen optimization engines
CRISPR-omics tracking (EditTrace)	dCas9-based epigenetic recording in livestock lineages	Tracking nutritional interventions' multi-generational effects	CRISPR barcode arrays	1000 embryos/batch	Longitudinal multi-omics monitoring	Off-target recording events	ML-guided sgRNA design (CRISPR-AI)	Genome-wide epigenetic dashboards
Digital twin integration (BovineTwin 2.0)	Physics-informed neural networks simulating omics cascades	Predicting heat stress responses in cattle	Dynamic digital twin models	Real-time simulation	Enables preventive husbandry strategies	Model drift over time	Continuous learning via IoT sensor data	Autonomous climate adaptation systems
Edge computing omics kits (Farm-Seq Mobile)	Field-deployable nanopore and mass spectrometry integration	On-site disease biomarker detection	Raw multi-omics streams	50 samples/field-day	Rapid disease outbreak management	Power supply limitations in remote areas	Solar-powered omics workstations	Drone-based omics surveillance networks
Ethical AI governance frameworks (OmicsGuard)	Blockchain-audited AI models for bias detection	Ensuring fairness in genomic selection algorithms	Model explainability reports	100 models/day	Maintains ethical breeding practices	Regulatory fragmentation across regions	Global livestock AI ethics consortium	Decentralized omics governance DAOs

2.11 Phenomics: High-throughput phenotyping

Through the systematic, scalable capture of biological traits across spatiotemporal and organizational scales, phenomics emerges as the observational cornerstone of livestock omics, bridging the gap between genotypic potential and phenotypic expression. Fundamentally, the field reinterprets phenotypic analysis by moving away from isolated trait measurement and toward continuous, multifaceted characterization. This paradigm shift is made possible by AI's ability to integrate disparate data streams into useful biological narratives. Phenotypes are positioned as dynamic interfaces where genetic architecture, environmental inputs, and management practices converge, going beyond conventional morphometric or production-focused evaluations. From molecular phenomics mapping protein-metabolite interactions to organismal-level quantification of behavioral thermoregulation patterns, this scope includes hierarchical observation tiers that are all situated within the social and environmental dynamics of herds (Shan et al., 2025; Klingström et al., 2024).

The analytical prowess of this discipline emerges from its integrative breadth, where AI integrates data from hyperspectral imaging, wearable biosensors, and behavioral tracking to develop probabilistic models of biological function. ML architectures decouple changeable phenotypic traits from genetically-determined ones by correlating multidimensional trait variability with molecular signatures derived from the omics, demonstrating that epigenetic changes or exposure to the microbiome drive observable phenotypes. More importantly, phenomics emphasizes that phenotypic traits are dynamic processes rather than static states, operationalized via recurrent neural networks modeled on growth curves, lactation curves, or stress adaptation cycles as continuous functions rather than fixed moments. This temporal aspect enables predictive phenotyping, the use of early-life phenotypic traits to predict lifetime production, or disease susceptibility, reconstituting livestock management as a predictive science (Upadhyay et al., 2024; Yu et al., 2024).

With geospatial AI combining satellite-derived environmental data with animal-level phenotypic records to decode genotype-by-environment interactions at landscape scales, spatiotemporal expansion further expands the scope of phenomics. The quantification of subtle behavioral phenotypes that are difficult for humans to observe, such as predator vigilance postures, social hierarchy dynamics, or grazing selectivity, is automated by DL models trained on terabyte-scale video datasets. Through AI-mediated feature selection, molecular phenomics simultaneously interacts with the proteomic and metabolomic layers to identify biomarker combinations that resolve ambiguous phenotypes into mechanistically distinct subtypes. Through this multifaceted integration, phenomics is positioned as a knowledge generator and data integrator, converting unprocessed sensor outputs into biological insights that guide welfare optimization, precision breeding, and climate resilience tactics (Steibel, 2023; Koldasbayeva et al., 2024).

The scope of phenomics requires careful consideration of its ethical and epistemological aspects. By introducing metrics for cognitive states or affective responses derived from gait analysis or vocalization patterns, AI-driven phenotyping is upending conventional definitions of animal welfare as it permeates livestock systems (Ghavi Hossein-Zadeh, 2025). Explainable AI frameworks that check phenotypic classifications for biological plausibility while avoiding anthropomorphic projection help the field navigate this complexity. Phenomics simultaneously addresses the ontological conflict between holistic organismal understanding and reductionist trait quantification; this conflict is resolved by hybrid AI architectures that allow for targeted intervention at the individual animal level while maintaining emergent properties in herd-level phenotypic networks (Vlaicu et al., 2024).

Phenomics will probably move from descriptive analytics to predictive systems biology in the future, where federated learning models will use global phenotypic datasets to model how livestock will react to new genomic or environmental perturbations. Quantum computing could help reveal nonlinear phenotypic interactions that are currently hidden by traditional algorithmic constraints, and synthetic phenomics could create virtual livestock populations to test management techniques under stress before they are put into practice in the real world (Xiao et al., 2025). The scope of phenomics ultimately goes beyond simple measurement technology; it is a fundamental reimagining of biological observation, where the latent order governing the interaction between life and environment in animal production systems is revealed through AI-mediated synthesis of phenotypic data across scales and modalities. The latest phenomics technologies in livestock production are set out in Table 2.11.

TABLE 2.11 Recent phenomics technologies in livestock production.

Key technologies	Main characteristics	Primary applications	Data type	Throughput	Major benefits	Major challenges	Potential solutions	Future perspectives
AI-enhanced 3D biometric imaging	Photonic chips capturing 5000 body points/second	Precision monitoring of swine growth trajectories	3D point clouds	1000 animals/hour	Non-invasive carcass quality prediction	Data storage requirements	Edge computing compression (BioZip)	Real-time genetic merit estimation
Quantum thermal sensors (Q-Therm)	12-qubit system detecting 0.01°C variations	Early mastitis detection in dairy herds	Quantum heat signatures	200 cows/minute	48-hour early disease warning	Environmental noise interference	Photonic noise cancellation	Whole-herd health forecasting systems
CRISPR-integrated PhenoBiosensors	dCas9-based reporters tracking muscle development	Real-time marbling progression in cattle	Optical density gradients	500 samples/day	Molecular-level phenotype correlation	Tissue-specific delivery challenges	Lipid nanoparticle carriers (BioNav)	Automated premium beef certification
Autonomous PhenoDrones	Swarm intelligence mapping 100ha pastures/hour	Assessing pasture utilization efficiency	Spectral vegetation indices	5000 acres/day	Optimizes rotational grazing	Regulatory airspace limitations	AI-powered collision avoidance	Fully autonomous grazing management
Blockchain phenome ledger	ZK-STARKs verifying genetic-phenotype correlations	Certifying heritage breed characteristics	Cryptographic trait proofs	10K transactions/hour	Prevents fraudulent breeding claims	Computational overhead	Lightweight recursive proofs	DAO-managed breed registries
Spatial transcripto-Phenomics	10 μm resolution MERFISH imaging	Mapping muscle fiber-type distribution in lamb loins	Spatial gene expression maps	8 tissues/week	Links genetics to meat quality	Sample preparation complexity	Microfluidic tissue processing	3D meat structure prediction models
Livestock wearable arrays	50-parameter IoT sensors with federated learning	Continuous welfare monitoring in free-range poultry	Multimodal biometric streams	10K birds/hour	Detects stress 72-hour pre-clinical	Battery life limitations	Kinetic energy harvesting	Self-powered welfare assurance systems
Edge AI PhenoAnalyzers	TensorRT-accelerated vision systems in milking parlors	Automated body condition scoring during milking	Compressed feature vectors	300 cows/hour	Enables real-time nutritional adjustment	Limited model generalization	Continual learning frameworks	Autonomous feed dispensing robots
Ethical phenome editing framework	AI governance for CRISPR-phenotype modifications	Developing heat-tolerant cattle breeds	Risk-benefit matrices	100 designs/week	Ensures ethical trait selection	Regulatory fragmentation	Global livestock ethics consortium	Climate-adaptive breed blueprints

References

Adetunji, A.O., Owusu, H., Adewale, E.F., Adesina, P.A., Xedzro, C., Saliu, T.P., Islam, S., Zhu, Z., Morenikeji, O.B., 2025. DNA methylation: a key regulator in male and female reproductive outcomes. Life 15 (7). https://doi.org/10.3390/life15071109.

Al-Remawi, M., Aburub, F., Al-Akayleh, F., Abdel-Rahem, R.A., Ali Agha, A.S.A., 2025. Artificial intelligence in lipidomics: advancing biomarker discovery, pathway analysis, and precision medicine. In: 1st International Conference on Computational Intelligence Approaches and Applications, ICCIAA 2025 - Proceedings. Institute of Electrical and Electronics Engineers Inc., Jordan https://doi.org/10.1109/ICCIAA65327.2025.11013531. http://ieeexplore.ieee.org/xpl/mostRecentIssue.jsp?punumber=11012670.

Alem, D., Aklilu, H., 2025. Balancing genetic improvement with conservation of biodiversity a review. Bioprocess Eng. 9 (1), 36–43. https://doi.org/10.11648/j.be.20250901.14.

Alhamoud, Y., Abudumijiti, T., Wu, J., Lu, L., Zhao, M., Luo, X., Feng, F., Wang, J., 2024. Stimulation of non-shivering thermogenesis by bioactive compounds: a focus on gut microbiota-mediated mechanisms. Trends Food Sci. Technol. 154. https://doi.org/10.1016/j.tifs.2024.104779.

Alparthi, K., Kumar Panigrahy, S., 2024. Design of an efficient model for detecting diseases using genomics with graph relationship networks. In: 4th International Conference on Artificial Intelligence and Signal Processing, AISP 2024. Institute of Electrical and Electronics Engineers Inc., India https://doi.org/10.1109/AISP61711.2024.10870820. http://ieeexplore.ieee.org/xpl/mostRecentIssue.jsp?punumber=10870579.

Alsayaydeh, D.A.J., Narayanamurthy, V., Futane, A.S., Alsayaydeh, J.A.J., Najib, S.B.M., 2025. Advances of artificial intelligence-enabled epigenetics. Health Technol. 15 (4), 663–676. https://doi.org/10.1007/s12553-025-00998-0.

Annan, R., Noland, J., Perkins, K., Yuan, X., Roy, K., Qingge, L., 2025. Genomic privacy and security in the era of artificial intelligence and quantum computing. Discov. Comput. 28 (1). https://doi.org/10.1007/s10791-025-09627-w.

Anteghini, M., Gualdi, F., Oliva, B., 2025. How did we get there? AI applications to biological networks and sequences. Comput. Biol. Med. 190, 110064. https://doi.org/10.1016/j.compbiomed.2025.110064.

Antonazzo, G., Gaudet, P., Lovering, R.C., Attrill, H., 2024. Representation of non-coding RNA-mediated regulation of gene expression using the Gene Ontology. RNA Biol. 21 (1), 981–993. https://doi.org/10.1080/15476286.2024.2408523.

Bovo, S., Ribani, A., Fanelli, F., Galimberti, G., Martelli, P.L., Trevisi, P., Bertolini, F., Bolner, M., Casadio, R., Dall'Olio, S., Gallo, M., Luise, D., Mazzoni, G., Schiavo, G., Taurisano, V., Zambonelli, P., Bosi, P., Pagotto, U., Fontanesi, L., 2025. Merging metabolomics and genomics provides a catalog of genetic factors that influence molecular phenotypes in pigs linking relevant metabolic pathways. Genet. Sel. Evol. 57 (1). https://doi.org/10.1186/s12711-025-00960-8.

Campo, M.del, Montossi, F., Manuel Soares de Lima, J., Brito, G., 2025. Future cattle production: animal welfare as a critical component of sustainability and beef quality, a South American perspective. Meat Sci. 219, 109672. https://doi.org/10.1016/j.meatsci.2024.109672.

Carabaño, M.J., Díaz, C., Ramón, M., 2025. Breeding for thermotolerance in dairy cattle: production versus fertility traits. J. Dairy Sci. https://doi.org/10.3168/jds.2025-26258.

Cerantonio, A., Greco, B.M., Citrigno, L., De Benedittis, S., Qualtieri, A., Maletta, R., Montesanto, A., Passarino, G., Spadafora, P., Cavalcanti, F., 2025. Epigenetic clocks and their prospective application in the complex landscape of aging and Alzheimer's disease. Genes 16 (6), 679. https://doi.org/10.3390/genes16060679.

Chu, L.X., Wang, W.J., Gu, X.P., Wu, P., Gao, C., Zhang, Q., Wu, J., Jiang, D.W., Huang, J.Q., Ying, X.W., Shen, J.M., Jiang, Y., Luo, L.H., Xu, J.P., Ying, Y.B., Chen, H.M., Fang, A., Feng, Z.Y., An, S.H., Li, X.K., Wang, Z.G., 2024. Spatiotemporal multi-omics: exploring molecular landscapes in aging and regenerative medicine. Milit. Med. Res. 11 (1). https://doi.org/10.1186/s40779-024-00537-4.

Crossa, J., Montesinos-Lopez, O.A., Costa-Neto, G., Vitale, P., Martini, J.W.R., Runcie, D., Fritsche-Neto, R., Montesinos-Lopez, A., Pérez-Rodríguez, P., Gerard, G., Dreisigacker, S., Crespo-Herrera, L., Pierre, C.S., Lillemo, M., Cuevas, J., Bentley, A., Ortiz, R., 2025. Machine learning algorithms translate big data into predictive breeding accuracy. Trends Plant Sci. 30 (2), 167–184. https://doi.org/10.1016/j.tplants.2024.09.011.

Dagasso, G., Wilms, M., Souza, R., Forkert, N.D., 2025. Accounting for population structure in deep learning models for genomic analysis. J. Biomed. Inf. 169, 104873. https://doi.org/10.1016/j.jbi.2025.104873.

De Filippis, G.M., Amalfitano, D., Russo, C., Tommasino, C., Rinaldi, A.M., 2025. A systematic mapping study of semantic technologies in multi-omics data integration. J. Biomed. Inf. 165. https://doi.org/10.1016/j.jbi.2025.104809.

Dublino, R., Ercolano, M., 2025. Artificial intelligence redefines agricultural genetics by unlocking the enigma of genomic complexity. Crop J. https://doi.org/10.1016/j.cj.2025.05.008.

Ghavi Hossein-Zadeh, N., 2024. An overview of recent technological developments in bovine genomics. Vet. Anim. Sci. 25, 100382. https://doi.org/10.1016/j.vas.2024.100382.

Hassanine, N.N.A.M., Saleh, A.A., Essa, M.O.A., Adam, S.Y., Din, R.M.U., Rehman, S.U., Ali, R., Husien, H.M., Wang, M., 2025. Candidate genes, markers, signatures of selection, and quantitative trait loci (QTLs) and their association with economic traits in livestock: genomic insights and selection. Int. J. Mol. Sci. 26 (16), 7688. https://doi.org/10.3390/ijms26167688.

Hossein-Zadeh, N.G., 2025. Artificial intelligence in veterinary and animal science: applications, challenges, and future prospects. Comput. Electron. Agric. 235, 110395. https://doi.org/10.1016/j.compag.2025.110395.

Husien, H.M., Saleh, A.A., Hassanine, N.N.A.M., Rashad, A.M.A., Sharaby, M.A., Mohamed, A.Z., Abdelhalim, H., Hafez, E.E., Essa, M.O.A., Adam, S.Y., Chen, N., Wang, M., 2024. The evolution and role of molecular tools in measuring diversity and genomic selection in livestock populations (traditional and up-to-date insights): a comprehensive exploration. Vet. Sci. 11 (12). https://doi.org/10.3390/vetsci11120627.

Ishida, S., Sawai, T., 2024. From CRISPR to conscience: ethical dilemmas in gene editing and genetic selection. Am. J. Bioeth. 24 (8), 67–70. https://doi.org/10.1080/15265161.2024.2361900.

Jitjumnong, J., Taweechaipaisankul, A., Lin, J.C., Wongchanla, S., Chuwatthanakhajorn, S., Lin, C.J., Khang, L.T.P., Linh, N.V., Sangsawad, P., Dinh-Hung, N., Tang, P.C., Moonmanee, T., 2025. An overview of advancements in proteomic approaches to enhance livestock production and aquaculture. Animals 15 (13). https://doi.org/10.3390/ani15131946.

John, A., Nallusamy, S., Natesan, S., Nathan, B., Venugopal, A., Chinnaswamy, H.B., 2025. An overview of spatial transcriptomics methodologies in traversing the biological system. Curr. Bioinf. https://doi.org/10.2174/0115748936352261241224053340.

Kang, M.G., Kwak, M.J., Kang, A., Park, J., Lee, D.J., Mun, J., Kim, S., Mun, D., Lee, W., Choi, H., Seo, E., Choi, Y., Jeong, K.C., Oh, S., Kim, J., Kim, Y., 2025. Metagenome-based microbial metabolic strategies to mitigate ruminal methane emissions using Komagataeibacter-based symbiotics. Sci. Total Environ. 987. https://doi.org/10.1016/j.scitotenv.2025.179793.

Karakas, E., Bulut, M., Fernie, A., 2025. Metabolome guided treasure hunt - learning from metabolic diversity. J. Plant Physiol. 309, 154494. https://doi.org/10.1016/j.jplph.2025.154494.

Kate, M., Neethirajan, S., 2025. Decoding Bovine Communication with AI and Multimodal Systems ~ Advancing Sustainable Livestock Management and Precision Agriculture. bioRxiv, Canada. https://doi.org/10.1101/2025.03.03.641174.

Khoshnam, A., Chafamo, D., Tokcan, N., 2025. Tensor-based approaches for omics data analysis: applications, challenges, and future directions. La Matematica 4 (2), 486–520. https://doi.org/10.1007/s44007-025-00149-x.

Klingström, T., Zonabend König, E., Zwane, A.A., 2024. Beyond the hype: using AI, big data, wearable devices, and the internet of things for high-throughput livestock phenotyping. Brief. Funct. Genom. 24. https://doi.org/10.1093/bfgp/elae032.

Koldasbayeva, D., Tregubova, P., Gasanov, M., Zaytsev, A., Petrovskaia, A., Burnaev, E., 2024. Challenges in data-driven geospatial modeling for environmental research and practice. Nat. Commun. 15 (1). https://doi.org/10.1038/s41467-024-55240-8.

Koluman, N., Paksoy, Y., Göncü, S., 2025. Innovative approaches and solutions for climate resilient and sustainable livestock systems. Arch. Curr. Res. Int. 25 (6), 165–179. https://doi.org/10.9734/acri/2025/v25i61260.

Kyriazos, T., Poga, M., 2024. Application of machine learning models in social sciences: managing nonlinear relationships. Encyclopedia 4 (4), 1790–1805. https://doi.org/10.3390/encyclopedia4040118.

Le, H., Simmons, C.H., Zhong, X., 2025. Functions and mechanisms of histone modifications in plants. Annu. Rev. Plant Biol. 76 (1), 551–578. https://doi.org/10.1146/annurev-arplant-083123-070919.

Li, W., Sun, J., Wu, Q., Kwok, L.Y., Dong, G., Sun, Z., 2025. Global genomics of Lactococcus lactis: horizontal gene transfer and intergenic variation drive multiple domestication and dairy adaptation. J. Adv. Res. https://doi.org/10.1016/j.jare.2025.07.053.

Lu, Y., Li, M., Gao, Z., Ma, H., Chong, Y., Hong, J., Wu, J., Wu, D., Xi, D., Deng, W., 2025. Advances in whole genome sequencing: methods, tools, and applications in population genomics. Int. J. Mol. Sci. 26 (1), 372. https://doi.org/10.3390/ijms26010372.

Lv, T., Zhang, Y., Liu, J., Kang, Q., Liu, L., 2024. Multi-omics integration for both single-cell and spatially resolved data based on dual-path graph attention auto-encoder. Briefings Bioinf. 25 (5). https://doi.org/10.1093/bib/bbae450.

Madhu, H., Rocha, J.F., Huang, T., Viswanath, S., Krishnaswamy, S., Ying, R., 2025. HEIST: A Graph Foundation Model for Spatial Transcriptomics and Proteomics Data (Version 1). https://doi.org/10.48550/ARXIV.2506.11152.

Maghdid, S.S., Rashid, T.A., Askar, S.K., 2025. Graph neural networks: historical backgrounds, present revolutions, and conventionalization for the. Int. J. Data Sci. Anal. https://doi.org/10.1007/s41060-025-00797-w.

Mahala, A., Ranjan, A., Priyadarshini, R., Vikram, R., Dansena, P., 2025. A fast (CNN + MCWS-transformer) based architecture for protein function prediction. Stat. Appl. Genet. Mol. Biol. 24 (1). https://doi.org/10.1515/sagmb-2024-0027.

Malarvannan, M., Naik Sabavath, B.T., Gaddam, V., Paul, D., 2025. Transformative potentials, challenges and innovative solutions of lipidomics in multiple clinical applications. Talanta 291, 127855. https://doi.org/10.1016/j.talanta.2025.127855.

Mollandin, F., Acloque, H., Ballester, M., Bink, M., Calus, M., Crespo-Piazuelo, D., Croiseau, P., Djebali, S., Foissac, S., Gilbert, H., Giuffra, E., Guyomar, C., Madsen, O., Mercat, M.J., da Costa Perez, B., de Vos, J., Rau, A., 2025. Guiding eQTL Mapping and Genomic Prediction of Gene Expression in Three Pig Breeds with Tissue-Specific Epigenetic Annotations from Early Development. bioRxiv, France. https://doi.org/10.1101/2025.03.05.641618.

Montesinos-Lopez, O.A., Montesinos-López, A., Brandon, A., Delgado-Enciso, I., Chavira-Flores, M., Crossa, J., Breseghelo, F., Dreisigacker, S., Sun, J., Gerard, G.S., Vitale, P., Ortiz, R., 2025. Genomic prediction powered by multi-omics data. Front. Genet. https://doi.org/10.3389/fgene.2025.1636438.

Monzó, C., Liu, T., Conesa, A., 2025. Transcriptomics in the era of long-read sequencing. Nat. Rev. Genet. https://doi.org/10.1038/s41576-025-00828-z.

Morabito, A., De Simone, G., Pastorelli, R., Brunelli, L., Ferrario, M., 2025. Algorithms and tools for data-driven omics integration to achieve multilayer biological insights: a narrative review. J. Transl. Med. 23 (1). https://doi.org/10.1186/s12967-025-06446-x.

Muloi, D.M., Caron, A., Berkley, J.A., Hassell, J.M., Brito, I.L., King, K., Moodley, A., Fèvre, E.M., 2025. Using the microbiota to study connectivity at human–animal interfaces. Trends Microbiol. https://doi.org/10.1016/j.tim.2025.05.006.

Naumova, E.N., Hsieh, A., Ran-Ressler, R.R., Jang, C.W., Huey, S.L., Dionisi, F., 2025. Application of artificial intelligence to lipid nutrition: a narrative review. Lipids. https://doi.org/10.1002/lipd.70000.

Nirmal, A.K., Mishra, N.T.P., Mishra, S., Talreja, S., Fatima, B., Srivastava, S., Nigam, C., Misra, S., 2024. Transforming transcriptomics: the impact of RNA sequencing technology. Asian J. Med. Health 22 (10), 40–57. https://doi.org/10.9734/ajmah/2024/v22i101103.

Okani-Onyejiaka, M.C., Etuk, E.B., Ogundu, U.E., 2024. Nutrigenomics in livestock research and production: principles, applications and challenges. Niger. J. Anim. Prod. 155–158. https://doi.org/10.51791/njap.vi.4328.

Qian, L., Sun, R., Aebersold, R., Bühlmann, P., Sander, C., Guo, T., 2024. AI-empowered perturbation proteomics for complex biological systems. Cell Genom. 4 (11). https://doi.org/10.1016/j.xgen.2024.100691.

Qin, W., Zheng, S., Zhou, L., Liu, X., Chen, T., Wang, X., Qi, L., Zhao, Y., Wang, D., Xu, G., 2025. High-coverage metabolomics reveals gut microbiota-related metabolic traits of type-2 diabetes in serum. J. Proteome Res. 24 (4), 1649–1661. https://doi.org/10.1021/acs.jproteome.4c00507.

Rahaman, M.M., Islam, M.R., Bhuiyan, M.M.R., Noman, I.R., Aziz, M.M., Das, K., 2025. Harnessing big data in biotechnology: a machine learning approach to multi-omics. In: 2nd International Conference on Machine Learning and Autonomous Systems, ICMLAS 2025 - Proceedings. Institute of Electrical and Electronics Engineers Inc., United States, pp. 1391–1401. https://doi.org/10.1109/ICMLAS64557.2025.10967731. http://ieeexplore.ieee.org/xpl/mostRecentIssue.jsp?punumber=10967566.

Richter, S., Scherer, L., Hegwood, M., Bartlett, H., Bossert, L.N., Frehner, A., Schader, C., 2024. Conceptual framework for considering animal welfare in sustainability assessments of foods. Sustain. Prod. Consum. 52, 179–209. https://doi.org/10.1016/j.spc.2024.10.004.

Rout, A.K., Rout, S.S., Panda, A., Tripathy, P.S., Kumar, N., Parida, S.N., Dey, S., Dash, S.S., Behera, B.K., Kumar Pandey, P., 2025. Potential applications and future prospects of metagenomics in aquatic ecosystems. Gene, 149720. https://doi.org/10.1016/j.gene.2025.149720.

Rybina, O.Y., Pasyukova, E.G., 2024. Multi-omic rejuvenation: a new strategy for lifespan extension. Mol. Biol. 58 (5), 805–813. https://doi.org/10.1134/s0026893324700377.

Sabaria, S., 2024. Harnessing machine learning for metagenomics: discovering the invisible microbial world. Commun. Appl. Nonlinear Anal. 31 (8s), 147–164. https://doi.org/10.52783/cana.v31.1462.

Sajjanar, B., Krishnaswamy, N., Saxena, V.K., Dhara, S.K., 2025. Stress responses to changing environmental factors in the domestic animals: an epigenetic perspective. J. Anim. Physiol. Anim. Nutr. 109 (4), 1000–1016. https://doi.org/10.1111/jpn.14115.

Sanches, P.H.G., de Melo, N.C., Porcari, A.M., de Carvalho, L.M., 2024. Integrating molecular perspectives: strategies for comprehensive multi-omics integrative data analysis and machine learning applications in transcriptomics, proteomics, and metabolomics. Biology 13 (11). https://doi.org/10.3390/biology13110848.

Santana, T.C., Guiselini, C., Pandorfi, H., Vigoderis, R.B., Barbosa Filho, J.A.D., Soares, R.G.F., Araújo, M.d.F., Gomes, N.F., Lima, L.D.d., Santos, P.C.d.S., 2025. Ethics, animal welfare, and artificial intelligence in livestock: a bibliometric review. AgriEngineering 7 (7). https://doi.org/10.3390/agriengineering7070202.

Shan, K., Zhou, G., Li, C., phenomics, M., 2025. A novel concern linking meat across the whole chain. Meat Sci. 229, 109913. https://doi.org/10.1016/j.meatsci.2025.109913.

Sindhu, P., Magotra, A., Sindhu, V., Chaudhary, P., 2024. Unravelling the impact of epigenetic mechanisms on offspring growth, production, reproduction and disease susceptibility. Zygote 32 (3), 190–206. https://doi.org/10.1017/s0967199424000224.

Sonstegard, T., Bostrom, J., Martins, K., Kim, E.-S., Correia, C., MacHugh, D., Larson, S., Carlson, D., 2024. Genome-edited livestock to secure sustainability. Reprod. Fertil. Dev. 37 (1). https://doi.org/10.1071/rd24145.

Sossidou, E.N., Banias, G.F., Batsioula, M., Termatzidou, S.A., Simitzis, P., Patsios, S.I., Broom, D.M., 2025. Modern pig production: aspects of animal welfare, sustainability and circular bioeconomy. Sustainability 17 (11). https://doi.org/10.3390/su17115184.

Steibel, J.P., 2023. 431 awardee talk: phenomics and phenotyping in animal breeding. J. Anim. Sci. 101 (Suppl. 3), 17. https://doi.org/10.1093/jas/skad281.021.

Struhl, K., 2024. The distinction between epigenetics and epigenomics. Trends Genet. 40 (12), 995–997. https://doi.org/10.1016/j.tig.2024.10.002.

Su, H., Hao, N., 2025. Computational systems biology approaches to cellular aging—integrating network maps and dynamical models. Quant. Biol. 13 (4). https://doi.org/10.1002/qub2.70007.

Sztandarski, P., Marchewka, J., Jaszczyk, A., Solka, M., Michnowska, H., Pogorzelski, G., Gotxi, J.R., Damaziak, K., Wójcik, W., Siwiec, D., Seidavi, A., Horbańczuk, J.O., 2025. Scientific literature on AI technologies to enhance animal welfare, health and productivity. Anim. Sci. Pap. Rep. 43 (2), 115–132. https://doi.org/10.2478/aspr-2025-0009.

Tahir, M., Norouzi, M., Khan, S.S., Davie, J.R., Yamanaka, S., Ashraf, A., 2025. Artificial Intelligence and Deep Learning Algorithms for Epigenetic Sequence Analysis: A Review for Epigeneticists and AI Experts. arXiv, Canada. https://doi.org/10.48550/arXiv.2504.03733.

Tang, Q., Yin, X., Wen, G., Luo, Z., Zhang, L., Tan, S., 2025. Unraveling the composition and function of pig gut microbiome from metagenomics. Anim. Microbiome 7 (1). https://doi.org/10.1186/s42523-025-00419-7.

Titus, S.D., Emmanuel, A., Stephen, E.C., Tatah, S.V., Adebisi Arowora, K., 2025. Recent advances in lipid metabolism and regulations: a review. J. Multidiscip. Sci. MIKAILALSYS 3 (1), 208–224. https://doi.org/10.58578/mikailalsys.v3i1.5044.

Tomiyasu, N., Izumi, Y., Heraviadeh, O., Takahashi, M., Bamba, T., 2025. Evaluation of separation performance and quantification accuracy in lipidomics methods. J. Chromatogr. A 1758, 466165. https://doi.org/10.1016/j.chroma.2025.466165.

Tripathi, T., Uversky, V.N., Giuliani, A., 2025. 'Intelligent' proteins. Cell. Mol. Life Sci. 82 (1). https://doi.org/10.1007/s00018-025-05770-1.

Ullah, K., Waheed, A., 2025. Review: genetic improvement in livestock: a journey from conventional breeding to genomic precision. Vet. Biomed. Clin. J. 7 (1), 30–44. https://doi.org/10.21776/ub.vetbioclinj.2025.007.01.3.

Upadhyay, V.R., Ramesh, V., Kumar, H., Somagond, Y.M., Priyadarsini, S., Kuniyal, A., Prakash, V., Sahoo, A., 2024. Phenomics in livestock research: bottlenecks and promises of digital phenotyping and other quantification techniques on a global scale. OMICS A J. Integr. Biol. 28 (8), 380–393. https://doi.org/10.1089/omi.2024.0109.

van der Meer, D., Rahman, Z., Ottas, A., Parekh, P., Kutrolli, G., Stinson, S.E., Koromina, M., Rokicki, J., Sønderby, I.E., Parker, N., Tesfaye, M., Hindley, G., Rødevand, L.N., Koch, E., Võsa, U., Kronberg, J., Abner, E., Taba, N., Palta, P., Hudjashov, G., Nelis, M., Metspalu, M., Mägi, R., Esko, T., Metspalu, A., Steen, N.E., Berg, J.P., O'Connell, K.S., Smeland, O.B., Frei, O., Dale, A.M., Djurovic, S., Lehto, K., Alver, M., Milani, L., Shadrin, A.A., Andreassen, O.A., 2025. Pleiotropic and sex-specific genetic mechanisms of circulating metabolic markers. Nat. Commun. 16 (1). https://doi.org/10.1038/s41467-025-60058-z.

Vlaicu, P.A., Gras, M.A., Untea, A.E., Lefter, N.A., Rotar, M.C., 2024. Advancing livestock technology: intelligent systemization for enhanced productivity, welfare, and sustainability. AgriEngineering 6 (2), 1479–1496. https://doi.org/10.3390/agriengineering6020084.

Wadood, A.A., Bordbar, F., Zhang, X., 2025. Integrating omics approaches in livestock biotechnology: innovations in production and reproductive efficiency. Front. Anim. Sci. 6. https://doi.org/10.3389/fanim.2025.1551244.

Wang, J., Ye, W., He, J., Zhang, L., Huang, G., Yu, Z., Liang, Z., 2026. Integrating biological and machine intelligence: Attention mechanisms in brain–computer interfaces. Inf. Fusion. 125, 103417. https://doi.org/10.1016/j.inffus.2025.103417.

Xiao, Y., Zhao, D., Li, X., Li, T., Wang, R., Wang, G., 2025. A federated learning-based data augmentation method for privacy preservation under heterogeneous data. IEEE Trans. Mobile Comput. 24 (9), 8105–8118. https://doi.org/10.1109/tmc.2025.3553501.

Xiong, Z., Lin, L., Ouyang, K., Qu, M., Qiu, Q., 2025. Deciphering heat stress mechanisms and developing mitigation strategies in dairy cattle: a multi-omics perspective. Agriculture 15 (14), 1477. https://doi.org/10.3390/agriculture15141477.

Xu, S., Akhatayeva, Z., Liu, J., Feng, X., Yu, Y., Badaoui, B., Esmailizadeh, A., Kantanen, J., Amills, M., Lenstra, J.A., Johansson, A.M., Coltman, D. W., Liu, G.E., Curik, I., Orozco-terWengel, P., Paiva, S.R., Zinovieva, N.A., Zhang, L., Yang, J., Li, M., 2024. Genetic advancements and future directions in ruminant livestock breeding: from reference genomes to multiomics innovations. Sci. China Life Sci. 68 (4), 934–960. https://doi.org/10.1007/s11427-024-2744-4.

Xu, S., Akhatayeva, Z., Liu, J., Feng, X., Yu, Y., Badaoui, B., Esmailizadeh, A., Kantanen, J., Amills, M., Lenstra, J.A., Johansson, A.M., Coltman, D. W., Liu, G.E., Curik, I., Orozco-terWengel, P., Paiva, S.R., Zinovieva, N.A., Zhang, L., Yang, J., Liu, Z., Wang, Y., Yu, Y., Li, M., 2025. Genetic advancements and future directions in ruminant livestock breeding: from reference genomes to multiomics innovations. Sci. China Life Sci. 68 (4), 934–960. https://doi.org/10.1007/s11427-024-2744-4.

Yang, W., Wang, Y., Wang, Y., 2025. Deep Learning-Driven Protein Structure Prediction and Design: Key Model Developments by Nobel Laureates and Multi-Domain Applications. arXiv, China. https://doi.org/10.48550/arXiv.2504.01490.

Ye, X., Sahana, G., Lund, M.S., Cai, Z., 2025. The crosstalk between host and rumen microbiome in cattle: insights from multi-omics approaches and genome-wide association studies. World J. Microbiol. Biotechnol. 41 (8). https://doi.org/10.1007/s11274-025-04504-6.

Yu, Z., Han, Y., Cha, L., Chen, S., Wang, Z., Zhang, Y., 2024. Design of an intelligent wearable device for real-time cattle health monitoring. Front. Robot. AI 11. https://doi.org/10.3389/frobt.2024.1441960.

Zack, M., Stupichev, D.N., Moore, A.J., Slobodchikov, I.D., Sokolov, D.G., Trifonov, I.F., Gobbs, A., 2025. Artificial intelligence and multi-omics in pharmacogenomics: a new era of precision medicine. Mayo Clin. Proc.: Digit. Health 3 (3). https://doi.org/10.1016/j.mcpdig.2025.100246.

Zhang, H., Pang, Z., Wang, W., Qiao, L., Liu, W., 2025. Impact of selection signature on genomic prediction and heritability estimation in livestock. Animals 15 (10), 1383. https://doi.org/10.3390/ani15101383.

Zhou, H., Clark, E., Guan, D., Lagarrigue, S., Fang, L., Cheng, H., Tuggle, C.K., Kapoor, M., Wang, Y., Giuffra, E., Egidy, G., 2025. Comparative genomics and epigenomics of transcriptional regulation. Ann. Rev. Anim. Biosci. 13 (1), 73–98. https://doi.org/10.1146/annurev-animal-111523-102217.

Chapter 3

Fundamentals of artificial intelligence in livestock management

Artificial intelligence (AI) in animal husbandry is a game-changer, transforming an entire sector from a traditional husbandry industry to one based on data-driven precision agriculture. With demand for sustainable protein growing at record rates across the globe, AI is the technology making that vital connection between biological complexity and computational power, harnessing terabytes of omics data to drive insights and excellence in modern cultivation.

This chapter lays the algorithmic foundation for AI technologies that are transforming how we monitor, analyze, and optimize living systems at scale. Modern livestock systems produce multi-dimensional information streams, from genomics at one end to day-to-day scores with changing scores. The intersection of machine learning (ML) with sophisticated sensor networks now makes it possible to track animal physiology at an unprecedented resolution, from the dynamics of the rumen microbiome to stress-induced epigenetic changes. Yet, this data deluge comes with its own set of challenges: mining these sparse high-dimensional omics datasets requires tailored pre-processing pipelines, and livestock growth has a temporal nature that cannot be easily decomposed into proteins and genes.

We start by breaking down more established key ML paradigms, which have been used for animal biosystems. AI-based models for livestock are not like typical AI-based systems; they need to be adjusted for biological variation between breeds, environmental stochasticity, and the ethical obligations of the management of sentient beings. Subsequent parts also highlight the potential use of deep neural networks in learning a latent representation of patterns embedded in multi-omics data layers that allow predicting disease outbreaks weeks in advance of clinical presentation.

The chapter questions the infrastructure needs for AI at the agricultural level. Blockchain-secured data lakes maintain lineage integrity across global supply chains, and edge computing architectures can now process video feeds from automated phenotyping stations with a latency of less than 50 ms. Novel approaches such as federated learning make collaborative model training possible on farms spread across the globe without ever moving the data, a development that will be key in the fight against anti-microbial resistance (AMR) via distributed surveillance. Even as we chart this terrain, ethical considerations are still key. Finally, the chapter discusses the interpretability of genomic selection algorithms to breeders according to explainable AI (XAI) writ small, as well as models of human-AI collaboration that guarantee farmer agency in high-stakes decisions. Via a broad and deep collection of subjects, including very specific applications, throughout, the reader will have access to a range of theories and practical applications of AI and analytical tools as applied to animal science, in one convenient place.

3.1 Conceptual foundations of AI-driven livestock systems

The AI within farming, especially precision agriculture, is a transformative intersection between farming and computation that enables adaptive systems to balance bio-complexity with computational precision. At its core, this paradigm functions through interconnected layers of sensor-mediated data capture, machine-learning-based optimization, and ethical governance frameworks that inscribe reciprocal, dynamic feedback loops between digital analysis and living matter. Sensors networks incessantly translate biological signals, from fluctuations in hormones to behavior, into data streams, with predictive algorithms digesting this information to create management interventions that maximize production, welfare, and environmental sustainability. This synthesis makes it possible the real-time responsive to animals, leaving the realm of observation and becoming an active part of the system (Ghavi Hossein-Zadeh, 2025; Mohammed et al., 2025).

Artificial Intelligence in Livestock Omics. https://doi.org/10.1016/B978-0-443-45663-3.00015-4

At the core of this integration is the multi-dimensional analysis of biological data, where genomic profiling, rumen microbiome dynamics, and behavioral biometrics come together to create this new understanding of animal systems. Complex neural networks can filter out non-linear relationships between expression markers and phenotypic as well as pharmacological effects, and federated learning models retain data privacy across different agriculture sites. The computational strategy goes beyond static analysis and uses reinforcement learning strategies to navigate between conflicting objectives through ethically weighted parameters informed by veterinary science and welfare guidance. These systems are intrinsically temporal, and the fast cycles of the iterations used to train ML models must match the long time scales of biological processes like gestation or growth, and require the development of new ways of training algorithms that respect natural phases (Yetgin, 2025; Kumar et al., 2024).

Ethical questions infuse each level of architectural design, requiring technical enactments that meet biological facts. Welfare monitoring systems use transformer-based ethological data analysis to non-invasively reveal stress, and resource optimization algorithms take rumen fermentation kinetics into account to formulate feed. The nascent field of CRISPR-interpretable neural networks raises the possibility of flexible genetic culling in the presence of essential epistasis. These developments highlight the need for XAI systems that can retain transparency in automated decision-making processes affecting sentient beings, to keep computational intentions tethered to veterinary ethics, as well as animal-centered values (Biswas, 2025; Manikandan and Neethirajan, 2025).

As the field matures, it faces basic problems posed by the mutual and deepening interweaving of biological and digital systems. That the opaque architectures of AI could be permanently relied on for routine husbandry decisions hangs in the balance between densening time (via trends toward ever-accelerating technological expiration and replacement) and the multi-generational breeding cycles. Measuring welfare, even specifically animal welfare, in terms of algorithmic loss functions raises philosophical and technical challenges that call for methods avoiding anthropomorphic bias, while accounting for the subtleties of animal experience. Development will likely evolve toward a hybrid system that couples quantum-inspired optimization and real-time control of such hypercomplex variables, but this will require corresponding advocacy of regulatory frameworks to avoid value-drift between computational efficiency and life. The way forward relies on holding to the standards of scientific excellence while also encouraging dialog among computer scientists, ethologists, and ethicists, a joint exploration aimed at ensuring that technological development never forgets that it is a stewarding of life itself (Sztandarski et al., 2025; van der Staay et al., 2025; Long et al., 2025). The AI-driven livestock management process is shown in Fig. 3.1.

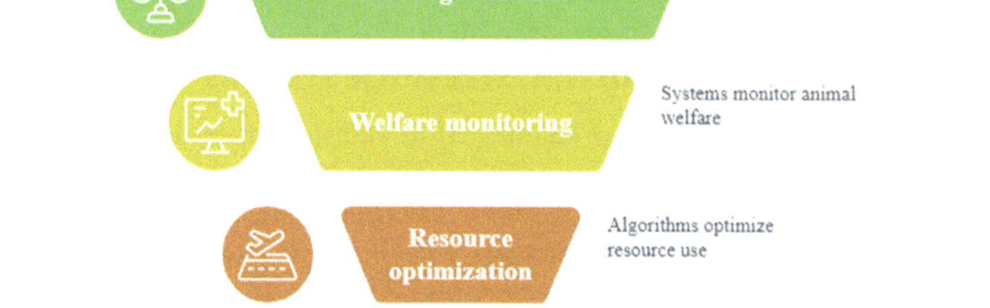

FIGURE 3.1 AI-driven livestock management process. This figure shows a system for managing livestock that is powered by AI and is made up of five parts that all work together. The first step is data processing, which gathers and gets biological data ready for analysis. Then, machine learning algorithms find patterns, like links between genetic markers and climate resilience, which can be used to improve breeding or decide how to divide up resources. These algorithmic results are controlled by ethical standards that put biocultural conservation and epistemic justice at the top of the list. This makes sure that decisions protect Indigenous livestock breeds and don't lead to genetic homogenization. Welfare monitoring systems keep an eye on stress biomarkers and other real-time indicators. They feed this information back into the system to improve predictions based on AI suggestions, ensuring a balance between productivity and sustainability. The closed-loop design makes sure that things are always getting better, and ethical governance makes sure that the results are in line with what people believe in.

3.2 Core ML paradigms in animal omics analysis

ML techniques transform animal omics analysis into a reconceptualization of biological interpretation, wherein the structures of ML algorithms become intermediaries between molecular data and zootechnical knowledge. This paradigm shift addresses traditional bioinformatics by being able to explore high-dimensional manifolds of genomic, proteomic, and metabolomic features where conventional statistical techniques fail due to the combinatorial explosion of interacting features. At its heart, the livestock ML ecosystem follows a composition of supervised and unsupervised structures, the orchestration of which is dependent on the ontological aspect of the scientific questions being asked. Supervised learning models, in particular those utilizing deep neural architectures, have become indispensable tools for representing non-linear links between polygenic signatures and phenotypic manifestations, translating single-nucleotide polymorphism (SNP) arrays and RNA-seq results into predictions for feed efficiency or disease resistance. In such environments, well-defined optimization targets guide the system between the panacea phase and complex behavior. The hidden layers of this behavior act as computational petri dishes for in silico experimentation with virtual genetic combinations (Ghavi Hossein-Zadeh, 2025; Mia et al., 2025; Crossa et al., 2025).

Beyond these application-driven methodologies, unsupervised learning frameworks are capable of uncovering hidden biological narratives encoded in omics datasets through manifolds and/or topology. Not covered here, clustering techniques are used to deconvolute intricate combinations of epigenetic marks and environmental stress cues, and variational autoencoders are applied to resurrect elemental biological dimensions along which such populations naturally live. This ability is especially important for revealing emerging properties in metagenomics microbiome data, in which the complex crosstalk between host genetics and gut-microbial composition resists reductionist analysis. For that matter, the real power of these unsupervised systems is not to be able to reproduce known biological facts, but rather to be able to formulate new biological hypotheses; in a manner of speaking, to invert the "direction" of scientific inquiry that is typically outlined in animal science (Larin and Karabelsky, 2025; Simpson et al., 2025).

The most promising profiles are currently found in hybrid architectures that combine the flexibility of the unsupervisory approach with the predictive advantage of the supervisory role. Semi-supervised methods make use of small amounts of labeled data that are typical within the realm of agricultural studies, allowing for knowledge to be transferred through well-connected graph neural networks that honor the internal connectivity of biological systems. Multi-modal transformers also redefine multi-omics integration, handling sequences emanating from genomic, transcriptomic, and metabolomic pipelines via attention mechanisms that adhere to the biological hierarchy. Such architectures especially show much potential when it comes to modeling temporal-biological processes, with their self-attention mechanisms learning the cyclic interdependent relationships of gene expression patterns among developmental stages and breeding cycles (Baião et al., 2025; Duy Nhat, 2025; Yetgin, 2025).

Both ethical and practical issues are related to the deployment of such ML paradigms, especially in connection to the interpretability-transparency paradox. Though deep learning (DL) models can predict methane emission reactivity scores, potential values, and marbling subjective scores at a level superior to those obtained by classical approaches, their black-box natures will directly contradict the expected requirement of understanding the causality of the predicted traits for the breeding sector. This tension has spurred the development of XAI frameworks, which bring together the representational abilities of neural networks and the symbolic reasoning of symbolic reasoning systems or glass-box models, these models that adhere to biological curiosity demands and regulatory scrutiny. It is also facing the challenge, resulting from globalized livestock genomics, of data sovereignty in which federated learning architectures need to weigh the collaborative benefits of shared omics repositories with the ethical rights of individual breeding companies (Kulaklıoğlu, 2024; David et al., 2025; Mia et al., 2025).

The future of education is moving in the direction of self-emergent learning systems that change and adapt to ongoing biological realities. The meta-learning architectures are expected to facilitate model development across livestock species by transferring data from omics-rich to sparse organisms. Quantum ML methods are on the horizon, promising to revolutionize predictions of protein folding and metabolic network simulations by exponential acceleration of combinatorial operations. However, these technological potentials are tempered by basic biological limitations, irreducible noise in the measurement of gene expression, epigenetic memory that persists through generations, the context-dependent genotype-phenotype relationships, and the dynamic nature of the genotype-environment interactions. ML paradigms discussed here are not simply technical tools, but frame a new epistemological standpoint on the comprehension of biological complexity, which requires a ceaseless interplay between algorithmic innovation and deep zootechnical wisdom for its responsible use in redesigning animal husbandry of the future (Mahawar and Rattan, 2025; Mu et al., 2025). A comparative matrix of ML algorithms in animal omics analysis is shown in Table 3.1.

TABLE 3.1 Comparative matrix of ML algorithms in animal omics analysis.

Algorithm class	Data profile	Core capabilities	Technical constraints	Animal omics applications	Biological value proposition
Supervised learning	▸ Labeled phenotypes ▸ Annotated genotypes	▸ Trait prediction ▸ Genomic selection accuracy	▸ Requires large labeled datasets ▸ Sensitive to class imbalance	▸ Breeding value estimation ▸ Disease susceptibility mapping	▸ Direct genotype-phenotype mapping ▸ Quantitative trait locus identification
Unsupervised learning	▸ Unlabeled omics data ▸ Sensor streams	▸ Population structure discovery ▸ Feature relationship modeling	▸ Interpretation challenges ▸ Sensitive to distance metrics	▸ Microbiome community typing ▸ Metabolite pattern discovery	▸ Hypothesis-free biological insight generation ▸ Latent biological process discovery
Semi-supervised learning	▸ Sparse labels ▸ Abundant unlabeled data	▸ Leverage cheap unlabeled data ▸ Improve model generalizability	▸ Requires careful pseudo-labeling ▸ Domain shift risks	▸ Phenotype imputation ▸ Cross-species knowledge transfer	▸ Cost-effective annotation ▸ Biological knowledge propagation
Ensemble learning	▸ Heterogeneous data sources ▸ Multi-modal features	▸ Error reduction through consensus ▸ Robust to noisy farm data	▸ Computational overhead ▸ Model interpretability issues	▸ Complex trait prediction ▸ Multi-omics data fusion	▸ Biological system complexity capture ▸ Compensate for individual model biases
Deep learning (DL)	▸ High-dimensional data ▸ Temporal sequences	▸ Automatic feature learning ▸ Complex pattern recognition	▸ Graphics Processing Unit (GPU) dependency ▸ Black-box limitations	▸ Histopathology analysis ▸ Whole-genome sequence modeling	▸ Hierarchical biological representation learning ▸ Cross-modal relationship extraction
Self-supervised learning	▸ Unannotated omics data ▸ Auxiliary tasks	▸ Pre-training without labels ▸ Context-aware representations	▸ Task design complexity ▸ Domain adaptation needs	▸ Pre-training on reference genomes ▸ Cross-breed feature learning	▸ Biological prior incorporation ▸ Foundation model development
Multi-task learning	▸ Correlated traits ▸ Shared biological mechanisms	▸ Joint feature learning ▸ Data efficiency improvement	▸ Task conflict risks ▸ Complex optimization landscape	▸ Pleiotropic effect analysis ▸ Multi-trait genomic prediction	▸ Biological system modeling ▸ Shared genetic architecture exploration
Explainable AI (XAI)	▸ Regulated applications ▸ Stakeholder-facing systems	▸ Model decision tracing ▸ Biological rationale extraction	▸ Performance-accuracy tradeoff ▸ Interpretation validation	▸ Regulatory compliance ▸ Farmer-facing decision systems	▸ Biologically plausible model auditing ▸ Causal relationship verification
Federated learning	▸ Distributed farm data ▸ Privacy-sensitive records	▸ Collaborative model training ▸ Data privacy preservation	▸ Communication overhead ▸ Heterogeneous data alignment	▸ Multi-farm disease detection ▸ Breed-specific model personalization	▸ Privacy-compliant knowledge sharing ▸ Distributed biological pattern discovery

Hybrid models	▶ Multi-omics data ▶ Heterogeneous data types	▶ Complementary strength integration ▶ Cross-domain knowledge transfer	▶ Integration complexity ▶ Training coordination challenges	▶ Genotype-environment interaction analysis ▶ Epistatic network modeling	▶ Holistic biological system modeling ▶ Multi-scale biological process integration
Generative AI	▶ Data-scarce scenarios ▶ Rare event modeling	▶ Synthetic data generation ▶ Biological process simulation	▶ Mode collapse risks ▶ Biological plausibility validation	▶ Rare disease modeling ▶ Experimental outcome prediction	▶ Biological mechanism exploration ▶ In silico hypothesis testing
Quantum ML	▶ High-complexity problems ▶ Combinatorial optimization	▶ Exponential speedup potential ▶ Quantum feature mapping	▶ Hardware limitations ▶ Algorithm maturity	▶ Protein folding prediction ▶ Optimal breeding scheme design	▶ Complex biological system simulation ▶ Large-scale genomic optimization
Causal ML	▶ Intervention analysis ▶ Confounded relationships	▶ Causal effect estimation ▶ Counterfactual reasoning	▶ Strong assumptions required ▶ Validation challenges	▶ Feed intervention impact analysis ▶ Breeding program optimization	▶ Actionable biological insights ▶ Mechanistic understanding
Swarm intelligence	▶ Dynamic optimization ▶ Non-linear landscapes	▶ Distributed problem solving ▶ Adaptive search strategies	▶ Parameter sensitivity ▶ Convergence guarantees	▶ Breeding strategy optimization ▶ Vaccine design optimization	▶ Biological-inspired optimization ▶ Complex system adaptation
Neuro-symbolic AI	▶ Knowledge-rich domains ▶ Rule-based systems	▶ Symbolic reasoning integration ▶ Biological knowledge infusion	▶ Knowledge engineering costs ▶ Hybrid system tuning	▶ Metabolic pathway analysis ▶ Gene regulatory network inference	▶ Biologically-grounded reasoning ▶ Domain knowledge preservation
Transfer learning	▶ Cross-species data ▶ Limited target data	▶ Knowledge reuse across domains ▶ Rapid model adaptation	▶ Negative transfer risks ▶ Domain shift challenges	▶ Model sharing across breeds ▶ Wildlife disease spillover prediction	▶ Cross-species biological knowledge transfer ▶ Conservation biology applications

3.3 DL architectures for multi-omics data integration

The combination of multi-omics data by DL is a game-changing concept in livestock systems biology that facilitates the translation of genomic, transcriptomic, proteomic, and metabolomic information into a single biological story that choreographs life itself. This computational problem goes beyond mere data fusion and requires the construction of architectures capable of disentangling hierarchical, nonlinear relationships that underlie biological systems across scales of resolution ranging from the atomic specificity of enzyme-substrate interactions to the emergence of complex behaviors in the whole ecosystems of microbial populations in the digestive system of ruminants. At its heart, multi-omics integration requires models that capture the idiosyncratic statistical properties of each layer, from the sparse pulsations of epigenetic modification to the cascading rhythms of gene expression waves and the stoichiometric balance of metabolic networks, as well as the collective spatiotemporal order that emerges when layers interact. Traditional ML methods, restricted by linear assumptions and an inability to learn a complex feature hierarchy, collapse in the face of this multi-dimensional complexity in the same way a cartographer trying to map quantum fluctuations with Renaissance-era tools does (Ballard et al., 2024; Sartori et al., 2025; Wadood et al., 2025).

Such complexity can be tackled by deep neural structures, which are tailored for distributed representation learning and can learn naturally to represent complex biological patterns in terms of sequences of nonlinear transformation layers that resemble the nested organization of organisms. These quotes are an inescapable evidence of how autoencoder-related formats, such as multimodal deep belief networks, have been successful in the task of mapping omics data of different types of omics data in a constrained way into a shared representation and preserving the modality-specific features through well-designed constraint operation. These architectures are particularly effective at learning cross-omics regulatory signatures (such as the intricate dance between DNA methylation patterns that silence transposable elements and the cascading effects throughout the lipid metabolism pathways, e.g.) by uncovering joint representations that encode the conditional dependencies of disparate data types through repeated cycles of compression and reconstruction. The real magic is in their capacity to capture pleiotropy (single genetic polymorphisms influencing multiple phenotypic traits through chain reaction molecular pathways) employing attention mechanisms that dynamically assign weights to inter-omics connections dependent on contextually-dependent biological significance, in the same way that cellular signaling networks prioritize essential communication lines during immunological challenges (Hussein et al., 2024; Yetgin, 2025; Zhou et al., 2024).

Convolutional neural networks are originally used on spatial data and have been recast for the integration of omics data using topological data analysis to represent nucleotide sequences as mathematical manifolds, leading to the emergence of epigenetic landmarks as persistent homology features. Such architectures treat the genomic sequences as a 1D manifold curved by the evolutionary selection pressures, versus protein interaction networks that are modeled as the graph-structured data representing allosteric communication pathways, and localize the patterns while preserving the global biological context by resorting to the dilated convolution operators with exponentially increasing receptive fields. This approach is particularly useful in studying host-microbiome interactions, where the spatial arrangement of microbial genes and their metabolites has to be jointly examined with host immune gene expression profiles, which are induced in a circadian and feeding cycle pattern. The use of residual connections and depthwise separable convolutions effectively alleviates the curse of dimensionality of multi-omics datasets, a phenomenon that is exaggerated in livestock species with high levels of genome duplication, enabling models to scale across species and breeding populations while reusing features in a parameter-efficient manner (Limbeck and Rieck, 2024; Alharbi et al., 2025; Tan et al., 2025).

In this landscape, transformers, with their self-attention mechanism, are revolutionizing how we integrate our omics data streams in time through semantic sequences of biological activity, ready for contextual semantics. We demonstrate their utility to handle longitudinal data structures, such as developmental transcriptomic trajectories across embryonic organogenesis or dynamic metabolic responses to dietary interventions, by encoding long-range dependencies across time points in both chronological order and developmental phase transitions through positional encoding strategies. The inherent nature of transformer architectures supports the chronological order of biological processes, such as epigenetic reprogramming events in early embryogenesis or metabolic adaptations of lactation cycles, and the multi-head attention layers can find conserved regulatory motifs among evolutionary divergent livestock species in a comparative analysis of attention heatmaps. This power is transforming the field of comparative genomics, allowing the insights from model species to be extended to agriculturally important species using cross-species attention mapping methods to align the orthologous gene loci while retaining the species-specific control features that regulate gene expression, a digital analog to molecular clock analyses implemented in extremely high-dimensional feature spaces (Camacho et al., 2024; Liu and Park, 2024).

Drugs, on the other hand, have traditionally been considered in a reductionist, single-gene-centric manner, largely neglecting network-level interactions between genes, proteins, and pathways. Through the node representation of genes using epigenetic states, and weighted gene-gene interaction by their chromatin accessibility scores, graph neural networks (GNNs) propagate information in molecular networks and maintain their topological structure using message-passing algorithms, which hold the directionality of the transcriptional regulations. This architecture does particularly well in pharmacogenomics applications where drug-target interaction predictions need to be made that take into account whole metabolic pathways that meander around the inside of the cell, and in breeding programs where polygenic patterns of inheritance need to be simulated across pedigree networks that run generations back and forwards in time. The message-passing mechanism in GNNs reflects a reality in biology in which cellular decision making results from the collective activities of interacting biomolecules engaged in an ongoing molecular conversation, in contrast to old models that depicted genes as lonely actors singing monologues on the stage of a cell (Zhao et al., 2024; Ma et al., 2025; Shi et al., 2024).

Careful thought should be given to the ethical considerations these architectures entail, especially concerning the interpretability-transparency paradox that appears to complicate biological discovery in the way that the uncertainty principle trails quantum measurement. While DL holds the promise of predicting complex traits from multi-omics data at an unprecedented level, predicting susceptibility to mastitis from milk exosome miRNA profiles or feed efficiency from rumen metagenome assemblies, its black-box nature complicates the identification of causal biological mechanisms and instead leaves researchers to traverse a maze of correlated features without an Ariadne's thread. Emerging neurosymbolic approaches seek to narrow this epistemic gap by combining differentiable reasoners with biological knowledge graphs to induce models not only capable of phenotypic prediction but also hypothesis generation about the molecular pathways underlying those phenotypes via abductive inference engines. At the same time, federated learning frameworks are tackling data sovereignty issues in international livestock genomics programs by using an encryption-based approach to collaborative model training without centralized data collection, which is crucial due to the proprietary nature of elite breeding lines and the regional disease prevalence patterns that make the conventional model of data sharing untenable and ethically fraught (Ouhmouk et al., 2025; Sartori et al., 2025; Şahin et al., 2024).

Future advances in this field include self-supervised architectures that exploit unlabeled multi-omics information at continental scales, which may even address the data scarcity for rare livestock breeds and understudied production environments through contrastive learning paradigms that help to draw biological insights from data shadows. The incorporation of physics-informed neural networks guarantees that omics predictions are rooted in biophysical constraints, guaranteeing that simulated metabolic networks satisfy thermodynamic principles and enzyme kinetics laws conditioned by Hamiltonian regularization terms. However, these technological achievements are still limited by basic biological challenges, the non-linear dynamics of gene regulatory networks that cannot be comprehensively explained through linear perturbation analyses, the context-dependent nature of epigenetic memory, which is adjusted by the type of environmental stress, and the multiscale interactions between molecular biology and herd-level management practices, which mix on cellular processes and agricultural decision-making. As DL algorithms grow ever more complex in an endless race to match the human brain's architectural glory, we can only hope that this dialectic of computational ambition and deep zootechnical wisdom will keep AI on a leash, so that the algorithms drain brain gain out of the technological infrastructure of animal farming without turning into black boxes indifferent to the earthly substrates of barnyards and pastures (Dradjat et al., 2025; Muthivhi and van Zyl, 2025). Architectural specializations for multi-omics integration challenges in livestock production are outlined in Table 3.2.

3.4 Data acquisition and preprocessing in livestock AI pipelines

The process of AI adoption in livestock management systems involves the systematic collection and processing of multimodal biological data, mediating the empirical knowledge of animal husbandry with computational abstraction. Data collection starts from connected networks of biosensors, IoT-based monitoring tools, and digital veterinary record systems, which generate constituted digital inputs from biological signals, including but not limited to physiological variables and behaviors. Embedded accelerometers and biometric sensors collect fine-grained measurements such as rumen motility, heart rate variability, and grazing dynamics, whereas computer vision systems infer welfare indicators from thermal imaging and postural analysis. These automated data feeds are then joined with historical husbandry records through optical character recognition and natural language processing (NLP) pipelines to create temporally aligned data sets that preserve the contextual richness of farming operations but lend themselves to computationally tractable analysis (Taiwo et al., 2024; Tangorra et al., 2024).

TABLE 3.2 Architectural specializations for multi-omics integration challenges in livestock production.

Architecture type	Livestock-specific challenge	Key components	Multi-omics processing mechanism	Production considerations	Biological validation use cases
Hybrid CNN	▶ Muscle development patterns ▶ Placental efficiency mapping	▶ 3D convolutional layers ▶ Tissue-specific kernel initialization	▶ Multi-scale feature fusion across histology/transcriptomics ▶ Spatial registration modules	▶ Requires farm-collected imaging data ▶ GPU-optimized for field deployment	▶ Meat quality prediction ▶ Reproductive efficiency optimization
Transformer-based	▶ Cross-breed genomic variation ▶ Feed efficiency patterns	▶ Breed-specific tokenization ▶ Attention masking for missing phenotypes	▶ Self-attention across SNP/transcriptome/proteome ▶ Adaptive context windows	▶ Benefits from distributed learning across farms ▶ Prone to overfitting small herds	▶ Selective breeding models ▶ Methane emission reduction strategies
Graph neural networks	▶ Gut microbiome-host interactions ▶ Metabolic pathway analysis	▶ Species-specific knowledge graphs ▶ Edge-weighted message passing	▶ Multi-omics node features ▶ Dynamic graph updating from longitudinal data	▶ Requires curated biological networks ▶ Sensitive to missing edge relationships	▶ Rumen function optimization ▶ Disease resistance network modeling
Autoencoder variants	▶ Missing microbiome data ▶ Sparse phenotypic records	▶ Robust encoders for noisy farm data ▶ Multi-task decoders	▶ Cross-modal reconstruction ▶ Adversarial domain adaptation for different breeds	▶ Handles uneven data quality ▶ Farm-to-farm transfer learning capabilities	▶ Gut health monitoring ▶ Stress response imputation
Attention mechanisms	▶ Priority trait identification ▶ Epistatic interaction detection	▶ Sparse attention mechanisms ▶ Biological prior-guided attention	▶ Dynamic feature weighting across omics layers ▶ Hierarchical attention stacking	▶ Memory-efficient designs needed ▶ Requires trait-specific pretraining	▶ Heat stress resilience markers ▶ Feed conversion ratio optimization
Multimodal fusion	▶ Wearable sensor integration ▶ Environmental adaptation analysis	▶ Time-sync fusion layers ▶ Sensor-omics alignment modules	▶ Adaptive fusion gates ▶ Residual cross-modal connections	▶ Real-time processing requirements ▶ Edge computing compatibility needs	▶ Precision nutrition systems ▶ Climate adaptation breeding programs
Memory-augmented	▶ Disease outbreak patterns ▶ Genetic lineage tracking	▶ Differentiable neural dictionaries ▶ Episodic memory recall	▶ Long-term pattern storage ▶ Context-aware retrieval mechanisms	▶ Requires compressed memory architectures ▶ Benefits from federated learning setups	▶ Zoonotic disease prediction ▶ Breeding value estimation
Capsule networks	▶ Pathogen structure analysis ▶ Muscle fiber organization	▶ Livestock-specific capsule dimensions ▶ Dynamic routing protocols	▶ Hierarchical part-whole relationships ▶ Pose-invariant feature detection	▶ Computationally intensive ▶ Requires high-resolution imaging data	▶ Vaccine development support ▶ Meat texture quality assessment
Diffusion models	▶ Rare breed genomics ▶ Low-coverage sequencing data	▶ Species-conditional denoising ▶ Ancestral state guidance	▶ Progressive data imputation ▶ Uncertainty-aware generation	▶ Slow sampling requires preprocessing ▶ Benefits from pretrained livestock models	▶ Genetic diversity preservation ▶ Ancient breed characterization
Neural ODEs	▶ Growth trajectory modeling ▶ Vaccine response dynamics	▶ Adaptive step solvers ▶ Latent space regularizers	▶ Continuous-time feature evolution ▶ Multiscale dynamics capture	▶ Memory-efficient for long-term studies ▶ Requires temporal data alignment	▶ Feed efficiency trajectory prediction ▶ Immune development modeling

The technical backbone of data collection fits the varying conditions of livestock production systems. IoT sensors are networked across grazing lands and use self-calibration techniques to adjust measurements and ensure they remain accurate during changing climatic conditions, and edge-based computing structures allow raw sensor data to be processed on-site to minimize delays in remote and low-bandwidth environments. Airborne surveillance technologies support the ground-based approaches from the power of hyperspectral image collection to develop models of pasture biomass, and from remotely sensed forage quality driven by ML that deliver tangible geotagged datasets that link nutritional amenity with animal movement. Power limitations typical in large-scale farming operations are overcome via energy generation and capture solutions, which transform kinetic energy from livestock activity into green electricity for 24/7 monitoring without impacting animal welfare. This cycle of iterative feedback between engineering constraints and biological needs parallels the adaptive interactions between domesticated organisms and their manipulated habitats (Ghavi Hossein-Zadeh, 2025; Losacco et al., 2025).

Preprocessing pipelines grapple with the intrinsic complexity of biological data because they involve layered computational approaches that are aimed at increasing the fidelity of the signal, while maintaining ecological validity. Adaptive filtering algorithms distinguish bio-markers of interest from environmental noise, such as stress-induced vocalizations against acoustic interference, using spectral analysis and context-sensitive pattern recognition. Imputation of missing data is used in the methodology by means of GNNs, which exploit the dependence between the states of physiological factors, social activities, and microclimate parameters in order to rebuild incomplete data on the basis of learned biological associations without using averages. Normalization pipelines normalize across potential sources of inter-breed or inter-species variation with dynamic-scaling algorithms in the context of population-specific metabolic rates, growth rates, and behavioral repertoires, allowing for cross-system comparison without denying genetic differences. Generative adversarial networks are used to enrich sparse pathological datasets with synthetic ones, producing virtual disease profiles consistent with biosecurity practices and extending the range of representational samples used to train diagnostic models (König et al., 2025; Murthy and Kumar, 2025).

Continued difficulties in data normalization and governance require novel solutions at the nexus of computation and agriculture. Diverse data standards with proprietary farm management systems require blockchain financial networks (BFNs) to include blockchain-regulated interoperable protocols to create common metadata standards while maintaining farm-level data control. In privacy-preserving federated learning frameworks, collective model training workflows are possible across decentralized agricultural networks without revealing sensitive operational content, providing a trade-off between shared intelligence on the one hand and competitive commercial interests on the other. The biological variability typical of mixed-production systems, ranging from intensive feedlot operations to free-range grazing, necessitates fair-aware preprocessing procedures that avoid algorithmic bias toward prevalent husbandry systems and thus ensure equitable model performance in distinct agricultural settings. Despite such challenges, new regulatory frameworks will continue to promote the development of explanatory preprocessing workflows, requiring organized audit trails of data processing operations that influence welfare-critical decisions to ensure some degree of human control in an automated husbandry setting (Maraveas et al., 2024; Dembani et al., 2025).

The trustworthiness of these data practices is what will ultimately determine the translational validity of AI applications in livestock management, whether computational models act as brittle abstractions or robust decision-support tools. ML systems can interpret biological signals in their ecological and ethical contexts thanks to robust preprocessing techniques. For example, they can identify that a change in vocalization patterns could be a sign of social stress rather than nutritional deficiencies, or that a spike in milk conductivity could be a sign of subclinical mastitis. In the context of increasing climate volatility and zoonotic disease risks, the ability to process multimodal stream data in near real-time is essential for predictive modeling of emerging health threats and space-induced resource allocation problems. The development of these pipelines suggests the future of autonomous systems responsive to reinforcement learning to self-optimize a quality metric for acquired data, changing acquisition parameters and preprocessing techniques in response to the changing environmental conditions and production objectives. Over time, as this data infrastructure is continuously improved, agricultural AI evolves from a grab-bag of analytic components into an adaptable system for sustainable livestock management, one that respects and encourages the biological complexity of managed ecosystems, but leverages computational precision to tackle the food-security problems of the 21st century (Ghavi Hossein-Zadeh, 2025; Kumar et al., 2025; Ma et al., 2025).

3.5 Edge AI and IoT integration in precision livestock networks

Combining edge AI with IoT frameworks ushers in a deep reconfiguration of precision livestock management by building distributed computing ecosystems tightly coupled with biological processes. Such a paradigm shift moves data analytics capabilities from centralized cloud infrastructures to the edge of agricultural settings, allowing the in-situ processing of diverse data streams, biometric, behavioral, and environmental, in real time, through context-aware actuation loops integrated in the operational layer of the farms. Edge devices, including subdermal biosensors that track metabolic dynamics and agricultural drones that monitor pasture conditions, constitute decentralized networks that perform on-site processing and thus retain the timeliness and location awareness of data that diminishes in transit to distant servers. The architecture is inspired by the structure of the brain and the nervous system, and so embodies the kind of distributed intelligence of a natural ecological system with processing hierarchies that interpret, filter, and respond to data at increasingly abstraction levels. Operational noise-reduction filters, such as those found in the microcontrollers, and attention-based architectures, such as the gateway nodes, ensure that the network focuses on important signals such as changes in respiration or feeding, and together execute a computational metabolism that is phase-locked to the natural rhythms of animal physiology (Ghavi Hossein-Zadeh, 2025; Kumar et al., 2025; Losacco et al., 2025).

The spatiotemporal compression achieved through edge computing reduces the constraint on cloud-based and cloud-restricted architectures for latency-sensitive interventions. These federated learning protocols allow for ongoing custom learning of models across device networks to utilize regional disease prevalence or nutrient suboptimal patterns to lift collective intelligence without undermining farm-level data sovereignty via cryptographic knowledge distillation. This resembles the adaptive dynamics of ruminant digestions with dataspaces offloading specialized analytical processing to different tiers in an energy-efficient way that can be implemented for specific precision in computations. Real-time applications are possible through closed-loop systems where edge-based reinforcement learning algorithms fine-tune environmental controls and food formulations using instant feedback from methane sensors, thermal imaging arrays, and acoustic monitors that investigate vocal stress signals. Economic, social, ethical, and cultural dimensions of such universal surveillance raise new frameworks that must deal with the ontological tension between livestock as ontogenetic agents and as data-generating network nodes, promoting encrypted edge processing and differential privacy technologies to prevent more biological insights while keeping the possibility of tailored welfare interventions. The sustainability issues that stem from the decentralized nature of IoT infrastructures push the frontiers of innovation in the realm of energy harvesting technologies that mimic ecological circularity. The bioenergy solutions convert kinetic energy harvested from animal movement using piezoelectrics and metabolic byproducts using microbial fuel cells into self-sustaining power cycles that minimize dependency on external energy grids. These systems implement the concept of regenerative agriculture in the realm of technology by tying maintenance of the sensor network to the metabolic outputs of the observed organisms. As edge systems independently carry out minute modifications to ventilation and feeding schedules, the cognitive revolution of farm management is taking place. This could replace traditional husbandry skills and shift human attention toward strategic welfare optimization and ethical supervision of automated decision streams (Alzu'bi et al., 2025; Modak et al., 2025; Raha et al., 2025; Rajawat et al., 2025).

Future biohybrid constructs may involve livestock playing an active role in data ecosystems through their interaction with neuromorphic interfaces tailored to species neural processing. While self-organizing mesh networks could dynamically change communication topologies in response to herd movement patterns, simulating biological collective intelligence seen in flocking behaviors, spiking neural networks might interpret behavioral patterns using temporal encoding schemes that mimic animal cognition timescales. Quantum-inspired tensor networks shall break the nonlinear crosstalk between genomic predisposition and environmental stressors that could lead to pre-symptomatic markers for disease prediction, as they correlate epigenetic re-programming and microbiome perturbations. These developments require interdisciplinary coherence, inspiring new theoretical approaches in animal-computer interaction research to bridge ethological requirements with the data collection needs (Singh et al., 2024; Li et al., 2025a).

This technological integration disrupts, at a deep level, the epistemology of animal husbandry itself, leading to the emergence of hybrid knowledge systems where ML outputs and traditional wisdom inform each other iteratively as habitus, while traditional and algorithmic priors for farmed animals' lives are mutually entangled. This tripartite cognitive ecology of human operators, animals, and artificial agents disrupts traditional hierarchies of agricultural decision-making, calling for new shapes of distributed responsibility and multispecies consent. The metrics of success transcend technical sophistication, prioritizing improvement of ecological balance through targeted interventions that minimize pharmaceutical dependence and optimize welfare outcomes. As edge intelligence species sinks into the agricultural system, it should cast off the fetters of technological determinism and uphold its support of eco-ethics and sustainable stewardship. Finally, the largest challenge is to develop systems that do not treat AI as an external authority, but as an embedded agent whose role is to maintain the fine balance between productivity and biological integrity, respecting the co-evolutionary agreement between humans and their domesticates in every act of computation (Ghavi Hossein-Zadeh, 2025; Santana et al., 2025). A convergence framework for edge computing components and livestock monitoring functions is presented in Table 3.3.

TABLE 3.3 Convergence framework for edge computing components and livestock monitoring functions.

Edge component	Technical specifications	Monitoring function	Data processing workflow	Value proposition
Edge sensors	▸ Multi-modal sensing (thermal/3D accelerometer/audio) ▸ Ultra-low power (<1 mW sleep) ▸ IP67 waterproof rating	▸ Real-time biometric tracking ▸ Environmental microclimate measurement ▸ Behavioral pattern recognition	▸ Adaptive sampling rates ▸ Local anomaly detection ▸ Compressed data packaging	▸ Early lameness detection ▸ Heat stress mitigation ▸ Feed efficiency optimization
Edge gateways	▸ Multi-protocol support (LoRaWAN/Zigbee) ▸ 256-bit encryption ▸ Edge-node management	▸ Herd-level aggregation ▸ Cross-sensor correlation ▸ Alert prioritization	▸ Temporal data alignment ▸ Context-aware filtering ▸ Protocol translation	▸ Distributed compute load balancing ▸ Cyber-physical system security ▸ Network resilience
AI accelerators	▸ 8-TOPS NPU performance ▸ Model quantization support ▸ Dynamic voltage scaling	▸ On-device lameness scoring ▸ Vocalization analysis ▸ Feed intake estimation	▸ Parallel model execution ▸ Attention mechanism optimization ▸ Gradient pruning	▸ Sub-100 ms inference latency ▸ Energy-efficient processing ▸ Model personalization
Edge databases	▸ Time-series optimization ▸ Federated learning support ▸ General Data Protection Regulation (GDPR)-compliant anonymization	▸ Longitudinal record keeping ▸ Cross-batch comparison ▸ Trend analysis	▸ Rolling window aggregation ▸ Differential privacy ▸ Spatial indexing	▸ Lifetime productivity tracking ▸ Breeding program support ▸ Regulatory compliance
Edge controllers	▸ Real-time OS (RTOS) ▸ PLC integration ▸ Fail-safe mechanisms	▸ Automated environment adjustment ▸ Robotic milking coordination ▸ Feed dispensing	▸ Closed-loop control algorithms ▸ Safety constraint enforcement ▸ Priority queuing	▸ Precision resource allocation ▸ Labor cost reduction ▸ Animal welfare compliance
Edge analytics	▸ Streaming SQL engine ▸ Federated learning orchestration ▸ Concept drift detection	▸ Cross-herd benchmarking ▸ Productivity forecasting ▸ Anomaly root cause	▸ Online feature engineering ▸ Model versioning ▸ Explainability scoring	▸ Operational intelligence ▸ Predictive maintenance ▸ Continuous improvement
Edge visualization	▸ WebGL-accelerated rendering ▸ AR overlay support ▸ Multi-tenant dashboards	▸ Heatmap generation ▸ 3D movement tracking ▸ Facility layout optimization	▸ Real-time rasterization ▸ Spatial data fusion ▸ Interactive query processing	▸ Intuitive system monitoring ▸ Rapid decision support ▸ Collaborative troubleshooting
Edge security	▸ Hardware root of trust ▸ Behavioral biometrics ▸ Zero-trust architecture	▸ Device authentication ▸ Data provenance ▸ Cyber-physical threat detection	▸ Continuous attestation ▸ Anomaly pattern learning ▸ Secure over-the-air updates	▸ Supply chain integrity ▸ Intellectual property protection ▸ Operational continuity

Continued

TABLE 3.3 Convergence framework for edge computing components and livestock monitoring functions.—cont'd

Edge component	Technical specifications	Monitoring function	Data processing workflow	Value proposition
Edge actuators	▶ Industrial IoT protocols (OPC UA, MQTT) ▶ Fail-safe mechanical design ▶ Precision servo controls	▶ Automated feed dispensing ▶ Climate system regulation ▶ Gate control	▶ Real-time command execution ▶ Safety interlocks ▶ Predictive maintenance triggers	▶ Closed-loop system automation ▶ Energy-optimized operations ▶ Immediate environmental response
Edge drones	▶ Autonomous navigation (RTK-GPS) ▶ Multispectral imaging ▶ Swarm coordination	▶ Pasture quality assessment ▶ Herd localization ▶ Predator detection	▶ Onboard NDVI calculation ▶ Obstacle mapping ▶ Thermal anomaly detection	▶ Large-scale terrain monitoring ▶ Rapid emergency response ▶ Growth monitoring
Edge mesh network	▶ Self-healing topology ▶ Dynamic spectrum access ▶ Multi-hop communication	▶ Network coverage optimization ▶ Device connectivity health ▶ Data backhaul	▶ Adaptive channel selection ▶ Packet prioritization ▶ Congestion control	▶ Farm-wide connectivity assurance ▶ Fault-tolerant communications ▶ Scalable infrastructure
Edge fog nodes	▶ Distributed computing fabric ▶ Latency-sensitive orchestration ▶ Edge-cloud synchronization	▶ Regional data aggregation ▶ Cross-farm benchmarking ▶ Disaster recovery	▶ Federated learning coordination ▶ Geo-distributed analytics ▶ Version control	▶ Hybrid intelligence deployment ▶ Multi-enterprise collaboration ▶ Cloud burst capability
Edge blockchain	▶ Lightweight consensus (PoA) ▶ Smart contract engine ▶ Immutable audit trails	▶ Supply chain provenance ▶ Data integrity verification ▶ Access control	▶ Cryptographic hashing ▶ Tokenized asset tracking ▶ Permissioned ledger updates	▶ Trusted data sharing ▶ Regulatory audit compliance ▶ Anti-counterfeiting protections
Edge digital twins	▶ Physics-based modeling ▶ Real-time data fusion ▶ Scenario simulation	▶ Virtual herd monitoring ▶ Facility stress testing ▶ Growth prediction	▶ Multiscale system emulation ▶ Parameter optimization ▶ Failure mode analysis	▶ Risk-free operational planning ▶ Predictive what-if analysis ▶ Training simulator development
Edge AR interfaces	▶ Spatial computing SDKs ▶ Hand gesture recognition ▶ Ruggedized HMD support	▶ Equipment maintenance guidance ▶ Virtual fencing setup ▶ Training simulations	▶ 3D registration ▶ Context-aware overlays ▶ Collaborative annotation	▶ Enhanced human-animal interaction ▶ Reduced cognitive load ▶ Remote expert assistance

3.6 Predictive analytics for disease surveillance and herd health optimization

The rise of predictive analytics for livestock disease control spells a new era of proactive governance as opposed to reactive intervention, in which AI synthesizes multi-omics signatures and environmental vectors into probabilistic frameworks that can model biological cascades across spatial and temporal scales. This computational revolution utilizes transformer architectures to analyze epigenetic drift patterns, volatile organic compound emissions, and gut microbiome dynamics in unison with geospatial pathogen distribution maps to generate four-dimensional manifolds that model disease propagation as emergent phenomena within complex adaptive systems. The coevolutionary dance between pathogen mutation rates and host immunogenetics is modeled by neural ordinary differential equations, and the coupled dynamics of microbial evolution and animal behavior are quantified by graph attention networks, which depict transmission pathways through livestock social hierarchies. At the analytical core lie Bayesian frameworks that capture uncertainty using random sampling to construct Hamiltonian Monte Carlo models and render them susceptible to probabilistic forecasts of antibiotic resistance emergence or heat stress-induced viral recombination events, together with interpretability enabled through layer-wise relevance propagation methods (James Ezeh et al., 2024; Eckhardt et al., 2025; Lin et al., 2025).

At the core of this development is the re-framing of the management of herd health as a dynamic optimization problem over intertwining biological and operational axes. Deep reinforcement learning platforms coordinate targeted interventions by interpreting streaming salivary biomarker and respiratory aerosol data through edge computation nodes, monitoring adaptive ventilation protocols or micronutrient supplementation in function of envisioned infection threats. The application of epigenetic aging clocks and immune senescence profiles contributes temporal depth to culling strategies, weighting genetic potential against longevity through multi-objective functions reflecting productivity against welfare. Rising above static epidemiological models, this work involves a rendering of livestock as part of digital twins of agricultural ecosystems, virtual metaherds on which climate change simulations and emergent pathogen scenarios can be played out that stress-test intervention strategies in the face of the Intergovernmental Panel on Climate Change (IPCC)-grade environmental projections (Ismail et al., 2024; Hassan et al., 2025; Ghavi Hossein-Zadeh, 2026).

Ethical issues grow in articulation from predictive resolution as it approaches individual-animal prognostication, resulting in moral dilemmas in algorithmic prioritization of preventative resource allocation. Polygonal risk scores deriving from polygenic susceptibility indices and social network centrality metrics complicate traditional veterinary ethical clarity with triage circumstantial contexts, which are predicated on herd prioritization of outcomes. The ontological positioning of the probabilistic ramblings of machines raises philosophical debates about how to treat the qualitative risk assessment of animals for early medical intervention, while avoiding the reduction of animals to fungible human identities (an actuarial dusk or value). Simultaneous algorithmic equity dilemmas arise as algorithmic original training data were developed from an intensive agricultural setting, and subsequently manifest inherent biases that could multiply inequities in global governance to improve herd health, where external setting conditions and resource allocation cannot be generalized to livestock ecosystem (pastoralist) systems. Hybrid architectures that couple XAI with ethological decision trees lie between these two extremes. These architectures result in auditable programmatic pathways that ground future orientation predictions arising from veterinary first principles, while maintaining the capability of the model to be flexible through federated learning layering (Ghavi Hossein-Zadeh, 2025; Gonzalez-Argote et al., 2025; Naripeddy and Thotakura, 2025).

Technology is progressing toward symbiotic predictive systems that feature co-evolving biological and computational systems utilizing reflexive feedback to facilitate states of coevolution. Synthetic biology interfaces come in the form of engineered phage vectors, which use pathogen loads to create a bioluminescent signal captured by hyperspectral imaging drones. Neuromorphic chips deliver disease pattern distributions across spectrums of spiking neural networks utilizing livestock neurophysiology. Quantum annealing systems facilitate combinatorial optimization processes for vaccination scheduling, taking into account the impact of immune imprinting in the viral lineage based on qubit-based antigenic distance matrices. Blockchain-based health systems provide immutable chains of custody for predictive insights, sustaining ongoing model development through cryptographic commitment to the relationship between forecasts and outcomes. These systems conduct a cybernetic conversation between algorithmic projections and biological reality, generating collectively learned behavior within farming networks (Max et al., 2025; Lu and Xiao, 2024).

Yet, these advances have fundamental limitations determined by biological complexity at baseline (what could be referred to as biological complexity at baseline) because you cannot derive fundamental emergent properties from the interplay of functions that take place within the microbiome-immune axis external to the livestock system itself, and also, epigenetic memory and its transgenerational significance that cannot be reduced to simplified modeling. These fields are maturing through epistemic humility processes that measure the uncertainty in models, hybrid models that juxtapose canonical anecdotal ethnoveterinary knowledge against the sensor data, and human oversight protocols that leave the mechanism open to say yes or no to life-denying predictive models (Wang et al., 2024; Mittal et al., 2025).

By developing a more organic approach to systematizing, we can create resilient processes where AI acts alongside to support (rather than replace) the necessary depth of thinking needed for sustainable husbandry. In this just version of the future of agriculture, then, there are no baselines that can be declared because there is no absence of disease, only growing adaptive capacity that expands the abilities of livestock populations to co-evolve with threats from the environment so that livestock populations, and all the relations, seek to support and protect livestock health, which can thrive in a more mutualistic future that collaboratively creates health narratives to displace entrenched norms through ongoing mutual adaptation, where computational foresight meets biological plasticity. The components of the early warning system in relation to the pathogen transmission characteristics in herds are described in Table 3.4.

3.7 Computer vision in automated phenotypic trait analysis

Computer vision in livestock phenomics is revolutionizing agricultural science by re-engineering standardized farm settings into advanced analytics. We are now working with existing systems that incorporate traditional analytics, hyperspectral imaging arrays, LiDAR-based depth sensing, and thermal imaging to measure multidimensional phylogenetic datasets of anatomical and physiological traits. This activity is all being guided by a geometric DL construct to map animal morphology into invariant feature spaces to account for noise in analysis across spatial and temporal variations. In addition, we will apply convolutional neural networks with spatiotemporal attention to decode complicated behaviors manifested in livestock, such as muscle flexion (i.e., muscle temperature related to specific flexion behaviors), and small micro-expressions associated with stress to form a computational basis for understanding the phenotypic response (Fatoki et al., 2024; Ghavi Hossein-Zadeh, 2025; Yang et al., 2025).

This technological innovation presents phenotypic analysis as a continuous data-driven process. Vision systems record developmental changes in real time, connecting biological deviations in bone growth to genomic markers of skeletal health or linking the development of feather follicles with epigenetic markers of aging. In addition to capturing data for phenotypic recordings, self-supervised learning architectures can identify minute phenotypic invariants that escape human observation, such as the variations in udder morphology predictive of mastitis. Physics-informed neural networks synthesize known laws of biomechanics to estimate internal tissue properties from changes in external deformations due to engagement in locomotion. These algorithms enable true non-invasive measures that discern muscle mass or can help identify early signs of lameness through analysis of gait. These analyses contribute to the establishment of dynamic digital twins that evolve with live biological subjects and can provide real-time data regarding the ongoing health and productivity associated with the artificial being (Böttcher et al., 2025; Ren et al., 2025).

Considerations of ethics arise as vision systems reach practically unlimited observational resolution. For example, due to continuous biometric monitoring, such as pupil dilation metrics that detect stress and vulvar colorimetry to predict estrus, the ethical ramifications of constant surveillance linger regarding animal privacy. The risk of algorithm bias is only beginning to be recognized, in particular concerning models trained on a specific breed that fail to generalize to genetically different populations, pointing again not only to a need for federated learning frameworks, but also to sovereign privacy protections affording localized model training. Environmental sustainability will also continue to spur the development of eco-efficient computing resources, including bio-inspired vision chips and event cameras simulating retinal ganglion composition, given orders of magnitude of computational simplicity (Tahir, 2024; Katoch, 2025).

New technologies will allow computer vision in molecular-scale phenotyping. For instance, some integrated systems contain optical sensing platforms coupled with Raman spectroscopy to allow external and internal metabolite levels to be related in ways to phenomenologically derive holistic health assessments. Some enhanced imaging technologies exploit quantum mechanics to bypass optical diffraction boundaries, allowing analysis of feather barbule or hair follicle microbiomes in nanometer resolution, without damaging sampling. Moreover, using AI, typically in the form of generative adversarial networks, phenotypic model generation can be done in entirely virtual environments, allowing researchers to design experiments on genetic crosses where they can explore potential outcomes in a model before testing with actual crosses. Of great significance for machine physiological or behavioral responses specifically are new neuromorphic vision systems using spiking neural networks, and exploiting biological sensory latencies, to allow detecting real-time physiological events like heat stress or fear reflex responses (Lu et al., 2024; Zhou et al., 2025).

Notwithstanding these strides, there remains a basic difference in biological complexity. Fractals in animal coats, chaotic trajectories of breathing in fluid systems, and quantum biological phenomena do not allow for full computational modeling. Future growth will depend on hybrid approaches that combine ML with expertise from the domain, along with uncertainty quantification to identify ambiguous observations for human ratification, while sharing interactive interfaces

TABLE 3.4 The components of the early warning system in relation to the pathogen transmission characteristics in herds.

System component	Technical specifications	Pathogen transmission characteristics addressed	Detection methodology	Mitigation alignment
IoT bio-sensors	▶ Multi-parameter sensing (temp/ECG/respiration) ▶ 0.1°C thermal accuracy ▶ Subcutaneous implantation compatible	▶ Host susceptibility variations ▶ Pre-symptomatic shedding ▶ Stress-induced immunosuppression	▶ Continuous vital sign streaming ▶ Behavioral baseline comparison ▶ Fever pattern recognition	▶ Individual quarantine protocols ▶ Automated ventilation adjustments ▶ Precision antibiotic dosing
Pathogen genomics	▶ Nanopore sequencing ▶ CRISPR-based detection ▶ Cloud-based strain libraries	▶ Antigenic drift/drift ▶ Zoonotic crossover potential ▶ Antibiotic resistance markers	▶ Wastewater metagenomic analysis ▶ Airborne particle capture ▶ Milk somatic cell profiling	▶ Vaccine matching optimization ▶ Transmission chain mapping ▶ Biosecurity protocol updates
ML prediction models	▶ Transformer architectures ▶ Federated learning ▶ SHAP explainability	▶ Non-linear transmission dynamics ▶ Super-spreader potential ▶ Environmental persistence	▶ Spatio-temporal cluster detection ▶ Network contagion modeling ▶ Weather pattern correlation	▶ Dynamic vaccination scheduling ▶ Movement restriction zoning ▶ Feed additive optimization
Environmental monitors	▶ Ammonia/VOC sensors ▶ Particulate counters ▶ UV radiation detection	▶ Aerosol transmission range ▶ Fomite survival duration ▶ Vector breeding conditions	▶ Air quality index calculations ▶ Surface contamination mapping ▶ Microclimate modeling	▶ UV-C disinfection cycles ▶ Bedding material rotation ▶ Natural ventilation optimization
Wearable trackers	▶ UWB positioning (<10 cm accuracy) ▶ Social graph mapping ▶ Activity classifiers	▶ Direct contact networks ▶ Density-dependent transmission ▶ Maternal-neonatal transfer	▶ Proximity duration logging ▶ Interaction frequency analysis ▶ Maternal bonding monitoring	▶ Automated cohort regrouping ▶ Farrowing pen isolation ▶ Colostrum management systems
Drone surveillance	▶ Multispectral imaging ▶ Thermal anomaly detection ▶ Autonomous patrolling	▶ Wildlife-livestock interfaces ▶ Carcass transmission risks ▶ Waterborne spread vectors	▶ Pasture contamination scanning ▶ Wildlife intrusion alerts ▶ Water source quality imaging	▶ Perimeter fencing automation ▶ Carcass removal prioritization ▶ Water treatment activation
Blockchain traceability	▶ Smart contract triggers ▶ GS1-compliant tagging ▶ Immutable movement logs	▶ Supply chain amplification ▶ Cross-herd transmission ▶ Market-day exposure risks	▶ Livestock passport verification ▶ Transportation route auditing ▶ Market contact tracing	▶ Dynamic risk-based trading restrictions ▶ Transport vehicle sanitization mandates ▶ Breeder certification systems
Digital twins	▶ Agent-based modeling ▶ Fluid dynamics simulation ▶ Reinforcement learning	▶ Ventilation-driven spread ▶ Equipment-mediated transfer ▶ Staff-mediated transmission	▶ Airflow pattern analysis ▶ Equipment usage simulation ▶ Human-animal interaction modeling	▶ Facility redesign recommendations ▶ Equipment sterilization protocols ▶ Staff rotation schedule optimization

to visualize and reflect on neural network decision-making processes with ethologists. By balancing the capabilities of technology and an epistemological humility, computer vision will become a collaborative approach to allow for a broader human understanding of animal biology, while recognizing the scope and idiosyncratic nature of living systems. The nature of this transition underscores the historical transition from knowledge as a mode of surveillance to one that reveals our ethical responsibilities as stewards of nature, acknowledging the need to maintain farm productivity with animal welfare imperatives (Pyrkov et al., 2024; Neethirajan, 2025). Phenotypic feature extraction techniques across livestock species are listed in Table 3.5.

3.8 Intelligent adaptive feeding systems

Intelligent adaptive feeding systems illustrate the intersection of computational biology and precision agriculture, creating a cyber-physical system where real-time physiological measurements contribute to adaptive feed solutions. The systems integrate multiple situational data streams, including biometric indicators of behavior, biochemical telemetry, physiological parameters, and environmental measures, and are modeled on multilevel neural networks focused on metabolic equilibrium. By deconstructing spectral analyses of rumen fluid and utilizing key measures of grazing behavior from Inertial Measurement Units, these systems generate multi-faceted nutrient requirement profiles. The key advantage of intelligent adaptive feeding systems is self-optimizing reinforcement learning, which develops feed formulations and optimizes the stoichiometric ratios of amino acids against epigenetic indicators of muscle hypertrophy at circadian metabolic rates. Integrated biosensor arrays that utilize surface-enhanced Raman spectroscopy allow the measurement of nutrient absorption from intestinal villi with nanometer resolution, creating closed-loop-feedback feeding systems that can respond to preclinical levels of acidosis through probabilistic adjustments to the ratio of fiber in the feedstock (Ghavi Hossein-Zadeh, 2025; Kalbhor, 2025).

The operating framework goes beyond operational adjustments with reactive systems to develop predictive metabolic forecasting models using transformer-based models on longitudinal metagenomic databases. Each individual would build on a cumulative record of all the proteomic responses associated with hepatic gluconeogenesis, along with the metagenomic responses of the methane-producing bacteria associated with enteral dynamics, thus providing counterfactual modeling-based simulations of nutrient partitioning within each production scenario. A substantial step forward is represented in the systems' modeling ability to detect latent biomarkers (e.g., minor differences in mastication acoustics, or differences in infrared thermography vascular perfusion) that can presage overt metabolic distress. The adoption of federated learning setups further enhances robustness by providing predictive modeling with privacy preservation from distributed farms or farms with local data, as well, so they encode geo-specific variability in the mycotoxin profile of forage or water salinities, which alter nutrient bioavailability (Khawaja et al., 2024).

Future developments in intelligent feeding systems will leverage advancing synthetic microbial consortia engineering, combined with developments in neuromorphic computing materials and testable and programmable probiotic vectors capable of safely delivering vitamins in situ, which are likely to interface with AI-controlled microfluidic dispensing to provide dynamic reservoirs of nutrients hosted in the gastrointestinal tract. As with AI, applying topological data analysis to the digestion process may also lead to the ability to dynamically adjust the feed particle concerning geometry to optimize the topological conditions about enzyme access to the digestion matrix, and those same models using quantum annealing architectures could provide solutions to solving combinatorial optimization problems with respect to least-cost rations for changing commodity markets. In addition, the ethical commitment to avoiding movement outside the state of ecological disequilibrium when using hyper-precision feeding to practice livestock animal husbandry will also necessitate a commitment to developing new multi-objective optimization practices that focus on optimizing animal welfare, resource efficiency, and circularity in the environment. The new intelligent feeding systems will form a part of an ecosystem, from feeding optimizers to holistic livestock management systems to integrate genomic selection platforms and carbon footprint trackers, while sustainability in agriculture is being redefined based on the impacts of humans on the earth since the start of the Anthropocene (Bishwas et al., 2024; Tulpan, 2024; Fushai et al., 2025; Safra Zaabar et al., 2025).

3.9 NLP for veterinary clinical text mining

The use of NLP in veterinary clinical text mining is an emerging discipline that gives rise to a new computational livestock management paradigm, linking unstructured, raw clinical language to biomedical data-informed interpretations and intentions. Today's systems utilize transformer-based architectures to decipher the semantic complexity of veterinary clinical data. For example, veterinary records comprise an array of irregular terminology; using veterinary-set programming languages, records will include vernacular descriptions of clinical symptoms, abbreviated codes for clinical

TABLE 3.5 Phenotypic feature extraction techniques across livestock species.

Livestock species	Computer vision technique	Phenotypic features extracted	Technical specifications	Data processing pipeline	Value proposition
Cattle	▶ 3D depth sensing ▶ Thermal imaging ▶ Gait analysis algorithms	▶ Body condition scoring ▶ Muscle volume estimation ▶ Udder morphology ▶ Hoof angle measurement	▶ 0.5 mm spatial resolution ▶ 30fps capture rate ▶ Dual-spectrum (VIS + NIR) imaging ▶ Azure Kinect DK integration	▶ Point cloud registration ▶ Skeletonization models ▶ Dynamic time warping for gait ▶ Volumetric reconstruction	▶ Precision breeding selection ▶ Early lameness detection ▶ Milk yield prediction ▶ Feed efficiency optimization
Swine	▶ Top-view RGB-D cameras ▶ Infrared thermography ▶ Depth-based segmentation	▶ Backfat thickness ▶ Loin eye area ▶ Tail posture analysis ▶ Vulva development staging	▶ 4K @ 60fps resolution ▶ ±1°C thermal accuracy ▶ Multi-hog tracking algorithms ▶ Stain-resistant housing	▶ Depth thresholding ▶ Thermal pattern recognition ▶ Growth curve alignment ▶ Farrowing readiness scoring	▶ Carcass quality prediction ▶ Estrus cycle monitoring ▶ Tail biting prevention ▶ Lean meat percentage estimation
Poultry	▶ High-speed cameras ▶ Feather coverage analysis ▶ Beak morphology detection	▶ Gait scoring ▶ Breast muscle conformation ▶ Comb color analysis ▶ Footpad dermatitis	▶ 1000fps motion capture ▶ UV reflectance imaging ▶ Hyperspectral line scanning ▶ Dust-resistant optics	▶ Optical flow analysis ▶ Feather density mapping ▶ Colorimetric calibration ▶ Lesion segmentation networks	▶ Broiler welfare assessment ▶ Egg production potential ▶ Disease resistance scoring ▶ Heat stress adaptation
Sheep/Goats	▶ LiDAR-based scanning ▶ Wool fiber analysis ▶ Facial recognition	▶ Fleece density ▶ Horn development ▶ Body condition scoring ▶ Udder attachment	▶ 0.1 mm LiDAR precision ▶ Polarized light microscopy ▶ Iris pattern recognition ▶ Outdoor lighting compensation	▶ Fiber diameter measurement ▶ 3D horn modeling ▶ BCS neural networks ▶ Multi-animal tracking	▶ Wool quality grading ▶ Genetic merit assessment ▶ Parasite load estimation ▶ Twin lamb disease prevention
Aquaculture	▶ Underwater multispectral imaging ▶ Stereo-video systems ▶ Hyperspectral analysis	▶ Body symmetry ▶ Fin erosion ▶ Scale reflectance ▶ Operculum movement	▶ 10-bar pressure housing ▶ 470–900 nm spectral range ▶ Turbidity compensation ▶ Schooling behavior tracking	▶ Morphometric landmarking ▶ Water quality correlation ▶ Stress response quantification ▶ Biomass estimation	▶ Growth rate prediction ▶ Disease outbreak early warning ▶ Stock density optimization ▶ Feed conversion efficiency

diagnoses, and clinical observations contingent on various contexts. Underpinning these models are continuous conceptual models or "embeddings", which include a hierarchical attention mechanism, in addition to knowledge embedding layers that are domain-adapted, that untangle hidden patterns in textual data, such as chronological tracking of subclinical infections or idiosyncratic behavioral symptomatology, into structural units in an actionable fashion. The groundwork for canine health ontologies, along with additional veterinary-relevant ontologies, allows for the knowledge representation of initial free text entries into measurable "phenotype ontologies", providing for knowledge to cross institutional boundaries while enabling resolutions to populate the original essence of clinical language representations (Davies et al., 2024; Elvas et al., 2025; Ghavi Hossein-Zadeh, 2025).

Improvements in few-shot learning approaches have increased the applicability of veterinary NLP systems to the unique data limitation issues of small-scale annotated datasets for specialized domains. Prototypical networks trained using multimodal clinical corpora, such as radiology reports, postmortem reports, and records of treatment over time, learn how to generalize from a limited number of examples, and they will identify implicit linguistic features that are predictive of new zoonotic diseases or patterns of resistance to antibiotics. Importantly, cross-modal alignment techniques allow these systems to align spoken language with other unstructured and complementary data streams, such as associations of descriptive phrases about coughing or respiratory distress in clinical notes with synchronized accelerometer data about the frequency of coughing. Multimodal reasoning naturally allows these systems to identify previously hidden associations between husbandry practices and (i) the respiratory disease susceptibility in poultry husbandry, and (ii) polygenic confounding factors associated with non-infectious diseases related to genes and environments (Balakrishnan et al., 2025; Li et al., 2025b).

The implementation of veterinary NLP requires new approaches to domain adaptation that consider the unique lexico-syntactic characteristics of clinical text in the livestock context. Curriculum learning, which allows models to learn progressively more specialized vocabulary, from veterinary vernacular to local dialects and herd-specific slang, produces strong outcomes in diverse farming contexts. At the same time, federated learning paradigms enabled by privacy preservation will foster collaborative retraining of models across veterinary networks while keeping the identifiable information of clients private, allowing for the integration of the latest lexicons on new diseases while meeting rigid data governance policies. One key development is explainable attention interfaces that allow the model's decisions to be represented visually and related to indexed text in the clinical narrative. This will engender more trust by demystifying the model's algorithmic inferences about disease progression or treatment (Kesgin and Amasyali, 2024; Khan et al., 2024).

Advances in veterinary NLP will have the potential to align with advances in causal inference and knowledge graph creation. For instance, the use of dynamic temporal graph networks that capture patient state change via clinical narratives could support real-time prediction of disease complications, whereas neurosymbolic systems that combine neural language models with rule-based reasoning may allow for automating the generation of evidence-based treatment recommendations by summarizing the preceding multiple case reports/searches. Furthermore, there will be fascinating opportunities to combine large language models with veterinary omics databases to frame hypotheses about the underlying molecular mechanisms associated with textual representations of clinical phenotypes. Over time, as these systems are developed, we will see them transition from passive analytical systems into more active partners in veterinary practice. For example, they will help identify inconsistencies in diagnosis, provide differential diagnoses based on epidemiological information, and ultimately, play an important role in the development of personalized approaches to veterinary medicine focused on improving the welfare of individual animals while ensuring farm profitability (Davies et al., 2024; Nalongo Bina, 2025; Ghavi Hossein-Zadeh, 2025; Gao et al., 2024).

3.10 Federated learning frameworks for multi-farm data collaboration

The emergence of federated learning systems has transformed collaborative intelligence in livestock management by allowing decentralized knowledge amalgamation with data sovereignty across various agricultural ecosystems. In these systems, knowledge emerges from practice into a secure storage place: the federated learning framework uses cryptographic protocols and a distributed optimization to query incomplete datasets (held locally on the farm) to train ML models, solving the ethical, logistical, and technical issues of centralized data, which results in data sovereignty. The federated learning system can thus allow participant farms to pool their "learnings" from (i) experiments, and (ii) examinations/analyses of historical factors, such as husbandry approaches, illness prevalence, and types of livestock (with distinguished genetics), while leaving genomic data and management strategies proprietary, including the years of experimentation to exhibit learning. The federated learning framework employs a key innovation of biologically relevant global design patterns and contextually relevant local design patterns, where research may find epigenetic adaptations to microclimates or metabolisms related to the breeds of the animals. Encryption protocols can enable privacy-preserving gradient sharing types of schemes to maintain ownership, even in the event of model inversion attacks (Dwarampudi and Yogi, 2024; Parekh and Shinde, 2024; Zhang et al., 2025).

It is worth highlighting that advanced adaptations include adaptive weighting methods to manage the intrinsic data heterogeneity and adaptively adjust the contribution of participatory nodes according to any representativeness and quality measures of the datasets. This is especially important when dealing with the reconciliation of omics-scale data from traceable precision farming practices with objective livestock husbandry records, as the transformer-based feature extractors are capable of normalizing disparate data types into compatible latent representations. The addition of homomorphic encryption will help further secure model aggregation, permitting more flexibility for individual entities contributing sensitive phenotypic data (i.e., subclinical diseases and reproductive performance data) to co-created intelligence models, in light of rapid data governance changes. The most impactful marker of progress will be the emergence of blockchain as a verification layer that audibly tracks the evolution of model trajectories and provides insights into how the information accumulated in models has been used; for example, when a group of collaborators produced a predictive algorithm for forecasting a new epidemic or optimizing nutritional strategies (Carraro et al., 2024; Pan et al., 2024; Rane et al., 2024).

The realization of these frameworks faces distinctive challenges related to the temporal asynchrony associated with farm data streams and the non-stationary biological systems associated with those data streams. New methods of synchronization based on principles of neuro-synaptic plasticity permit confident convergence of models with variable joint participation frequencies, while attention-based aggregation filters minimize the influence of stale or seasonally biased local data updates. Both frameworks show their true potential for transforming agricultural bioinformatics through the possibility of transferring knowledge between species, where learned patterns from poultry microbiome data may be used to inform bovine metabolic disorder predictive models through intentional domain adaptation layers. Most importantly, this cross-pollination accelerates the identification of common biomarkers among species for stress resilience and feed efficiency, while balancing the data constraints on species and maintaining their isolated data structures (Khezri et al., 2024; Komorniczak, 2025).

Future advancements in federated learning systems for animal husbandry are likely to align with developments in swarm intelligence and edge computing architectures. Self-organizing neural ensembles with dynamic reconfigurable topologies could manage a trade-off between exploring new ways of managing animal production and exploiting existing best practices of husbandry at participating nodes. As genomic data are collected with greater resolution and commercial sensitivity, it will be necessary to implement quantum-resistant encryption algorithms. Federated reinforcement learning should allow for a concurrent exploration of unrealized vaccine schedules and breeding programs through decentralized trial-and-error learning. Ethical frameworks also need to be updated to address new challenges in algorithmic accountability, specifically, for ensuring equitable distribution of benefits for collaborative entities while avoiding the evil of incompetent, inequitable predictive power sharing, whilst operating under a monopoly established by agricultural technology companies. Eventually, these systems will build toward a new normal for data ecosystems in animal husbandry that converts independent farms into nodes of collective intelligence, engendering space for privacy-conscious and collaborative learning for sustainability and animal production welfare (Bharathi et al., 2024; Janaki and Umanandhini, 2024). Data sovereignty protocols in the multi-stakeholder model training for livestock production are presented in Table 3.6.

3.11 XAI in genomic prediction systems

The capacity to embed XAI in genomic prediction systems is a necessary advancement in livestock omics, which addresses two underlying principles: algorithmic transparency and biological interpretability, especially in the context of complex and consequential breeding decisions. Current XAI frameworks use a combination of hierarchical attention methods and perturbation sensitivity analyses to disentangle the black-box nature of DL algorithms, interpreting contributions or impacts from specific haplotype blocks or epistatic (also known as "epigenetic") interactions as they relate to polygenic trait predictions and associations. Saliency maps are relative measures of the impact of each of the detected non-coding regulatory regions on the resulting phenotype in the livestock omics' decision-making predictions to account for the learned associations and quantitative genetics dogma that may be used when validating model predictions. Coupling the gradient-based class activation and attribution with the genome browser allows the researcher a powerful method to remap model decisions to the chromosomal coordinates used to test against the theoretically defined known quantitative trait loci, while simultaneously identifying untested candidate regions for future functional validation (Hussein et al., 2024; Imam et al., 2024; Ajay Vardhan et al., 2025).

An important innovation is the establishment of counterfactual explanation engine models of plausible genomic scenarios that represent a way for breeders to begin to understand how intended edits to single-nucleotide polymorphisms

TABLE 3.6 Data sovereignty protocols in multi-stakeholder model training for livestock production.

Protocol component	Technical implementation	Data governance features	Stakeholder rights enforcement	Value proposition
Federated learning architecture	▸ Horizontal federation with weighted averaging ▸ Secure aggregation protocols ▸ Model versioning with a Git-like system	▸ Raw data never leaves the source ▸ Local differential privacy ▸ Data usage auditing	▸ Granular data contribution consent ▸ Dynamic participation control ▸ Model ownership tracking	▸ Collaborative intelligence without data pooling ▸ Regulatory compliance ▸ IP protection
Differential privacy	▸ Rényi differential privacy ▸ Adaptive noise injection ▸ Per-parameter clipping thresholds	▸ Privacy-loss accounting ▸ (ε, δ)-guarantees ▸ Cross-jurisdictional compliance	▸ Farm-specific privacy budgets ▸ Opt-out mechanisms ▸ Data contribution visibility	▸ Quantifiable privacy protection ▸ Attack resilience ▸ Legal risk mitigation
Homomorphic encryption	▸ Lattice-based cryptography ▸ Partial homomorphism ▸ Multi-key support	▸ Encrypted model updates ▸ Secure parameter aggregation ▸ Quantum-resistant algorithms	▸ Data sovereignty during computation ▸ Multi-party computation governance ▸ Decryption control	▸ End-to-end encrypted workflows ▸ Protection against model inversion ▸ Secure cross-cloud operations
Data provenance tracking	▸ Blockchain-anchored hashing ▸ Zero-knowledge proofs ▸ Temporal metadata embedding	▸ Immutable lineage records ▸ Data pedigree verification ▸ Usage restriction enforcement	▸ Attribution guarantees ▸ Data misuse detection ▸ Granular access revocation	▸ Audit-compliant recordkeeping ▸ Fraud prevention ▸ Dispute resolution mechanisms
Model watermarking	▸ Backdoor-based techniques ▸ Parameter permutation ▸ Federated fingerprinting	▸ Ownership verification ▸ Model provenance tracking ▸ Tamper-evident artifacts	▸ IP rights protection ▸ Unauthorized model sharing prevention ▸ License compliance monitoring	▸ Counterfeit detection ▸ Royalty distribution ▸ Model marketplace security
Smart contracts	▸ Hyperledger-based execution ▸ Automated compliance checks ▸ Multi-signature approvals	▸ Data usage agreements ▸ Model licensing terms ▸ Revenue sharing mechanisms	▸ Automated policy enforcement ▸ Dispute resolution protocols ▸ Dynamic contract amendments	▸ Trustless collaboration ▸ Automated governance ▸ Transparent value distribution
Edge-based filtering	▸ On-device data masking ▸ Feature hashing ▸ Synthetic data generation	▸ Sensitive information protection ▸ Data minimization ▸ Context-aware anonymization	▸ Field-level redaction control ▸ Data utility preservation ▸ Re-identification risk mitigation	▸ Privacy-by-design implementation ▸ Regulatory burden reduction ▸ Ethical AI compliance
Data residency controls	▸ Geo-fenced model sharding ▸ Sovereign cloud gateways ▸ Data localization policies	▸ Jurisdiction-specific storage ▸ Cross-border transfer restrictions ▸ Encryption-at-rest	▸ Territorial compliance ▸ Data sovereignty mapping ▸ Legal hold capabilities	▸ Multi-national regulation adherence ▸ Data localization ▸ Sovereignty conflict resolution
Federated analytics	▸ Secure multi-party computation ▸ Privacy-preserving SQL ▸ Encrypted statistics	▸ Aggregate insights without raw data ▸ Differential private queries ▸ Role-based access	▸ Collaborative benchmarking ▸ Anonymized trend analysis ▸ Data donation mechanisms	▸ Industry-wide insights ▸ Competitive confidentiality ▸ Collective intelligence
Audit trails	▸ Immutable logging ▸ Three-way reconciliation ▸ Automated compliance scoring	▸ Model development transparency ▸ Data flow mapping ▸ Real-time compliance monitoring	▸ Stakeholder access to audit records ▸ Automated reporting ▸ Regulatory submission automation	▸ Accountability assurance ▸ Continuous compliance ▸ Reduced audit costs

could and will alter their estimated breeding values for complex traits (e.g., feed efficiency or methane emissions). These systems utilize variational autoencoders to explore the latent space of all genomic variations, and hence, the number and sophistication of scenarios produced can be detailed to the level needed to disentangle pleiotropic sources of variation as well as locus-specific variation of interest. Their implementation in practical contexts would require new hybrid systems that encompass symbolic reasoning components that can be blended with neural networks, while also encoding reasoning consistent with Mendelian rules of inheritance as well as the non-linear effects that accompany polygenic adaptation (Gao and Zhao, 2024; Pegios et al., 2024).

The distinct complexities of livestock genomic XAI arise from the scale and heterogeneity of multi-omics data streams, and attention-based transformer models with biological prior constraints will soon contribute to the hierarchical attribution of prediction importance across genomic, transcriptomic, and metabolomic feature spaces. This multi-modal explainability becomes especially necessary when interpreting genotype-environment interactions, where the model explanations must balance the contextually dependent expression of genetic "potential" with various management practice variables. Privacy-preserving explanation models help to allay industry concerns further by developing synthetic feature importance profiles that respect the confidentiality of specific animals' data while maintaining population-level descriptions of marker-trait associations (Maqsood et al., 2024; Toussaint et al., 2023).

Neural-symbolic methods that challenge XAI in genomic selection are emerging, which combine biologically-informed knowledge graphs of biological pathways with predictions made using deep learners to incorporate causally based explanations. These systems provide automatic hypotheses about an explanatory state mechanism that can, through the stage of biological mechanisms as identifiers, map machine-based evidence of marker effects on canonical gene networks. Biologically referenced explanations would improve stakeholder confidence by situating AI-supported explanations alongside historical domain knowledge, while also providing evidence of epistatic network relationships that were previously obscured from existing rationalist breeding models (Himabindu et al., 2023; Zhu and Sun, 2024).

Looking into the future, the trajectory of XAI in livestock genomics is heading toward dynamic explanation systems that are capable of adjusting their interpretive frameworks as more biological evidence emerges. Explanation regulators based on reinforcement learning will optimize the trade-off between interpretive simplicity and biological completeness, while offering unique, personalized explanation granularity for every target end user, from molecular geneticists to production managers. With quantum computing architectures beginning to address genomic optimization problems, brand new XAI approaches will need to be developed to interpret entanglement-based feature interactions in quantum neural networks. These developments will not only identify XAI as a validation tool but also as a discovery tool in its own right, while systematically interpreting AI's identified patterns of prediction in the complex bioinformatics arena of livestock genomics to reveal the underlying biological principles (Fukuchi and Yamada, 2024; Nancy Ambritta et al., 2025; Ajay Vardhan et al., 2025). Interpretability techniques mapped to genomic prediction confidence levels in livestock are presented in Table 3.7.

3.12 Ethical AI frameworks in livestock genomic selection

To responsibly implement AI in livestock genomic selection, we require complex frameworks that reconcile the technological possibility of distinct breeding options with the social and cultural responsibility inherent in the production of lifeforms, which requires an understanding of the interplay between the loss of infrastructure through the optimization of technology in production and the complexity of a spectrum of ethical issues about lifeform engineering. Now, ethical AI frameworks are moving beyond traditional bio-ethical frameworks to incorporate principles of epistemic justice directly into genomic prediction models, allowing the algorithmic decisions made to recognize that genetic diversity has intrinsic value beyond the metrics of productivity that breeders normally define. These frameworks generate models that employ multi-objective optimization, balancing goals of commercial breeding with conservation objectives, and that can change the selection indices on a case-by-case basis to avoid narrowing the allelic richness of adaptive genes that provide organisms resilience to climate change or disease. If indigenous knowledge systems have been incorporated, as part of the ontological alignment layers, not ontologies or epistemologies, the AI system will be able to recognize indigenous cultural heritage as potential biocultural heritage within livestock genomics contexts, preserving livestock breeds and types that associate cultural identity without the limiting biases and homogenizing effects of conventional livestock improvement that predominate pure efficiency-based algorithmic selection (Egon et al., 2023; Ghavi Hossein-Zadeh, 2025; Hofmann, 2025).

A pivotal innovation is the realization of formal architectures that promote transparency by design, which allows for the traceability of genomic influence networks throughout the breeding pipeline. These systems use blockchain-anchored audit trails to track the extent to which specific AI recommendations change population genetic structures over generations, creating accountability structures for the lasting effects of selection decisions. The frameworks use anticipatory

TABLE 3.7 Interpretability techniques mapped to genomic prediction confidence levels in livestock.

XAI technique	Confidence level range (%)	Genomic application	Technical specifications	Validation metrics	Stakeholder impact
SHAP values	70–80	▶ Trait prioritization ▶ SNP effect quantification ▶ Additive variance decomposition	▶ GPU-accelerated kernel SHAP ▶ Whole-genome sampling ▶ Epistasis interaction weighting	▶ Consistency across breeds ▶ Additive variance explained ▶ Directional accuracy	▶ Breeder decision support ▶ Genomic selection training ▶ QTL discovery prioritization
LIME explanations	60–75	▶ Local model behavior ▶ Environmental interaction effects ▶ Non-linear effect mapping	▶ Superpixel segmentation ▶ Perturbation sampling ▶ Ensemble stability analysis	▶ Fidelity scores ▶ Instance-level consistency ▶ Cross-validation stability	▶ Veterinarian diagnostics ▶ Field trial analysis ▶ GxE interaction education
Attention mechanisms	85–92	▶ Haplotype block identification ▶ Cis-regulatory element detection ▶ Imprinting effects	▶ Transformer architectures ▶ Positional encoding ▶ Multi-head attention layers	▶ Attribution congruence ▶ Biological plausibility ▶ Conservation scores	▶ Geneticist research tools ▶ Breeding program design ▶ Regulatory element discovery
Saliency maps	75–85	▶ Genomic region importance ▶ Pleiotropy detection ▶ Pathway activation patterns	▶ Integrated gradients ▶ SmoothGrad filtering ▶ Class activation mapping	▶ Pixel-wise perturbation tests ▶ Biological pathway alignment ▶ ROC-AUC	▶ Genomic counseling ▶ Research hypothesis generation ▶ Educational visualizations
Counterfactual analysis	65–78	▶ Alternative allele effects ▶ Genomic editing simulations ▶ Threshold trait modification	▶ Genetic algorithm search ▶ Latent space traversal ▶ Biological constraint modeling	▶ Plausibility scores ▶ Causal validity ▶ Domain expert evaluation	▶ Precision breeding planning ▶ Gene editing impact assessment ▶ Risk communication
Rule extraction	80–88	▶ Dominance/ recessiveness patterns ▶ Epistatic rules ▶ Threshold effect identification	▶ Decision tree distillation ▶ Fuzzy logic systems ▶ Association rule mining	▶ Rule consistency ■ Biological validity index ▶ Coverage metrics	▶ Producer decision tools ▶ Genetic defect avoidance ▶ Selection index optimization
Feature interaction	88–95	▶ Epistatic network visualization ▶ Genomic × environment effects ▶ Pleiotropic pathways	▶ N-order interaction detection ▶ Graph neural networks ▶ Multivariate dependency analysis	▶ Interaction strength metrics ▶ Biological network congruence ▶ Stability scores	▶ Complex trait dissection ▶ Genomic architecture studies ▶ Breeding value refinement
Uncertainty quantification	90–98	▶ Model confidence intervals ▶ Breeding value reliability ▶ Genomic prediction risk	▶ Bayesian neural networks ▶ Monte Carlo dropout ▶ Ensemble variance analysis	▶ Calibration curves ▶ Uncertainty-error correlation ▶ Reliability diagrams	▶ Risk-informed selection ▶ Genomic investment strategies ▶ Insurance product development

ethics modules that model multi-generational consequences of breeding strategies based on evolutionary algorithms, where trajectories that could undermine animal welfare or disrupt other ecosystem services through unintended pleiotropic effects are flagged. Finally, privacy-preserving federated learning approaches have been proposed in the context of previously announced collaborative data use agreements that, at least ideally, protect the knowledge of traditional breeders and prevent encroachment by corporations of the collectively carried out genome-based genetic improvement strategies developed by groups of animal breeders and scientists (Dakshayini et al., 2024; Pérez-Enciso et al., 2025).

The implementation of ethical AI in this setting presents its challenges due to the different values held by the parties involved. Adaptive mechanisms for weighting moral values allow for temporary prioritization of the ethical principles relevant for a particular breeding scenario, such as animal autonomy and maintaining the environment. Neuro-symbolic architectures combine DL's predictive capacity with logic-based ethical reasoning engines that can maintain feasible compliance with essential international governance agreements about the management of genetic resources, while also acknowledging local agro-ecological realities. Some solutions use counterfactual fairness testing to observe the sources of bias in existing genomic evaluations, where official recognition may unintentionally exclude breeds with heritage status or policy consideration for access to smallholder farming systems (Iyer et al., 2025; Neethirajan, 2025).

The ethical frameworks of the future must address the ontological uncertainties that gene-editing technologies will bring when combined with ML genomic predictions. Quantum moral simulation environments are being constructed to provide moral simulations of the ethical dimensions of CRISPR-based enhancements suggested by algorithms, but these tools will be capable of predicting ethically consequential outcomes for all dimensions of alternative futures simultaneously. This may promote One Health ethics within AI frameworks, which offers a way of aligning livestock genomic improvement plans with the health of the ecosystem, especially as the well-being of animals, humans, and environments is interconnected. Likewise, while synthetic biology is beginning to redefine the parameters of genomic-based modification, ethical AI frameworks will need to build in biosafety thresholds within algorithms and models that automatically restrict dangerous gain-of-function predictions while simultaneously allowing biological pathways to innovation for desirable traits (Ishida and Sawai, 2024; Ajay Vardhan et al., 2025;Dixit et al., 2024; Sagar et al., 2025).

The development of ethical AI for livestock genomics signals models of polycentric governance, whereby the algorithmic decision-making process becomes a site of continuous moral consideration. Embedded within these systems will be decentralized ethical oversight functions, where actors can actively participate in and leverage democratic AI architectures. We will have the capacity to give proportional consideration to a range of stakeholder perspectives, the corporate environment, and a pastoralist context. The development of generative AI for self-reflexive, meta-ethical reasoning regarding its genomic selections is not only an opportunity with respect to technical sophistication; in terms of philosophy, it will demand that the governance of the power to rearrange living organisms is through wisdom, and not just the optimization of computation. As livestock genomics embraces these types of systems, the community has not only the capacity to manage the challenges of the Anthropocene but also maintain its commitment to stewarding the co-evolution of humans and other animals, like livestock (Bikkasani, 2024; Ghavi Hossein-Zadeh, 2025; Kiran Vuyyuru, 2025).

3.13 Blockchain-enabled traceability in AI-managed supply chains

The intersection of AI and blockchain technology has created a transformational opportunity for livestock systems to provide transparent supply chains that have created a persistent record of biological and operational data flows from genomic origins to consumer endpoints. Blockchain architectures facilitate a distributed ledger of transactions and biological events managed by AI across the supply chain to ensure that there exists a permanent and auditable trail that is cryptographically secured. Such a trial can verify not only the genomic source of genetic material, but also the health interventions needed throughout the production period, and whether husbandry practices were ethical. Distributed ledgers provide a consumer verification mechanism through smart contract protocols, allowing for the automation of quality assurance protocols and confirming compliance, all while maintaining the autonomy of data exchange among multiple parties. This provision ensures that traditional records will be maintained on-chain alongside the values from these schemes, based on the verified outcomes delivered through the use of on-chain records. The tamper-evidence nature of blockchain records, combined with machine-learning decision making, has the potential to create a self-regulating environment in the livestock-human-AI interaction and goals, including breeding stock selection, feed resource consumption, and disease management, which can therefore be linked to actual evidence on-chain, based on biological outcomes rather than solely to a decision made by an individual or automated process (Yekeen et al., 2024; Rahman et al., 2025; Ravikumar et al., 2025).

One of the key innovations is the tokenization of livestock genomic profiles that incorporate blockchain-verified, non-fungible tokens (NFTs), which represent the individual genetic identities but share access to connected omics data. This data will encapsulate many very comprehensive longitudinal datasets, covering all epigenetic, microbiome, and metabolomic variations, which will be available to AI models; however, these data types will not give away any of the genetic data from the animal being monitored. Similarly, the systems will use zero-knowledge proofs to authenticate an advancement in algorithms if the algorithms provide evidence of phenotypes (i.e., traits in animals) such as antimicrobial resistance or heat tolerance; here, the stakeholders can trust the AI predictions and recommendations even though the genetic data would not be disclosed. This type of architecture is crucial in terms of verifying standards in streamlined global supply chains. The zero-boundary characteristics of blockchain can help link conflicting national standards for animal welfare and biosecurity (veterinary) standards with machine-readable regulatory smart contracts (Varma et al., 2024; Krishnan and Ganesan, 2025).

The establishment of these hybrid systems, which combine a robust blockchain with predetermined biological baselines, tackles long-standing issues associated with supply chain opacity through the immutability of temporal, geolocated data. Anomaly detection algorithms powered by AI and ML will continuously observe the blockchain and supply chain pathways for changes from strict biological baselines, essentially alerting researchers to potentially emerging disease outbreaks and/or unethical husbandry practices in the ecosystem as the incidents occur. The closed-loop feedback of these systems implies that researchers will be able to write interdisciplinary commands as predictive interventions in the blockchain ledger (to replace whatever feed formulation on the farm, impose quarantine, and so on), using conditional statements as the time-stamped biosecurity indicators are assessed (e.g., associated individual livestock, breeds/herds). Blockchain security protocols, which are essential to the integrity of the supply chain, can also integrate quantum-resistant encryption protocols, which future-proof the supply chain against quantum computers and emerging computational threats and protect genetic intellectual property derived from animal genomics and breeding algorithms, as composite genomics of livestock increasingly becomes an exascale data science (Ghavi Hossein-Zadeh, 2024; Misra et al., 2024; Sweta Dixit et al., 2024; Yekeen et al., 2024).

The ethical aspects of blockchain-AI convergence can be understood in their ability to enable democratized supply chain governance through decentralized autonomous organizations (DAOs). These now stakeholder-governed workspaces take the form of DAOs, using token-weighted voting to help stakeholders negotiate shared AI learning priorities. Typically, these priorities balance ownership-driven objectives with social motivations such as sustainability and animal welfare. The distributed trust in blockchain forms the basis for the new level of accountability in these AI-management systems. Each decision is recorded transparently in a ledger, making it possible to identify how increases or decreases in the data used to design AI parameters conform to levels of genetic diversity and ecosystem responses through breeding generations. Neuro-symbolic AI systems, emerging architectures that construe AI layers as separate, yet interdependent systems solutions, contribute to the ethics of AI by including content-based compliance with standards of international governance of livestock stewardship into blockchain consensus rules. Such a design could ensure automated compliance with evolving conventions on biodiversity conservation and climate-responsible practices in animal agriculture (Hazarika and Shah, 2024; Yadav, 2024; Neulinger et al., 2025).

In the future, as blockchain-enabled AI systems continue to mature, circular bioeconomies can become commonplace in livestock supply chains. Self-executing smart contracts will offer agricultural by-products upcycling to resource nutrient cycles, and AI models will use real-time blockchain data on benefits to soil health and emissions to optimize material flows, along with material flows. The establishment of federated learning consortia based on permissioned blockchains will enable the development of collective intelligence models while allowing firms to maintain their competitive advantages offered by proprietary breeding algorithms. As augmented reality interfaces learn more about blockchain-verified insights from AI, stakeholders at multiple levels of the supply chain will be able to almost reflexively access validated data about the complete lifespan of the animal (the genomic potential of the fetus at embryonic developmental stages to metabolic legacy after death). This evolution will mean that a consumer will gain access to the food and agricultural system with radical transparency, changing the way humanity engages in the domestication of other species. For the livestock production system, distributed ledger technology and machine intelligence converge to move into a future where every algorithmic decision can always be traced back to a biological paradigm of truth, ethical responsibility, and ecological consideration (Papa et al., 2024; Vinay et al., 2024; Venturini, 2025). Immutable record types across supply chain echelons are presented in Table 3.8.

3.14 Real-time biometric monitoring systems

The incorporation of real-time biometric monitoring technologies has changed the paradigm of precision livestock management from being a linear one to a dynamic system, where there is constant feedback/loop between physiological states and algorithmic decision making that can provide even real-time interventions when needed at the intersection of animal

TABLE 3.8 Immutable record types across supply chain echelons.

Supply chain echelon	Core data types	Blockchain implementation	AI integration	Compliance features
Genetic breeding	▶ Genomic sequencing records ▶ Hereditary disease markers ▶ Breeding performance metrics	▶ Smart contracts for genetic rights management ▶ Cryptographic hashing of DNA profiles	▶ Predictive trait selection models ▶ Inbreeding coefficient monitoring	▶ CITES compliance ▶ Biosecurity protocol adherence ▶ Genetic resource treaties
Feed production	▶ Ingredient provenance logs ▶ Nutritional analysis reports ▶ Antibiotic-free certifications	▶ IoT sensor data anchoring ▶ Batch-specific QR code tracing	▶ Feed efficiency optimization ▶ Contaminant pattern detection	▶ FDA GRAS compliance ▶ HACCP documentation ▶ Organic certification validation
Farm operations	▶ Animal health monitoring ▶ Welfare assessment logs ▶ Environmental impact metrics	▶ Edge device data oracles ▶ Geotagged welfare evidence storage	▶ Disease outbreak prediction ▶ Carbon footprint optimization	▶ OIE terrestrial code ▶ GLOBAL G.A.P. standards ▶ GHG protocol tracking
Processing facilities	▶ Slaughterhouse hygiene logs ▶ Carcass grading data ▶ Cold chain integrity records	▶ Temperature-proof hashes ▶ Worker biometric authentication	▶ Yield optimization algorithms ▶ Quality defect in computer vision	▶ USDA FSIS regulations ▶ HALAL/Kosher audits ▶ Worker safety compliance
Distribution networks	▶ Customs clearance docs ▶ Refrigerated transport logs ▶ Cross-border phytosanitary certificates	▶ Multi-chain interoperability ▶ Automated customs smart contracts	▶ Route optimization engines ▶ Spoilage risk prediction models	▶ WTO-TBT agreements ▶ IATA LAR regulations ▶ Export health certificate validation
Retail operations	▶ Shelf-life monitoring ▶ Consumer complaint tracking ▶ Recall management systems	▶ Dynamic NFT product passports ▶ Decentralized recall coordination	▶ Demand forecasting models ▶ Sentiment analysis integration	▶ Country of origin labeling ▶ Allergen disclosure ▶ Fair trade certification
Consumer endpoints	▶ Product authentication ▶ Ethical sourcing proofs ▶ Nutritional impact tracking	▶ Mobile-optimized verification ▶ Tokenized sustainability claims	▶ Personalized nutrition insights ▶ Carbon credit calculation	▶ GDPR ▶ Green claims verification ▶ Animal welfare labeling

biology and computing intelligence. Real-time telemetry systems rely on multimodal sensors that an animal wears, and these sensor arrays can capture high-dimensional biometric signatures that range from subcutaneous temperature gradients and arrhythmia, rumen fermentation, and cortisol levels induced by stress, thus producing stream datasets that AI models can feed into temporal convolutional networks. The ontological value of creating and using the streams of these heterogeneous data creates a living digital twin for each animal, which replaces itself over time in concert with biological rhythms to predict metabolic crises or subclinical infections before the animal demonstrates phenotypic changes. The new edge computing architecture enables local processing within the sensor itself, and uses federated learning to maintain privacy when creating insights at the herd level based on the distributed neural networks that train on biometric features from multiple animals using encoders that keep their data private (Siegford, 2024; Losacco et al., 2025; Mahato and Neethirajan, 2025).

One significant advancement relates to the creation of self-calibrating biosensors, which adaptively configure their measuring protocols based on environmental context, compensating for variables like ambient humidity or activity level through adversarial neural networks. Thereafter, the adaptive sensing systems can leverage blockchain-based health records to create immutable provenance chains for livestock biometric data, from raw sensor inputs recorded by the wearable device through to decision-making and management by AI, leveraging AI, leveraging optimization algorithms to find the trade-off between the frequency of measurement and energy efficiency, which increases the potential incremental lifetime of a wearable sensor system but can maintain real-time responsiveness to physiological events on the nanosecond scale. Incorporating XAI modules into the firmware of the sensor means that the rationale for alerting can be transparently communicated to the person wearing the device, and the importance of the alert can be communicated to the human as part of the story of the biometric data reliably ascertained through the sensor, without the ambiguities present in interpreting machine-processed health data (Akkaş et al., 2025; Thute and Gudadhe, 2025).

The operationalization of these systems relies on unique challenges associated with biological variability in livestock populations. Evolutionary optimization strategies have recently been introduced to deep neural models to develop individual animal health baselines while accounting for breed-associated metabolic profiles and developmental ontogeny. Cross-modal attention models decouple correlated biometric signals, specifically regarding intra-class variability in thermal stress in extreme heat and early detection of pathogenic stress with hierarchical feature attribution. Complementing this with biometric information and prior genomic predictions serves to create an analytic framework in which real-time physiological monitoring supports long-term breeding value estimates in a closed-loop genetic improvement context that can respond to emergent environmental pressures. Furthermore, privacy-preserving synthetic data generation algorithms will address the ethics of sharing biometric data generated from individual livestock by allowing joint learning to occur across multiple farms and minimizing individual animal response fingerprints (Qazi et al., 2024; Solona and Lisovyi, 2024; Wasudeorao, 2024).

The ethical implications of massive biometric surveillance require new governance frameworks that will trade off the improvement of an individual's welfare against the preservation of their behavioral autonomy. Neuro-symbolic AI systems now interpret ethical boundaries on how biometrics are used, and prevent owners from exploiting biometric knowledge of their animals and driving particular behaviors for hyper-productivity to the detriment of animal welfare. Biometric consent mechanisms, built on the analysis of animals' behavioral patterns, now allow animals to have social agency that influences the parameters of the collection of biometric data with sensor-apparatus-based environments, which is an early ethical framework for human-digital coexistence with livestock. These systems increasingly utilize One Health frameworks and relate livestock biometric developments to environmental epidemiological signals to allow digital early-warning systems to reduce the risk of zoonotic spillover and recognize disease transmission patterns across mammalian and vector species (Ghavi Hossein-Zadeh, 2025; Teja Reddy Maryala, 2025).

Future opportunities for biometric monitoring indicate the potential use of non-invasive quantum sensor arrays capable of resolving physiological processes at the molecular scale in real-time, from mitochondrial energy production to neurotransmission. The development of biometric AI and genetic expression profiling will enable closed-loop epigenetic modulation systems capable of changing animal physiology to cope with environmental tensions without intervention from the animal's status through specifically targeted food and nutrition. As neuromorphic computers evolve, integrated biometric processors will achieve brain-like efficiencies in pattern recognition and apply their recognition and decision-making at the edge with respect for the latency-sensitive biological timescale. The human-livestock-AI ecosystem envisioned will exemplify the use of continuous biometric dialog to inform humane and ethically mindful husbandry practices. The journey will traverse livestock management from a retrospective discipline to an anticipatory science that seeks a predictive future without distorting the irreducible complexity of living systems. In this technological worldview, real-time biometric monitoring is less of a management solution and more like the lingua franca of cross-species communication in an algorithmic age (Oh et al., 2024; Tafadzwa Mpofu and Mthunzi-Kufa, 2024; Yu et al., 2025).

3.15 Human-AI collaboration in farm decision support systems

The advancement of human-AI cooperation in livestock management has moved beyond a traditional decision-support model to form a coupled cognitive architecture where algorithmic precision and human knowledge combine into flexible networks for problem-solving. These systems use neuro-symbolic approaches that mesh DL's ability to recognize patterns with the coordination of stated knowledge based on the ontological principles of agriculture to form joint conditions for reasoning that both leverage data-derived understandings and farmers' tacit knowledge. The AI systems are continually learning through NLP dialog interfaces that allow real-time contextualization of metrics, suggestions, and actions within the farmer's world. The farmer is likewise contextualizing the recommendations by presenting staged feedback of the metrics through the situated knowledge of their experience. They are accordingly adjusting the algorithmic weights of prioritization. Together, these processes are also building a distributed intelligence phenomenon, wherein neither party simply assists the other, nor does it mean they are simply swapping one method of decision-making for the other. It means there is a co-evolution of decision-making heuristics, in terms of knowledge representations in a joint state and mutual concept (Kazembe and Mkandawire, 2024; Mate et al., 2024).

At the core of this collaboration is the creation of cognitive ergonomic interfaces that convert between machine-learned feature spaces and human-perceptible agricultural phenomena. Leading-edge visualization devices and learning systems already exist, which allow farmers to view an AI's predicted herd health or feed efficiency as augmented reality overlays to a real-world livestock setting, enabling them to interrogate the algorithm-driven conclusions by gestural interaction with the hologram. XAI modules distill complex, model-made inferences into worthwhile "cause" stories that conform to the pathways of historical husbandry decisions. This maintains human agency, but takes advantage of the future visibility of AI. These systems include a level of affective computing, so they can respond to farmer stress, uncertainty, or similar signals with alternatives to their extraction from a high-stakes decision environment. With this information, the presentation of material can be dynamically and situationally adjusted to reduce cognitive load and increase confidence in decisions (Ibrahim and Danmaigoro, 2024; Hilmi et al., 2024).

The ethical framework of these collaborative systems is demonstrated through value-sensitive design principles that are embedded in the design, which incorporates socioecological interests, ensuring that algorithmic optimization does not supersede them. Hybrid governance models leverage DAOs to facilitate democratic value prioritization of stakeholders between animal welfare proposals or indigenous land management practices and decisions on the AI training agendas. Through multi-party computation, there is an intersection where decisions are made that involve the ethical expectations of humans, and this output can be audited and provides accountability trails to decisions on selecting genetic material or resource allocation. The systems are implementing anticipatory ethics modules, which are systems to anticipate or simulate the collective consequences of collaborative decisions in biological, economic, and sociocultural domains, allowing farmers to design their decision-making to navigate in a pre-determined choice architecture that can contextualize trade-offs for productivity against their sustainability aspirations (Lustenberger et al., 2025; Long et al., 2024).

Human-AI collaboration is fundamentally beset by operational challenges, centered on finding the right balance between automation and human oversight. Self-aware AI systems have begun to inherently use meta-reasoning features to identify contexts of decision making wherein human judgment is necessary, and automatically defer in instances where risks arise from new pathogens or the onslaught of climate unpredictability. Autonomy and trust in AI are fostered through adaptive competency models that assess farmers' trust in and understanding of AI recommendations to modulate system autonomy levels, ensuring skills do not erode while introducing and developing their fund of knowledge afforded through technology. As digital twins technology enters the hybrid practice of direct and indirect mediation, it provides a space to innovate and experiment by allowing time and space to co-design and test together, thus providing human-AI teams a space to innovate intervention options before executing them, and allows predictive analytics to be melded with experiential learning in a risk-free simulation of farm ecosystems (Ampatzidis et al., 2025; Samdani et al., 2025).

Future developments suggest quantum-supplemented collective intelligence systems that will allow instantaneous consensus-making across scattered agricultural networks through entangled AI-human decision states. A level of cooperation that was previously impossible will be made possible by next-generation neuromorphic AI chips integrated into agricultural assets and infrastructure. These chips will also enable systems to recognize neural patterns and infer possible human decision pathways while maintaining the boundaries of cognitive sovereignty. As collaborative systems evolve, they will increasingly adopt the One Health tenets and integrate livestock management decision-making with metrics related to ecosystem health and public health goals through multi-species welfare optimization algorithms (the engineering of the trajectory of individual and collective action for the most favorable outcome). A post-anthropocentric approach to decision-making, where AI and human systems work together within current and future stewardship frameworks, is what this trajectory represents. The ecological management of interactions between global food systems, agricultural ecosystems, and domesticated species is the main focus. Under ethical guidance, this transition seeks to transform farms from simple production locations into centers of computationally enhanced biocultural innovation (Blessing Akinnagbe, 2024; Milian Gómez and Byttebier, 2025; Usman et al., 2025).

References

Ajay Vardhan, B., Rushil Phanindra, K.P., Sumith, K., Sirisha, T., Kakulapati, V., 2025. Explainable AI for livestock disease detection: an integrated ML/DL framework. Asian J. Res. Comput. Sci. 18 (6), 147–153. https://doi.org/10.9734/ajrcos/2025/v18i6686.

Akkaş, T., Reshadsedghi, M., Şen, M., Kılıç, V., Horzum, N., 2025. The role of artificial intelligence in advancing biosensor technology: past, present, and future perspectives. Adv. Mater. https://doi.org/10.1002/adma.202504796.

Alharbi, F., Vakanski, A., Zhang, B., Elbashir, M.K., Mohammed, M., 2025. Comparative analysis of multi-omics integration using graph neural networks for cancer classification. IEEE Access 13, 37724–37736. https://doi.org/10.1109/ACCESS.2025.3540769.

Alzu'bi, S., Kanan, T., Elbes, M., Kanaan, G., Trrad, I., 2025. Energy-efficient edge deployment of generative AI models using federated learning. Clust. Comput. 28 (5). https://doi.org/10.1007/s10586-025-05263-7.

Ampatzidis, Y., Bhandari, M., Ferreyra, A., Gentimis, T., McReynolds, E., Murray, S.C., Peterson, M., Rodriguez Lopez, C., Strong, R., Tedeschi, L., Vitale, J., Ye, X., 2025. AI in agriculture: opportunities, challenges, and recommendations. Counc. Agric. Sci. Technol. https://doi.org/10.62300/iaag042514.

Baião, A.R., Cai, Z., Poulos, R.C., Robinson, P.J., Reddel, R.R., Zhong, Q., Vinga, S., Gonçalves, E., 2025. A technical review of multi-omics data integration methods: from classical statistical to deep generative approaches. Briefings Bioinf. 26 (4). https://doi.org/10.1093/bib/bbaf355.

Balakrishnan, T.S., Gururama Senthilvel, P., Ebenezar, U.S., Karthikeyan, L., Kishan, B.S., 2025. Exploring few-shot learning to enhance NLP's cross-domain capabilities. In: Proceedings of 2025 International Conference on Computing for Sustainability and Intelligent Future, COMP-SIF 2025. Institute of Electrical and Electronics Engineers Inc. https://doi.org/10.1109/COMP-SIF65618.2025.10969937

Ballard, J., Wang, Z., Li, W., Shen, L., Long, Q., 2024. Deep learning-based approaches for multi-omics data integration and analysis. BioData Min. 17 (1). https://doi.org/10.1186/s13040-024-00391-z.

Bharathi, M., Aditya Sai Srinivas, T., Bhuvaneswari, M., 2024. Federated learning: from origins to modern applications and challenges. J. Inform. Technol. Cryptogr. 1 (2), 29–38. https://doi.org/10.48001/joitc.2024.1229-38.

Bikkasani, D.C., 2024. Navigating artificial general intelligence (AGI): societal implications, ethical considerations, and governance strategies. https://doi.org/10.20944/preprints202407.1573.v3.

Bishwas, A.K., Som, A., Choudhary, S., 2024. Investigation into the potential of parallel quantum annealing for simultaneous optimization of multiple problems: a comprehensive study. doi: 10.48550/arXiv.2403.05764.

Biswas, I., 2025. Ethical dimensions and societal implications: ensuring the social responsibility of CRISPR technology. Front. Genome Ed. 7. https://doi.org/10.3389/fgeed.2025.1593172.

Blessing Akinnagbe, O., 2024. Human-AI collaboration: enhancing productivity and decision-making. Int. J. Educ. Manag. Technol. 2 (3), 387–417. https://doi.org/10.58578/ijemt.v2i3.4209.

Böttcher, L., Fonseca, L.L., Laubenbacher, R.C., 2025. Control of medical digital twins with artificial neural networks. Phil. Trans. R. Soc. A Math. Phys. Eng. Sci. 383 (2292). https://doi.org/10.1098/rsta.2024.0228.

Camacho, J., Armstrong, M.S., García-Martínez, L., Díaz, C., Gómez-Llorente, C., 2024. Single-cell spatial (scs) omics recent developments in data analysis. doi: 10.48550/arXiv.2412.13591.

Carraro, A., Saurio, G., Marinello, F., 2024. Towards rigorous dataset quality standards for deep learning tasks in precision agriculture: a case study exploration. SSRN. https://doi.org/10.2139/ssrn.4789228.

Crossa, J., Montesinos-Lopez, O.A., Costa-Neto, G., Vitale, P., Martini, J.W.R., Runcie, D., Fritsche-Neto, R., Montesinos-Lopez, A., Pérez-Rodríguez, P., Gerard, G., Dreisigacker, S., Crespo-Herrera, L., Pierre, C.S., Lillemo, M., Cuevas, J., Bentley, A., Ortiz, R., 2025. Machine learning algorithms translate big data into predictive breeding accuracy. Trends Plant Sci. 30 (2), 167–184. https://doi.org/10.1016/j.tplants.2024.09.011.

Dakshayini, M., Kavalleswari, A., Suhasini, H.S., 2024. Blockchain-NFT enabled privacy and ownership protection system for genomics[POPS-G]. In: International Conference on Emerging Technologies in Computer Science for Interdisciplinary Applications, ICETCS 2024. Institute of Electrical and Electronics Engineers Inc. https://doi.org/10.1109/ICETCS61022.2024.10544201

David, R., Shankar, H., Kura, P., Kowtarapu, K., Uma Maheswari, S.K.S., 2025. Advancement in explainable AI: bringing transparency and interpretability to machine learning models for use in high-stakes decisions. In: International Conference on Emerging Smart Computing and Informatics, ESCI 2025. Institute of Electrical and Electronics Engineers Inc. https://doi.org/10.1109/ESCI63694.2025.10988079

Davies, H., Nenadic, G., Alfattni, G., Arguello Casteleiro, M., Al Moubayed, N., Farrell, S.O., Radford, A.D., Noble, P.J.M., 2024. Text mining for disease surveillance in veterinary clinical data: part one, the language of veterinary clinical records and searching for words. Front. Vet. Sci. 11. https://doi.org/10.3389/fvets.2024.1352239.

Dembani, R., Karvelas, I., Akbar, N.A., Rizou, S., Tegolo, D., Fountas, S., 2025. Agricultural data privacy and federated learning: a review of challenges and opportunities. Comput. Electron. Agric. 232, 110048. https://doi.org/10.1016/j.compag.2025.110048.

Dixit, S., Kumar, A., Srinivasan, K., Vincent, P.M.D.R., Krishnan, N.R., 2024. Advancing genome editing with artificial intelligence: opportunities, challenges, and future directions. Front. Bioeng. Biotechnol. https://doi.org/10.3389/fbioe.2023.1335901.

Dixit, S., Maurya, M., Jain, V., Subramaniam, G., 2024. Blockchain in Supply Chain Management Artificial Intelligence-Enabled Businesses: How to Develop Strategies for Innovation. Wiley, pp. 57–79. https://doi.org/10.1002/9781394234028.ch4.

Dradjat, K., Hamidi, M., Bartet, P., Hanczar, B., 2025. Self-supervised learning on gene expression data (Version 1). https://doi.org/10.48550/ARXIV.2507.13912.

Duy Nhat, D., 2025. Combining supervised and semi-supervised models to enhance personalized education. J. Basic Appl. Res. Int. 31 (1), 44–49. https://doi.org/10.56557/jobari/2025/v31i19101.

Dwarampudi, A., Yogi, M.K., 2024. Application of federated learning for smart agriculture system. Int. J. Inform. Technol. Comput. Eng. 43, 36–47. https://doi.org/10.55529/ijitc.43.36.48.

Eckhardt, R., Arablouei, R., Ingham, A., McCosker, K., Bernhardt, H., 2025. Livestock behaviour forecasting via generative artificial intelligence. Smart Agric. Technol. 11, 100987. https://doi.org/10.1016/j.atech.2025.100987.

Egon, K., Karl, L., Eugene, R., 2023. Animal Welfare and Ethical Considerations in Precision Livestock Farming (PLF). https://doi.org/10.31219/osf.io/dh984.

Elvas, L.B., Almeida, A., Ferreira, J.C., 2025. Natural language processing in medical text processing: a scoping literature review. Int. J. Med. Inf. 204. https://doi.org/10.1016/j.ijmedinf.2025.106049.

Fatoki, O., Tu, C., Hans, R., Bello, R.W., 2024. Role of computer vision and deep learning algorithms in livestock behavioural recognition: a state-of-the-art-review. Edelweiss Appl. Sci. Technol. 8 (6), 6416–6430. https://doi.org/10.55214/25768484.v8i6.3396.

Fukuchi, Y., Yamada, S., 2024. User decision guidance with selective explanation presentation from explainable-AI. doi: 10.48550/arXiv.2402.18016.

Fushai, F., Chitura, T., Oke, O.E., 2025. Climate-smart livestock nutrition in semi-arid Southern African agricultural systems. Front. Vet. Sci. 12. https://doi.org/10.3389/fvets.2025.1507152.

Gao, Z., Zhao, Q., 2024. Counterfactual explainability of black-box prediction models. doi: 10.48550/arXiv.2411.01625.

Gao, Y., Li, R., Croxford, E., Caskey, J., Patterson, B.W., Churpek, M.M., Miller, T.A., Dligach, D., Afshar, M., 2024. Leveraging a medical knowledge graph into large language models for diagnosis prediction (Preprint). https://doi.org/10.2196/preprints.58670.

Ghavi Hossein-Zadeh, N., 2024. An overview of recent technological developments in bovine genomics. Vet. Anim. Sci. 25, 100382. https://doi.org/10.1016/j.vas.2024.100382.

Ghavi Hossein-Zadeh, N., 2025. Artificial intelligence in veterinary and animal science: applications, challenges, and future prospects. Comput. Electron. Agric. 235, 110395. https://doi.org/10.1016/j.compag.2025.110395.

Ghavi Hossein-Zadeh, N., 2026. Advancing climate-resilient livestock systems: next-generation emission mitigation strategies and integrated technological innovations. Vet. Anim. Sci. 31, 100588. https://doi.org/10.1016/j.vas.2026.100588.

Gonzalez-Argote, J., Maldonado, E., Maldonado, K., 2025. Algorithmic bias and data justice: ethical challenges in artificial intelligence systems. EthAIca 4, 159. https://doi.org/10.56294/ai2025159.

Hassan, M.M., Mohamad Ramadan, G., Karthikeyan, R., Shkola, V., Järvis, M., Mostova, A., 2025. Harnessing digital twin technology for detailed simulation and analysis of climate impact on agricultural systems. SHS Web Conf. 216, 01070. https://doi.org/10.1051/shsconf/202521601070.

Hazarika, A.V., Shah, M., 2024. Blockchain-based distributed AI models: trust in AI model sharing. Int. J. Sci. Res. Arch. 13 (2), 3493–3498. https://doi.org/10.30574/ijsra.2024.13.2.2598.

Hilmi, A.H., Abdul Hamid, A.R., Wan Ibrahim, W.A.R.A., 2024. Advancements in cognitive ergonomics: integration with human-robot collaboration, workload management, and industrial applications. Malays. J. Ergon. 6, 39–51. https://doi.org/10.58915/mjer.v6.2024.1368.

Himabindu, M., Revathi, V., Gupta, M., Rana, A., Chandra, P.K., Abdulaali, H.S., 2023. Neuro-symbolic AI: integrating symbolic reasoning with deep learning. In: 10th IEEE Uttar Pradesh Section International Conference on Electrical, Electronics and Computer Engineering, UPCON 2023. Institute of Electrical and Electronics Engineers Inc., pp. 1587–1592. https://doi.org/10.1109/UPCON59197.2023.10434380

Hofmann, B., 2025. Biases in AI: acknowledging and addressing the inevitable ethical issues. Front. Digit. Health 7. https://doi.org/10.3389/fdgth.2025.1614105.

Hussein, A., Prasad, M., Braytee, A., 2024. Explainable AI methods for multi-omics analysis: a survey. doi: 10.48550/arXiv.2410.11910.

Ibrahim, U., Danmaigoro, A., 2024. Human-computer interaction in agricultural user interfaces. Int. J. Appl. Sci. Res. 2 (2), 187–198. https://doi.org/10.59890/ijasr.v2i2.1381.

Imam, N.M., Ibrahim, A., Tiwari, M., 2024. Explainable artificial intelligence (XAI) techniques to enhance transparency in deep learning models. IOSR J. Comput. Eng. 26 (6), 29–36. https://doi.org/10.9790/0661-2606012936.

Ishida, S., Sawai, T., 2024. From CRISPR to conscience: ethical dilemmas in gene editing and genetic selection. Am. J. Bioeth. 24 (8), 67–70. https://doi.org/10.1080/15265161.2024.2361900.

Ismail, S., Diaz, M., Carmona-Duarte, C., Vilar, J.M., Ferrer, M.A., 2024. CowScreeningDB: a public benchmark database for lameness detection in dairy cows. Comput. Electron. Agric. 216. https://doi.org/10.1016/j.compag.2023.108500.

Iyer, V., Manshad, M., Brannon, D., 2025. A value-based approach to AI ethics: accountability, transparency, explainability, and usability. Mercados y Negocios 2025 (54), 3–12. https://doi.org/10.32870/myn.vi54.7815.

James Ezeh, C., Chioma Anioke, S., Oyewole, S., Garnvwa David, M., 2024. The role of predictive analytics in enhancing public health surveillance: proactive and data-driven interventions. World J. Adv. Res. Rev. 24 (3), 3059–3077. https://doi.org/10.30574/wjarr.2024.24.3.3909.

Janaki, G., Umanandhini, D., 2024. Federated learning approaches for decentralized data processing in edge computing. In: Proceedings of the 5th International Conference on Smart Electronics and Communication, ICOSEC 2024. Institute of Electrical and Electronics Engineers Inc., pp. 513–519. https://doi.org/10.1109/ICOSEC61587.2024.10722731

Kalbhor, A., 2025. AI and machine learning in precision agriculture: the future of agricultural precision agriculture. Int. J. Res. Appl. Sci. Eng. Technol. 13 (2), 648–654. https://doi.org/10.22214/ijraset.2025.66920.

Katoch, N., 2025. Addressing bias in AI: ethical concerns, challenges, and mitigation strategies. Int. J. Sci. Res. Eng. Manag. 09 (04), 1–9. https://doi.org/10.55041/ijsrem45370.

Kazembe, I., Mkandawire, M., 2024. AI driven animal farming and livestock management system. Int. J. Adv. Res. Sci. Commun. Technol. 23–27. https://doi.org/10.48175/ijarsct-22604.

Kesgin, H.T., Amasyali, M.F., 2024. Advancing NLP models with strategic text augmentation: a comprehensive study of augmentation methods and curriculum strategies. doi: 10.48550/arXiv.2402.09141.

Khan, Y., Sánchez, D., Domingo-Ferrer, J., 2024. Federated learning-based natural language processing: a systematic literature review. Artif. Intell. Rev. 57 (12). https://doi.org/10.1007/s10462-024-10970-5.

Khawaja, S.A., Alturise, F., Alkhalifah, T., Khan, S.A., Khan, Y.D., 2024. Gluconeogenesis unraveled: a proteomic Odyssey with machine learning. Methods. https://doi.org/10.1016/j.ymeth.2024.09.002.

Khezri, S., Tanha, J., Samadi, N., 2024. An experimental review of the ensemble-based data stream classification algorithms in non-stationary environments. Comput. Electr. Eng. 118, 109420. https://doi.org/10.1016/j.compeleceng.2024.109420.

Kiran Vuyyuru, H., 2025. Navigating ethical dilemmas in generative AI development: a framework for responsible innovation. Int. J. Sci. Res. Arch. 14 (1), 1416–1423. https://doi.org/10.30574/ijsra.2025.14.1.0166.

Komorniczak, J., 2025. Synthetic non-stationary data streams for recognition of the unknown. doi: 10.48550/arXiv.2505.13745.

König, T., Wagner, F., Liebschner, M., Kley, M., 2025. Generative adversarial networks in fault diagnosis and an implementation approach: a review. IEEE Sens. J. 25 (14), 26289–26299. https://doi.org/10.1109/jsen.2025.3575284.

Krishnan, S., Ganesan, L.P., 2025. Integrated dairy production and cattle healthcare management using blockchain NFTs and smart contracts. Systems 13 (1). https://doi.org/10.3390/systems13010065.

Kulaklıoğlu, D., 2024. Explainable AI: enhancing interpretability of machine learning models. Hum. Comput. Interact. 8 (1), 91. https://doi.org/10.62802/z3pde490.

Kumar, R., Ruhel, R., van Wijnen, A.J., 2024. Unlocking biological complexity: the role of machine learning in integrative multi-omics. Acad. Biol. 2 (4). https://doi.org/10.20935/acadbiol7428.

Kumar, L., Nainwal, K., Sanjay, Nitharwal, P., Shukla, Y.K., Shukla, R., Chugh, P., 2025. A review on the role of artificial intelligence in livestock health management. Arch. Curr. Res. Int. 25 (5), 365–370. https://doi.org/10.9734/acri/2025/v25i51216.

Larin, I., Karabelsky, A., 2025. Riemannian manifolds for biological imaging applications based on unsupervised learning. J. Imag. 11 (4), 103. https://doi.org/10.3390/jimaging11040103.

Li, S., Liu, K., Wang, H., Yang, R., Li, X., Sun, Y., Zhong, R., Wang, W., Li, Y., Sun, Y., Wang, G., 2025a. Pose estimation and tracking dataset for multi-animal behavior analysis on the China Space Station. Sci. Data 12 (1). https://doi.org/10.1038/s41597-025-05111-8.

Li, J., Li, M., Subramaniya, K., 2025b. Few-shot learning: methods and applications. ITM Web Conf. 70, 02012. https://doi.org/10.1051/itmconf/20257002012.

Limbeck, K., Rieck, B., 2024. Detecting spatial dependence in transcriptomics data using vectorised persistence diagrams. doi: 10.48550/arXiv.2409.03575.

Lin, M., Guo, J., Gu, Z., Tang, W., Tao, H., You, S., Jia, D., Sun, Y., Jia, P., 2025. Machine learning and multi-omics integration: advancing cardiovascular translational research and clinical practice. J. Transl. Med. 23 (1). https://doi.org/10.1186/s12967-025-06425-2.

Liu, Z., Park, T., 2024. DMOIT: denoised multi-omics integration approach based on transformer multi-head self-attention mechanism. Front. Genet. 15. https://doi.org/10.3389/fgene.2024.1488683.

Long, Y., Novak, L., Walsh, C.G., 2024. Searching for value sensitive design in applied health AI: a narrative review. Yearb. Med. Inform. 33 (01), 075–082. https://doi.org/10.1055/s-0044-1800723.

Long, R., Sebo, J., Sims, T., 2025. Is there a tension between AI safety and AI welfare? Philos. Stud. 182 (7), 2005–2033. https://doi.org/10.1007/s11098-025-02302-2.

Losacco, C., Pugliese, G., Forte, L., Tufarelli, V., Maggiolino, A., De Palo, P., 2025. Digital transition as a driver for sustainable tailor-made farm management: an up-to-date overview on precision livestock farming. Agriculture 15 (13), 1383. https://doi.org/10.3390/agriculture15131383.

Lu, S., Xiao, X., 2024. Neuromorphic computing for smart agriculture. Agriculture 14 (11), 1977–2024.

Lu, S., Huang, Y., Shen, W.X., Cao, Y.L., Cai, M., Chen, Y., Tan, Y., Jiang, Y.Y., Chen, Y.Z., 2024. Raman spectroscopic deep learning with signal aggregated representations for enhanced cell phenotype and signature identification. PNAS Nexus 3 (8). https://doi.org/10.1093/pnasnexus/pgae268.

Lustenberger, M., Wollenschläger, S., Küng, L., 2025. DAO as digital governance tool for collaborative housing. Front. Blockchain 8. https://doi.org/10.3389/fbloc.2025.1523951.

Ma, J., Zhao, Z., Liu, Y., Li, T., Zhang, R., 2025. Designing an adaptive learning framework for predicting drug-target affinity using reinforcement learning and graph neural networks. Eng. Appl. Artif. Intell. 139, 109472. https://doi.org/10.1016/j.engappai.2024.109472.

Mahato, S., Neethirajan, S., 2025. Dairy DigiD: an edge-cloud framework for real-time cattle biometrics and health classification. AI 6 (9), 196. https://doi.org/10.3390/ai6090196.

Mahawar, K., Rattan, P., 2025. Empowering education: harnessing ensemble machine learning approach and ACO-DT classifier for early student academic performance prediction. Educ. Inf. Technol. 30 (4), 4639–4667. https://doi.org/10.1007/s10639-024-12976-6.

Manikandan, V., Neethirajan, S., 2025. AI-powered vocalization analysis in poultry: systematic review of health, behavior, and welfare monitoring. Sensors 25 (13), 4058. https://doi.org/10.3390/s25134058.

Maqsood, K., Hagras, H., Zabet, N.R., 2024. An overview of artificial intelligence in the field of genomics. Disc. Artif. Intell. 4 (1). https://doi.org/10.1007/s44163-024-00103-w.

Maraveas, C., Rajarajan, M., Arvanitis, K.G., Vatsanidou, A., 2024. Cybersecurity threats and mitigation measures in agriculture 4.0 and 5.0. Smart Agric. Technol. 9, 100616. https://doi.org/10.1016/j.atech.2024.100616.

Mate, S., Somani, V., Dahiwale, P., 2024. Applications of machine learning to address complex problems in livestock. In: 3rd International Conference for Innovation in Technology, INOCON 2024. Institute of Electrical and Electronics Engineers Inc. https://doi.org/10.1109/INOCON60754.2024.10511471

Max, K., Sames, L., Ye, S., Steinkühler, J., Corradi, F., 2025. Synthetic biology meets neuromorphic computing: towards a bio-inspired olfactory perception system. Neuromorph. Comput. Eng. 5 (3), 034010. https://doi.org/10.1088/2634-4386/aded2d.

Mia, N., Sarker, T., Halim, M.A., Alam, A.M.M.N., Ali, M.S., Rahman, M.M., Hashem, M.A., 2025. Machine learning overview and its application in the livestock industry. Meat Res. 5 (1). https://doi.org/10.55002/mr.5.1.109.

Milian Gómez, J.F., Byttebier, K., 2025. Agroecological sustainability: exploring the intersection of digital agriculture, ethics and the right to food. Disc. Agric. 3 (1). https://doi.org/10.1007/s44279-025-00228-1.

Misra, M.K., Mukherjee, P., Mahawar, S., Anusha, K., Vamshi Krishna, P., Charan Modak, K., Umejesi, I., 2024. Blockchain and AI Integration for Efficient and Secure Supply Chain Management. Informa UK Limited, pp. 44–48. https://doi.org/10.1201/9781003596721-11.

Mittal, M., Juneja, S., Mittal, R., 2025. Unraveling complex interactions between microbiota and immune system. Microb. Immun. 4790. https://doi.org/10.36922/mi.4790.

Modak, M., Pritom, M.M., Banik, S.C., Rabbi, M.S., 2025. Internet of things-based health surveillance systems for livestock: a review of recent advances and challenges. IET Wirel. Sens. Syst. 15 (1). https://doi.org/10.1049/wss2.70013.

Mohammed, S.P., Deepika, J., Sritharan, N., Ravichandran, V., Prasanthrajan, M., Kannan, P., 2025. A systematic literature review on artificial intelligence in transforming precision agriculture for sustainable farming: current status and future directions. Plant Sci. Today 12 (2). https://doi.org/10.14719/pst.6175.

Mu, M., Jiang, H., Wang, X., Dong, Y., 2025. Adaptive model-agnostic meta-learning network for cross-machine fault diagnosis with limited samples. Eng. Appl. Artif. Intell. 141, 109748. https://doi.org/10.1016/j.engappai.2024.109748.

Murthy, V.S.V.S., Kumar, J.N.V.R.S., 2025. Missing data imputation: a systematic comparison of traditional, machine learning and deep learning approaches. In: 2025 International Conference on Inventive Computation Technologies (ICICT), pp. 1209–1215. https://doi.org/10.1109/icict64420.2025.11005380.

Muthivhi, M., van Zyl, T.L., 2025. Wildlife target re-identification using self-supervised learning in non-urban settings. doi: 10.48550/arXiv.2507.02403.

Nalongo Bina, K., 2025. The role of natural language processing in medical documentation. Res. Output J. Biol. Appl. Sci. 5 (1), 11–14. https://doi.org/10.59298/rojbas/2025/511114.

Nancy Ambritta, P., Mahalle, P.N., Bhapkar, H.R., Shinde, G.R., Sable, N.P., 2025. Improving explainable AI interpretability with mathematical models for evaluating explanation methods. Int. J. Inf. Technol. https://doi.org/10.1007/s41870-025-02444-w.

Naripeddy, S.C.B., Thotakura, V.R., 2025. Ethical challenges and solutions in the deployment of artificial intelligence sys-tems. J. Artif. Intell. Gen. Sci. 8 (02), 81–110. https://doi.org/10.60087/jaigs.v8i02.392.

Neethirajan, S., 2025. Agency in livestock farming—a perspective on human–animal–computer interactions. Hum. Centr. Intell. Syst. https://doi.org/10.1007/s44230-025-00108-3.

Neulinger, A., Sparer, L., Roshanaei, M., Ostojić, D., Kakka, J., Ramljak, D., 2025. Is blockchain the future of AI alignment? Developing a framework and a research agenda based on a systematic literature review. J. Cybersecur. Priv. 5 (3), 50. https://doi.org/10.3390/jcp5030050.

Oh, S., Jekal, J., Liu, J., Kim, J., Park, J.-U., Lee, T., Jang, K.-I., 2024. Bioelectronic implantable devices for physiological signal recording and closed-loop neuromodulation. Adv. Funct. Mater. 34 (41). https://doi.org/10.1002/adfm.202403562.

Ouhmouk, M., Baichoo, S., Abik, M., 2025. Challenges in AI-driven multi-omics data analysis for oncology: addressing dimensionality, sparsity, transparency and ethical considerations. Inform. Med. Unlocked 57, 101679. https://doi.org/10.1016/j.imu.2025.101679.

Pan, Y., Sun, J., Yu, H., Bai, G., Ge, Y., Luck, J.D., Awada, T., 2024. Transforming Agriculture with Intelligent Data Management and Insights. https://doi.org/10.48550/arxiv.2401.13672.

Papa, M., de Medeiros Oliveira, S.R., Bergier, I., 2024. Technologies in cattle traceability: a bibliometric analysis. Comput. Electron. Agric. 227, 109459. https://doi.org/10.1016/j.compag.2024.109459.

Parekh, N.H., Shinde, M.V., 2024. Federated learning: a paradigm shift in collaborative machine learning. Int. J. Sci. Res. Eng. Manag. 08 (11), 1–6. https://doi.org/10.55041/ijsrem38501.

Pegios, P., Feragen, A., Hansen, A.A., Arvanitidis, G., 2024. Counterfactual explanations via Riemannian latent space traversal. doi: 10.48550/arXiv.2411.02259.

Pérez-Enciso, M., Zingaretti, L.M., de los Campos, G., 2025. Generative AI for predictive breeding: hopes and caveats. Theor. Appl. Gene. 138 (7). https://doi.org/10.1007/s00122-025-04942-8.

Pyrkov, A., Aliper, A., Bezrukov, D., Podolskiy, D., Ren, F., Zhavoronkov, A., 2024. Complexity of life sciences in quantum and AI era. WIREs Comput. Mol. Sci. 14 (1). https://doi.org/10.1002/wcms.1701.

Qazi, A., Razzaq, T., Iqbal, A., 2024. AnimalFormer: multimodal vision framework for behavior-based precision livestock farming. In: IEEE Computer Society Conference on Computer Vision and Pattern Recognition Workshops. IEEE Computer Society, pp. 7973–7982. https://doi.org/10.1109/CVPRW63382.2024.00795.

Raha, A.D., Adhikary, A., Gain, M., Hong, C.S., 2025. A federated learning framework for optimizing edge computing with semantic offloading. In: International Conference on Advanced Communication Technology, ICACT. Institute of Electrical and Electronics Engineers Inc., pp. 142–149. https://doi.org/10.23919/ICACT63878.2025.10936753

Rahman, M.S., Hossain, M.S., Rahman, M.K., Islam, M.R., Sumon, M.F.I., Siam, M.A., Debnath, P., 2025. Enhancing supply chain transparency with blockchain: a data-driven analysis of distributed ledger applications. J. Bus. Manag. Stud. 7 (3), 59–77. https://doi.org/10.32996/jbms.2025.7.3.7.

Rajawat, A.S., Arora, R.K., Muqeem, M., Faisal, S.M., Jamal, M.K., 2025. Enhancing IoT efficiency by minimizing latency and boosting real-time processing with edge computing. SN Comput. Sci. 6 (6). https://doi.org/10.1007/s42979-025-04192-x.

Rane, J., Kaya, Ö., Kumar Mallick, S., Liladhar Rane, N., 2024. Smart Farming Using Artificial Intelligence, Machine Learning, Deep Learning, and ChatGPT: Applications, Opportunities, Challenges, and Future Directions. Deep Science Publishing. https://doi.org/10.70593/978-81-981271-7-4_6.

Ravikumar, H., Ramachandran, A., Pandithurai, P., Vasanthan, Poovan, S., 2025. Artificial intelligence (AI) and blockchain integration: synergies, applications, and future prospects. ShodhKosh J. Vis. Perform. Arts 6 (1). https://doi.org/10.29121/shodhkosh.v6.i1.2025.4682.

Ren, Z., Zhou, S., Liu, D., Liu, Q., 2025. Physics-informed neural networks: a review of methodological evolution, theoretical foundations, and interdisciplinary frontiers toward next-generation scientific computing. Appl. Sci. 15 (14), 8092. https://doi.org/10.3390/app15148092.

Safra Zaabar, L., Yacob, A.A., Nathan, D.K., Yip, E.W.H., Razali, N.A.M., 2025. A review of livestock smart farming for sustainable food security. JOIV Int. J. Inf. Visual. 9 (2), 728. https://doi.org/10.62527/joiv.9.2.2794.

Sagar, G., Roshni Akarshana, H., Chakravardhan, R., 2025. Revolutionizing genetic manipulation: artificial intelligence-powered precision in CRISPR-Cas9 gene editing. In: 2025 3rd International Conference on Disruptive Technologies (ICDT), pp. 1549–1554. https://doi.org/10.1109/icdt63985.2025.10986581.

Şahin, E., Arslan, N.N., Özdemir, D., 2024. Unlocking the black box: an in-depth review on interpretability, explainability, and reliability in deep learning. Neural Comput. Appl. 37 (2), 859–965. https://doi.org/10.1007/s00521-024-10437-2.

Samdani, G., Viswanathan, G., Dasu Jegadeesh, A., 2025. Human-AI collaboration: balancing agentic AI and autonomy in hybrid systems. Int. J. Cloud Comput. Serv. Arch. 15 (1), 01–15. https://doi.org/10.5121/ijccsa.2025.15101.

Santana, T.C., Guiselini, C., Pandorfi, H., Vigoderis, R.B., Barbosa Filho, J.A.D., Soares, R.G.F., Araújo, M.D.F., Gomes, N.F., Lima, L.D.D., Santos, P.C.D.S., 2025. Ethics, animal welfare, and artificial intelligence in livestock: a bibliometric review. AgriEngineering 7 (7). https://doi.org/10.3390/agriengineering7070202.

Sartori, F., Codicè, F., Caranzano, I., Rollo, C., Birolo, G., Fariselli, P., Pancotti, C., 2025. A comprehensive review of deep learning applications with multi-omics data in cancer research. Genes 16 (6), 648. https://doi.org/10.3390/genes16060648.

Shi, D., Han, A., Lin, L., Guo, Y., Wang, Z., Gao, J., 2024. Design your own universe: a physics-informed agnostic method for enhancing graph neural networks. Int. J. Mach. Learn. Cybern. 16 (2), 1129–1144. https://doi.org/10.1007/s13042-024-02326-w.

Siegford, J.M., 2024. Precision Livestock Farming and Technology in Pig Husbandry. Elsevier BV, pp. 449–469. https://doi.org/10.1016/b978-0-323-85676-8.00015-8.

Simpson, C., Tabatsky, E., Rahil, Z., Eddins, D.J., Tkachev, S., Georgescauld, F., Papalegis, D., Culka, M., Levy, T., Gregoretti, I., Meehan, C., Schiller, C., Bestak, K., Schapiro, D., Chernyshev, A., Walther, G., Ghosn, E.E.B., Orlova, D., 2025. Lifting the curse from high-dimensional data: automated projection pursuit clustering for a variety of biological data modalities. GigaScience 14. https://doi.org/10.1093/gigascience/giaf052.

Singh, S., Sabharwal, I., Kushwaha, S., Jain, D.S., Jain, D.M., 2024. Enhancing human-machine interaction: leveraging neuromorphic chips for adaptive learning and control in neural prosthetics and artificial intelligence. Int. J. Sci. Res. Comput. Sci. Eng. Inform. Technol. 10 (6), 933–940. https://doi.org/10.32628/cseit241061135.

Solona, O., Lisovyi, D., 2024. Prediction and monitoring of animal health using artificial intelligence. Vibrat. Eng. Technol 2 (113), 33–42. https://doi.org/10.37128/2306-8744-2024-2-4.

Sztandarski, P., Marchewka, J., Jaszczyk, A., Solka, M., Michnowska, H., Pogorzelski, G., Gotxi, J.R., Damaziak, K., Wójcik, W., Siwiec, D., Seidavi, A., Horbańczuk, J.O., 2025. Scientific literature on AI technologies to enhance animal welfare, health and productivity. Anim. Sci. Pap. Rep. 43 (2), 115–132. https://doi.org/10.2478/aspr-2025-0009.

Tafadzwa Mpofu, K., Mthunzi-Kufa, P., 2024. Recent Advances in Quantum Biosensing Technologies. IntechOpen. https://doi.org/10.5772/intechopen.1007858.

Tahir, G.A., 2024. Ethical challenges in computer vision: ensuring privacy and mitigating bias in publicly available datasets. doi: 10.48550/arXiv.2409.10533.

Taiwo, G.A., Ali, A., Mansouri, T., 2024. Review of farmer-centered AI systems technologies in livestock operations. CABI Rev. 19 (1). https://doi.org/10.1079/cabireviews.2024.0038.

Tan, C.Y., Ong, H.F., Lim, C.H., Tan, M.S., Ooi, E.H., Wong, K.S., 2025. Amogel: a multi-omics classification framework using associative graph neural networks with prior knowledge for biomarker identification. BMC Bioinf. 26 (1). https://doi.org/10.1186/s12859-025-06111-6.

Tangorra, F.M., Buoio, E., Calcante, A., Bassi, A., Costa, A., 2024. Internet of Things (IoT): sensors application in dairy cattle farming. Animals 14 (21), 3071. https://doi.org/10.3390/ani14213071.

Teja Reddy Maryala, B., 2025. AI governance framework for health data and sensitive domains: a comprehensive approach to ethical data utilization. Eur. J. Comput. Sci. Inform. Technol. 13 (35), 76–85. https://doi.org/10.37745/ejcsit.2013/vol13n357685.

Thute, T., Gudadhe, A., 2025. Wearable biosensors: transformative innovations and future pathways for continuous health monitoring. In: 2nd International Conference on Machine Learning and Autonomous Systems, ICMLAS 2025—Proceedings. Institute of Electrical and Electronics Engineers Inc., pp. 1203–1207. https://doi.org/10.1109/ICMLAS64557.2025.10968261

Toussaint, P.A., Leiser, F., Thiebes, S., Schlesner, M., Brors, B., Sunyaev, A., 2023. Explainable artificial intelligence for omics data: a systematic mapping study. Briefings Bioinf. 25 (1). https://doi.org/10.1093/bib/bbad453.

Tulpan, D., 2024. Advanced computing and information technology to address challenges in livestock production. J. Anim. Sci. 102 (Suppl. 3), 295–296. https://doi.org/10.1093/jas/skae234.337.

Usman, M., Hamza, A.U., Abdullahi, S., Ja'afar, S., Hassan, A., Musa, H., 2025. The implementation of artificial intelligence in quantum technology. J. Sci. Innov. Technol. Res. https://doi.org/10.70382/ajsitr.v8i9.029.

van der Staay, F.J., Goerlich, V.C., Meijboom, F.L.B., Arndt, S.S., 2025. Animal welfare definitions, frameworks, and assessment tools: advancing the measurement and laying the foundation for improved animal welfare through a three-step approach. Anim. Welf. 34. https://doi.org/10.1017/awf.2025.23.

Varma, A., Dixit, N., Ray, S., Kaur, J., 2024. Blockchain technology for sustainable supply chains: a comprehensive review and future prospects. World J. Adv. Res. Rev. 21 (3), 980–994. https://doi.org/10.30574/wjarr.2024.21.3.0804.

Venturini, R.E., 2025. Technological innovations in agriculture: the application of blockchain and artificial intelligence for grain traceability and protection. Braz. J. Dev. 11 (3), e78100. https://doi.org/10.34117/bjdv11n3-007.

Vinay, S.Y., Kumar, A., Renu Narwal, P., 2024. Synergizing intelligence: revolutionizing supply chains with blockchain and AI. Int. J. Adv. Res. Sci. Commun. Technol. 56–60. https://doi.org/10.48175/ijarsct-17610.

Wadood, A.A., Bordbar, F., Zhang, X., 2025. Integrating omics approaches in livestock biotechnology: innovations in production and reproductive efficiency. Front. Anim. Sci. 6. https://doi.org/10.3389/fanim.2025.1551244.

Wang, L., Hu, C., Fan, M.Z., 2024. Editorial: the role of nutritional strategies in the regulation of gut microbiota and host immune system. Front. Microbiol. 15. https://doi.org/10.3389/fmicb.2024.1501596.

Wasudeorao, K.A., 2024. Evolutionary optimization of deep learning models: enhance performance through hyperparameter and structural tuning with evolutionary algorithms. Adv. Nonlinear Var. Inequalities 27 (2), 632–649. https://doi.org/10.52783/anvi.v27.1351.

Yadav, A., 2024. Future of blockchain governance: exploring decentralized democracy. J. Unique Law Stud. III (I). https://doi.org/10.59126/v3i1a8.

Yang, S.X., Han, Y., Ma, W., Tulpan, D., Li, J., Li, J., Yue, Y., 2025. Review of computer vision for livestock body conformation assessment. Agric. Commun., 100099 https://doi.org/10.1016/j.agrcom.2025.100099.

Yekeen, A.O., Paul-Mikki Ewim, C., Sam-Bulya, N.J., 2024. Enhancing supply chain security and transparency with AI and blockchain integration for future-proof solutions. Int. J. Front. Res. Sci. Technol. 4 (1), 043–062. https://doi.org/10.56355/ijfrst.2024.4.1.0052.

Yetgin, A., 2025. Revolutionizing multi-omics analysis with artificial intelligence and data processing. Quant. Biol. 13 (3). https://doi.org/10.1002/qub2.70002.

Yu, H., Ma, M., Zhang, B., Wang, A., Zhong, G., Zhou, Z., Liu, C., Li, C., Fang, J., He, Y., Ren, D., Deng, F., Hong, Q., Zhao, Y., Guo, X., 2025. Bionic sensors for biometric acquisition and monitoring: challenges and opportunities. Sensors 25 (13), 3981. https://doi.org/10.3390/s25133981.

Zhang, Y., Liu, J., Wang, J., Dai, L., Guo, F., Cai, G., 2025. Federated learning for cross-domain data privacy: a distributed approach to secure collaboration. doi: 10.48550/arXiv.2504.00282.

Zhao, Y., Li, H., Zhou, H., Attar, H.R., Pfaff, T., Li, N., 2024. A review of graph neural network applications in mechanics-related domains. Artif. Intell. Rev. 57 (11). https://doi.org/10.1007/s10462-024-10931-y.

Zhou, Y., Geng, P., Zhang, S., Xiao, F., Cai, G., Chen, L., Lu, Q., 2024. Multimodal functional deep learning for multiomics data. Briefings Bioinf. 25 (5). https://doi.org/10.1093/bib/bbae448.

Zhou, Y., Tang, X., Zhang, D., Lee, H.J., 2025. Machine learning empowered coherent Raman imaging and analysis for biomedical applications. Commun. Eng. 4 (1). https://doi.org/10.1038/s44172-025-00345-1.

Zhu, S., Sun, S., 2024. A short survey: exploring knowledge graph-based neural-symbolic system from application perspective. https://doi.org/10.48550/arXiv.2405.03524.

Chapter 4

Artificial intelligence integration in livestock genomics

Artificial intelligence (AI) is being used to transform the genomics of livestock. The future system is a dynamic ecosystem with recommendation goals, biological and epigenetic environment, and genomic variation in perpetual optimization instead of a linear pipeline of marker detection with a predictive function. AI systems make use of heterogeneous, dynamic, and partially annotated data and can find hidden structure and complex genotype-phenotype relationships. On the contrary, quantitative genetics involves the use of structured information and the development of models to derive correspondence. The new representation learning, informed uncertainty, and choice maximization enabled with AI will bring about new theoretical descriptions of genomic prediction beyond only offering a small contribution to prediction and enable AI to perform as an integrative layer in genomic studies of livestock that incorporates long-standing constructs of heritability, selection response, and heritable variation stewardship (Džermeikaitė et al., 2025).

The substrate of the genome is in itself coadapting. The fact that one-dimensional variation will be substituted with richer structural variation (e.g., copy number), rare variation (e.g., mutations and/or structural variation), and population-specific variation (both in epigenetics and population genetics) will expand the functional signals and computational abstraction required to detect them. The AI techniques are innovations spanning a range of abstraction spaces, with the ability to train on flexible representations for innumerable applications that demand flexible consideration of intricate mutant structures, detangling uninformative population structure, and overcoming sparse sampling in patchy datasets. It presents the possible implications of temporal multi-omics and of increasing streams of microbiomes, environments, and high-throughput phenomics to realistic, core genomic layers and proposes AI as a translational scaffold to generate coherence on some aspects of molecular resolutions with actionable breeding and management implications, rather than as an and/or replacement of mechanistic understanding (Dara et al., 2025; Mardinoglu et al., 2025).

To effectively integrate, it will be necessary to be disciplined in transparency in the data foundation. The selection of the sampling design (e.g., multi-desampling and multimodal sources), the completeness of the metadata, the adoption of the interoperability standards, quality management, provenance tracking, and the approach to governance will all lead to downstream model faithfulness, equity, and reproducibility. Although multi-core, multi-parameter, multi-regional, multi-layered, and multi-sector algorithmic complexity may be useful, it is not capable of addressing systemic bias, disparate and underrepresented genetic make-up, and data provenance obscurity. It is then possible to decouple ethically informed design and technically informed design and incorporate welfare as an active requirement, equity of benefit, privacy, and transparency requirements on the initial design choices, and not an ex-post facto audit (Taddese et al., 2025).

The breadth of methods is reflective of the assumptions on how things work. These mental models of governmental lifecycle governance, benchmarking rigor, domain shift detection, and pipeline maintenance need to be considered and provided as infrastructure and not maintenance labor. The section shows how the imperative to quantify uncertainty, generate explainability interfaces, and finally formulate multi-objective optimization structures can reap benefits when connecting trust, relating, as well as embedding methodologies when stakeholder groups are heterogeneous. With that said, even computational workflows, perhaps tentatively looking ahead, need to consider energy consumption and environmental footprint in the context of larger mandates related to sustainability, resilience, or being a good steward of resources that otherwise mimic the biological urge towards desiring resilience or managing resources (Souza et al., 2024).

The chapter focuses on future directions that may further reconstruct the field, such as simulations on digital twins where they can be used to explore scenarios using changing climatic, epidemiological, and economic pressures, causally explicit reasoning engines, foundation-level genomic representations, and federal coordinate systems. AI-enabled livestock genomics is formed with the help of statistical genomics, machine intelligence, systems biology, and responsible

Artificial Intelligence in Livestock Omics. https://doi.org/10.1016/B978-0-443-45663-3.00012-9

innovation. The field is mapping out accurate, adaptive, transparent, and equitable genomic decision ecosystems through the conceptualization of ethical governance, governance methodological rigour, infrastructure of the future, and principled data stewardship.

4.1 Introduction and conceptual foundations

AI is well-positioned to initiate the path of livestock genomics and showcase new goals, directions, and organizational forms and practices (Fig. 4.1).

Genomic selection techniques were initiated with well-specified statistical guesses, in a reasonably stable atmosphere, with the anticipation of creating a bit effective genetic additive documents to be effectively utilized, by means of sampling outlined marker panels. The coming and new genomic uses are systems in which we will be able to model many dynamical environmental factors, as well as heterogeneous determinants of phenotype, biological annotation at many scales, and the mapping of a rapidly diverging catalog of variants. AI offers some level of flexibility by enabling models to compute hierarchical abstractions that provide connections between the essentially raw molecule depictions and the metrics that pertain to performance, health, well-being, and sustainability. The AI development is a less-additional and more codesigning process of data collection, representation learning, and decision support downstream, rather than the mere addition of new datasets increasing the accuracy. To its predictive pipelines, AI replaces a genomic evaluation as an iterative process of recalibration to an adaptive loop with models that have the ability to quantify uncertainty, perform fairness auditing, and multi-objective optimization. This way, the strategic advantage turns out to be a governance and stewardship strategic advantage. Models with drift and recalibration capabilities have now become an opportunity for resilient innovation in the face of change in markets, regulation, and climate (Ghavi Hossein-Zadeh, 2024; Džermeikaitė et al., 2025; Nawaz et al., 2025).

The most important element in converting the representational substrate is the shift between sparse arrays of markers to much richer and more pluralistic assemblies of genomic informatics. Linear reference genomes are pragmatically utilitarian, but they pull the variation of a population into one coordinate reference frame, unlike critical genomic aspects (e.g., structural rearrangements, copy number variations, presence-absence variation, and context-dependent regulatory features), which are thus challenging to observe. The pangenome and graph-based reference frameworks are more relaxed reference constraints, directly including alternative haplotypes and structures, which allow algorithms to traverse a reference landscape that is better modeled by population-level genomic topology. Embedding (based on either learning or graph neural) AI approaches can more effectively learn complex variation neighborhoods, position dependencies, and higher-order interactions with minimal feature engineering. Giving the representational substrate more dimensions provides a broader scope of lens through which to see functional signal, but injects more dimensionality, sparsity, and undesired signal. Integration back then was the challenge of identification of adaptive regularization, disentangling of causal-link artefacts and signal of interest, and principled noninclusion of misalignment because of missingness. Moreover, the conversion into pangenomic representation makes equity a matter of concern; explicitly inclusive construction of graph references assists here, to minimize inequity in performance of minimal performed lineages aligned to a graph reference tilted in favor of commercially significant groups. Representation, then, constitutes an intersection in the scale of technical design and ethical control (Chikhi et al., 2024; Cui et al., 2025; Lehmann et al., 2025).

Quantitative genetic theory in combination with machine intelligence leads to radically new problems in genotype-phenotype modeling. The traditional methods involve the division of phenotypic variance into additive, nonadditive, and environmental terms and approximate the terms in linear mixed models, which are directed by relationship matrices. AI provides an opportunity to learn nonlinear manifolds, in which the representation may represent epistatic modulation, dominance compression, pleiotropic architecture, and genomic background contexts and covariates. Machine learning does not displace the foundational concepts, but it may be used to apply them in another context, and the possibilities of extending the inferences made based on them. As an example, the allocation of representational weight to parts of the regulatory space or groups of variants that covary can be done through attention-based mechanisms, whereas the estimation of posterior uncertainty to inform calibrated decision thresholds can be done by probabilistic layers. The use of causal discovery models offers a systematic way of delineating predictive correlates using putative intervention targets. This difference could be critical in the process of prioritization of gene editing or management interventions. A greater complexity (capabilities) is associated with its own problems: It gets more difficult to overfit to

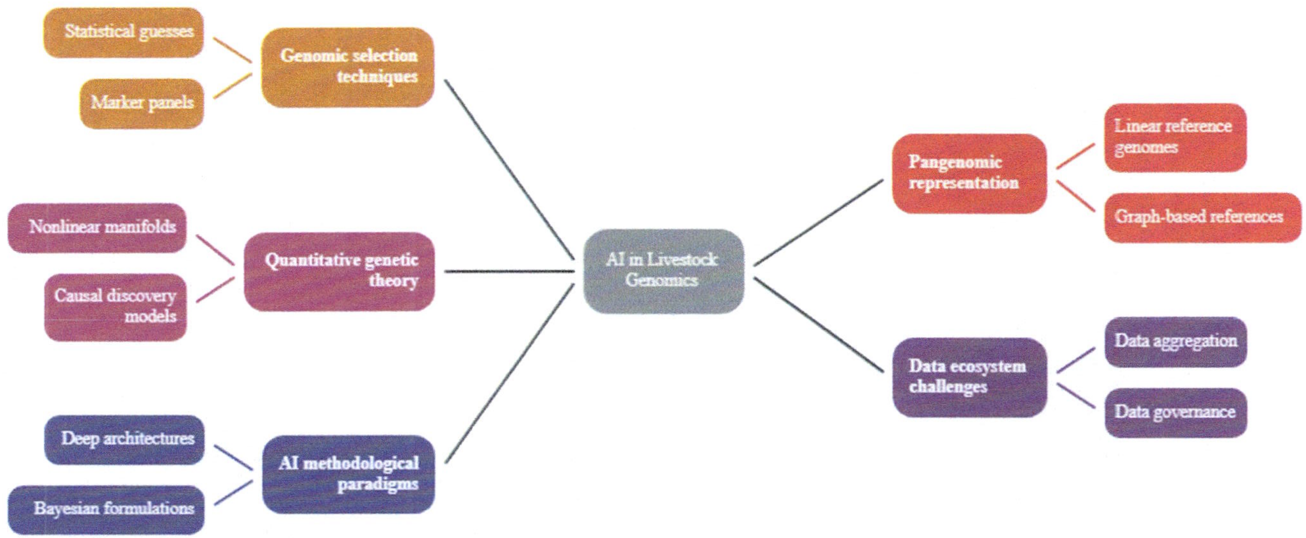

FIGURE 4.1 **An overview of AI in livestock genomics.** A conceptual map of the main domains and components that underpin the integration of AI into livestock genomics.

a latent structure that can remain unobservable; it becomes compounded, thereby ratifying ascertainment biases; and it mixes up population stratification with the signal of interest. Hence, to overcome this difficulty, builders are encouraged to do cross-population validation, sensitivity analysis should start on possible confounding covariates, and transparency of reporting on architectural decisions, loss function, and regularization regimes should be promoted. The theoretical synthesis uses AI as an appendix to supplement but not substitute the explanatory scaffolding of the quantitative genetics through the marineering of an inferential rigor with an adaptive flexibility (Forrest et al., 2025; Han, 2025; Rijal et al., 2025).

The bigger data ecosystem that models are embedded in is actually heterogeneous, fragmented, and nonrepresentative. Genomic resources are also available as a by-product of a phenomic instrument, environmental monitoring, or pedigree records. The variations in the sampling intensity, technological platforms, the completeness and quality of metadata are all causes of batch effects, missingness patterns, and bias. This is found to lack in the geographic regions, production systems, and locally adapted breeds, whose scope of representation limits the possibility of generalizing the studies to provide equity in genetic gain and resiliency. To address these gaps, pipelines will need the creation of data aggregation techniques aimed at putting diversity into the first rank, the development of ontologies to have harmonized trait semantics, and the application of provenance tracking to guide the creation of strong downstream models. Limits of privacy, sovereignty, and commercial ownership keep the integration of the data between the institutions restricted in such a way that it cannot be used outside the institutions. They also need to address cyberbiosecurity threats as a protection mechanism against data contamination, model inversion, and data exfiltration. The next pillar, thereby, that will carry equal significance to the complexity of the algorithms in the evolutionary cascade of adaptive genomic intelligence, will be data governance (Ramanan et al., 2025; Rao, 2025).

In such an infrastructural and conceptual setting, the complementary advantages of a broad set of AI methodological paradigms exist. The classical linear mixed models and their genomic analogs have continued to be valuable/useful because of their statistical transparency, computational tractability, and congruence with established paradigms of interpretation. The extension of the nonlinearity of the kernel and ensemble extensions is made available without compromising on some interpretability. Hierarchical features can be learned on sequence, haplotype, and graph-encoded representations directly using deep architectures consisting of convolutional, recurrent, attention-oriented, and transformer-based components, and learn local motifs and long-range interactions. Graph neural networks employ relational inductive bias, which represents the population structure, the network of variants' cooccurrence, and biological interaction graphs, and signal propagation by topological neighborhoods. Bayesian and probabilistic formulations put prior biological knowledge on the table, propagate the uncertainty, and allow posterior inference, which are critical to risk-sensitive decision making. Transfer, multi-task, meta, and few-shot approaches improve data-sparse situation behavior through the process of cross-breed, cross-trait, cross-environment, and continual and online learning, modifying models to changing times. Computation is redistributed by federated and privacy-preserving schemes, and it reduces the barriers associated with centralized aggregation. Generative models produce synthetic surrogates or synthetic genomic or phenotypic data to add to them, which are cautioned about in terms of governance structures countering amplification of bias or privacy invasion. Causal discovery and structural modeling attempt to assist in screening between associational and intervention-relevant indicators (Feng et al., 2025; Pasupuleti, 2025; Ghavi Hossein-Zadeh, 2026a). Multi-objective optimizers integrate productivity, welfare, diversity conservation, resilience, and emissions goals within unified allocation strategies. Model compression, pruning, quantization, and specialized hardware acceleration contribute to energy efficiency and scalability, thereby aligning computational practices with broader sustainability imperatives. A comparative assessment of these paradigms requires multidimensional evaluation criteria spanning predictive accuracy, calibration, fairness across subpopulations, robustness to distributional shifts, interpretability sufficient to gain stakeholder trust, resource efficiency, and ease of lifecycle maintenance (Chakraborty et al., 2025; Ghavi Hossein-Zadeh, 2026b).

These conceptual foundations all lead to a coherent vision in which data architecture, representational expansiveness, statistical genetics, and machine intelligence evolve progressively, transparently, and justly into equitable genomic decision ecosystems. Acquisition pipelines, model portfolios, governance standards, and interfaces to the stakeholders must be coordinated across strategic deployment, as it is understood that the failure of any of the layers can fail the entire system. The shift to pangenomic, multi-omics, and situational modeling opens opportunities to embrace biological complexity at the cost of methodological stringency and reproducibility, and have ethical foresight. Later chapters will continue this outline to discuss application domain areas of specialization, multi-layer fusion strategies, validation, lifecycle management processes, and innovations that will further reorganize the limits of capability. The further discussion is based on the following principles, which help to align the technical progress with the reasonable management and the value creation in the long term. All the additional strata of analysis are informed in broad terms by this background.

4.2 Data foundations and infrastructure

Strong AI application to livestock genomics can start with purposive experimental and sample design to preempt downstream algorithm requirements instead of retrofitting old collection paradigms. Design decisions can determine the statistical efficiency of the estimation of the parameters, the ability to deconfound the genetic signal and confounding structure, as well as the equitability of the predictive generalization between production systems and breed lineages. Representation of allelic diversity, environmental gradients, management regimes, and temporal cohorts is achieved using stratification strategies based on the principles of population genetics to ensure that the models learn invariant and context-sensitive components without being overfit to narrow ecological or business niches. Longitudinal sampling improves the ability to model developmental career paths, genotype-age interactions, and time-varying health conditions, and nested family or cohort designs provide data needed to disaggregate additive, dominance, and epistatic effects. Power analyses are no longer limited to the conventional heritability targets to include predicted effect sizes, distributions of rare structural variants, cross-modal correlations in multi-omics fusion, and uncertainty tolerances of risk-sensitive decision layers. Early codesign of domain geneticists, data engineers, and machine learning architectures matches sequencing depth, platform uniformity, phenotyping cadence, and instrumentation of environmental sensors to architectural expectations of missingness mechanisms, temporal context windows, and variant embedding resolution. The resulting datasets are therefore not just curated to be rich in biology but also to be computationally tractable, allowing models to use information-dense, ethically inclusive, and balanced training corpora (Van Hilten et al., 2024; Lehmann et al., 2025; Mardinoglu et al., 2025).

The connective tissue that makes heterogeneous datasets machine-actionable is metadata infrastructure and semantic interoperability. In the absence of standard ontological anchors, trait descriptors, management interventions, health observations, and environmental annotations reduce to idiosyncratic vocabularies that make aggregation, federated learning, and reproducible benchmarking difficult. Implementation of community schema and exchange frameworks that are partners of global genome data sharing efforts enables consistent representation of sample characteristics, consent limitations, and transformation of the analysis. Semantic resolution is enhanced by breed ontologies, trait hierarchies, anatomical and growth stage taxonomies, pathological state name sets, and controlled vocabularies of nutritional and climatic variables; and ontology-driven feature expansion is made possible by logical inference. Interoperability is based on harmonized identifier systems, versioned controlled terms, machine-readable provenance, and programmatic application programming interfaces that allow incremental synchronization. Conformance to FAIR principles provides findability, access with controlled conditions, interoperability in the context of analytic stacks, and reusability through express licensing and consent metadata. These semantic layers enable query expansion, automatic quality and validation rules, and dynamic schema evolution, enabling AI pipelines to consume, reconcile, and reason across heterogeneous modalities without fragile, human-crafted mapping tables (Sinaci et al., 2024; Vogt et al., 2025).

Data quality control, error profiling, and bias mitigation are not a one-time preprocessing step but a multi-stage, continuous process. Platform-specific artefact detection, statistical normalization, and cross-sample concordance assessment are applied to raw sequence reads, genotype calls, expression quantifications, and sensor-derived phenotypes, and the error models are propagated but not discarded to support downstream quantification of uncertainty. Replicate concordance, Mendelian consistency, and context-aware sequencing depth distributions are used to calibrate variant calling thresholds and imputation frameworks that harmonize reference panels of different ancestry to limit the biases of differential accuracy, which can amplify equity gaps. Mixed modeling, adversarial representation learning, or covariate balancing algorithms are used to separate batch effects that might be introduced by laboratory experiments, library preparation kits, or device software updates and maintain biological variance at the cost of technical noise. As a surface representation of representational imbalance across axes of stratification, bias audits review allele frequency spectra, haplotype diversity, tail of phenotype distribution, and missingness patterns on the basis of breed category, production system, climatic zone, and sex. The conventional mitigation methods include reweighting, specific supplemental sampling, synthetic augmentation subject to governance constraints, domain adaptation, and fairness regularization on model loss functions. Constant observation identifies data-generating process drift, and recalibration is done before the predictive integrity or trust of the stakeholders is compromised by the buildup of insidious error (Aninze, 2024; Mansilla et al., 2025).

The choice of reference model has a far-reaching impact on downstream analysis that defines the alignment of reads, the representation of variants, and the reconstruction of haplotypes. Single linear assemblies are simpler, more toolable, and have a fixed coordinate space, but condense structural polymorphism and lineage-specific sequences to a canonical scaffold, introducing mapping bias, coverage dropout, and false representation of complex rearrangements. Pangenome graphs expand representational fidelity by encoding alternative insertions, inversions, translocations, and presence-absence regions, so that alignment algorithms can explore topological neighborhoods, which are more reflective of population genomic architecture. Computational scalability requires the use of graph-based indexing and succinct data structures that preserve query responsiveness. Complexity subgraph areas also have more uncertainty in alignment; probabilistic alignment scoring frameworks and downstream variant calling frameworks that can model path ambiguity are required. Interoperability between graph-derived variant catalogs and the legacy linear-based annotation ecosystems is mediated by coordinate translation layers that require a must-be version-controlled semantic mapping to guarantee interpretive consistency. The shift to pluralistic references is also a socio-technical project, and this necessitates clear inclusion rules to haplotype integration, fair representation of minor and geographically isolated groups, and mechanisms to update iteratively without disrupting historical analytical comparability (Liu et al., 2024; Chikhi et al., 2024).

Combining structural, copy number, rare, and mobile element variation expands the analytical horizon past single-nucleotide polymorphisms to genomic events with disproportionate genetic effect on gene dosage, regulatory topology, and chromatin architecture. The detection pipelines using paired-end signature, split read evidence, read depth segmentation, assembly-based reconstruction, and mobile element anchoring have to cope with nonuniform sensitivity and specificity characteristics depending on the complexity of genomes, repetitive content, and local GC composition. Representation of variants, breakpoint uncertainty ranges, and genotype likelihood models can be harmonized to allow downstream AI architectures to take advantage of these features alongside small variant matrices without inflating induced sparsity. The feature engineering approaches use length-normalized encodings, type-specific embeddings, graph-derived adjacency representations, and frequency-sensitive regularization to ensure that common structural classes do not over-dominate but still signal ultra-rare events. There is no easy way to impute structural variants, and hybrid methods that combine haplotype clustering with extensive generative models calibrated on empirical copy number distributions are required. Mobile element polymorphisms require special consideration because of the potential for insertional homoplasy and regulatory disruption. Finally, extensive variant integration increases the ability to address causal architecture, refine polygenic models, and enhance cross-population portability in methods that maintain equal treatment of variant classes and sound uncertainty propagation (Harris et al., 2024; Chen et al., 2025b).

Ethical, legal, and operational legitimacy is based on secure data stewardship, access governance, and provenance tracking. Role and permissions, permission constraints, embargo duration, and benefit sharing requirements are modeled in governance structures, implemented as fine-grained access control lists, attribute-based authorization, and auditable policy enforcement engines. Data partitioning methods isolate identifiable or sensitive layers between generalized analytical derivatives and identifiable or sensitive layers, and cryptographic hashing, digital signatures, and immutable ledger technologies offer tamper-evident confidence to the integrity of datasets across transformation steps. Provenance ontologies record lineage over raw ingestion, quality control processes, normalization processes, imputation processes, annotation versions, and model training processes, such that they can be reproducible, legally required, and forensically studied in case of anomalous results. Differential privacy, secure multi-party computation, homomorphic encryption, and trusted execution environments add more confidentiality to those areas where jurisdictional or commercial concerns forbid traditional centralization. Stewardship is also applied to lifecycle deidentification, retention policies consistent with the rights of consent revocation, and standardized machine-readable data use agreements that encode purpose constraints that are permissible. Organizations that incorporate these controls in the context of automated pipelines, as opposed to ad hoc governance, develop scalable trust infrastructures that support collaborative innovation (Kuthuru, 2025; Taddese et al., 2025).

Cyberbiosecurity and threat mitigation have become necessary complements to technical optimization, acknowledging that adversarial manipulation may dehouse model performance, drain sensitive information, or compromise strategic assets of breeding. Threat models include data poisoning, label flipping, adversarial example injection, membership inference, model inversion to reconstruct genotype surrogates, and the compromise of analytical containers by adversaries, ransomware against high-value genomic repositories, and stealing proprietary selection indices. Defensive architectures use staged validation sandboxes, cryptographic authenticity checking of input streams, anomaly detection based on statistical and machine learning detectors, and strong training paradigms that make them less vulnerable to malicious gradient manipulation. Attack surface is minimized by segmentation of computing environments, least privilege credential management, continuous dependency chain vulnerability scanning, and playbooks for incident response. Attribution and remediation are enabled by secure logging with immutable timestamping, and red teaming exercises approximate attacker behavior to find latent weaknesses. The combination of privacy-preserving learning primitives makes recovery by adversarial reconstruction methods more challenging, but with appropriate calibration, privacy budgets, and predictive utility can be balanced. Cyberbiosecurity operationalizes resilience in this way so that the integrity, availability, and confidentiality of genomic AI pipelines are maintained in changing threat environments (Annan et al., 2025; Clement et al., 2025).

The integration of these underlying building blocks results in an infrastructure level that is technically resilient and ethically principled, and with sufficient capability to support higher-order methodological innovation without sacrificing reliability or equity. An experimental design is needed so as to be sure that the raw informational substrate is suitably diverse and statistically effective. Semantic interoperability makes it possible to integrate heterogeneous modalities. Signal fidelity and fairness are preserved by quality control and bias mitigation. Enhanced reference and variant representation models expand available genomic architecture, but require close algorithmic customization. Trust anchors afforded by secure stewardship and cyberbiosecurity help to protect continuity and legitimacy. Transparency in the lineage and accountability are guaranteed by provenance tracking and governance orchestration. They can constitute the scaffolding within which later sections on the advanced modeling and areas of application, verification guidelines, and implementation plans can be constructed, and the idea that sustainable development of AI-enabled livestock genomics is driven as much by infrastructural excellence and stewardship as it is by creative algorithm design (Eyinade et al., 2025; Han, 2025). Foundational livestock genomics data types are shown in Table 4.1.

4.3 Advanced AI methodologies in livestock genomics

The current methodology in the field of livestock genomics is advancing along a spectrum that begins with basic statistical genetics and then proceeds to adaptive, multimodal, and semantically enriched AI ecosystems. The conceptual primacy of mixed models, which separate phenotypic variance into systematic and random components based on relationship structures, was established by baseline frameworks such as best linear unbiased prediction. The genomic extension replaced pedigree-derived expectations with the actually realized genomic relationships based on dense marker panels, partitioning of the additive signal more effectively, and facilitating a faster selection process. Bayesian formulations extended this fundamental with heterogeneous shrinkage priors to allow sparse large effects that can be generated within a polygenic background, with shrinkage intensity being tempered based on empirical evidence as opposed to a uniform penalty. These methodologies are operationally central due to interpretability, statistical transparency, and congruence with long-established decision rules that have long been enshrined in breeding programs. However, their linear models, reliance on precomputed kernels, and limitations in modelling complex nonadditive architectures drive the incorporation with machine learning paradigms able to find higher-order representations and maintain calibrated inference (Husien et al., 2024; Ullah and Waheed, 2025).

The deep learning architectures were created to solve representational bottlenecks of manually engineered features. Convolutional networks discuss genomic sequence or haplotype windows as spatially-organized signals, and learn motif-level filters, which can combine to hierarchical abstractions that represent regulatory, structural, or linkage patterns. Dilated, depthwise separable, and residual designs increase the area of receptivity in a manner that does not grow in parameter space without providing the capacity to represent long-range interactions of interest to cis–trans regulatory interaction and long haplotype interactions. Recurrent and gating mechanisms, less common with the predominance of attention operators, help model temporal or ordered modality streams, including longitudinal phenotypes consistent with sequential genomic or epigenomic states. Autoencoder families reduce high-dimensional variant matrices, disentangling shared and trait-specific sources of variation, allowing multi-trait prediction and dimensionality reduction without the heavy dependence on prior kernels. On genomic data of livestock, these architectures face the challenge of extreme sparsity, imbalance in classes of rare variant patterns, and scale mismatches between sequence, structural, and contextual data, leading to a hybrid architecture that integrates convolutional filters on local motif detecting with global attention or graph components on distal integrating (Lan, 2024; Mao, 2025).

Attention-based mechanisms and transformer architectures provide representational plasticity by enabling an adaptable assignment of attention to variable-length tokenized genomic inputs. Tokenization approaches can divide a sequence into k-mer units, haplotype blocks, structural variant symbols, or ontology-grounded trait descriptors, allowing a common embedding space. Scaled dot product attention calculates contextualized representations that learn to adapt to the combinatorial interaction space as opposed to using fixed convolutional footprints. Sinusoidal, learned, or relative positional encodings retain information about ordering that is important to interpret phased haplotypes and regulatory adjacency. Objectives informed by masked reconstruction, contrastive denoising, permutation recovery, or next segment prediction generate general-purpose genomic embeddings that can be transferred to downstream tasks, including predicting genomic estimated breeding values for multiple traits, environmental interaction models, or functional annotation imputation. Multi-breed, multi-species, or multi-omics trained foundation models can capture cross-context invariants and still support fine-tuning to local production settings. Application in livestock areas requires adjustments to the lack of balance in representation of breeds, nonoverfitting to lineage unique sequence idiosyncrasies, and the incorporation of

TABLE 4.1 Foundational livestock genomics data types.

Data type	Source	Format	Applications	Limitations
Raw whole-genome sequencing (high-coverage)	- Commercial sequencing providers (Illumina, BGI, and PacBio HiFi) - Research consortia (e.g., 1000 bull genomes and national breeding programs) - In-house core facilities	- FASTQ (paired-end short reads) - BAM/CRAM (aligned reads) - Quality metrics (QC reports)	- Variant discovery (SNP, indel) training datasets - Building reference variant panels for imputation - Training deep variant callers and quality prediction models - Detecting rare/novel alleles impacting traits	- High cost limits sample breadth (breed diversity) - Batch/platform heterogeneity introduces bias - Storage and I/O intensive for large cohorts - Coverage variability impacts downstream reliability
Low-pass whole-genome sequencing	- Commercial low-pass services - Population-scale consortium projects - Pilot genomic surveillance programs	- FASTQ (reduced depth) - Postimputation VCF/BCF - Coverage summary files	- Cost-efficient genotype capture for large AI training cohorts - Input to genotype imputation models (statistical or deep learning) - Rapid surveillance of emerging alleles	- Higher genotype uncertainty preimputation - Bias toward common variants after imputation - Accuracy drops in underrepresented breeds
Whole-genome long-read data	- PacBio HiFi/ONT community projects - Breed-specific assembly initiatives - Structural variant discovery studies	- FASTQ (long reads) - BAM/CRAM (aligned long reads) - Raw signal (FAST5 for ONT)	- Structural variant (SV) and complex indel detection - Phasing haplotypes for AI haplotype embeddings - Improving reference/pangenome graphs - Training SV classification neural networks	- Higher per-base cost - Platform-specific error profiles - Computationally heavy assembly/polishing steps
Genotype SNP array data	- National breeding organizations (e.g., USDA and Dairy recording bodies) - Commercial array vendors (e.g., Illumina, Affymetrix/Thermo Fisher, etc.) - Breed improvement programs	- PLINK (BED/BIM/FAM) - VCF (post-QC export) - Allele frequency reports	- Baseline input for genomic prediction (GBLUP) - Training set for imputation to the sequence level - Feature matrices for classical and ML trait models - QC filtering reference for sequencing validation	- Ascertainment bias (marker selection) - Limited rare variant representation - Cross-array harmonization challenges
Imputed genotypes	- Imputation pipelines (Beagle, Minimac, Eagle, and GLIMPSE) - Breed-specific reference panels - Multi-breed merged panels	- VCF/BCF with genotype probabilities (GP and DS) - INFO metrics per variant - Imputation quality summary tables	- Expanding feature density for AI genomic prediction - Rare variant association augmentation - Enhanced haplotype reconstruction for selection decision tools	- Accuracy dependent on reference panel composition - Propagates reference biases (underrepresented breeds) - Uncertainty is not always modeled by downstream AI

Variant call sets (SNP and small indels)	- Joint calling pipelines (GATK, DeepVariant, and FreeBayes) - Consortium releases - Breed-focused variant surveys	- Multi-sample VCF/BCF - Annotation-enhanced VCF (VEP/ANN fields) - Frequency and functional effect tables	- Training genotype-to-phenotype predictive models - Functional prioritization (missense, LOF) in AI feature selection - Constructing variant embeddings for deep learning	- False positives in low-complexity regions - Annotation updates change functional interpretation - Multi-tool discrepancies require reconciliation
Structural variants (SVs)	- Long-read and optical mapping studies - Dedicated SV callers (Sniffles, SVIM, and Manta) - Pangenome comparative analyses	- VCF (SVTYPE, END, and CIPOS annotations) - BEDPE/BED intervals - Graph components (GFA)	- Inclusion of large-effect variants in genomic prediction - AI feature engineering for copy-number dosage - Detection of breed-specific adaptation signatures	- Lower sensitivity in repetitive regions - Inconsistent breakpoints across callers - Difficult integration into SNP-centric pipelines
Copy number variants (CNVs)	- Read-depth tools (CNVnator and qCNV) - Array intensity analyses - Long-read depth normalization	- Segmented BED files - VCF with CN (copy number fields) - Log_2 ratio matrices	- Augmenting genomic prediction with dosage features - Identifying gene amplification/deletion impacting production traits - Biomarker discovery for disease susceptibility	- High false discovery rate at low coverage - Poor standardization of CN annotations - Hard to phase with SNP data
Haplotype and phased data	- Phasing tools (Eagle, SHAPEIT, and WhatsHap) - Trio-based phasing datasets - Reference panel-based phasing	- VCF with phased GT (symbol)	- Haplotypes block files - Genetic map (cM) files	- Haplotype-based genomic prediction (captures LD structure) - Founder lineage tracing and selection signature AI - Incorporation into recombination-aware models
Pangenome graph representations	- Breed-diverse assembly consortia - Graph construction pipelines (minigraph, vg) - Comparative genomics studies	- GFA graph files - GBZ/XG (graph indexes) - Augmented VCF (graph-aware)	- Improved mapping accuracy for underrepresented haplotypes - Discovering novel sequence variation for AI feature expansion - Graph-based variant embedding for deep models	- Tooling ecosystem still maturing - Higher compute requirements for mapping - Complex integration with legacy pipelines
Reference genome assemblies (primary and alternate haplotigs)	- Ensembl/RefSeq releases - Breed-specific high-contiguity assemblies - HiFi and Hi-C scaffolding projects	- FASTA (primary, alternate contigs) - Assembly reports (AGP, stats) - Index files (FAI)	- Stable coordinate framework for variant indexing - Anchor for annotation and feature mapping in AI datasets - Comparative assembly quality modeling	- Assembly gaps reduce variant detection - Reference bias skews read alignment - Version drift complicates longitudinal studies
Genome annotation (genes, transcripts, and regulatory)	- Ensembl, NCBI, RefSeq - FAANG regulatory tracks - Community-curated lncRNA catalogs	- GFF3/GTF annotation files - BED promoter/enhancer sets	- Mapping variants to functional regions for AI importance weighting	- Incomplete UTR/lncRNA annotations - Regulatory region boundaries are ambiguous

Continued

TABLE 4.1 Foundational livestock genomics data types.—cont'd

Data type	Source	Format	Applications	Limitations
		- Feature ID mapping tables	- Building regulatory-context feature matrices - Training models linking sequence to gene expression (eQTL)	- Frequent ID/version updates
Epigenomic profiles (ATAC-seq, ChIP-seq, and histone marks)	- FAANG epigenome datasets - Tissue-specific regulatory projects - Public functional studies	- FASTQ (raw reads) - Peak BED/BroadPeak files - Signal bigWig tracks	- Enhancing variant prioritization through accessibility context - Training multi-modal sequence and epigenetic AI models - Defining promoter/enhancer sets for network inference	- Sparse tissue/breed coverage—peak caller variability affects reproducibility - Batch and antibody quality issues
DNA methylation (WGBS/RRBS)	- Targeted epigenetic studies - Developmental and stress response projects - Commercial methylome profiling	- FASTQ (bisulfite converted) - CpG methylation tables (bedGraph) - Aggregated DMR BED files	- Integrating methylation as a regulatory feature in genomic prediction - Identifying epigenetic biomarkers for fertility/health - Modeling genotype–epigenotype–phenotype causal chains	- High cost for whole-genome coverage - Interpretation complexity (context-dependent effects) - Incomplete temporal resolution (snapshot bias)
Chromatin conformation (Hi-C/Omni-C)	- Genome architecture mapping projects - Reference assembly scaffolding efforts - Developmental regulation studies	- Contact matrices (cooler/.hic) - Compartment and TAD BED files - Normalized interaction frequency tables	- Incorporating 3D genome context for enhancer–gene AI linking - Prioritizing structural regulatory variants - Building graph features for GNN models	- Resolution limited by depth and bin size - TAD boundary calling inconsistency - High storage for dense matrices
Expression QTL (eQTL) and other molecular QTL summary statistics	- Integrated genotype-expression studies - Public QTL repositories - Breed/tissue-specific investigations	- Summary stats (TSV: SNP, effect, and *P*-value) - Fine-mapping credible sets (JSON/TSV) - Colocalization result tables	- Linking variants to gene regulation in genomic prediction features - Informing causal variant prioritization with AI models - Building regulatory mediation layers (genotype → expression → trait)	- Tissue- and condition-specificity limits transfer - Power insufficient for rare variants - Heterogeneous pipelines complicate meta-analysis
GWAS summary statistics (trait associations)	- Published studies (production, health, and fertility) - Breeding company internal analyses - Public domain trait archives	- Summary tables (SNP, beta, SE, and *P*-value) - Manhattan plot data (TSV) - LD reference panels (binary)	- Training polygenic risk/genomic estimated breeding value (GEBV) AI models - Sensitivity analysis of variant importance weighting	- Winner's curse inflates effect estimates - LD differences across breeds reduce portability - Incomplete reporting of covariates

			- Multi-trait integration for pleiotropy modeling	
Pedigree and relationship records	- Breed registries - Herd management software - National genetic evaluation centers	- Pedigree text (ID, sire, dam) - Relationship matrices (A-matrix) - SNP-based GRM (binary matrix)	- Incorporating additive relationships in hybrid AI and BLUP models - Validation of inferred kinship from genotypes - Bias correction for population structure	- Pedigree errors (mis-parentage) - Missing depth for crossbred lines - Incomplete capture of Mendelian sampling variance
Genomic relationship matrices (GRM) and inbreeding metrics	- Computed from SNP/sequence data - Breeding evaluation pipelines - Imputed genotype sets	- Dense matrices (binary/HDF5) - Inbreeding coefficient reports - Eigen decomposition outputs	- Feature augmentation (relationship-aware embeddings) - Regularization in genomic prediction networks - Monitoring inbreeding trends through AI dashboards	- High memory footprint for large populations - Sensitive to marker filtering strategies - Not directly interpretable by stakeholders
Phenotypic and production trait databases	- On-farm records (milk yield, growth, FCR, etc.) - Centralized breeding value repositories—research trial phenotyping	- Relational DB tables (CSV/SQL dumps) - Time-stamped longitudinal records - Derived EBV/ reliability scores	- Supervised targets for genomic prediction models - Multi-trait AI modeling (correlated outputs) - G×E interaction discovery when linked to environments	- Measurement error and inconsistent protocols - Missingness and censoring biases - Confidentiality limits public sharing
Environmental and management metadata	- Farm sensors (temperature, humidity, and THI) - Feed formulation logs - Housing/mobility systems	- Time-series CSV/ JSON - Sensor streams (IoT formats) - Ontology-tagged categorical fields	- Modeling genotype × environment interactions - Domain adaptation/covariate adjustment in AI models - Early warning systems integrating genomic risk and environment	- Heterogeneous formats and vocabularies - Temporal misalignment with genomic sampling - Data gaps/sensor failures
Microbiome (metagenomic shotgun/16S) linked to host genotypes	- Rumen/fecal sampling surveys - Feed efficiency microbiome projects - Metagenomic consortia	- FASTQ (raw reads) - Taxonomic/functional abundance tables (BIOM/TSV) - MAG genome FASTA sets	- Integrative host–microbiome genomic prediction (multi-source features) - AI models for methane emission and efficiency - Causal inference (genotype → microbiome → trait pathways)	- High interindividual variability requires a large N - Compositional data constraints complicate integration - Temporal instability versus static host genotype
Quality control metrics and reports	- Pipeline outputs (FastQC, VerifyBamID, and Picard) - Internal QC dashboards - Consortium data validation steps	- HTML/JSON/TSV metrics - Sample-level flag summaries - Batch QC aggregation tables	- Filtering low-quality samples before AI training - Covariate generation (coverage, duplication) for model weighting - Drift detection across sequencing batches	- Fragmented formats across tools - Manual interpretation needed for some thresholds - Overfiltering risks data sparsity
Data access and security logs	- Institutional storage systems - Cloud access logs (AWS CloudTrail and GCP audit)	- JSON log streams - Audit CSV exports	- Compliance monitoring for sensitive genotype/phenotype data	- Not directly biologically informative

Continued

TABLE 4.1 Foundational livestock genomics data types.—cont'd

Data type	Source	Format	Applications	Limitations
	- Federated learning orchestration	- Access control policy files	- Auditing federated AI training events - Detecting anomalous data usage patterns	- High volume requires aggregation/AI anomaly detection - Privacy constraints on sharing logs
Data integration manifests and metadata schemas	- Data lakes/LIMS systems - Consortium standard templates (MIxS and FAANG metadata) - Ontology-driven curation	- Structured spreadsheets (CSV/TSV) - JSON/YAML metadata descriptors - Ontology term ID mappings	- Harmonizing multi-source datasets for unified feature matrices - Automating ingestion pipelines feeding AI training - Enabling reproducibility and provenance tracking	- Incomplete or inconsistent fields hinder automation - Ontology evolution requires updates - Schema drift over the project lifecycle

uncertainty heads that moderate overconfident extrapolation in underrepresented genomic regimes (Lindsey et al., 2025; Zhang et al., 2025).

Graph neural networks add an inductive bias of a distinctly relational nature that is especially applicable to structured dependencies that pervade livestock genomics. On the one hand, animals may be modeled as nodes linked by the edges that indicate the genomic relationship coefficient, pedigree connection, or similarity scores in multi-omics profiles. The information is passed on by message passing updates that execute prediction on sparsely phenotyped individuals by aggregate neighborhood signal using this relational scaffold. On still a higher level, variant interaction networks may be built, with all nodes denoting markers, structural events, or regulatory elements but interlinked by an edge indicating physical proximity, linkage disequilibrium, chromatin contact, coexpression, or inferred epistatic synergy. Graph convolutional, attention, spectral, and neural diffusion networks learn to map local neighborhoods to contextual embeddings that characterize polygenic interactions. Heterogeneous graphs represent several types of nodes and edges and allow supporting concurrent reasoning with animals, variants, traits, pathways, microbial taxa, and environmental nodes. Dynamic graph formulations are models that include the time evolution of relationships when under selection pressure or management interventions. These graph paradigms have to face scalability, sampling bias in edge construction, and propagation of noise across densely connected subgraphs, which are solved by sparsification, edge uncertainty modeling, hierarchical pooling, and curriculum training regimes (Malla and Banka, 2025; Li et al., 2025).

The robustness of uncertain and probabilistic policies cuts across all architectural classes, since decision-making in the risky context demands the quantification of both the epistemic and aleatoric aspects in a manner that is calibrated. Variants of Bayesian neural networks can use variational inference to introduce priors of weights, or provide function space approximations (Monte Carlo dropout, stochastic gradient Langevin dynamics, or Laplace approximations), giving posterior predictive distributions instead of point estimates. Deep ensembles are a pragmatic alternative that learns a variety of local minima, which together approximate the posterior spread. The density ratio estimation and conformal prediction provide distribution-free coverage guarantees, which strengthen the selection index thresholds. The hierarchical probabilistic models combine multi-level modeling: population, breed, herd, and individual effects, and include a balance of global sharing and local specificity in partial pooling. Upstream measurement uncertainty and imputation, and variant call likelihood propagated to downstream genomic predictions, avoid unnecessary shrinkage of credible intervals. By allowing downside risk penalty or value at risk constraints to be added to these frameworks, risk-sensitive multi-objective optimization is also supported, based on breeder risk tolerance profiles (Ajirak et al., 2025; Gillis et al., 2025).

The strategies of transfer, multi-task, meta, and few-shot learning are in response to the issues of data stratification and scarcity of data that are inherent to less intensively characterized breeds or emergent characteristics or novel environmental conditions. The sharing of parameters across tasks exploits pleiotropic correlation structures and allows the predictive fidelity of low heritability or sparsely phenotyped traits to be enhanced by induction transfer between associated endpoints. The tension between positive sharing and adverse transfer is controlled by hard and soft parameter sharing architectures, cross-stitch mechanisms, and task routing networks. Metalearning algorithms can be trained to quickly change environments or breed without substantial extra gradient steps by keeping the starting parameter states adaptable. Few-shot adaptation uses prototype representations or metric learning to classify or predict in new cohorts. Domain adaptation aligns latent distributions between production systems, reducing the performance decline in the event of changes in management practices or environmental exposures. Continual learning mitigates selection-induced temporal drift, climate variability, and changing disease pressures, using flexibility buffers, elastic regularization, or dynamic architectural growth to avoid catastrophic forgetting and incorporation of novel signals (Upadhyay et al., 2024; Agrawal, 2025).

Generative models bring in complementary abilities, such as data augmentation to mechanistic hypothesis exploration. Variational autoencoders are trained to learn latent genomic spaces in which interpolation is realistic in terms of plausible haplotypic transitions, and synthetic individuals can be trained to inherit allele frequency spectra and linkage structure within governance constraints. Adversarial generative models are trained to be realistic, but also need special attention to prevent mode collapse and incidental memorization of rare sensitive haplotypes. Diffusion models are more fidelity- and controllability-controlled, and gradually denoise latent representations to sample complex genotype or multi-omics distributions. Conditional variants produce data that is conditional on the values of traits, environmental characteristics, or the values of breed identifiers, and serve to both explore counterfactual scenarios and to balance underrepresented strata. The overlays of governance are important: differentiating privacy, auditing membership inference, and bias evaluation pipelines are essential measures to ensure that synthetic data do not reveal identifiable patterns or increase bias. Synthetic augmentation can best be justified in situations where it is applied to clean up decision boundaries in low-density areas, as opposed to correcting systemic underrepresentation in explicit targeted real sampling (Li et al., 2025; Salmè et al., 2025; Ghavi Hossein-Zadeh, 2026a).

Federated and privacy-preserving learning models address the trade-off between collaborative model improvement and limitations on the centralization of raw data. Instead of genomic records, federated averaging coordinates decentralized updates of gradients over institutions, breeds, or geographical locations, and averages model parameters. Non-homogeneity of local data distributions requires adaptive algorithms to weight updates based on quality, divergence, or fairness purposes, so that large high-resource nodes do not dominate. Secure aggregation algorithms, homomorphic encryption, and trusted execution models protect the aggregate updates against adversarial audit, and differential privacy introduces some noise to the gradients to ensure that the risk of leakage is kept within bounds. Personalization layers may be added to cross-silo federation to customize global backbones to local breed-specific quirks. Scalability of communication also necessitates gradient sparsification, quantization, and periodic update methods that are cooptimized with convergence stability. Governance contracts specify what models can be used, audit logs accept contributions, and incentive systems reward participants based on the utility benefits they obtain through their participation (Lazaros et al., 2024; Jimenez-Gutierrez et al., 2025).

The field of knowledge graph and neuro-symbolic integration deals with the issue of integrating curated biological knowledge into data-driven pipelines. Knowledge graphs are the representation of a gene, variant, pathway, traits, taxa of microbes, drugs, management practices, and environmental factors, and are connected through typed relations based on literature curation, pathway data, and ontologies. Graph embedding methods are used to map this structured semantic space into representations in vector form that are compatible with learned genomic feature representations and allow joint reasoning to satisfy known biological plausibility constraints. The neuro-symbolic hybrids permit rule-based constraints, logical entailments, or causal graph structures to tune or regularize neural predictions to minimize biologically implausible extrapolations. Causal discovery algorithms, with the support of constraint-based search, score-based optimization, invariant risk minimization, or functional causal modeling, are designed to distinguish structural and associative dependencies, which can inform prioritization to gene editing or targeted management interventions. A combination of interventional data, temporal perturbations, and multi-environment contrasts enhances the identifiability of causal edges. These structures need to cope with confounding, selection bias, and measurement error, and extrapolate to high-dimensional genotype spaces; hybrid approaches that confine causal search to latent embeddings or pathway-constrained subspaces provide manageable trade-offs (Li et al., 2025; Pojer et al., 2025).

The importance of model compression, acceleration, and energy efficiency is on the rise with the complexity of architectures and the size of the datasets. Pruning eliminates redundant weights found by saliency analysis or structured sparsity induction, and inference latency is cut without noticeable predictive fidelity loss when retraining is used to counteract capacity reduction. Quantization sacrifices numerical accuracy of parameters and activations, and uses integer arithmetic acceleration, but uses calibration or mixed precision schemes to maintain stability in the more sensitive layers (e.g., normalization or attention scoring). Knowledge distillation transfers are a transfer of representational competence between large teacher ensembles or foundation models to lean student architectures that can be deployed on on-farm edge environments, where bandwidth and energy budget are limited. Multi-objective optimization-driven neural architecture search finds trade-off optimal architectures, such as those that trade off accuracy, latency, memory footprint, and energy consumption. Resource intensity is further reduced by compiler-level optimization, operator fusion, and use of specialized hardware accelerators, such as low-precision tensor cores and field programmable logic. Carbon accounting metrics, overall energy-conscious training programs, the early termination based on marginal utility curves, and energy-conscious training programs incorporate the environmental stewardship aspect into the model development life-cycle governance that is congruent with the principles of sustainability and operational efficiency (Agrawal et al., 2025; Soundararajan, 2025).

The integration of these levels of methodological rigor at cross-cutting levels gives rise to a unified ecosystem where classical statistical rigor, deep representation learning, relational reasoning, uncertainty quantification, adaptability, and ethical governance have a positive influence on each other. Baseline mixed models define calibrated baselines and provide interpretable components of variance; graph and knowledge infused models handle relational and semantic complexity, probabilistic layers provide reliable confidence intervals, transfer and meta learning propagates innovation to data sparse domains, generative models promote robustness under safe conditions, federated infrastructure facilitates fair collaborative development, neuro symbolic and causal models steer models towards mechanistic relevance, and compression with energy optimization make them realistic and viable to practice. The unique requirements of livestock genomics, such as multi-trait selection in the presence of resource and welfare limitations, interactions between the environment of heterogeneous nature, accelerating climatic changes, the emergence of pathogens, and the necessity of equitable representation of breeds, compound the worth of this multi-paradigm synthesis. The choice of the method is then a game of portfolio design, aligning job aspects with the capabilities of the architect, with transparency, equity, and robustness requirements encoded into assessment matrices (Ghavi Hossein-Zadeh, 2025; Tedeschi et al., 2025).

To maximize the potential of modern AI approaches, it is necessary to implement a careful lifecycle management that includes data curation to align with architectural assumptions, an approach to systematically ablate to ensure that performance improvements can be attributed to meaningful modeling innovation and not to accidental data leakage, and a protocol of cross-population validation that stresses test generalization. Interpretability tooling designed to support complex architecture, such as attribution of attention weights, counterfactual explanation, concept activation, and graph saliency mapping, spans the epistemic gap between algorithmic inference and expert domain analysis. A continuous monitoring system follows calibration drift, fairness indicators, computational costs curves, and cyberbiosecurity indicators and responds adaptively to mitigate degradation, which undermines strategic performance. Within this unified framework, innovation is not characterized only by foreseeable benefits but by the harmonious development of accuracy, reliability, cost-effectiveness, and moral congruency, which preconditions a sustainable basis for the further application-oriented sections that turn the methodological ability into the actual genetic enhancement and sustainability effects (Coté and Aires, 2025). Advanced AI methodologies in livestock genomics are presented in Table 4.2.

4.4 Multi-layer and contextual integration

Future developments in the direction of multi-layered biological and contextual information in livestock genomics are dependent on the ability to combine multi-layered biological and contextual data into integrated computational representations that are operationally tractable and that maintain hierarchical causality. Multi-omics fusion is the backbone, which goes beyond static catalogs of variations into epigenomic marks that index chromatin accessibility, methylation dynamics, and histone modification landscapes, transcriptomic outputs of bulk tissue snapshots, single cell resolution heterogeneity, and spatially resolved expression gradients, proteomic profiles of post-translational modulation and interaction stoichiometry, and metabolomic end points of pathway flux, and environmental biochemical exchange. The layers represent a variety of noise profiles, temporal half-lives, and sparsity regimes, which force fusion architectures to consider encoders that are modality specific to strip technical artefact off biological signal and project latent factors to a common manifold that matches downstream prediction goals. Variational models of hierarchies are offered, cross-attention fusion transformers, and graph-aligned multi-view embeddings offer ways to support partially overlapping sets of samples, asynchronous acquisition periods, and missing modalities without collapsing integration to the lowest common denominator (Baião et al., 2025). The propagation of uncertainty is essential in a disciplined manner; epigenomic peak calling uncertainty, single cell expression matrix dropouts, peptide identification false discovery, and ambiguity of metabolite annotations should be propagated in posterior distributions so that composite genomic predictions show evidence of compounded evidence, and not unjustified deterministic consolidation. An additional interpretation is enhanced by temporal coupling, where relationships between epigenomic remodeling and transcriptional reprogramming occur, where transcriptional reprogramming changes proteomic abundance and metabolic fluxes with phenotypic outcome in production or disease resistance (Balsa-Canto et al., 2025) (Fig. 4.2).

The axis of integration is extrapolated by host genomic and microbiome interaction modeling to ecological collaborators that can coregulate metabolic efficiency, immune calibration, and nutrient transformation. The microbiome presents constraints of compositional data, zero inflation, degree of strain polymorphism, and horizontal gene flow dynamics, which complicate naive feature concatenation. Compositional aware transforms, phylogeny-informed graph informatics, and strain resolved k-mer embeddings permit host genotype to interface with microbial functional potential under models that consider mutualistic, antagonistic, and context contingent relationships. The role of causal disentanglement is central, as the host genetic variation may cause microbial community organization by immune or physiological intermediates and vice versa, where microbial metabolic products cause epigenetic marks and metabolic phenotypes that, in turn, provide feedback to selection indices. Directional hypotheses can be expressed using joint probabilistic graphical models and structural equation hybrids, and spurious correlation inflation can be reduced by minimizing risk invariance across different dietary or management cohorts. The portability is increased by integration of microbial gene functionality but not taxonomic labels due to community compositional turnover among environments. To balance sampling depth divergences and heterogeneity of sequencing platforms, it is necessary to harmonize the samples by layers, such that abundance distributions appear on a standard latent scale, where technical confounding does not present as a biologically significant host microbe epistasis. The resulting composite embeddings broaden the predictive capabilities of traits related to feed conversion, methane emissions, and resistance to disease, and provide intervention levers that integrate genomic selection and microbiome-modulating intervention strategies (Ferretti et al., 2025; Mishra et al., 2025).

Genomic inference is further based on high-throughput phenomics and sensor data fusion, which place genotype expression in the context of continuous behavioral, physiological, environmental, and welfare signals. Multi-rate, irregularly sampled, and possibly drifting time series are generated by wearable accelerometry, ingestion behavior

TABLE 4.2 Advanced AI methodologies in livestock genomics.

AI technique	Potential applications	Description	Advantages	Disadvantages
Transformer architectures	- Sequence-to-trait prediction (regulatory and splicing) - Long-range haplotype and epistasis modeling - Multi-omics (genome–transcriptome–epigenome) fusion - Variant effect contextualization—cross-species foundation model adaptation	- Attention-based architectures capturing global dependencies - Token or patch encodings for sequence, graph paths, or modality embeddings - Pretraining through masked sequence tasks	- Models long-range regulatory interactions - Scales to heterogeneous modalities - Effective self-supervised pretraining - Strong transfer across breeds/species	- High compute/energy demand - Risk of overfitting to stratification - Attention ≠ causality - Complex tokenization and positional encoding design
Graph neural networks (GNNs)	- Pangenome/variation graph analytics - Pedigree, genomic, and microbiome integration - Gene/protein/pathway interaction modeling - Knowledge graph trait reasoning - Population structure-informed prediction	- Message passing on nodes (variants, genes, and animals) and edges (relations and interactions) - Heterogeneous/multiplex graph support	- Relational inductive bias - Integrates structured prior knowledge - Reduces manual feature crafting - Flexible multimodal graph fusion	- Oversmoothing/oversquashing issues - Graph construction bias propagates - Scaling very large complex graphs - Harder interpretability of embeddings
Variational autoencoders (VAEs)	- Latent genotype–environment (G×E) factor discovery - Multi-omics compression - Phenotype/omics missing data imputation - Rare trait profile synthesis - Noise disentanglement	- Probabilistic encoder–decoder learning latent distributions - Reconstruction and KL regularization	- Uncertainty-aware latent factors - Data denoising and imputation - Augments scarce trait combinations - Regularizes downstream models	- Latent axes may lack biological meaning - Posterior collapse risk - Blurs sharp class boundaries - Sensitive to prior/architecture choices
Generative adversarial networks (GANs)	- Synthetic phenotypes for rare breeds - Class imbalance mitigation (disease states) - Image phenotyping (e.g., body conformation, carcass, etc.) - Cross-domain translation (sensor → image) - Stress scenario augmentation	- Adversarial generator–discriminator training - Conditional, cycle, and style variants for control/domain shift	- High-fidelity sample generation - Balances skewed datasets - Enables modality translation - Enhances data diversity	- Mode collapse/instability - Hard biological validity assurance - Lacks calibrated uncertainty - May encode hidden biases

Diffusion models	- High-fidelity multi-omics synthetic data - Progressive phenotype simulation under constraints - Data augmentation for rare environmental contexts - Denoising of noisy sensor/genomic signals	- Iterative denoising generative process learning data distribution through forward (noise) and reverse (denoise) chains	- Stable training versus GANs - Better coverage (less mode dropping) - Supports conditional generation - Intrinsic denoising capability	- Computationally expensive sampling (many steps) - Large training resources needed - Potential to fabricate unrealistic rare biology - Hallucinated structure risks misuse
Self-supervised contrastive learning	- Unlabeled sequence/sensor pretraining - Cross-modal embedding alignment (genome–transcriptome) - Breed-general representation learning - Early anomaly detection	- Learns invariances by contrasting positive (augmentations/biological links) versus negative pairs - Projection layers refine latent space	- Reduces labeled data reliance - Robust, generalizable embeddings - Modular for multi-task fine-tuning - Encourages modality alignment	- Negative sampling sensitivity - Latent space opacity - Poor augmentations induce artifacts - Large batch/memory needs
Structural causal models (SCMs) and causal discovery	- Distinguish causal versus linked loci - Counterfactual gene edit/diet/ management evaluation - Microbiome–host causal mediation - Climate resilience intervention design	- Graph-based structural equations - Leverage independence tests, invariance, and interventions to infer directionality	- Actionable insight (not just correlation) - Supports intervention prioritization - Clarifies confounding pathways - Enhances trust in editing targets	- Limited by unmeasured confounders - Computationally intensive at scale - Model misspecification risk - Requires expert biological validation
Reinforcement learning (RL) and multi-objective RL	- Sequential mating/selection policy optimization - Adaptive feeding and environmental control - Disease surveillance scheduling - Balancing gain, diversity, welfare, and emissions	- Policy learning maximizing cumulative (possibly vector) rewards - Integrates simulators/digital twins	- Long-horizon optimization - Explicit trade-off navigation - Real-time adaptive control - Scenario stress testing	- Reward design complexity - Sample inefficiency (needs simulation) - Transferability challenges - Ethical oversight imperative
Neuro-symbolic integration	- Rule/ontology-constrained selection - Pathway-informed variant interpretation	- Combines neural embeddings with symbolic logic/ontologies - Hybrid inference ensures constraint satisfaction	- Improved interpretability and compliance	- Knowledge base incompleteness - Engineering integration overhead

Continued

TABLE 4.2 Advanced AI methodologies in livestock genomics.—cont'd

AI technique	Potential applications	Description	Advantages	Disadvantages
	- Constraint enforcement (welfare/diversity) - Explainable trait reasoning		- Reduces biologically implausible outputs - Enables traceable rationales - Facilitates regulatory acceptance	- Risk of rigidity if rules stale - Conflict resolution challenges
Federated learning and privacy-preserving analytics	- Cross-organization GEBV modeling - Rare breed inclusion under sovereignty - Multi-country disease resistance analytics - Secure sensor and omics fusion	- Decentralized training with parameter/gradient aggregation - Can combine secure aggregation and DP	- Preserves data locality and confidentiality - Increases cohort diversity - Reduces central breach risk - Enables collaborative innovation	- Non-IID data hampers convergence - Communication overhead - Gradient leakage risks w/o safeguards - Complex legal/governance setup
Differential privacy mechanisms	- Privacy-safe publication of model outputs - Protecting individual animal genomes - Regulatory-compliant data sharing - Secure synthetic data release	- Adds calibrated noise (e.g., DP-SGD) to gradients/queries to bound reidentification risk	- Formal privacy guarantees - Encourages broader data access - Mitigates membership inference attacks - Composable privacy accounting	- Utility loss with high privacy budgets - Hyperparameter tuning complexity - Cumulative noise in multi-round training - May obscure rare variant signals
Secure multiparty computation (MPC) and homomorphic encryption (HE)	- Joint genomic evaluation without plaintext disclosure - Secure cross-institution variant effect estimation - Confidential federated aggregation - Encrypted GWAS computations	- Cryptographic protocols enabling computation over encrypted or secret-shared data	- Strong cryptographic guarantees - Avoids raw data pooling - Complements federated learning - Reduces trust assumptions	- High computational overhead - Limited supported ops (especially HE) - Latency and bandwidth burdens - Complex implementation expertise needed

Bayesian deep learning	- Uncertainty-calibrated GEBVs - Risk-aware disease susceptibility stratification - Active acquisition (which animals to sequence) - Decision threshold tuning	- Neural networks with probabilistic weights/outputs (variational layers, MC dropout, and ensembles)	- Explicit epistemic and aleatoric uncertainty - Supports risk-sensitive selection - Improved robustness to shift - Informs active learning	- Computational cost increase - Approximate inference quality varies - Uncertainty source disentanglement is hard - Deployment complexity
Gaussian processes and kernel methods	- Small-sample trait prediction - Nonlinear genotype–phenotype mapping - Environmental response surface modeling - Uncertainty-aware spatial predictions	- Kernel-based nonparametric Bayesian models (e.g., RBF and genomic kernels)	- Strong uncertainty quantification - Effective with limited data - Flexible nonlinear modeling - Interpretable kernel contributions	- Poor scalability (O(n3)) - Kernel choice critical - Memory-intensive for large cohorts - Limited high-dimensional raw sequence handling
Gradient boosted decision trees (GBDT)	- Trait prediction with tabular omics aggregates - Feature importance screening - Rapid prototyping versus deep models - Integrating environmental and management covariates	- Sequential tree ensembles (e.g., XGBoost and LightGBM) minimizing additive loss	- Strong performance on structured data - Fast training and tuning - Built-in handling of missingness - Interpretable feature importances	- Limited raw sequence modeling - Struggles with very high-dimensional sparse variants - Captures limited complex interactions - Potential overfitting without regularization
Support vector machines (SVM) and large margin methods	- Binary/rare trait classification - High-dimensional marker-based discrimination - Early-stage variant filtering - Robust small-sample modeling	- Margin maximization with kernel trick for nonlinear separation	- Effective in high-dimensional small-n regimes - Clear decision boundary principle - Kernel flexibility - Resistant to some overfitting	- Scaling issues with large datasets - Parameter and kernel tuning are sensitive - Probabilistic calibration is often needed - Less suited for multi-output modeling
Convolutional neural networks (CNNs)	- Image phenotyping (body shape, lesions) - 1D convolution for local motif detection in sequence - Spectral/sensor signal feature extraction - Spatial microbiome abundance patterns	- Local receptive field filters capturing translation-invariant features - Adaptable to 1D/2D/3D data	- Efficient local pattern learning - Parameter sharing reduces overfit risk - Mature tooling and transfer learning - Good for imaging and structured signals	- Limited long-range dependency capture - Requires architecture tuning for genomics - Potential to miss global epistasis - Data augmentation is critical

Continued

TABLE 4.2 Advanced AI methodologies in livestock genomics.—cont'd

AI technique	Potential applications	Description	Advantages	Disadvantages
Recurrent/temporal models (RNN, LSTM, and temporal transformers)	- Longitudinal health and growth modeling - Time-varying environmental response trajectories - Sensor-based behavior sequence analysis - Temporal anomaly detection	- Sequential models capturing ordered dependencies - Variants incorporate attention or gating	- Models temporal dynamics directly - Handles variable-length sequences - Captures lagged effects - Complements static genomic features	- Vanishing/exploding gradient risks (classic RNN) - Less parallelizable (RNN/LSTM) - Data hungry for complex patterns - May underperform modern attention models
Multi-task learning (MTL)	- Joint prediction of correlated traits - Shared representation for health, production, and fertility - Cross-breed transfer with auxiliary tasks - Simultaneous GEBV and risk factor modeling	- Shared backbone with task-specific heads leveraging inductive transfer across tasks	- Gains for low-data traits through shared signal - Regularization through auxiliary tasks - Encourages holistic selection indices - Reduces redundant training	- Negative transfer risk between dissimilar tasks - Task weighting complexity - Conflicting gradient directions - Requires careful architecture partitioning
Transfer learning and domain adaptation	- Cross-breed or species model reuse—adapting lab to field phenotyping - Climate/environmental shift adjustment - Low-resource trait bootstrap	- Reuse of pretrained representations - Alignment through fine-tuning, adversarial, or discrepancy minimization strategies	- Reduces labeled data needs - Accelerates deployment in new populations - Mitigates distribution shift effects - Leverages foundation models	- Risk of negative transfer - Source bias propagation - Requires shift detection and validation - Fine-tuning instability
Meta-learning (few-shot)	- Rapid adaptation to emerging traits/diseases - Breed-specific calibration with a few samples - Personalized management recommendations—fast task onboarding	- Learns initialization or adaptation rules enabling quick learning with minimal new data (e.g., MAML and prototypical networks)	- Efficient few-shot generalization - Reduces retraining overhead - Supports rapid trait onboarding - Complements scarce data regimes	- Meta-overfitting risk - Task distribution mismatch degrades performance - Complex training loops - Limited interpretability

Active learning	- Prioritizing animals for genotyping/sequencing - Targeted phenotype collection (expensive traits) - Efficient microbiome sampling strategies - Reducing labeling costs for novel traits	- Iteratively selects the most informative samples using uncertainty, diversity, or expected model change criteria	- Reduces data acquisition cost - Accelerates model improvement - Focuses on high-value individuals - Synergizes with uncertainty-aware models	- Acquisition bias if the strategy is narrow - Requires reliable uncertainty estimates - Operational complexity in fast cycles - Diminishing returns later
AutoML and neural architecture search (NAS)	- Automated model selection for trait prediction - Hyperparameter optimization across breeds - Tailoring architectures to multi-omics fusion - Rapid benchmarking pipelines	- Search over model architectures/parameters through Bayesian optimization, evolutionary, or gradient-based search	- Reduces manual tuning burden - Potentially superior bespoke architectures - Standardizes benchmarking - Accelerates experimentation	- High compute cost - Risk of overfitting on validation splits - Opaque architecture rationale - Reproducibility challenges
Digital twin simulation platforms	- Scenario testing (climate, feed, disease) - Multi-generation selection strategy evaluation - Diversity and inbreeding trajectory forecasting - RL policy training sandbox	- Dynamic virtual replicas integrating genomic, phenotypic, environmental, management, and economic layers	- Safe counterfactual exploration - Reduces costly real trials - Stress-test resilience strategies - Enables closed-loop RL training	- Fidelity constraints limit transferability - High integration and maintenance cost - Risk of overreliance on simulation - Data drift erodes accuracy
Hybrid mechanistic–data-driven models	- Integrating nutritional physiology with ML growth prediction - Disease spread and genomic resistance modeling - Thermoregulation and environmental sensor fusion - Gene regulatory network priors and deep learning	- Combines mechanistic equations/compartments with ML residual or coupling components	- Improved extrapolation under shift - Biological interpretability preserved - Reduces data needs through priors - Constrains implausible outputs	- Model coupling complexity - Calibration of mechanistic parameters needed - Potential identifiability issues - Longer development cycles
Ensemble and hybrid modeling	- Robust prediction across heterogeneous breeds - Multi-trait integration through diverse learners	- Aggregates multiple models (stacking, blending, voting, and hierarchical ensembles)	- Improved generalization and resilience	- Interpretation complexity - Higher compute/runtime overhead

Continued

TABLE 4.2 Advanced AI methodologies in livestock genomics.—cont'd

AI technique	Potential applications	Description	Advantages	Disadvantages
	- Stability under environmental shift - Combining statistical, deep, and causal models		- Mitigates individual model bias - Leverages legacy pipelines - Incremental modular expansion	- Risk of redundancy and overfitting - Attribution of gains is difficult
Tensor/factorization and multi-omics data fusion	- Joint decomposition of genotype–transcript–metabolite arrays - Cross-environment trait factor discovery - Integrative biomarker identification - Missing block completion (omics layers)	- Matrix/tensor factorization (CP, Tucker), coupled NMF, and multi-view latent variable fusion	- Captures shared and modality-specific factors - Reduces dimensionality with structure - Handles partially observed blocks - Interpretable latent components (sometimes)	- Assumes low-rank structure—sensitive to scaling and normalization - May miss nonlinear interactions - Component interpretability not guaranteed
Sparse modeling and feature selection (Lasso, elastic net, and group sparsity)	- Selecting informative SNP subsets - Multi-trait marker prioritization - Reducing model complexity for deployment - Biomarker panel design	- Penalized regression enforcing sparsity or structured sparsity at variant/group levels	- Enhances interpretability - Reduces overfitting risk - Enables cost-effective genotyping panels - Efficient optimization	- May miss polygenic diffuse effects - Sensitive to correlated predictors - Requires careful penalty tuning - Limited nonlinear capture
Explainable AI (XAI) techniques (SHAP, integrated gradients, and counterfactuals)	- Variant/region importance attribution - Trust building for breeders and regulators - Bias and fairness audits - Counterfactual GEBV scenarios	- Post-hoc or intrinsic methods attributing predictions to inputs or concepts - Counterfactual generation for what-if analysis	- Improves transparency and adoption - Aids debugging and bias detection - Supports regulatory reporting - Enables rational communication	- Potentially inconsistent explanations - Computational overhead (e.g., SHAP) - Susceptible to spurious attributions - Not inherently causal
Edge AI and on-device inference	- Real-time barn sensor anomaly detection	- Model compression, quantization, and pruning for deployment on	- Low-latency decisions	- Accuracy loss from compression

	- On-farm imaging (lameness, body condition) - Latency-sensitive decision support (feeding) - Privacy-preserving local analytics	constrained hardware (edge gateways and IoT devices)	- Reduced bandwidth and cloud dependence - Enhanced privacy (local processing) - Energy efficiency potential	- Maintenance and update complexity - Hardware heterogeneity challenges - Limited capacity for large models
Energy and carbon-aware optimization (Green AI)	- Carbon footprint tracking of genomic pipelines - Energy-efficient hyperparameter search - Scheduling compute by renewable availability - Model architecture energy-performance trade-offs	- Methodologies for measuring and optimizing energy/carbon of training/inference - Integrates lifecycle sustainability metrics	- Reduces environmental impact - Aligns with sustainability mandates - Encourages efficient model design - Cost savings through resource optimization	- Added monitoring overhead - May constrain model complexity - Limited standardized reporting frameworks - Potential trade-off with peak accuracy
Diffusion/Graph contrastive hybrid	- Robust pangenome embedding pretraining - Cross-graph alignment (breed graphs) - Denoising variant neighborhood representations - Low-label trait transfer	- Combines diffusion corruption-denoising with graph contrastive augmentation for structured genomic graphs	- Enhances robustness to noise - Improves low-label performance - Captures structural and contextual signals - Flexible augmentation pipeline	- Increased training complexity - Hyperparameter sensitivity (noise schedule) - Limited mature tooling - Interpretability challenges

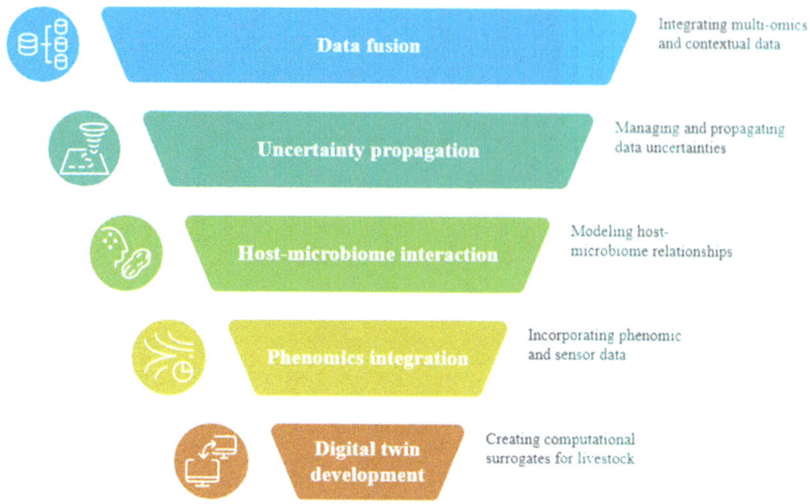

FIGURE 4.2 An overview of genomic AI development. This figure illustrates a layered schematic of key emerging domains in AI-driven livestock genomics, showing how they build upon one another to support more holistic, predictive models. Starting from data fusion (combining contextual and multi-omics information), it moves through uncertainty propagation, host-microbiome interaction, phenomics integration, and culminates in digital twin development (virtual surrogates of livestock), highlighting progressive complexity and integration in system design.

tracking, thermal imaging, computer vision posture detection, and ambient environmental sensing streams. Temporal attention sequence-to-sequence encoders, neural controlled differential equation frameworks that are irregularly sampled, and multimodal temporal graph networks transform these streams into latent trajectories, which can be aligned with genomic and multi-omics embeddings. In order to match asynchronous phenotyping with developmental milestones, health events, or management interventions, time warping of dynamic time alignment requires the mastery of learned temporal anchoring events or differentiable time warping modules. Sensor drift, calibration offsets, and missing windows require strong imputation strategies with physics-constrained priors, where feasible, which manage to avoid hallucination at long gaps in data. Decision time fusion may be staged, and early fusion must be used to integrate closely coupled modalities such as locomotion and thermal stress, and late fusion to maintain modality specificity. The integration of welfare indicators into the same representational space as productivity traits makes the multi-objective optimization structures feasible, in which ethical and sustainability imperatives are taken into account by punishing model choices that deteriorate welfare embeddings despite increases in productivity projections. Online adaptation can also be facilitated by continuous phenomic feedback, in which the implied predicted genomic breeding values are updated in response to observed new patterns of behavioral or physiological deviation, indicative of genuine interaction realizations between genes and their environment that are not fully recaptured in fixed baseline models (Le et al., 2024; Essien and Neethirajan, 2025).

The compounding of these merged information layers into computational surrogates simulating single animals and aggregate herd systems under hypothetical conditions is summed up as the development of digital twins. A livestock digital twin learns the residual dynamics or nonlinear interactions unmodelled by a model, which is coupled between mechanistic submodels of growth, metabolism, immune response, and reproductive physiology, and data-driven components. Genomic and multi-omics embeddings learn latent physiological parameter priors, and sensor-derived phenomics update state estimates by either Bayesian filtering or differentiable state space inference, preserving trajectory coherence. Control inputs may include management interventions, nutritional interventions, vaccination interventions, and environmental interventions, and the twin will project phenotypic responses, risk responses, and resource utilization responses. The interaction-mediated processes that can be simulated at the herd level using agent-based extensions include pathogen transmission or social stress propagation, and the network structures are informed with proximity sensing and housing configuration data. Calibration requires stringent validation cycles in which predictive fidelity is validated against withheld longitudinal datasets across diverse environments and guarantees generalization. Leverage points identified in sensitivity analysis and counterfactual interrogation of twin dynamics show where genetic, microbiome-modulation, environmental-modification, or restructuring management of performance or welfare can result in disproportionate performance or welfare returns. Layers of ethical governance make sure that any recommendations to the decisions have taken into consideration uncertainty levels and do not overly rely on the extrapolation model beyond empirical training distributions (Moškon and Režen, 2024; Nawaz et al., 2025).

The implementation of climate, environmental, and management history into genomic AI completes the circuit between biological potential and actual performance under changing and more changeable external pressures. Climatic variables such as thermal loading, patterns of humidity, solar radiation, and seasonal variability sympathetically tune gene expression networks, epigenetic stability, and metabolic allocation strategies, yielding genotype by environment interaction landscapes, which may dilute the transferability of selection gains across geographies. Context-aware models introduce environmental covariates in the form of conditioning vectors in attention, gating, or hypernetwork modules that parameterize predictive layers based on contextual state, allowing genotype effect surfaces instead of fixed coefficients. Continuous as opposed to categorical characterization of the environment is made possible by spatially explicit embeddings of geostatistical encoders or remote sensing composition, improving the ability to find gradients of interaction. This raises management practices such as feeding strategy, housing density, health practice, and reproductive timing, which introduce extra contextual dimensions; neuro-symbolic constraints can be coded as organized ontological features, so that neuro-symbolic interaction constraints can provide plausible interaction modeling in relation to domain knowledge. Digital twin/predictive pipelines can be injected with climate projection scenarios to stress test the robustness of existing genetic portfolios, to find allelic combinations susceptible to future heat stress regimes or disease emergence zones. Traits are then weighted by adaptive selection indices in response to the projected changes in environmental distribution through risk-adjusted optimization, which balances between the projected expected gain and resilience variance indicators (Bussiman et al., 2025; Džermeikaitė et al., 2025).

Taken together, this multi-layered and contextual integration paradigm puts livestock genomic AI in a different context of readiness to translate genotype into phenotype into dynamic systems models in which molecular, ecological, behavioral, environmental, and managerial cues intersect. The keys to success are principled harmonization of data, uncertainty-aware fusion, causal sensitivity, and governance restricting overreach and making actionable insight feasible. The resulting architectures should balance representativity and interpretability, and the embedded integrated system needs to be decomposable into contributory factors that can be interrogated by the stakeholders. With downstream chapters turning to application domains and deployment, the integrative bases here will offer the scaffold upon which resilient, equitable, and sustainable progress is achievable, with selection and management strategies taking into account the entire richness of biological potential interacting with a dynamic world.

4.5 Core application domains

The development of AI is changing the derivation of genomic estimated breeding values, which is substituting hand-designed kernels with adaptive representation learning, which comprehends multilayered genetic architecture. Deep sequence encoders, haplotype transformers, graph neural networks over linkage disequilibrium and functional interaction graphs, and multi-omics fusion embeddings produce latent factors that disaggregate additive, dominance, epistatic, and regulatory effects and subdue error due to imputation uncertainty or sequencing artefacts. These latent abstractions enable nonlinear genotype to phenotype mappings without loss of calibration, coupled with Bayesian or conformal uncertainty layers to convert internal variance to actionable selection confidence intervals. Attention modalities identify contributory genomic parts in trait-specific situations, which allow post hoc understandability anchoring choices in known biological pathways. Multi-breed foundation style pretraining on multi-breed corpora addresses the issue of sparsity in smaller or indigenous breed datasets and later fine-tuning to local trait distributions and management regimes. Incremental updating models maintain model relevance in a changing set of allele frequencies generated during directional selection, hence stabilizing the predictive base on which to force cumulative genetic gain (Džermeikaitė et al., 2025; Xing et al., 2025).

Extending from the concept of single-trait focus toward the multi-trait and indicator-trait genomic prediction, AI architectures leverage similar genetic and biological relationships to increase performance when tackling difficult endpoints. Multi-task networks that use disentangled shared and private subspaces make use of pleiotropy, whereas hierarchical latent variable models allow partial pooling to defend against negative transfer in case trait correlations depend on the context. Predictive surfaces of late expressing, low heritability, or sex limited traits are then refined with integration of early measurable indicator traits, longitudinal sensor-derived phenomics, or intermediate molecular endophenotypes by providing surrogate constraints during training. It can be demonstrated through cross-attention fusion of temporally staged phenotypes with fixed genomic embeddings, which captures changes in development, showing temporal windows within which genetic effects have the greatest effect. The optimization of selection indexes is made subtler when models are provided with full covariance structures with uncertainty bands, allowing portfolio style balancing between production, welfare, adaptation, and environmental goals. Elastic weighting programs dynamically reweigh the contribution of each latent trait representation in order to guard against the dominance of high-prevalence or high heritability traits that might overpower minority goals of welfare or resilience (Cooper et al., 2025; Fan and Waldmann, 2025).

The AI-powered search is beneficial to mating optimization and inbreeding and genetic diversity management by providing genomic relationship representations to guide combinatorial parental pairings. Simulation of progeny genotype distributions under alternative matings is made possible by graph embeddings of realized relationship matrices, which are augmented with projections of segment-level homozygosity risk and runs of homozygosity projections. The trade-offs between projected genetic merit increase and future inbreeding trajectory constraints are assessed in evolutionary optimization, reinforcement learning, and differentiable surrogate modeling, which incorporate stochastic projections of allele frequency changes and effective population size at various levels of selection. The concept of diversity preservation is redefined as a multi-objective optimization, in which a set of measures of genomic uniqueness, haplotype diversity retention, and functional variant rarity balance against short-term gain. Constraint regularization is a form of mating plan penalty that rewards plans that increase the level of coancestry along genomic loci involved in the functioning of immunity, fertility, or stress resistance in order to protect the adaptability of the long horizon. Scenario analysis tools reveal the vulnerability of the recommendations to uncertainty in estimated breeding values, in the sense that they are robust and not fragile to the exploitation of close prediction peaks (Forneris et al., 2025; Pook et al., 2025).

Early life screening and embryo screening strategies are strategies that leverage AI to speed up the selection process by front-loading decision-making with predictive signals based on embryonic genotypes, low-depth sequence reads, methylation signals, metabolite signatures, or microimaging phenotypes. Low coverage sequence denoising models recreate high fidelity genotype likelihoods that can be further used in prediction without the expense of full depth sequencing. Multimodal encoders combine nongenetic early biomarkers correlating to later performance into a multimodal system that combines both with genomic latent spaces, making predictive sharpness more predictive before the availability of traditional phenotyping. Active learning models are used to choose the embryos or neonates to be subjected to the costlier tests, maximizing the information acquisition at a fixed cost. Scheduling of an embryo transfer and culling decisions use probabilistic forecasts of multi-trait merit using risk penalized on variance explanatory of sparse early data. The ethics governance modules address systematic bias, which may focus the selection to limited genomic backgrounds in cases where early predictions have more uncertainty, and therefore maintain the genetic basis of long-term advancement (Chen et al., 2025a; Moysis et al., 2025).

Disease risk genomics is a combination of polygenic risk scoring and host-pathogen interaction modelling used to inform prevention and remedial strategies. The progress of AI improvements extends beyond the summation of polygenic risks to the interaction of conscious risk surfaces that mimic sequence variation in immune genes, accessibility of regulatory regions, microbiome operational patterns, and metadata of historic exposures. Host gene networks and maps of pathogen effector interactions are encoded with graph neural networks to permit message passing through to identify sensitized subnetworks with collective perturbation that promotes susceptibility. Temporal models consume longitudinal health event logs and sensor-generated behavioral anomalies to revise dynamic risk predictions to form a feedback loop between surveillance and genomic predisposition evaluation. Calibration is important: Bayesian deep ensembles or evidential regression models guarantee that the risk probabilities are always conservative in different environments with poorly represented genotypes. Outputs are input to the targeting of vaccination priorities, biosecurity zoning, and differentiated management practices, which can match the distribution of preventive resources to stratified risk landscapes (Ozdemir et al., 2024; Sahragard et al., 2025).

Genomic surveillance and emerging disease early warning systems are disease modeling at the population scale, incorporating distributed sequencing, metagenomic pathogen detection, environmental sampling, and mobility or trade metadata. Stream processing architectures can be used to perform anomaly detection on rolling genotype frequency distribution, detecting abnormal changes in haplotype, suggesting the presence of a selection pressure by some emerging pathogen or a change in management. Phylogenetic embedding models insert new pathogen sequences into latent evolutionary spaces, estimating the velocity of divergence and potential host shift. Coupled host genomic tracking allows for assessing whether pathogenic expansions of interest overlap with clusters of vulnerability determined by host risk embeddings, allowing preemptive breeding or culling efforts. Federated analytics and differential privacy are able to preserve the custodians of data but allow the aggregation of signals across regions. Multi-criteria scoring is included in alert systems, and this combines statistical deviation magnitude, relevance of affected loci in the functional aspect, and epidemiological context, decreasing the false positives that may diminish stakeholder trust (Ahmed, 2025; Li et al., 2025).

Feed efficiency, nutrient utilization, and gene-diet interaction modeling use integrative AI to model the complexity of metabolism between genotype and rumen or gut microbiome functionality, feed composition, and phenomic ingestion patterns. Learning conditional representations enhances the genomic effect estimates with diet composition vectors, or seasonal forage quality embeddings, to generate interaction-sensitive breeding values. Multi-scale models integrate mechanistic nutrient partitioning equations with neural residual learners, which explain unmodeled nonlinearities, including differences in microbiome-mediated volatile fatty acid production. Counterfactual simulation modules can determine the anticipated changes in efficiency with different feed formulations of individual genotypes, informing the design of precision feeding programs to complement those of selection. Uncertainty propagation provides that pairing recommendations of diet genotypes have explicit confidence intervals and that they do not prematurely truncate the interactions on sparse strata. Joint loss functions incorporate the environmental coobjectives such as the intensity of methane or nitrogen emission to ensure efficiency and footprint reduction requirements are balanced (Deng et al., 2025; Lakamp et al., 2025; Ghavi Hossein-Zadeh, 2026b).

Reproductive genomics and fertility exploit AI to break down complicated physiological cascades underlying conception, gestation maintenance, and parturition outcomes. The predictive models fuse sequence and epigenetic features in reproductive tissues, the time series of hormone profiles, as well as high-resolution behavioral estrus signatures into predictive, more reliable, and earlier clues to fertility phenotypes. The causal graph discovery algorithms are useful in identifying the mediating factors, which transform the genomic predisposition to apparent fertility signs, guiding management strategies that supplement selection. Digital twin reproductive models recapitulate follicular dynamics and embryonic development based on the individual genomic and phenomic state, which allows optimization of the time of intervention. Integrating alignment between sexes is helpful to maintain consistency of selection on common reproductive traits as well as to model sex specific expression adjustments. Fertility prediction risks are incorporated in mating and culling decisions and guard against unintentional progression of infertility threat by aggressive choice on associated production characteristics (Lantzi et al., 2025; Orovou et al., 2025).

Climate stress-related resilience and adaptation characteristics require context-sensitive AI, which simulates genotype performance surfaces along gradients of thermal, humidity, and pathogen burden. Conditional modulation layers or hypernetworks reparameterize core prediction networks using climatic embeddings, which allows genotype by environment interaction mapping explicitly. Latent resilience indices combine measures of stability, recovery rates of stress perturbations, and strength of deviation in performance, providing goals in selection in addition to average productivity. Profiles of dynamic adaptation are enhanced by longitudinal sensor data that records heat load responses, changes in activities, and changes in feeding patterns. Climate model outputs as a form of scenario projections can be used to test the resilience of existing genetic portfolios, revealing a set of vulnerabilities that have potentially not been realized in the past. Adaptive selection indices use future environmental change probabilities to defensively, as opposed to reactively, weight resilience characters once performance deterioration becomes a reality (Chen and Dong, 2025; Naik, 2025).

Genomic selection is a method of environmental footprint and emissions reduction, which joins sustainability accounting and trait prediction (Ghavi Hossein-Zadeh, 2026b). Methane emission, nitrogen excretion, water use intensity, and carbon opportunity cost proxies are combined as either a direct phenotype or derived composite measure based on feed intake, microbiome functional gene patterns, and metabolic biomarkers. Multi-objective optimization models represent environmental externalities as shadow prices, resulting in efficient frontiers where a marginal sacrifice in production gain receives an unwarranted disproportionate dividend of emission reduction. Informed priors based on life cycle analysis limit model extrapolations such that forecasted footprint gains are biophysically realistic. Attribution methods determine regions of the genome that support cobeneficial pleiotropy that can either encourage productivity and mitigate emissions or are antagonistic and require compensatory management or index weighting corrections (Džermeikaitė et al., 2025; Sepulveda et al., 2025).

Prioritization of gene editing targets uses AI-based causal inference, network centrality measurement, and a measure of uncertainty about effect size to prioritize candidate loci whose modulation may do the best at the minimal cost of adverse pleiotropy. Structural causal models give estimates of the expected downstream system-wide perturbations of simulated edits, whereas counterfactual generative models suggest multi-locus combinations of edits that are constrained to satisfy welfare, diversity conservation, and regulatory constraints. The models used in off-target risk assessment are transformer-based sequence specificity models and chromatin accessibility informed context embeddings to estimate unintended edit probability spectra across the genome, and combine such models with functional consequence predictions to generate composite risk scores. Ethical governance levels implement screening levels that do not violate social and regulatory standards so that prioritization processes are consistent with social acceptability and ecological integrity (Kosaraju, 2022; Matthay et al., 2025).

Application of genomics in holistic decision platforms completes the application spectrum by integrating predictive, optimization, simulation, and surveillance modules into unified interfaces that inform strategic and operational decisions. Genomic predictions, management actions, economic parameters, welfare indicators, and environmental outcomes are linked in knowledge graphs and can be explored, resulting in the generation of queries and explanations. Multi-horizon planning is built into decision support systems, enabling adaptation of short-term mating and culling decisions with long-term diversity and sustainability goals in case of uncertainty. Adaptive intervention sequences proposed by reinforcement learning agents incorporate penalties on risk, resilience, and externality in their value functions, and human-in-the-loop oversight checks are plausible and ethically sound. On-going monitoring dashboards expose drift in surface calibration, imbalance in fairness between breeds, and signals of risk, which lead to retraining or governance review processes. Such an integrative fabric can enable genomic AI to cease being a peripheral analytic role and become the core nervous system of data-driven, ethically based, and future resilient livestock improvement (Dara et al., 2025). Core application domains in livestock genomics are presented in Table 4.3.

4.6 Multi-objective and decision optimization

The multi-objective and decision optimization of livestock genomics reconceptualizes the problem of breeding as a unidimensional search to optimize productivity, instead of a negotiation between partially competing objectives on welfare, genetic diversity, resistance to stressors, and environmental emission mitigation. AI provides the algorithmic machinery to model, learn, and navigate these elaborate trade spaces by building high-fidelity predictive surfaces on each of these objectives and linking them to optimization engines that search through combinatorial decision variables such as selection candidates, mating allocations, culling schedules, and management interventions. At the heart of this reframing lies a clear recognition that goals in heterogeneous mathematical forms: productivity characteristics can have the structure of a continuous response surface with moderately smooth genotype landscape structure, welfare measures can be complex indices of behavioral and physiological embeddings, diversity can be characterized in nonlinear functions of genomic relationship matrices and haplotype frequency spectra, resilience can be characterized in terms of stability measures and recovery processes under disturbed environmental covariates, and emissions can be additively and interactionally related to metabolic, microbiome and feed intake representations. AI-based surrogate models combine such modalities to form a latent decision space, allowing effective gradient-based or sample-efficient black box exploration with biological plausibility constraints through constraints imposed by prior knowledge graphs and mechanistic submodels. The balancing act is implemented mathematically by Pareto frontier estimation, adaptive scalarization, or evolutionary multi-objective heuristics, keeping provider sets spread on the frontier instead of being drawn too soon to a small area that is biased in terms of historically overweight economic payoffs. Dynamic reweighting strategies informed by tracking of objective attainment paths change selection emphasis in response to leading indicators of welfare standards erosion or genomic diversity contraction beyond preset safety thresholds, and thus, provide closed-loop governance (Ehrgott et al., 2025; Franco et al., 2025; Ding et al., 2026).

Layers of optimization within economic, ethical, and risk-constrained normative structure map versions of the technical balancing process, versions of the societal expectations, and enterprise sustainability goals into formal constraint, penalty functions, or risk-sensitive objective augmentations. Weights in the selection indices have long been given with economic valuation; AI gives weights by making probabilistic forecasts of market and input cost volatility and policy modulations, and so converts deterministic economic weights into distributions to drive robust or stochastic optimization programs. Ethical considerations such as welfare minimums, lack of extreme phenotypes that cause discomfort, and equal treatment of minor or indigenous breeds are expressed in hard constraints or fairness regularizers that limit the disparity in the anticipated returns that various strata of the population can produce. Risk enters through some channels: uncertainty of parameters in genomic predictions, model specification, environmental uncertainty, and tail risk in infrequent, but high-intensity, disease or climate events. More recent approaches that internalize downside protection are the conditional value at risk, distributionally robust optimization, and Bayesian decision theory. Examples of reinforcement learning agents, which offer sequential decision support, are the use of risk shaping in reward functions, and the punishment of strategies that offer low dependency on a constrained genetic base, or the high emission profiles in the face of likely future regulation measures. Interpretability ensures that optimization paths are auditable, and attribution can be used to disaggregate marginal contributions of genomic segments, and management can leverage the flow of composite goals, enabling stakeholder groups to ethically assess and communicate (Tariq et al., 2025; Toderas, 2025).

Scenario and sensitivity analysis are strategic tools that may be employed to challenge the capacity and the malleability of the proposed breeding portfolios to pose questions of nonreversible commitment before they are implemented. The scenario generation pipelines synthesize climate forecasts, feed resource variability curves, disease outbreak probabilities, and market demand variations, and combine sets of viable future scenarios to which candidate sets of decisions are stress-tested. Multi-scale simulation frameworks and digital twin systems simulate all scenarios at biological, economic, and environmental scales and resort to the distributions of objective results, revealing latent correlations capable of increasing systemic vulnerability. The global sensitivity analysis methods are applied as the standard variance of the composite objective output is split into driving inputs and into the genomic diversity metrics, resilience index, or emission reduction lever, which is ranked according to its impact and is used to establish priorities in surveillance and allocate the resources to demonstrate data quality. Local perturbation and elasticities are used to measure marginal responsiveness, and Shapley-style coalition analyses are used to measure synergistic and antagonistic combinations of decisions. Putting these learnings together, through sound portfolio building, to select solutions with palatable performance envelopes across

TABLE 4.3 Core application domains in livestock genomics.

Application domain	Description	AI techniques used	Examples
Genomic selection and breeding value prediction	Estimating genomic breeding values (GEBVs) for economically important traits by integrating high-density genotype, pedigree, and phenotypic records to accelerate genetic gain	GBLUP/ssGBLUP, Bayesian regression (BayesA/B/C, BayesR), ensemble and hybrid models, transformers, GNNs (pedigree and genomic graphs), multi-task learning, Bayesian deep learning (uncertainty), federated learning (cross-herd)	- Predicting milk yield, fertility, and somatic cell score in dairy cattle - Across-breed beef growth GEBVs - Decentralized resistance prediction between countries through federated models
Trait discovery and variant prioritization	Identifying causal loci, regulatory elements, and functional variants influencing traits using integrative statistical and machine learning pipelines	GWAS and fine-mapping with sparse models, transformers (regulatory sequence modeling), VAEs (latent variant effect grouping), GNNs (pangenome/LD graphs), causal discovery (SCMs), XAI (SHAP/IG), diffusion/contrastive pretraining (sequence embeddings)	- Prioritizing mastitis resistance SNPs - Identifying heat tolerance enhancers in *Bos Taurus* - Functional ranking of splice-altering variants for fertility
Disease resistance and health surveillance	Modeling susceptibility, early detection, and progression of infectious and production diseases through genomic, sensor, and clinical data streams	CNN/RNN/Temporal transformers (sensor time series), graph models (contact/pen networks), anomaly detection (autoencoders), Bayesian deep learning (risk stratification), contrastive pretraining, federated learning, RL for surveillance scheduling	- Early mastitis detection from milking parlor conductivity and accelerometer data - Genomic prediction of bovine respiratory disease resistance - Adaptive sampling for parasite monitoring
Reproductive management and fertility optimization	Improving reproductive performance (conception, calving interval, and semen quality) through predictive modeling and decision support	Multi-task learning (fertility and production), Survival models with deep features, transformers (hormonal time series), causal inference (nutrition versus fertility), RL (insemination timing), edge AI (real-time estrus detection)	- Predicting estrus onset from activity and temperature sensors - Semen dose quality classification - Optimizing the insemination window using RL policies
Nutritional optimization and feed efficiency	Personalizing diets and predicting feed conversion efficiency using genomics, metabolomics, and intake/behavior data	Hybrid mechanistic–ML models, ensemble regressors (GBDT and DL), VAEs (nutritional latent factors), RL (dynamic ration adjustment), causal models (diet–microbiome–performance), GNNs (ingredient interaction graphs)	- Optimizing feed conversion ratio in beef cattle - Predicting methane-adjusted feed efficiency - Adaptive precision feeding in swine grow-finish phases
Animal welfare and behavior monitoring	Continuous assessment of welfare indicators (lameness, stress, and aggression) using multimodal sensing integrated with genomic predisposition	CNNs (video/image gait), pose estimation and transformers, RNN/LSTM (activity sequences), anomaly detection (autoencoders), edge AI, multimodal fusion (sensor and genotype), explainable AI dashboards	- Lameness scoring from barn camera streams - Aggression detection in group-housed pigs - Stress susceptibility genomic risk layering with behavioral alerts
Environmental adaptation and climate resilience	Identifying and selecting animals resilient to heat, humidity, altitude, or other environmental stressors; modeling genotype × environment interactions	G×E reaction norm models, multi-task and domain adaptation, transformers (climate and genomic fusion), causal mediation (environmental pathways), GNNs (environment-location graphs), Bayesian hierarchical models	- Heat tolerance selection in dairy cattle under rising THI - Adaptation modeling for sheep in arid regions - Predicting resilience to fluctuating pasture quality

Continued

TABLE 4.3 Core application domains in livestock genomics.—cont'd

Application domain	Description	AI techniques used	Examples
Microbiome–host integration	Incorporating rumen/gut microbiome composition with host genomics to explain and predict complex traits (efficiency, emissions, and health)	Multi-view learning, Coembedding (contrastive), GNNs (microbe–metabolite–host networks), sparse modeling (microbial feature selection), VAEs (joint latent factors), causal mediation analysis	- Linking rumen microbial taxa to methane emissions - Predicting feed efficiency from host SNP and microbiome synergy - Microbiome-assisted mastitis risk modeling
Precision phenotyping and imaging	Automated extraction of phenotypes from images, video, 3D scans, wearables, and acoustic or thermal devices	CNNs/Vision transformers, 3D point cloud networks, Self-supervised contrastive vision pretraining, edge inference, active learning (annotation efficiency), and diffusion models (synthetic augmentation)	- Body condition scoring from side-view images - Carcass trait estimation through hyperspectral imaging - Thermal camera-based early fever detection
Multi-omics integration and systems biology	Joint modeling of genomics, transcriptomics, epigenomics, proteomics, and metabolomics to elucidate regulatory architecture and predict complex traits	Tensor/matrix factorization, multi-modal transformers, graph/multiplex networks, VAEs (variational multi-omics), kernel fusion, neuro-symbolic pathway constraints	- Integrative prediction of milk composition - Pathway-centric interpretation of fertility traits - Metabolome-transcriptome fusion for disease resilience
Genetic diversity and inbreeding management	Monitoring and optimizing diversity while achieving selection goals to avoid inbreeding depression and maintain adaptive potential	Optimization (multi-objective RL/evolutionary algorithms), graph/pedigree GNNs, causal modeling (diversity–trait trade-offs), Bayesian risk modeling, and scenario simulation (digital twins)	- Dynamic mating allocation balancing GEBV and inbreeding - Conservation strategy for endangered cattle lines - Forecasting diversity loss under intensified selection
Gene editing target prioritization and validation	Selecting high-confidence causal targets for CRISPR or allele introgression to improve traits while mitigating off-target risks	Causal inference (SCMs), functional sequence transformers, GNNs (regulatory/gene networks), XAI for variant attribution, and in silico off-target predictive models	- Prioritizing edits for hornlessness (polled trait) introgression - Ranking heat tolerance regulatory elements - Evaluating pleiotropic risk of fertility targets
Supply chain traceability and product quality	Linking genomic and phenotypic data through the production chain to assure provenance, quality, and compliance	Blockchain and federated analytics, anomaly detection (consistency checks), graph models (supply chain nodes), classification (quality grading), and edge AI (on-site verification)	- Traceable genomic-driven beef origin labeling - Milk quality anomaly detection in cooperative networks - Authenticity verification for breed-labeled products
Biosecurity and epidemiological modeling	Predicting, simulating, and mitigating disease spread within and across herds and regions using genomics and movement/sensor data	Graph epidemiological models, GNNs (contact networks), agent-based and RL intervention optimization, Bayesian hierarchical models, Spatio-temporal transformers, and digital twins for outbreak scenarios	- Foot-and-mouth disease spread simulation - Optimal vaccination scheduling with RL - Genomic-informed tracing of bovine TB clusters

Sustainability and emissions mitigation	Reducing environmental footprint (methane, nitrogen, and carbon) through selection, management, and diet optimization	Multi-objective optimization, causal pathway analysis (feed → microbiome → emissions), hybrid mechanistic–ML (rumen fermentation), ensemble prediction, and RL (diet-emission trade-offs)	- Selecting low-methane dairy sires - Optimizing manure management strategies - Ration design minimizing CH_4 while preserving production
Economic decision support and multi-objective optimization	Balancing profitability, welfare, sustainability, and genetic progress under uncertain market and environmental conditions	Multi-objective RL, Bayesian decision analysis, scenario simulation (digital twins), evolutionary algorithms, and stochastic programming with ML predictors	- Breeding index simulation integrating emissions and profit - Adaptive culling strategy under feed price volatility - Dynamic selection intensity adjustment
Rare breed conservation	Preserving genetic uniqueness and adaptive traits of minority or heritage breeds through informed genomic strategies	Diversity-aware optimization, graph/pedigree analysis, Bayesian effective population size modeling, representation learning (to capture unique haplotypes), and active sampling prioritization	- Identifying priority individuals for cryopreservation - Minimizing kinship in mating plans for endangered sheep - Targeted sequencing of underrepresented alleles
Antimicrobial stewardship and resistance monitoring	Tracking resistance markers and guiding prudent antimicrobial use to mitigate AMR emergence	Genomic resistance classifier models, temporal anomaly detection, causal analysis (treatment → resistance dynamics), dashboard XAI, and federated data sharing (privacy)	- Predicting resistance probability in mastitis pathogens - Early detection of emerging resistance haplotypes - Optimizing treatment protocols, reducing AMR risk
Digital twins and simulation-based policy testing	Virtual replicas of herds integrating genetics, health, nutrition, environment, and economics for safe experimentation and policy evaluation	Integrated simulation and RL, hybrid mechanistic–ML models, causal scenario evaluation, and Bayesian updating, generative models (synthetic trajectories)	- Testing long-term inbreeding control policies - Evaluating heat stress mitigation investments - Comparing vaccination strategies preoutbreak
Data governance, privacy, and compliance	Ensuring secure, ethical, and regulation-aligned usage of genomic and phenotypic data across stakeholders	Federated learning, differential privacy, secure multiparty computation, homomorphic encryption, audit trail analytics, and access anomaly detection	- Multi-country dairy genomic evaluation without raw data movement - Privacy-preserving release of synthetic genotype panels - Secure cross-farm disease risk collaboration
Edge and real-time on-farm analytics	Local, low-latency inference for immediate decision support using compressed or specialized models	Model compression (quantization, pruning), Tiny CNNs/temporal models, on-device anomaly detection, streaming feature extraction, and incremental learning	- Real-time feeding adjustment based on rumination sensors - Wearable-based calving alert system; on-site carcass quality grading with mobile vision models

scenario ensembles, rather than finding a solution that is optimal to a perceived single path. Adaptive planning horizons allow a reevaluation periodically because the posterior plans about scenarios are revised as real-world information becomes available, and Bayesian model averaging is biased downwards by either one story about the world. Participatory governance is also a part of the optimization loop, and scenario narratives can also be used as a way of making abstract optimization trade-offs concrete futures that stakeholders can consider (Føre et al., 2024; Wei et al., 2025).

Superior quality of data curation and harmonization is the key to implementing multi-objective balancing, constraint-aware optimization, and scenario-based sensitivity assessment; biases in welfare phenotyping, underrepresentation of climatic extremes, or insufficient measurements of emissions are prone to bias the estimation of the frontier and to misallocate the optimization effort. AI-guided data quality diagnostics identify distributional change, pattern of missingness, and underlying latent subpopulation organization, which may jeopardise fairness or overstate the perceived severity of trade-offs. The quantification of uncertainty is still found everywhere, and predictive distributions are provided with every objective and propagated both analytically and by Monte Carlo as part of optimization and scenario pipelines to ensure that epistemic gaps do not lead to intrinsic biological conflict. Decision dashboards combine frontier visualizations, risk-adjusted utility surfaces, and rankings on sensitivity into summaries that are cognitively tractable but do not overwhelm the user with underlying model granularity. Operational discipline is codified in governance frameworks that react to thresholds that lead to automatic reoptimization, such as drops in effective population size, increases in the number of indicators of welfare deviation, or increases in the intensity of emission in relation to regulatory targets (Griffiths et al., 2025; Panda et al., 2025).

Overall, the multi-objective and decision optimization paradigm would shift livestock genomic enhancement from an implicit, economically distorted exercise to an explicit systems engineering initiative based on open-mindedness, ethical consistency, and risk-conscious computing. AI provides the representational fidelity to describe intricate biological and environmental interactions, the algorithmic variety to search high-dimensional trade space, and the analytical rigor to evaluate robustness in the face of uncertainty. The resulting strategic portfolios not only seek productivity benefits but also manage genetic resources, boost animal welfare, strengthen resilience to increasing climatic and animal disease volatility, and reduce environmental externalities. The integrative decision fabric provides the basis of further deployment chapters, where operationalization and sustained governance guarantee the manifestation of theoretical multi-objective commitments into lifelong, adaptive practice.

4.7 Ethical, legal, and societal dimensions

The ethical, legal, and societal aspects of AI meant that livestock genomics must be framed in an integrated sense to view technological competence and moral responsibility as complementary and not successive. The timescale of predictive accuracy, selection intensity, and intervention granularity is increasing the space of potential actions much more rapidly than the traditional governance doctrines were suitable to support, and requires proactive integration of normative guardrails into computational architectures. Animal welfare is not an offshoot axis but a prerequisite axis since algorithmic optimization is simply not directionally selected with the explicit terms of welfare due to its risky, exploitative directional selection, which enhances production correlated burdens. Modern indices are thus demanding formal welfare protections being instantiated as layers of constraints, penalty terms, or multi-goal cuts that ensure that morphological integrity, physiological resilience, behavioral expression, and health resilience remain intact as economic characteristics evolve. The contributions of AI are that it quantifies the latent welfare constructs by multimodal sensor fusion and physiological signal modeling and converts the soft qualitative measures into reproducible and auditable embeddings. These embeddings feed selection models that can distinguish between correlated productivity gains, which are either neutral or welfare benefiting, and those that involve hidden compromise. Ongoing feedback control loops that identify drift determine if achieved phenotypic paths do not conform to the ethical purpose, causing recalibration. In this way, welfare has been internalized in the predictive optimization pipeline, and ethical compliance is a property of the algorithmic process and not a posterior audit (Neethirajan, 2024; Trotsyuk et al., 2024).

The second critical dimension is fairness and representational equity, as data-driven genomic models are inherently unequal in breed representation, as well as capable of increasing health disparities by favoring heterogeneity of the management frameworks and environmental coverage throughout the area. Bias may occur based on systematic over- or underestimation of breeding values of undersampled populations, unfair increase in variance; therefore, decimation of selection opportunity, and unintentional exclusion of adaptive or culturally important genotypes. To overcome this, it will be necessary to have bias detection protocols that examine error distributions by population cluster, ecological niche, and data provenance attributes. Some of the ways algorithmic fairness interventions can be applied are reweighting, domain adaptation, transfer learning, and synthetic augmentation, because they balance predictive performance across strata while maintaining biological signal fidelity. Representation audits look at the coverage of the genomic feature space to highlight areas where haplotype diversity is underrepresented, which might cause the algorithmic myopia. Equity precepts also apply to control of resource distribution for phenotyping and sequencing, in that the investments made should not be more biased towards already advantaged production systems. The measurement of ethical success is then based on the decrease in performance inequality periods and the sustainability of working population size proxies within the segments of the population. Notably, the concept of fairness is presented with a different angle, based not on statistical equality but rather on the maintenance of functional diversity that supports long-term resilience and stability of the food system (Lehmann et al., 2025; Sucharita and Shaw, 2025).

The control of gene editing and the accelerated genetic gain, ethically, is dependent on the definition of responsible limits of the intensity of intervention, the choice of targets, and the speed of the trait change. Causal inference, network perturbation simulation, and off-target risk modeling are AI-based methods to prioritize targets, which, when not put under control, can incentivize aggressive multiplex editing patterns with unknown systemic impacts. The governance structures should be made to express assessment standards that combine the welfare effects, environmental externalities, pleiotropic uncertainty, intergenerational equity, and the acceptability of society. The risk assessment is now multidimensional, as both probabilistic models of unintended genomic changes and interaction effects with preexisting polymorphisms are now needed, and genetic diversity may be narrowed when swept at high speed with edited alleles. Ethical veto layers can be incorporated in oversight architecture as the suggested sets of editing within models go through rule-based and deliberative review that queries necessity, proportionality, and reversibility. One way is to temporally regulate the speed of deployment to enable the intentional modulation of phenotypic validation under diverse conditions, to intercept the emergence of latent adverse effects before fixation. Openness in reported uncertainty procedures about anticipated edit results provides counterfeiting overconfidence scripts and the maintenance of trust. Acceleration in this framing can only be ethically justified to the extent that it is accompanied by compensating safeguards to ensure optionality to future adaptation, and safeguards of animal well-being (Sagar et al., 2025; Wang, 2025).

Issues of data privacy, sovereignty, benefit sharing, and intellectual property come together when genomic, phenomic, environmental, and management data streams are combined into AI training corpora with high derived value. There is often genomic data of livestock that crosses jurisdictional boundaries, and this involves different legal regimes of the biological resources and the digital assets. Sovereignty states that the original communities or countries maintain negotiated interest in the secondary applications, aggregation, and commercialization, which questions the unidirectional data extraction models. Although privacy is dissimilar to the human personal data contexts, it nevertheless includes confidential enterprise data and possibly sensitive geospatial disease indicators, as well. Technical applications to leveraged federated learning, secure multiparty computation, and differential privacy that are directly differentiated to retain the utility of the model and limit reconstruction risks or unauthorized inference of proprietary characteristics. The benefit-sharing models establish fair sharing of value created as a result of shared data, and this might be tiered access licensing, performance-based royalties, or capacity-building commitments to enable the contributors to develop local analytical acumen. The intellectual property systems should strike a balance between the need to encourage innovation and the need to avoid enclosing the basic genomic predictions, which act as infrastructure. Open interoperability standards of models and traceability metadata of data lineage facilitate accountability and dispute resolution, and keep the legal frameworks aligned with computational practices (Monteiro et al., 2024; Mandayam, 2025).

The communicative foundation of the ethically aligned AI genomic systems is made of transparency, clarification, proper explanation, and appropriate communication with stakeholders involved. Causal intuitions may be lost behind the complexity of deep representation models and multi-objective optimization, and cause skepticism or misunderstanding. Explainability strategies thus generate stratified accounts tuned to the literacy level of the stakeholders: molecular level attributions allowing researchers to understand which variance is contributed to which genomic region, welfare impact summaries allowing ethicists to understand the risks, risk-adjusted selection advice allowing producers to understand how to profit, and societal outcome framing allowing policy makers to understand the effects. Model cards, datasheets, and decision provenance logs are used to document training data composition, validation procedures, known limitations, bias audits, and update cadence. Counterfactual reasoning interfaces can enable users to simulate how different trait weightings or constraint thresholds would have affected the choice they would have been recommended to make, making it seem within control. Communicative transparency also involves revelation of uncertainty, with confidence intervals and epistemic gaps being emphasized instead of the deterministic outputs being given. Ethically, the concept of opacity is regarded as a hidden risk that can lead to a loss of trust in the field of collaborative governance, hinder its detection, and postpone the identification of emergent harms. The dedication to interpretability does not risk reductive over-simplification, and this applies to foster data literacy and participatory evaluation frameworks that combine experiential knowledge with computational evidence (Hooper and Lunn, 2024; Androutsopoulou et al., 2024).

Socio-economic impact and global equity in technology diffusion talks about the distributive effects of AI made genomic advances on scale, geography, and cultural systems. Unless designed with deliberate intent, unequal distribution of access to sequencing infrastructure, phenotyping sensors, computational capacity, and expertise may cause increasing productivity and resilience disparities, which strengthen structural inequities. Equity-oriented diffusion models are capable of having capacity-building features, open educational resources, and modular toolchains that can be compatible with different resource environments. Economic studies do not just focus on aggregate efficiency benefits but on livelihood stability, diversification of risk exposure, and dependency relationships that might arise when forecasting platforms get monopolized by centralized platforms. Ethical analysis involves analysis of the potential displacement impacts of accelerated genetic gain, altering the market conditions to the degree of pressuring less capitalized producers, potentially decreasing biodiversity as niche genotypes become economically ineffective. Some approaches might consist of collaborative data sharing alliances pooling resources to scale, but maintaining local control, or adaptive licensing schemes that reduce barriers of access in low-resource settings. International ethics also question cross-border externalities, so that the option to cut down on emissions or to make welfare gains in one country does not involve uncompensated costs being expropriated away in other countries in terms of feed resource adjustments or the susceptibility of genetic uniformity (Calzada, 2024; Darii, 2025).

The interconnection within these areas reinforces the idea that welfare protection, fairness, gene editing regulations, privacy controls, transparency discipline, and equity spreading plans are synergetic and not a set of checklists that are followed, combining control regardless of their activity level. In any privacy architecture where marginalized groups do not participate in the conversation of data, marginalized groups may be inadvertently disadvantaged, because the model calibration would suffer due to their missing inclusion, and vice versa, any privacy architecture that lacks welfare, or welfare embedded in its data, would expedite production at the cost of welfare. This, therefore, achieves ethical coherence within a systemic design wherein every element provides reinforcement feedback: transparency improves accountability towards both welfare and fairness indicators; equitable benefit sharing incentivizes the diversity in participation that enhances model generalizability; strong editing supervision secures welfare and long-term adaptive capacity that supports socio-economic resilience. The position of AI changes by not being a passive analytical implementer, but a proactive implementer who uses ethical tools once the model goals, loss functions, and governance elicit internal normative commitments. The literature is even moving in favor of ethics by architecture, making constraint satisfaction, bias reduction, and welfare optimization part of intrinsic learning instead of postprocessing (Mittal et al., 2024; Ortega-Bolaños et al., 2024).

The pathways of ethical, legal, and societal governance of AI-enhanced livestock genomics in the future are likely to take the direction of adaptive governance through consistent monitoring and deliberative review. Policy instrumentation can recalibrate tolerable levels of risk, or the tightness of welfare constraints, according to empirical trend evidence, scientific discovery, or changing values in society. Proactive intervention can be made possible by interdisciplinary ethics boards with real-time dashboards of diversity measures, welfare measures, bias audits, and model update logs, which could prevent path dependency, which would be expensive to reverse. The international harmonization could aim at establishing minimum standards in the scope of gene editing, sovereignty of data, and transparency standards as a way to reduce regulatory arbitrage and encourage the diffusion of innovation. Incorporating reflexivity, systems thinking, and inclusivity into the development of methods enhances legitimacy and trust, such that acceleration in the genomic accuracy of precision translates to the ethical distribution of benefits. In this way, the transformative possibilities of AI in livestock genomics can be achieved without disregarding ethical duties to animals, ecosystems, communities, and future generations. Ethical, legal, and societal dimensions of AI integration in livestock genomics are presented in Table 4.4.

4.8 Validation, benchmarking, and lifecycle management

The structural backbone of determining whether AI integration in livestock genomics becomes a reliable decision infrastructure or a pile of vaguely defined and flimsy prototypes is validation, benchmarking, and lifecycle management. The unique multi-layered nature of livestock genomic prediction with its longitudinal environmental modulation, multi-objective optimization, ethical limits, and selection of its highly complex trait structure requires a philosophy of validation beyond the snapshot of accuracy. Lifecycle management in this case involves a spectrum that starts with standardizing the protocol design of the benchmarking, moves onto the robustness testing and articulating of uncertainties, and reaches a deployment stage where the model has evolved with controlled and manageable changes and accountable audits. All of them need to be interconnected in a manner that the assertions in the model are kept evidentiary as biological population shifts, climatic natural environments, nutritional supplies, control systems, and policy expectations change (Nawaz et al., 2025; Tedeschi et al., 2025).

TABLE 4.4 Ethical, legal, and societal dimensions of AI integration in livestock genomics.

Dimension	Description	Implications	Examples
Data privacy and sovereignty	Data scope: genomic; phenotypic; sensor; supply chain; data classes: personal IDs; herd aggregates; environmental/welfare; proprietary strategies; sovereignty: localization and export controls; Reidentification risks: unique haplotypes, linkage attacks; protection: federated learning, differential privacy, SMPC, homomorphic encryption, encryption in transit/at rest; stewardship: purpose limitation, minimization, retention, tiered access, audit logs	Legal: cross-border transfer limits; controller versus processor duties; compliance duties; ethical: Prevent exploitative value extraction; respect autonomy; operational: granular access control, key rotation, classification taxonomy; economic: reluctance if no benefit sharing; possible competitive lockout; social: trust erosion through opaque secondary use; risk: breeding strategy inference; targeted manipulation	- Federated genomic evaluation (share gradients, not SNPs) - Differential privacy on breeding value summaries - Region-locked (EU-only) genotype processing - Attribute-based access for welfare video streams
Informed consent and secondary use	Consent models: broad; tiered; dynamic; purpose-specific; secondary use enumeration: welfare, carbon, insurance exclusions; revocation: propagation to derivatives; retraining exclusion; provenance: immutable consent logs (hash/blockchain); function creep checks: policy and - technical guardrails	Legal: invalid/overbroad consent voids downstream use; withdrawal right variability; ethical: autonomy; perceived fairness; governance: central consent registry before model updates; auditable trails; operational: automated exclusion pipelines; dataset version tagging; risk: mass withdrawals reduce generalizability; reputational harm	- Blockchain consent tokens controlling genotype access - Farmer dashboard toggling carbon analytics - Retraining, excluding revoked IDs - Exportable consent ledger for auditors
Transparency and explainability	Layers: technical (attribution), procedural (governance/update cadence), outcome (uncertainty); artifacts: model cards, lineage graphs, change logs, calibration plots; communication: plain-language, no genetic determinism simplification; uncertainty: intervals, reliability, coverage stats	Legal: explainability mandates possible; evidentiary support in disputes; ethical: reduces asymmetry; builds trust; operational: aids debugging drift/bias/data errors; economic: faster adoption; social: public reassurance; risk: misinterpreted attributions; overconfidence if uncertainty hidden	- SHAP plots for methane predictors - Model card noting excluded breeds/ time span - Calibration curve for lameness model - Release notes detailing feature/ performance changes
Bias, fairness, and distributional equity	Bias types: sampling (elite herds), label (phenotype scoring inconsistency), algorithmic (objective skew); fairness targets: accuracy parity (breed/system/geography/herd size); monitoring: stratified metrics, disparity indices, drift checks; genetic concern: diversity narrowing risk	Ethical: avoid disadvantaging underresourced producers; genetic: mitigate inbreeding risk; legal: Trade/competition scrutiny; economic: widening productivity gap; social: perceived favoritism harms trust; risk: misallocated breeding investment; skewed sustainability claims	- Fairness report with MAE per breed - Reweighting underrepresented smallholder data - Constraint for minimum minority breed accuracy - Data acquisition in tropical/arid systems
Animal welfare and ethical monitoring	Enhancement: early detection (lameness, mastitis, heat stress); surveillance limitation: minimize intrusive imaging/invasive sensing; Differentiation: welfare-benefiting versus purely productivity surveillance; alert governance: prioritization, throttling, escalation paths	Ethical: balance oversight and natural behavior; regulatory: certification alignment; social license: Demonstrated welfare impact; operational: threshold tuning, false positive control; risk: productivity optimization masking subtle welfare decline; alert fatigue	- Vision gait scoring triggering early hoof care - Multi-sensor stress index validated by vets - Alert suppression after confirmed treatment - Welfare KPI linking alert to recovery outcomes

Continued

TABLE 4.4 Ethical, legal, and societal dimensions of AI integration in livestock genomics.—cont'd

Dimension	Description	Implications	Examples
Environmental responsibility and climate claims	Metrics: absolute versus intensity reduction; modeled versus measured; traceability: documented assumptions (feed/emission factors); proxy validation: MIR spectra methane correlation; *Rebound safeguards: prevent expansion negating per-unit gains*	Legal: antigreenwashing; carbon credit compliance; ethical: authentic net reduction; market: access to eco-premiums/finance; operational: reproducible calculation pipelines; risk: overcrediting from extrapolation; rebound effects	- Versioned carbon parameter repository - Chamber validation subset - Counterfactual herd emission simulations - Emission label with CI and model version
Genetic diversity preservation	Metrics: Ne; ROH; allelic richness; heterozygosity; optimization: multi-objective (gain vs. inbreeding penalty); monitoring: temporal diversity dashboards; early warning: thresholds for ROH/Ne	Ethical: stewardship for future adaptability; biological: resilience to disease/climate stress; economic: sustained long-run productivity; operational: complexity of multi-constraint optimization; risk: silent homozygosity buildup	- Selection algorithm capping genomic inbreeding - Simulations of diversity under alternative mating - ROH heatmaps guiding mating exclusions - KPI tying bonuses to Ne maintenance
Intellectual property and benefit sharing	Asset layers: raw data, features/embeddings, model architectures, trained weights, and composite indices; rights allocation: licensing and derivatives; value return: revenue share, credits, and discounts; interoperability: open formats and standard ontologies	Legal: ambiguity triggers disputes; ethical: fair compensation drives participation; innovation: interoperability accelerates progress; economic: Incentivizes sustained contribution; risk: vendor lock-in; reduced transparency	- Cooperative revenue dividend model - Open trait ontology for portability - Dual licensing (research vs. commercial) - Smart contract royalty per API call
Accountability and liability	Responsibility: data quality (farm), model dev (vendor), and deployment interpretation (advisor); provenance: log model version, inputs, and rationale; incident: escalation and root cause workflow; distinction: model error versus misuse versus corrupted input	Legal: liability allocation impacts insurance; ethical: transparent postmortems; operational: monitoring for defensibility; economic: rapid resolution reduces downtime; risk: blame diffusion delays fixes	- Incident log with model hash and dataset checksum - SLA for critical false negatives - Automatic rollback on drift detection - Root cause template categorizing failure
Regulatory compliance and standards	Frameworks: data protection, welfare statutes, gene editing, and environmental reporting; standards: FAIR, GA4GH, and ISO security; horizon scanning: multi-jurisdiction monitoring; evidence automation: generated compliance dossiers	Legal: avoid fines/export limits; strategic: early alignment shapes norms; operational: automation lowers audit burden; risk: fragmentation raises integration complexity	- Compliance engine location-check pretraining - FAIR metadata enrichment pipeline - Regulatory change alert feed - Automated retention/adherence report
Dual use and misuse prevention	Misuse potentials: unauthorized gene target discovery, biothreat profiling, and cloning facilitation; access stratification: tiered raw versus aggregated; monitoring: anomaly detection (query volume and trait extremes); preapproval: ethics/biosecurity review	Ethical: anticipate harm; security: reduce exposure; legal: biosecurity/export control alignment; risk: backlash/policy overcorrection postincident	- Restricted API (no bulk raw sequence dumps) - Alerts on repeated rare resistance locus queries - Red-team adversarial query drills - Mandatory ethics review checklist
Labor and workforce transformation	Task shifts: observation → interpretation; data entry → exception handling; phenotyping → sensor oversight; displacement versus	Economic: retraining cost versus efficiency gain; social: digital divide risk; ethical: duty for equitable upskilling; operational: workflow design	- Training module on confidence intervals for breeding - Shadow phase before full AI reliance

	augmentation: remove repetitive tasks retain judgment; skills: data literacy, uncertainty handling, and ethical oversight; Reskilling: micro-credentials and cooperative hubs	influences adoption; risk: misinterpretation by undertrained staff	- Reskilling subsidy for smallholders - AR guidance for interventions
Smallholder inclusion and equity	Barriers: connectivity, capital, and language/digital literacy; inclusion: edge inference, tiered pricing, local language UI, cooperative pooling; data *equity: avoid dominance by large datasets*	Ethical: prevent widening inequity; economic: broader genetic progress; social: stronger community resilience; risk: model bias if smallholder *data is absent*	- Offline-capable mobile sync - Microsubscription with genotype credits - *Cooperative sampling campaigns* - Multilingual dashboard with pictorial aids
Consumer trust and social license	Trust drivers: AI transparency, gene edit labeling, and proven welfare/sustainability outcomes; communication: balanced risk-benefit narratives; Engagement: participatory forums; traceability: product-to-metric linkage	Market: influences willingness to pay; legal: labeling and claim substantiation; ethical: honest disclosure; risk: misinformation rejection of innovation	- QR code linking product to methane reduction and uncertainty - Public welfare outcome dashboard - Stakeholder codesign transparency KPIs - Plain-language gene edit explainer
Validation, reliability, and model integrity	Validation layers: internal CV; external benchmarking; temporal drift; prospective trials; robustness: stress tests (distribution shift and adversarial perturbations); calibration: predicted versus observed monitoring; security: poisoning detection and input sanity checks	Legal: certification may need independent validation; ethical: prevent harmful errors; operational: early drift detection lowers remediation cost; risk: silent degradation; adversarial manipulation	- Quarterly external validation report - Drift dashboard (calibration trending) - Adversarial perturbation suite (vision lameness) - Alert when Brier score exceeds threshold, triggering recalibration
Antimicrobial stewardship integration	Decision support: nonantibiotic alternatives; use differentiation: prophylactic versus therapeutic; resistance surveillance: allele trend deviation alerts; feedback: link recommendations to usage metrics	Public health: AMR reduction; legal: stewardship reporting compliance; ethical: preserve drug efficacy; operational: integrate prescription/treatment logs; risk: undertreatment from overcaution; overreliance on imperfect predictions	- Alternative husbandry recommendation before antibiotics - Resistance anomaly alert prompting lab confirmation - Stewardship dashboard correlating guidance with usage reduction - Causal analysis comparing outcomes versus alternatives

Benchmarking protocols begin by setting trait-aligned metrics and objective-aligned metrics, which need to capture biological and operational value propositions of the predictive system. Although required, conventional measures of correlation or mean squared error are not enough since selection decisions rely on rank stability, tail fidelity of genotypes of rare but high merit, calibration of probabilistic results, and cross-trait covariance coherence in the event of multi-trait or multi-objective indices. The key is calibration: any probability, such as output, reporting a risk-conscious breeding or health-based selection, has to show reliability between breeds, between levels of management, and between phenotype frequency distributions. In addition to benchmarking suites of rank-based criteria that measure the presence or absence of relative ordering by stratification of the environment, calibration curves, sharpness measures, and appropriate scoring functions are also included. They are external validation panels, including temporarily distinct cohorts, geographically scattered herds, and genetic lineages that are underrepresented, which are stress tests that display overfitting to unique linkage disequilibrium trends or training regimes inherent in the training data. These panels need strict curation using standardized phenotyping standards, harmonized environmental covariates, and missing data should be transparently managed so that estimates of performance can be compared across modeling methods. Cache repositories composed of fixed dataset hashes, written preprocessing pipelines, and consensus metric computation programs minimize ambiguity and make cumulative progress (Pan et al., 2024; Lane, 2025).

Benchmarking is even richer when multi-objective contexts are taken into account. Predictive quality of productivity traits needs to be evaluated alongside welfare-pertinent phenotype fidelity, resilience measures, emission proxy, and genomic diversity maintenance measures. Composite evaluation schemas do not naively aggregate, and instead they record surfaces of trade-off, such that even a model in which a slight increase in the primary trait accuracy comes at a cost of steep welfare calibration deterioration is not mischaracterized as being better. The preservation of cross-trait genetic correlation is examined to ensure that learned representations are not an unintentional bias propagator against the underlying covariance pattern, which would transfer bias to the construction of selection indexes. Moreover, time-stratified validation schemes differentiate between biological drift and measurement process drift, which decouples whether performance erosion is due to actual changes in the genomic architecture or some phenotyping bias. This will support specific remediation priorities (Vlaicu et al., 2024; Mbebi et al., 2025).

The concept of robustness and generalizability deals with the potential of models to be able to maintain reliable inference in conditions of domain shift. Domain shift in the case of livestock genomics has been observed to be exhibited by climatic abnormalities, pressures by new pathogens, changes in the composition of feed, or demographic changes that shift allele frequencies. Robustness testing integrates perturbation tests that systematically perturb covariates of the environment, mimic changes in genotype frequencies, and add random noise to phenotypic functions, and test how the stability of predictive distributions and rank ordering remains. Generalization audits test the performance on low-relatedness validation individuals in order to measure how much performance depends on realized relatedness matrices as opposed to actual learning of marker effects. Domain shift detection entails statistical monitoring of a stream of incoming data, which uses distributional divergence measures among genotype major elements, environment feature embeddings, and phenotype residual distributions. Recognized divergence activates conditional recalibration channels or safeguarded adaptation methods. A concept sensitivity boundary and a covariate sensitivity principle are primarily upheld: The former may be addressed by importance weighting or aligning the features, but the latter may mandate training a new model or architecture. Stress resistance measures are a quantification of widening of prediction intervals with respect to shift, and with this measure being guaranteed, such that an increasing uncertainty is reported, not hidden behind brittle point projections (Mai et al., 2025; Javed et al., 2024).

The risk-aware decision integration is based upon the quantification of uncertainty. The differences between aleatoric uncertainty experienced due to natural biological variability and epistemic uncertainty created by lack of data or model insufficiency guide the resources to be devoted to further data gathering or improvement of models. Bayesian hierarchical mixed models, deep ensembles, stochastic variational approximations, and conformal prediction models are also exploited to generate calibrated intervals around breeding value predictions, genomic risk scores, and welfare-associated trait predictions. Interval evaluation is both nominal and conditional evaluation, which encompasses conditional coverage analysis at the rarest genotype strata, environmental regimes, and across the deciles of trait magnitudes. Propagation of uncertainty along multi-trait indices is a direct calculation that explicitly links covariance of predictive errors, such that downstream optimization can make use of joint risk contours, instead of naive assumptions of independent variance. Communication layers are used to convert uncertainty into actionable advice, with high expected merit having broad epistemic ranges that should be allowed to undergo deferred irreversible selection and moderate merit having narrow ones that should be enabled to proceed with implementation. The decision-theoretic expected utility frameworks combine uncertainty in order to adaptively adjust selection thresholds and to make them less susceptible to regret in case of negative realizations (Ajirak et al., 2025; Walha et al., 2025).

The explainable uncertainty attribution complements confidence reporting: interval width contributions are broken down into data sparsity, model disagreement, and shift-induced extrapolation, which is nitpicking to reduce. The quantification of uncertainty improvements with additional phenotyping or sequencing validates the marginal value of utilizing data acquisition strategies, establishing evidence-based feedback loops on which decisions on investments can be made. In the case of safety-critical welfare constraints, conservative uncertainty adjustment with risk bounds that are calibrated gives a guarantee that the probability of constraint violation, under hidden shift conditions, is tolerably low (Fan, 2025).

The formalization of reproducibility, version control, and model lineage is based on an MLOps framework specific to a genomic data lifecycle. Data provenance is recorded during ingestion through cryptographic hashing of raw sequence files, phenotypic records, and environmental covariates source, and is accompanied by semantic metadata of information about collection instruments, curation procedures, and imputation approaches. Engineering pipelines that are declaratively defined can be regenerated across computational environments deterministically. Model artifacts have a version and are linked to commits in training code, dependency manifests, hyperparameter configurations, random seeds, and snapshots of training data that are defined by uniquely identifiable immutable identifiers. Lineage graphs graphically represent derivations of base models, using fine-tuning variants to fit each breed or production system, enabling audit queries to track any prediction back through transformation layers to the evidence of original data. Reproducibility tests automatically resubmit benchmark suites when dependencies are changed, or when parameters are changed, and report statistically significant variation in performance. Ingestion, preprocessing, training, evaluation, packaging, deployment, and monitoring are modular with idempotent execution, which workflow orchestrates, speeding up safe iteration and making regulatory inspection straightforward (Kampezidou et al., 2024; Chevuri, 2025).

The deployment is an evolving service and not an object through continuous tuning and drift management as well. Accuracy surrogates, calibration, uncertainty coverage, and fairness disparity indicators in population strata are displayed in dashboards using real-time/batch latency-aligned time-based monitoring of metric trajectories. Population stability index, change point detectors, and embedding similarity scores are used in drift detection algorithms to distinguish between unstable noise and systematic changes. Response policies are regularized: lightweight recalibration layers (Platt scaling or isotonic regression updates) can be run when minor covariate drift occurs, and concept shift causes retraining on augmented datasets with holdout validation to ensure the absence of regression on past satisfied welfare or diversity constraints. Continuous learning pipelines use active learning modules that focus on training labels on phenotypes in areas of high model uncertainty or sparse representation, which helps the convergence of epistemic gaps in a faster way. The performance of previous cohorts has safeguards against catastrophic forgetting, either through replay buffers or regularization constraints. Major model alterations are subjected to human sign-off at governance review gates, and the simulated counterfactual analysis of historical data will ensure that the ethical constraints are not violated (Mannapur, 2025).

The oversight is institutionalized in the auditing, traceability, and accountability frameworks. Detailed audit logs capture all prediction requests, model version identifiers, input feature hash, output values, and associated uncertainty, downstream decision action, and aggregate, along with evidence of this stored information that can be retrospectively analyzed about adverse outcomes or challenged selection decisions. Traceability can go down to the level of the data subject, where it can delete or correct erroneous records with automated repropagation with dependent models and indices. Responsibilities are determined by accountability structures. The data stewards ensure the integrity of the data by keeping the models and information up to date. Ethics officers ensure compliance by carrying out well-being and equity checks. Independent auditors of data carry out independent external surveys. Audit processes evaluate compliance to defined minimum metrics, that retraining has taken place because of what has been documented, that uncertainty calibration has not exceeded tolerance, and that lineage metadata is complete. The incident management protocols lay out a specific system of detecting, escalating, removing, and reporting performance degeneration, the emergence of bias, or even the violation of constraints once it has been identified. Formal assurance reports summarize the findings of an audit to a handful of stakeholders in the form of summaries and hold on to detailed technical annexes to be reviewed by experts. Standardized reporting schemas also contribute to integration with legal and regulatory expectations because model documentation has to be aligned with a compliance framework to solve the issue of data sovereignty, gene editing regulation, and welfare regulation. Measurement of accountability maturity is based not on the mere lack of incidents, but on the responsibility and openness to deal with them (Busch et al., 2024; Lam et al., 2024).

Lifecycle synthesis is the result of balancing benchmarking, uncertainty, reproducibility, monitoring, and auditing into a systematic feedback system. Results of benchmarking inform initial deployment gating criteria, and reproducibility infrastructure ensures future improvement is traceably attributed, monitoring provides empirical information leading to recalibration or retraining, and uncertainty analytics help address focused data acquisition, minimizing the risks of epistemism, and auditing ensures that every adaptation honors ethical, legal, and performance commitments. This paradigm lifecycle lessens technical debt, risk of unintentional performance erosion, and ensures trust of leaders in AI-adorned genomic decision making. Within this strong culture, some innovation in architecture modeling or data modeling can be absorbed, without disrupting operational stability, so that the increases in the sophistication of algorithms can be represented in long-term, responsible gains in livestock genomic performance.

4.9 Deployment, adoption, and impact realization

The process of deploying, adopting, and realizing the impact of AI in livestock genomics shifts the discussion of algorithmic possibilities to operational change, which must be carefully coordinated through the organization of structures, socio-technical interfaces, economic logic, human agency, and regulation. The translational arc is not linear: predictive models that do well in controlled benchmarking do not deliver their best when faced with fragmented data custodianship, inhomogeneous farm management cultures, varying market incentives, and changing policy cues. For an effective integration, AI systems are rethought as part of breeding pipelines, decision feedback loops, and strategic planning resonances, which redefine how genetic gain, welfare protection, sustainability goals, and economic sustainability are collectively addressed (Taddese et al., 2025).

The first step in integration into breeding organizations and on-farm workflows is the harmonization of data infrastructures, including pedigree repositories, genomic variants, longitudinal phenotypes, environmental telemetry, health and welfare measurements, and economic performance measures. Interoperable data schemas and ontology matching allow an easy ingestion into model serving layers, which can generate dynamic genomic prediction of values, indexing updates, and risk-annotated recommendations. The deployment architecture makes a distinction between having centralized model governance in breeding cooperatives or nucleus herd versus having decentralized inference endpoints that are closer to commercial production environments, where latency-sensitive decisions are made. Embedding of workflow means that AI outputs are not in parallel silos to the current deliberations of the selection committee, mating allocation tools, culling criteria, nutritional strategy adjustments, and health surveillance activations. Decision provenance capture records every action implemented to versions of underlying models, and the rationale behind weighting traits and uncertainty profiles, which are used to retrospectively analyze achievable and projected genetic and economic performance. The interface affordances get refined with repeated codesign with breeders and farm managers to make sure that the granularity, temporal rhythm, and explanatory labels of outputs are based on operational cycles such as the breeding season, calving schedules, or the periodic revision of feed formulations. The observational deviations, emergent phenotypic anomalies, and contextual nuances are fed back into the model refinement pipelines through feedback loops, producing a socio-technical learning process in a way that increasingly adapts AI behavior to the domain realities (Jayaram, 2025; Kancharla, 2025).

The issue of human factors, trust calibration, and change management determines whether stakeholders use AI insights in consequential decisions or perceive them as noise. The amplification of confidence that is unconditional does not enhance trust, and instead, selective amplification of transparency between stable, data-rich trait domains and emergent, data-sparse modalities, where caution is advisable, can enhance trust. The psychological acceptance is enhanced when the users have early-stage advisory deployments that operate in shadow mode with their current heuristics, showing a comparative advantage without a commitment that cannot be reversed. Change management plans present a story defining AI implementation as an enhancement of professional knowledge and speeding up ethical and sustainability objectives as opposed to a replacement for human judgment. There is a perceived-opaqueness resistance phenomenon; so, overlaying descriptions, counterfactual exploration tools, and simulation sandboxes in which users can experiment with different weightings of traits promotes a sense of agency. Trust calibration involves methodical monitoring of consistency between model-recommended and human-implemented decisions, as well as downstream outcome variance analysis that can indicate an excessive bias in automation or systematic override patterns, which can be indicative of miscalibrated model framing (McGrath et al., 2025).

Economic analysis and model of returns is the quantitative framework of strategic investment and scaling. Under conditions of environmental and market volatility, cost-effectiveness analyses segment value streams due to better and more precise choice of better germplasm, shorter generation interval, better welfare-related productivity strength, lower costs on health intervention based on risk-stratified preventive control, and better resource allocation. Types of expenditure include sequencing and high-resolution expansion, modernization of data infrastructure, operation of computational pipeline, model governance staffing, workforce training, and change management. Dynamic ROI simulations include the timing of diffusion, adoption saturation curves per segment, and lagged acquisition of the gain of genetic improvement as enhanced phenotypes are expressed in the next generation. The scenario analysis examines net benefit trajectory sensitivity to the assumptions of market price change, sustainability metrics, regulatory incentives, welfare compliance costs, and climate-induced environmental stressors. Real option valuation models staged investment under uncertainty, because the first use of deployment itself provides the values of learning and flexibility before the full economic benefits crystallize. Examples of externalities include the conservation of genetic diversity and reductions in emissions intensity, which are internalized as extended cost-benefit models in either shadow pricing or sustainability-consistent performance indices, to ensure that there is not a narrow focus on short-term profitability indicators at the expense of long-term resilience benefits (Aidoo and Kwofie, 2025; Huikku et al., 2025).

The enablers underlying the translation of algorithmic capability to the development of institutional competence include capacity building, training pathways, and cross-disciplinary workforce development. By integrating quantitative genetics principles, trait architecture complexities, and breeding program design constraints, genomicists need to know how to evaluate machine learning models, interpret uncertainty, and consider ethics in data governance, and data scientists and AI engineers need to internalize these principles and complexities. Interdisciplinary curriculum encompasses statistical genomics, causal inference, constrained optimization, welfare science, and sustainability measures and regulatory compliance. Applied training focuses on reproducible pipeline building, effective feature engineering of mixed data modalities, the detection of bias procedures, and the creation of interpretability artifacts. Mentorship programs that match domain experts with specialists in technical domains are helpful in hastening the transfer of tacit knowledge. Capacity growth at farm and regional cooperative scales encompasses the capacity to compete in quality assurance of data capture, sensor calibration, annotation procedures, and safe data sharing behaviors to maintain sovereignty and attain the scale of analysis. Skill currency is upheld by credentialing models and continuing education programs as a modeling paradigm and regulating expectations changes. Open educational resource investment and collaborative sandbox data democratize participation and reduce the concentration of expertise inside of a few large organizations, as well as provide diversity in innovation (Chiekezie et al., 2024; Shi, 2025).

The evolutionary forces of policy and regulation have a directional effect on the deployment paths by determining a boundary condition, incentive arrangement, and compliance requirement related to AI-enabled genomic practices. Data sovereignty regulatory advice defines allowable cross-border data aggregation, defining federated learning or privacy-preserving analytics implementation. New welfare laws that incorporate quantitative welfare outcome surveillance cement the incorporation of welfare constraint elements in the computation of a selection index that links ethical and economic motivations. Gene editing and accelerated genetic gain oversight mechanisms must also consist of clear records of AI-assisted target prioritization, off-target risk modelling, and phenotypic validation guidelines, and algorithmic accountability must be encompassed in approval mechanisms. The policy of environmental sustainability that compensates for the intensity of emission adds other objective dimensions to the index of genomic selection and encourages investment in the infrastructure of phenotyping that measures methane proxies or feed efficiency. Standard-setting bodies can issue documentation templates on model cards, lineage tracking, and calibration reports, reducing the cost of compliance due to harmonized reporting. Incorporation of adaptive regulation, including periodic consultation of the stakeholders, empirical evaluation of impact, and refinement of the innovation, promotes innovation and ensures protection against the emergent risks, but prescriptive regulation that is rigid runs the risk of obsolescence in the fast pace of methodology development. Regulatory fragmentation that would otherwise preclude cross-jurisdictional breeding collaboration and pooling of data can be reduced by international harmonization efforts, aiding in the fair distribution of AI benefits (Bahgat, 2025; Sedeeq and Arman, 2025).

The achievement of this change depends on the ultimate reconciliation of these aspects with an implementation ecology where organizational adoption, human trust, economic justification, capacity, and policy scaffolding feed off each other. The sustained assessment of achieved genetic gain programs, stability of welfare performance, sustainability performance, and economic resiliency are pathways of assuring the materialization of promised improvements within the limits of operational capabilities. Difficulties are between projected and actual stimuli of impacting force: model recalibration, retraining, redesigning incentives, or lobbying for a regulatory change. Socio-technical resilience occurs in situations where the design of a system predicts entropy: the departure of staff, old-fashioned technology, the drift of data, or the change in social expectations. That way, deployment is repackaged as a continuous adaptive governance model instead of a singular endpoint, which keeps AI implementation in livestock genomics abreast with ethical requirements, environmental issues, and market forces, and thus provides long-term value to producers, animals, consumers, and other food system stakeholders.

4.10 Future horizons

A set of technological, methodological, and sustainability imperatives that are converging to redraw the boundaries of how biological variation is represented, interpreted, and operationalised is defining future horizons in AI integration in livestock genomics. The movement is not away from individual predictive tasks but into integrated layers of intelligence between molecular and ecological scales, with the ability to reason across modalities and have adaptive responses to real-time production environments and constraints on the planet. According to this new paradigm, genomics is no longer seen as an off-the-shelf documentation of allelic differences but as a kind of computational master file in which dynamic models negotiate productivity, resilience, welfare, diversity, and carbon intensity trade-offs. While the following advancements in underlying model architecture, distributed analytics infrastructure, hardware acceleration, graph-based genome representations, causal simulation engines, and sustainability-aware, up to the embedded optimization, are entwined, the realization of such an ecosystem depends on them (Vohra, 2024).

The congruence of genomic, transcriptomic, and epigenomic foundational models will change the emphasis on analysis towards inquiring into narrow control of prediction to deep representation learning, representing regulatory logic and development contexts. Big-scale transformer-based and diffusion models are trained on heterogeneous corpora of raw sequence fragments, multi-tissue expression time series, chromatin accessibility maps, histone modification maps, methylation mosaics, and three-dimensional approximations of contact measures on different breeds and environmental regimes. Objective cross-modality alignments promote latent spaces, which contain sequence motifs, regulatory element accessibility, and induced expression responses on structured manifolds that represent causal adjacency. This latent organization allows minimal adaptations of a few shots to new characteristics, fills in gaps in phenotyping, and makes them better interpolated by being able to use trait-related embeddings to back-project onto regulatory circuits. Cross-species transfer learning based on conservation of regulatory grammar helps in the acceleration of model development in underresourced livestock populations, whereas cautious domain adaptation helps to prevent the misuse of patterns learned in well-studied reference breeds in genetically differentiated or locally adapted populations. Generative tasks can also be achieved with foundational multi-omics encoders, such as the imputation of unknown epigenetic states or the simulation of changes in expression under hypothetical edits of a variant, enabling the anticipation of selection or editing decision-making before investment of resources (Sharma and Jabin, 2025).

Real-time edge cloud hybrid genomic analytics that reflect progress toward prompt decisions have been obtained. Portable sequencing and sensor systems result in variant calls in real-time, signature calls to microbial communities, physiological biomarkers, and environmental telemetry at the interface of the herd. Smaller foundation models, shrunk down to edge inference modules, offer low-latency quality control, local phasing, pathogen variant screening, and preliminary approximation of breeding values when connections are intermittent and delegate computationally expensive multi-trait assimilation, uncertainty refinement, and federated model updates to cloud orchestration layers. Hierarchical compression and adaptive sampling reduce bandwidth by giving priority to the transmission of innovative haplotypes, abnormal sensor patterns, or regions of high uncertainty in genomes. To address privacy, sovereignty, and biosecurity, federated learning protocols address these issues by consolidating gradient updates, but not raw data, and by tuning the amount of injected noise differentially between desirable performance metrics and adverse selection of the relevant signal. The scheduling algorithms dispatch edge compute cycles between alignment, real-time health surveillance inference, and decision support rendering, based on utility estimates, time sensitivity, and information gain expectations. This dispersed continuum minimizes latency between the appearance of biological signals and intervention, provides resistance in the face of network failures, and increases the range of producers (both remote geographically and of smaller scale) to contribute to collective genomic development (Shah, 2025).

The prospects in quantum-inspired and specialized hardware acceleration consider the rapidly growing number of calculations of multi-omics foundation training, pangenome graph traversal, causal simulation, and on-device inference. Although fault-tolerant quantum machines are far off, hybrid algorithms based on tensile network factoring, variational quantum circuit representation, and annealing majority of the area of activity, classical optimization optimizers making use of haplotype phasing, genotype likelihood marginalization, and combinatorial mating distribution with multi-objective optimization. Simultaneously, specific accelerator designs on reconfigurable logic or application-specific integrated circuits have been designed with high-throughput k-mer counting, seed extension alignment, succinct de Bruijn and variation graph navigation, and sparse attention primitives needed to support long context molecular transformers. Near-memory processing architectures reduce penalties of data mobility incurred by multiple scans of large sequences and epigenomic matrices. Although neuromorphic paradigms are still young to genomics, they hold promise of delivering ultra-low power neural impetus detection of anomalies in edge-deployed biosurveillance through spiking representations of streamed sensor modality fusion. Scheduling frameworks can dynamically assign tasks to different hardware tiers based on factors such as precision tolerance, latency demands, and energy budgets. These frameworks can also embed carbon-aware orchestration, deferring non-urgent retraining to periods when there is a surplus of renewable energy. This aligns computational scaling with sustainability commitments (Liao et al., 2024; Silvano et al., 2025).

New generation pangenome graph systems and dynamic update platforms replace the linear reference dependence and allow equal representation of structural, copy number, and presence-absence variation between breeds and geographically adapted lineages. Graph-based assemblies of references incorporate contigs that have been resolved using haplotype and structural variants breakpoints, as well as annotated regulatory elements, into highly structured nodes and edges. Dynamic incorporation pipelines confirm individuals or environmental adaptation haplotypes that have been newly sequenced using consensus support metrics, conflict resolution heuristics, and provenance tagging recording sequencing technology, assembly procedure, and curation choice. Positioned between incremental indexing algorithms based on encoding using near real-time augmentation, and versioning semantics based on downstream model reconciliation with predictions across changing graph states, feature extraction can be anchored to stable coordinate abstractions. The epigenomic anchor integration links sequence variability to patterns of temporal accessibility, which facilitates scoring of regulatory impacts that fluctuate with the change of environmental or management contexts. Effective compression and streaming algorithms present subgraphs specific to breeds or genomic intervals to the edge devices, which saves memory footprint and speeds up variant effect inference. The types of governance maintain oversight and management of a community, promoting donation of underrepresented lineage sequences and removing the ability of a single person to control the direction of reference evolution (Chikhi et al., 2024; Nyaga et al., 2025).

The more advanced causal and counterfactual simulators project predictive utility into explanatory and prospective contexts in which decisions depend on critical knowledge of the intervention pathways and resilience to uncertainty. Structural causal models articulate layered relationships among genotype, regulatory states, expression, intermediate physiological phenotypes, microbiome dynamics, management interventions, environmental stressors, and production or welfare outcomes. Identification strategies make use of natural experiments, breed introgression events, temporal shocks, and instrumental variants to break the confounding and quantify mediation using molecular intermediates. Based on these causal graphs and grounded in foundation model latent representations, counterfactual generators are used to simulate phenotypic trajectories when breeding index values are changed hypothetically, genetic diversity is conserved to specific threshold levels, nutritional regimes are altered, or subject to climate stress conditions. Reinforcement learning overlays treat selection and management as sequential policies constrained by inbreeding, emission budgets, and welfare minima, optimizing long-term utility rather than single cycle gain. Simulation ensembles transmit epistemic uncertainty in causal structure, avoiding excessive reliance on estimates of intervention impacts and targeting appropriate data acquisitions to separate competing ways of finding the world in terms of mechanistic theory. Ethical guardrails stop exploration by not going into welfare undermining or environmentally unfriendly policy regimes, and fine-tune algorithmic search to normative promises (Kosaraju, 2022; Poinsot et al., 2024).

Model goals that incorporate sustainability measures and carbon accounting close the loop between genomic progress and planetary stewardship. Multi-objective optimization captures the intensity of greenhouse gas emissions, nitrogen and phosphorus excretion indices, land use efficiency, water footprint indices, biodiversity support indices, and the productivity and robustness characteristics. Life cycle inventory data, through dynamic emission factors, can respond to changes in diet, management of manure, and region-specific climatic conditions, lessening the discrepancy between the environmental impacts anticipated and those achieved. Carbon accounting models assign credit to downstream emission abatement to breedomic-induced feed efficiency or health resilience, assign credit to breeding organizations, producers, and technology providers using open attribution formulas that address the problem of possible duplication. Shadow pricing of carbon and nutrient discharge incorporates externalities into the construction of the selection index, and risk-adjusted optimization hedges against the uncertainty of scenarios in the future regulation stringency or volatility of carbon market prices. Sustainability phenotype models that account for measurement error extend the confidence interval of sustainability measures to selection response models, avoiding placing unwanted emphasis on proxies with high noise. Adaptive weighting plans rereadjust objective trade-offs as monitoring infrastructure develops, microbial control technologies develop, or policy incentives change to maintain consistency relative to changing societal views on ecosystem services (Costa et al., 2024; Gu et al., 2025).

Crosses between such areas of the horizon accentuate the power of transformation. The foundation models provide coherent latent representations that are easy to integrate with causal graphs, sustainability signals, and pangenome structural variation. The temporal granularity required to causally validate interventions that happen and environmental perturbations that take place is provided by the edge cloud orchestration. Computational intensity is matched with emission reduction commitment by special hardware and carbon-conscious scheduling. Pangenome dynamism is vital in ensuring an unbiased causal and sustainability impact inference in various genetic architectures to promote a fair benefit allocation. Counterfactual simulation engines, in their turn, capitalize on rich multi-omics embeddings and updating graphs to refine scenario planning, which allows them to plan the ways to adjust breeding and management during the projected climate pathways (Wu and Xie, 2024). The ecology of this realization requires the simultaneous development of

TABLE 4.5 Future horizons of AI integration in livestock genomics.

Aspect of AI integration	Current applications	Potential future developments	Challenges and considerations
Multi-omics and systems genomics integration	- Genomic SNP and basic phenotypes in indices - Targeted RNA-seq variant prioritization - Pilot metabolomics panels - Emerging proteomic biomarkers - Static regulatory network analyses	- Full genome–epigenome–transcriptome–proteome–metabolome–microbiome fusion - Temporal multi-omics tracking - Causal graph inference - Automated functional annotation and pathway scoring - Individual intervention optimization	- High dimensionality versus sample size - Batch effects and ontology gaps - Longitudinal cost - Overfitting and spurious edges - Consent scope governance - Equity concentration risk
Real-time edge genomic analytics	- Compressed SNP panels (cloud EBVs) - Basic on-device alerts - Periodic offline EBV sync	- Streaming local evaluations - Edge multimodal anomaly detection - Local generative imputation - Context-aware decision scoring - Peer-to-peer model deltas	- Computational and energy limits - Tamper/security of updates - Drift versus the central model - Intermittent connectivity - Portable model format needed
Digital twins and simulation optimization	- Herd-level feed/repro simulators - Static genetic what-if projections - Separate nongenomic epidemiology	- Individual physiologic-genomic-behavioral-microbiome twins - RL multi-year optimization - Stress scenario testing—multi-objective (profit/welfare/emissions/diversity) - Capital planning scenarios	- Dense data requirements - Computational intensity - Twin validation challenges—RL interpretability - Ethical automation guardrails
AI-guided gene editing and synthetic biology	- Monogenic variant prioritization - CRISPR target ranking - Basic combinatorial feasibility scoring	- Polygenic minimal edit sets - Epistasis-aware multiplex planning - Regulatory element modulation - Host–microbiome synthetic constructs - Probabilistic safety simulation	- Regulatory divergence - Enhancement ethics - Diversity narrowing risk - Off-target and pleiotropy uncertainty - Public acceptance
Adaptive breeding objective optimization	- Static economic indices - Uniform trait weights - Minimal sustainability signals	- Streaming recalibration - Bayesian local personalization - Externalities integration - Climate scenario adaptation - Consumer premium responsiveness	- Volatility and overfitting risk - Transparency and explainability - Smallholder fairness - Governance of triggers - Shock overreaction
Precision animal health and predictive epidemiology	- Threshold sensor alerts - Regional outbreak models - Select host susceptibility markers	- Individual risk stratification - Adaptive transmission modeling - Multi-pathogen syndrome indices - Rapid on-site AI sequencing triage - Intervention efficacy versus stewardship optimization	- Privacy (health and location) - Alert fatigue - System interoperability gaps

			- Cross-border biosecurity compliance - Access equity
Microbiome-informed selection and intervention	- Correlation studies (e.g., efficiency, methane, etc.) - Experimental model features - Generic probiotics	- Routine sequencing in selection - AI-designed consortia - Diet–microbiome–genome optimization - Functional metagenomics inference - Precision prebiotics	- Temporal/diet variability - Causality ambiguity - Longitudinal cost - Regulation of engineered consortia - Ecological displacement
Climate resilience and sustainability modeling	- GHG intensity predictions - Heat stress indices - Partial and decoupled LCA	- Integrated resilience scores - Dynamic LCA and decisions - Microbiome-informed methane modeling - Genotype × future climate mapping - Emission strategy optimization	- Sparse GxE datasets - Attribution complexity - Emission factor inconsistency - Greenwashing risk - Access equity
Autonomous phenotyping and computer vision	- Lameness video scoring - 2D weight estimation - CNN body/condition scoring	- 3D multi-spectral tracking - Behavioral micropattern mining - Continual self-supervised edge vision - Genomic-health-vision fusion indices - Explainable spatiotemporal heatmaps	- Label subjectivity - Cross-site variability - Storage/bandwidth load - Privacy (facility/personnel) - Breed/color bias
Federated and privacy-preserving collaboration	- Federated averaging - TLS parameter exchange - Occasional DP noise	- Hierarchical adaptive federation - Secure multiparty computation - Homomorphic encrypted gradients - Dynamic privacy budgets - Geo-fenced compliance nodes	- Latency and overhead - Privacy–utility trade-offs - Non-IID skew - Verifiable consent/contribution logs - Incentive design
Explainable and responsible AI frameworks	- Model cards and feature importance - Ad hoc fairness checks - Manual incident reviews	- Continuous fairness dashboards - Counterfactual/causal explanations - Automated adversarial red-teaming—ethical risk gating - Self-documenting compliance pipelines	- Oversimplified explanations risk - Metric overload - Governance artifact upkeep - Regulatory evolution - Operational monitoring cost
Genomics-linked supply chain traceability and value differentiation	- Batch-level RFID traceability - Limited genomic merit label linkage - Sustainability claims (no uncertainty)	- Immutable genotype-to-product provenance chains - Real-time sustainability/welfare credential embedding - Dynamic premium pricing engines - Consumer transparency portals - Smart contract certification triggers	- Data integrity and tamper resistance - Standards fragmentation - Farm data privacy and commercial sensitivity - Overclaim (green/ethical washing) risk - Smallholder cost burden

governance, standardization, and participatory design. Graph streaming, uncertainty annotation, sustainability metric schema, as well as causal model serialization have open-interface specifications, which encourage interoperability as well as reproducibility. The incentive systems reward data, algorithmic novelty, curation, and validation expertise so that no value capture asymmetries occur. Constant ethical assessment questions the new actions of generative and simulation elements, even in a situation where automated optimization may unwillingly reduce genetic variation or obscure welfare trade-offs through composite measures. Pipelines in education become broader, now to include graph bioinformatics, quantum-inspired optimization literacy, rigor in causal inference, and sophistication in carbon accounting, developing a workforce with the skills to manage AI genomic infrastructures of greater complexity (Amen, 2024).

Overall, IoT futures in AI led livestock genomics to an integrated, adaptive, ethically based computational fabric, able to learn, simulate, and cooptimize biological systems at molecular depth and ecological breadth. Production of multimodal fundamental representations, distributed real-time analytics, faster computation, dynamic pangenome knowledge bases, causal reasoning engines, and internalization of sustainability prepares stakeholders to move toward genetic advancement that is not only speedy but wiser, robust, transparent, and consistent with the environmental and societal demands defining the future of food systems. Future horizons of AI integration in livestock genomics are presented in Table 4.5.

References

Agrawal, R., 2025. Adaptive few-shot learning (AFSL): tackling data scarcity with stability, robustness, and versatility. https://doi.org/10.48550/arXiv.2501.13479.

Agrawal, R., Kumar, H., Lnu, S.R., 2025. Efficient LLMs for edge devices: pruning, quantization, and distillation techniques. In: 2nd International Conference on Machine Learning and Autonomous Systems, ICMLAS 2025—Proceedings. Institute of Electrical and Electronics Engineers Inc, pp. 1413–1418. https://doi.org/10.1109/ICMLAS64557.2025.10968787.

Ahmed, R., 2025. Global surveillance systems for emerging infectious diseases: a critical review of infrastructure, policy, and data sharing. Premier J. Data Sci. https://doi.org/10.70389/pjds.100004.

Aidoo, R., Kwofie, E.M., 2025. Circular bioeconomy accounting tool (CBEAT): a holistic framework for agriculture and agri-food system circularity practice. Sustain. Sci. 20 (1), 33–53. https://doi.org/10.1007/s11625-024-01578-3.

Ajirak, M., Ravishankar, A., Djurić, P.M., 2025. Uncertainty quantification in probabilistic machine learning models: theory, methods, and insights. doi: 10.48550/arXiv.2509.05877.

Amen, M.A., 2024. AI-driven sustainable habitat design: key policy frameworks and ethical safeguards. Smart Des. Policies 1 (1), 23–32. https://doi.org/10.38027/smart-v1n1-4.

Androutsopoulou, M., Askounis, D., Carayannis, E.G., Zotas, N., 2024. Leveraging AI for enhanced eGovernment: optimizing the use of open governmental data. J. Knowl. Econ. 16 (3), 12998–13033. https://doi.org/10.1007/s13132-024-02317-w.

Aninze, A., 2024. Artificial intelligence life cycle: the detection and mitigation of bias. Int. Conf. AI Res. 4 (1), 40–49. https://doi.org/10.34190/icair.4.1.3131.

Annan, R., Noland, J., Perkins, K., Yuan, X., Roy, K., Qingge, L., 2025. Genomic privacy and security in the era of artificial intelligence and quantum computing. Disc. Comput. 28 (1). https://doi.org/10.1007/s10791-025-09627-w.

Bahgat, A., 2025. AI governance and data privacy in cross-border contexts: a comparative analysis of regulatory frameworks. J. Data Protect. Priv. 8 (1), 8. https://doi.org/10.69554/bojl3033.

Baião, A.R., Cai, Z., Poulos, R.C., Robinson, P.J., Reddel, R.R., Zhong, Q., Vinga, S., Gonçalves, E., 2025. A technical review of multi-omics data integration methods: from classical statistical to deep generative approaches. Briefings Bioinf. 26 (4). https://doi.org/10.1093/bib/bbaf355.

Balsa-Canto, E., Campo-Manzanares, N., Moimenta, A.R., Roudaut, G., Troitiño-Jordedo, D., 2025. Quantifying and managing uncertainty in systems biology: mechanistic and data-driven models. Curr. Opin. Syst. Biol. 42. https://doi.org/10.1016/j.coisb.2025.100557.

Busch, F., Kather, J.N., Johner, C., Moser, M., Truhn, D., Adams, L.C., Bressem, K.K., 2024. Navigating the European Union artificial intelligence act for healthcare. Npj Digit. Med. 7 (1). https://doi.org/10.1038/s41746-024-01213-6.

Bussiman, F., Lourenco, D., Hidalgo, J., Chen, C.Y., Holl, J., Misztal, I., Vitezica, Z.G., 2025. Genotype-by-environment interaction with high-dimensional environmental data: an example in pigs. Genet. Sel. Evol. 57 (1). https://doi.org/10.1186/s12711-025-00974-2.

Calzada, I., 2024. Artificial intelligence for social innovation: beyond the noise of algorithms and datafication. Sustainability 16 (19), 8638. https://doi.org/10.3390/su16198638.

Chakraborty, A., Ghosh, S., Kumar, B., Mandal, S., Paul, S., 2025. Resource-aware deep learning: neural network optimization for edge devices: a review. Int. J. Environ. Sci. 3859–3863. https://doi.org/10.64252/yc79fn98.

Chen, C., Dong, J., 2025. Deep learning approaches for time series prediction in climate resilience applications. Front. Environ. Sci. 13. https://doi.org/10.3389/fenvs.2025.1574981.

Chen, X., Wei, S., Sun, C., Yi, Z., Wang, Z., Wu, Y., Xu, J., Tao, J., Chen, H., Zhang, M., Jiang, Y., Lv, H., Huang, C., 2025b. Computational tools for studying genome structural variation. OMICS 29 (2), 36–48. https://doi.org/10.1089/omi.2024.0200.

Chen, K., Zuo, J., Han, W., Guo, J., 2025a. Intelligent assisted reproduction: Innovative applications of artificial intelligence in embryo health assessment. LabMed Discov. 2 (2), 100075. https://doi.org/10.1016/j.lmd.2025.100075.

Chevuri, R.R., 2025. Demystifying MLOps: core principles for scalable machine learning. Int. J. Adv. Eng. Manag. 7 (3), 884–890. https://doi.org/10.35629/5252-0703884890.

Chiekezie, N.R., Obiki-Osafiele, A.N., Ebele Agu, E., 2024. Preparing the workforce for AI technologies through training and professional development for future readiness. World J. Eng. Technol. Res. 3 (1), 001–018. https://doi.org/10.53346/wjetr.2024.3.1.0050.

Chikhi, R., Dufresne, Y., Medvedev, P., 2024. Constructing and personalizing population pangenome graphs. Nat. Methods 21 (11), 1980–1981. https://doi.org/10.1038/s41592-024-02402-7.

Clement, T., Gbaja, C., Onayemi, H., 2025. Adversarial machine learning: defense mechanisms against poisoning attacks in cybersecurity models. Int. J. Eng. Comput. Sci. 14 (06), 27286–27308. https://doi.org/10.18535/ijecs.v14i06.5156.

Cooper, M., Tomura, S., Wilkinson, M.J., Powell, O., Messina, C.D., 2025. Breeding perspectives on tackling trait genome-to-phenome (G2P) dimensionality using ensemble-based genomic prediction. Theor. Appl. Genet. 138 (7). https://doi.org/10.1007/s00122-025-04960-6.

Costa, G.de S., da Silva Ferko, G.P., Dimy Barbosa Bitar, S., Lopes, J., 2024. Otimizando ganhos no sistema de integração lavoura-pecuária-floresta: aplicação da otimização multiobjetivo. Contribuciones a Las Ciencias Sociales 17 (6), e7197. https://doi.org/10.55905/revconv.17n.6-002.

Coté, M., Aires, S., 2025. Futurity as infrastructure: a techno-philosophical interpretation of the AI lifecycle. doi: 10.48550/arXiv.2508.15680.

Cui, Y., Peng, C., Xia, Z., Yang, C., Guo, Y., 2025. A survey of sequence-to-graph mapping algorithms in the pangenome era. Genome Biol. 26 (1). https://doi.org/10.1186/s13059-025-03606-6.

Dara, M., Dianatpour, M., Azarpira, N., Tanideh, N., 2025. The transformative role of artificial intelligence in genomics: opportunities and challenges. Gene Rep. 41, 102314. https://doi.org/10.1016/j.genrep.2025.102314.

Darii, C., 2025. The socio-economic impact of artificial intelligence. Simpozion Ştiinţific al Tinerilor Cercetători. Ediţia a 22-a. Culegere de Lucrări Ştiinţifice 2, 207–210. https://doi.org/10.53486/sstc2024.v2.48.

Deng, Z.C., Liu, M., Cao, K.X., Khalil, M.M., Guan, L.L., Sun, L.H., 2025. Gut microbiome and postbiotics: bridging the dietary nutrition and feed efficiency in food-producing animals. Sci. China Life Sci. https://doi.org/10.1007/s11427-025-3063-5.

Ding, W., Liu, J., Dong, W., Luo, F., Gu, C., 2026. MPCMO: an improved multi-population co-evolutionary algorithm for many-objective optimization. Inf. Sci. 723, 122671. https://doi.org/10.1016/j.ins.2025.122671.

Džermeikaitė, K., Šidlauskaitė, M., Antanaitis, R., Anskienė, L., 2025. Enhancing genomic selection in dairy cattle through artificial intelligence: integrating advanced phenotyping and predictive models to advance health, climate resilience, and sustainability. Dairy 6 (5), 50. https://doi.org/10.3390/dairy6050050.

Ehrgott, M., Köksalan, M., Kadziński, M., Deb, K., 2025. Fifty years of multi-objective optimization and decision-making: from mathematical programming to evolutionary computation. Eur. J. Oper. Res. https://doi.org/10.1016/j.ejor.2025.06.012.

Essien, D., Neethirajan, S., 2025. Multimodal AI systems for enhanced laying hen welfare assessment and productivity optimization. doi: 10.48550/arXiv.2508.07628.

Eyinade, W., Jacqueline Ezeilo, O., Abiodun Ogundeji, I., 2025. Navigating the complexities of ethical AI and algorithmic accountability in modern technological practices. Comput. Sci. IT Res. J. 6 (6), 371–381. https://doi.org/10.51594/csitrj.v6i6.1962.

Fan, X., 2025. Position paper: integrating explainability and uncertainty estimation in medical AI (Version 1). https://doi.org/10.48550/ARXIV.2509.18132.

Fan, Y., Waldmann, P., 2025. Multi-task genomic prediction using gated residual variable selection neural networks. BMC Bioinf. 26 (1). https://doi.org/10.1186/s12859-025-06188-z.

Feng, Y., Zhang, Y., Ying, S., Du, S., Gao, Y., 2025. Kernelized hypergraph neural networks. IEEE Trans. Pattern Anal. Mach. Intell. 47 (10), 8938–8954. https://doi.org/10.1109/tpami.2025.3585179.

Ferretti, P., Johnson, K., Priya, S., Blekhman, R., 2025. Genomics of host–microbiome interactions in humans. Nat. Rev. Genet. https://doi.org/10.1038/s41576-025-00849-8.

Føre, M., Alver, M.O., Alfredsen, J.A., Rasheed, A., Hukkelås, T., Bjelland, H.V., Su, B., Ohrem, S.J., Kelasidi, E., Norton, T., Papandroulakis, N., 2024. Digital Twins in intensive aquaculture—challenges, opportunities and future prospects. Comput. Electron. Agric. 218. https://doi.org/10.1016/j.compag.2024.108676.

Forneris, N.S., Bosse, M., Gautier, M., Druet, T., 2025. Genomic prediction of individual inbreeding levels for the management of genetic diversity in populations with small effective size. Mol. Ecol. Resour. 25 (4). https://doi.org/10.1111/1755-0998.14068.

Forrest, I.S., Vy, H.M.T., Rocheleau, G., Jordan, D.M., Petrazzini, B.O., Nadkarni, G.N., Cho, J.H., Ganapathi, M., Huang, K.L., Chung, W.K., Do, R., 2025. Machine learning–based penetrance of genetic variants. Science 389 (6763). https://doi.org/10.1126/science.adm7066.

Franco, P., Mullot, R., Owczarek, V., 2025. Pareto multi-objective optimization for high locality-preserving space-filling curve identification. Swarm Evol. Comput. 92, 101797. https://doi.org/10.1016/j.swevo.2024.101797.

Ghavi Hossein-Zadeh, N., 2024. An overview of recent technological developments in bovine genomics. Vet. Anim. Sci. 25, 100382. https://doi.org/10.1016/j.vas.2024.100382.

Ghavi Hossein-Zadeh, N., 2025. Artificial intelligence in veterinary and animal science: applications, challenges, and future prospects. Comput. Electron. Agric. 235, 110395. https://doi.org/10.1016/j.compag.2025.110395.

Ghavi Hossein-Zadeh, N., 2026a. Generative artificial intelligence in animal genomics for smart agriculture: applications, challenges, and future prospects. Vet. Anim. Sci. 33, 100702. https://doi.org/10.1016/J.VAS.2026.100702.

Ghavi Hossein-Zadeh, N., 2026b. Advancing climate-resilient livestock systems: next-generation emission mitigation strategies and integrated technological innovations. Vet. Anim. Sci. 31, 100588. https://doi.org/10.1016/j.vas.2026.100588.

Gillis, H.M., Xu, I., Trappenberg, T., 2025. Uncertainty estimation using variance-gated distributions (Version 1). https://doi.org/10.48550/ARXIV.2509.08846.

Griffiths, E.J., Jurga, E., Wajnberg, G., Shay, J.A., Cameron, R., Barclay, C., Sehar, A., Dooley, D., John, N.S., Scott, A., Johnson, L.A., Robertson, J., Schonfeld, J., Bastedo, D.P., Tang, J., Yin, X., Rehman, A., Wallace, R.L., Thomas, K., Eagle, S.H.C., McAllister, T., Diarra, M.S., Nash, J.H.E., Topp, E., Domselaar, G.V., Taboada, E., Tamber, S., Kess, T., Broadbent, J., Poulin-Laprade, D., Smith, D.D.N., Reid-Smith, R., Zaheer, R., Laing, C.R., Carrillo, C.D., Hsiao, W.W.L., 2025. Crossing the streams: improving data quality and integration across the one health genomics continuum with data standards and implementation strategies. Can. J. Microbiol. 71. https://doi.org/10.1139/cjm-2024-0203.

Gu, H., Liu, S., Huan, C., Cheng, M., Dong, X., Sun, H., 2025. Multi-objective land use optimization based on NSGA-II and PLUS models: balancing economic development and carbon neutrality goals. Land 14 (8), 1585. https://doi.org/10.3390/land14081585.

Han, H., 2025. Challenges of reproducible AI in biomedical data science. BMC Med. Genom. 18 (S1). https://doi.org/10.1186/s12920-024-02072-6.

Harris, L., McDonagh, E.M., Zhang, X., Fawcett, K., Foreman, A., Daneck, P., Sergouniotis, P.I., Parkinson, H., Mazzarotto, F., Inouye, M., Hollox, E. J., Birney, E., Fitzgerald, T., 2024. Genome-wide association testing beyond SNPs. Nat. Rev. Genet. 26 (3), 156–170. https://doi.org/10.1038/s41576-024-00778-y.

Hooper, K., Lunn, S., 2024. A scoping review of transparency and explainability in AI ethics guidelines. Int. FLAIRS Conf. Proc. 37. https://doi.org/10.32473/flairs.37.1.135594.

Huikku, J., Karjalainen, J., Seppälä, T., 2025. Strategic factors and the use of risk analysis techniques in strategic investment decisions. Account. Bus. Res. 1–29. https://doi.org/10.1080/00014788.2025.2486753.

Husien, H.M., Saleh, A.A., Hassanine, N.N.A.M., Rashad, A.M.A., Sharaby, M.A., Mohamed, A.Z., Abdelhalim, H., Hafez, E.E., Essa, M.O.A., Adam, S.Y., Chen, N., Wang, M., 2024. The evolution and role of molecular tools in measuring diversity and genomic selection in livestock populations (traditional and up-to-date insights): a comprehensive exploration. Vet. Sci. 11 (12). https://doi.org/10.3390/vetsci11120627.

Javed, H., El-Sappagh, S., Abuhmed, T., 2024. Robustness in deep learning models for medical diagnostics: security and adversarial challenges towards robust AI applications. Artif. Intell. Rev. 58 (1). https://doi.org/10.1007/s10462-024-11005-9.

Jayaram, D.D., 2025. AI-augmented decision making: a framework for enterprise workflow transformation. Int. J. Sci. Res. Comput. Sci. Eng. Inform. Technol. 11 (1), 1950–1959. https://doi.org/10.32628/cseit251112184.

Jimenez-Gutierrez, D.M., Falkouskaya, Y., Hernandez-Ramos, J.L., Anagnostopoulos, A., Chatzigiannakis, I., Vitaletti, A., 2025. On the security and privacy of federated learning: a survey with attacks, defenses, frameworks, applications, and future directions. https://doi.org/10.48550/arXiv.2508.13730.

Kampezidou, S.I., Tikayat Ray, A., Bhat, A.P., Pinon Fischer, O.J., Mavris, D.N., 2024. Fundamental components and principles of supervised machine learning workflows with numerical and categorical data. Eng 5 (1), 384–416. https://doi.org/10.3390/eng5010021.

Kancharla, S.S.R., 2025. Data quality and integration: the AI-driven evolution. J. Comput. Sci. Technol. Stud. 7 (7), 882–888. https://doi.org/10.32996/jcsts.2025.7.7.97.

Kosaraju, D., 2022. Causal inference and AI: elevating decision-making in uncertain environments. Galore Int. J. Appl. Sci. Human. 6 (3), 179–184. https://doi.org/10.52403/gijash.20220720.

Kuthuru, A., 2025. Responsible AI in database systems: governance frameworks for generative AI data access. World J. Adv. Res. Rev. 26 (2), 3017–3026. https://doi.org/10.30574/wjarr.2025.26.2.1942.

Lakamp, D., Adams, S., Wells, J.E., Snelling, W.M., Kuehn, L.A., Fernando, S., Spangler, M.L., 2025. Predicting feed efficiency traits in beef cattle using genomics and rumen microbiomics. J. Anim. Sci. 103 (Suppl. 1), 18–19. https://doi.org/10.1093/jas/skaf102.020.

Lam, K., Lange, B., Blili-Hamelin, B., Davidovic, J., Brown, S., Hasan, A., 2024. A framework for assurance audits of algorithmic systems. https://doi.org/10.48550/arXiv.2401.14908.

Lan, B., 2024. Deep learning-based functional prediction model for genome sequence. In: Proceedings—2024 3rd International Conference on Data Analytics, Computing and Artificial Intelligence, ICDACAI 2024. Institute of Electrical and Electronics Engineers Inc, pp. 207–212. https://doi.org/10.1109/ICDACAI65086.2024.00045.

Lane, R.O., 2025. A comprehensive review of classifier probability calibration metrics. https://doi.org/10.48550/arXiv.2504.18278.

Lantzi, M.A., Papakonstantinou, E., Vlachakis, D., 2025. Bioinformatic analysis of complex in vitro fertilization data and predictive model design based on machine learning: the age paradox in reproductive health. Biology 14 (5), 556. https://doi.org/10.3390/biology14050556.

Lazaros, K., Koumadorakis, D.E., Vrahatis, A.G., Kotsiantis, S., 2024. Federated learning: navigating the landscape of collaborative intelligence. Electronics 13 (23), 4744. https://doi.org/10.3390/electronics13234744.

Le, L.P., Nguyen Thi, X.H., Nguyen, T., Riegler, M.A., Halvorsen, P., Nguyen, B.T., 2024. Missing data imputation for noisy time-series data and applications in healthcare. https://doi.org/10.48550/arXiv.2412.11164.

Lehmann, B., Bräuninger, L., Cho, Y., Falck, F., Jayadeva, S., Katell, M., Nguyen, T., Perini, A., Tallman, S., Mackintosh, M., Silver, M., Kuchenbäcker, K., Leslie, D., Chatterjee, N., Holmes, C., 2025. Methodological opportunities in genomic data analysis to advance health equity. Nat. Rev. Genet. 26 (9), 635–649. https://doi.org/10.1038/s41576-025-00839-w.

Li, D., Huang, Y., Li, M., Zhou, T., Zhang, X., Liu, H., 2025. Generative Models for Synthetic Data: Transforming Data Mining in the GenAI Era. https://doi.org/10.48550/arXiv.2508.19570.

Liao, X.F., Zhao, W.J., Jin, H., Yao, P.C., Huang, Y., Wang, Q.G., Zhao, J., Zheng, L., Zhang, Y., Shao, Z.Y., 2024. Towards high-performance graph processing: from a hardware/software co-design perspective. J. Comput. Sci. Technol. 39 (2), 245–266. https://doi.org/10.1007/s11390-024-4150-0.

Lindsey, L.A.M., Pershing, N.L., Habib, A., Dufault-Thompson, K., Stephens, W.Z., Blaschke, A.J., Jiang, X., Sundar, H., 2025. The impact of tokenizer selection in genomic language models. Bioinformatics 41 (9). https://doi.org/10.1093/bioinformatics/btaf456.

Liu, Z., Wang, N., Su, Y., Long, Q., Peng, Y., Shangguan, L., Zhang, F., Cao, S., Wang, X., Ge, M., Xue, H., Ma, Z., Liu, W., Xu, X., Li, C., Cao, X., Ahmad, B., Su, X., Liu, Y., Huang, G., Du, M., Liu, Z., Gan, Y., Sun, L., Fan, X., Zhang, C., Zhong, H., Leng, X., Ren, Y., Dong, T., Pei, D., Wu, X., Jin, Z., Wang, Y., Liu, C., Chen, J., Gaut, B., Huang, S., Fang, J., Xiao, H., Zhou, Y., 2024. Grapevine pangenome facilitates trait genetics and genomic breeding. Nat. Genet. 56 (12), 2804–2814. https://doi.org/10.1038/s41588-024-01967-5.

Mai, J., Gao, C., Bao, J., 2025. Domain generalization through data augmentation: a survey of methods, applications, and challenges. Mathematics 13 (5), 824. https://doi.org/10.3390/math13050824.

Malla, A.M., Banka, A.A., 2025. Comprehensive review of graph neural networks: challenges, classification, architectures, applications, and potential utility in bioinformatics. Expert Syst. 42 (8). https://doi.org/10.1111/exsy.70091.

Mandayam, R., 2025. AI and personal data: the art of balancing privacy. Int. J. Sci. Res. Eng. Manag. 09 (03), 1–9. https://doi.org/10.55041/ijsrem42363.

Mannapur, S.B., 2025. Understanding data drift and concept drift in machine learning systems. Int. J. Sci. Res. Comput. Sci. Eng. Inform. Technol. 11 (1), 318–330. https://doi.org/10.32628/cseit25111239.

Mansilla, L., Echeveste, R., Gonzalez, C., Milone, D.H., Ferrante, E., 2025. BM-CL: bias mitigation through the lens of continual learning (Version 1). https://doi.org/10.48550/ARXIV.2509.01730.

Mao, Z., 2025. Lightweight network architecture optimization of deep learning models. Appl. Comput. Eng. 175 (1), 169–176. https://doi.org/10.54254/2755-2721/2025.ast26509.

Mardinoglu, A., Turkez, H., Shong, M., Srinivasulu, V.P., Nielsen, J., Palsson, B.O., Hood, L., Uhlen, M., 2025. Longitudinal big biological data in the AI era. Mol. Syst. Biol. 21 (9), 1147–1165. https://doi.org/10.1038/s44320-025-00134-0.

Matthay, E.C., Neill, D.B., Titus, A.R., Desai, S., Troxel, A.B., Cerdá, M., Díaz, I., Michele, S., Thorpe, L.E., 2025. Integrating artificial intelligence into causal research in epidemiology. Curr. Epidemiol. Rep. 12 (1). https://doi.org/10.1007/s40471-025-00359-5.

Mbebi, A.J., Mercado, F., Hobby, D., Tong, H., Nikoloski, Z., 2025. Advances in multi-trait genomic prediction approaches: classification, comparative analysis, and perspectives. Briefings Bioinf. 26 (3). https://doi.org/10.1093/bib/bbaf211.

McGrath, M.J., Duenser, A., Lacey, J., Paris, C., 2025. Collaborative human-AI trust (CHAI-T): a process framework for active management of trust in human-AI collaboration. Comput. Hum. Behav. Artif. Hum. 6, 100200. https://doi.org/10.1016/j.chbah.2025.100200.

Mishra, V.H., Bankar, N.J., Sani, T., Noman, O., 2025. Microbial metabolism: gut microbiota's role in host genetics and metabolic health. J. Cell. Biotechnol. 11 (1), 31–34. https://doi.org/10.1177/23523689251334037.

Mittal, S., Thakral, K., Singh, R., Vatsa, M., Glaser, T., Ferrer, C.C., Hassner, T., 2024. On responsible machine learning datasets emphasizing fairness, privacy and regulatory norms with examples in biometrics and healthcare. Nat. Mach. Intell. 6 (8), 936–949. https://doi.org/10.1038/s42256-024-00874-y.

Monteiro, D., Mavinkurve, I., Kambli, P., Surve, S., 2024. Federated learning for privacy-preserving machine learning: decentralized model training with enhanced data security. Int. J. Res. Appl. Sci. Eng. Technol. 12 (11), 355–361. https://doi.org/10.22214/ijraset.2024.65062.

Moškon, M., Režen, T., 2024. Integration and analysis of omics data using genome-scale metabolic models. Metabolites 14 (11), 595. https://doi.org/10.3390/metabo14110595.

Moysis, L., Iliadis, L.A., Vergos, G., Sotiroudis, S.P., Boursianis, A.D., Papatheodorou, A., Kokkinidis, K.I.D., Abdul Matin, M., Sarigiannidis, P., Siniosoglou, I., Argyriou, V., Goudos, S.K., 2025. Artificial intelligence-empowered embryo selection for IVF applications: a methodological review. Mach. Learn. Knowl. Extract. 7 (2). https://doi.org/10.3390/make7020056.

Naik, M., 2025. AI-driven climate modelling and forecasting enhancing environmental resilience through predictive analytics. Int. J. Environ. Sci. 1762–1768. https://doi.org/10.64252/0hqedz44.

Nawaz, U., Zaheer, M.Z., Khan, F.S., Cholakkal, H., Khan, S., Anwer, R.M., 2025. AI in agriculture: a survey of deep learning techniques for crops, fisheries and livestock. https://doi.org/10.48550/arXiv.2507.22101.

Nawaz Khan, M., Ahmad, I., 2025. Harnessing digital twins: advancing virtual replicas for optimized system performance and sustainable innovation. Babylonian J. Mech. Eng. 2025, 18–33. https://doi.org/10.58496/bjme/2025/002.

Neethirajan, S., 2024. Metaverse for enhancing animal welfare—leveraging sensor technology and ethical considerations. J. Emerg. Comput. Technol. 4 (1), 6–14. https://doi.org/10.57020/ject.1460995.

Nyaga, D.M., Zaied, R.E., Silander, O.K., Black, M.A., O'Sullivan, J.M., 2025. Beyond single references: pangenome graphs and the future of genomic medicine. Front. Genet. 16. https://doi.org/10.3389/fgene.2025.1679660.

Orovou, E., Tzimourta, K.D., Tzitiridou-Chatzopoulou, M., Kakatosi, A., Sarantaki, A., 2025. Artificial intelligence in assisted reproductive technology: a new era in fertility treatment. Cureus 17 (4). https://doi.org/10.7759/cureus.81568.

Ortega-Bolaños, R., Bernal-Salcedo, J., Germán Ortiz, M., Galeano Sarmiento, J., Ruz, G.A., Tabares-Soto, R., 2024. Applying the ethics of AI: a systematic review of tools for developing and assessing AI-based systems. Artif. Intell. Rev. 57 (5). https://doi.org/10.1007/s10462-024-10740-3.

Ozdemir, O.B., Chen, R., Li, R., 2024. A Deep Ensemble Encoder Network Method for Improved Polygenic Risk Score Prediction. https://doi.org/10.1101/2024.07.31.24311311.

Pan, S., Liu, Z., Han, Y., Zhang, D., Zhao, X., Li, J., Wang, K., 2024. Using the Pearson's correlation coefficient as the sole metric to measure the accuracy of quantitative trait prediction: is it sufficient? Front. Plant Sci. 15. https://doi.org/10.3389/fpls.2024.1480463.

Panda, C., Chaturvedi, R., Mukherjee, A., Kaur, N., Goyal, R., Rodrigues, A., 2025. Learning tax-aware allocation: an AI-infused framework for predictive and prescriptive mutual fund strategy. Int. J. Environ. Sci. 3617–3623. https://doi.org/10.64252/sn4n7474.

Pasupuleti, M.K., 2025. Neural Computation and Learning Theory: Expressivity, Dynamics, and Biologically Inspired AI. https://doi.org/10.62311/nesx/rriv425.

Poinsot, A., Leite, A., Chesneau, N., Sébag, M., Schoenauer, M., 2024. Learning Structural Causal Models through Deep Generative Models: Methods, Guarantees, and Challenges. https://doi.org/10.48550/arXiv.2405.05025.

Pojer, R., Passerini, A., Larsen, K.G., Jaeger, M., 2025. A Neuro-Symbolic Approach for Probabilistic Reasoning on Graph Data. https://doi.org/10.48550/arXiv.2507.21873.

Pook, T., Hassanpour, A., Niehoff, T.A.M., Calus, M.P.L., 2025. Strategies to improve selection compared to selection based on estimated breeding values. https://doi.org/10.1101/2025.01.09.632197.

Ramanan, V., Vinod, R., Williams, C., Ramachandran, S., Venkatasubramanian, S., 2025. Principles and Policy Recommendations for Comprehensive Genetic Data Governance. https://doi.org/10.48550/arXiv.2502.09716.

Rao, A.K., 2025. Big data in genomics: managing and analyzing large-scale clinical and proteomic data. Int. J. Res. Appl. Sci. Eng. Technol. 13 (3), 3283–3291. https://doi.org/10.22214/ijraset.2025.68053.

Rijal, K., Holmes, C.M., Petti, S., Reddy, G., Desai, M.M., Mehta, P., 2025. Inferring genotype-phenotype maps using attention models. https://doi.org/10.48550/arXiv.2504.10388.

Sagar, G.L., Roshni Akarshana, H., Chakravardhan, R., 2025. Revolutionizing genetic manipulation: artificial intelligence-powered precision in CRISPR-Cas9 gene editing. In: ICDT 2025—3rd International Conference on Disruptive Technologies. Institute of Electrical and Electronics Engineers Inc., pp. 1549–1554. https://doi.org/10.1109/ICDT63985.2025.10986581

Sahragard, R., Arabfard, M., Najafi, A., 2025. Predicting host-pathogen interactions with machine learning algorithms: a scoping review. Infect. Genet. Evol. 130, 105751. https://doi.org/10.1016/j.meegid.2025.105751.

Salmè, M., Tronchin, L., Sicilia, R., Soda, P., Guarrasi, V., 2025. Beyond the generative learning trilemma: generative model assessment in data scarcity domains. SSRN. https://doi.org/10.2139/ssrn.5196687.

Sedeeq, F.S., Arman, P., 2025. Human-centric AI governance: an adaptive public international law framework for ethical and inclusive AI regulation in public health. J. Law Med. Ethics 53 (4), 563–574. https://doi.org/10.1017/jme.2025.10175.

Sepulveda, B.J., González-Recio, O., Chamberlain, A.J., Khansefid, M., Cocks, B.G., Wang, J., Prowse-Wilkins, C.P., Marett, L.C., Williams, S.R.O., Jacobs, J.L., García-Rodríguez, A., Jiménez-Montero, J.A., Pryce, J.E., 2025. Rumen metagenome as a genomic selection target to reduce enteric methane emissions. J. Dairy Sci. 108 (8), 8619–8636. https://doi.org/10.3168/jds.2024-25436.

Shah, M.R.V., 2025. Quantum-edge synergy: a novel framework for real-time IoT analytics beyond cloud and edge computing. J. Inform. Syst. Eng. Manag. 10 (37s), 925–935. https://doi.org/10.52783/jisem.v10i37s.6746.

Sharma, D., Jabin, S., 2025. Review of large language models for genomic data and medical text. Int. J. Bioinform. Intell. Comput. 10–27. https://doi.org/10.61797/ijbic.v4i2.483.

Shi, L., 2025. Global perspectives on AI competence development: analyzing national AI strategies in education and workforce policies. Hum. Resour. Dev. Rev. 24 (4). https://doi.org/10.1177/15344843251332360.

Silvano, C., Ielmini, D., Ferrandi, F., Fiorin, L., Curzel, S., Benini, L., Conti, F., Garofalo, A., Zambelli, C., Calore, E., Schifano, S., Palesi, M., Ascia, G., Patti, D., Petra, N., De Caro, D., Lavagno, L., Urso, T., Cardellini, V., Cardarilli, G.C., Birke, R., Perri, S., 2025. A survey on deep learning hardware accelerators for heterogeneous HPC platforms. ACM Comput. Surv. 57 (11), 1–39. https://doi.org/10.1145/3729215.

Sinaci, A.A., Gencturk, M., Alvarez-Romero, C., Laleci Erturkmen, G.B., Martinez-Garcia, A., Escalona-Cuaresma, M.J., Parra-Calderon, C.L., 2024. Privacy-preserving federated machine learning on FAIR health data: a real-world application. Comput. Struct. Biotechnol. J. 24, 136–145. https://doi.org/10.1016/j.csbj.2024.02.014.

Soundararajan, B., 2025. Developing new AI model compression techniques. Int. J. Sci. Res. Eng. Manag. 09 (03), 1–8. https://doi.org/10.55041/ijsrem43474.

Souza, R., Caino-Lores, S., Coletti, M., Skluzacek, T.J., Costan, A., Suter, F., Mattoso, M., Da Silva, R.F., 2024. Workflow provenance in the computing continuum for responsible, trustworthy, and energy-efficient AI. In: Proceedings—2024 IEEE 20th International Conference on E-Science, E-Science 2024. Institute of Electrical and Electronics Engineers Inc. https://doi.org/10.1109/e-Science62913.2024.10678731.

Sucharita, S., Shaw, K., 2025. Bias detection and mitigation in AI: a framework for fair and inclusive machine learning models. Int. J. Sci. Res. Eng. Manag. 09 (07), 1–9. https://doi.org/10.55041/ijsrem51300.

Taddese, A.A., Addis, A.C., Tam, B.T., 2025. Data stewardship and curation practices in AI-based genomics and automated microscopy image analysis for high-throughput screening studies: promoting robust and ethical AI applications. Hum. Genom. 19 (1). https://doi.org/10.1186/s40246-025-00716-x.

Tariq, B., Ashraf, M.R., Rashid, U., 2025. Ethical imperatives in AI design: a comprehensive framework for risk mitigation and responsible innovation. Ubiquitous Technol. J. 1 (2), 61–73. https://doi.org/10.71346/utj.v1i2.23.

Tedeschi, L.O., Guarnido Lopez, P., Menendez III, H.M., Seo, S., 2025. Advancing precision livestock farming: integrating artificial intelligence and emerging technologies for sustainable livestock management. Anim. Biosci. https://doi.org/10.5713/ab.25.0289.

Toderas, M., 2025. Artificial intelligence for sustainability: a systematic review and critical analysis of AI applications, challenges, and future directions. Sustainability 17 (17), 8049. https://doi.org/10.3390/su17178049.

Trotsyuk, A.A., Waeiss, Q., Bhatia, R.T., Aponte, B.J., Heffernan, I.M.L., Madgavkar, D., Felder, R.M., Lehmann, L.S., Palmer, M.J., Greely, H., Wald, R., Goetz, L., Trengove, M., Vandersluis, R., Lin, H., Cho, M.K., Altman, R.B., Endy, D., Relman, D.A., Levi, M., Satz, D., Magnus, D., 2024. Toward a framework for risk mitigation of potential misuse of artificial intelligence in biomedical research. Nat. Mach. Intell. 6 (12), 1435–1442. https://doi.org/10.1038/s42256-024-00926-3.

Ullah, K., Waheed, A., 2025. Review: genetic improvement in livestock: a journey from conventional breeding to genomic precision. Vet. Biomed. Clin. J. 7 (1), 30–44. https://doi.org/10.21776/ub.vetbioclinj.2025.007.01.3.

Upadhyay, R., Phlypo, R., Saini, R., Liwicki, M., 2024. Sharing to learn and learning to share; fitting together meta, multi-task, and transfer learning: a meta review. IEEE Access 12, 148553–148576. https://doi.org/10.1109/access.2024.3478805.

Van Hilten, A., Katz, S., Saccenti, E., Niessen, W.J., Roshchupkin, G.V., 2024. Designing interpretable deep learning applications for functional genomics: a quantitative analysis. Briefings Bioinf. 25 (5). https://doi.org/10.1093/bib/bbae449.

Vlaicu, P.A., Gras, M.A., Untea, A.E., Lefter, N.A., Rotar, M.C., 2024. Advancing livestock technology: intelligent systemization for enhanced productivity, welfare, and sustainability. AgriEngineering 6 (2), 1479–1496. https://doi.org/10.3390/agriengineering6020084.

Vogt, L., Strömert, P., Matentzoglu, N., Karam, N., Konrad, M., Prinz, M., Baum, R., 2025. Suggestions for extending the FAIR Principles based on a linguistic perspective on semantic interoperability. Sci. Data 12 (1). https://doi.org/10.1038/s41597-025-05011-x.

Vohra, D.S., 2024. Big data analytics in bioinformatics: transforming genomics and proteomics for advanced biological insights. In: International Conference on Big Data Analytics in Bioinformatics, DABCon 2024. Institute of Electrical and Electronics Engineers Inc. https://doi.org/10.1109/DABCon63472.2024.10919359.

Walha, N., Gruber, S.G., Decker, T., Yang, Y., Javanmardi, A., Hüllermeier, E., Buettner, F., 2025. Fine-Grained Uncertainty Decomposition in Large Language Models: A Spectral Approach (Version 1). https://doi.org/10.48550/ARXIV.2509.22272.

Wang, J., 2025. Machine learning-based risk assessment of gene editing: degeneration during gene editing possibility of target effect. Theor. Nat. Sci. 82 (1), 33–38. https://doi.org/10.54254/2753-8818/2024.ka20158.

Wei, Y.-M., Liu, L.-J., Zhang, S.-X., Liu, J.-X., Zhao, L., Cong, R.-G., Liang, Q.-M., Han, T., Yuan, X.-C., Chen, Y.-M., Hu, Y.-X., Xu, Z.-Q., Chen, H.-B., Xiao, Y.-X., Wang, P., Yan, S.-Y., Huang, X.-L., Wang, T.-Y., Li, X.-Q., Cui, R., 2025. Global Scenario Reference Datasets for Climate Change Integrated Assessment with Machine Learning. Copernicus GmbH. https://doi.org/10.5194/essd-2025-299.

Wu, Y., Xie, L., 2024. AI-driven multi-omics integration for multi-scale predictive modeling of causal genotype-environment-phenotype relationships. https://doi.org/10.48550/arXiv.2407.06405.

Xing, L., Wang, H., Yang, Y., Mei, B., 2025. An integrative multi-omics framework for genomic prediction in livestock using Bayesian deep neural networks. Front. Epigenet. Epigenom. https://doi.org/10.3389/freae.2025.1664757.

Zhang, H., Shi, Y., Wang, Y., Yang, X., Li, K., Im, S.-K., Han, Y., 2025. Biological sequence representation methods and recent advances: a review. Biology 14 (9), 1137. https://doi.org/10.3390/biology14091137.

Chapter 5

Artificial intelligence integration in livestock transcriptomics

Artificial intelligence (AI) is positioned in this chapter as an initiating layer that will change the transcriptomics of livestock in a descriptive snapshot of gene activity into a dynamic, predictive, and decision-oriented field. It introduces the transcriptome as a responsive molecular interface at which genotype, environment, management, physiology, and health status intersect, and suggests that AI techniques have become necessary to unravel the context-dependent structure that is fluid and that is the interface. The story begins with a statement of the conceptual gap between the raw RNA profiles and the operable intelligence, defining the transition to the model focused on the lifecycle, in which the process of sample acquisition, the modeled computational representation, and the interface of intervention planning in breeding and management pipelines are connected.

Principal attention is paid to the data skeleton according to which plausible AI systems operate. The chapter focuses on strict experimental design, longitudinal and multicontext sampling, new modalities of sequencing, metadata standardization, and harmonization practices that can open up interoperability and reuse. Then it follows the process of going through raw reads and analysis-ready matrices, starting with quality control, normalization, mitigating batch effects, imputation, and feature engineering, highlighting how preprocessing decisions are recoded (as of) model behavior. On this foundation, a paradigm survey of modeling extends to encompass classical baselines, deep architecture, representation and generative learning, multimodal fusion, causal and uncertainty-aware frameworks, and adaptive design strategies as a toolkit to extract robust biological signals out of the high-dimensional, noisy, and heterogeneous realms of expression.

Themes, such as health, welfare, efficiency in production, and genetic enhancement, are tailored to the application domains, with transcriptomics as an intermediary layer enhancing predictive values of breeding, precision in selection, and proactive intervention of resilience and resource management. Integration sections demonstrate how transcript data are brought into agreement with genomics, epigenomics, proteomics, metabolomics, and information about the microbiome to provide multiscale insight and how cellular regulation is brought into agreement with herd-level success. Sharing ethical, legal, and socioeconomic issues is done in cross actions to consider data ownership, privacy, equity, transparency, animal welfare, and regulation alignment, and, therefore, its governance structures are based on a premise where accountability, explainability, and inclusivity are inherent and not a back-seat addition.

The chapter addresses technical and operational issues such as data heterogeneity, sparsity, domain shifts, computational costs, additional interpretive tensions, integration complexities, and lifecycles. It also provides monitoring and validation practices to maintain model fidelity in the face of changing biological and production situations. Workforce development emerges as a strategic enabler, advocating a combination of molecular literacy and computational proficiency, collaborative cultures, and continuous education. The concept of sustainability is considered in both molecular and digital terms: the former is examined through transcript indicators of adaptation and environmental impact, while the latter is assessed through energy-aware modeling and infrastructure choices.

Looking forward, the chapter provides an overview of emergent trends such as large-scale foundation models, multimodal single-cell and spatial integration, real-time edge analytics, gene editing guidance, privacy-sensitive synthetic data, neuro-symbolic and causally explicit reasoning, interoperable knowledge graph ecosystems, and exploratory hardware horizons. It ends with a list of future suggestions that place standardization, ethical foresight, diversity of biological representation, multiscale model coupling, rigorous validation, responsible yet open innovation, sustainable computation, and global capacity building as the foundation of resilient, ethical, and efficient livestock systems, based on AI-empowered transcriptomics.

Artificial Intelligence in Livestock Omics. https://doi.org/10.1016/B978-0-443-45663-3.00010-5

5.1 Introduction to AI in livestock transcriptomics

Livestock transcriptomics may be interpreted as the methodical description of the dynamic RNA regulation portrayal among tissues, cell types, physiological conditions, progression phases, and control or environmental circumstances in domesticated fauna, with a two-fold aim of clarifying the regulatory action and applying those findings to better health, well-being, productivity, resistance, and viability results. The transcriptome, as opposed to constituting just a static genomic variance, is a temporally and contextually sensitive molecular interface, whereby genotype interacts with nutrition, housing, microbial exposures, climatic stressors, health difficulties, and management interventions. It covers breadth, therefore, between foundational level quantification of genes via splicing, allele-specific expression, noncoding RNA layers, and emergent single-cell and resistively resolved design to systems-level abstractions such as coexpression modules, regulatory networks, and latent embeddings that synthesize regulatory state. In this context, livestock transcriptomics is both a discovery science studying functional regulation and an enabling technology that catalyzes translation between breeding, health management, and resource stewardship action decisions and molecular biology (Zambon et al., 2025).

The biological and production relevance of this domain lies in the distinct ability of transcriptional readouts to record proximate regulatory responses that get antecedent or concomitant fluxions to phenotypic expression, providing earlier, more detailed, and mechanistically grounded signals than traditional performance or health protocols. The expression profiles communicate multidimensional data regarding genetic setup, epigenetic training, metabolic activity, immune preparedness, stress adaptation, and developmental pathway. Through questioning patterns of differentiation, regulatory consistency, and network perturbation, transcriptomics provides intermediate phenotypes, which narrow down the explanation of causal activity as well as separate complex, polygenic/environmentally modulated qualities. Within a production context where efficiency, robustness, reproductive success, quality of products, and a lesser environmental impact are sought across increasingly elevated ethical and sustainability demands, the ability to imagine the underlying regulatory state enhances refined control and selective improvement measures. Further, biomarkers obtained through transcriptomes increase the content of phenotypic differences beyond the conventional measures of measurement, increasing the viewable characteristic landscape that guides selection indices, health surveillance, or adaptive control intervention development (Kamble et al., 2025; Zeitlinger et al., 2024).

Traditionally, livestock transcriptomics developed out of low-throughput expression assays with a gene-centered orientation to more and more dimensional platforms until eventual comprehensive sequencing-based profiling. The initial descriptive cataloging of transcripts in individual conditions was replaced by comparative analyses, followed by integrative statistical modeling of coexpression and activity of pathways. This trend was increased in terms of sequencing depth, read accuracy, versatility in library preparation, and computational alignment and counting pipelines. It has since shifted the field of retrospective differential expression contrasts to predictive and prescriptive modeling paradigms where streams of expression data are fed into machine learning models predictive of phenotype, are predictive of environmental variability because of resilience, and offer actions to optimization or selection problems. AI is, therefore, not a window dressing, but a structural point of demarcation that allows one to cross the post hoc bio description boundary and make forward-looking, adaptive decisions based on mechanistic signals (Aburub et al., 2025).

In the larger livestock omics spectrum, the transcriptome exists as an intermediate that is both affected upstream by genomic, structural, and epigenomic variation and downstream by effects on proteomic composition, metabolic abundances, parameters of physiological performance, and sensor-derived data of behavior or environmental interaction. This intermediate position makes transcriptomics an integrative interface to multiscale modeling and causal chaining and is able to work out the differences between potential outcomes on the basis of genotype and actual phenotypes under conditions of variable management. Congruence with genomic selection platforms enables expression-restrained optimism of trait plans, whereas congruence with epigenomic maps puts into perspective regulatory adaptability. When combined with proteomic and metabolomic data layers, fusion of diverse layers can increase attributive functionality, and integration of microbiome and environmental telemetry data can position host regulatory state within ecological and management kidneys. Proper positioning of transcriptomics means, in part, acknowledging its dual ability to provide predictive features and act as a relational layer to reflexivize both semantics and time dynamics of the various omics and phenomic modalities (Nasir et al., 2024).

The motivation of using AI methods in this space is dictated by data-specific properties inherent to the data: high data dimensionality (compared to sample size), widespread sparsity or zero inflation in some modalities, the presence of nonlinear and hierarchical regulation influence, confounding batch and environmental noise, expression architecture-specific to a breed or line, and time- or space-varying. Traditionally, linear or univariate tests do not utilize the latent structure efficiently and have trouble with the explosion of interactions, as well as have no methods of uncertainty quantification or causal exploration under distributional change. The AI methodologies, including representation learning, deep architectures, probabilistic graphical formulations, attention-based sequence models, graph neural methods, and hybrid symbolic models, provide increased abilities to reduce dimension and still ensure functional signal, disentangle variation, predict regulatory network topology, characterize temporal progression, understand predictive uncertainty, and transfer learned representations across breeds/environment or even related traits. Moreover, generative and simulation-oriented models allow the ability to augment data and perform counterfactual reasoning as well as exploration of the impact of interventions, whereas interpretability advances aim to explain in biologically meaningful terms feature attributions as well as pathway levels of relevance (Pasupuleti, 2025; Ur-Rahaman and Sudheer, 2025).

The conceptual model of AI-enabled livestock transcriptomics can be defined as a flow that runs between the acquisition, curation, transformation, modeling, interpretation, deployment, and feedback. Acquisition is the design of experiments that is biologically replicated, mindful, longitudinal in structure, or at least in a multifactorial format, with tissue and cell type resolution, and controlled manipulation to stabilize preanalytical variables. Curation entails strict standardization of metadata, ontological annotation, and integrity validation so as to achieve interoperability and reusable datasets. Transform provides sequence quality control, alignment or pseudo-alignment, quantification, normalization, mitigation of batch effects, and noise modeling, as well as feature engineering, such as aggregated module construction, pathway scores, or learned embeddings. Modeling layers convert the transformed data into predictive or generative functions optimized to perform well, with calibration and generalizability, including regularization, transfer learning, or multitask structures where needed. Interpretation is used to transform internal representations of the models to biologically and operationally coherent stories, including attribution, enrichment, and network reconstruction, to create transparency and trust. In breeding models are deployed, a decision ecosystem is placed, indicating priorities, health, resources, or welfare via models, governance, drift monitoring, or lineage traceability. The postdeployment outcomes, experimental validation results, and changes in distributions of feedback loops are stored to recalibrate the models, add more training corpora, and optimize the design assumptions, and this creates a vicious cycle that brings empirical biology into the cycle with adaptive algorithmic evolution (Aburub et al., 2025; Abdulla et al., 2024).

With these components consolidated, the introductory part presents that the intermediation of livestock transcriptomics and AI represents a paradigm shift area that transcends the nonadaptive cataloging toward an adaptive and integrative infrastructure that is ethically conscious in iterating the process of molecularly informed decision-making. The pledge does not lie only in approach specificity and doggedness of quality of data or methodological stability, biologic interpretiveness, stewardship prudence, and translation adherence to production, welfare, and sustainability aims. An already existent implementation vision preempts equitable access, transparency, exploiting a wide variety of genetic resources and production systems, and ongoing capacity building, such that AI-empowered transcriptomics enhances scientific knowledge and provides concrete, responsible benefits throughout the livestock sector.

5.2 Data foundations for AI-enabled transcriptomics

The quality, depth, representativeness, and semantic richness of data used are a strength that limits or enables robust AI in livestock transcriptomics, instead of the ability of downstream algorithms to achieve higher sophistication. The data foundation thus goes beyond simple compilation of sequencing outputs to include rigorous experimental design, rigorous sample management, expansive and standardized metadata acquisition, multiscale biological coverage, principled ethical stewardship, and architectural preparedness of an integrative, longitudinal, and adaptive model. The central idea here is the appreciation that transcriptional measurements constitute context-sensitive outputs of genotype in interaction with the environment, management, physiological condition, microbiome, and time course, the outcome of which any omission, distortion, or unrecorded variability in acquisition is instilled into the preprocessing and modeling phase as biased inference, weakened generalizability, and poorly calibrated decision support (Caufield et al., 2025).

Experimental design creates the statistical identifiability of biological contrasts and predicts the limitations of high-dimensional modeling in low sample sizes. Balanced representation of breeds, lines, production systems, developmental stages, sexes, physiological or health states, and environmental regimes eliminates the possibility of a threat where the AI models spurious correlations regarding unmodeled structure. Close stratification and randomization between sequencing batches, library preparation lots, technicians, and temporal collection windows minimize confounding, which subsequently would need extreme efforts in corrective measures that could potentially depress genuine biological signals. Instead of univariate considerations of power, this space should have its power considerations in a more multivariate way, which foreshadows the interaction of sample size, expression variance structure, compression in feature space, and the capacity of a model class to avoid overfitting latent noise. Longitudinal and multicontext designs are both operationally challenging but provide both time continuity and cross-environment contrast, which is highly useful in disentangling transient suppression and oscillatory regulation baselines, and allow AI models to generalize both dynamic trajectories as opposed to snapshots (Victorseelan et al., 2025).

The handling and procurement of samples influence the RNA integrity, degradation artifacts, and compositional changes, particularly in tissues that have high endogenous nuclease activity or have high postmortem transcriptomic dynamics. Preanalytical metadata detailing the time of collection with respect to interventions, stabilization steps, storage time, freeze–thaw events, and extraction chemistry is not only necessary as quality control triage information but also in covariate modeling that separates the technical and biological components of variation. The additional sensitivity of increasing resolution by single-cell and spatial transcriptomic modalities implies more sensitivity to dissociation bias, cell type-selective lysis, and ambient RNA contamination, which requires additional metadata information regarding capture chemistry, sequencing depth targets, imaging parameters, and cellular annotation confidence to model the subsequent probabilistic structure. This scaling out to these modalities also requires a close match between bulk and cellular scale designs to allow multilevel integration without compositional confounding (Arora et al., 2025).

The other axis of foundation is the reference resources. The applicability of assemblies of genomes, the completeness of genome annotation of coding and noncoding transcripts, splice junction catalogs, and variant-sensitive transcript models directly affect the accuracy of quantification, especially in genetically diverse or admixed livestock populations, where the use of a single linear reference can produce mapping bias and allele-specific distortion. New representations of graphs and pangenomes provide a structural approach to reduce the bias in references and improve fair representation of the variation of expression between breeds. The versions of the annotation and open provenance are essential to ensure that reproducibility and interpretation are possible in case of recreating the historic datasets or combining the cohorts assembled using the work under diverse annotation standards (Nyaga et al., 2025).

Semantic interoperability is based on metadata architecture. Harmonization and machine-readable integration take place more quickly with the adoption of controlled vocabularies and ontology-based descriptors of traits, anatomical sites, cell types, developmental stages, management practices, environmental exposures, and health status. Ontology alignment also facilitates hierarchical abstraction, which facilitates multigranular modeling where AI systems can easily traverse between fine-grained cell state features and functional summaries of an organ or whole animal functions. Rich phenotypic metadata increases the label space of supervised or semisupervised learning, whereas detailed environmental and management descriptors allow domain adaptation, covariate adjustment, and causal disentanglement. Contextual framing, which can be achieved by the inclusion of microbiome compositional agglomers and nutritional or sensor-derived behavioral or microclimatic data, can render transcriptomic agglomers more intelligible as nodes in larger ecological and production networks (Hamdan et al., 2024).

Quality control pipelines put quality control metrics into practice, including read quality, mapping performance, duplication, contamination, library complexity, uniformity of uniform gene body coverage, and clustering of outlier samples. Clear boundaries, combined with soft flagging as opposed to indiscriminate top-cutting, make it possible to perform delicate decision-making and downstream modeling, which can bias or readjust marginal data, as opposed to top-cutting it. Design stage balancing enhances the prediction of batch effect, but systematic variation remains that requires prediction by explicit linear, empirical Bayesian, factor analytic, or deep latent variable models. The normalization strategy selection recurs with the downstream modeling hypothesis concerning distributional form, zero inflation, and variance mean dependence, which highlights the necessity to align preprocessing changes with the statistical characteristics anticipated by AI architectures of interest (Zhou et al., 2024).

Complete data and balance of classes issues (of certain breeds, of rare health conditions, or of extreme environmental conditions) plague livestock transcriptomic corpora. Missingness is often organized, which is associated with logistical constraints or performance results, and naive imputation carries with it synthetic bias. Guided imputation, probabilistic imputation, graph-based imputation, or latent embedding imputation methods can recover partially observed expression or phenotype fields without losing uncertainty quantification to encourage well-calibrated uncertainty estimates. Other complementary approaches include active learning-guided supplemental sampling, which iteratively reduces epistemic uncertainty by prioritizing the acquisition of the most informative samples. Generative modeling can have a beneficial effect on synthetic data generation, though only with a very rigorous validation process to prevent the amplification of training distribution artifacts and loss of lineage traceability (Benhamza et al., 2025).

Data foundations are associated with ethical and governance aspects. The legal and social considerations of trust overlap with ownership, permission to access, respect for privacy, and fair representation of various production systems. Federated and privacy-preserving analytic models enable the training of models collaboratively without concentrating the raw data, reducing the risk of exposure, and expanding diversity. Transparent data usage agreements, provenance logs, version-controlled processing workflows, and auditable lineage records promote accountability and reproducibility, thereby fostering stakeholder confidence in the outputs of downstream AI processes (Shu, 2025).

Scalability, sustainability, and adaptability are defined depending on infrastructure and lifecycle stewardship. The storage structures should support raw signal data, processed count matrices, derivations of embeddings, and rich metadata graphs and support low-latency access to repeated modeling. Modular integration of preprocessing, model training, evaluation, and deployment pipelines is facilitated by standardized file formats, schema validation, and application programming interfaces. Resource allocation plans that are energy and cost-conscious, compression methods that are specific to sparse expression tensors, and caching of the intermediate representations have been used to minimize environmental and economic costs without affecting the analytical fidelity. The process of continuous integration, preprocessing pipeline regressions using automated regression tests on benchmark data, and drift detection on baseline quality metrics are used to develop a dynamically growing backbone of data with adaptive innovation of methods as experimental methods and biological scale are increased (Maddipudi, 2024).

Together, these interwoven elements constitute a data foundation, which is not fixed but rather reflexive by incorporating feedback of model performance diagnostics, biological validation experiments, and changing production challenges. A deliberate, governance-inspired, and semantically informed model of data governance turns uncooked transcriptomic gauges into a robust base over which dependable, extensible, and biologically sound AI decision-supportive structures can be solidly formed and enhanced via reiterative assimilation. Foundational data modalities underpinning AI-enabled livestock transcriptomics are presented in Table 5.1.

5.3 Preprocessing, quality control, and normalization pipelines

Preprocessing belongs to the message veins between uncoded input with sequencing data and representations interpretable to analytically feignable derivatives, and defensibility sets the limit with respect to possible biological understanding and predictive integrity. The pipeline should not be viewed as a sequence of what numerous checklists or casters have accomplished, but should be viewed as an irreducibly complex system where decisions made in the early model can limit or invite modeling tactics in the later models, where decisions can alter the way the new learning models are interpreted and transferred to other settings. The fundamental idea of this system is the maintenance of natural biological structure and modeling, attenuating, or flagellating, in a systematic manner, technical artifacts, stochastic noise, and confounding factors of undesired variation without causing distortions to causal interpretability or inflated spurious correlations (Lee et al., 2025).

Rudimentary quality control starts with the inspection of the raw read integrity, complexity, and contamination reports, and then progresses to adapter and low-quality sequence processing with an explicit philosophy of minimum but adequate intervention to avoid the spread of insignificant errors. Decisions on trimming aggressiveness, ambiguous retention of base case depth, and treatment of polynucleotide stretches balance between downstream mapping bias and fragment length distributions and have effects on normalization assumptions. The combination of reference representation and alignment (or quasialignment) strategies dictates the sensitivity to splice junction identification, allele-specific expression, and isoform separation, which is especially high in genetically diverse livestock populations in which structural polymorphism may preclude consistent read positioning. Mapping variant-aware/graph-based versions will eliminate allele-biased mapping, but of course, can create more complexity when tracing and mapping to legacy linear coordinates, and additional effort is needed in maintaining provenance records to carefully document annotation version morphos to ensure compatibility with the past (Koptekin et al., 2025; Jain, 2025).

Quantification transitions align or pseudo-align fragments into matrices of gene, transcript, exon, junction, or custom feature counts or abundances. At this stage, evaluative metrics such as library complexity, distributional sparsity, and saturation plateaus inform filtering and normalization strategies. A key epistemic principle is that filtering criteria should relate to evidence of unreliable measurement rather than absolute expression magnitude. Otherwise, low-abundance transcripts that are pivotal for regulation may be disproportionately removed. Strategies that employ detectability thresholds based on per-feature prevalence across samples can balance the retention of rare but consistently measured elements with the removal of stochastic single-sample noise. Considering mitochondrial, ribosomal, or contamination-associated fractions in parallel helps to distinguish biologically meaningful shifts in metabolic or translational activity from technical enrichment due to degradation or preparation anomalies (Cole et al., 2025).

Normalization is used to systematically adjust the bias of the depth of sequencing, changes in compositions, and, in some instances, transcript length and attempts to preserve the relative differentiability of the biological signal. The diversity of livestock tissues, cell types, and experimental designs complicates the ability to scale simply due to the

TABLE 5.1 Foundational data modalities underpinning AI-enabled livestock transcriptomics.

Data type	Source	Relevance to AI-enabled transcriptomics	Key considerations
Raw RNA-seq reads	• On-farm tissue or blood biopsies • Commercial/academic sequencing cores • Public repositories (NCBI SRA, ENA, and FAANG) • Experimental challenge studies	• Enables end-to-end deep learning from raw signal • Supports adapter/error/artifact detection models • Variant-aware quantification (SNP/indel integration) • Domain adaptation across breeds/cohorts • Quality metrics feed sample weighting	• Read quality variance introduces noise • Depth heterogeneity biases representation • Library prep (polyA vs. rRNA) shifts transcript classes • Strandedness needed for orientation-aware models • Cost versus broader cohort diversity trade-off • Ethical/logistical constraints for invasive sampling
Processed expression matrices (counts/TPM/CPM)	• In-house standardized pipelines (STAR and RSEM, Salmon/Kallisto) • FAANG processed sets • LIMS exports • Consortium harmonized matrices	• Primary feature space for prediction and clustering • Input to embedding/autoencoder architectures • Rapid prototyping without raw reprocessing • Cross-species transfer via ortholog mapping • Integration hub for phenotypes and metadata	• Normalization choice affects comparability • Zero inflation and sparsity handling are critical • Residual batch effects mislead attribution • Gene set mismatch across pipelines • Versioned gene IDs needed to prevent drift
Single-cell/single-nucleus RNA-seq	• Dissociated tissues (immune, mammary, and muscle) • Nuclei preps (brain and adipose) • 10x/SMART-seq platforms • Early livestock cell atlases	• Cell-type deconvolution for bulk models • Train cell-state trajectory and pseudotime predictors • Weak supervision labels for bulk samples • Rare cell type discovery impacting traits • Marker validation for feature selection	• Dropout sparsity challenges modeling • Subjective cell-type annotation; ontology alignment needed • Platform chemistry differences hinder harmonization • High per-sample cost reduces breed diversity • Nucleus versus whole-cell capture bias
Spatial transcriptomics	• Fresh-frozen slides (muscle, liver, and placenta) • 10x Visium and emerging platforms • Histology-integrated datasets	• Supplies spatial priors for GNNs • Fusion of morphology (image) and expression • Local niche (immune/metabolic) inference • Enhances interpretation of bulk differential results	• Spot-level mixtures require deconvolution • Stain variability impacts alignment • Limited replicates risk of overfitting location artifacts • Slide-to-slide chemistry variability • Large image and matrix storage demands
Long-read isoform sequencing	• PacBio Iso-Seq/Nanopore targeted tissues • Breed-specific transcript diversity surveys • Public pilot Iso-Seq datasets • Custom capture panels	• Isoform-aware deep modeling (TSS/polyA usage) • Improves short-read quantification accuracy • Splice/isoform pattern prediction under conditions • Allele-specific isoform discovery for causal inference	• Higher raw error rates need polishing • Depth bias toward abundant transcripts • Chemistry updates shift error/length profiles • Computational cost of isoform clustering • Need stable isoform identifiers
Alternative splicing and junction counts	• STAR/HISAT2 junction outputs • rMATS/SUPPA events • Long-read validated junction catalogs	• Features for splicing QTL models • Inputs to splice prediction neural nets • Improves functional variant interpretation • Enables condition-specific splice graph GNNs	• Event class imbalance biases classifiers • Low support junction false positives • Incomplete annotations inflate novel calls • Cross-pipeline event definition variance • Need breed-specific splice references

Epigenomic accessibility and histone marks (ATAC, ChIP)	• Chromatin assays (tissue/immune cells) • FAANG epigenome tracks • In-house regulatory maps • Public ChIP-seq sets	• Regulatory priors for expression prediction • Guide feature selection (open chromatin genes) • Multimodal attention (sequence and accessibility) • Enhancer–gene linkage for causal graphs	• Sparse species coverage limits generalization • Batch differences (transposition efficiency) • Peak caller variability • Limited cell-type resolution masks specificity • Coordinate/version harmonization required
Genetic variants and haplotypes	• WGS panels • SNP arrays (Illumina/Affymetrix) • Breed association data • dbSNP/EVA repositories	• Genotype-informed expression (eQTL) modeling • Allele-specific expression and cis-regulation • Multimodal causal variant ranking • Haplotype embeddings aid breed transfer	• Array ascertainment bias versus WGS • Population structure confounds signals • Incomplete structural variant catalogs • Phasing errors degrade haplotype models • Privacy constraints on elite lines
Reference genome and annotation	• Ensembl/RefSeq releases • FAANG curated sets • Breed-specific assemblies • Lift-over mappings	• Stable feature indexing for pipelines • Mapping model outputs to interpretable entities • Cross-version transfer learning alignment • Incorporation of novel transcripts iteratively	• Version drift causes feature mismatch • Incomplete gene models (UTRs and lncRNAs) • Paralog/pseudogene ambiguity • Need controlled vocab for feature lineage • Assembly gaps bias regional analyses
Noncoding RNA catalogs	• miRBase, NONCODE • circRNA detection (CIRI, find_circ) • lncRNA assemblies (long-read) • Specialized ncRNA studies	• Expands regulatory feature space • Posttranscriptional control modeling (miRNA–mRNA) • ncRNA biomarker discovery (health/performance) • Multitask modeling (coding and noncoding)	• Fragmented annotations and redundancy • circRNA false positives (junction noise) • Low abundance complicates modeling • Poor conservation limits transferability • Naming inconsistency causes duplication
Phenotypic and production trait records	• On-farm performance DBs (milk, ADG, and FCR) • Breed associations • Herd management systems • Research trial measurements	• Targets for expression-to-trait prediction • Multilabel models for complex traits • Causal mediation (expression as intermediate) • Multitask architectures share representations	• Inconsistent measurement protocols • Temporal misalignment with sampling • Missingness and censoring bias • Manual entry errors • Commercial confidentiality
Environmental and management metadata	• Climate sensors (temperature and humidity) • Housing system logs • Season/parity/lactation stage records • Stressor exposure logs	• Context-aware environment × expression models • Domain adaptation via covariates • Robustness improvement through stratification • Disentangles gene–environment modulation	• Temporal granularity mismatch • Sensor calibration variance • Missing intervals cause leakage • Multicollinearity among factors • Need standardized vocabularies
Health and clinical/disease status records	• Veterinary EHRs • Lab diagnostics • Vaccination and treatment logs • Challenge trial outcomes	• Disease risk prediction from expression • Early warning classifiers (subclinical shifts) • Host–pathogen prognosis models • Survival/longitudinal modeling	• Heterogeneous diagnostics/test kits • Delayed or incomplete entries • Privacy and regulatory constraints • Class imbalance for rare diseases • Treatment confounding (antibiotics/vaccines)
Sensor and wearable physiology/behavior	• Accelerometers, rumination collars • Temperature/HR boluses • Camera-derived activity metrics • Real-time location systems	• Multimodal temporal models (behavior and expression) • Context for dynamic transcript changes • Anomaly detection linking physiology and omics • Supports causal mediation analyses	• Frequency mismatch versus transcript snapshots • Signal noise and device drift • Storage and bandwidth costs • Device downtime/missing segments • Vendor format heterogeneity

Continued

TABLE 5.1 Foundational data modalities underpinning AI-enabled livestock transcriptomics.—cont'd

Data type	Source	Relevance to AI-enabled transcriptomics	Key considerations
Dietary and nutritional composition data	• Feed lab analyses (NIR) • Ration formulation exports • Ingredient batch certificates • Experimental diet logs	• Diet–transcript interaction (nutrigenomics) • AI diet adjustment recommendations • Disentangles nutrient versus genetic regulation • Enhances methane/emission expression predictors	• Analytical precision variability • Mid-trial reformulations create hidden confounders • Missing micronutrient detail limits resolution • Undocumented ingredient substitutions • Proprietary formula secrecy
Breed, pedigree, and relationship matrices	• Breed registries • Pedigree software • SNP-based GRMs • Crossbred classification files	• Population structure correction layers • Breed-aware domain adaptation • Heritability-informed feature attribution • Transfer learning via relationship embeddings	• Mis-parentage/pedigree errors • Unequal representation biases models • Crossbred ambiguity in labeling • High-dim matrices raise memory cost • Privacy over elite genealogies
Proteomics (quantitative)	• LC-MS/MS experiments • Pilot proteome datasets • Targeted assays (ELISA) • Tissue-specific atlases	• Translation efficiency inference with transcripts • Multi-omics fusion boosts trait prediction • Validates pathway activation inferred from RNA • Detects transcript–protein discordance (posttranscriptional control)	• Limited sample overlap with transcriptomics • Missingness and dynamic range issues • Batch/instrument drift • Peptide-to-protein inference ambiguity • Higher cost restricts replication
Metabolomics (targeted/untargeted)	• GC-MS/LC-MS biofluids • NMR panels • Rumen fermentation assays • Reference metabolite libraries	• Downstream phenotype proxy features • Pathway-centric feature engineering • Causal modeling (metabolite mediators) • Multimodal health status classifiers	• Feature identification ambiguity (untargeted) • Batch drift and ion suppression artifacts • Rapid turnover, temporal instability • Sparse cross-study overlap • Normalization/transform choice impacts comparability
Microbiome profiles	• Rumen/gut/fecal sampling • 16S/shotgun sequencing labs • Public rumen microbiome datasets • Intervention trials	• Host–microbiome functional network integration • Feed efficiency/methane joint embeddings • Causal inference (microbiome → expression) • Resilience and health risk augmentation	• Compositional data constraints • High interindividual variability requires a large N • Contamination risks (low biomass) • Annotation inconsistency (taxa/functions) • Temporal instability versus single RNA time points
Pathogen genomics/Dual RNA-seq	• Infection challenge experiments • Field outbreak sequencing • Pathogen reference genomes • Dual host–pathogen RNA-seq studies	• Joint host–pathogen transcriptional modeling • Early infection detection • Virulence–host interaction networks • Intervention trajectory optimization	• Biosafety and ethical constraints limit sampling • Critical temporal sampling windows • Variable pathogen load affects signal • Mixed infections complicate labels • Pathogen reference quality variability
Imaging and histopathology	• H&E/immunostained slides • Digital slide scanners • Ultrasound/radiography archives • Automated lesion scoring	• Vision–transcript multimodal learning • Predict tissue remodeling and cell states • Weak labeling of spatial gene patterns • Explainability via linking ROIs to gene modules	• Stain variability and color normalization needs • Labor-intensive annotations • Large file size throughput challenges • Device/domain shift effects

			• Privacy (facility/identifiable markers)
Sampling and temporal metadata	• Time-of-day logs • Developmental stage records • Lactation/parity stage • Postfeeding or treatment intervals	• Time-aware recurrent/attention models • Separate temporal versus treatment effects • Circadian-adjusted normalization • Forecasting longitudinal expression trajectories	• Missing/coarse timestamps reduce resolution • Confounding events near sampling • Seasonal daylight variation shifts baselines • Irregular interval modeling difficulty • Need standardized temporal ontology
Batch/technical metadata	• Library prep kit versions • Sequencing lane/run IDs • RIN scores • Operator/lab identifiers • Flowcell chemistry versions	• Batch-effect correction (adversarial/conditional VAEs) • Dynamic quality-based sample weighting • Technical drift detection in continuous learning • Reproducibility and audit traceability	• Incomplete capture leaves hidden confounders • Overcorrection may remove biology • Rapid tech updates cause shifting distributions • Naming inconsistency hinders automation • Privacy about operator/lab data
Ontologies and knowledge graphs (GO, KEGG, trait, and disease)	• GO/KEGG pathway sets • Animal trait/disease ontologies • Veterinary term sets • Custom curated knowledge graphs	• Ontology-informed embeddings enhance interpretability • Pathway-level predictive modeling (set attention) • Constraint-based multitask learning linking genes to traits • Concept-level explainable AI versus raw gene lists	• Version drift alters membership • Incomplete livestock-specific coverage • Overlapping/redundant pathways complicate attribution • Mapping errors from outdated gene IDs • Governance needed for graph evolution

compositional methodology of empirical studies, based on lively statistical hubs or ratio-based transformations, which have strength when the small number of features that store lots of counts are skewed. In cases where the length of transcripts affects interpretive interests on isoform utilization, as opposed to overall expression of a gene, length-adjusted scales may provide parity in cross-feature analyses, but would need to be balanced with model structures that presuppose count scale distributions. Mean–variance relationships that would stabilize variance and regularized log-like transformations can be used to support distance-based clustering, latent embedding learning, and gradient-based optimization, but their use should be chosen with respect to distributional assumptions in the later modeling layers to prevent incompatibilities between the transformations that swell reconstruction loss or damp predictive calibration (Nixon et al., 2024).

The characterization and mitigation of batch effects is an overwhelming imperative with multisite, multiinstrument, and time-staggered livestock studies. The earlier experimental balance accounts for less retrospective correction, although latent structure is usually left over. Methods of correction include linear covariate regression, empirical Bayes frameworks of shrinkage, and matrix factorization of unwanted variation, as well as deep generative or adversarial representations that factor out both technical and biological sources of variation. The critical evaluation criterion goes further than visual homogenization to maintenance of authentic biological variations; the overcorrection will arise in reduced fold change, failure to deflate fold, or reduced downstream predictive results. Quantitative adjudication of adequacy of correction is supported by the incorporation of replicate or control samples, spike-in references, and orthogonal phenotypic anchors. Notably, workflows of AI that incorporate batch identifiers as explicit conditioning variables can be trained to factor (instead of erasing) technical structure, an alternative to previous aggressive removal (Yu et al., 2024b; John and Alicia, 2025).

Outlier detection is related to quality control and robust modeling. Multivariate diagnostics that utilize distance measures in latent spaces, influence measures, or probabilistic reconstruction errors can discriminate between samples high in global anomalies and those representing true extremes of biological spectra. When it comes to dealing with cases of this sort, policies based on handling them should instead promote graded weighting or isolation in model training as opposed to wholehearted exclusion to retain awareness of phenotypic tail cases that were of interest to resilience or stress modeling. Exogenous species and barcode swapping, as well as endogenous introgenous probe contamination in single-cell scenarios, require probabilistic deconvolution methods, which approximate the contamination rates and readjust the expression profile and propagate uncertainty. This subtle use avoids the unnecessary boost of confidence in corrected values, which may confuse feature attribution in the future (Mayrhofer et al., 2024).

The process of feature engineering starts in the preprocessing of features by processing the raw features into biologically meaningful modules, pathway activity scores, coexpression eigenvectors, or network centrality measures and optimizes signal-to-noise ratios in the high-dimensional regimes. At the same time, unsupervised dimensionality reduction using matrix factorization, autoencodings, or manifold learning may generate small-sized embeddings preserving significant regulatory axes where all significant levels of technical scatter are eliminated. The danger of premature compression is entanglement of subtle but functionally relevant variation; therefore, repetitive assessment of embeddings that delineates their association with independent phenotypic, environmental, or genotype-derived covariates is mandatory to the adoption of such embeddings as primary AI inputs. In any case of multimodal extensions with single cell, spatial, microbiome, or sensor mode of operation, alignment procedures, like mutual nearest neighbor anchoring or shared latent variable modeling, need to be incorporated as steps to not impose artificial concordance and thereby obscure valid modality-specific divergences (Karthik, 2025).

Missingness in the expression matrices and other metadata is frequently in a nonsensory mode in association with the quality of the samples, the availability of the tissue, or the severity of the condition. Simple imputation may create an illusion of spurious coherence; probabilistic methods that explicitly model the mechanisms of missingness or that use latent factor-conditioned sampling will preserve amounts of uncertainty that downstream AI models can encourage flow into prediction intervals as opposed to collapsing to false precision. However, incorporating caution is justified: selective dropout of features with highly prevalent and informative missingness can outperform vigorous imputation, particularly when sets of models are strong enough to factor in the form of sparse allowances or permit masking indications as being members of their feature groups (Prakash et al., 2024).

The aspect of data leakage prevention is an underestimated part of preprocessing. Leakage may be caused by estimating normalization parameters, a batch correction transformation, decisions about feature selection, or even imputation models using pooled datasets with one or more of the subsequent steps, including validation or test partitions, accidentally passing on information. A strict pipeline performs strict partition-aware fitting of preprocessing transformations and then transparently applies these parameter objects that are specific to the training data to held-out partitions. To support reproducibility, provenance logging every transformation step is used and powered by checksum verification and declarative workflow descriptions to make the workflow application auditable with potential leakage paths.

The result of such operations is the expression matrices, derived sets of features, and metadata schemas that are computationally compatible with sophisticated AI systems and statistically faithful as well as biologically faithful. However, preprocessing must be considered a dynamic, feedback-informed staging. When there are violations in model calibration, shifts in data quality, or biases revealed through fairness audits, it is essential to thoughtfully reevaluate certain elements of the pipeline rather than simply applying quick fixes. Preprocessing, quality assurance, and normalization should not be seen as static checkpoints but as dynamic, evolving systems. These systems must adapt alongside the expanding landscape of livestock transcriptomics, technology, and ethical considerations to ensure AI-generated insights remain trustworthy, transparent, and fair.

5.4 AI techniques and modeling paradigms in livestock transcriptomics

The AI methods of livestock transcriptomics have come to form a networked ecosystem of methodological paradigms of predictive, generative, causal, integrative, and adaptive modeling, each with different layers of biological complexity and translational requirements. Livestock production biology features, such as polygenic regulation of complex traits, genotype-by-environment interaction, management heterogeneity, and the ethical requirement of sustainability and maximization of animal welfare, place modeling demands and opportunities that are not necessarily reflected in biomedical or plant transcriptomic settings. It has thus been developed that the field focuses on high-dimensional molecular measurements that can be reconciled with sparse or noisy phenotypic datasets; technology has been developed to utilize structural priors based on regulatory biology, can be trained to adapt under distributional shifts between breeds and production systems, and generates outputs that are amenable to integration into decision support pipelines, such as breeding, health monitoring, nutritional optimization, and environmental stewardship. Fundamentally, the modeling effort is not bound to a listing or labeling of algorithm novelty through enumeration of such synergies among representation learning, bias-conscious prediction, mechanistic interpretability, and governance-responsive implementation (Li et al., 2024; Xiang et al., 2025).

Supervised learning is also an indicator and remains a fundamental modality with transcriptomic expression patterns, which may be enhanced with genotype-derived variant summaries, epigenomic phenotypes, microbiome descriptors, or environmental covariates, being mapped into quantitative performance indicators, categorical health status, resilience measures, or even latent trait surrogates. Linear and additive models historically offered baseline interpretability, although they cannot opine on nonlinear regulatory epistasis or the hierarchical structure of coexpression. Deep feedforward architecture and residual formulations yoke capacity to the interaction spectra, whereas regularization structures reconcile parameter elicitness in small sample regimes that are typical of livestock groups. Multitask learning provides efficiency benefits by simultaneously learning related characteristics through cross-tissue expression representations, meaning that they share lower-level representations that model invariant regulatory grammar but allow task specialization. Vertical information flow by a hierarchical design is based on biological stratification of cells, tissues, organs, and the whole animal, and noise attenuation is provided by structured sharing of parameters. Semisupervised extensions exploit the abundance of unlabelled transcriptomes by coupling supervised loss functions with auxiliary objectives such as reconstruction, consistency, or contrastive learning. These objectives refine the embeddings, even when phenotypic labels are scarce. This reduces the impact of the annotation bottleneck, which is particularly acute for welfare and stress phenotypes, for which routine ground truth is lacking (Liang et al., 2024; Yu et al., 2024a).

Representation learning has become the key to decomposing composite regulatory readings imprinted in high-dimensional, sparse, and context-specific expression matrices. Dimensionality compaction at the initial stages is offered by matrix factorization and probabilistic latent variable models; however, the deep autoencoding frameworks can extend expressivity abilities by modeling nonlinear manifolds that can be representative of conditions that are regulated by regimes. Variational formulations provide principled quantification of uncertainty in latent space, which can be used to calibrate downstream and perform stochastic exploration of plausible regulatory states on which a scenario is being analyzed. Identity paradigms apply contrastive embeddings of biologically identical pairs between conditions, tissues, or modalities and reject nonequivalent counterparts, enhancing discrimination of delicate state changes. Self-supervised goals mimicking or corrupting sets of genes and training supernets to produce lost content without using hand-designed feature engineering pressure the learning of distributed dependencies. Centric architectures obtained in the course of sequence modeling are reused to process gene expression vectors as unordered sets in which adaptive weighting conditions context-specific regulatory effects, and enable interpretable saliency patterns to be extracted. Self-supervised pretraining on large amalgamated multibreed transcriptomic corpora creates foundation embeddings that can be applied downstream to fine-tune marketing models under a limited amount of data and in generalization to underrepresented genetic backgrounds (Yang et al., 2024; Tatyana, 2025).

Generative modeling is becoming a more strategic part of augmenting data efficiency, counterfactual investigation, and creating integrated multi-omics views. Latent regulatory coded probabilistic decoders can predict plausible transcriptomic responses to proposed environmental stressors, nutritional modifications, or genomic modifications, which are useful in prioritization before the expensive in vivo experiments. The generative adversarial or diffusion-oriented frameworks support distribution-faithful augmentation capable of alleviating class imbalance of rare resilience phenotype or health perturbation and must be carefully validated to ensure entrenchment of latent biases. The connection between coupling layers of generation and mechanistic modules that do not decompose pathway characteristics or even economic balance promotes biological plausibility and contends with a belief in the results produced through simulation. Digital twin models are built upon multiscale generative components that connect transcriptomic dynamics with metabolic flux, physiological measurements, and management intervention in order to support iterative in silico experimentation in alternative breeding or nutritional dynamics that inform adaptive management strategies consistent with sustainability goals. These simulated systems are calibrated using longitudinal observational cohorts so that they converge to fidelity as opposed to diverging toward synthetic artifacts (Cusworth et al., 2024; Rosales-Morales et al., 2024).

Elucidation of regulatory and interaction maps is an equal paradigm in which the objective changes to no longer focus on direct phenotype prediction but rather reestablish graph-structured dependence between genes, transcription factors, mediators of signaling, or external modulators like microbiome-derived metabolites or climatic indices. The variants of graphical lassoes, Bayesian network structures, dynamic Bayesian formulations, and neural message passing architectures all offer avenues to the approximate topology of regulatory influence latents. The entity of the graph neural networks (GNNs) is that prior knowledge graphs are fed into, which are further trained with data refinements of the edges, and the resultant node projection of the GNNs is a representation of the context-specific influence patterns that can be exported into supervised tasks. The volatility of dynamic graph modeling of temporal and longitudinal transcriptomes could be used to differentiate between core regulatory scaffolds that are stable and peripheral rewiring in response to stress or developmental history. Causal discovery methods seek to leave behind the associational coherence and incorporate intervention data or genetic variation as instrumental anchors or invariant risk-minimizing principles across environmental strata to deduce directional effects that may be useful in breeding or management. Counterfactual modeling superimposes the causal graphs with generative simulators that allow what-if analysis of perturbation of a regulatory node, manipulation of management variables, or translocation of genetic backgrounds and hence converts network abstraction into decision-relevant information (Cutello et al., 2024; Vélez-Cruz and Papandreou-Suppappola, 2024).

Domain adaptation, transfer learning, and meta learning can tackle the endemic problem of distributional shift across breed production systems, geographies, and temporal scales. Different covariate shifts and label shift divergences indicate that naive model portability is compromised and thus requires alignment strategies, which involve reweighting or transforming representations to obtain conditional consistency. Adversarial domain discriminators that are consulted to obfuscate the origin of embeddings induce the derivation of invariant regulatory cores, yet conserve discriminative characteristics, but careful balance to prevent homogenization is necessary to shield the braver breed's particular adaptive signatures. Few-shot adaptation protocols can quickly adapt foundation models to niche herds with little to no fine-tuning samples, and metalearning paradigms can be trained to learn how to reuse their dynamics to adapt quickly to common variations of tasks, enhancing their resilience to emergent health threats or climatic volatility. Continuous learning methods overcome disastrous forgetting with models updated sequentially with new transcriptomics batches, using rehearsal buffers, regularization penalties against parameter drift, or modular expansions to isolate new knowledge with legacy competency. Federated learning is a technique of distributed training that deploys data silos like separate breeding organizations or regional cooperatives, which retain data possession and privacy and aggregate worldwide parameter improvement; privacy-maintaining gradients and safe aggregation deal with the issues of exposures. Nevertheless, heterogeneity in the cohort structure and sequencing methods requires strong aggregation weighting and drift detection measures (Agrawal, 2025; Wang et al., 2025a).

These paradigms are filled with robustness and fairness concerns. Adversarial vulnerability may occur, whereby even minor changes in expression vectors cause disproportionate changes in model outputs, casting doubt on their reliability in automated decision-making. Resilience is improved by the use of strong optimization and smoothing models and the detection of out-of-distribution samples based on distance-based and likelihood-based criteria. Fairness auditing is used to assess performance equality among breeds, sexes, production systems, or levels of management that contain systemic error inflation that may give way to selection or allocation of intervention. Correction can either reweight losses, debias adversarial training, or make post hoc adjustments to calibrate losses in order to balance the distribution of errors and keep the overall accuracy the same. The quantification of uncertainty by way of Bayesian neural layers, ensembles, or evidential outputs provides all necessary calibrated prediction intervals to guide risk-conscious decision-making from one's breeding or health surveillance activity to avoid the overconfident extrapolation of sparse data regimes (Pradipta and Gambetta, 2024; Zhou, 2024).

Identifiable attributes, suited to the interface between complex model internals and a domain expert's viewpoint, entail interpretability and explainability. It is possible to get partial lenses of model reasoning using saliency attribution, feature permutation influence, integrated path methods, and attention weight extraction, but in each specific context, they need to be seen through the correlated structure of transcriptomic features lest they be used to misassign causal primacy. Attribution aggregation to pathways, modules, or network community scales enhances cognitive correspondence to biological reasoning frameworks. Causal surrogate modeling is able to model and estimate black-box decision boundaries using structured, mechanistically annotated models, providing two levels of performance and intelligibility. The interpretability outputs can be improved in terms of reliability by analyzing stability to perturbed inputs, cross-model consensus scoring, and benchmarking the outputs against manually curated regulatory knowledge repositories. Integrated explanations that incorporate uncertainty express not only what features seem to be influential, but also the degree of confidence in such decisions, which allows subtle expert acceptance (Goel and Weber, 2025).

The use of model uncertainty and expected information gain measures to optimize experimental design is used in active learning and experimental design optimization to prioritize future transcriptomic sampling that decreases epistemic uncertainty as much as possible or eliminates ambiguity between competing mechanistic hypotheses. Sequential selection of animals, tissues, time, or environmental context, all guided by acquisition functions, is known to increase the rate at which knowledge is acquired within resource constraints. Reinforcement learning builds upon this principle with policies that are modeled by reinforcement schedules, for example, adaptive sampling schedules or management modifications that use reinforcement schedules which reward long-term decrease in predictive error or enhancement in resilience measures in simulated production seasons. Coupling to the digital twin worlds allows exploring policy space without on-the-job risk, and close matching of the simulated and measured dynamics eliminates the issue of policy fragility in practice (Thomas and Houssineau, 2024).

Paradigms of evaluation extend past the traditional measures of accuracy that focus on accuracy-centric measures to include calibration, predictive interval sharpness, resampling stability, performance on external validation groups, fairness parity, robustness to synthetic perturbations, and alignment with biological plausibility. Multiobjective analysis redefines success as an efficient frontier consisting of predictive utility, interpretability richness, cost of computation, energy use, and compliance with governance. Awareness of energy and carbon footprint drives architectural pruning, low-rank factorization, quantification, and knowledge distillation into producing less weighty models, which can be deployed in on-farm edge infrastructures or low-connectivity settings to support near real-time inference combined with sensor streams sufficient to induce adaptive management signaling (Burden et al., 2025).

Lifecycle governance manages the performance drift, data shift, and emergent bias through continuous monitoring, and retraining pipelines are launched automatically with a constraint on provenance and reproducibility. The version of the models, data, and preprocessing transformations and interpretability reports form an auditable lineage, which will meet regulatory, ethical, and stakeholders' expectations of transparency. The risk assessment models define model applications based on the severity of the potential impact and reduce validation rigor, documentation level, and level of oversight accordingly. Integrating an ethical contemplation of modeling decisions promotes the prevention of obscurity that might diminish stakeholder trust or spread spillover unease pressures that may harm genetic diversity or welfare results (Paleti, 2025).

These modeling paradigms are moving toward a convergent integration to an integrative fabric with representation learning providing a stable multilingual biological embedding, generative and causal modules making scenario planning, and supervised and multitask predictors transforming embeddings to actionable indicators; adaptive mechanisms of governance provide the reliability of the paradigm and distribute opportunities fairly to reduce the ecological footprint. The commercial success of livestock transcriptomic AI will not rely on any single case of secluded algorithmic innovation but rather on synthesis where data stewardship, methodological purity, interpretability, faithfulness, and practicality translation will all coemerge. Within this synthesis, models are not points in time but dynamic and collaborative tools that enhance biological understanding, informed innovation, and resilient and sustainable livestock systems. AI techniques and modeling paradigms for livestock transcriptomics are presented in Table 5.2.

5.5 Tooling and platforms

The AI-enabled livestock transcriptomics tooling and platform architecture comprise the operational substrate that enables operational scalability, reproducibility, interpretability, and velocity in translational research. The unique requirements of

TABLE 5.2 AI techniques and modeling paradigms for livestock transcriptomics.

AI technique	Modeling paradigm	Application in livestock transcriptomics	Benefits	Limitations
Regularized linear models (Lasso/Elastic net/Ridge)	• Parametric linear regression with penalty terms • Sparse (L1) versus shrinkage (L2/elastic) priors • Convex optimization ensuring global optimum • Handles multicollinearity via regularization	• Baseline gene expression for trait prediction • Feature selection for biomarkers (sparse panels) • Adjusting for covariates (batch and breed) • Interpretable baseline for comparing complex models	• High interpretability • Fast training on large matrices • Provides ranked features • Low overfitting risk versus deep models • Easy integration into pipelines	• Assumes linear relationships • Limited capture of high-order interactions • Performance drops on highly nonlinear gene networks • Sensitive to correlated features (unstable selection)
Tree-based ensembles (random forest, gradient boosted trees)	• Bagging (RF) versus boosting (GBT) paradigms • Nonlinear, hierarchical partitions • Implicit feature interaction modeling • Out-of-bag internal validation (RF)	• Trait prediction from expression matrices • Prioritizing influential genes for panel reduction • Handling mixed feature types (expression and metadata) • Robust baseline for health risk models	• Captures nonlinearities and interactions • Handles noise/outliers relatively well • Provides feature importance metrics • Requires minimal preprocessing	• Limited performance on extremely high-dimensional sparse matrices without pruning • Interpretability of interactions is opaque • Memory-heavy with many trees • Can be biased toward high-variance features
Support vector machines (linear/kernel)	• Large-margin classifier/ regressor • Kernel trick maps to high-dimensional feature space • Convex optimization (primal/ dual forms)	• Disease state classification from expression • Subclinical condition detection with limited samples • Transcriptomic response separation under stressors	• Effective in small to medium sample, high-dimension regimes • Robust decision boundary • Flexible kernels (RBF and polynomial)	• Kernel scaling is poor with very large datasets • Parameter tuning (C and gamma) sensitive • Outputs less probabilistic without calibration
k-Nearest neighbors (kNN)	• Instance-based lazy learning • Distance metric driven (Euclidean/cosine) • Nonparametric decision surface	• Similar expression profile retrieval (case-based reasoning) • Local reference set for anomaly detection • Baseline clustering sanity checks	• Simple and transparent • No training time overhead • Adapts instantly to updated data	• Curse of dimensionality in high-gene space • Sensitive to scaling and noise • Memory and latency cost at inference
Ensemble stacking/blending	• Metalearning combining diverse base learners • Weighted or learned aggregation • Reduces model-specific bias	• Improving trait prediction by merging linear, tree, and deep outputs • Robust consensus disease risk scoring • Stability across breeds/ environments	• Boosts predictive robustness • Mitigates single-model overfitting • Leverages complementary inductive biases	• Increased complexity and maintenance • Risk of overfitting metalearner • Harder attribution to biological rationale

Matrix factorization (PCA, NMF, ICA)	• Low-rank latent space decomposition • Additive (NMF) versus orthogonal (PCA) versus independent (ICA) constraints • Dimensionality reduction for denoising	• Removing batch/technical variation components • Latent factor extraction for pathway modules • Input compression for downstream deep models	• Noise reduction and visualization • Faster downstream model training • Identifies coexpression structures	• Linear assumptions miss nonlinear pathways • Component biological meaning sometimes ambiguous • Overremoval may discard signal
Denoising autoencoders	• Neural encoder–decoder with noise corruption • Learns robust latent embedding • Reconstruction loss driven	• De-noising expression matrices (dropout, batch noise) • Feature compression for trait models • Pretraining backbone for multitask nets	• Captures nonlinear structure • Improves signal-to-noise ratio • Enables transfer of multiple tasks	• Risk of learning technical artifacts if noise is mis-modeled • Latent space not inherently interpretable • Needs sufficient data volume
Variational autoencoders (VAEs)	• Probabilistic latent variable model • KL divergence regularized latent prior • Generative reconstruction paradigm	• Latent generative embedding for clustering cell states (pseudo-bulk) • Data augmentation for rare condition balancing • Multi-omics integration via joint latent spaces	• Uncertainty-aware representation • Smooth latent manifold for interpolation • Supports semisupervised extensions	• Posterior collapse risk • Hyperparameter sensitivity (beta and capacity) • Blurry reconstructions reduce fine-grained fidelity
Self-supervised contrastive learning	• Instance discrimination using similarity maximization • Negative sampling/InfoNCE objectives • Augmentation-driven invariance learning	• Pretraining on large unlabeled expression cohorts • Cross-breed invariant embedding generation • Aligning modalities (gene expression vs. microbiome)	• Reduces labeled data need • Learns robust, transferable features • Improves downstream task generalization	• Choice of augmentations critical • Negative sampling biases possible • Computationally intensive with large batches
Masked modeling (masked gene/token prediction)	• Mask-and-reconstruct paradigm (BERT-like) • Contextual embedding learning • Denoising objective	• Recover masked gene expression values • Impute low-depth transcriptomic regions • Pretrain transformer encoders for downstream tasks	• Captures contextual gene coregulation • Scales with large corpora of expression matrices • Improves few-shot task performance	• Mask strategy impacts representation quality • Potential to learn dataset-specific artifacts • Requires substantial computing for large vocabularies
Convolutional neural networks (1D/2D)	• Local receptive fields and weight sharing • Hierarchical feature extraction • Translation invariance (adapted to ordered signals)	• Modeling ordered genomic regions (promoter accessibility and expression) • Spatial spot transcriptomics image fusion • Temporal convolution over sliding transcript windows	• Efficient parameter usage • Captures local motifs/ patterns • Fast inference once trained	• Gene expression vectors lack natural spatial order • Requires engineered ordering or reshaping • Less effective for long-range dependencies
Recurrent neural networks (LSTM/GRU)	• Sequence modeling with gated memory • Temporal state propagation • Handles variable-length inputs	• Longitudinal expression trajectory prediction (time-course challenge studies) • Time-aligned health deterioration forecasting • Dynamic response modeling to interventions	• Captures temporal dependencies • Handles irregular sequence padding via masking • Interpretable via attention overlays	• Vanishing gradient risk (mitigated but not eliminated) • Slower training versus parallel architectures • Less scalable than transformers for very long sequences

Continued

TABLE 5.2 AI techniques and modeling paradigms for livestock transcriptomics.—cont'd

AI technique	Modeling paradigm	Application in livestock transcriptomics	Benefits	Limitations
Transformers	• Self-attention mechanism globally modeling dependencies • Positional encoding for order • Highly parallelizable	• Global gene–gene interaction inference • Multimodal fusion (sequence, expression, and environment) • Cross-sample attention for cohort-level pattern discovery	• Models complex long-range relationships • Scales to large datasets and multihead interpretability • Adaptable (masking, multitasking)	• Data-hungry and compute-intensive • Risk of learning spurious correlations • Attention weights are not always causal explanations
Graph neural networks (GNNs)	• Message passing over graph nodes/edges • Structured relational embedding • Integrates topology and node features	• Gene coexpression/regulatory network modeling • Spatial transcriptomics spot graph inference • Multi-omics integration via heterogeneous graphs	• Leverages known biological networks • Encourages structured generalization • Supports inductive inference on new nodes	• Graph construction quality critical • Oversmoothing with many layers • Computational overhead for large dense graphs
Knowledge graph embeddings	• Translational/tensor factorization (TransE, DistMult) • Triple scoring for relation prediction • Latent representation of entities and relations	• Integrating gene–trait–pathway–disease relationships for annotation propagation • Enhancing interpretation of latent transcript features	• Adds semantic context to models • Improves zero-shot inference for rare traits • Facilitates explainability via relation chains	• Requires curated, up-to-date graph • Sparse or biased relations degrade performance • Risk of propagating ontology errors
Multimodal fusion (early/intermediate/late)	• Early concatenation of raw features • Shared latent fusion (intermediate) • Ensemble decision fusion (late)	• Combining expression, microbiome, and proteomics • Integrating image histopathology and spatial expression • Merging environmental metadata with transcriptomics	• Captures complementary signals • Increased robustness across conditions • Flexible architecture selection	• Risk of modality dominance • Missing modality handling complexity • Higher engineering and computing costs
Multitask learning (MTL)	• Shared backbone with task-specific heads • Inductive transfer via parameter sharing • Joint optimization balancing losses	• Predicting multiple traits (production and health) simultaneously • Joint expression, splicing, and isoform abundance modeling • Copredicting disease onset and severity	• Efficiency through shared representations • Improved generalization on low-sample tasks • Encourages biologically coherent features	• Loss balancing delicate tasks (task interference) • Negative transfer risk • Harder error attribution per task
Transfer and domain adaptation	• Pretrain–finetune paradigm • Domain-invariant feature learning (adversarial) • Weight freezing/partial adaptation	• Adapting models across breeds or populations • Cross-species transfer (cattle → sheep ortholog expression) • Environment shift correction (farm-to-farm)	• Reduces labeled data requirements • Accelerates deployment in new cohorts • Mitigates performance collapse under shift	• Source-target mismatch risk • Catastrophic forgetting without strategies • Requires careful validation to avoid leakage

Metalearning/few-shot (e.g., MAML and prototypical networks)	• Learn initialization enabling rapid adaptation • Metric-based embedding comparison • Gradient-based task adaptation loops	• Rapid tuning for rare disease expression signatures • Personalized farm-level model adaptation • Fast specialization to novel stressor conditions	• Efficient with very small labeled sets • Improves responsiveness to emerging phenotypes • Reduces retraining time	• Complex training loops • Sensitive to task sampling design • May underperform on large, stable datasets
Active learning	• Query strategies (uncertainty, diversity) • Human-in-the-loop labeling loops • Iterative model refinement	• Prioritized sequencing/labeling of ambiguous samples • Optimizing phenotyping budget allocation • Curating validation sets for new traits	• Reduces labeling cost • Accelerates model improvement • Focuses effort on high-value data	• Requires reliable uncertainty estimates • Potential sampling bias if the strategy is flawed • Operational complexity for iterative labs
Semisupervised learning (pseudo-label, consistency)	• Leverages unlabeled and labeled data jointly • Consistency regularization under perturbations • Pseudo-label refinement loops	• Exploiting large unlabeled expression repositories • Extending disease classifiers with limited confirmed cases • Enhancing rare tissue models with bulk unlabeled sets	• Improves performance with scarce labels • Reduces overfitting to a small labeled subset • Scales with available unlabeled corpora	• Propagates early pseudo-label errors • Sensitivity to augmentation choices • Risk of confirmation bias
Causal inference and causal graph/SCM methods	• Directed acyclic graph modeling • Intervention and counterfactual estimation • Do-calculus/causal discovery algorithms	• Identifying expression mediators between genotype and trait • Prioritizing intervention targets (diet and management) • Evaluating putative causal pathways in stress responses	• Distinguishes correlation versus causation • Enhances trust in actionable biomarkers • Supports policy/intervention simulation	• Requires strong assumptions (no hidden confounders) • Limited by observational biases • Causal discovery unstable with noisy omics
Bayesian hierarchical modeling	• Probabilistic multilevel parameter sharing • Posterior inference (MCMC, variational) • Explicit uncertainty quantification	• Modeling breed-level versus individual-level expression variance • Integrating prior knowledge (pathways) into trait models • Shrinkage for low-expression genes	• Interpretable uncertainty estimates • Robustness to small subgroup sizes • Flexible incorporation of priors	• Computationally intensive at scale • Model specification complexity • Convergence diagnostics required
Generative adversarial networks (GANs)	• Generator–discriminator minimax training • Implicit distribution learning • Adversarial signal refinement	• Synthetic expression sample augmentation for minority classes • Batch effect style transfer (domain translation) • Simulation of stress response profiles	• Enhances data balance and robustness • Can model complex joint distributions • Enables domain generalization experiments	• Training instability and mode collapse • Hard to assess biological plausibility • Potential to introduce artifactual patterns

Continued

TABLE 5.2 AI techniques and modeling paradigms for livestock transcriptomics.—cont'd

AI technique	Modeling paradigm	Application in livestock transcriptomics	Benefits	Limitations
Diffusion probabilistic models	• Iterative noise addition/removal (denoising steps) • Learned reverse Markov chain • Score-based generative modeling	• High-fidelity synthetic transcriptome generation • Imputation of missing gene expression patterns • Scenario simulation under hypothetical interventions	• Stable training versus GAN • Captures complex multimodal distributions • Provides a controllable generative process	• Computationally expensive (many steps) • Risk of generating unrealistic rare patterns • Limited adoption of tooling in livestock yet
Reinforcement learning (RL)	• Policy optimization via reward signals • Sequential decision processes • Exploration–exploitation trade-offs	• Adaptive sampling schedules (when to collect transcriptomics) • Dynamic intervention sequencing (nutritional changes guided by expression feedback) • Multiyear breeding–expression strategy simulation	• Optimizes long-term reward (e.g., cumulative trait gain) • Learns adaptive policies • Simulates counterfactual strategies	• Reward design is challenging • Data scarcity for real trajectories • Interpretability of learned policies is limited
Survival/time-to-event deep models (DeepSurv, Cox NN)	• Hazard function parameterization • Partial likelihood or ranking loss • Incorporates censored data	• Predicting time to disease onset from baseline expression • Longevity or productive lifespan forecasting • Time-to-relapse in infection challenges	• Handles censored records appropriately • Provides risk stratification curves • Integrates continuous covariates flexibly	• Requires sufficient event numbers • Assumes proportional hazards (classical variants) • Calibration can drift over time
Temporal convolutional networks (TCNs)	• Dilated causal convolutions • Long receptive fields with fewer layers • Parallel sequence processing	• Modeling dynamic transcriptomic responses postvaccination • Time-resolved expression-based health deterioration detection • Temporal integration of wearable and periodic RNA data	• Efficient long-range dependency capture • Stable gradients versus RNNs • Fast training with large sequences	• Needs evenly spaced or resampled time points • Less intuitive biologically than mechanistic models • Sensitive to padding/truncation choices
Anomaly and outlier detection (isolation forest, one-class SVM, AE reconstruction)	• Unsupervised density/isolation paradigms • Reconstruction error thresholds • Boundary estimation for the normal class	• Early detection of atypical expression indicative of emerging disease • QC flagging of low-quality libraries • Identifying novel stress phenotypes	• Works without labeled anomalies • Enhances quality control pipelines • Supports biosurveillance signals	• High false positive risk if distribution shifts • Hard to interpret the cause of the anomaly • Performance sensitive to contamination in training data

Privacy-preserving AI (federated learning, differential privacy, secure MPC)	• Decentralized gradient/model aggregation • Noise injection for privacy budgets • Cryptographically secure aggregation protocols	• Collaborative multifarm model training without raw data pooling • Protecting proprietary breeding line expression profiles • Regulatory-compliant health biomarker development	• Enables data sharing under constraints • Reduces central breach risk • Expands training cohort diversity	• Communication overhead • Utility–privacy trade-off (noise degrades accuracy) • Complex orchestration and compliance auditing
Explainable AI (SHAP, LIME, counterfactuals, saliency)	• Post hoc local/global attribution • Perturbation or path-integrated gradients • Counterfactual outcome generation	• Interpreting gene contributions to trait predictions • Validating disease classifier rationale • Communicating biomarkers to stakeholders	• Increases trust and adoption • Aids biological hypothesis generation • Identifies spurious shortcut features	• Explanations can be unstable • Computational overhead for large models • Risk of overtrust in approximations
Bayesian optimization/ AutoML	• Surrogate modeling (Gaussian Process/TPE) • Sequential acquisition (EI, UCB) • Automated hyperparameter search	• Optimizing architecture/ hyperparameters for expression → trait models • Tuning preprocessing (normalization, feature selection) • Efficient search for multimodal fusion strategies	• Reduces manual trial-and-error • Improves model performance reproducibly • Handles expensive evaluation budgets efficiently	• Surrogate mis-specification risk • Runtime cost is still high for large deep models • May overfit validation splits if mismanaged
Compositional data modeling (Dirichlet-multinomial, Aitchison geometry)	• Constrained simplex-aware probability models • Log-ratio transformations (CLR, ALR) • Accounts for compositional dependencies	• Integrating microbiome relative abundances with host expression • Correcting compositional bias when fusing multi-omics • Improving feed efficiency or methane prediction embeddings	• Proper handling of compositional constraints • Reduces spurious correlation risk • Enhances joint modeling interpretability	• Requires careful transformation choices • Sensitive to zeros (pseudocount necessity) • Limited off-the-shelf deep integration tooling
Hybrid mechanistic and ML models	• Coupling differential equation/ kinetic modules with learned residuals • Physics/biology-informed constraints • Gray-box modeling	• Enhancing metabolic pathway expression simulators • Constraining gene regulatory predictions with known kinetics • Predicting nutrient–expression dynamics under changing diets	• Improves biological plausibility • Reduces data demands via priors • Facilitates extrapolation beyond observed range	• Model integration complexity is high • Potential mismatch between mechanistic assumptions and real data • Hard debugging of hybrid errors
Mixture-of-experts (MoE)/ Ensemble of specialists	• Gating network routes inputs to experts • Conditional computation • Expert specialization (breed/ task/domain)	• Breed-specific or tissue-specific expression predictors • Specialized experts for stress versus baseline states • Scaling large models with sparse activation	• Efficient scaling (compute on active experts only) • Improves domain specialization performance • Mitigates negative transfer across divergent cohorts	• Load imbalance across experts • Complex training stability (expert collapse) • Harder interpretability of gating decisions

livestock systems, which include heterogeneous data provenance, distributed acquisition contexts at the farm level, issues of privacy with proprietary genotypes or management practices, and the requirements of timely feedback loops of decisions, supply design constraints that are less than those of more central biomedical research environments. Good platform strategies thus combine layered computing infrastructure, semantically rich information fabrics, standardized workflow orchestration, life cycle manageable model tracing, federated work, adaptable visualization, as well as interpretation interfaces all over a governance framework that instills compliance, ethical custody, and sustainability consciousness (Rosati, 2025).

On the infrastructure layer, a mixed high-performance cluster mesh of elastic cloud and edge-level compute is used to support differentiated workload classes at the infrastructure layer. High-bandwidth storage and parallelizable accelerators are an advantage to intensive preprocessing and large-scale representation learning, whereas latency-sensitive inference of health or management alerts requires lightweight model containers that can be deployed to locations where connectivity can be intermittent. Active workload scheduling based on telemetry of queue depth, energy usage, and cost indicators can be used to perform responsive allocation to avoid bottlenecks without overallocation. Data locality schemes minimize the unwanted data transfer through colocating compute with storage shards or using in situ query primitives on huge transcriptomic and metadata tables. The compression coders and columnar or graph storage engines, optimized with sparse high-dimensional matrices and semantic triples, reduce storage footprints without compromising access time (Wang et al., 2024).

Data management systems are becoming more than flat repositories and are being, in their turn, transformed into knowledge-centric fabrics with ontological identifiers, lineage relationships, provenance assertions, and access control policies. Each read, processed quantification, derived embedding, regulatory network construct, phenotypic annotation, environmental condition variable, and downstream decision output has a semantic model connected in a layered way, allowing traversable queries to support the generation of hypotheses and debugging of the model. Versioning semantics use each transformation as a nonmutable node in a directed lineage graph, making it straightforward to roll back or audit comparatively and certify that the transformation was also reproducible. The access governance implements the principle of least privilege and allows different privacy perspectives or encrypted enclaves of sensitive attributes. The engines are policy implementations that combine consent metadata and usage controls, preventing or pointing out analytic flows that would otherwise violate data sharing agreements (Borgelt et al., 2025).

The workflow orchestration models provide a declarative specification of the pipeline of multistages of the ingestion, quality control, normalization, feature construction, model training, evaluation, packaging, and deployment. Best practice is contained in parameterized templates, which allow contextual overrides of the organism or tissue, modality, or modeling objective. The caching and incremental recomputation heuristics identify unchanged intermediate states, reducing the length of an iteration cycle and minimizing the energy consumption. The introspection of workflow presents performance metrics, data drift indicators, and artifact checksums and allows detection of anomalies at an early stage. Containerized execution is used to attain cross-platform portability, a fact described by abstract resource descriptors without linking the design of logical pipelines with the underlying hardware heterogeneity. Workflows have policy gates to impose mandatory quality thresholds, generate interpretability reports, and conduct fairness audits before models are promoted to the levels of production.

Model lifecycle management is often implemented in an operations paradigm that combines the concepts of continuous integration and delivery with domain-specific validation. Not only the weights of parameters but also the schema signature of anticipated inputs, dependency manifests, training hyperparameters, and evaluation dashboards are stored in repository structures. Performance traces offered by automated benchmarking against curated gold standard panels and external validation cohorts are longitudinal performance traces, providing the opportunity to identify regressions occurring postcode or data updates. The agents of drift monitoring compare the statistical profiles of the incoming batches of inference with the historical baselines. When the statistical profile of the batched inferences does not match the adaptive threshold, the retraining or human examination is initiated. The use of ensemble registries enables the use of controlled stacking or gating policies, which take advantage of the complementary capabilities of heterogeneous architectures, and governance policies that ensure that ensemble complexity is interpretable and maintainable (Mateo-Casalí et al., 2025).

Federated learning systems and privacy-preserving learning systems deal with the distributed data custodianship reality. The raw data is safeguarded by means of secure aggregation protocols, the calibration of the noise of privileged privacy, and encrypted gradient exchange, as well as allowing parameters to be refined jointly. Heterogeneity-aware optimization balances the size of the variable cohort, prevalence of labels, and sequencing platform peculiarities, avoiding overpowering by large and/or noisy nodes. To protect equity issues and facilitate incentive schemes to compensate for the quality of data stewardship, audit trails record contribution magnitudes. The compliance modules are used to trigger the exclusion of jurisdictions or nodes that do not pass the validation test, and the collaborative fabric is kept intact (Zhang et al., 2025).

Digital twins' layers provide the connection between transcriptomic state space models and mechanistic simulators of physiology, environmental progression modules, and management intervention policy. Tooling coordinates the consistency of observational feeds of empirical data to simulated experience by means of an assimilation algorithm that modifies latent states to match newly received observed measurements. Scenario experimentation consoles enable the desire to investigate different breeding strategies, diets, or mitigating stress conditions with a quantitative comparison of the estimated resilience, productivity, and sustainability indicators. Uncertainty propagation engines make certain beliefs known across simulated horizons, which discourages the extrapolation of beliefs in lightly observed regimes (Khan et al., 2024).

Issues of sustainability and stewardship of resources are permeating platform design. Energy telemetry and carbon attribution estimators are configured to work with workload schedulers in a way that causes the work to be biased in either direction, such as a lower emission time window or a hardware profile. Model compression toolchains compress inference in the presence of pruning, low-rank decomposition, quantization, and distillation with no worse than unacceptable loss of calibration or interpretability. Policies in archives distinguish between high-value irreversible raw materials and regenerable intermediate objects, which makes it possible to selectively retain in order to meet long-term scientific value and ecological responsibility. Green benchmarking registries report both efficiency measurements and predictive performance and bring the community to the level of balanced optimization goals (Reddy Pinnapareddy, 2025).

The security and resilience models ensure the integrity of the platforms against corruption of the data, unauthorized access, and tampering with the model. A defense in depth posture consists of multifactor authentication, detection of anomalies over access patterns, cryptographic signing of models and workflows, and irrevocable ledgering of key governance activities. Playbooks of incident response and autonomous quarantine are employed to minimize the spread of compromised components. Continuity of essential inference services during hardware failures or network fragmentation is guaranteed by regular chaos testing of failover pathways, which are needed in the deployment of time-sensitive health monitoring (Jaiswal, 2025).

The tooling ecosystem is achieved through human factors and capacity building. Adaptive training portals customize the educational material concerning pipeline design, interpretability assessment, fairness auditing, and ethical considerations based on the user's proficiency, decreasing the time to onboarding and minimizing the occurrence of manual errors. Judgment consensualization using collaborative annotation interfaces can be done over ambiguous phenotypic labels or interpretations of regulatory events, where adjudication algorithms can be used to weigh up contributor reliability to arrive at high-confidence ground truth expansions. The visualization layers have feedback capture, which enables practitioners to report suspicious model behavior that initiates iterative refinement cycles that view the user as an active cocreator and not as a consumer (Ray, 2024).

All of these strata of interoperable tools and platforms combine fragmented transcriptomic data and single experiments in analysis into a unified, continuously learning infrastructure. The platform layer enhances AI by incorporating semantic richness, rigorous provenance, adaptive governance, federated collaboration, interpretability fidelity, and sustainability mindfulness by enabling it to exist as a continuum of knowledge generation and deployment. This type of integrated ecosystem forms the basis of achieving credible, fair, and meaningful livestock transcriptomic innovation that attaches genetic enhancement, welfare, and environmental protection goals to a robust operational structure. An ecosystem of computational tools and platforms enabling AI-driven livestock transcriptomics is presented in Table 5.3.

5.6 Applications

5.6.1 Enhancing animal health and welfare

The use of AI-based transcriptomics on animal health and welfare has transformed the field of retrospective pathology to advance precision-driven stewardship that combines molecular sentinels of physiological disruption with environmental, behavioral, and genetic history. The core of this shift is the ability of transcriptome-wide profiles, which are analyzed using a flexible representation learning map and causal inference map, to show signs of deviations out of homeostatic regulatory equilibria of cells long before macroscopic clinical symptoms can be observed. The basis of early warning is based on the detection of coordinated dysregulation within the context of immune surveillance, mitigation of oxidative stress, metabolic allocation, and neuroendocrine signaling modules, the fine patterns of coactivation of which are often invisible when considered individually based on the biomarker heuristics. Individualized molecular baselines and dynamic ranges are also learned by AI models trained on longitudinally sampled cohorts, which means that deviation scoring can

TABLE 5.3 Ecosystem of computational tools and platforms enabling AI-driven livestock transcriptomics.

Tool or platform name	Description	Key features	Applications in livestock transcriptomics	Advantages	Limitations
FAANG data portal	• Consortium repository for functional annotation of animal genomes • Curated multi-omics (RNA-seq, ATAC-seq, and ChIP-seq) for livestock species • Standardized metadata and quality metrics	• Harmonized metadata schemas • Programmatic API access • Cross-species comparative datasets • Quality-controlled submissions	• Sourcing baseline tissue expression profiles • Benchmarking AI normalization and batch correction models • Training cross-species transfer learning embeddings • Validating regulatory element–expression AI predictions	• High data quality and standardization • Reduces preprocessing burden • Enables reproducible benchmarking • Facilitates multimodal integration starting points	• Coverage still incomplete for rare tissues/breeds • Update lag versus new experiments • Limited phenotype diversity in early releases • Requires careful filtering for study design heterogeneity
NCBI sequence read archive (SRA)	• Largest public raw sequencing repository • Broad species and experimental scope • Contains raw FASTQ, metadata, and sometimes partial annotations	• Massive scale for pretraining models • Accession-based retrieval (prefetch APIs) • Cloud-enabled access (AWS/GCP mirrors) • Supports controlled access (dbGaP)	• Unlabeled corpora for self-supervised gene embedding learning • Data augmentation for rare condition modeling • Multicohort robustness evaluation • Domain shift studies (farm/lab protocols)	• Scale accelerates representation learning • Diverse protocols simulate real-world noise • Enables metaanalyses across studies	• Metadata inconsistency/ incompleteness • Variable quality; need stringent QC • Batch confounding is pervasive • Download/storage cost for large-scale use
Ensembl/BioMart	• Reference genome and annotation provider • BioMart enables bulk feature retrieval • Regular release cycles with version control	• Stable gene/transcript IDs • Orthology mappings for cross-species transfer • Variant effect annotations • REST and GraphQL APIs	• Building consistent feature dictionaries for models • Mapping orthologs for cross-species AI adaptation (cattle ↔ sheep) • Augmenting features with biotype/function tags • Variant-aware transcript quantification pipelines	• Centralized, versioned reference reduces drift • Facilitates explainability (map outputs to gene names) • Supports integration with ontology IDs	• Version changes can break pipelines if unmanaged • Lag in incorporating novel transcripts from long-read data • Pseudogene/paralog ambiguity complicates attribution
Nextflow and nf-core	• Workflow engine for scalable, reproducible pipelines • nf-core provides curated community RNA-seq, ATAC, and variant pipelines • Container and cloud-native	• Portable execution (local, HPC, and cloud) • Continuous integration testing • Version pinning and provenance tracking • DSL2 modular pipeline reuse	• Standardized expression processing for AI feature generation • Reproducible multibreed data harmonization • Automated pretraining dataset assembly • Scalable reruns for model retraining	• Strong reproducibility and scalability • Reduces custom scripting errors • Modular design accelerates iteration	• Learning curve for DSL2 modules • Resource tuning needed for cost efficiency • Depends on community maintenance cadence

Snakemake	• Declarative workflow system using Pythonic rules • Focus on readability and incremental builds • Popular for academic-scale pipelines	• Automatic DAG generation • Integrated conda/env and container support • Checkpoint and partial rerun efficiency • Cluster profile templates	• Custom exploratory transcriptomics preprocessing • Rapid prototyping of AI-specific feature extraction (e.g., splice events) • Integrating bespoke QC and model training steps	• Low barrier to entry for Python users • Fine-grained control over pipeline steps • Efficient for iterative development	• Less standardized than the nf-core ecosystem • Scaling very large cohorts may need tuning • Complex multicloud orchestration is less native
Docker/Singularity (Apptainer)	• Containerization platforms for reproducible environments • Singularity suited for HPC without root access • Encapsulate dependencies for pipelines and ML	• Immutable environments • Layered caching for build efficiency • Cross-platform portability • Security (Singularity user-space execution)	• Packaging AI training environments (PyTorch and bio libs) • Ensuring pipeline reproducibility across farms/institutions • Long-term archival of analysis contexts	• Eliminates "it works on my machine" issues • Facilitates pipeline sharing/publication • Speeds continuous integration testing	• Large image sizes impact bandwidth • Requires maintenance to update security patches • GPU driver compatibility pitfalls
Data version control (DVC)	• Data and model artifact tracking layer atop Git • Handles large files via remote storage • Reproducibility metadata lineage	• Versioned datasets and models • Pipeline dependency graph (dvc.yaml) • Remote backends (S3, GCS, and SSH) • Metrics and params tracking	• Tracking the evolution of expression matrices across normalization strategies • Managing iterative model checkpoints for trait prediction • Rolling back to previous feature engineering states	• Improves auditability and provenance • Enables collaborative experimentation • Lightweight integration with Git workflows	• Added complexity versus plain Git • Requires discipline in committing to stages • Potential storage cost for numerous versions
FastQC	• Per-sample raw sequencing QC tool • Generates summary quality metrics and plots • De facto first-pass QC	• Adapter content detection • Per-base quality distribution • Sequence duplication and GC content • Overrepresented sequence flags	• Screening livestock RNA-seq libraries pre-AI ingestion • Filtering low-quality samples to prevent model noise inflation • Baseline quality benchmarking across breeds	• Rapid, easy visualization • Low compute overhead • Widely integrated into pipelines	• Limited multisample aggregation • Not predictive of downstream bias alone • Requires manual interpretation of flags
MultiQC	• Aggregates outputs from many QC tools • Produces unified dashboard reports • Extensible plugin architecture	• Cross-sample QC summaries • Trend and outlier detection plots • Support for numerous bioinformatics tools • Exportable reports	• Cohort-level RNA-seq quality harmonization • Identifying systemic batch issues before model training • Documenting training dataset quality for regulatory reporting	• Centralizes QC for decision-making • Saves manual collation time • Enhances transparency in model inputs	• Relies on upstream tool completeness • Custom tool outputs may need plugin development • Large cohorts can produce heavy reports

Continued

TABLE 5.3 Ecosystem of computational tools and platforms enabling AI-driven livestock transcriptomics.—cont'd

Tool or platform name	Description	Key features	Applications in livestock transcriptomics	Advantages	Limitations
Seurat/Scanpy	• Single-cell/single-nucleus analysis frameworks (R/Python) • Provide preprocessing, clustering, and integration workflows • Widely adopted in scRNA communities	• Normalization and scaling • Batch integration (CCA, Harmony, BBKNN) • Differential expression and marker detection • Dimensional embeddings (UMAP and t-SNE)	• Deriving cell-type markers to deconvolve bulk livestock tissues • Training weak supervision labels for bulk AI models • Exploring immune cell heterogeneity under disease	• Rich, battle-tested pipelines • Community tutorials and extension • Interoperable with downstream ML embeddings	• High memory usage for large cells × genes • Parameter choices influence cluster stability • Batch correction may remove subtle breed signals
WGCNA	• Weighted gene coexpression network analysis in R • Constructs scale-free networks and modules • Correlates modules with traits	• Soft thresholding adjacency • Module eigengene summarization • Hub gene identification • Trait association statistics	• Feature reduction for AI (module eigengenes) • Prior network scaffolds for GNN initialization • Cross-breed conservation of module evaluation	• Reduces dimensionality meaningfully • Biologically interpretable modules • Enhances explainability of AI outputs	• Assumes approximate scale-free topology • Sensitive to outliers and batch effects • Not optimized for extremely large datasets without tuning
MOFA+	• Multi-omics factor analysis framework • Bayesian latent variable model for integration • Captures shared and modality-specific factors	• Probabilistic factor inference • Handling missing modalities • Variational inference scalability • Factor variance decomposition	• Integrating transcriptomics with proteome/metabolome in feed efficiency models • Generating low-dimensional inputs for predictive deep nets • Disentangling environment versus genetic factors	• Handles incomplete multi-omics gracefully • Provides interpretable latent factors • Reduces overfitting in downstream tasks	• Requires careful hyperparameter tuning • Factor interpretability is still subjective • Computational cost increases with modalities
PyTorch	• General-purpose deep learning framework • Dynamic computation graphs • Extensive ecosystem (Lightning and TorchGeo)	• GPU/TPU acceleration • Autograd flexibility • Distributed data parallel training • Custom operator extensibility	• Building transformer and GNN architectures for gene expression • Implementing self-supervised livestock transcript embeddings • Federated prototypes via add-on libraries	• High flexibility for novel architectures • Strong community support • Rich debugging and profiling tools	• Steeper learning curve for nonprogrammers • Need to implement domain-specific layers manually • Risk of reproducibility issues without seeding
TensorFlow/Keras	• Deep learning framework with static/dynamic graph modes • Keras high-level API simplifies model prototyping • Strong production deployment tooling	• TF Serving/TFLite export • AutoGraph optimization • Integrated distributed strategies • Model serialization standardization	• Rapid prototyping of expression → trait prediction models • Deploying inference services for on-farm decision tools • Implementing multitask breeding trait predictors	• Mature deployment pipeline • High-level APIs accelerate iteration • Broad ecosystem integrations	• Verbosity in low-level customization • Version incompatibilities are occasionally disruptive • Debugging static graphs can be less intuitive

SHAP/Captum	• Model interpretability libraries (global and local explanations) • SHAP: Shapley-value-based; Captum: PyTorch attribution suite • Support multiple explanation algorithms	• Feature attribution (Shapley and Integrated Gradients) • Layer and neuron-level insights • Visualization utilities • Multiinput support	• Explaining gene contributions to predicted health traits • Auditing bias toward batch or environment covariates • Prioritizing candidate biomarkers for validation	• Increases trust and transparency • Guides feature panel reduction • Identifies spurious shortcuts	• Computationally expensive for large models • Explanations not strictly causal • Variability across methods complicates consensus
Flower (federated learning)	• Framework for federated model training across silos • Python-first integration with major DL stacks • Pluggable strategies (FedAvg variants)	• Secure parameter aggregation • Custom client selection policies • Simulation and real deployment modes • Extensible strategy API	• Cross-farm model training without sharing raw expression • Breed-specific adaptation while preserving privacy • Federated fine-tuning of disease early-warning models	• Preserves proprietary data locality • Scales collaborative efforts • Reduces central data governance burden	• Network and synchronization overhead • Non-IID client data slows convergence • Requires careful privacy threat modeling
Terra/DNAnexus	• Cloud genomic analysis workspaces • Integrated data management and workflow execution • Collaboration and access control	• GUI and API pipeline launches • Data provenance tracking • Built-in security/ compliance (HIPAA, etc.) • Workspace-level sharing	• Hosting reproducible, multi-omics livestock pipelines • Sharing cohorts with controlled-access phenotypes • Scalable AI training with managed compute	• Low DevOps overhead • Compliance frameworks accelerate collaborations • Central audit trails	• Compute/storage costs accumulate • Vendor lock-in risk • Less flexible for highly custom GPU orchestration
AWS Batch/ ParallelCluster	• Managed batch scheduling and HPC provisioning in the cloud • Scales containerized or MPI workloads • Automates resource elasticity	• Auto-scaling compute environments • Spot instance cost optimization • Heterogeneous job queue policies • CloudWatch monitoring	• Scaling large self-supervised embedding training • High-throughput realignment or quantification • Distributed hyperparameter searches	• Elastic resource scaling reduces queue times • Pay-as-you-go efficiency • Integrates with object storage pipelines	• Requires cloud cost governance • Setup complexity for optimal throughput • Data egress charges for large downloads
Cytoscape	• Network visualization and analysis platform • Plugin ecosystem for functional enrichment • Interactive graph exploration	• Layout algorithms and styling • App store (enrichment and clustering) • Session reproducibility • Import/export multiformat	• Visualizing gene coexpression or regulatory networks from AI models • Presenting hub genes and module architectures to stakeholders • Merging trait associations onto network nodes	• Enhances interpretability and communication • Rich plugin ecosystem extends analytics • Low barrier for exploratory analysis	• Manual curation not easily automated at scale • Large networks become visually cluttered • Limited native support for real-time streaming data

Continued

TABLE 5.3 Ecosystem of computational tools and platforms enabling AI-driven livestock transcriptomics.—cont'd

Tool or platform name	Description	Key features	Applications in livestock transcriptomics	Advantages	Limitations
Neo4j (graph database)	• Native property graph database • Query language for complex relationships • Optimized for connected data traversal	• Index-free adjacency for speed • Graph algorithms library (centrality and community) • APOC procedures for ETL • Scalable clustering options	• Storing gene–trait–pathway–variant knowledge graphs powering AI embeddings • Executing subgraph queries for feature engineering • Supporting GNN training with dynamic graph retrieval	• Efficient complex relationship querying • Flexible schema evolution • Integrates with GNN pipelines via adapters	• Requires data modeling expertise • Memory-intensive for very large graphs • Licensing considerations for enterprise features
GENIE3/GRNBoost	• Machine learning methods for gene regulatory network inference • Tree-based importance decomposition • Predict regulator–target edges	• Handles nonlinear interactions • Scalable to thousands of genes • Produces ranked edge importance • Parallelization support	• Generating prior regulatory graphs for GNN initialization • Identifying key transcription factors in production traits • Feature selection via regulator hub extraction	• Captures complex dependencies beyond correlation • Provides interpretable edge rankings • Extensible to multicondition analyses	• False positives without experimental validation • Computationally heavy for full genome scale • Sensitive to expression normalization choices
JupyterLab	• Interactive computational notebook environment • Multilanguage kernel support (Python/R/Julia) • Rich UI for prototyping	• Inline visualization and widgets • Extension management (Git and terminals) • Reproducible narrative blending code/text • Remote kernel integration	• Rapid iteration on model architectures and feature engineering • Interactive QC and exploratory data analysis • Demonstrating model interpretability to domain experts	• Enhances collaboration and transparency • Low barrier for experimentation • Supports literate, reproducible analysis	• Notebook sprawl/version drift risk • Hidden state can cause irreproducibility • Not ideal for large-scale production pipelines
GitHub/GitLab	• Distributed version control hosting platforms • CI/CD automation and collaboration tools • Issue and project tracking	• Branching and pull requests • Actions/Runners for automated tests • Release and tagging workflows • Access control and auditing	• Managing AI pipeline code and model definitions • Automating QC/test suites for preprocessing scripts • Tracking issues for model performance regressions	• Structured collaborative development • Automated continuous integration ensures stability • Transparent change history aids reproducibility	• Requires governance to avoid code fragmentation • Large data storage limited (needs LFS/DVC) • Public repos risk IP exposure if misconfigured
Ontology portals (OBO Foundry/AgroPortal)	• Repositories for standardized biomedical and agricultural ontologies • Provide machine-readable term hierarchies • Support semantic interoperability	• Stable term identifiers and hierarchies • Cross-ontology mappings • API access for annotation pipelines • Versioned releases	• Harmonizing phenotype and tissue metadata labels for AI models • Enriching model outputs with pathway/trait ontology terms • Building knowledge graphs for explainable embeddings	• Improves data integration consistency • Enables semantic feature engineering • Facilitates cross-dataset alignment	• Gaps in livestock-specific coverage • Version drift requires synchronization • Mismatched term granularity can complicate modeling

be customized and not generic to the entire population, and that small-amplitude preclinical shifts can be detected. This type of individualized modeling assists the welfare by facilitating low-impact intervention at the right time to prevent the invasion to the point of intensive pharmacological reactions (Rappaport et al., 2025).

Disease resilience modeling elaborates this paradigm by distinguishing between molecular signatures of resistance, which is containment and speedy resolution of pathogen challenge with minimal collateral tissue disruption, and tolerance, which is attenuation of damage pathways despite pathogen persistence. Transcriptomic trajectories, which are causally unbiased using AIs that combine transcriptomic trajectories with measures of pathogen loads and physiological readouts, separate the causal directionality between the intensity of immune activation and productive performance tradeoffs. Resultant embeddings would facilitate stratification of the animals along the resilience axes that would guide breeding and management allocation strategies in pursuit of maximizing the level of robustness in the population without reducing efficiency. These lessons can be linked to the goal of antimicrobial stewardship, as these animals possess native regulatory orchestration that lessens the need for prophylactic or metaphylactic antimicrobial use, decreasing selective pressure toward the development of resistance and being congruent with the tenet of health to the population (Choudhary et al., 2024).

Transcriptomic clocks of multitissue sensitivity to nutritional imbalance, thermal load, disruption of social hierarchy, transport, and housing density are used to implement stress biology applications. Latent factor models of AI are isolated in order to separate conserved stress response motifs with orthogonal context-specific modifiers, which allow one to distinguish between adaptive and maladaptive regulatory adjustments. Combination with steady environmental sensor data and behavioral telemetry makes it possible to incorporate multimodal fusion models, which attribute probabilistic credits to particular stressors to make specific environmental adaptations rather than generalized mitigation, which can consume resources unnecessarily. The chronic subclinical level of stress observance is especially effective with respect to welfare because the chronic disturbance of inflammatory and metabolic networks predetermines the susceptibility to secondary diseases and shortened lifespan in animals, which in some cases remain unnoticed under the traditional system of monitoring, oriented on the acute events (Yalawar et al., 2025).

Transcriptomic deconvolution of cell-type-specific dynamics of activation and regulatory cytokine milieu is useful in immune repertoire and optimization of vaccination response. Signaling cascade prior knowledge graph-based AI models reveal network bottlenecks to prevent effective adaptive responses or excessive inflammatory amplification. This guides the improvement of the scheduling and adjuvant formulation to maximize immunogenicity and reduce welfare-compromising reactogenicity. Hypothetical intervention sequences simulated by coupled causal modeling can predict downstream transcriptional and phenotypic outcomes, which can be used to make rational decisions in immunomodulatory interventions to be used against breed-specific or region-specific pathogen ecological communities. The identical frameworks play a role in zoonotic spillover vigilance by raising red flags on abnormal host response capabilities that are suggestive of emergent or shifting pathogen interactions, which facilitates One Health surveillance integration (Yazdani et al., 2025).

The concept of metabolic welfare overlaps with the field of precision nutrition, where transcriptome-based characterization of nutrient sensing, mitochondrial performance, and anabolic/catabolic allocation of fluxes generates refined adjustments of the ration and helps to stabilize energy balance and mitigate the metabolic disorders that degrade welfare and productivity. Latent metabolic phenotypes identified by AI-driven clustering of transcriptomic states have nutritional needs unrelated to generalized feeding group predictions and thus can be fed stratified, reducing susceptibility to underfeeding-induced and inflammatory and oxidative stress burdens inherent to generalized feeding group models. Adaptive management cycles of dietary adjustment to transcriptomic reevaluation of performance outcomes and the mechanisms of feedback between the dietary adjustment and the transcriptomic reevaluation and the performance outcomes are a part of adaptive management that is guided by the convergence criteria encoded in the model's monitored stability thresholds (White, 2025).

Reproductive health applications entail the usage of an endometrial, ovarian, and embryonic transcriptomic exponent to close the fertility prediction interval, define the processes of subfertility, and understand the most suitable moment to intervene. Mapping AI sequences to outcomes across developmental stages can detect the abnormal timing of gene activation modules that compromise embryonic competence or placental development. This enables early management decisions that preserve animal welfare. Insights into transcriptomic reproductive immunotolerance and inflammatory balance reduce reliance on hormonal synchronisation protocols by facilitating naturally aligned breeding that respects physiological rhythms (Stokholm et al., 2024).

Genomic prediction using transcriptomes is becoming more significant in genetic selection as concerns welfare, such as robustness, relaxed temperament, soundness of the hoof, and disease resistance. The AI models are trained on the genotype to transcriptome and phenotype chain of mediation, which can be used to partition the genetic variance into mechanistically interpretable regulatory components. This facilitates indices of selection that scale molecularly based resilience metrics and manufacturing properties to provide a buffer against antagonistically related reactions, which might harm welfare. Uncertainty estimates impose constraints on the overconfident attempt to reach into the qualities of partial definition in underrepresented breeds using transcriptomics, which is expected to protect the evolution of genetic variation and adaptive latency (Xu et al., 2024).

Digital twins' systems can be used to model virtual cohorts, the dynamics of which can be simulated in reaction to environmental perturbations, management interventions, and genetic manipulations, and provide information about the distribution of the welfare outcomes without necessarily carrying out the experiment in vivo. The fidelity can be conserved on the real-world sensor-enriched longitudinal data, and the new observational data can be brought into the fray by the Bayesian updating processes to make the predictions truer to the world. The informed use of investment in facility design changes is done by use of scenario analysis to determine the expected welfare improvement by realigning prophylaxis or breeding program by quantifying the benefits and costs associated with the action (Castiglioni et al., 2025).

The ethical control of AI requires transparency and interpretability modalities, which convert complex latent representations into biologically understandable justifications that can be easily read and interpreted by veterinarians, nutritionists, and breeders. Actionable-based pathway-aggregated attributions and counterfactual explanations promote informed consent to intervention adoption and protection against misuse. Fairness auditing ensures that there are no lopsided distributions of predictive sensitivity and false alarm rates by breed, sex classes, and management systems that might be used to reinforce the welfare inequities. The postdeployment method of continuously monitoring model calibration and drifting with structured feedback by ingesting feedback in the form of field practitioners can quickly counteract the performance degradation caused by changes in seasonal pathogen shifts or changing management practices (Ning et al., 2025).

Sustainability synergies are created in which enhanced health and welfare lead to a decrease in mortality, morbidity, and inefficiencies that add to the environmental footprint per unit of production. The AI transcriptomic applications enhance the resilience, optimize the nutritional alignment, and prevent the disease outbreak, allowing the reduction of waste, the decrease of antimicrobial discharge, and the increase of resource conversion, which supports the integrative vision where welfare improvement and environmental stewardship are coemergent results. In this way, the application space demonstrates a trend of moving beyond the solitary diagnostic ambition to a systemic ideal (where molecular intelligence is a supplement to human know-how), proactive care (as a coordinated strategy), and welfare (as a quantitatively optimized drive) in the overall ecology of the livestock systems. Transcriptomics-driven AI application areas enhancing animal health and welfare are presented in Table 5.4.

5.6.2 Improving production efficiency

AI-enhanced transcriptomics applications to efficiency of production focus on decoding and controlling the molecular parameters of resource conversion, growth dynamics, reproduction output, and product quality, and, at the same time, restricting environmental externalities. The livestock systems are emerging as efficient due to the multilayered biological processes that involve regulation of feed intake, digestive and absorptive capacity, metabolic partitioning between maintenance and productive synthesis, resilience to stress, and the utilization of micronutrients. Transcriptome-wide measurements give time- and context-resolved snapshots of regulatory states operating such processes, but their intrinsic dimensionality and context sensitivity cannot be directly operationalized except with sophisticated modeling. The advantages of AI architecture are closing this gap with the learning of latent representations representing coexpression modules that are sensitive to metabolic pathways, endocrine signaling axes, and mitochondrial bioenergetics and translating high-dimensional signals into predictable and prescriptable constructs in nutrition, breeding, and management pipelines (Gao et al., 2025).

Feed efficiency is a critical performance where minor changes in energy sensing, ruminal microbial interaction, intestinal barrier integrity, and postabsorptive metabolic prioritization interact with each other to have an effect on conversion efficiency. Host transcriptomic models that combine host transcriptomes with metatranscriptomic or suggested microbial functional signatures disentangle host intrinsic and microbiome-mediated effects to permit more precise stratification of animals based on mechanistic efficiency archetypes. This type of stratification guides the formulation of a precision ration to match the inferred genomic and transcriptomic metabolic capacity to reduce systemic wastage of energy and subclinical inflammatory leakage that diverts nutrients to immune activation. The transcriptional responses to ration adjustments are modeled dynamically to enable the optimization loop of iterative optimization, which has an endpoint at the individualized feeding strategy that minimizes the variation in the conversion performance (Deng et al., 2025). The same is true of growth and muscle accretion: transcriptomic surveillance of myogenic differentiation, protein

TABLE 5.4 Transcriptomics-driven AI application areas enhancing animal health and welfare.

Application area	AI technique used	Benefits to animal health	Examples
Early systemic disease detection	• Supervised classifiers (Random Forest and Gradient Boosted Trees) on peripheral blood RNA-seq profiles • Regularized logistic regression for sparse *biomarker panels* • Ensemble stacking of linear, tree, and shallow neural nets • Feature selection via differential expression and SHAP filtering	• Detects disease before clinical signs • Enables earlier intervention, lowering morbidity • Reduces treatment costs and mortality • *Improves welfare through preemptive care*	• Cattle: early detection of bovine respiratory disease (BRD) from circulating leukocyte transcriptional signatures • Swine: preclinical identification of viral *outbreaks (PRRS) via blood transcript panels*
Subclinical mastitis/ udder health monitoring	• Support vector machines on milk somatic cell expression signatures • Multitask learning combining inflammation and pathogen load predictions • Sparse Lasso panels for deployable qPCR assays • Bayesian hierarchical models incorporating herd-level effects	• Identifies inflammation before yield loss • Reduces unnecessary blanket antibiotic use • Minimizes chronic tissue damage • Supports selective dry cow therapy strategies	• Dairy cattle: predicting subclinical mastitis 1–2 weeks before SCC spike • Buffalo: early immune gene signature indicating staphylococcal infection risk
Respiratory pathogen surveillance and severity stratification	• Transformer-based embedding of host response gene sets • Semisupervised contrastive learning on mixed-labeled cohorts • Multiclass gradient boosting for pathogen-type discrimination • Causal inference to distinguish coinfection versus primary agent	• Stratifies severity, enabling targeted treatment • Limits spread through faster isolation decisions • Reduces broad-spectrum antimicrobial reliance • Improves prognostic accuracy	• Calves: differentiating viral versus bacterial BRD etiologies from nasal swab host RNA • Sheep: classifying severity tiers in pneumonia-associated pathogen exposure
Parasitic and helminth burden assessment	• Graph neural networks integrating host gene coexpression and parasite life-cycle knowledge graph • Random Forest models on blood/intestinal mucosa transcripts • Domain adaptation to transfer signatures across breeds • Explainable AI (SHAP) to isolate immune pathway drivers	• Enables timely targeted deworming versus routine mass treatment • Slows anthelmintic resistance emergence • Improves growth and feed efficiency • Enhances immune resilience monitoring	• Small ruminants: predicting Haemonchus contortus burden from PBMC transcriptomes • Cattle: early detection of liver fluke-associated hepatic stress transcriptional patterns
Vaccine response prediction and optimization	• Variational autoencoders (latent immunomodulatory factors) • Temporal convolutional networks for postvaccination time-course RNA • Multitask learning (antibody titer and cell-mediated signature) • Bayesian optimization for adjuvant-responsive gene module selection	• Personalizes booster scheduling • Identifies nonresponders for alternative strategies • Enhances the durability of immunity • Lowers outbreak susceptibility	• Swine: predicting PRRS vaccine responder status from day 0 baseline blood transcriptome • Poultry: adjuvant selection guided by early cytokine gene module activation

Continued

TABLE 5.4 Transcriptomics-driven AI application areas enhancing animal health and welfare.—cont'd

Application area	AI technique used	Benefits to animal health	Examples
Antimicrobial stewardship and treatment necessity prediction	• Probabilistic classifiers integrating gene expression and minimal clinical metadata • Ensemble stacking with calibrated uncertainty (Platt scaling) • Active learning to refine ambiguous infection cases • Causal modeling to isolate inflammation not requiring antibiotics	• Reduces unnecessary antibiotic usage • Lowers resistance selection pressure • Improves targeted therapy decisions • Enhances stewardship compliance	• Dairy cows: distinguishing sterile inflammation versus bacterial mastitis • Feedlot cattle: flagging viral-only BRD cases unlikely to benefit from antibiotics
Heat stress resilience and thermal load monitoring	• Recurrent neural networks (LSTM) combining longitudinal blood transcriptomics and temperature-humidity index (THI) • Attention-based models highlighting heat-shock and oxidative genes • Domain adaptation across climatic regions • Latent factor models linking expression to physiological stress markers	• Early warning before a drop in intake or fertility • Maintains productivity under heat waves • Guides cooling and nutritional interventions • Reduces mortality in severe events	• Dairy cattle: predicting decline in milk yield 72 h ahead using leukocyte transcripts • Broilers: identifying heat-resilient flocks via stress pathway expression profiling
Nutritional and feed efficiency optimization	• Multimodal fusion (transcriptome and rumen fermentation metadata) • Elastic Net for sparse metabolically relevant gene panels • Reinforcement learning simulation using metabolic pathway-informed embeddings • Causal mediation (diet → gene expression → feed conversion ratio)	• Improves feed conversion and reduces waste • Minimizes metabolic disorders linked to diet shifts • Supports precision ration formulation • Lowers methane intensity indirectly	• Beef cattle: transcript-based residual feed intake (RFI) prediction models • Dairy goats: selecting diet formulations to reduce inflammatory gene activation
Metabolic disorder early prediction	• Gradient boosted survival models incorporating transcriptomic risk scores • Denoising autoencoders for latent metabolic dysregulation signatures • Time-to-event deep Cox networks with censored health records • k-NN anomaly detection for abrupt metabolic transcript shifts	• Prevents severe energy imbalance episodes • Reduces culling and fertility setbacks • Lowers veterinary intervention costs • Improves welfare via proactive supplementation	• Dairy cows: predicting subclinical ketosis from prepartum liver/blood RNA • Feedlot cattle: early acidosis risk flag via rumen epithelium transcriptional markers
Reproductive health and fertility support	• Multitask deep learning (pregnancy outcome, embryo viability, and estrus timing) • Feature selection via stability selection on reproductive tract transcripts • Transfer learning from human reproductive datasets (ortholog mapping) • Bayesian hierarchical modeling for parity-adjusted predictions	• Enhances conception rates with optimized AI (artificial insemination) timing • Reduces embryonic loss through early supportive care • Improves selection of fertile sires/dams • Shortens calving intervals	• Cattle: early pregnancy detection using day 18–21 blood gene panels • Swine: predicting litter size potentials from prebreeding endometrial expression

Immune profiling and inflammation differentiation	• Unsupervised clustering (NMF/PCA) to derive immune modules • Explainable transformers highlighting cytokine regulatory interactions • Semisupervised label propagation using limited immunophenotyping • Causal graph models partitioning innate versus adaptive activation	• Distinguishes pathogenic versus sterile inflammation • Enables targeted antiinflammatory interventions • Limits the overuse of broad immunomodulators • Improves disease resilience selection	• Sheep: separating parasite-induced eosinophilic response from bacterial inflammation • Cattle: profiling chronic low-grade inflammation affecting fertility
Stress and welfare monitoring (transport, handling, and crowding)	• Contrastive self-supervised embeddings of acute versus baseline stress states • SVM classification on blood cortisol-associated gene sets • Temporal anomaly detection using sliding window transcript scores • Domain adaptation across transport durations	• Mitigates prolonged stress exposure • Informs adjustments to handling protocols • Reduces immunosuppression and disease susceptibility • Enhances behavioral welfare assessments	• Beef cattle: identifying high-stress transport events from postarrival gene expression • Poultry: crowding stress biomarker panel guiding stocking density adjustments
Gastrointestinal integrity and rumen health	• Random Forest models on rumen papillae/mucosal transcripts • Multi-omics fusion (microbiome and host RNA) with graph embeddings • Sparse partial least squares (sPLS) for digestibility-linked gene sets • Composition-aware neural networks incorporating microbiome ratios	• Prevents subacute rumen acidosis and mucosal damage • Improves nutrient absorption efficiency • Reduces the incidence of digestive disorders • Supports methane mitigation strategies	• Dairy cows: linking epithelial gene expression and microbiome to acidosis risk • Sheep: host–microbe transcript interplay predicting pasture bloat susceptibility
Precision genetic selection for robustness and health traits	• Genomic prediction augmented with transcriptomic-derived genomic feature weights • Multitask models (health and production) sharing expression-informed latent layers • Graph neural networks integrating gene regulatory networks and SNPs • Mixture-of-experts specialized by breed	• Accelerates gains in disease resilience • Balances production with longevity and robustness • Reduces reliance on pharmaceuticals • Enhances heritable health trait accuracy	• Cattle: integrating leukocyte expression modules into GEBV for mastitis resistance • Pigs: adding immune gene expression eigenvectors to robustness selection indices
Zoonotic and emerging pathogen spillover risk assessment	• Anomaly detection on host antiviral transcriptional signatures • Semisupervised learning incorporating limited labeled outbreaks • Network propagation over host–pathogen interaction graphs • Diffusion models simulating hypothetical host response patterns	• Early flagging of atypical infection patterns • Reduces cross-species transmission risk • Informs surveillance targeting • Supports One Health strategies	• Swine: detecting atypical influenza-like host transcript responses • Poultry: identifying novel avian pathogen exposure via interferon-stimulated gene surges
Personalized management and digital twin health modeling	• Recurrent/transformer sequence models on longitudinal transcriptomics and sensor covariates • Variational state-space models estimating latent health trajectories • Active learning to prioritize the next sampling time point • Reinforcement learning for intervention policy simulation	• Tailors interventions (nutrition, vaccination timing) per individual • Anticipates deviations from the healthy baseline early • Reduces blanket interventions • Enhances lifetime productivity and welfare	• Dairy cows: individualized metabolic risk trajectory forecasting • Beef cattle: adaptive supplementation scheduling guided by transcriptomic state shifts

Continued

TABLE 5.4 Transcriptomics-driven AI application areas enhancing animal health and welfare.—cont'd

Application area	AI technique used	Benefits to animal health	Examples
Environmental toxicant/mycotoxin exposure detection	• One-class SVM/isolation forest on baseline-adjusted detoxification gene expression • Pathway-enriched Random Forest focusing on xenobiotic metabolism genes • Attention-based models weighting oxidative stress transcripts • Causal mediation distinguishing feed versus pathogen stress responses	• Rapidly identifies contaminated feed exposure • Prevents chronic organ damage and performance losses • Supports targeted feed batch recalls • Protects reproductive and immune function	• Dairy: aflatoxin-associated liver transcriptomic perturbation screening • Swine: fumonisin exposure detection via sphingolipid metabolism gene signature
Lameness/musculoskeletal inflammation risk screening	• Sparse logistic regression on inflammation and cartilage turnover transcripts (serum/blood-derived) • Ensemble models combining gait sensor features and transcriptomic biomarkers • Multimodal fusion of expression and acute phase proteins • SHAP explanations to validate biomarker panel	• Enables earlier corrective hoof care • Reduces chronic pain and secondary infections • Maintains mobility and feed intake • Lowers culling due to locomotion issues	• Dairy cattle: detecting subclinical laminitis inflammation before locomotion score change • Broiler chickens: early tibial dyschondroplasia stress gene panel monitoring
Pain and welfare biomarker identification	• Differential expression ranking and metalearning across stress paradigms • Denoising autoencoders extracting latent nociceptive signatures • Transfer learning from human/rodent nociception ortholog datasets • Integrated gradients highlighting key neuro-immune genes	• Enables objective pain assessment • Improves analgesic deployment precision • Reduces chronic suffering unrecognized clinically • Enhances ethical compliance reporting	• Cattle: postoperative pain transcript panel postdisbudding • Small ruminants: acute handling stress versus inflammatory pain signature separation
Multipathway antioxidant and oxidative stress management	• Multitask networks predicting oxidative damage markers and health outcomes • Gene set enrichment–guided feature engineering with regularized models • Temporal convolutional networks modeling antioxidant response kinetics • Causal graph modeling linking diet supplements → gene response → oxidative biomarkers	• Prevents cellular damage under heat or metabolic load • Supports longevity and reproductive capacity • Enhances resilience to immune challenges • Guides antioxidant supplementation timing	• Dairy cows: periparturient oxidative stress prediction from leukocyte transcripts • Poultry: vitamin E/selenium supplementation effect modeled via liver oxidative gene shifts

balance of turnover, oxidative and glycolytic fiber specification, and vascular support pathways. AI-based temporal alignment of expression patterns throughout development recognizes deviations of efficient growth patterns, thus early intervention is made to the developmental schedule or customized supplementation that stabilizes further development to the attractable compositional phenotypes without overly compensatory feed. Latent factor models predicting expression patterns onto downstream carcass quality proxies permit multiobjective optimization, in which efficiency improvements do not impact product attribute standards. These systems can be controlled by quantifying uncertainty, which mitigates the risk of the decision when exploring an underrepresented genetic background or new feeding regimes (Haas et al., 2025).

The downstream effect of reproductive efficiency on aggregate production efficiency is via the effect of reproductive efficiency on the generational interval, replacement rate, and lifecycle productivity. Reproductive readiness, implantation capability, and endocrine synchrony are markers found upon transcriptomic profiling of reproductive tissues, peripheral immunity, and metabolic interfaces. With the help of AI, trained on combined molecular and phenological data, probability distributions of successful reproductive outcomes are predicted, which can be used to schedule adaptively to minimize nonproductive days and resource use based on unsuccessful cycles. Coupling causal inference layers to separate regulatory bottlenecks, the modulation of which by nutritional, environmental, or selective breeding interventions should have multiplicative efficiency returns; in that regard, it stabilizes reproductive throughput (Emsen et al., 2025).

Nutrient partitioning and metabolic adaptive processes are facilitated by transcriptomic clarifying regulatory nodes coordinating the process of lipid mobilization, amino acid sparing, gluconeogenic robustness, and mitigation of oxidative stress. AI systems develop personalized scores of metabolic efficiency, which indicate the potential to sustain productive output when exposed to changing energy supply or temporary stressors without excessive catabolic loss. Decision support systems take in these scores with real-time information about the environment and feeding systems and suggest small corrective actions in the composition of the ration or thermal mitigation strategies to maintain metabolic stability and eliminate the expensive variation of performance. The archive of intervention outcomes optimizes policy networks over time, which eventually become anticipatory modulation and not reactive correction (Jasbi et al., 2025).

The transcriptomic intermediates can be used to gain resolution in genomic selection pipelines by introducing mechanistically proximal phenotypes, which can be transcriptomic intermediates. Mediation modeling separates genetic effects that are transmitted by fixed regulatory modules and diffuse background pathways and increases the accuracy of estimated breeding values of efficiency-related traits and decreases antagonistic correlated responses. Transfer learning makes it possible to reuse the underlying transcriptomic embeddings on related efficiency endpoints and reduces the data needs on emergent attributes, including resilience-adjusted feed conversion or emissions intensity normalized productivity. Federated modeling functions also enable distributed breeding agents to cooperatively increase efficiency prediction without losing sovereignty over data and more efficiently spread the improvement without the need for centralization (Haas et al., 2025).

Production efficiency is disrupted by the environmental efficiency on the grounds of shared molecular mediators of nutrient retention, regulation of nitrogen and methane, and oxidative stability. AI-improved transcriptomic analytics identify regulatory circumstances to predict the decline in waste release or the reduction in the strength of emissions at a particular productivity rate, which enables the synthesis of selection and control strategies that reduce environmental footprint per unit production. These molecular predictors are combined with model biophysical processes using digital twins to predict the long-horizon sustainability paths endowed with various breeding and feeding conditions and assess the tradeoffs and are used to optimize the tradeoffs between productivity and stewardship (Xing et al., 2025).

The efficiency loss caused by health conditions takes place when the metabolic load is diverted by a subclinical inflammatory or stress load. The combination of efficiency-based transcriptomic signature with welfare surveillance systems differentiates between primary inefficiency phenotypes and secondary inflammatory inefficiency and is applied to select the specific remediation to be implemented that will not artificially adapt the nutritional composition in case the rate-limiting factor is immune control. This segregation keeps the cost-effectiveness of the feed intake and reduces the long-term exposure to direct action interventions of potentially wide-spectrum reactivity that may upset the microbiome homeostasis or trigger resistance development (de Oliveira et al., 2025).

The operationalization of insights into transcriptomic efficiency depends on interpretable communication layers that translate shifts in latent factors into actionable levers, such as reformulating rations, altering grouping strategies, adjusting environmental controls, or recalibrating selection indices. Explainability modules summarize gene-level influences to create pathway narratives that are contextualised within metabolic or endocrine frameworks that are familiar to nutritionists and geneticists, thereby ensuring their adoption. Continuous monitoring of model calibration prevents the erosion of recommendation reliability caused by drift as production systems evolve or novel feed ingredients emerge. Fairness auditing also safeguards equitable efficiency gains across diverse breeds and management strata (Gao et al., 2025; Haas et al., 2025).

Lastly, the application spectrum of AI-enhanced transcriptomics to make production more efficient is suggestive of a paradigm shift in the systemic focus of the coarse-aggregate productivity measures to a fine, molecularly informed orchestration of efficiency. By threading transcriptomic intelligence through the production system in breeding, nutrition, environmental management, and health resiliency, the production systems are transformed into adaptive networks with anticipatory capabilities to match the biological capacity and resource provisioning. Economic sustainability is enhanced, environmental stewardship and ethical responsibility are strengthened, and resilience to production volatility is improved with this convergence, which is among the pillars of sustainable development of livestock production in an ever-limited resource and climatic environment. AI-enabled transcriptomic application areas driving production efficiency are shown in Table 5.5.

5.6.3 Genetic selection and breeding strategies

The use of AI to integrate transcriptomic data into genetic selection models is transforming breeding practices by interfering with mechanisms of genetic regulation, with mechanistic layers between genotype and multifaceted phenotypic behaviors. Conventional genomic selection uses statistical shrinkage or machine learning to compress vast marker panels into predictive breeding values, but remains partially agnostic to causal regulatory circuitry. In contrast, transcriptomics provides intermediate molecular phenotypes that capture dynamic gene regulation, splicing, and pathway-level coordination. This furnishes AI models with a structured signal that improves the attribution of variance to biologically coherent modules. By modelling the paths of mediation from genomic variation through transcript abundance to trait expression, algorithmic architectures can separate direct additive genetic effects from those routed through context-dependent regulatory cascades. This improves the estimation of breeding values and reduces the noise introduced by linkage disequilibrium with non-causal polymorphisms (Ghavi Hossein-Zadeh, 2024; Haas et al., 2025; Wang et al., 2025b).

Deep representation learning augmented by expression quantitative trait mapping enhances the discovery and prioritization of regulatory variants, such as distal modulators, the effect of which is either condition- or combinatorial-dependent. Multitask networks trained with cross-tissue and developmental window supervision are a way to learn common embeddings revealing pleiotropic regulatory centers without context specificity required to accurately characterize genotype-by-environment interaction. Additional incorporation of allele-specific expression signals further increases the ability to detect cis-acting elements and parent-of-origin effects that can be used to guide imprinting-aware selection policies. A shift in single trait mapping to combined network inference facilitates the detection of master regulators, with the addition of which to selection indices can spread desirable shifts in clustered traits without compelling many peripheral loci to cumulative inbreeding pressure (Strober et al., 2024).

It is necessary to use principled feature distillation when integrating transcriptomic profiles into composite selection indices to prevent overfitting and redundancy. Graph-constrained autoencoders and variational encoders learn the coexpression structure by encoding structure into low-dimensional latent factors, which are associated with biological pathways, at reduced dimensionality without affecting predictive sufficiency. These latent regulatory scores turn into candidate indicator characteristics that have greater heritability than unprocessed complex phenotypes since they eject environmental noise and put aside heritability-orchestrated components of variance. AI-based multiobjective optimization equalizes weight distributions in the conventional performance metrics, resilience metrics, welfare-related regulatory signatures, and environmental footprint proxies, creating indices that internalize long-term sustainability and productivity (Bohrer and Dorn, 2024; Ghavi Hossein-Zadeh, 2026a).

The importance of accounting for antagonistic genetic correlations arises when raising molecular indicators to the level of selection criteria. The causal graph learning approach can draw a distinction between correlation due to the presence of common upstream regulators and actual tradeoffs in which genetic benefit in one process inherently limits another. This allows strategic buffering whereby a selection pressure on a central regulatory module is balanced or counterbalanced with the weighting of antagonistic modules. The dynamic reestimation of covariance structures through the rolling integration of newly sequenced cohorts and updated transcriptomic baselines ensures the continued relevance of the index as production environments change, thereby preventing the system from becoming locked into outdated regulatory architectures (Job et al., 2025).

The preservation of genetic diversity and an increase in molecularly informed selection are enabled by AI-assisted mating plan optimization akin to the expected progeny merit, inbreeding coefficients, and complementarity of regulatory haplotypes. The constraint-aware evolutionary algorithms are used to assess the simulated progeny transcriptomic potential in alternative mating networks and select crosses that construct combinations of synergistic regulatory modules without spreading allele frequencies and undergoing drift to erode adaptive capacity. Quantification of uncertainty

TABLE 5.5 AI-enabled transcriptomic application areas driving production efficiency.

Application area	AI technique used	Benefits to production efficiency	Examples
Feed efficiency (residual feed intake and RFI) optimization	• Multitask deep neural networks linking gene expression modules to RFI and average daily gain (ADG) • Elastic Net/Lasso for sparse biomarker panels deployable via qPCR • Graph neural networks (GNN) integrating coexpression and metabolic pathway topology • Transfer learning across breeds to stabilize prediction variance • Model explainability (SHAP, Integrated Gradients) to refine panels	• Lowers feed cost per unit gain • Improves dry matter intake (DMI) to gain ratio • Enables early selection of efficient juveniles • Reduces overfeeding waste • Accelerates genetic gain in feed efficiency indices	• Beef cattle: blood/leukocyte transcript panel predicting low RFI class prefeedlot • Dairy heifers: juvenile expression signatures forecasting mature efficiency • Sheep: rumen epithelium expression ranking pasture efficiency
Milk yield and composition efficiency	• Recurrent neural network (RNN) modeling of lactation curve and temporal expression shifts • Multi-omics fusion (RNA and milk MIR spectra) with attention layers • Bayesian hierarchical models adjusting for parity and days-in-milk • Reinforcement learning simulating diet adjustments based on predicted transcript states	• Optimizes energy partitioning for sustained peak • Improves milk solids per unit feed • Reduces inefficiency from extended negative energy balance • Informs the ration timing to smooth the curve persistence • Increases return on feed inputs	• Dairy cows: early postpartum blood transcripts forecasting peak milk-solids ratio • Goats: mammary tissue expression guiding protein:fat ratio tuning
Growth rate and lean muscle accretion	• Variational autoencoders capturing latent anabolic regulation • Semisupervised learners leveraging partially labeled gain datasets • Convolutional models integrating regulatory motif scores → expression → growth traits • GNN connecting myogenic transcription factors to downstream effectors	• Increases lean gain per day • Reduces days to market weight • Improves uniformity across cohorts (lower variance) • Enhances feed conversion via reduced nonproductive adiposity • Lowers the cost of gain	• Swine: muscle biopsy RNA predicting ADG before finishing • Broilers: early chick blood transcript score stratifying growth classes
Carcass quality and meat composition (marbling, tenderness and IMF%)	• Ensemble models (gradient boosting and deep MLP) on muscle transcriptomes • Stability selection isolating lipid deposition pathways • Multitask learning predicting marbling, tenderness, and shear force • Domain adaptation (grass vs. grain finishing)	• Aligns finishing duration with optimal carcass endpoint • Improves grade (marbling score) consistency • Reduces overfattening feed wastage • Enhances tenderness predictability, lowering variability penalties • Supports premium market segmentation	• Beef: longissimus dorsi RNA panel predicting marbling EBV improvement • Lamb: transcript-based tenderness index for harvest timing decisions

Continued

TABLE 5.5 AI-enabled transcriptomic application areas driving production efficiency.—cont'd

Application area	AI technique used	Benefits to production efficiency	Examples
Reproductive turnaround and shortening of nonproductive days	• Deep survival (Cox neural nets) integrating luteal-phase expression • Multitask models (conception, calving interval, and progeny vigor) • Transfer learning across parity groups • Ensemble feature attribution vetting biomarkers	• Reduces days open, improving lifetime milk per cow-year • Lowers replacement heifer requirement (cost) • Increases calves/weaned progeny per female per year • Improves synchronization success rates • Enhances reproductive resource allocation efficiency	• Dairy: days 21–28 blood expression predicting conception outcome • Sows: uterine transcript panels forecasting litter efficiency metrics
Accelerated genetic gain (efficiency-weighted selection indices)	• Genomic prediction augmented with expression-derived pathway weights • Multiobjective optimization (Pareto front: production vs. robustness) • Federated learning across herds without raw data pooling • Explainable AI ensuring balanced trait weighting	• Increases accuracy of breeding values for efficiency traits • Shortens generation interval for efficiency-focused sires/dams • Prevents selection of metabolically costly alleles • Enhances cumulative feed efficiency gains per year • Improves index stability across environments	• Dairy: integrating metabolic gene module eigenvectors into Net Merit variants • Pigs: composite EBV using feed efficiency and robustness transcript features
Methane intensity reduction aligned with performance	• Multimodal deep fusion (rumen tissue RNA, microbiome, and diet) • Sparse PLS isolating methanogenesis pathway signatures • Causal mediation (diet → transcript → CH_4) modeling • Clustering for low-emission high-output phenotypes	• Lowers energy loss as methane, improving gross feed efficiency • Supports carbon intensity reduction targets (market premiums) • Identifies animals combining low emissions and high yield • Reduces variability in methane per kg of product • Guides diet formulation for both efficiency and emissions	• Dairy: rumen epithelium transcripts predicting methane per kg ECM (energy-corrected milk) • Sheep: blood expression biomarkers for low-CH_4 high-growth selection
Precision nutrient partitioning (protein and energy utilization)	• Dynamic Bayesian networks modeling nitrogen metabolism gene regulation • Attention temporal models on serial liver/blood RNA • Reinforcement learning simulating rational scenarios • Optimization algorithms minimizing predicted N excretion	• Improves amino acid utilization efficiency • Reduces costly protein oversupplementation • Lowers nitrogen waste (environmental compliance) • Enhances net energy retained for growth or milk • Stabilizes production during diet changes	• Dairy: mid-lactation liver transcript panel guiding MP (metabolizable protein) adjustments • Beef: urea cycle expression informing precision protein supplementation

Heat resilience with maintained output	• LSTM, attention combining THI, and transcript trajectories • Domain adaptation across climatic zones • Latent state-space disentangling heat versus production signals • Early warning classifiers for yield drop thresholds	• Preserves milk/ADG under thermal stress (higher output per hot day) • Reduces production losses requiring compensatory feeding • Improves persistence in high-heat seasons • Optimizes timing of cooling interventions (cost-efficient) • Stabilizes intake patterns	• Dairy: predictive blood RNA alert 72 h before heat-induced yield decline • Broilers: transcript-based resilience scoring supporting stocking adjustments
Weaning and transition efficiency (growth continuity)	• Temporal anomaly detection on peri-weaning expression series • Contrastive learning distinguishing smooth versus poor transitions • Clustering of immune-metabolic adaptation modules • Risk scoring for postweaning growth dip	• Minimizes growth setbacks, reducing extra days to target weight • Improves feed conversion in the early postweaning phase • Reduces need for remedial high-cost diets • Enhances uniformity entering grower/finisher stages • Lowers mortality-related production loss	• Piglets: preweaning blood transcript index predicting postweaning ADG stability • Calves: expression markers forecasting successful starter adoption
Precision supplementation and additive ROI optimization	• Causal inference linking supplement → pathway activation → performance KPI • Bayesian optimization for minimal effective dose discovery • Few-shot learning for evaluating new additives with limited data • Ensemble regression estimating diminishing returns curves	• Lowers feed additive cost per kg gain or kg milk • Prevents oversupplementation inefficiency • Accelerates evaluation of novel additives (faster adoption cycles) • Increases nutrient conversion efficiency • Improves the rationcost:output ratio	• Dairy: selenium/vitamin E dose refinement from oxidative stress transcripts • Beef: ionophore response estimation via rumen epithelial expression shifts
Slaughter timing and finishing optimization	• Regression models on muscle/blood transcript panels predicting optimal finish date • Survival/threshold models detecting plateau in marginal carcass value gain • Reinforcement learning exploring days-on-feed versus value curves • Cost-aware AI integrating feed cost and carcass forecast	• Reduces unnecessary feed days (cost savings) • Optimizes carcass yield grade versus input cost ratio • Improves uniformity in finished groups • Decreases overconditioning penalties • Enhances scheduling logistics	• Beef: blood RNA predicting marbling plateau for early harvest decision • Lamb: transcript-informed reduction in days on concentrate without yield loss
Muscle fiber type and meat quality modulation	• NMF clustering of fiber-type expression modules • GNN linking regulatory transcription factors to metabolic shifts • Semisupervised classification of oxidative versus glycolytic bias • Generative intervention simulation (expected quality outcomes)	• Aligns muscle metabolism with desired carcass traits • Improves consistency in pH/tenderness, reducing downgrades • Reduces inefficient energy use linked to imbalanced fiber profiles • Supports targeted management (exercise/nutrition) for quality • Enhances predictability of processing yields	• Pigs: longissimus dorsi modules predicting ultimate pH and drip loss • Broilers: expression-based selection for balanced fiber oxidative capacity

Continued

TABLE 5.5 AI-enabled transcriptomic application areas driving production efficiency.—cont'd

Application area	AI technique used	Benefits to production efficiency	Examples
Persistence of lactation/egg production longevity	• Multitask temporal transformers (yield, persistency, and decline slope) • Survival neural networks modeling time to production threshold drop • Regularized feature selection stabilizing biomarker panels • Longitudinal embedding of expression trajectories	• Extends productive lifespan, reducing replacement cost • Increases total lifetime yield per unit rearing investment • Smooths late-stage production, preventing feed inefficiency • Improves capital efficiency (fewer replacements) • Enhances predictability for supply planning	• Dairy: mid-lactation transcript modules forecasting persistency curve • Layers: gene signatures predicting extended peak-lay duration
Crossbreeding and heterosis optimization	• Admixture-aware deep models combining breed composition and expression • Transfer learning aligning breed-specific expression spaces • Bayesian decomposition of additive versus dominance expression effects • Graph models linking allelic variants to heterosis modules	• Identifies cross-combinations maximizing feed efficiency and growth • Captures hybrid vigor without excessive trial matings • Enhances robustness, reducing nonproductive days • Improves multitrait efficiency balance • Speeds heterosis exploitation in commercial lines	• Beef composites: the expression heterosis index predicting gain and efficiency • Dairy crossbreds: blood transcripts anticipating hybrid feed efficiency uplift
Digital twin production optimization (individual animals)	• Variational recurrent state-space models fusing sensor and transcript data • Reinforcement learning recommending individualized feeding or grouping • Active learning prioritizing next sampling time (cost-efficient) • Multiobjective optimization balancing output versus resource use	• Increases per-animal yield precision versus blanket strategies • Reduces overallocation of high-density feed • Improves forecasting accuracy for inventory planning • Enhances lifetime output per animal • Lowers intervention cost per productivity unit	• Dairy: digital twin predicting energy balance to preempt yield dips • Swine: adaptive grouping decisions guided by transcriptomic state forecasts
Nutrient waste (nitrogen/phosphorus) minimization with output retention	• Systems biology neural nets linking expression to N/P excretion metrics • Multiobjective optimization (output vs. waste) • Penalized regression isolating urea cycle and transporter genes • Scenario simulation via generative surrogate models	• Reduces feed protein/mineral cost while maintaining output • Improves feed conversion by lowering unutilized nutrient loss • Supports compliance (environmental constraints), sustaining operations • Enhances the ration efficiency margins • Increases net income per kg of product	• Dairy: liver expression guiding protein ration cuts without milk loss • Poultry: intestinal transporter gene panel predicting phosphorus utilization

Rapid diet transition adaptation (forage ↔ concentrate)	• Domain adaptation correcting batch/diet expression shifts • Temporal convolutional models tracking adaptation kinetics • Anomaly detection flagging maladaptive inflammatory gene responses • Causal pathway modeling separating diet effect from health confounders	• Shortens adaptation period, reducing lost gain or yield • Maintains intake consistency, improving feed conversion • Minimizes setbacks requiring extra transition feed phases • Reduces variability across cohorts posttransition • Increases throughput efficiency in finishing systems	• Dairy: rumen epithelium transcripts forecast safe escalation to high-concentrate diets • Sheep: expression-guided pasture-to-feedlot transition protocols
Precision culling and retention (efficiency-based)	• Ensemble gradient boosting predicting remaining productive value • Deep survival models estimating time to efficiency decline threshold • Counterfactual modeling (retain vs. cull scenarios) • Explainable dashboards ranking drivers	• Removes chronically inefficient animals early (feed savings) • Retains late-maturing but efficient animals • Improves overall herd average efficiency metrics • Optimizes resource allocation (housing, feed) • Increases cumulative output per housing space	• Dairy: mid-lactation transcript and yield persistency scoring informing retention • Beef: identifying low-efficiency feeders for earlier marketing
Integrated multitrait economic index enhancement (efficiency-centric)	• Multioutput deep learning predicting suite of economic KPIs from expression modules • Hierarchical Bayesian models modeling correlated trait structure • Pareto front analysis visualizing trade-offs (yield vs. feed, and vs. emissions) • Federated learning enabling cross-region model generalization	• Balances multiple efficiency traits simultaneously • Increases economic return per animal by optimized trait weighting • Reduces antagonistic selection outcomes • Increases stability of the index under environmental variation • Speeds deployment of updated indices with new data	• Dairy: integrating feed efficiency, methane intensity, and solids yield transcript drivers • Swine: composite index combining ADG, FCR, robustness gene modules

incorporated in breeding value prediction discourages too much focus on the candidates with a mediated but loosely supported transcriptomic advantage, thus limiting unintended reduction in the genetic source (Pook et al., 2025).

When transcriptomic plasticity measures are included, genotype-by-environment interaction management is resolved. The models that encode reaction norms on the regulatory module level predict stability or responsiveness of gene networks to thermal, nutritional, or pathogen pressure gradients. Selection can then discriminate between robustness, in which there is maintenance of optimal regulatory homeostasis across environments, and adaptive plasticity, where controlled modulation enhances fitness in a given stressor. Trading these profiles over a breeding population can help maintain resilience to climate volatility and not to reduce efficiency in conventional settings, and AI scenario simulation can help to predict long-horizon implications of alternative weighting plans in the face of plausible environmental dynamics (Waters et al., 2024; Ghavi Hossein-Zadeh, 2026b).

The addition of the epigenetically mediated transcriptional variation in the selection strategies adds further layers. Although the heritability of epigenomic marks can be biased or situation-specific, AI models can be trained to understand probabilistic associations between sustained chromatin accessibility states and patterns of transcription factor occupancy, along with motif-enriched expression modules and phenotypic benefits that are stable. This type of coupling enables indirect selection on regulatory settings that can hasten response to selection when traditional additive consequences reach a plateau. Nevertheless, the existence of layers of governance imposes a conservative usage whereby it is necessary to replicate epigenetically mediated relationships between cohorts and prove that inferred marks are not simply transient management artifacts that are easily degraded (Tahir et al., 2024).

Polygenic resilience scoring builds upon conventional risk prediction by averting attention toward metrics of susceptibility to risk prediction in favor of the ability to recover regulation quickly after perturbation. Dynamical response parameters, for example, metabolic or immune modules' speed of restabilization after imposed challenge, are trained on recurrent architectures and stored as resilience properties within selection models. These attributes make up the static efficiency or production indices by encoding the kinetic characteristics of biological regulation, hence balancing selection and long-term system stability. Digital twin simulations extrapolate the cumulative population-level effect of the gradual changes in the resilience scores to facilitate strategic investment in slower accruing and compounding increases in robustness (Alves da Mata, 2024).

Ethical and socioeconomic aspects are also intertwined because AI enhances molecularly guided selection. Transparency mechanisms convert the hidden factors with complexity into storytelling based on familiar biological pathways, enabling the stakeholders to assess the welfare and sustainability consequences of suggested index changes. Fairness audits ensure that predictive accuracy and error patterns of transcriptomic-mediated breeding values are held in balance by subpopulations, without any systemic underestimation of possible underrepresented genetic lines in the past. The data governance models facilitate consent compliance and fair data sharing in cases of collaboration with transcriptomic datasets on the basis of accuracy gains to promote trust and further data flow necessary to recalibrate models (Shafik, 2024).

Postdeployment monitoring is a continuous measure that bridges the gap between the estimated genetic development and the actual phenotypic development. Detection algorithms for drift identify a difference between expected and observed overall transcriptomic changes, and root cause analysis can be used to identify the root cause of an emergent environmental pressure, unmodeled epistasis, or inadvertent selection of proxy artifacts. Retraining of AI using such signals helps to maintain the relationship between the models and changing biological and operational realities. This integrative, reflexive process results in genetic selection being more of a statistical venture than a mechanistically based, ethically aware, and adaptively controlled system, which speeds up balanced genetic accretion, while preserving diversity, resilience, and social permission to exist through transcriptomics-implemented AI. An overview of genetic selection and breeding strategies utilizing AI and transcriptomics is presented in Table 5.6.

5.7 Integration with other data modalities

Combining transcriptomic data with other data modalities using AI defines a multiscale evidentiary continuum between variations in genes, regulation, physiological condition, behavioral expression, environmental contact, and management situation of livestock systems. The conceptual change between remote molecular layers to holistic digital phenotypes is based on algorithms that can be used to bring heterogeneous structures into harmony, reconcile different time granularities, and find causal but not merely correlative cross-modal relationships. Genomic variation gives the genomic template underlying transcriptional regulation, but untranslated polymorphisms are opaque until they are placed in context in the landscape of chromatin accessibility, methylation patterning, and transcription factor occupancy as the product of epigenomic assays. These regulatory topologies are coded in AI-mediated graph representations, which can be path-based,

TABLE 5.6 Overview of genetic selection and breeding strategies utilizing AI and transcriptomics.

Application area	AI technique used	Benefits	Examples
Transcriptome-augmented genomic prediction (improving EBV/GEBV accuracy)	• Multikernel GBLUP (genomic and expression-derived kernels) • Bayesian multi-omics prediction (BayesRC/BayesB with expression priors) • Deep learning (multilayer attention over SNP blocks and gene expression modules) • Stacked ensemble (GBLUP, gradient boosting, and transformer embeddings)	• Higher prediction accuracy for complex traits • Faster genetic gain per generation • Better early-life selection decisions • Improved reliability for low-heritability traits	• Dairy: adding leukocyte gene modules to mastitis resistance GEBV • Beef: feed efficiency EBV uplift using muscle expression kernels • Poultry: egg persistency prediction improved with ovarian transcript patterns
Expression-informed variant and SNP prioritization	• Deep sequence-to-expression CNNs predicting regulatory impact • Feature attribution (SHAP) ranking SNP influence on key gene modules • Graph neural networks linking variants → regulatory network nodes • Active learning to iteratively validate uncertain variants	• Focused genotyping panels (cost reduction) • Enhanced causal variant discovery • Reduced noise in genomic models • Accelerated functional follow-up	• Swine: promoter variant prioritization for growth genes • Cattle: regulatory SNP ranking in immunity loci • Sheep: selection of variants modulating wool fiber expression
Multitrait AI-optimized selection indices	• Multiobjective evolutionary algorithms (Pareto fronts) • Multitask deep nets predicting correlated trait outcomes • Economic weighting optimization with reinforcement learning simulators • Hierarchical Bayesian modeling integrating trait covariance with expression	• Balanced genetic progress without negative trade-offs • Reduced antagonism between production and robustness traits • Improved lifetime profitability • Transparent trade-off visualization	• Dairy: custom index integrating milk solids, fertility, somatic cell score (SCS) expression modules • Beef: index balancing growth, marbling, and resilience transcripts • Pigs: feed efficiency, robustness, and survivability index
Disease resistance polygenic risk scoring with expression weighting	• Weighted PRS using eQTL effect sizes • Causal mediation models (genotype → expression → phenotype) • Sparse Bayesian regression emphasizing immune gene modules • Transfer learning across breeds for resistance trait sharing	• Improved early resistance classification • Reduced morbidity and treatment cost • Supports targeted vaccination strategy • Increases selection intensity for resilience	• Cattle: bovine respiratory disease resistance PRS and leukocyte expression weights • Poultry: coccidiosis resistance risk stratification • Sheep: parasite (Haemonchus) burden resistance scores
Accelerated eQTL and regulatory architecture discovery	• Matrix factorization (NMF) to compress expression for eQTL scans • Bayesian fine-mapping with deep prior on regulatory regions • Graph-based colocalization (variant–gene–trait) networks • Cloud-parallelized sparse regression (distributed AI scheduling)	• Faster identification of regulatory hotspots • Improved causal inference for trait pathways • Prioritized targets for selection or editing • Reduced computational cost per discovery	• Dairy: lactation stage-specific eQTL for milk components • Swine: muscle growth eQTL acceleration • Chickens: immune tissue eQTL for vaccine response

Continued

TABLE 5.6 Overview of genetic selection and breeding strategies utilizing AI and transcriptomics.—cont'd

Application area	AI technique used	Benefits	Examples
Gene regulatory network (GRN) reconstruction for target trait pathways	• Dynamic Bayesian networks (time-course expression) • GNN inferring edge weights (TF–target) from multitissue data • Causal discovery (NOTEARS/DAG learning) • Ensemble GRN inference (ARACNe, regression, and attention fusion)	• Identification of master regulators • Refined biomarker panels for selection • Enables pathway-focused breeding strategies • Supports CRISPR target nomination	• Beef: myogenesis GRN highlighting key TF hubs • Dairy: lactation persistency regulatory modules • Poultry: heat-stress response GRN nodes
Crossbreeding and heterosis prediction (expression-enabled)	• Admixture-aware deep models (breed proportion embeddings and expression) • Dominance and epistasis modeling via tensor factorization • Bayesian decomposition of additive versus nonadditive expression effects • Metalearning transferring across crossbred populations	• Predicts hybrid vigor before large-scale matings • Maximizes performance × robustness synergy • Reduces trial-and-error cross-design cost • Maintains diversity while boosting output	• Beef composites: predicted heterosis for gain and fertility • Dairy crossbreds: feed efficiency uplift forecasts • Swine: cross-evaluation for litter size and robustness
Adaptive introgression and breed composition optimization	• Optimization algorithms (genetic algorithms) simulating allele introgression trajectories • Local ancestry inference with deep CNNs • Fitness prediction models combining genotype and expression plasticity • Scenario simulation (RL) for composition over generations	• Faster incorporation of adaptive alleles (e.g., heat tolerance) • Maintains core production genetics • Reduced time to stable composite lines • Enhanced environment-specific performance	• Cattle: slick coat heat-tolerance introgression modeling • Sheep: parasite resilience allele blending • Goats: arid-climate adaptation incorporation
Embryo/oocyte viability and developmental potential selection	• CNN and transformer on single-cell or bulk embryo transcriptomes • Probabilistic survival models for implantation success • Transfer learning from human/rodent embryogenesis datasets • Feature selection (stability selection) for minimal qPCR panels	• Higher implantation and pregnancy rates • Reduced cost per successful transfer • Early discard of low-probability embryos (ethical/resource gains) • Improved genetic gain via more viable selections	• Dairy: blastocyst RNA viability scoring • Beef: in vitro embryo competence prediction • Swine: oocyte maturation transcript markers
Gamete and sire fertility prediction (semen transcriptomics)	• Ensemble classifiers (gradient boosting and SVM) on sperm RNA profiles • Autoencoders capturing latent fertility signatures • Multitask models (motility, morphology, conception rate) • Explainable AI highlighting key fertility gene modules	• Improved sire selection accuracy • Reduced failed inseminations (cost/time) • Enhanced reproductive efficiency metrics • Supports cryopreservation prioritization	• Bulls: sperm RNA predicting conception success • Boars: semen transcript markers for litter size • Roosters: fertility retention under heat stress

Selection for environmental resilience (heat, disease, and nutritional stress)	• Domain adaptation aligning multienvironment expression signatures • Reaction norm modeling with expression covariates • Metalearning for rapid adaptation to new environments • Multitask resilience scoring (heat, pathogen, and feed restriction)	• Maintains productivity under stress • Reduces culling in extreme climates • Broadens geographical deployment of lines • Enhances sustainability credentials	• Dairy: heat resilience EBV boosted by leukocyte heat-shock transcripts • Poultry: combined heat-pathogen tolerance scoring • Sheep: drought resilience selection using metabolic expression
Longevity and functional herd life prediction	• Survival deep learning (Deep Cox/RNN) on longitudinal expression and pedigree • Recurrent state-space models tracking aging pathways • Multimodal fusion (health records and transcripts) • Feature attribution isolating aging biomarkers	• Extended productive lifespan • Lower replacement rearing costs • Improved cumulative yield per animal • Reduced involuntary culling	• Dairy: mid-lactation transcript predictors of herd life • Beef cows: longevity risk scoring from immune-metabolic expression • Layers: persistence gene modules linked to lay duration
Early-life immune competence and robustness selection	• Neonatal blood transcript classifiers (Random Forest and XGBoost) • Semisupervised learning on sparse challenge data • Transfer learning from adult immune panels • Causal modeling distinguishing maturation versus pathology genes	• Early identification of resilient stock • Reduced juvenile morbidity/ vaccination failures • Higher survival and uniformity in the market • Enhanced long-term performance foundation	• Calves: week-1 immune module score predicting disease incidence • Piglets: preweaning expression forecasting postweaning growth robustness • Chicks: early innate immune transcript scoring
Genetic diversity and inbreeding risk monitoring via expression variability	• Variance component modeling linking expression variance to the inbreeding coefficient • Clustering to detect loss of expression heterogeneity • Autoencoder reconstruction error as diversity proxy • Simulation of drift impacts on expression networks	• Early warning of diversity erosion • Informed mating to mitigate inbreeding depression • Preservation of adaptive potential • Supports compliance with conservation goals	• Dairy: expression variance tracking in elite bull lines • Heritage sheep: monitoring adaptive immune diversity • Poultry: layer line diversity safeguard program
Polygenic editing and CRISPR target prioritization	• Causal graph modeling to rank intervention nodes • In silico perturbation via generative models (simulate knockouts) • Off-target risk prediction with transformer DNA models • Multiobjective optimization (effect size vs. pleiotropy risk)	• Prioritizes high-impact low-risk edit targets • Reduces experimental validation cycles • Minimizes unintended trait trade-offs • Accelerates functional confirmation	• Beef: editing candidate myostatin regulatory enhancers • Dairy: lipid metabolism regulator prioritization • Pigs: resilience pathway editing shortlist
Rare and structural variant functional impact prediction	• Deep CNN/RNN models on sequence, chromatin, and expression context • Attention models integrating structural variant breakpoints with gene modules • Graph embeddings of variant–gene–pathway relationships • Semisupervised anomaly detection for rare variant effect inference	• Unlocks hidden heritability components • Improves accuracy for elite donors • Enables informed inclusion/exclusion of carriers • Guides bespoke mating plans	• Dairy: structural variant effect on milk fat genes • Swine: rare promoter variant impacting feed efficiency • Sheep: CNV influencing parasite resistance expression

Continued

TABLE 5.6 Overview of genetic selection and breeding strategies utilizing AI and transcriptomics.—cont'd

Application area	AI technique used	Benefits	Examples
Epigenetic–transcriptomic integration for heritable plasticity	• Multiview learning (canonical correlation/deep fusion of methylome and RNA) • Latent factor models capturing stable versus transient expression components • Causal mediation (epigenetic mark → expression → phenotype) • Temporal alignment algorithms for developmental windows	• Distinguishes stable selectable regulation from transient noise • Informs selection for adaptive plasticity traits • Enhances prediction of G×E performance • Refines biomarker panels to heritable components	• Dairy: methylation–expression fusion for fertility persistence • Beef: epigenetic modulation of feed efficiency modules • Poultry: early-life epigenetic marks predicting growth robustness
Digital twin breeding program simulation	• Agent-based RL simulating mating and selection decisions • Generative surrogate models predicting trait response to selection pressure • Scenario optimization (economic, genetic, and risk constraints) • Uncertainty quantification (Bayesian model ensembles)	• Strategic foresight before real-world implementation • Reduced costly trial generations • Optimized selection intensity versus diversity retention • Accelerated return on genetic investment	• Dairy: simulation of alternative fertility and production weighting strategies • Swine: evaluating robustness-inclusive indices • Beef: forecast of heat tolerance introgression timeline
Maternal line and uterine environment selection	• Transcriptomic profiling of uterine/placental tissues with predictive models • Multitask modeling (gestation success and progeny vigor) • Feature selection isolating maternal effect gene sets • Cross-generational modeling linking dam expression to offspring performance	• Improved neonatal survival and uniformity • Enhanced early growth trajectory • Reduced dystocia-associated losses • Strengthened maternal lineage value	• Beef: placental expression predicting calf vigor • Dairy: uterine transcript markers for conception retention • Sheep: maternal efficiency gene panel
Integrated AI-driven composite health–production–efficiency index development	• Multioutput deep nets linking expression modules to broad trait suite • Hierarchical Bayesian integration (health, production, and efficiency) • Pareto frontier mapping for stakeholder weighting scenarios • Federated learning across regions/breeds for generalizability	• Holistic improvement without sacrificing resilience • Faster deployment of updated indices • Enhanced stakeholder transparency • Robust performance across environments	• Dairy: composite index integrating mastitis, fertility, solids yield transcripts • Pigs: ADG, FCR, and robustness multimodule index • Poultry: disease resistance, egg persistency, and feed efficiency combination

attributed to thread sequence level perturbations through cis and trans control architectures into emergent coexpression modules and downstream complex traits. The higher-order graph edits of structural variants and copy number changes are included, and based on message passing operations, their propagated influence can be integrated with functional variation prioritization (Mardinoglu et al., 2025).

Extending transcriptomic analyses to incorporate proteomic and metabolomic data adds phenomena of translation efficiency, posttranslational modification, turnover, enzymatic flux, and network compensation of metabolic networks that overturn naive expectations of linear relationships between mRNA concentration and phenotype. Multimodal encoders use modality-specific feature extractors over common latent spaces constrained to be biophysical (e.g., stoichiometric or thermodynamic), which reduces the semantic disparity between transcriptional potential and realized biochemical throughput. Disentanglement objectives disentangle latent dimensions to include the core physiological constitutions, capturing conserved core physiology and context-adaptive responses that enhance generalization in the face of changing nutritional or climatic environments. Metabolomic time series, which have rapid kinetics compared to slower transcript reprogramming, measure temporal alignment modules, which in turn provide lag structures, and therefore arm predictive systems with a sense of anticipation, as opposed to a reactionary response (Li et al., 2024; Xu et al., 2025).

Microbiome and metatranscriptome integration introduces the cross-organismal level of regulatory interaction that microorganism community composition, functional gene expression, and metabolic secretion can regulate the host epithelial, immune, and metabolic transcriptomic settings. Coembedding AI methods together places host and microbial pathways in the same manifold, restricted by established biochemical exchange relations, which allows the identification of influence loop directions. The algorithms of causal discovery that use interventions as implicit as administration changes distinguish between direct signaling effects of microbial causation and indirect covariance with the same diet or housing setup. These findings guide the choice of host genotypes, the regulatory plasticity of which interacts with advantageous microbial functional guilds in promoting the coadapted efficiency of holobionts and resiliency strategies (Srinak et al., 2025).

Imaging-based, accelerometry-based, acoustic monitoring-based, and precision feeding platform-based high-dimensional phenomics bring along the continuous behavioral and anatomical descriptors that represent health, welfare, and performance regimes. The computer vision and sequence modeling architectures reduce these streams of sensors to time-based representations consistent with transcriptomic sampling windows through attention-based image alignment to deal with arbitrary sparsity in the acquisition of molecules. The fused representations make both early warning signals of incipient deviations more sensitive since they use the behaviorally mediated early warning signals that predict transcriptomic changes in some pathways and follow-on in others, and this increases the predictive robustness in either direction. Layers of environmental and climatic information, such as thermal load, humidity, air quality, and photoperiod, allow exogenous drivers whose interplay with regulatory networks determines adaptive or maladaptive transcriptional reactions. Spatiotemporal fusion models incorporate geolocated environmental context with molecular profiles, which allows the genotype-by-environment interaction to be quantified on a fine-grained scale and breeding values to be approximated and conditional on the environment (Brito et al., 2025).

The biological processes are placed in operational constraint landscapes by their management records, nutritional formulations, therapeutic interventions, and economic metadata. Knowledge graph models represent entities involving genes, pathways, feed, husbandry behavior, and outcome measures, where the relationship type may be regulatory activation, inhibition, consumption, supplementation, or economy of scale. GNNs explore these heterogeneous relationships to make context-directed embeddings, which guide multiobjective optimization, where one considers a balance between biological performance and resource efficiency and compliance parameters. Interoperability is dependent on the ontologies that are in harmony and the metadata schema that is standardized; ontology alignment with the help of AI-based applications is used to reconcile local terminologies to common vocabularies to minimize semantic fragmentation that otherwise would dismantle cross-farm generalizability (Lin et al., 2025).

Pervasive missingness and asynchronous sampling are also pillars of its approach to dealing with missingness. Variational temporal imputation models make use of intermodality redundancy to impute unobserved transcriptomic nucleotide stretches conditional on existing sensor values or environmental sequence, with uncertainty quantification defining uncertainty limits that offer a conservative downstream decision-making. In the federated learning model, the training of the model involves the agents that possess sensitive genomic and operational data, with differentially private noise added to minimize the chances of reidentification and achieve a meaningful signal. This type of governance preserves collaborative accuracy advantages and equal distribution of benefits, which is particularly essential in under-resourced production settings, where the datasets that are disconnected lack a lot of breadth to enable multimodal modeling (Hatfaludi and Serban, 2025).

Causal inference is lifted with counterfactual simulation engines of causal models of structures into the learned latent spaces, providing the ability to interrogate hypothetical interventions, for example, dietary redesign or environmental damage, with their integrated transcriptomic, metabolic, and phenotypic propagation effects. The instantiated models of digital twins are customized dynamic models whose state vectors encode genomic predisposition, regulatory configuration, physiological response, and external forcing factors, which respond to new measurements as they are received. The experiment sandboxes are these twins that are going to be used to experiment with the order of interventions and system stability and to identify when a deviation of adaptive trajectories first appears (Gkountouras et al., 2024).

Cross-species transfer learning may be efficient when it offers such benefits as data deficiency to match modality to less-studied breeds or ecotypes. Domain adaptation algorithms are in correspondence to latent distributions that do not change species-specific regulatory peculiarities, but instead allow selection of divergence on peripheral modules in specialized contexts of production. Adversarial training also promotes resistance to dataset shift and discourages a dependence on facility-specific batch artifacts or a signature of sequencing platform when it is applied in novel working conditions (Satou and Mitkiy, 2025).

Trust of the stakeholders is largely concerned with interpretability. The layers of post hoc attribution partially cross-modally contribute to pathway-oriented narratives and reflect those predictive weight contributions that are attributable to a genetic predisposition or an exposure to acute environmental stressors, microbiome functional changes, or behavior change. The approaches of contrastive explanation articulate small cross-modal interventions that require manipulating the predicted outcomes and indicate areas of leverage where the management can intervene. The calibration drift, covariate shift, and emergent bias by demographic or breed stratum are determined through constant monitoring of the models, and retraining can be reinitiated or recalibration amended in response to performance deviations, which are over preset tolerance levels (Mehdiyev et al., 2025).

Coupled with a demanding combination of transcriptomic, genomic, epigenomic, proteomic, metabolomic, microbial, phenotypic, environmental, and managerial modalities, AI becomes a nonisolated layer of analytics into systems intelligence to comprehend the dynamic biology of livestock. This convergence offers the decision makers mechanistically determined, temporally receptive, and ethically guiding insight streams that boost accuracy, reliability, sustainability, and transparency in the regions of breeding, health control, and cost-effectiveness of production and makes structural forecasts of progressive innovation with livestock omics.

5.8 Ethical, legal, and socioeconomic considerations

The integration of AI-enhanced transcriptomics into livestock systems poses a correlated series of ethical, legal, and socioeconomic inquiries, which surpass technical performance, to legitimacy, fairness, and licensure of the society. The informational capital concentrated in those actors that can collect, curate, and model large-scale datasets through the capture and algorithmic transformation of high-dimensional molecular data reorganizes the realities of power by concentrating informational capital in fewer actors and imposes the requirement of governance structures to ensure equitable distribution of benefits and avoid the enclosure of foundational biological knowledge. The provenance and ownership of data, as well as the custodianship, are refuted when transcriptomic profiles of privately owned herds overlap with publicly funded reference atlases to produce hybrid informational products, whose derivative value has challenged the traditional paradigm of intellectual property. A clear understanding of rights to collect, model, commercialize, and recalibrate access to the composite datasets is the basis of trust, whereas the possibility of uncertainty endangers the decline of data sharing that is critical to a solid model generalization (Ghavi Hossein-Zadeh, 2025; Rosati, 2025).

Although it has traditionally been the concern of human subjects, privacy finds its parallel in safeguarding strategic information of production, where the unwanted release of the information as metadata or as inverted models may undermine competitive advantage or reveal vulnerability in biosecurity. Differential privacy, secure multiparty computation, and federated learning models reduce the risk of reidentifications but require parameterization with an ethical tradeoff of accuracy versus confidentiality, which the stakeholders need to freely negotiate. Data contribution consent procedures must be clearly articulated with the analytical directions to be followed, the potential future secondary use of data, and exit procedures, so that it is not obscured and may subsequently lead to criticism of misuse or exploitation (Shenoy et al., 2025).

The substantive issue of algorithmic bias, that is, in the case of corpora being trained to be underrepresentative of certain breeds, production settings, or management models, is that systematic miscalibration is generated that undermines genetic potential measurement or predicts health risks in marginalized groups. This discrimination can cancel inequality in resource allocation through the allocation of genomic, nutritional, and breeding resources to already advantaged groups. Stratified performance reporting, fairness auditing, and adaptive reweighting strategies constitute an obligation and not a refinement that developers may or may not do, but must constantly monitor distributive impacts. The inability to deal with these imbalances threatens the loss of socioethical underpinnings of the precision livestock technologies and could see regulatory measures based on antidiscrimination ethics borrowed into the field (Agnes, 2025).

Using transcriptomic indicators to inform selection indices and management decision systems has ethical implications for the way genetic diversity changes in the long-term resilience. Immature optimization, which focuses on small molecular signatures, could speed up homogenization and reduce the adaptive potential to climatic volatility and compromise intergenerational stewardship responsibilities. Diversity conservancy limits, simulation of the scenarios of allelic richness at varying intensities of selection, and reporting the effects of diversity in a transparent manner operationalize precautionary ethics that trade off short-term productivity benefits with the sustainability of biosystems. Ethical assessment should also consider possible externalities of welfare in which it is unintentionally the case that choosing in favor of metabolic efficiency or faster growth binds itself to a set of regulatory modules that increase exposure to stress, illness, or reproductive loss. Welfare-oriented transcriptomic markers as multidimensional indices are the countermeasures that bring the state of genomic advancement in line with humanitarian treatment idealism (Ashango and Abtew, 2025).

Legal regulations are behind technological convergence, so compliance remains unclear as to cross-border transfer of genomic and transcriptomic data, the regulation of composite predictive models as regulated decision support systems, and who should be held liable when AI-directed interventions cause adverse events. The harmonization initiatives are necessary to avoid regulatory arbitrage that may disintegrate data ecosystems or encourage moving the hubs of analytical activity to jurisdictions with limited regulation. The allocation of liability should decompose causal chains in which a combination of model recommendations, human interpretation, management implementation, and latent environmental factors results in outcomes, and therefore, the provenance of the models, versioning, performance properties, and known limitations should be adequately documented. These transparency artifacts are both ethical tools helping to promote accountability and legal tools helping to promote evidentiary clarity (Zhuang et al., 2024).

The governance of AI models through intellectual property policies based on the training of models on multi-stakeholder datasets would involve striking a balance between incentives to innovate and protection against monopolistic bottlenecks in critical predictive infrastructure. Open licensing of base representations and allowing localized datasets to be fine-tuned is a democratic solution to allow baseline capability to be democratized and still allow differentiated value generation along the line. The presence of ethical tensions can be seen in the case of exclusive control over high-performing models, leading to rents extracted that are disproportionate to contributing data value, which can be reduced by benefit sharing, cooperative data trusts, and royalty distributions based on quantified data provenance (Wang, 2025).

The socioeconomic effects are not evenly spread, and capital-intensive activities are placed at the location where the initial benefits of efficiency and risk reduction can be obtained and thus may create discrepancies in productivity as compared to smallholder or resource-limited producers. Distributive justice is facilitated by inclusion programs that focus on capacity building, subsidization of access to core analytic services, data pipelines with interoperable low bandwidth, and simplification of interpretive interfaces to fit context. The lack of resourcing inclusivity to the challenge may lead to the internalization of structural inequity and the suppression of overall industry capacity to withstand it by marginalizing genetic reserves and tacit knowledge vested in various production systems. The south side of overtime labor dynamics adds another layer since any form of automation of the diagnostic or selection process can reorganize the demand to capacities, effectively removing the old functions and increasing the need to create data stewardship, biomatter expertise, and ethics expertise. The transition management of ethics entails retraining pathways, participatory design that keeps human control of value-laden judgments, and expression of the new professional standards that combine domain husbandry learning and data literacy (Denashurya et al., 2025).

Animal welfare takes a main place on the ethical axis. Transcriptomic surveillance enhances the ability to identify subclinical stress or immunological challenge earlier than standard phenotypic surveillance but should not be used to instrumentalize the outcome of the negligence of welfare protections in the name of increasing the pressure of production. Governance tools can require welfare biomarkers as cooptimalizing goals and not support indicators and entrench ethical pluralism in algorithmic objective functions. Honesty to consumers and wider civil society on the use of molecular information in breeding, nutrition, and welfare judgments enhances social license and is a means of overcoming the suspicion of ulterior techno-managerialism (Córdoba et al., 2025).

The biosecurity and dual-use issues arise whereby the sophisticated modeling may be reused to design interactions of pathogenicity or capitalize on weaknesses of supply continuity. The strategy of risk assessment protocols, multitiered access, and multidisciplinary representation ethics review boards offers anticipatory control. The connections between environmental ethics and predictive models are that the allocation of resources, mitigation of emissions, and land-use decisions are affected by predictive models. The localized efficiency gains will not trickle into augmented aggregate environmental burdens via rebound effects due to life cycle-aware modeling, which internalizes externalities (Lu and Lewis, 2025).

The questions of cultural and indigenous knowledge are raised when genomic and transcriptomic resources turn to breeds located in the heritage practices. The idea of ethical integration requires recognition of any custodial relationships and negotiation of terms of benefit sharing, and respect toward epistemic plurality that could oppose a purely utilitarian conceptualization of genetic resources. The models of responsible innovation focus on reflexivity, inclusivity, anticipation, and responsiveness and institutionalize a process of iterative ethical evaluation instead of post hoc remediation. The ongoing monitoring, participatory audit, and adaptive loops of governance mean that they can be recalibrated as the emergent consequences emerge to ensure that there is a constant alignment between the technological development and the societal values (Bernal-Hernández et al., 2025).

When we put these dimensions together, ethical, legal, and socioeconomic stewardship does not come as an ancillary component to technical excellence but represents a foundational part of the process of sustainable AI-facilitated transcriptomic integration. Trust is based on a planned system of transparency, equity, maintenance of diversity, focus on welfare, equal access, transparency in liability, and culturally sensitive benefit sharing, which enables the contribution of data in collaboration and predictable innovation directions. By using this form of integrative governance, the transformative potential of AI in livestock transcriptomics may be achieved without compromising other social responsibilities and intergenerational needs. Ethical, legal, and socioeconomic considerations in AI-driven livestock transcriptomics are presented in Table 5.7.

5.9 Technical challenges and limitations

The combination of AI and livestock transcriptomics is faced with a nexus of technical limitations that mitigate the effect of translation and define the limits of prevailing methodological maturity. The limitations of these include biological complexity, structural constraints, frictions of data governance, and epistemic voids of multiscale molecular signals into actionable predictions. Although sequencing, sensing, and machine learning architectures have rapidly advanced to broaden the limits of analysis, current limitations on the quality of data, representations of data, interpretations, and even reproducibility are hampering full confidence deployment to heterogeneous production environments (Wang et al., 2025c).

The underlying issues of data acquisition are heterogeneous sampling regimes, whereby less represented states of time, tissue, and physiology create latent confounding factors that complicates confounding between real regulatory variability and artifacts of the procedures. Batch effects are spreading between library preparations, chemistries, sequencing technologies, and in the laboratory, despite normalization methods that fail in cases of covariates being correlated with biological covariates. Bulk transcriptomics blurs variation in cellular composition, and deconvolution models are based on reference signatures, which are seldom specific to different breeds or environmental stress factors, leading to systematic misrepresentation of expression to pseudo-cell types. The isoform level quantification has limitations on read length and hazy mapping, which weakens the power to correct alternative splicing events, which could be the source of trait differences. Single-cell techniques have fine granularity but are limited by cost, sparsity caused by dropout, and scalability to large cohorts that are needed to do genetic association, limiting the resolution continuum (Wolfram-Schauerte et al., 2025).

Incomplete reference genomes and annotation subject all further modeling to downstream workflows to uncharted areas, as they are unmarked transcripts, structural variants, or unprocessed noncoding regulatory elements that are never captured during feature engineering pipelines. Pan-genomic variation across breeds has introduced alignment biases when using a single linear reference and inflates apparent changes in expression of the region that has an insertion, deletion, or copy number variation. Albeit conceptually better in reducing reference bias, reference frameworks based on graphs are computationally expensive and do not have a standardized toolkit to integrate smoothly with a standard machine learning workflow. Regulatory polymorphisms are also uncertain in the prediction of variability when their context-specific effects cannot be predicted by the models that have been trained on a small set of functional genomic assays, and the reliance on correlation but not mechanistic attribution continues (Nyaga et al., 2025).

Multimodality integration increases the complexity with discordant scales, noise distributions, and missingness patterns of genomic, epigenomic, proteomic, metabolomic, microbiome, sensor, and environmental datasets. When normalization strategies are harmonized, the distortion of the signal geometry of a different modality by transformations suitable for one would hinder harmonization. Cross-modality correspondence usually assumes synchronous sampling, which is operationally infeasible, and results in timing misregistration, which reduces the fidelity of sequence models.

TABLE 5.7 Ethical, legal, and socioeconomic considerations in AI-driven livestock transcriptomics.

Ethical considerations	Legal framework	Socioeconomic impact	Recommendations
Data governance and ownership: unclear control; function creep risk; inequitable extraction; power asymmetry	• Data protection (GDPR/CCPA) for linked metadata • Contracts/license terms • IP/trade secret • Emerging farm data charters	• Value aggregation by large platforms • Smallholder leverage loss • Potential new cooperative revenue • Participation hesitancy	• Plain-language data-use terms • Producer representation boards • Federated/edge modeling • Tiered benefit and revenue sharing
Privacy and reidentification: residual identifiability; overcollection; ethical duty of minimization	• Privacy statutes (human-linked info) • Confidentiality clauses • Biosurveillance overlap • Audit obligations	• Trust erosion reduces data supply • Compliance cost burden • Innovation slowdown	• Apply differential privacy/k-anonymity • Limit granular geo-temporal metadata • Formal re-ID risk audits • Publish privacy impact summaries
Animal welfare and sampling burden: invasive biopsies; frequency versus necessity; 3Rs adaptation	• National welfare acts • IACUC/ethics approvals • Transport/procedure compliance	• Welfare perception affects social license • Refined methods increase unit cost • Premium branding potential	• Prefer non-/minimally invasive matrices • Adaptive (uncertainty-triggered) sampling • Disclose welfare KPIs • Integrate welfare constraints in models
Bias and representativeness: elite breed overfit; neglect of indigenous genetics; risk of genetic homogenization	• Conservation funding conditions • Public grant diversity mandates • Competition law (indirect)	• Marginalization of local breeds • Resilience decline • Market dominance consolidation	• Dataset diversity audits • Weighted loss functions • Incentives for minor breed inclusion • Open repositories for underrepresented cohorts
Transparency and explainability: black-box decisions; automation bias; traceability need	• EU AI Act (risk documentation) • ISO/IEC AI management drafts • Procurement audit clauses	• Higher adoption with clarity • Opaque models increase lock-in • Trust accelerates diffusion	• Publish model cards and data lineage • Deploy explainable dashboards • Periodic third-party audits • User interpretability training
Accountability and liability: responsibility diffusion; ethical ambiguity in adverse outcomes	• Contract liability clauses • Tort/product liability • Regulatory sandboxes • Insurance alignment	• Risk aversion in SMEs • Large actors absorb shocks • Insurance premium stratification	• Define RACI matrices • Human-in-loop gates for high-impact actions • Sector liability guidelines • Performance-linked insurance products
IP and benefit sharing: enclosure of shared biology; inequitable downstream value; biomarker privatization	• Patents (methods/panels) • Database rights (EU) • Nagoya Protocol adjacency • Licensing/EULAs	• Royalties bypass producers • Paywalled innovation gaps • Slowed diffusion in low-resource regions	• Tiered (research vs. commercial) licensing • Transparent revenue allocation • Open core and proprietary extensions • Annual benefit distribution reporting
Genetic diversity and inbreeding: intensified selection, narrowing gene pool, latent deleterious fixation risk	• Breed registry rules • Effective population size guidelines • Conservation program frameworks	• Systemic vulnerability to disease/climate shifts • Future rescue costs • Volatility in productivity	• Optimal contribution selection constraints • Real-time genomic inbreeding dashboards • Diversity KPIs in indices • Early warning thresholds
Dual-use and biosecurity: misuse of expression data; pathway exploitation; agro-terror risk	• Biosecurity and select agent rules • Export control • Cybersecurity standards (ISO 27001) • Data access vetting	• Breach impacts trust and markets • Compliance cost disparities • Jurisdictional leakage risks	• Dual-use risk screening prerelease • Tiered credentialed access • Zero-trust architecture • Incident response drills and playbooks
Environmental sustainability alignment: optimizing output at ecological cost; ignoring emissions/N losses	• GHG inventory standards (ISO 14064) • Nutrient management regulations • Carbon market MRV • Sustainability certification	• Access to green finance/premiums • Risk of excluding smallholders • Widened sustainability performance gap	• Embed CH_4/N KPIs in objective functions • Low-cost proxy measurement kits • Standardized lifecycle protocols • Shared carbon/eco-credit models

Continued

TABLE 5.7 Ethical, legal, and socioeconomic considerations in AI-driven livestock transcriptomics.—cont'd

Ethical considerations	Legal framework	Socioeconomic impact	Recommendations
Equity and digital divide: access disparity; proprietary ecosystem lock-in; informed consent gaps	• Antitrust/competition oversight • Public procurement fairness • Funding conditionality for openness	• Rural stratification • Underutilized productivity potential • Dependence on dominant platforms	• Interoperable open APIs • Subsidized training and capacity building • Cooperative data trusts • Lightweight edge-ready model variants
Labor displacement and skill transition: automation of observational husbandry; deskilling; just transition duty	• Labor and training regulations • Vocational/extension policy • National AI workforce strategies	• Short-term displacement • Emergence of data stewardship roles • Wage polarization risk	• Codesign human–AI task matrices • Funded reskilling/microcredentials • Labor impact monitoring • Preserve experiential husbandry inputs
Public trust and social license: perception of unnatural "algorithmic breeding"; misinformation amplification	• Advertising/labeling truth standards • Public consultation stipulations • Consumer protection laws	• Market acceptance influences premium capture • Negative narratives slow approvals • Investment sensitivity	• Stakeholder fora and citizen panels • Plain-language impact summaries • Independent ethics advisory reporting • Proactive myth-busting campaigns
Fair compensation and value distribution: risk of "data colonialism"; undervaluing local knowledge	• Contractual revenue-sharing • International trade norms • ABS principles (context-dependent)	• Development gaps widen • Lower participation willingness • Geopolitical tension over data flows	• Clear benefit-sharing formulas • Producer cooperatives/data trusts • Publish value distribution KPIs • Multilateral equity frameworks
Model drift and lifecycle ethics: silent degradation; bias reintroduction; outdated assumptions	• Quality management (ISO 9001) • AI lifecycle governance drafts • SLA update clauses • Audit trails	• Efficiency loss and hidden error cost • Confidence erosion • Disadvantage to actors without refresh capacity	• Continuous drift monitoring metrics • Scheduled ethical and technical revalidation • Versioned model registry • Deprecation/sunset policies
Sampling and resource inequality: high-cost omics pipelines favor large integrators; exclusion of low-resource regions	• Grant equity requirements • Technology transfer provisions • Public–private partnership guidelines	• Innovation concentration • Slower global trait improvement • Dependency loops	• Tiered assay cost models • Pooled regional sequencing hubs • Open protocol standardization • Capacity-building grants
Consent and secondary use: broad consent vagueness; function creep into unrelated analytics	• Contractual consent scope clauses • Data reuse governance • Ethical review for secondary analyses	• Participant withdrawal risk • Reduced willingness to share prospective samples • Erosion of trust capital	• Layered (specific and broad) consent forms • Dynamic consent platforms • Transparent secondary use registry • Routine reconsent checkpoints
Algorithmic monoculture: convergence on a few dominant models; systemic failure risk	• Competition oversight • Interoperability mandates • Standards bodies fostering plurality	• System-wide vulnerability to shared blind spots • Reduced innovation diversity • Lock-in switching costs	• Encourage model plurality and benchmarking • Open reference datasets • Stress-test scenarios (adversarial and rare events) • Procurement diversity criteria
Cross-border data flows: jurisdictional mismatch in protections; uneven enforcement	• Data localization laws • Cross-border transfer agreements (SCCs) • Trade and biosecurity controls	• Fragmented innovation networks • Increased compliance overhead • Exclusion of developing markets	• Harmonize contractual clauses • Federated/locally trained models • Shared international governance frameworks • Differential access tiers respecting local law
Ethical use of predictive culling signals: premature culling based on probabilistic outputs; welfare versus efficiency tension	• Animal welfare statutes • Liability for unjustified culling • Documentation/audit requirements	• Potential productivity gains offset by reputational risk • Market scrutiny on decision criteria	• Confidence thresholds and human validation • Dual-metric (welfare and efficiency) dashboards • Audit logs for culling decisions • Periodic ethical review panels

Naive approaches based on concatenation find compositional constraints in microbiome data, with mixtures having variations with different depth scales. High-dimensional context, such as microclimatic fluctuations and management interventions injected into the environment, results in attenuation of fine genetic regulatory effects unless the model is hierarchically structured (Nunes et al., 2025).

The state of model development is faced with an endemic imbalance of sample size against parameter richness of deep architectures, which gives rise to overfitting and fragile generalization with domain shift. Genotype-by-environment interactions introduce nonstationarity, which is not represented by the simpler models, and reaction norm learning is complicated by sparse sampling along stress gradients. Ambitions of causal inference are beyond the available interventional information, and observational confounding removes identifiability of regulatory edges in inferred networks. The architecture in sequence-based transformers takes in raw k-mer representations but is likely to encode spurious patterns of linkage over multiple cells, not biological motifs, and thus may be hard to interpret biologically. Latent factor models feed down coexpression information but have the risk of confounding independent pathways when regularization forces against factor proliferation, resulting in less precise downstream intervention targeting (Castro et al., 2025).

Saliency attribution is not fully interpretable, and uncertainty quantification is not fully determined and is sensitive to input perturbations and likely to point to high-variance components instead of causally effective ones. The black box transparency impedes compliance and credibility among regulators and stakeholders, whereas post hoc descriptions may be inconsistent between methods, which compromises epistemic coherence. The calibration drift occurs as the production systems change, yet the monitoring frameworks and statistics on when to retrain the models have not been properly established, and thus, the subtle erosion of the predictive reliability is silent. Benchmarking flaws are caused by a lack of universally accepted reference datasets that cover a variety of breeds, tissues, and environmental conditions, making it impossible to make sound comparative testing of the algorithmic innovations and promoting the overstatement of performance (Majlatow et al., 2025).

Limitations in computational infrastructure manifest as storage, memory bandwidth, and energy constraints associated with large-scale, multi-omics tensors and high-resolution, temporal sensor streams. The cost of training and inference restricts iteration velocity and biases experimentation towards resource-privileged actors, thereby reinforcing disparities in reproducibility. If not carefully validated, model compression techniques can degrade biological fidelity, while distributed training introduces synchronisation overhead and gradient staleness that exacerbate convergence instability. Privacy-preserving paradigms, such as federated learning and secure aggregation, introduce statistical heterogeneity (client drift) and communication bottlenecks, which slow optimization and reduce accuracy relative to centralized baselines. This challenges their practical utility when regulatory or competitive pressures mandate data localization (Cui et al., 2024).

Biases disseminate based on breed overrepresentation, undersampling of fringe, and happy capture of high-performing operations, which lead to latent shifts in domains that corrupt fairness and exalted returns on deployment to lower characterized settings. The subjectivity and late demonstration of phenotypic tags of welfare, resilience, and complicated health outcomes introduce disturbance to transcriptomic sampling and diminish the quality of signals operated by supervision, improving the quality of signals used in supervised signal quality. The lack of labels is partially addressed in the semisupervised and weak supervision models, which can also cause the propagation of an annotation bias by accident. The jitter and dropout, and nonuniform sampling frequency also contribute to sensor integration, and make it hard to fuse time, and incite modes of failure where models have continued to optimize continuously streaming, comparatively weakly informative proxies and underutilization of comparatively sparse but rich molecular snapshots. Digital twin simulations (often known as decision sandboxes) are founded on structural assumptions and parameter prior knowledge not always right in emergent disease behavior and climate anomalies, leading to overconfident intervention recommendations (Zhioua et al., 2025).

A lack of solutions to problems in variant-to-function mapping, dynamic regulation circuit elucidation, and cross-tissue coordination impedes the ability to proceed to mechanistic design beyond associative prediction. The biases of integration of epigenetic context are that the heritability and context specificity of marks are not complete, which limits the transferability of observed regulatory states across different management regimes. Network inference is also facing the combinatorial outburst of potential interactions when perturbational variety is restricted, and a regularizing heuristic setup might be killing actual but insidious edges needed to comprehend polygenic trait structure. The tension between computational efficiency and the aims of environmental sustainability creates a new constraint between energy-intensive training and larger-scale missions of minimized production system footprints, which provokes the need to develop algorithmic efficiency improvements that preserve biology (Fang, 2025).

All these technical constraints highlight the reality that the present-day AI-powered transcriptomic systems, despite their strength, are working in frail capsules of validity that are determined by the representativeness of the data, the levels of annotation of the data, the transparency of the model, and structural resiliency. To go further, coherent standardization of metadata schemas, the creation of open, but privacy-conscious consortia, the phenomenalization of causal and probabilistic modeling systems sensitive to agricultural nonstationarity, and the combination of energy-efficient computational paradigms are required. Awareness of these limitations leads to more reasonable expectations, a focus on investing in core resources that facilitate algorithmic innovation, and the expectation that future increases in the complexity of livestock production contexts will be matched by equivalent improvements in reliability and interpretability. Key technical challenges and limitations in AI-enabled livestock transcriptomics are presented in Table 5.8.

5.10 Monitoring, validation, and governance of AI models

AI models of livestock transcriptomics should be monitored, validated, and governed, and this is an integrated lifecycle practice that ensures reliability, biological fidelity, and ethical legitimacy with the change toward operational decision influence. The unique heterogeneity, temporal variability, and perturbations applied by managers to livestock production environments accentuate threats of silent model degradation, which complicates miscalibration, unless designed surveillance architectures are realized at the outset. Validation has been turned into a repetitive show of evidence by which predictive performance, calibration, and mechanistic plausibility are periodically redetermined in the context of the life cycle of molecular baselines, changing genotypic compositions, changing environmental stress factors on a seasonal basis, and the new health issues. This active framing presupposes governance of an adaptive scaffold and perfect alignment of technical development and regulatory anticipations, bioethical, and trust criteria among stakeholders (Kate and Neethirajan, 2025).

Basic validation is initiated by extreme partitioning approaches where animals, herds, farms, and time blocks are hierarchically partitioned to avoid information leakage to artificially overestimate generalization capacity. The capacity to cross-context validate across divergent breeds, management systems, climatic zones, and different regimes of exposure to pathogens creates robustness envelopes and reveals failure modes of interaction whereby otherwise apparently good overall performance is found to be systematically poor in underrepresented strata. Discrimination, calibration, sharpness of uncertainty, feature attribution stability, and coherence in biological pathways are multiobjective analyses to ensure that optimization does not occur on short-term measures that lack sufficient negotiability for downstream decision utility. Adding negative controls, permutation-based baselines, and synthetic perturbation probes makes available negative signals to spurious correlation, and additionally verifies that salient features correspond plausibly with known regulatory structures (Džermeikaitė et al., 2025).

The postdeployment surveillance presents ongoing data quality auditing, which profiles distributional properties of incoming transcript counts, isoform ratios, quality flags, and related contextual covariates to identify a shift with a divergence threshold that has been set. Drift characterization is used to differentiate between benign seasonal modulation and pathological domain shift that develops due to the change in assay protocols, instrument recalibration, breeding innovations, or epidemiological phenomena. The early warning indicators exploit the ensemble disagreement, increasing predictive entropy, decreasing slope of calibration, and reducing sparsity of pathway-level attribution as compound signals of developing degradation. Lag structure checkers use modules of temporal alignment to evaluate the reliability of stability between transcriptomic signals and other auxiliary phenotypic or sensorized frameworks and raise warnings about deviations in coordination design, which might cause bias in causal determination (Roschewitz et al., 2024).

Governance models professionalize positions that define the model stewardship, data custodianship, judgment of the biological domain, and ethical management to eliminate the acculturation of unrestrained power within algorithm creation teams. Artifacts of documentation, such as provenance-enhanced model cards, data sheets explaining sampling frames and preprocessing transformations, and decision impact statements defining the boundaries of intended use, form clear interfaces to the auditors and other external parties. The practices of version control ensure that updates to parameters or changes in the structure of the feature schema or retraining can be traced to an initial version, and the semantic versioning practices provide a way to communicate the extent of behavioral change. Change management procedures require predeployment shadowing, where the new models run concurrently with the current systems to build up comparative evidence before they are promoted and hence reduce the disruption probability (Leslie et al., 2024).

Uncertainty management is used as a governance pillar, which converts probabilistic outputs into risk stratification that is measured to fit intervention cost–benefit frameworks and welfare protection limits. Epistemic and aleatoric uncertainty decomposition leads to prioritization of data augmentation over model refinement, and abstention mechanisms allow one to defer to human expertise when the level of confidence is less than the minimum defined by the governance. Human control goes even further, with review gateways on high-stakes deliverables like selection index adjustments or early disease interventions to incorporate accountability loops and give experiential feedback to guide active learning query choice. The ingestion pipelines' feedback participates in adjudicated disagreements between model advice and expert

TABLE 5.8 Key technical challenges and limitations in AI-enabled livestock transcriptomics.

Challenge/Limitation	Description	Impact on transcriptomics	Proposed solutions
Data quality and noise	Transcriptomics data often suffer from low signal-to-noise ratio, dropout events (in single-cell datasets), technical noise, batch effects, and inconsistency in sample preparation and sequencing depth	Misleading gene expression measurements; false positives/negatives; lower reproducibility; reduced power for detecting subtle or rare transcripts; degraded model performance	Standardize library preparation protocols; use quality control filters; imputation methods for dropout; normalization techniques; replicate samples; use of spike-ins; cross-batch correction
Annotation and reference genome limitations	In livestock species, the reference genomes may be incomplete, misassembled, or lacking fine gene models; annotations of noncoding transcripts and isoforms are often sparse	Misalignment of reads; errors in quantification of alternative splicing or isoform expression; difficulty in interpreting noncoding RNAs; reduced accuracy of transcript-level predictions	Improve genome assemblies (use long-reads); update and curate annotation databases; integrate de novo transcriptome assembly; include species- or breed-specific annotations; cross-validate with proteomics
High computational and storage demands	Large datasets (bulk RNA-seq, single-cell RNA-seq, spatial transcriptomics) generate very large raw and processed data; AI/ML models require compute for training and inference; requirements grow with scale	Resource bottlenecks in generating, storing, and processing data; delays in model training or inference; limited adoption, especially in resource-constrained settings; potential compromises in model complexity	Use cloud and HPC (high-performance computing); data compression; streaming or edge processing; efficient algorithms; federated learning; optimize pipelines; hardware accelerators (GPUs and TPUs)
Overfitting and poor generalizability	AI/ML models, especially complex deep learning ones, may fit too closely to specific training data (e.g., specific breeds, labs, and environments) and fail when applied elsewhere; cross-validation issues	Poor performance in new data; limited transferability across breeds, environments; risk that models work well in controlled experiments but poorly in field settings; reduced trust	Collect diverse training datasets; cross-breed, cross-environment validation; regularization methods; use of transfer learning; domain adaptation techniques
Interpretability and explainability	Many powerful AI models are "black boxes," making it hard to understand why they make particular predictions; this reduces trust, hinders debugging, and may violate regulatory or ethical norms	Users (breeders, regulators, scientists) may be reluctant to adopt AI tools; difficulty in identifying biological meaning; hampered use for discovery (versus just prediction)	Incorporate Explainable AI (XAI) methods; use interpretable models where possible; feature importance, attention, SHAP/LIME; visualization of internal model representations; engage domain experts to validate model outputs
Sampling bias and representativeness	Data may overrepresent certain breeds, ages, tissues, or conditions; underrepresentation of rare breeds or challenging environments; a limited number of replicates; cell subtype bias in single-cell transcriptomics	Models are biased toward overrepresented conditions; predictive models are less effective for underrepresented breeds; missed biological variation; unfair performance	Design studies with balanced sampling; include underrepresented groups; ensure adequate replicate numbers; add metadata about tissue, breed, environment; consider federated datasets
Temporal and spatial resolution limitations	Many transcriptomic studies sample at one time point, or bulk samples that mask cell-level spatial heterogeneity, loss of tissue context, and inability to capture dynamic processes	Dynamics (e.g., stress response, infection) may be missed; spatial heterogeneity (e.g., within organs) may not be captured; model predictions may be less relevant to real-world conditions	Use time-series designs; single-cell and spatial transcriptomics; paired sampling; integrate time/stage metadata; develop longitudinal AI models
Ethical, privacy, and data sharing constraints	Concerns over sharing of genetic/transcriptomic data, consent, ownership, especially for indigenous or proprietary livestock genetic resources; lack of standardized policies	Slower accumulation of large, shared datasets; fragmented data; some data may remain inaccessible; reduced ability to build large, robust AI models; legal or ethical risks	Develop clear data-sharing policies; use controlled access repositories; ensure consent and benefit-sharing; anonymization or de-identification; international collaboration for governance
Cost and accessibility	Sequencing (especially single-cell or spatial), sample processing, compute infrastructure, and AI expertise are expensive; many regions or institutions lack capacity	Limits scale and diversity of studies; slower adoption; potential widening of "omics gap" between well-funded and less-resourced settings; underrepresentation of certain livestock species or geographic areas	Promote technology cost reduction; establish shared infrastructures; training and capacity building; cloud-based analytics platforms; scalable and low-cost protocols

choices. The difference in model advice and expert choices in curated relabeling events facilitates the model adaptation without escalating noise (Bickford Smith et al., 2024).

Protection of security and integrity holds adversarial or unintentional perturbation of transcriptomic inputs or auxiliary contextual information that can contaminate transcriptomic output. The reduction of surface area to manipulate is minimized with the use of anomaly checking of improbable expression vector configurations and isolation of training pipelines of operational decision channels, unlike hash-based integrity verification. The belief in privacy-sensitive aggregation techniques in federated learning settings is vetted to possess respectable noise budgets and amplification of prejudices, and ought not to contradict the data protection principles by surreptitiously compromising the equities or predictive equilibrium among the contributors that are dispersed over the grid. Accountability and prevention of abuse through providing an accountable access control and logging system in terms of forensic transparency by enabling one to reconstruct the sequence of events led to the creation of anomalous outputs (Shafik et al., 2024).

The element of fairness monitoring in the concept of ethical governance introduces the component of stratification of the performance and calibration of the breeds, management levels, geographical locations, and types of welfare, which activate the mitigation processes in cases where the disparity measures are beyond the predetermined tolerance ranges. With constrained diversity preservation, a choice is made with respect to cumulative selection in relation to the output of a model, which identifies the oncoming loss of measures of genetic variability before irreversible homogenization pathways would become a reality. Welfare protection is achieved when sentinel transcriptomic markers of stress, inflammation, or metabolic overload are incorporated in model interpretive layers, with rules of governance stipulating that a conservative decision modulation be taken under welfare risk indicators independent of benefits oriented toward production (Kamil and Al-askari, 2025).

The collective validation is founded on benchmarking consortia, and common sets of reference challenges would exclude duplicated effort and promote proprietary obscurity, which has the impact of compromising comparability. Interoperability is also imposed through harmonized metadata ontologies and minimal information reporting standards, thereby permitting the repetition and independent reprocessing, normalization, and model training process. Periodic reproducibility auditing reconstructs published performance claims in containerized environments after locking them down, the outcome of which is reported as variable and attributable either to stochastic variability in initiation, the version drift of the library, or unknown parameter settings. The disclosure of discrepancies procedures will promote the integrity of reporting the replication shortfalls and changing them into system learning events instead of reputational risk (Antunes and Hill, 2024).

Lifecycle governance anticipates the retirement and archiving of models, specifying the criteria for decommissioning those whose domain assumptions are no longer valid or whose maintenance costs exceed their marginal utility. Archival practices preserve parameter states, training data hashes, and validation reports, facilitating retrospective analysis of historical decision-making processes and supporting the longitudinal attribution of breeding or management outcomes to specific algorithmic influences. Sustainability considerations include energy profiling of training cycles and inference throughput. This incentivises enhancements to algorithmic efficiency and informs environmental impact accounting within broader sustainability reporting frameworks (Chen et al., 2025).

By using integrative methods of rigorous multilevel validation, proactive drift and integrity control, transparent record keeping, systematic human control, equity and wellbeing surveillance, security hardening, and joint benchmarking of AI models in livestock transcriptomics, it moves to anticipatory rather than responsive governance. This unified device will keep the predictive systems biologically, contextually, ethically, and publicly responsible to the rising complexity. A long-lasting investment in governance infrastructure, therefore, becomes indivisible with technical innovation and sets the stage for trust and scalability, sustainability, and accountability of transcriptomic intelligence applications in the context of various livestock production ecosystems.

5.11 Emerging trends and future directions

The futures involving AI integration into livestock transcriptomics are convergent in the sense that they aim to liquidate the remnant boundary between molecular resolution, system-level cognition, adaptive decision-making, and planetary sustainability mandates (Fig. 5.1). The subsequent stage is marked by the growth of static, assay-centric modeling to dynamic, continuous, and context-aware intelligence, internalizing the time, space, environmental, and socioecological dimensions. The multi-omics fusion that is emerging in the center here is that transcript abundance becomes intermixed with chromatin accessibility, methylation status, proteoform distribution patterns, metabolic flux compartmentalizations, community dynamics in the context of the microbiome, flows of physiological sensors, and microenvironmental satisfying telemetry. The change in representation learning architectures is moving toward modality-aligned latent manifolds that

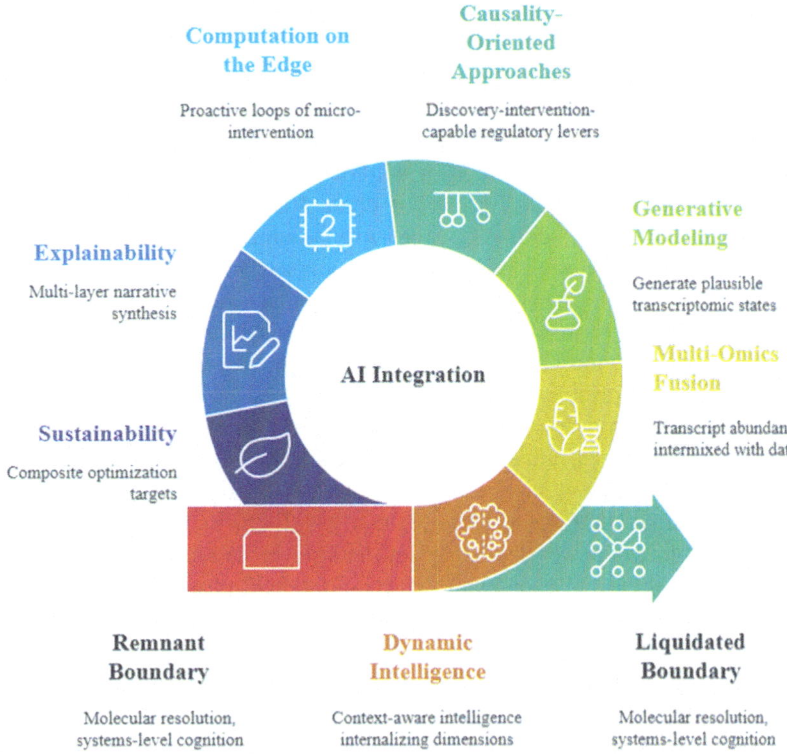

FIGURE 5.1 AI integration into livestock transcriptomics. Conceptual framework for AI integration into livestock transcriptomics, illustrating key components, including computation on the edge, causality-oriented modeling, generative modeling, multi-omics fusion, sustainability objectives, and explainability, that progressively collapse the boundaries between molecular detail and systems-level intelligence in context-aware livestock systems.

retain modality-specific stochasticity while allowing cross-domain transfer, self-supervised learning goals, and extracting invariant structure of unlabeled corpora, being used to reduce the chronic shortage of phenotype annotations (Syeda, 2025).

Spatially resolved and single-cell-based transcriptomics are driving a new definition of trait architecture in which cellular ecology has been transformed, in which lineage paths, microenvironmental gradients, immune niches, and cell–stroma interactions are predictive substrates. New models learn the combination of spatial coordinates, morphological embeddings, and subcellular transcript distributions without requiring training on any extra data, and only graph-based operators and attention-mediated operators are tuned to the topology of neighborhoods and sparsity-induced noise. It has been envisaged that this intratissue fine structure may report selection strategies that are aware of cellular composition stability, stress buffering capacity, and resilience signaling plasticity beyond blunt expression summaries to microanatomical systems engineering. Simultaneous construction of pan-genomic graph references offers a framework which allele-specific expression, structural variant modulation, and breed-adaptive regulatory rewiring can be combined, eliminating reference bias, such that context-sensitive inference of transcriptomic implications of underrepresented genomic embodiments is possible (Gulati et al., 2024).

Generative modeling represents a rapidly growing field in which variational sequence decoders, diffusion models, and energy-based formulations are used to generate plausible transcriptomic states under the condition of genotype, microbiome composition, nutritional condition, or climatic condition. These synthetic cohorts are both used as augmentation substrates for low-frequency trait assemblies and stress situations that are underrepresented in empirical evidence and are also useful to stress test model robustness to extreme perturbations. Mechanistically plausible prior inclinations, structural constraints of pathways, and thermodynamic plausibility constraints encourage the use of mechanistic priors, biologically implausible mosaics of expression, and anticipate a hybrid modeling paradigm where data-motivated expressivity is constrained by biochemical plausibility (Kenneweg et al., 2024).

Causality-oriented approaches are developing into more effective tools for discovery and intervention compared to associative coexpression methods. Counterfactual inference and invariant risk minimization frameworks using time-ordered data streams, perturbational datasets, allele-specific expression contrasts, and structural prior graphs based on comparative functional genomics find counterfactual prediction relationships across environmental strata. Connecting causal candidates to multiscale simulation spaces aids in creating translators to digital twin ecosystems where virtual herds are simulated as instantiations of physiological, immunological, and metabolic submodels to which transcriptomic embeddings are presented as dynamic state vectors. These digital twins help explore in silico scenarios of management adjustments and selection policy, avoiding the burden of empirical trials and providing an interpretive laboratory for discovering emergent nonlinear feedback loops (Taylor-King et al., 2024).

Computation on the edge and near data sources is expected to counter bandwidth, latency, and privacy limitations related to centralized processing of the high-velocity sensor-coupled molecular diagnostics. Model compression, quantization-aware training, sparse attention, and neuromorphic-inspired event-driven inference are supposed to reach low-energy deployments at distributed farm nodes capable of performing initial searches of anomalies, selective data triage, and adaptive sampling orchestration. This changes the paradigm of periodic retrospective analysis to proactive loops of microintervention whereby molecular sampling times, nutritional modulation, or welfare checks are redefined adaptively depending on the emergent transcriptomic divergence clues. Federated and swarm learning schemas build on top of decentralization and allow parameter refinement to jointly consider and fine-tune the operations of geographically spread operations without the pooling of raw data, utilizing secure aggregation, drift-aware optimization, and incentive fairness to avoid domination by large contributors (Adu-Manu et al., 2025).

Explainability is changing from feature-level saliency into multilayer narrative synthesis of alignment of molecular signatures, inferred pathways, and predicted phenotypic effects in the field of uncertainty quantification. Counterfactual reasoning modules produce sets of transcript perturbations needed to modify predicted risk strata, which form the foundation for communicative transparency. The treatment of uncertainty is expanding out of scalar calibration and into more structured forms of epistemic maps that localize the ignorance in genomic regions, the compartments of cells, or the local environment, and inform explicit data acquisition plans in order of the amount of information that is likely to be gained. Active learning pipelines combine these maps to use them to dynamically schedule the allocation of sequencing depth, tissue selection, and temporal sampling intensities so as to maximize marginal improvement in decision-critical model dimensions and limit resource consumption (Iversen et al., 2025).

The priorities of sustainability and climate resilience are driving the consideration of the parameters of life cycle assessment, the proxies of emissions, heat stress biomarkers, and adaptive immunity signatures as composite optimization targets. It is estimated that AI-mediated transcriptomic analysis will be used to inform breeding portfolios that will internalize the mitigation of methane intensity, nutrient conversion efficiency, and resilience in the face of climatic volatility, which are supported by predictive scenario ensembles quantifying tradeoffs between short-term productivity and long-term ecological stability. Crossing over to circular bioeconomy models promotes the valorization of molecular know-how in waste stream upcycling, nutrient recycling optimization, and precision feed formulation that are based on soil health and biodiversity conservation indicators, transcending animal-centric performance of transcriptomic value chains (Shashank et al., 2024).

The regulatory codevelopment and standards harmonization is a metatrend recognizing the need to have an interoperable metadata scheme, assay quality ontology, model documentation template, and auditing protocol to enable cross-jurisdictional trust and portability. Open science consortia are incubating common benchmark corpora, challenge assessments, and reproducibility infrastructure based on containerized workflows and cryptographic provenance. However, to strike a balance between openness and proprietary innovation and biosafety, it is necessary to invest in governance innovations like licensing the access to tiers and the release of raw versus abstracted representations and ethical vetting processes of high-power generative or predictive systems. The objective of anticipatory policy engagement is to thwart fragmentation by expressing mutual principles of accountability in algorithms, genetic diversity conservation, protection of welfare, and equitable distribution of benefits (Rosen et al., 2025).

The future state of computational substrates may be the exploration of quantum-inspired optimization heuristics and photonic or memristive computation of matrices and attention kernels, which is likely to run shorter high-dimensional inference times with the capability to potentially substantially lower the energy intensity. The focus on algorithmic sustainability is giving rise to green model scorecards benchmarking carbon footprint, parameter efficiency, and performance trade curves to provide incentives to architectural parsimony without biologically compromising resolution. The lifecycle carbon accounting on the governance dashboards will presumably move the design choices in the direction of modular incremental fine-tuning instead of retraining the whole system on a large scale with universal regulatory grammars through self-supervised goals (Pimenow et al., 2024).

The proposed field of cross-domain integration between human and wildlife health under a One Health conceptual umbrella would aid in generating early warning of zoonotic interface risks by comparing transcriptomics of host response activation profiles across host and species boundaries. These translational bridges have the potential to hasten the process of discovering resilience alleles and immunomodulatory therapy that would strengthen animal health, decrease the use of antimicrobials, and decrease the risk of spillover. These scientific perspectives are interspersed with ethical and cultural considerations that provide the justification of participatory codesigning, respect for indigenous knowledge, and assessment of social impacts that would ensure that technological pathways do not become inconsistent with pluralistic value landscapes (Raza et al., 2025).

All these new streams of developments outline the future where AI-enhanced livestock transcriptomics becomes an emerging adaptive, transparently managed, causally conscious, sustainability-oriented, and fair knowledge infrastructural system. This vision is dependent on the careful management of methodological innovation and investment in the common resource, moral guardrails, energy-conscious computation, and ecosystems of collaborative governance. With a combination of granular molecular intelligence and systems-level foresight, the discipline is set to overcome complexity and provide resilient productivity and welfare improvements and environmental cobenefits, which entrench the future of livestock systems in the wider societal and ecological stewardship commitments.

5.12 Recommendations for future research and practice

To move closer to incorporating AI into livestock transcriptomics, it is necessary to intentionally shift the focus of effort toward the moonshot of enhancing the accuracy of the incremental model improvements and building foundational data, annotation, and governance assets to raise the innovation frontier in its entirety. It is necessary to prioritize the broadening of existing transcriptomic reference corpora as well as the in-depth studies of breeds, development, physiological conditions, tissues, and environmental conditions under harmonized conditions with machine-actionable metadata. Capturing of sampling intent, temporal coordinates, management interventions, and assay provenance systematically will minimize confounding latent variables and allow strong domain adaptation. Similar investment in full functional annotation, resolution of alternative isoforms, noncoding elements, regulatory variants, and context-specific use of enhancers will expand the biologically interesting feature space that can be used by representation learners. Pan-genomic graph frameworks should be made practical to perform routine alignment and expression quantification to alleviate reference bias and facilitate allele-specific inference, and benchmarking initiatives should be used to realize the efficiency of computations and reproducibility in production pipelines (Aburub et al., 2025).

There should be a strengthening of paradigms of self-supervised and weakly supervised methodology, which can take advantage of large volumes of unlabeled sequence and expression time series data and sensor time series data, and pretrain multimodal foundation models that learn generalizable regulatory grammars. These models may then be fine-tuned to emergent tasks with sparsely labeled phenotypes in a resource-efficient fashion, without as much resource wastage or the threat of overfitting. Work on architectures that inherently encode hierarchical biological structures—genes within pathways, cells within tissues, animals within herds, and herds within agroecological zones will allow the disentanglement of components of variance and allow interscale transfer. Research into causal discovery frameworks that combine a time-based ordering, genetic perturbation proxies, and cross-environment invariance constraints should continue to go beyond associative prediction to actionable insights that identify key regulatory leverage points. It is possible to incorporate mechanistic priors into deep generative and discriminative models, which means restricting hypothesis space and enhancing extrapolative fidelity in novel stressors (Dradjat et al., 2025).

Standardized requirements of model validation and monitoring require formalization in community-approved model protocols that set minimum evidence requirements to be used in breeding decision support, disease surveillance, welfare analysis, and resource optimization. Validation should be carried out systematically and should include calibration, uncertainty decomposition, attribution stability, and fairness between underrepresented strata and should be able to disclose the performance envelopes and failure typologies transparently. Studies need to come up with statistically principled drift detection methods that can distinguish benign cyclical variation and material, consequential distributional shifts with adaptive retraining policies that maintain historical continuity yet adopt emergent patterns. Accountability will be enhanced by probabilistic auditing tools with the ability to trace the presence of influential training cases and feature subsets that form the basis of a particular prediction to aid in remediation in case data quality problems are detected (Stafford et al., 2025).

The principles of data governance and ethical stewardship must change their form from fixed compliance checklists to dynamic stewardship systems with built-in risk evaluation, welfare protection, genetic diversity conservation, and equitable benefit sharing. The suggested practice involves the development of federated consortia that would facilitate decentralized model refinement without compromising data sovereignty and privacy, backed by the development of communication-efficient aggregation, reduction of bias in a heterogeneous client environment, and ensuring through provenance monitoring. Responsible utilization of generative models with the capability to generate transcriptomic profiles to generate transcriptomic profiles ought to be codified into principles to circumvent misuse and to permit the use of generative models in legitimate applications, such as scenario stress testing and data augmentation. The integration of social science into the research of adoption will clarify the determinants of trust, as it will design an explainability interface that is understandable to the stakeholders without excessive biological simplicity (Abdul, 2025).

Active learning and adaptive experimental design need to be given more attention to maximize sequencing and phenotyping resource allocation to observations that are likely to give the most information. There should be development of methods that aim to include decision-theoretic models incorporating uncertainty maps, economic cost models, animal welfare concerns, and genetic diversity goals to suggest sampling schedules and tissue choices that can hasten convergence of model reliability. A combination of these strategies and edge computing infrastructure can be used to support real-time feedback loops where initial analyses are used to drive real-time refinement of sampling intensity given the field conditions and minimize the time delay between the appearance of biological signals and their assimilation into the analysis. The dynamical systems methods to resolve the incompatibility between asynchronous sampling rates and dissimilar noise mechanisms without loss of causal interpretation will be useful in research into vigorous sensor transcriptomics fusion (Balsa-Canto et al., 2025).

The sustainability needs promote the energy and carbon efficiency of the computational workflows to the first-class research agenda. Guidelines based on comparative lifecycle analyses of model architectures, training regimens, and inference deployment modalities should be used to encourage the use of parameter-efficient methods, modular fine-tuning, and the use of sparsity without compromising biological depth. Providing a common framework for environmental impact reporting of analytical pipelines will enable accountability and promote innovation in designing green algorithms. At the same time, the model goals must be extended in order to encode combined optimization of productivity, welfare, resilience, and environmental footprints and require multiobjective learning frameworks and a clear expression of tradeoffs in the face of uncertainty. Interdisciplinary cooperation with climate science, soil ecology, and microbiome ecology can be used to provide extra contextual sets of variables and enhance prediction consistency with ecosystem-scale performance (Malik, 2025).

Capacity building is an essential element that supports the process of sustainable progress. Combined curricula in quantitative machine learning, statistical genetics, molecular biology, and ethical governance should be increased, and readily available training environments with low barriers to entry should be provided to a wide variety of institutions. Competence will be democratized, and methodological fragmentation will be alleviated by open educational resources on end-to-end reproducible workflows between raw sequence processing and model interpretation and monitoring. Best practices can be transmitted in mentorship circles and joint fellowships along academic, industry, and producer boundaries and bring research agendas close to reality. Publication and funding systems must be redefined to provide incentives for high-quality dataset construction, replication experiments, negative outcome reporting, and addition to the common infrastructure at the same level as the new algorithmic performance assertions (AlSagri and Sohail, 2024).

One strategic enabler that will be required is interoperability, which requires the convergence of community-controlled ontologies, schema registries, and programmatic interfaces that will streamline connectivity of omics layers, sensor streams, and environmental telemetry. The suggestion is also applicable to the design of architecture for modular microservices that will permit replacement and instance of comparative benchmarking of preprocessing, normalization, feature engineering, and modeling modules without the need to rewrite the entire pipeline. Sufficiency criteria of metadata standards that will be embedded in the requirements of submissions to repositories will improve the utility of the baseline data. The development of minimal information guidelines to be used when reporting AI-enabled transcriptomic studies will be collaboratively defined, which will lead to reproducibility and comparability, with the consequent reduction of duplication and the possibility of metaanalyses to draw generalizable principles (Barret et al., 2025).

Lastly, a reflexive orientation must be developed where regular strategic reviews reconsider the assumptions regarding data representativeness, model suitability in the face of an emergent disease, ethical importance of the acceleration of selection pressure, and social expectations surrounding transparency and animal welfare. Coordinated futuristic activities and scenario planning, which involves the participation of producers, veterinarians, geneticists, ethicists, environmental scientists, and policymakers, will be used to guide the direction of research toward resiliency in systemic shock. Constant harmonization of technical innovation with pluralistic value systems will be used to ensure legitimacy and promote responsible dissemination of AI-augmented transcriptomic knowledge. By a concerted effort toward the realization of these overlapping guidelines, the discipline will move out of the technically stunning but disjointed models to be a mature, dependable, and equitable knowledge base that quantitatively changes productivity, adaptability, welfare, and sustainability across the global livestock systems.

References

Abdul, R., 2025. AI-driven ethical governance: balancing innovation and accountability in data-intensive enterprises. Int. J. Eng. Sci. Adv. Technol. 25 (10), 17–21. https://doi.org/10.64771/ijesat.2025.v25.i10.pp17-21.

Abdulla, S., Aevermann, B., Assis, P., Badajoz, S., Bell, S.M., Bezzi, E., Cakir, B., Chaffer, J., Chambers, S., Cherry, J.M., Chi, T., Chien, J., Dorman, L., Garcia-Nieto, P., Gloria, N., Hastie, M., Hegeman, D., Hilton, J., Carr, A., 2024. CZ CELLxGENE discover: a single-cell data platform for scalable exploration, analysis and modeling of aggregated data. Nucleic Acids Res. 53 (D1), D886–D900. https://doi.org/10.1093/nar/gkae1142.

Aburub, F., Al-Akayleh, F., Abdel-Rahem, R.A., Al-Remawi, M., Agha, A.S.A.A., 2025. AI-driven transcriptomics: advancing gene expression analysis and precision medicine. In: 1st International Conference on Computational Intelligence Approaches and Applications, ICCIAA 2025—Proceedings. Institute of Electrical and Electronics Engineers Inc. https://doi.org/10.1109/ICCIAA65327.2025.11013076

Adu-Manu, K.S., Kyeremanteng, P.S., Ofori, P., Adjetey, C., Lawson, P.K., Cobbah, E., 2025. Federated Learning at the Edge: Enhancing Privacy and Efficiency in Distributed AI Systems. https://doi.org/10.5772/intechopen.1011986.

Agnes, K., 2025. Addressing bias in AI algorithms for health applications. IAA J. Biol. Sci. 13 (1), 37–43. https://doi.org/10.59298/iaajb/2025/1313743.

Agrawal, R., 2025. Adaptive Few-Shot Learning (AFSL): Tackling Data Scarcity with Stability, Robustness, and Versatility. https://doi.org/10.48550/arXiv.2501.13479.

Ali Raza, M., Zahara, K., Zahra, L., Zahra, A., 2025. One health concept: an integrative approach to combat zoonotic diseases at the human-animal-environment interface. Middle East Res. J. Agric. Food Sci. 5 (03), 47–50. https://doi.org/10.36348/merjafs.2025.v05i03.006.

AlSagri, H.S., Sohail, S.S., 2024. Evaluating the role of Artificial Intelligence in sustainable development goals with an emphasis on "quality education". Discov. Sustain. 5 (1). https://doi.org/10.1007/s43621-024-00682-9.

Alves da Mata, K., 2024. Machine Learning for Predictive Resilience Modeling. https://doi.org/10.62791/20389.

Antunes, B., Hill, D.R.C., 2024. Reproducibility, replicability and repeatability: a survey of reproducible research with a focus on high performance computing. Comput. Sci. Rev. 53. https://doi.org/10.1016/j.cosrev.2024.100655.

Arora, J.K., James, L.K., Charoensawan, V., 2025. Understanding and mitigating the impact of ambient mRNA contamination in single-cell RNA-sequencing analysis. PLoS One 20 (9). https://doi.org/10.1371/journal.pone.0332440.

Ashango, Z., Abtew, W.G., 2025. Impacts of Evolutionary Forces on Allele and Genotype Frequency of Organisms. https://doi.org/10.20944/preprints202506.0510.v1.

Balsa-Canto, E., Campo-Manzanares, N., Moimenta, A.R., Roudaut, G., Troitiño-Jordedo, D., 2025. Quantifying and managing uncertainty in systems biology: mechanistic and data-driven models. Curr. Opin. Syst. Biol. 42, 100557. https://doi.org/10.1016/j.coisb.2025.100557.

Barret, N., Bernasconi, A., Bikbov, B., Pinoli, P., 2025. I-ETL: an interoperability-aware health (meta)data pipeline to enable federated analyses. BMC Med. Inf. Decis. Making 25 (1). https://doi.org/10.1186/s12911-025-03188-0.

Benhamza, K., Benselim, R., Naidja, H., Seridi, H., 2025. A comprehensive survey of imputation methods in medical missing data analysis. Appl. Intell. 55 (11). https://doi.org/10.1007/s10489-025-06602-2.

Bernal-Hernández, P., Boni, A., Schot, J., Velasco, D., 2025. Transformative innovation policy meets responsible research and innovation: a critical reflection. Sci. Publ. Pol. https://doi.org/10.1093/scipol/scaf051.

Bickford Smith, F., Kossen, J., Trollope, E., van der Wilk, M., Foster, A.S., Rainforth, T., 2024. Rethinking Aleatoric and Epistemic Uncertainty. https://doi.org/10.48550/arxiv.2412.20892.

Bohrer, J.da S., Dorn, M., 2024. Enhancing classification with hybrid feature selection: a multi-objective genetic algorithm for high-dimensional data. Expert Syst. Appl. 255, 124518. https://doi.org/10.1016/j.eswa.2024.124518.

Borgelt, H., Kitel, F.G., Kockmann, N., 2025. Ontology-based data pipeline for semantic reaction classification and research data management. Computers 14 (8). https://doi.org/10.3390/computers14080311.

Brito, L.F., Heringstad, B., Klaas, I.C., Schodl, K., Cabrera, V.E., Stygar, A., Iwersen, M., Haskell, M.J., Stock, K.F., Gengler, N., Bewley, J., Hostens, M., Vasseur, E., Egger-Danner, C., 2025. Invited review: using data from sensors and other precision farming technologies to enhance the sustainability of dairy cattle breeding programs. J. Dairy Sci. 108 (10), 10447–10474. https://doi.org/10.3168/jds.2025-26554.

Burden, J., Tešić, M., Pacchiardi, L., Hernández-Orallo, J., 2025. Paradigms of AI Evaluation: Mapping Goals, Methodologies and Culture. https://doi.org/10.48550/arXiv.2502.15620.

Castiglioni, V., Lanotte, R., Loreti, M., Tini, S., 2025. DT-Stark: a tool for evaluating the effectiveness of digital twins through feedback and perturbations. Int. J. Softw. Tool. Technol. Tran. https://doi.org/10.1007/s10009-025-00826-w.

Castro, M., Vida, R., Galeano, J., Cuesta, J.A., 2025. Scarce data, noisy inferences and overfitting: the hidden flaws in ecological dynamics modelling. J. R. Soc. Interface 22 (231). https://doi.org/10.1098/rsif.2025.0183.

Caufield, J.H., Ghosh, S., Kong, S.W., Parker, J., Sheffield, N., Patel, B., Williams, A., Clark, T., Munoz-Torres, M.C., 2025. Standards in the Preparation of Biomedical Research Metadata: A Bridge2AI Perspective. https://doi.org/10.48550/arXiv.2509.10432.

Chen, X., Zhai, X., Sun, X., Hong, Q., Hu, J., 2025. Intelligent lifecycle management of distribution networks: a machine learning framework for efficiency, resilience, and environmental sustainability. Int. J. Inf. Commun. Technol. 26 (22), 40–54. https://doi.org/10.1504/ijict.2025.146906.

Choudhary, R.K., Sunil Kumar, B.V., Mukhopadhyay, C.S., Kashyap, N., Sharma, V., Singh, N., Tazerji, S.S., Kalantari, R., Hajipour, P., Singh Malik, Y., Rabbani, I., 2024. Animal wellness: the power of multiomics and integrative strategies. Vet. Med. Int. 2024 (1). https://doi.org/10.1155/2024/4125118.

Cole, N., Wu, W., Head, S.T., Bresnahan, S.T., Bhattacharya, A., 2025. Quantification method a/ects replicability of eQTL analysis, colocalization, and TWAS. https://doi.org/10.1101/2025.08.20.671303.

Córdoba, A., Lazzarini, D., Vargas-Bello-Pérez, B., 2025. Consumer perceptions, knowledge and preferences on animal welfare in grazing and confined dairy systems. J. Dairy Res. 1–9. https://doi.org/10.1017/s0022029925101398.

Cui, L., Guo, G., Ng, M.K., Zou, Q., Qiu, Y., 2024. GSTRPCA: irregular tensor singular value decomposition for single-cell multi-omics data clustering. Briefings Bioinf. 26 (1). https://doi.org/10.1093/bib/bbae649.

Cusworth, S., Gkoutos, G.V., Acharjee, A., 2024. A novel generative adversarial networks modelling for the class imbalance problem in high dimensional omics data. BMC Med. Inf. Decis. Making 24 (1). https://doi.org/10.1186/s12911-024-02487-2.

Cutello, V., Pavone, M., Zito, F., 2024. Inferring a Gene Regulatory Network from Gene Expression Data. An Overview of Best Methods and a Reverse Engineering Approach, 14070. Springer Science and Business Media LLC, pp. 172–185. https://doi.org/10.1007/978-3-031-55248-9_9.

de Oliveira, P.S.N., Andrade, B.G.N., Cardoso, T.F., Conteville, L.C., Pena, G.A.C., Malago-Jr, W., Bruscadin, J.J., Pascoal, J.J., Almeida, L.F., Josahkian, L.A., Ventura, H.T., Maciel, G.A., Mourão, G.B., Coutinho, L.L., Reecy, J., Regitano, L.C.A., 2025. miRNA–microbiome correlations in *Bos indicus* feed efficiency. Sci. Rep. 15 (1). https://doi.org/10.1038/s41598-025-20408-9.

Denashurya, N.I., Tiwut Atmojo, S., Muhlas Sambodo, I., 2025. Model bisnis inklusif untuk pemberdayaan petani kecil di kabupaten sambas. Jurnal Pertanian Agros 27 (2), 250–256. https://doi.org/10.37159/jpa.v27i2.81.

Deng, Z.C., Liu, M., Cao, K.X., Khalil, M.M., Guan, L.L., Sun, L.H., 2025. Gut microbiome and postbiotics: bridging the dietary nutrition and feed efficiency in food-producing animals. Sci. China Life Sci. https://doi.org/10.1007/s11427-025-3063-5.

Dradjat, K., Hamidi, M., Bartet, P., Hanczar, B., 2025. Self-supervised learning on gene expression data. https://doi.org/10.48550/arXiv.2507.13912.

Džermeikaitė, K., Šidlauskaitė, M., Antanaitis, R., Anskienė, L., 2025. Enhancing genomic selection in dairy cattle through artificial intelligence: integrating advanced phenotyping and predictive models to advance health, climate resilience, and sustainability. Dairy 6 (5), 50. https://doi.org/10.3390/dairy6050050.

Emsen, E., Kutluca Korkmaz, M., Baran Odevci, B., 2025. Artificial intelligence-assisted selection strategies in sheep: linking reproductive traits with behavioral indicators. Animals 15 (14), 2110. https://doi.org/10.3390/ani15142110.

Fang, L., 2025. Context-specific regulatory variants in precision medicine and agriculture. Nat. Rev. Genet. 26 (2), 81. https://doi.org/10.1038/s41576-024-00798-8.

Gao, Z., Zheng, J., Xu, G., 2025. Molecular mechanisms and regulatory factors governing feed utilization efficiency in laying hens: insights for sustainable poultry production and breeding optimization. Int. J. Mol. Sci. 26 (13), 6389. https://doi.org/10.3390/ijms26136389.

Gkountouras, J., Lindemann, M., Lippe, P., Gavves, E., Titov, I., 2024. Language agents meet causality—bridging LLMS and causal world models. https://doi.org/10.48550/arXiv.2410.19923.

Ghavi Hossein-Zadeh, N., 2024. An overview of recent technological developments in bovine genomics. Vet. Anim. Sci. 25, 100382. https://doi.org/10.1016/j.vas.2024.100382.

Ghavi Hossein-Zadeh, N., 2026a. Generative artificial intelligence in animal genomics for smart agriculture: Applications, challenges, and future prospects. Vet. Anim. Sci. 33, 100702. https://doi.org/10.1016/J.VAS.2026.100702.

Ghavi Hossein-Zadeh, N., 2026b. Advancing climate-resilient livestock systems: Next-generation emission mitigation strategies and integrated technological innovations. Vet. Anim. Sci. 31, 100588. https://doi.org/10.1016/j.vas.2026.100588.

Goel, P., Weber, R.O., 2025. Measuring interpretability: an investigation of domain independent interpretability. In: 6th International Conference on Pattern Recognition and Machine Learning, PRML 2025. Institute of Electrical and Electronics Engineers Inc., pp. 276–284. https://doi.org/10.1109/PRML66062.2025.11160328

Gulati, G.S., D'Silva, J.P., Liu, Y., Wang, L., Newman, A.M., 2024. Profiling cell identity and tissue architecture with single-cell and spatial transcriptomics. Nat. Rev. Mol. Cell Biol. 26 (1), 11–31. https://doi.org/10.1038/s41580-024-00768-2.

Haas, W., Duenk, O., Ponsuksili, B., Calus, 2025. Incorporating transcriptomic data into genomic prediction models to improve the prediction accuracy of phenotypes of efficiency traits. Genet. Sel. Evol. 57. https://doi.org/10.1186/s12711-025-01008-7.

Hamdan, N.M., Admodisastro, N., Osman, H.B., Muhammad, M.S.B., 2024. Semantic interoperability in multi-cloud platforms: a reference architecture utilizing an ontology-based approach. Int. J. Adv. Sci. Eng. Inf. Technol. 14 (6), 1967–1975. https://doi.org/10.18517/ijaseit.14.6.19861.

Hatfaludi, C.A., Serban, A., 2025. Foundational models and federated learning: survey, taxonomy, challenges and practical insights. https://doi.org/10.48550/arXiv.2509.05142.

Hossein-Zadeh, N.G., 2025. Artificial intelligence in veterinary and animal science: applications, challenges, and future prospects. Comput. Electron. Agric. 235, 110395. https://doi.org/10.1016/j.compag.2025.110395.

Iversen, P., Witzke, S., Baum, K., Renard, B.Y., 2025. Identifying drivers of predictive aleatoric uncertainty, 5453–5462. https://doi.org/10.24963/ijcai.2025/607.

Jain, N., 2025. Enhancing Data Integrity through Provenance Tracking in Semantic Web Frameworks. https://doi.org/10.48550/arXiv.2501.09029.

Jaiswal, D., 2025. Zero-trust architecture for telecom API security: a framework for the communications economy. Eur Modern Stud. J. 9 (4), 11–21. https://doi.org/10.59573/emsj.9(4).2025.2.

Jasbi, P., Mohr, A.E., Murthy, M.H.S., Klein-Seetharaman, J., 2025. Understanding metabolic resilience by unraveling temporal dynamics of cellular responses. Trends Endocrinol. Metabol. https://doi.org/10.1016/j.tem.2025.04.006.

Job, S., Tao, X., Cai, T., Xie, H., Lin, L., Li, Q., Yong, J., 2025. Exploring causal learning through graph neural networks: an in-depth review. WIREs Data Min. Knowl. Discovery 15 (2). https://doi.org/10.1002/widm.70024.

John, M.F., Alicia, L., 2025. Covariate balanced allocation of samples to batches to mitigate the impacts of technical variability. https://doi.org/10.1101/2025.03.21.644523.

Kamble, M.A., Dhabarde, D.M., Ingole, A., 2025. Transcriptomics: Mapping the RNA Landscape. Springer Science and Business Media LLC, pp. 187–198. https://doi.org/10.1007/978-981-96-3648-8_10.

Kamil, I.A., AL-Askari, M.A., 2025. Mitigating bias in AI algorithms: techniques for fair and ethical AI systems. Aswan Sci. Technol. Bull. 77–92. https://doi.org/10.21608/astb.2025.390962.1025.

Karthik, T., 2025. Dynamic Feature Engineering Through Reinforcement and Prompt Based Learning. https://doi.org/10.20944/preprints202505.2193.v1.

Kate, M., Neethirajan, S., 2025. Giving cows a digital voice—AI-enabled bioacoustics and smart sensing in precision livestock management. Ann. Anim. Sci. https://doi.org/10.2478/aoas-2025-0091.

Kenneweg, P., Dandinasivara, R., Luo, X., Hammer, B., Schönhuth, A., 2024. Generating synthetic genotypes using diffusion models. https://doi.org/10.48550/arXiv.2412.03278.

Khan, T., de Koning, K., Endresen, D., Chala, D., Kusch, E., 2024. TwinEco: A Unified Framework for Dynamic Data-Driven Digital Twins in Ecology. https://doi.org/10.1101/2024.07.23.604592.

Koptekin, D., Yapar, E., Vural, K.B., Sağlıcan, E., Altınışık, N.E., Malaspinas, A.S., Alkan, C., Somel, M., 2025. Pre-processing of paleogenomes: mitigating reference bias and postmortem damage in ancient genome data. Genome Biol. 26 (1). https://doi.org/10.1186/s13059-024-03462-w.

Lee, K.W., Jung Min, H., Park, H.W., Manavalan, B., Jeon, Y.-J., 2025. Abstract 2427: optimization of preprocessing strategies for developing AI-based disease diagnosis model using whole transcriptomic data. Cancer Res. 85 (8_Suppl. 1), 2427. https://doi.org/10.1158/1538-7445.am2025-2427.

Leslie, D., Rincon, C., Briggs, M., Perini, A., Jayadeva, S., Borda, A., Bennett, S.J., Burr, C., Aitken, M., Katell, M., Fischer, C., 2024. AI Ethics and Governance in Practice: An Introduction. https://doi.org/10.48550/arXiv.2403.15403.

Li, M., Lu, Y., Gao, Z., Yue, D., Hong, J., Wu, J., Xi, D., Deng, W., Chong, Y., 2024. Pan-omics in sheep: unveiling genetic landscapes. Animals 14 (2), 273. https://doi.org/10.3390/ani14020273.

Liang, Q., Lin, G.-W., Ding, X., Zou, B., Liu, X., Li, J., Zhang, Y., Wen, X., Liang, L., Bei, J.-X., He, M., Jia, H., Lai, W., 2024. Multiomics integration unravels genotype-microbiome interactions shaping the conjunctival transcriptome. iMetaOmics 1 (2). https://doi.org/10.1002/imo2.37.

Lin, Y., He, J., Chen, J., Zhu, X., Zheng, J., Bo, T., Lu, Z., 2025. BioGraphFusion: graph knowledge embedding for biological completion and reasoning. Bioinformatics 41 (7). https://doi.org/10.1093/bioinformatics/btaf408.

Lu, A.B., Lewis, A.C.F., 2025. Governance strategies for biological AI: beyond the dual-use dilemma. Trends Biotechnol. https://doi.org/10.1016/j.tibtech.2025.09.012.

Maddipudi, S., 2024. Cloud security and data privacy: safeguarding critical infrastructure during application migration. Int. J. Sci. Res. Eng. Manag. 08 (12), 1–5. https://doi.org/10.55041/ijsrem17988.

Mahmood- Ur-Rahaman, S., Sudheer, S., 2025. Analyzing the efficiency of hybrid explainable AI models for feature extraction and pattern recognition in high-dimensional data mining tasks. Int. J. Innov. Sci. Res. Technol. 2514–2525. https://doi.org/10.38124/ijisrt/25jul1197.

Majlatow, M., Ahmed Shakil, F., Emrich, A., Mehdiyev, N., 2025. Uncertainty-aware predictive process monitoring in healthcare: explainable insights into probability calibration for conformal prediction. Appl. Sci. 15 (14), 7925. https://doi.org/10.3390/app15147925.

Malik, F., 2025. Carbon-conscious intelligence: life cycle assessment and green standards for generative AI. Int. J. Innov. Sci. Res. Technol. 1030–1037. https://doi.org/10.38124/ijisrt/25aug585.

Mardinoglu, A., Turkez, H., Shong, M., Srinivasulu, V.P., Nielsen, J., Palsson, B.O., Hood, L., Uhlen, M., 2025. Longitudinal big biological data in the AI era. Mol. Syst. Biol. 21 (9), 1147–1165. https://doi.org/10.1038/s44320-025-00134-0.

Mateo-Casalí, M.Á., Boza, A., Cândea, C., Fraile, F., Morand, L., Nahshon, Y., 2025. Continuous integration, deployment and validation: supporting scalable industrial ecosystems. 1–9. https://doi.org/10.1109/ice/itmc65658.2025.11106635.

Mayrhofer, M., Radojičić, U., Filzmoser, P., 2024. Robust covariance estimation and explainable outlier detection for matrix-valued data a preprint. https://doi.org/10.48550/arXiv.2403.03975.

Mehdiyev, N., Majlatow, M., Fettke, P., 2025. Interpretable and explainable machine learning methods for predictive process monitoring: a systematic literature review. Artif. Intell. Rev. 58 (12). https://doi.org/10.1007/s10462-025-11399-0.

Nasir, T., Rehman, A., Tariq, M., Kubra, K.T., Ashraf, N., Jamal Nasir, F., 2024. Role of multi-omics in functional genomic, transcriptomic, proteomic and metabolic mechanisms for milk production, growth, fertility and health in livestock animals. Zeugma Biol. Sci. 6, 38–46. https://doi.org/10.55549/zbs.1680746.

Ning, Y., Liu, M., Liu, N., 2025. Advancing ethical AI in healthcare through interpretability. Patterns 6 (6), 101290. https://doi.org/10.1016/j.patter.2025.101290.

Nixon, M.P., Gloor, G.B., Silverman, J.D., 2024. Beyond Normalization: Incorporating Scale Uncertainty in Microbiome and Gene Expression Analysis. https://doi.org/10.1101/2024.04.01.587602.

Nunes, F.V.M., Behrens, L.M.P., Weimer, R.D., Gonçalves, G.F., da Silva Fernandes, G., Dorn, M., 2025. Deep learning methods and applications in single-cell multimodal data integration. Mol. Omics. https://doi.org/10.1039/d5mo00062a.

Nyaga, D.M., Zaied, R.E., Silander, O.K., Black, M.A., O'Sullivan, J.M., 2025. Beyond single references: pangenome graphs and the future of genomic medicine. Front. Genet. 16. https://doi.org/10.3389/fgene.2025.1679660.

Paleti, S., 2025. Ensuring Data Governance, Lineage, Explainability, and Auditability in Artificial Intelligence-Driven Financial Models. Deep Science Publishing, pp. 156–173. https://doi.org/10.70593/978-93-49910-19-5_9.

Pasupuleti, M.K., 2025. Decision Theory and Model-Based AI: Probabilistic Learning, Inference, and Explainability. https://doi.org/10.62311/nesx/rriv525.

Pimenow, S., Pimenowa, O., Prus, P., 2024. Challenges of artificial intelligence development in the context of energy consumption and impact on climate change. Energies 17 (23), 5965. https://doi.org/10.3390/en17235965.

Pook, T., Hassanpour, A., Niehoff, T.A.M., Calus, M.P.L., 2025. Strategies to improve selection compared to selection based on estimated breeding values. https://doi.org/10.1101/2025.01.09.632197.

Pradipta, N., Gambetta, I.W., 2024. Analysis of in-processing algorithms meeting monotonic selective risk conditions for fairness and accuracy in machine learning. In: Proceedings of 2024 IEEE International Conference on Data and Software Engineering: Data-Driven Innovation: Transforming Industries and Societies, ICoDSE 2024. Institute of Electrical and Electronics Engineers Inc., pp. 43–48. https://doi.org/10.1109/ICODSE63307.2024.10829888

Prakash, S., Singh, S., Mankar, A., 2024. Bridging data gaps: a comparative study of different imputation methods for numeric datasets. In: 2nd IEEE International Conference on Data Science and Network Security, ICDSNS 2024. Institute of Electrical and Electronics Engineers Inc. https://doi.org/10.1109/ICDSNS62112.2024.10691111

Rappaport, N., Nogal, B., Perrott, K., Domina, V., Leroy, H., Price, N.D., 2025. Early detection of wellness-to-disease transitions in the AI era: implications for pharmacology and toxicology. Annu. Rev. Pharmacol. Toxicol. https://doi.org/10.1146/annurev-pharmtox-062124-013423.

Ray, P.K., 2024. Human Factors Engineering, Product Development and Sustainable Performance in Organizations: Issues and Challenges from an International Perspective. Springer Science and Business Media LLC, pp. 137–145. https://doi.org/10.1007/978-981-99-9122-8_10.

Reddy Pinnapareddy, N., 2025. Carbon conscious scheduling in Kubernetes to cut energy use and emissions. Int. J. Comput. Exp. Sci. Eng. 11 (3). https://doi.org/10.22399/ijcesen.3785.

Rosales-Morales, A.E., Gutiérrez-Alfaro, A., Ornelas-Rodríguez, M., Espinal, A., Rojas-Domínguez, A., Puga-Soberanes, H.J., Carpio, J.M., 2024. Generative models for class imbalance problem on BreakHis dataset: a case study. Stud. Comput. Intell. 1149, 105–119. https://doi.org/10.1007/978-3-031-55684-5_8.

Rosati, A., 2025. Invited paper Artificial intelligence and the future of animal husbandry: rethinking knowledge, practice, and governance. Arch. Latinoam. Prod. Anim. 33 (Suppl. 1), 567–572. https://doi.org/10.53588/alpa.330509.

Roschewitz, M., Mehta, R., Jones, C., Glocker, B., 2024. Automatic dataset shift identification to support root cause analysis of AI performance drift. https://doi.org/10.48550/arxiv.2411.07940.

Rosen, I.A., Sánchez-Núñez, P., Hartley, R., 2025. Open Science Governance: The Role of Persistent Identifiers and Metadata Standards. https://doi.org/10.31219/osf.io/9h564_v3.

Satou, H., Mitkiy, A., 2025. On the Mechanisms of Adversarial Data Augmentation for Robust and Adaptive Transfer Learning. https://doi.org/10.48550/arXiv.2505.12681.

Shafik, W., 2024. Toward a more ethical future of artificial intelligence and data science. IGI Global 362–388. https://doi.org/10.4018/979-8-3693-2964-1.ch022.

Shafik, W., Kalinaki, K., Fahim, K.E., Adam, M., 2024. Safeguarding Data Privacy and Security in Federated Learning Systems Federated Deep Learning for Healthcare: A Practical Guide with Challenges and Opportunities. CRC Press, Uganda, pp. 170–190. https://doi.org/10.1201/9781032694870-13.

Shashank, C.G., Sejian, V., Silpa, M.V., Devaraj, C., Madhusoodan, A.P., Rebez, E.B., Kalaignazhal, G., Sahoo, A., Dunshea, F.R., 2024. Climate resilience in farm animals: transcriptomics-based alterations in differentially expressed genes and stress pathways. BioTech 13 (4). https://doi.org/10.3390/biotech13040049.

Shenoy, D., Bhat, R., Krishna Prakasha, K., 2025. Exploring privacy mechanisms and metrics in federated learning. Artif. Intell. Rev. 58 (8). https://doi.org/10.1007/s10462-025-11170-5.

Shu, W., 2025. Legal logic of AI data governance based on federated learning: institutional evolution from privacy protection to rights distribution. Int. J. Ethical AI Appl. 1 (4), 7–12. https://doi.org/10.64229/efs37007.

Srinak, N., Krüger, F., Kaleta, C., Taubenheim, J., 2025. Metabolic modeling of host-microbe interactions. Comput. Struct. Biotechnol. J. 27, 4304–4319. https://doi.org/10.1016/j.csbj.2025.10.006.

Stafford, C., Provost, K., Jones, N., 2025. Model validation levels: an automatable framework for model validation. J. Simulat. https://doi.org/10.1080/17477778.2025.2536096.

Stokholm, M.N., Rabaglino, M.B., Kadarmideen, H.N., 2024. GEDI: an R package for integration of transcriptomic data from multiple platforms for bioinformatics applications. Curr. Protoc. 4 (10). https://doi.org/10.1002/cpz1.70046.

Strober, B.J., Tayeb, K., Popp, J., Qi, G., Gordon, M.G., Perez, R., Ye, C.J., Battle, A., 2024. SURGE: uncovering context-specific genetic-regulation of gene expression from single-cell RNA sequencing using latent-factor models. Genome Biol. 25 (1). https://doi.org/10.1186/s13059-023-03152-z.

Syeda, 2025. Harnessing multi-omics and genome-editing technologies for climate-resilient agriculture: bridging AI-driven insights with sustainable crop improvement. Plant Mol. Biol. 115 (6). https://doi.org/10.1007/s11103-025-01650-1.

Tahir, M., Norouzi, M., Khan, S.S., Davie, J.R., Yamanaka, S., Ahmed, A., 2024. Artificial intelligence and deep learning algorithms for epigenetic sequence analysis: a review for epigeneticists and AI experts. Comput. Biol. Med. 183, 109302. https://doi.org/10.1016/j.compbiomed.2024.109302.

Tatyana, A., 2025. Multi-Way Data Representation: A Comprehensive Survey on Tensor Decomposition in Machine Learning. https://doi.org/10.31224/4421.

Taylor-King, J.P., Bronstein, M., Roblin, D., 2024. The future of machine learning within target identification: causality, reversibility, and druggability. Clin. Pharmacol. Therap. 115 (4), 655–657. https://doi.org/10.1002/cpt.3158.

Thomas, J., Houssineau, J., 2024. Improving Active Learning with a Bayesian Representation of Epistemic Uncertainty. https://doi.org/10.48550/arXiv.2412.08225.

Vélez-Cruz, N., Papandreou-Suppappola, A., 2024. Bayesian learning of nonlinear gene regulatory networks with switching architectures. Front. Signal Process. 4. https://doi.org/10.3389/frsip.2024.1323538.

Victorseelan, D., Babitha, N., Babitha, N., Subramanyam, T., Zalki, M., Jadhav, P.N., Pawar, R.K., 2025. Experimental design and statistical analysis in biological sciences: best practices and pitfalls. J. Neonatal Surg. 14 (10S), 736–748. https://doi.org/10.52783/jns.v14.2922.

Wang, Y., 2025. Structural monopoly risk and regulatory governance of generative artificial intelligence. Trans. Soc. Sci. Educ. Hum. Res. 14, 331–342. https://doi.org/10.62051/syw9g748.

Wang, S., Zheng, H., Wen, X., Fu, S., 2024. Distributed high-performance computing methods for accelerating deep learning training. J. Knowl. Learn. Sci. Technol. 3 (3), 108–126. https://doi.org/10.60087/jklst.v3.n3.p108-126.

Wang, Q., Xiao, Z., Mao, Y., Qu, Y., Shen, J., Lv, Y., Ji, X., 2025a. Model predictive task sampling for efficient and robust adaptation. Res. Square. https://doi.org/10.21203/rs.3.rs-6700167/v1.

Wang, X., Si, J., Wang, Y., Fang, L., Zhang, Z., Zhang, Y., 2025b. Integrating gene expression data via weighted multiple kernel ridge regression improved accuracy of genomic prediction. Genet. Sel. Evol. 57 (1). https://doi.org/10.1186/s12711-025-00997-9.

Wang, M., Guo, A., Zhang, L., Wu, S., Du, J., Yu, T., 2025c. Technological innovations and applications of spatial transcriptomics in livestock and poultry research. Anim. Res. One Health. https://doi.org/10.1002/aro2.70016.

Waters, D.L., Clark, S.A., Brown, D.J., Walkom, S.F., van der Werf, J.H.J., 2024. Validation of reaction norm breeding values for robustness in Australian sheep. Genet. Sel. Evol. 56 (1). https://doi.org/10.1186/s12711-023-00872-5.

White, R.R., 2025. 262 Precision feeding technologies for health and welfare: practice, promise, and pie-in-the-sky. J. Anim. Sci. 103 (Suppl. 3), 13–14. https://doi.org/10.1093/jas/skaf300.016.

Wolfram-Schauerte, M., Vogel, T., Tuoken, H., Savitski, M.F., Simon, E., Nieselt, K., 2025. Approaching the holistic transcriptome—convolution and deconvolution in transcriptomics. Briefings Bioinf. 26 (4). https://doi.org/10.1093/bib/bbaf388.

Xiang, R., Fang, L., Liu, S., Liu, G.E., Tenesa, A., Gao, Y., Mason, B.A., Chamberlain, A.J., Goddard, M.E., Canela-Xandri, O., Wang, S., Yu, Y., Cai, W., Li, B., Pairo-Castineira, E., D'Mellow, K., Rawlik, K., Xia, C., Yao, Y., Navarro, P., Rocha, D., Li, X., Yan, Z., Li, C., Rosen, B.D., Van Tassell, C.P., Vanraden, P.M., Zhang, S., Ma, L., Cole, J.B., 2025. Genetic score omics regression and multitrait meta-analysis detect widespread cis-regulatory effects shaping bovine complex traits. PNAS Nexus 4 (7). https://doi.org/10.1093/pnasnexus/pgaf208.

Xing, R., Li, B., Dora, S., Whittaker, M., Mathie, J., 2025. Engineering-oriented explainable machine learning and digital twin framework for sustainable dairy production and environmental impact optimisation. Algorithms 18 (10), 670. https://doi.org/10.3390/a18100670.

Xu, S., Akhatayeva, Z., Liu, J., Feng, X., Yu, Y., Badaoui, B., Esmailizadeh, A., Kantanen, J., Amills, M., Lenstra, J.A., Johansson, A.M., Coltman, D. W., Liu, G.E., Curik, I., Orozco-terWengel, P., Paiva, S.R., Zinovieva, N.A., Zhang, L., Yang, J., Li, M., 2024. Genetic advancements and future directions in ruminant livestock breeding: from reference genomes to multiomics innovations. Sci. China Life Sci. 68 (4), 934–960. https://doi.org/10.1007/s11427-024-2744-4.

Xu, X., Li, Y., Liu, J., Lv, X., Liu, L., 2025. Rewiring microbial metabolism through post-translational switches. Trends Biotechnol. https://doi.org/10.1016/j.tibtech.2025.10.002.

Yalawar, D., Yadav, A., Lingampally, S., Jinnoji, S., 2025. Multimodal AI-enhanced robotics for real-time adaptive collaboration in dynamic environments. Int. J. Multidiscip. Res. 7 (5). https://doi.org/10.36948/ijfmr.2025.v07i05.54975.

Yang, X., Liu, G., Feng, G., Bu, D., Wang, P., Jiang, J., Chen, S., Yang, Q., Miao, H., Zhang, Y., Man, Z., Liang, Z., Wang, Z., Li, Y., Li, Z., Liu, Y., Tian, Y., Liu, W., Li, C., Li, A., Dong, J., Hu, Z., Fang, C., Cui, L., Deng, Z., Jiang, H., Cui, W., Zhang, J., Yang, Z., Li, H., He, X., Zhong, L., Zhou, J., Wang, Z., Long, Q., Xu, P., Shen, Q., Dong, K., Li, S., Zhang, S., Wang, Y., Huang, Y., Wang, T., Liu, Z., Luo, Q., Luo, Y., Liu, J., Yang, G., Li, Y., Huang, W., Qi, X., Zhang, Z., Xu, Z., Li, C., Qin, J., Qin, X., Zhao, Y., Chen, Y., Wang, Y., Xiao, M., Zhang, R., Wang, C., Zhang, C., Chen, Y., Wang, J., Zong, S., He, H., Wang, Z., Meng, Z., Li, F., Wang, X., Wang, Y., Zhou, Y., Li, Y., Guo, S., Hu, Y., He, C., Wu, H., Wang, H., Chen, Q., Ma, K., Jiang, S., Han, L., Xin, Y., Zhao, Y., Li, F., Ren, Y., Zou, Y., Lin, X., Wu, W., Yu, J., Feng, G., Zou, Z., Zhai, W., Gu, Q., Yu, L., Guo, J., Gao, F., Li, W., Hu, B., Wang, H., Li, X., Wang, H., Wang, Y., Guo, J., Zhao, Y., Zhou, Y., Li, F., Liu, J., Chen, Y., Yang, G., Li, X., 2024. GeneCompass: deciphering universal gene regulatory mechanisms with a knowledge-informed cross-species foundation model. Cell Res. 34 (12), 830–845. https://doi.org/10.1038/s41422-024-01034-y.

Yazdani, A., Mijakovac, A., Giulianini, F., Alotaibi, M., Gibson, H., Ammar, M., Hoshi, R., Cummings, R.D., Serhan, C.N., Lauc, G., Chasman, D., Mora, S., 2025. Immune and inflammatory resolution pathways through multi-omics using the AI-based Network Integration. Res. Square. https://doi.org/10.21203/rs.3.rs-7665795/v1.

Yu, J., Dai, Y., Liu, X., Huang, J., Shen, Y., Zhang, K., Zhou, R., Adhikarla, E., Ye, W., Liu, Y., Kong, Z., Zhang, K., Yin, Y., Namboodiri, V., Davison, B.D., Moore, J.H., Chen, Y., 2024a. Unleashing the Power of Multi-Task Learning: A Comprehensive Survey Spanning Traditional, Deep, and Pretrained Foundation Model Eras. https://doi.org/10.48550/arXiv.2404.18961.

Yu, Y., Ma, W., Zheng, Y., Shi, L., 2024b. Assessing and mitigating batch effects in large-scale omics studies. Genome Biol. 25 (1). https://doi.org/10.1186/s13059-024-03401-9.

Zambon, M., Mantica, F., Dias, M., Frazer, J., Irimia, M., 2025. Evolution of comparative transcriptomics: biological scales, phylogenetic spans, and modeling frameworks. Curr. Opin. Genet. Dev. 94, 102387. https://doi.org/10.1016/j.gde.2025.102387.

Zeitlinger, J., Roy, S., Ay, F., Mathelier, A., Medina-Rivera, A., Mahony, S., Sinha, S., Ernst, J., 2024. Perspective on recent developments and challenges in regulatory and systems genomics. Bioinform. Adv. 5 (1). https://doi.org/10.1093/bioadv/vbaf106.

Zhang, X., Luo, Y., Li, T., 2025. A review of research on secure aggregation for federated learning. Future Internet 17 (7), 308. https://doi.org/10.3390/fi17070308.

Zhioua, S., Binkyte, R., Ouni, A., Ktata, F.B., 2025. On the origins of sampling bias: implications on fairness measurement and mitigation. https://doi.org/10.48550/arXiv.2503.17956.

Zhou, Q., 2024. Optimisation Strategies for Ensuring Fairness in Machine Learning: With and Without Demographics. https://doi.org/10.48550/arxiv.2411.09056.

Zhou, J., Zhang, X., Wang, Y., Liang, H., Yang, Y., Huang, X., Deng, J., 2024. Contamination survey of insect genomic and transcriptomic data. Animals 14 (23), 3432. https://doi.org/10.3390/ani14233432.

Zhuang, Z., Lee, X., Wei, J., Fu, Y., Zhang, A., 2024. CBCMS: a compliance management system for cross-border data transfer. In: Proceedings—2024 IEEE International Conference on Big Data, BigData 2024. Institute of Electrical and Electronics Engineers Inc., pp. 4789–4798. https://doi.org/10.1109/BigData62323.2024.10826075

Chapter 6

Artificial intelligence integration in livestock proteomics

Artificial intelligence (AI) is becoming a driving interface within livestock proteomics that converts the changeable flux of proteins, peptides, and post-translational modifications to operational biological and management information. Whereas genomics is a relatively stable blueprint, the proteome is indicative of actual-time physiological condition, interaction with the environment, and regulatory modulation, and so is a fragile connection between hereditary possibility and delivered performance, health, welfare, and sustainability efficacies. A pillar of AI embodied in this paradigm not only changes the paradigm of proteomics as a series of almost discrete identification and quantification processes but also enables the analysis turned into the fabric of adaptive inferences that learn latent structure, decouple technical and biological variance, and continually redefine interpretation as new streams of data about their contexts flow in. In this chapter, proteomics is placed in the wider context of the omics continuum without undervaluing its temporal plasticity, structural diversity, and context dependency, which complicates traditional analytical pipelines but is compensated for by the advanced representation learning.

The story follows the progress in machine learning architecture, probabilistic modeling, and multimodal fusion to encode more faithfully a variety of heterogeneous proteomic signals, including but not limited to changes in abundance, modification state, interaction net, and novel spatial or structural labeling. It is stressed that the focus has shifted to the move toward handcrafted feature engineering and learned embeddings that can match peptide spectral data with biological functionality and downstream phenotypic significance. The uncertainty quantification and explainability are considered as mediators of trust, with the focus on having the predictive gains with the calibrated confidence and interpretive transparency to domain experts. Causal reasoning and counterfactual simulation are introduced as an edge that makes the difference between correlative biomarkers and putative intervention levers, which should be used to select, make management decisions, steward resources, and protect against overinterpreting associative patterns.

The notions of data governance can emerge on the notion of sampling design, metadata richness, provenance tracking, standard ontologies, privacy preservation, and fair representation across breeds and production settings, may all influence the fairness, generalizability, and ethical suitability of a model. This chapter introduces the concept of how federated and edge-aware infrastructures can boost the idea of participatory inclusiveness without violating the limits of sovereignty and biosecurity, and how stewardship lifecycle models (such as benchmarking, drift detection, recalibration, and versioning) can also be undertaken to offer sustainable reliability. The problems, such as high-dimensional sparsity, not at random intensity data, batch effects, annotation incompleteness in rare modifications, and integrative frictions with other layers, are not addressed as peripheral, but resolved in an architectural choice, and the possibility of implementing them in practice.

Some of the future trends that are being visualized include proteome-informed foundation models, graphical representation of interaction and pathway situations, generative improvement of low-density spectra, real-time streaming analytics, and sustainability-oriented optimization that internalizes environmental and welfare measures and productivity objectives. It is connected to ethical thinking that contains transparency standards, diversity preservation, reducing the effects of algorithmic bias, carbon-conscious computation, and responsible acceleration of innovation without damaging the trust of the stakeholders or the ecological resilience. Integrating both the depth of methodology and rigor of governance and translational purpose, the chapter talks of AI-enabled livestock proteomics as the foundation of an integrated, adaptive, and ethically-oriented omics intelligence ecosystem that will enhance livestock system resilience, efficiency, and sustainability in future livestock systems.

6.1 Introduction to AI in livestock proteomics

Proteomics has taken a strategic role in the current livestock systems in the sense that it embodies a dynamic molecular interface where genomic potential is regulated into actual physiological state, productivity, resilience, health, welfare, and

Artificial Intelligence in Livestock Omics. https://doi.org/10.1016/B978-0-443-45663-3.00013-0

environmental stewardship outcomes. As genomic variation represents a relatively stable pool of inherited ability, the proteome indicates the actual time-dependent assembly of the transcriptional regulation, translational efficiency, post-translational modification, protein folding, trafficking, interaction networking, and degradation relations in changing nutritional, microbial, management, and climatic conditions. This conditional and time-dependent plasticity promotes proteins and their modes of modification to the level of the evidentiary layer to interpret subclinical perturbation, adaptive stress response, metabolic allocation strategy, immunological preparedness, and quality attributes of product integrity and sustainability requirements. The promise of proteomic indicators to narrow down the selection goals by complementing the estimated breeding values of genomic data with a mechanistically based intermediate phenotype capable of increasing accuracy, speeding feedback mechanisms, and protecting genetic variation, also contributes to strategic significance by clarifying the pleiotropic tradeoffs and physiological limits beforehand in the innovation cycle. Enabling the incorporation of this type of information requires conceptual schemes, in which proteomic signals are no longer seen as a single biomarker, but rather as a node in multilayered biological and production knowledge networks, which drive ethical, economic, and ecological decision spaces (Jitjumnong et al., 2025; Zhong et al., 2025).

Placing proteomics in the continuum of omics needs to take into consideration and differentiate with respect to higher and lower strata of the molecular domain. The levels of transcripts, epigenetic states, landscape of chromatin accessibility, microbial community state, pools of metabolites, stream of digital phenotyping, and outputs of environmental sensing are all additional facets of the genotype-to-phenotype translation process. Proteins not only mediate much of the causal agency by conducting metabolic responses, structural assembly, signaling cascades, and immune effector functions, but also their level, isoform makeup, folding state, localization, and modification status all encode regulatory finesse that cannot be inferred alone based on nucleic acid-derived proxies. The proteome thus serves as an interface point at which differences in genomes, epigenetic modification, transcription, and exogenous effects are worked out by biochemical means. Simultaneously, proteomic readouts establish downstream-oriented layers of metabolism by giving shape to patterns of flux and turnover, establishing a dependent relationship in both directions, which disrupts linear analytical hierarchies. This positioning is further elaborated in detail by spatial and structural proteomics to map distributions of compartmentalized distribution-interaction interfaces and expand insight into bulk abundance information. In integrative livestock omics, the proteome becomes the actor, become-sentinel: the former by the fact that it is a mediator of the potential as that encoded in a genome; the latter by the fact that a change in composition or the landscape of modification can herald phenotypic change before it even becomes visible. Analytic methodologies should, therefore, honor their stratified relational context, retaining connections to their variant, transcript, metabolic, microbiome, and phenomic features and a sparsity, dynamic range, and heterogeneity of their form that is unique (Enitan et al., 2025; Happonen and Varjosalo, 2025).

The arguments that support the use of AI in this proteomic ecosystem are based on complexity, scale, and timeliness demands that surpass the ability of traditional deterministic or isolated statistical algorithms. Large-dimensional spectral observations, large peptide protein inference manifolds, stochastic and mechanism-specific missingness, large dynamic range spanning orders of magnitude, and the combinatorial explosion of potential post-translational modification states create challenges to analysis that require adaptive representation learning. Latent confounders due to batch effects, instrument drift, or protocol heterogeneity have the advantage of disentangling the effects of the latent confounder, but the disadvantage of losing subtle phenotype-connected variance components. The network dependencies between proteins, complexes, pathways, and regulatory modules are inviting relational inductive biases that can be operationalized on graph-oriented architectures. Proteomic profiles under longitudinal management regimes or environmental perturbations experience temporal evolution, which necessitates sequence and state space modeling that is capable of projecting the trajectories and identifying anomalous inflections at an early stage in time to inform intervention. The multimodal integration of genomic, transcriptomic, metabolomic, metagenomic, imaging, sensor, and management data requires flexible alignment mechanisms, attention-based cross-modal interaction capture, and hierarchical abstraction, which will retain interpretability in the face of compositional heterogeneity. The current constraints of timeliness of decisions provide an additional constraint: in real or near real-time inference pipelines, efficient edge to cloud orchestration, energy-conscious calculation, risk-calibration prediction of action thresholds, and trust scaffolds are essential, which can be sustained by practitioners and regulators. Causal inference integration further reasons why it is necessary since it distinguishes between predictive correlates and modifiable levers, to help prioritize resource-constrained situations and minimize inadvertent reinforcers of confounded relationships (Jadav and Mandanaka, 2025; Zeng et al., 2025).

The extent of the chapter is thus arranged in a manner that can be able to recasting of these integrated motivations into a logical, analytical, and operational blueprint. This is supported by the initial conceptual and terminology background that is expounded to reconcile the knowledge of proteoforms, modification taxonomies, acquisition modalities, spectral properties, peptide to protein inference issues, dynamic range considerations, missing data process, functional annotation ontologies, and cross-species mapping limitations. Principles of data engineering are subsequently preempted with a particular focus on the rigor of the sampling design, biobanking integrity, completeness of metadata, normalization policies, batch correction models, alternative feature representations, capturing provenance, quality measures, and reproducibility models, each of which has an impact on the downstream model fidelity, equity, and generalizability. Application-oriented discourse considers present and future applications across health monitoring, disease resilience, nutritional optimization, reproductive management, welfare assessment, enhanced quality of products, antimicrobial stewardship, and integration into selection indices, while maintaining predictive utility and ensuring compliance with ethical and sustainability principles. Deep learning is questioned by methodological exposition as a spectrally interpretable and de novo sequencer, embedding generator, interaction contextualizer, graph modeler, uncertainty propagator, probabilistic architecture, multi-task learner, multi-modal learner, causal model, and explainability toolchain, and computational scalability and energy efficiency.

High-dimensional sparsity, incomplete coverage, modification annotation gaps, cross-laboratory harmonization, low breed and production system representation, integration friction across omics timescales, model drift across varying environmental baselines, problems in interpretation, poor standards of validation, bad governance complications across privacy, sovereignty, auditability, and security are also interrogated in the chapter. Case-oriented storytelling integrates all of these methodological approaches into working pipelines of action that illuminate plausibility, translational operation, and lifecycle performance custodianship and does not overfit context-specific procedural niceties. The discourse of validation and benchmarking advocates the variety of measures, the external cohort examination, the uncertainty calibration, the drift monitoring, the protocol of continuous learning, and the audit trail preservation as inseparable to the accountable deployment. The directions sought by future oriented analysis include, but are not limited to, proteome informed basis models, real time streaming analytics, structural and spatial joining, generative augmentation approaches, federated collaboration ecosystem and digital twin connection, sustainability considered multi objective optimization, fair benefits allocation, and equitable benefits allocation, as an ethical reflection of reducing bias, openness, eco footprint, wellbeing, and fair benefits allocation.

Sustaining these aspects of coherence in a unified narrative, the chapter aims to provide a rigorous yet integrative framework, which will facilitate the use of AI by researchers, practitioners, and governance stakeholders in order to create proteomic insights through curative decision support and adaptive stewardship in livestock systems. The aim of indexing techniques is not only to conform, but also to align methodological choice with biological practicability, practical operational legal, or ethical, and sustainability demands, to facilitate a dependable, open-minded, and fair proteomic intelligence ecosystem that enhances genomic innovation or augments the broader livestock omics ecosystem.

6.2 Key concepts and terminology in proteomics

Proteomic research in livestock systems relies upon sophisticated understandings of the molecular entities being measured, starting with a distinction between forms representing canonical gene products and the myriad of proteoforms that are generated through alternative splicing, protein translation initiation variability, ribosomal frameshifting, sequence polymorphism, proteolytic actions, chemical derivatization, and the whole panoply of post-translational modifications. Isoforms denote amino acid sequence variations driven by splice variation, while the more synonymous term proteoform can be used to describe any molecular variant of a protein, including spatial configurations representing combinations of modifications that translate to charge distribution, stability, binding affinity, localization, rates of turnover, and ultimately, functional attributes of the protein's characteristics. Post-translational modifications bring layers of regulation by encoding conditional switches, timestamps, and/or integrative signatures that are inextricably linked to environmental, developmental, immune, and/or metabolic cues. The factor of site specificity, stoichiometry, crosstalk, and dynamic reversibility, accompanied by transient binding and structural changes to regulatory states and functional use of protein, increases the effective state space relative to the underlying gene count and increases the layers of biological complexity. Thus, AI must learn a form of representation that moves beyond a single abundance scalar per gene symbol, and utilize structure embeddings that represent patterns of modification, the positional context of modifications, and the interdependence of modifications among sites that allow for the reasoning of downstream mechanistic pathways and points for intervention leverage. Without this sort of detailed representation of the molecular landscape, AI's predictive capabilities fail to capture biologically informative heterogeneity that is consequential to constructs like interpretability and the generation of actionability related to gene products and/or proteins (Müller et al., 2025; Palabiyik and Palabiyik, 2025).

Quantitative paradigms have translated this conceptual richness into measurable signals via separate methodological philosophies; label-free strategies, multiplexed isobaric tagging strategies, and targeted assays for a priori hypothesis testing or targeted (high precision) monitoring. Label-free strategies estimate relative abundance based on spectral counting or integrating ion signal from chromatographic runs. This maximizes breadth of coverage, but it is valuable to consider that it can vary and is subject to alignment across runs. Isobaric tagging allows for multiplexing, whereby several samples are aggregated during acquisition; however, this has disadvantages due to the potential interference experienced with reporter ions, introducing potential inaccuracies if not controlled for. Targeted assays will restrict the analytical breadth to a set of predefined peptides in targets, which will lose exploratory opportunity value for improved accuracy, reliability, and quantitative linearity. Also, each sampling paradigm is subject to specific types of noise that include structured missing data, dynamic range, and batch effects; without proper modeling, these will carry downstream. AI pipelines will need to consider these paradigm-specific error structures and hence introduce paradigm-specific approaches to preprocessing, normalization, denoising, and uncertainty quantification, and such work should isolate paradigms, rather than employ naively homogenous approaches that rob themselves of the signal. In addition to bridging multiple paradigms, cross-scale calibration must also contend with intensity and domain adaptation strategies that harmonize the different quantitative scales and censoring behaviors, while respecting and not losing biological variance of interest for selection, health monitoring, and decision support (Ghoneim et al., 2025; Stoisser et al., 2025).

The characteristics of primary data, including the density of spectral data, fragmentation variability, allocation of duty cycle, and tradeoff between the depth of identification and extent of quantitative completeness, are impacted by the platform and acquisition mode of a mass spectrometer. Data-dependent acquisition modes stochastically sample high-intensity precursor features, which can result in richly informative fragmentation spectra, but this occurs at the expense of run-to-run variability in precursor sampling and non-randomly missing features. Data-independent acquisition modes systematically sample data from within the mass range of interest, generating rich fragmentary spectra providing greater breadth of recording at the expense of additional spectral congestion requiring computational approaches to deconvolution and attribution of peptides. Targeted acquisition modes focus instrument resources to monitor pre-determined transitions and/or product apex of precursor monitoring to increase the precision of quantification for a pre-determined panel of biological constituents. The operational distinctions associated with each acquisition mode lead to non-overlapping distributions of precursor interference and drift in retention time, mass accuracy, and fragmentation variability, which challenge the potential for equal treatment throughout downstream analytics. Models for deconvolution, denoising, retention alignment, and attribution of features based upon rule-based architectures informed by AI would also benefit from explicitly encoding the acquisition mode metadata, instrument evaluations associated with performance, and run contextual factors into constants during modeling to differentiate mode-specific technical variance from biological structure. Explicit encoding in multi-channel input tensor or graph formats would improve generalization over cohort, instruments, and temporal sampling, while also possibly allowing for recalibration based upon performance drift (Jebet et al., 2025; Phlairaharn et al., 2025).

The interpretive spine that creates biologically meaningful objects out of raw signals is made up of spectral libraries, feature detection algorithms, and inference of a peptide-to-protein relationship. The library-centric identification makes use of stored patterns of fragmentation and retention indices to speed up the search, but is limited in the cases where new sequence variants, unexpected changes, or species-specific peptides not found in the compendium emerge. The de novo sequencing expands into unknown sequence space; however, it has high error rates, which require probabilistic confidence modeling. Chromatographic peak groups, intensities, and charge state annotations are delineated by feature detection pipelines, but feature detectors face the problems of coelution, peak splitting, saturation artifacts, and shape perturbations. Subsequent parsimony or probabilistic protein inference overcomes shared peptide degeneracy, splice variability, and ambiguity of modification site localizations to generate hierarchical assignment structures. This is enhanced by AI: learned embeddings of fragment ion spectra can be used to search for similarities and expand libraries; a graph neural network can be used to model peptide-protein bipartite relationships, with uncertainty propagation; sequence-to-spectrum generative models can generate plausible libraries when there is a class imbalance; and Bayesian or evidential models can be used to scale peptide confidence scores. Handling ambiguity integratively, as opposed to straining to arrive at a definite result, maintains analytic flexibility downstream and multi-hypothesis reasoning, which is important in cases where decisions have economic or welfare consequences (Molla and Meseret, 2025).

There is also the issue of dynamic range and missingness mechanisms that make reliable quantification in livestock proteomics difficult. Protein and peptide concentrations range over large concentration differentials and result in censored low-intensity signals and the risk of saturation by high-concentration species. Missing values arise in a variety of ways, including stochastic undersampling of low-abundance precursors during competition with selection cycles; ion suppression and matrix effects; biological silencing in the deterministic absence; and acquisition strategy-specific coverage limitations. It is important to differentiate missing completely at random, missing at random, and missing not at random regimes since any imputation or modeling decisions that do not ignore the underlying mechanism may cause bias, inflated false discovery, or negated true variance. These difficulties are exacerbated by data imbalance, either a small fraction of proteins dominating the signal or the infrequent states of a modification site having a disproportionately large functional value. Such distortions are reduced by AI methods, which include mechanism-aware missingness modeling, zero-inflated and hurdle models, censored likelihood models, generative augmentation of rare classes, cost-sensitive optimization, and uncertainty decomposition that categorizes epistemic and aleatoric sources of variation. Effectively managing imbalance is useful in improving the detection of subtle phenotype-related signatures without excessive overfitting to the noise from sparsely observed peptide features (Mou et al., 2025).

Other conceptual terrains that require rigor are functional annotation, ontology integration, and cross-species mapping. Livestock species may have only partial or unbalanced annotation coverage with respect to model organisms, which are more comprehensively understood, thereby encouraging the use of orthology projection to the detriment of lineage-specific innovations, paralog divergence, and situation-specific functional translocation. Ontologies provide hierarchical organization to processes, components, and functions, which allow semantic aggregation, enrichment analysis, and constraint-based reasoning, although inconsistency, spreading of synonyms, and granularity disparities can help propagate ambiguity. AI can help in that it learns joint embeddings of sequence, structural, interaction, and literature on the basis of putative functions, disambiguation of paralogs, and prioritization of annotation curation targets. Transfer learning and domain adaptation methods reconfigure representations that have been trained with abundant annotation of species in the livestock environment and mitigate systematic bias. Probabilistic ontology mapping can encode uncertainty in annotation such that downstream decision systems can use evidence weights in the same way they would have in the absence of an ontology encoding. Cross-species integration should also consider genetic stratification at the breed level as well as selection history, conserving population uniqueness proteomic idiosyncrasies that can affect adaptation, resilience, and production phenotypes (Prieto-Baños et al., 2025).

Structural and spatial proteomics are able to increase interpretive granularity by questioning conformational states, interaction architecture, compartmental localization, and patterns of tissue or cell type resolved distributions. Confirmation sensitive measurements can identify the variants of the folding as well as the dynamics of complex assembly that can regulate functional output in a way not described by abundance alone. Spatial resolution profiling identifies the microenvironmental heterogeneity by mapping gradients, segregation, or co-localization patterns underlying signaling, metabolic specialization, or immune orchestration. The resulting multi-dimensional data on these modalities have spectral, coordinate, structural, and network properties with increased sparsity and context sensitivity. AI needs to thus exploit geometric, graph, and attention-based models capable of integrating Euclidean and relational cues, making predictions of interaction propensities and inferring missing spatial coverage, and implicating structural perturbations to phenotypic endpoints. There is the concept of multi-scale representation learning that connects the atomic or domain level embedding with the pathway and system scale abstractions to generate mechanistic hypotheses and reduce the overinterpretation of partially sampled tissue landscapes by uncertainty-aware spatial interpolation (Horvath and Coscia, 2025).

Both integrative proteogenomics and more general multi-omics alignment principles aim to transform proteomic measurements into corresponding genomic variants, transcriptomic shallow, epigenetic signatures, metabolite observations, microbiome structures, phenotypic sensor streams, and environmental metadata, and convert them all into consistent cause-and-effect explanations. Proteogenomics enhances annotation with unanticipated peptides of new splice junctions, sequence polymorphisms, or unannotated open reading frames and refines search space and sensitivity to improve identification. Coherence in modalities requires temporal coherence, scale coherence, and identity inconsistency resolution. Variations in measurement error structure, sampling frequencies, and missingness patterns are some of the problems that are problematic with naive concatenation methods. This fusion is achieved using AI by applying modality-specific encoders that project their output into common latent manifolds where cross-modal correlations, conditional regularities, and possible causal pathways can be explored. Graph-based knowledge integration uses existing biological networks with learned associations, and counterfactual simulation frameworks attempt to discover the effect of perturbation in candidate regulatory nodes on each layer. Multi-objective optimization is more accurate in prediction and interpretable at the same time, as it maintains fairness and is computationally efficient by guaranteeing that integrative models provide actionable information without obscuring the provenance or introducing unintentional bias. Translational reliability is protected by intensive testing across distributional shifts, such as the changing management practices or weather conditions (Oh and Li, 2024; De Filippis et al., 2025). An overview of proteomic data modalities and acquisition strategies relevant to AI applications is presented in Table 6.1.

6.3 Data engineering foundations for AI in livestock proteomics

The prerequisite to engineering the data underpinnings of AI in livestock proteomics precedes any interaction with the instrumentation, at the interface of a biological sample and the administration of related contextual information, which

TABLE 6.1 Overview of proteomic data modalities and acquisition strategies relevant to AI applications.

Data modality/acquisition strategy	Description	Relevance to AI-driven livestock proteomics	Key considerations for AI integration
Label-free quantification (LFQ, DDA/DIA)	Proteins or peptides are quantified without chemical labels; data-dependent acquisition (DDA) or data-independent acquisition (DIA) MS workflows assess relative or absolute abundance	Large-scale, cost-effective for multiple animals or conditions; generates rich quantitative features for AI models in livestock (e.g., biomarker identification and trait correlations)	• Greater sample throughput and fewer reagents needed. • Increased variability, missing values, and batch effects compared to some labeling methods. • AI models must handle sparse data, imputation, and non-uniform missingness
Isobaric labeling (TMT, iTRAQ) multiplexing	Chemical tags (e.g., TMT, iTRAQ) label peptides from different samples; multiplexing allows simultaneous analysis of many samples in one run and relative quantification	Enables high-throughput comparative studies across conditions (e.g., breeds and treatments) in livestock; AI models can exploit multiplex data for phenotype-prediction or responsive trait modeling	• Strong multiplexing and built-in internal normalization. • Tag-ratio compression, cost of labels, more complex workflows • AI must adjust for multiplex bias and interference artifacts
Targeted proteomics (SRM/MRM/PRM) and PTM-focused workflows	Selected reaction monitoring (SRM/MRM) or parallel reaction monitoring (PRM) target predefined peptides; workflows focus on post-translational modifications (PTMs) such as phosphorylation, ubiquitination	Ideal for validation of biomarker candidates or trait-linked proteins in livestock; AI models trained on validated panels can yield robust predictions of traits (e.g., health and growth)	• High sensitivity and reproducibility for known targets • Limited coverage of proteome; fewer discovery opportunities • AI models may over-fit small panels; need clear feature engineering and validation
Spatial/single-cell proteomics and low-input workflows	Advanced strategies enable proteomic profiling in individual cells or tissue subregions, often via micro-sampling, imaging-MS, or low-input MS methods	Highly relevant for livestock research when resolving tissue-specific responses (e.g., muscle and immune) or rare cell types; AI models can map fine-scale variation into breed or welfare traits	• Cutting-edge, complex, high cost; sparse data and more noise. • AI must accommodate ultra-high-dimensional, low-sample data; robust feature selection and dimension-reduction are crucial
Data-independent acquisition (DIA) proteomics	In DIA workflows, all precursor ions in defined m/z windows are fragmented, enabling broad coverage and consistent quantification across samples	Produces rich, consistent data across large sample sets in livestock populations; AI algorithms benefit from reduced missing-value rates and stronger consistency for modeling traits	• More complex spectra and heavy computational requirements • Requires comprehensive spectral libraries and sophisticated data processing • AI pipelines must integrate spectral-library generation, feature extraction, and robust training
Metaproteomics and complementary proteome/omics coupling	In livestock settings, proteomics may link with microbiome/host-proteome, exosome proteomics, or integrate with other omics layers (transcriptome and metabolome) to create richer data for systems modeling	Enables integrated AI models that link host proteome with microbiome, diet, or metabolomic context, crucial for livestock traits such as feed-efficiency and disease resilience	• Multi-modal data integration increases complexity, heterogeneity, and missingness • AI frameworks must perform feature fusion, reduce modality bias, and maintain interpretability

eventually preconditions interpretability and downstream modeling validity. Sampling procedures should balance temporal, physiological, and environmental state recording, the circadian effects, and the timing of feeding regimens, exposures to stress, disease progression periods, the reproductive status, and microclimatic change that may leave proteomic signatures that are not of interest. To address confounding between biological groups and technical sequencing, randomization and blocking designs are used, and the order of collection, operators, and time of handling is properly recorded to facilitate future disentanglement of latent nuisance variance. Biobanking infrastructure is molecular integrity maintained by an evidence-based cold chain stewardship, preservative minimization of thaws, aliquot strategy design, and consistent use of preservatives, the potential proteomics-specific artifacts of which are captured in structured metadata. Chain of custody systems use special persistent identifiers on each sample, and they are connected in two directions to animal identifiers, which have been confirmed by genomic or RFID (radio frequency identification)-based identity-checking to avoid phenotype misassignment. Integrity of phenotypic links denotes ontologically correct codification of traits, time-precision to allow alignment with longitudinal sensor streams, and alignment of farm management logs, veterinary operations, and environmental sensor readings under a temporally normalized event ledger. Pedigree and genomic relationship matrices are further combined with proteomic records to support mixed model adjustment and better learning of kinship awareness, whereas information systems tools delineate access management, consent congruence with both ethics and verifiable privacy degrading fashions when information swamps international borders with collaborative breeding or health surveillance centers (Tedeschi et al., 2025; Zhang et al., 2025).

Preprocessing pipelines convert raw spectral outputs in heterogeneous formats into structured matrices consumable by machine learning, with a trade-off between attenuation of noise and retention of small biologically pertinent variance values. Denoising includes the elimination of chemical and electronic base artifacts, deisotoping, charge state reconciliation, mitigation of precursor interference, and smooth chromatographic peaks using adaptive filters or learned denoising autoencoders that avoid reconstructing the variance components predicting orthogonally measured quality control references. Aligned retention time builds on warping algorithms or deep sequence and retention predictor, and culling false missingness across livestock purchases. Normalization methods should be mechanism-conscious: global scaling, variance stabilization, probabilistic quotient, median polish, and local regression adjustments are considered based on the criteria of batch variance suppression, preservation of known controlled ratios of spikes in cases where they are known, and preservation of covariance structures of phenotypes. The latter includes an idea of batch correction, which combines linear mixed modeling, empirical Bayes harmonization, unwanted variation factorization, adversarial domain adaptation, or conditional variational autoencoding to promote systematic shifts at the expense of explicit protection of biological contrasts by adding phenotype-anchored invariance constraints. The datasets of post-translational modification require enrichment efficiency variability and site occupancy properties that are not measured by total protein abundance, and which require specialized normalization. Sampling of quality control embedded into batches is used to provide anchors to drift modeling and recalibration signals, and adaptive feedback loops, which indicate deviation limits that signal instrument service or review of protocols before model training generates bias (Ma et al., 2025; Kumar & Kumar, 2025; Niloy et al., 2025).

The choice of feature representation arbitrates the mapping of molecular measurements into computational objects and has a significant impact on the ability of AI systems to use hierarchical biological organization. Peptide intensity matrices provide granular resolution and retain sensitivity of isoform and proteoform heterogeneity, but result in peptide-to-protein degeneracy and fragmentation of missing peptides. Protein-level rollups will decrease the number of dimensions and alleviate sparsity at the cost of potential obfuscation of regulatory subtlety at a particular modification site or alternative exon. Resolved post-translational modification signatures carry context-dependent regulatory markers. Understanding these markers requires considering their positions relative to domains, motifs, structural regions, disorder elements, and evolutionary conservation. Spectral level embeddings based on neural encoders that have been trained to reconstruct fragmentation patterns provide rich representations that can be used to drive clustering, novelty detection, or similarity search functions without early introduction of commitments to potentially erroneous peptide assignments. The formulations are graph-based, describing bipartite structures of protein-receptor, protein-protein interaction networks, and modification co-occurrence graphs, allowing graphic neural networks to spread information in the context and predict latent functional modules. Multi-instance learning models address heterogeneity by learning aggregation operators that are optimized to perform particular downstream prediction challenges on sets of peptides as bags of latent protein states. Exogenous features such as genomic variant burden, microbiome community summaries, environmental exposure indices, and results of management interventions are incorporated into multimodal architectures, which enhance representation to support integrative reasoning and mediate how the genotype and environment interact via proteomic pathways (Hein et al., 2025; Mo et al., 2025).

The reconstruction of missing values not by chance or of low abundance is a signal that requires principled statistical modeling rather than ad-hoc rules of imputation, which pervert variance structure and cause effects of bias to be estimated. Detection limits create left-censored intensities that necessitate censored likelihood models, Tobit-like models, or hurdle models that explicitly model the process of occurrence and magnitude. The benefit of the mechanism classification is that the supervised discriminators are conditioned on synthetic perturbations or simulator outputs of instruments able to label missingness causes, and downstream prediction of intervention plans. Generative probabilistic and diffusion-based imputers, which depend on local chromatographic neighborhood, fragmentation similarity, and network adjacency, are able to make coherent reconstructions with uncertainty intervals calibrated, allowing downstream models to use imputed entries based on their reliability. It is possible to use few-shot transfer learning to enhance the detection of low-abundance peptides with the help of the overall multi-project corpora embeddings, and open modification search with sequence to spectrum prediction can also be employed to recover modified low intensity species which would otherwise be missed. The problem of class imbalance is reduced without any artificial, implausible patterns through data augmentation techniques (noise injection using profiles of instrument variance, mixout techniques using high confidence analogs with low abundance spectra). Informational content of absence patterns themselves can be replaced by explicit modeling of their missingness indicators as auxiliary inputs by deep architectures that can derive biologically motivated suppression events in health or productivity phenotype (Azad et al., 2025; Lou et al., 2025).

The measures of the data quality, provenance, and data reproducibility schemes are all the basics of the credibility and regulatory acceptability of the proteomics-informed AI outputs. Quality measurement involves identification confidence distributions, adherence to false discovery rate at peptide, protein, and location level, mass accuracy stability, measuring drift at retention time, chromatographic peak form integrity, carryover evaluation, coefficient of variation between quality control replicates, dynamic range coverage, fraction of explained variance attributable to batch factors before and after correction and completeness measures stratified by functional categories to identify systematic dropout. Monitoring is a continuous process, and the emergent deviations are reported by the control charts and anomaly detection algorithms that may cause the contamination of longitudinal predictive models. Provenance tracking imposes schemas of organized metadata on the basis of controlled vocabularies of sample properties, processing choices, software releases, model hyperparameters, and data provenance transformations to permit fine-grained building of analytical paths. The cross-referentiality of institutions is ensured due to the conformity to the standards and compliance with the reporting standards, ontology mapping, and semantic versioning of spectral libraries. Reproducibility frameworks represent containerized deployment of applications, decorative workflow needs, immutable data snapshots, cryptographic hashes, and automated continuous integration claims of pipeline alterations on benchmark data of various breeds, tissues, and production systems. A transparent stewardship layer consists of exemplary artifacts, which have documentation that describes the intended scope of application, stratified performance depending on demographic or environmental subgroups, calibration of uncertainty diagnosis, and ethical risk evaluation (Jalbani et al., 2025; Kancharla, 2025).

This combination of engineering elements makes it possible to have a single ecosystem in which AI models can be trained, tested, deployed, and continually refined without sacrificing biological or operational reliability. Strict sample and metadata stewardship offers a high-integrity substrate with few possible avoidable confounders. Harmonization in control of the drowning of little physiological signals due to technical differences is guarded by advanced preprocessing, as well as batch harmonization. The rich hierarchy feature representations give the architecture access to multi-scale molecular logic encoding as well as paths to interpretability. In addition to improving sensitivity to early perturbations, mechanism-aware handling of missingness and low-abundance patterns avoids distortion of effect size estimation. The lengthy quality, provenance, and reproducibility regimes are playing a role in the fact that insights are checkable, transportable, and pliable in the view of the variable management, climatic, and hereditary landscape. This combination is what causes livestock proteomics not to be a descriptive type of modality, but an operational source of intelligence, one that can be utilized to manage health proactively, breed resiliency, develop nutritional interventions, maximize welfare, and manage sustainability. Having engineering rigor at each stage, the resultant continuum of analysis will be aligned with ethical demands, economic viability, and scientific capability, and the AI will become the exemplar of the transformative promise it holds to the multifaceted, dynamic, and responsibility-laden environment of modern livestock systems. Data engineering strategies for AI readiness in livestock proteomics are shown in Table 6.2.

6.4 Current applications of AI in livestock proteomics

Applications of AI in the field of livestock proteomics have shifted their focus from exploratory pattern recognition to operational decision support as part of health management, breeding, nutrition, and product quality optimization, welfare monitoring, and biosecurity systems. The overall concept that unifies these applications is the participation of proteomic

TABLE 6.2 Data engineering strategies for AI readiness in livestock proteomics.

Strategy category	Description	Purpose for AI readiness	Key methods or tools	Typical challenges
Data ingestion and harmonization	Collecting raw proteomics data from different instruments, formats, and labs, and consolidating into a unified schema	Ensures consistent input for AI models, reduces heterogeneity, and supports transferability	• Use of standard file formats (e.g., mzML, mzXML, and raw) • Metadata annotation with *ontology-backed fields* • ETL pipelines for import into central repositories	• Divergent instrument formats and vendor files • Missing or inconsistent metadata • *Lab-to-lab protocol* variation
Quality control (QC) and filtering	Assessing data quality (e.g., mass error, retention time drift, and peak width) and filtering out unreliable measurements or samples	Improves data reliability, reduces noise, and prevents garbage-in for AI models	• Peak alignment, outlier detection • Quantitative metrics (missed cleavages and dynamic range) • Automated QC dashboards	• Lack of standard QC thresholds across livestock species • High-throughput data volumes make manual QC impractical • Filter-induced bias (removing rare but valid signals)
Missing value imputation and normalization	Addressing missing protein-abundance values and normalizing across batches or runs to enable comparability	Prevents bias or instability in AI models; supports consistent feature patterns	• Imputation algorithms (k-NN and MICE) • Normalization techniques (median-based and quantile) • Batch-effect correction (ComBat and LIMMA)	• Highly sparse data in proteomics • Imputation may introduce artifacts • Batch correction may remove the biological signal
Feature engineering and dimensionality reduction	Transforming the high-dimensional proteome into meaningful features for AI models (e.g., ratio metrics, network scores, and embeddings)	Reduces model complexity, improves generalization, supports interpretability	• Principal component analysis (PCA), t-SNE, and UMAP • Protein co-expression network metrics • Embedding via autoencoders and feature selection	• Risk of losing rare but important signals • Over-reduction may hamper interpretability • Embedding training requires large sample sizes
Dataset labeling and cohort metadata linking	Associating proteomics samples with detailed phenotype, breed, treatment, and environment metadata	Ensures that AI models connect molecular features to meaningful biological outcomes	• Use of standardized metadata schema (breed, age, tissue, and environment) • Linking proteomic data to sensor/phenomic logs • Version control of data cohorts	• Incomplete or ambiguous metadata • Phenotype definitions may vary across farms/species • Linking challenges between molecular and field data

Continued

TABLE 6.2 Data engineering strategies for AI readiness in livestock proteomics.—cont'd

Strategy category	Description	Purpose for AI readiness	Key methods or tools	Typical challenges
Data storage, access, and governance	Architecture for secure, scalable storage and managed access to proteomic datasets in livestock settings	Facilitates large-scale AI training, reproducibility, and compliance (FAIR principles), and collaboration	• Cloud-based platforms and distributed databases • Secure access and audit logs • Data catalogs and APIs for downstream use	• Resource constraints in field settings • Data privacy, breed IP issues • Ensuring interoperability across systems
Data fusion and integration for multi-modal AI	Integrating proteomics with other omics (genomics and metabolomics), phenomics, or sensor data to build holistic AI models	Enables richer AI models with better predictive power and context awareness in livestock traits	• Multi-omics feature concatenation, graph-based fusion • Temporal/longitudinal alignment algorithms • Use of feature-level and model-level fusion strategies	• Differences in scale, resolution, and timing across modalities • Complex missingness patterns • Demands for advanced modeling and computing
Scalable pipeline automation and monitoring	Automating end-to-end data engineering pipelines from raw data ingestion to model-ready datasets, including monitoring and versioning	Supports large-scale livestock studies, multi-farm data flows, and reproducible AI workflows	• Workflow tools (Snakemake and Nextflow) • Pipeline logging, version control, and metadata tracking • Alerting and dashboarding for data drift	• Complexity of building field-scale automated pipelines • Versioning changes may affect downstream AI models • Need for domain expertise in pipeline design

data as a fixed catalog of molecular abundances in a dynamic, context-dependent expression of the processes of physiological state change, perturbations in regulatory networks, and host-environment interaction. Architectures have now become able to ingest high-dimensional peptide and post-translational modification signature signals as well as genomic variants, environment telemetry, behavior sensor streams, and longitudinal phenotype records and construct latent embeddings that conserve mechanistic signal and reduce nuisance variance. This integrative substrate makes it possible to estimate risks in a stratified way, identify early deviations, and plan the intervention individually. The focus has changed to explainable architectures, quantification of uncertainty, and heuristics of causal inference that increase trust and practical clarity, especially when economic and ethical interests include animal welfare, sustainability goals, and antimicrobial responsibility. The applications domain may be considered in terms of functional domains where proteomic intelligence is used to aid in biomarker discovery, metabolic efficiency modeling, reproductive management, stress resilience characterization, product quality improvement, pathogen interaction profiling, vaccination response analytics, precision nutrition, herd segmentation and genomic selection refinement, and each of these domains has data engineering requirements and modeling constraints to condition algorithm choice and assessment standards (Tedeschi et al., 2025; Zhang et al., 2025).

Multi-resolution pipeline proteomic representations based on sparse regularization, ensemble stability assessment, and network-informed constraint systems have re-energized biomarker discovery of health, welfare, and disease surveillance. The techniques of AI transform high-dimensional spectral or peptide-level manifolds onto discriminative subspaces in relation to subclinical states, inflammatory transitions, metabolic disruptions, or new infection signatures before apparent clinical dysfunction. Semi-supervised algorithms based on anomaly detection use extensive unlabeled longitudinal baselines to derive personalized molecular homeostatic envelopes, due to which patterns of deviations are identified with false alarm controls. Transfer learning and domain adaptation support breed, management system, tissue matrix, and instrument platform variability, which will lower the revalidation costs. Multi-instance and multiple kernel learning strategies are reconciled with heterogeneity in the peptide assignment confidence and complexities in proteomics, and probabilistic graphical models capture conditional dependencies between centric biomarker panels in the proteomic pathway, enhancing interpretability and mitigating single-feature dropout. Mechanistic validation, a post-discovery, takes advantage of causal structure learning that interrogates directionality between physiological indicators and proteomic candidates and eliminates correlational artifacts due to confounding variables like age, nutritional plane, or environmental stress. Online drift monitoring is added to the deployment layers to identify changes in the baseline protein expression distributions that may damage classifier calibration, which triggers adaptive retraining procedures that are based on continual learning models that maintain legacy knowledge and introduce novel molecular patterns (Adams and Bittremieux, 2025; Shao et al., 2025).

Predictive modeling of feed efficiency and metabolic acclimatization uses time-dependent proteomic patterns that are characteristic of nutrient allocation, ruminal or gut-related microbial cross-talk, mitochondrial biogenesis, lipid mobilization, and redox equilibrium. Recurrent neural architectures, temporal convolutional networks, and attention-based sequence models model lag formations between previous proteomic configurations and the resultant feed conversion or resilience on diet changes. Combination with metabolomic and metagenomic summits by multimodal fusion layers increases the ability to separate host intrinsic metabolic regulation and microbiome-mediated nutrient transformation. Representation learning focuses on the unfreezing of individual fixed features and dynamic plastic responses, using variational models that use the variability to separate latent spaces between fixed and adaptive parts in accordance with breeding or management lever categories. Counterfactual simulation modules approximate possible efficiency gains in the event of hypothetical nutritional interventions, and are used to suggest the calculated precise ration. Uncertainty-conscious predictions are based on heteroskedastic modeling reflecting varying degrees of confidence in variables across different phenotypic levels and assist in making risk-conscious decisions, balancing the cost of inputs and the predicted productivity with the consideration of welfare (Li et al., 2025).

Reproductive physiology and prediction of fertility uses a proteomic signature of gametogenesis quality, endocrine control, uterine receptivity, embryonic developmental competence, and maternal immunomodulatory tolerance. The sampling of multiple timepoints around periods of reproductive competitiveness drives survival analysis using advanced deep learning systems that can process the censored outcome data and situations of competing risks. Attribution of pathways across these embedded graphs that incorporate protein interaction networks, hormone signaling pathways, and structural matrix remodeling elements elucidates molecular groups that predict a conception, gestational maintenance, or postpartum recovery path. Transferable self-supervised large unlabeled proteome corpora pretrained in the form of embeddings decrease the requirements of labeled samples and improve generalization across physiological cycles. Causal mediation models examine the hypothesis of proteomic mediators of genomic variant effects on fertility phenotypes and, with this understanding, inform selection that maintains reproductive fitness in the quest for production benefits. Fairness auditing identifies whether subgroup calibration is done on parity, age, or physiology status to protect against systematic bias in prediction services that may unreasonably redistribute reproductive management interventions (Singh, 2025).

Proteomic signatures of stress, welfare, and environmental resilience inhabit a space where AI separates acute adaptive and chronic dysregulation. Multivariate change point detectors are applied in rolling proteomic windows to detect the rapid transitions between molecular states representing thermal load, social hierarchy breakages, transport stress, or subclinical inflammation before the manifestation of behavior. Latent trajectory clustering groups individual adaptation curves and demonstrates archetypal profiles of resilience that can be used to customize management. Causal discovery methods assist in the isolation of direct stress-induced proteomic modulation and secondary metabolic responses, boosting the specificity of intervention targeting. Composite welfare indices with better sensitivity and less noise invoked volatility are achieved by integrating proteomic capabilities with wearable sensor indices of activity, posture, or thermal gradients by hierarchical Bayesian fusion models. Interpretability in models focuses on the translation of influential proteomic to stress axis regulatory pathways, such as neuroendocrine, immune, and oxidative homeostasis modules, to mechanistically validate and gain trust among the stakeholders (Buchha, 2025).

This involves optimizing milk, meat, and product quality traits using predictive scaffolds between production proteomics and compositional properties, processing performance, organoleptic and shelf life determining features, and determinants of nutritional value. Structured sparsity multi-output regression models also identify shared and trait-specific proteomic drivers, which decreases redundancy and maintains distinct components of variance. Domain adaptation rectifies the change of processing batch or storage condition, and the prediction accuracy is stabilized in different segments of the supply chain. Proteoform conscious representations distinguish variants that are modulated by modification to function and regulate coagulation behavior, tenderness pathways, generation of flavor precursors, or lipid oxidation dynamics. Reinforcement learning models have arisen to emulate quality outcome dynamics in response to alternative management interventions with the aid of generative models to foresee proteomic changes with hypothetical feeding, housing, or handling changes. Multi-objective maximization presents a surface of balanced sets of decisions based on quantifying the anticipated quality improvements (as well as indicators of environmental footprint and welfare impact) (Durgun, 2024; Mathieu et al., 2025).

Monitoring of antimicrobial resistance and mapping of host-pathogen interactions involves AI to deconstruct complex mixtures of host and microbial proteomes, extracting signals of resistance mechanism activation, effector protein deployment, immune evasion strategies, and host defense mobilization. Mixed origin signals are separated into different clusters by deconvolution algorithms based on spectral library expansion, taxonomic priors, and peptide embedding clustering. Graph neural networks model bipartite host-pathogen interaction landscapes, and they use structural similarity, co-localization probability, and temporal co-expression to predict functional contact edges. Multi-instance learning solves the problem of sparse detection of low abundance of resistance proteins, whereas anomaly detection provides an indicator of an emerging composite pattern of resistance not previously defined in historical baselines. The causal relationship between antimicrobial-related events, resistance proteomic changes, and host inflammatory modulation is addressed by causal inference frameworks for the assessment of stewardship policies. Privacy-protected federated learning systems allow for cross-farm or regional model training on data protection surveillance without centralization of raw data, increasing statistical power while respecting the limits of data management (Wang et al., 2025a).

The temporal characterization of vaccine and immune responses by using temporal proteomics is based on the high-frequency sampling used to identify priming, expansion, deployment of effector, contraction, and memory formation activities. The sequence-based transformers are conditioned on peptide motif embeddings and structural antigen features, which can predict epitope-specific antibody or cellular response proteomic correlates. A mechanism of differential temporal attention is used that determines critical windows during which proteomic diversity or flux of modification is best used to distinguish between robust and suboptimal responders. The hybrid models are mechanistic, and data-driven residual learning is used to combine ordinary differential equation immunodynamics with prior knowledge on antigen presentation, cytokine signaling cascades, as well as clonal selection with observed proteomic variability. Uncertainty quantification gives confidence levels of the projected protective status to boosters that are used to personalize schedules. Cross-cohort transfer models allow quick adaptation of predictors of response to new vaccine formulations or variants of the pathogen by fine-tuning antigen-specific embedding layers whilst maintaining conserved immunoregulatory latent dimensions (Hui et al., 2025).

Precision nutrition advice based on proteomic flux patterns uses dynamic modeling of pathways that mediate amino acid usage, mitigation of nitrogen excretion, lipid mobilization, oxidative phosphorylation fine-tuning, and maintenance of the gut barrier. Models of latent flux inference combine the protein abundances with stoichiometric balances and learned memory coefficients that enable it to infer pathway throughput, such that the bottlenecks to the dietary regulation can be identified. Engineers recommended engines were modeled as constrained optimization to trade off performance gains, objectives on nutrient excretion reduction, cost, and risk of metabolic disorder. Continuous learning involves the feedback of actual production and health results, updating the policy of nutrient intervention, and calibration of the models. There are counterfactual reasoning modules simulating proteomic adjustments when other ration formulation is used, to enable proactive adaptation to changes in ingredient availability, or sustainability sourcing requirements. Mediated metabolite signaling representation through integration with microbiomes improves prediction of host proteomic changes caused by prebiotic or probiotic interventions (Zhu et al., 2025).

Herd stratification and risk segmentation models are data-driven, which use unsupervised and semi-supervised clustering, manifold learning, and probabilistic mixture modeling to identify subpopulations based on composite proteomic health, productivity potential, resilience capacity, and resource efficiency. Multi-view integration combines proteomic, genomic, behavioral, and environmental matrices in the same latent overlying matrices that do not collapse complementary variance structures. Dynamic segmentation monitors the changes in risk conditions, which allows prompt intensification of preventive activities or management changes in the focus. Interpretability routines project cluster defining proteomic features on biological processes, allowing stakeholders to understand and enabling them to see that segmentation has been done on meaningful physiological divergence, and not technical artifacts. Active learning focuses on labeling of the ambiguous individuals in order to perfect boundary demarcation, enhance the accuracy of triage, frequency allocation of monitoring, and resource allocation of interventions. Good segmentation is fair so that there is equal representation of breeds or management groups, and the data system fails to be biased in providing access to improved care pathways (Wang et al., 2025b).

Incorporating proteomics in selection indices and genomic decision support brings in a new intermediate molecular phenotype layer that narrows down the estimation of breeding values of complex traits that include robustness, fertility, efficiency, and quality aspects. Multi-trait genomic prediction models use proteomic principal components, pathway activity scores, or sparse biomarker panels as correlated traits, and this will enhance the predictive accuracy of phenotypes with low heritability or sparse measurement. Structural equation modeling provides mediation pathways in which genomic variants have effects via proteomic intermediates, distinguishing direct additive effects and modulatory network effects and providing the opportunity to prioritize causal candidate variants. Bayesian selection of variables makes proteomic variables substantially more stable to add and complexity more effective than simple methods, and cross-validation stratified to both cohort and temporal blocks reduces overfitting to transient environmental quality. Selection indices that include the economic weighting of projected downstream effects of proteomics inferred resilience or health attributes on lifetime productivity and avoidance of the treatment cost, produce composite scores, which may be used to balance genetic gain with sustainability and welfare goals. The decision support interfaces display the uncertainty broken down into genetic, proteomic mediation, and residual environmental factors, enabling the breeders to customize the risk tolerance level and operate on the diversification strategies that aim to mitigate the unexplained environmental changes (Ghavi Hossein-Zadeh, 2025; Xing et al., 2025).

In these functional areas, there are common methodological motifs that include resilience to batch effects and missingness, the ability to integrate longitudinally, interpretability based on biological networks, and recalibration to distributional shift. Application design is touched by ethical issues requiring that proteomic data be used in a transparent manner in informing welfare-related decisions, preventing overdependence on model products that may have concealed early welfare issues, and ensuring that data governance is considered in multi-stakeholder ecosystems. Recent implementations are increasingly taking advantage of hybrid modeling, which combines mechanistic priors with deep learning, resulting in systems that maintain falsifiability and mechanistic anchoring and learn nonlinear compositional interactions otherwise unreachable with traditional linear mixed models. Frameworks of validation integrate replicating cohorts externally, negative control tasks, and stress testing in simulated scenarios of the noise inflation or systematic drift, and provide resilience of deployed tools. Current issues consist of scaling of standardized spectral libraries to different breeds and physiological states, the inclusion of spatial and structural proteomics capabilities to make tissue-specific inferences, and the further development of models to be in line with changing regulatory and societal requirements of data use, welfare, and sustainability. Nevertheless, the interconnected portrayals made by the existing applications show that proteomics enabled by AI is moving away from descriptive post hoc analysis to prediction-based, precision-based, and ethically sensitive operational intelligence underpinning advancements in health stewardship, efficiency, product quality, resilience, and genetic enhancement in contemporary livestock production systems. Functional domains of AI applications in livestock proteomics are presented in Table 6.3.

6.5 AI methodological landscape for proteomic analysis

AI techniques for investigations in livestock proteomics have moved beyond narrowly defined boundaries to embrace a methodological spectrum from specific neural structures in processing raw spectra to causal integrated approaches,

TABLE 6.3 Functional domains of AI applications in livestock proteomics.

Application domain	AI techniques used	Key functions	Examples/outcomes
Biomarker discovery for production traits	Supervised machine learning (e.g., Random Forest and XGBoost); deep learning (autoencoders)	• Identification of protein signatures linked with production traits (growth, meat quality, and milk yield) • Selection of candidate proteins for prediction models	In livestock studies, AI methods applied to proteomics data identified protein panels associated with milk-production efficiency and disease-resistance traits
Protein structure and functional prediction	Graph neural networks (GNNs) and deep convolutional neural networks (CNNs) for sequence/structure data	• Predicting unknown protein functions, PTMs, protein-protein interactions (PPIs) • Enhancing proteome annotation in livestock species	AI models applied in proteomics can accelerate the annotation of regulatory proteins in livestock species lacking comprehensive databases
Trait prediction and selection modeling	Multi-omics integration with deep learning (e.g., multilayer perceptrons and attention models); transfer learning	• Mapping proteomic features to phenotypes (e.g., feed efficiency and fertility) • Incorporating proteomic data into genomic- and phenotypic-prediction models	In livestock, combining proteomics with AI improved predictive accuracy of trait models compared to single-omics approaches
Disease diagnosis and health surveillance	Unsupervised and supervised learning (clustering and classification), anomaly detection algorithms	• Detecting proteomic shifts indicative of early disease or stress responses • Real-time monitoring via biosensors or multiplex assays	Studies highlight AI's role in processing proteomic data for health outcomes, e.g., immune response profiling
Environmental and management response profiling	Reinforcement learning; time-series deep learning (RNNs and LSTM)	• Linking proteomic changes to environment (heat stress, diet, housing) • Supporting adaptive management of livestock via predictive proteome-based models	AI-driven proteome profiling supports the prediction of environmental impacts on livestock performance and welfare
Proteome data integration and systems modeling	Multi-modal deep learning; hybrid AI (graph convolution and ML); data-fusion methods	• Integrating proteomics with genomics, metabolomics, and phenomics • Building network models of proteome function in livestock systems	AI tools enabled the fusion of proteomics data with other omics, enabling holistic modeling of complex livestock traits
Real-time proteome-based monitoring and decision support	Edge-AI, embedded machine learning, and streaming analytics	• Continuous or near-real-time proteomic monitoring in livestock • Immediate decision-support for feed, health, and reproduction management	Emerging platforms show potential for field-deployable proteomic sensors with AI analytics in livestock systems
Rare breed and species adaptation via AI-proteomics	Transfer learning, few-shot learning, and domain adaptation algorithms	• Tailoring proteomic AI models for under-represented livestock species and local breeds • Accounting for breed-specific proteome signatures and limited sample availability	AI-adapted models have begun to show improved performance in niche livestock populations, promoting inclusivity and diversity in livestock omics
Explainable AI for proteomic insights in livestock	Explainable AI (XAI) frameworks: SHAP, LIME, and attention models integrated with proteomics	• Making AI-driven proteomic predictions transparent and actionable for breeders and veterinarians • Linking protein feature importance to biological mechanisms	Pilot studies in livestock proteomics show how XAI methods can identify critical proteins driving model predictions, improving trust and interpretability

espousing mechanistic explanatory power to relate protein network configurations to elaborate phenotypic levels of health, productivity, resilience, and quality of product. The methodological landscape has moved beyond stand-alone benchmarking to a system-oriented approach in which model design explicitly incorporates biological hierarchy, cross-species inference, and data sparseness, temporal dynamics, and cautions against the ethical and operational limitations of agricultural performance and ethical innovation. Layers in methodology, where deep learning methodology is complemented by transfer learning and foundation model strategies, graph- and pathway-aware representations, probabilistic and uncertainty modeling, multi-task and multi-modal integration, causal inference constructs, and advanced explainability tools, each provides methodological avenues to complement one another. The coherent orchestration of these practices defines the analytical envelope within which proteomics can be relied on to provide trusted, interpretable, and actionable intelligence for decision support and scientific discovery in livestock systems (Ma et al., 2025; Rahman et al., 2025).

Spectral interpretation and de novo sequencing architectures are now not restricted to pipelines based on early convolution models, such as heterogeneous ensembles, with sequence transformers, using graph-enriched peptide encoders and diffusion or autoregressive generative components that produce tandem mass spectra as structured probabilistic data, rather than flat collections of intensities. Modern models use raw or minimally processed spectra and provide ranked peptide candidates with confidence intervals, as well as approximate the distributions of post-translational modifications and charge-state refinements. Masked fragment modeling encourages contextual embeddings, enabling missing ion information to be inferred without the unnecessary reliance on instrument-specific artifacts, thereby facilitating domain transfer. Attention mechanisms accord dynamic emphasis to informational areas on m/z that enhance differentiation in those instances of poor signal recovery characteristic of low-abundance regulatory proteins. The additional inclusion of learned instrument response priors causes the cross-platform calibration and acquisition cross-method domain gaps to be small. Hybrid deterministic probabilistic decoders produce n-best lists with calibrated posterior probabilities, which are pipelines passed on to propagate uncertainty with protein inference, differential abundance analysis, and pathway activation scoring. Semi-supervised consistency regularization applies large reservoirs of unlabeled spectra and is invariant to recovery-influencing perturbations in the spectrum, such as synthetic noise injection or retention alignment shift. This reduces dependency on labeled annotation, which is a consideration in situations where assembled peptide libraries are not complete, in case the tissue or developmental stage under investigation is of interest. Aspects of de novo sequencing are progressively built upon the concept of an amino acid language model using multiple proteomes, which can generate realistic sequences even in cases where the database coverage of breed-specific isoforms or new splice junction peptides is limited and where the available detectable proteoform space within which the phenotype can be perturbed is large (Eloff et al., 2025; Sanders et al., 2025).

Transfer learning and foundation models of cross-species proteomics deal with the heterogeneity of proteomic landscapes among breeds, physiological states, and species in exploiting common biochemical regularities. Protein sequence, structure, or spectrum generation foundation models are trained on self-supervised tasks, like masked token recovery, contrastive alignment between predicted and empirical fragmentation patterns, or joint sequence-structure latent consistency. These pretraining regimes generate universal embeddings that learn physicochemical, structural, evolutionary, and ionization propensity attributes, which can be fine-tuned by downstream tasks with only a small labeled dataset. Resource-constrained agricultural research, low-rank adapters, and selective layer freezing are parameter-efficient adaptation methods that reduce overfitting and computational cost in resource-constrained system adaptation. Cross-species orthology-aware alignment brings in multi-task supervision that is characterized by homologous proteins in different taxa sharing representation subspaces and allows learned functional and spectral patterns to be transferred to less characterized proteomes. Domain adversarial elements promote the indistinguishability of embeddings both between sites of acquisition and between types of matrices and maintain discriminative contents in biological tasks, enhancing generalization in the presence of decentralized data curation. The use of multilingual style tokenization of peptide substructures can be used to support species-specific amino acid usage biases and common modification motifs, and the accurate identification of peptides in panels across production contexts. The constant learning methods prevent disastrous forgetting in the case where models sample new batches in sequential order, and rehearsal buffers are an optimal spectrum of examples or a distillation loss that maintains previous functional prediction accuracy (Ma et al., 2025; Truong and Bepler, 2025).

Graph-based models of protein interactions and pathways contextualization can use relational inductive biases to combine proteomic quantitative signals with network topology modeling of physical interactions, co-complexes, metabolic adjacency, regulatory effects, and spatial co-location. Multiplex architectures built upon static protein interaction graphs are augmented with canonical contact, predicted structural contact, co-expression similarity, and modification co-occurrence layers, which can propagate information and transform information across complementary relational perspectives implemented with graph neural networks. Conditional passing of attention-based messages distinguishes levels of importance of interaction dependent on task context, including the distinction between edges of stress adaptation and edges of maintenance of basal metabolism. Hypergraph formulations deal with higher-order complexes and pathway modules where the functional effect arises due to the action of multi-protein assemblies. Longitudinal proteomic trajectories are the input to temporal graph extensions, which model the dynamics of the shift in edge weights based on regulatory reconfiguration in response to nutritional, environmental, or health perturbations. By combining peptide-level evidence with integration, bipartite constructions where peptide measurements are combined with parent protein measurements are used to maintain uncertainty in cases where peptides can be distributed between isoforms. Embedding models based on these models are input to predictive tasks such as phenotype classification and risk stratification, and at the same time provide interpretability through subgraph extraction that reveals cohesive network motifs that add substantive predictive power. Graph contrastive learning takes advantage of augmentations that hide edges or permuted node features, which are better than incomplete or noisy interaction databases typical of non-model livestock proteomes (Moozhippurath & Natarajan, 2025; Wang, 2025).

Probabilistic modeling, uncertainty representation, and calibration are methods of modeling that lie in the methodology stratum and ensure that the forecasts offered to management or breeding decisions have an explicit characterization of confidence. The variational inference, stochastic gradient Monte Carlo, or ensemble diversification adds the weight/activation level uncertainty to the state of the art of Bayesian deep learning models to produce the posterior predictive distributions on the peptide identifications, diffusion abundance statistics, or phenotype risk scores. The heteroskedastic noise model takes into consideration the intensity-dependent variance and matrix-specific ion suppression effects without making incorrect interpretations of high-dispersion features as biologically volatile. Censored and zero-inflated likelihood models are better models of detection for restricted or sparse proteomic data, and there is a better parameter estimate of low-abundance regulators. Calibration techniques align nominal predictive probability with empirical frequency and are achieved by scaling temperature, by isotonic transformation, or by Dirichlet prior smoothing to maximize downstream decision reliability. To detect out-of-distribution (OOD) samples, out-of-distribution detection layers embed distance or latent density estimates and either signal to human operators when a spectrum, protein, or sample profile is out of the training manifold or detect silent failure when the proteomic state space changes due to novel pathogens, emergent changes, or other unexpected environmental stressors. The propagation of uncertainty from spectrum identification to protein inference enhances the assessment of network activities. This process ensures probabilistic coherence, allowing aggregated risk to be balanced with principled error control, thus replacing unreliable point estimates (Grefsrud et al., 2025; Weng Nyam et al., 2025).

The use of multi-task and multi-modal learning that considers genomic and metabolomic data implies the fact that proteomic variation is a part of a broader ecosystem of molecules, which is influenced by the genetic architecture, metabolic flux, interactions of the microbiome, and the environment. Concatenated or co-aligned modality representations are presented to shared encoders, and the task-specific decoders are trained on tasks like prediction of traits, inference of pathway activity, anomaly detection, or survival analysis. The hierarchical latent variable designs disentangle modality-specific and shared terms, and do not have much dominance of densely sampled modalities over sparsely sampled modalities, but can transfer cross-modal knowledge, which enhances the sensitivity of vulnerable proteomic cues. The cross-attention process finds a conditional association between genomic variant and proteomic intensities to be dynamically fused, which is regulated by metabolic context. The alignment losses facilitate similarity of the inferred patterns of the pathways used between the proteomics and metabolomics, which promotes biological plausibility. Known relationships between genotype and protein expression quantitative trait locus are used to make inferences regarding regularization of the latent mappings describing a set of interactions with the interpretation of plausible regulatory mediation, combining gradients about the similarities between genetic identity. Multi-task-based formulations can be simultaneously used to predict multiple phenotypes with auxiliary reconstruction tasks, which stabilize an imbalanced classwise representation. It is more convergent and less overfit because the curriculum sequencing of learning process tasks involves generic reconstruction and specialized phenotype prediction. Missing modalities in a generative imputer (conditioned by observed data) can also have their piping of data integrators in ways that transparently reflect the imputated contributions to downstream predictions to enable downstream predictions to exercise interpretive caution (Ballard et al., 2025; Benkirane et al., 2025).

Causal inference methods coupled between protein networks and phenotypic outcomes aim to be able to go beyond correlation to a mechanistic attribution that can be used in interventions, selection strategies, or management optimizations. Structural causal models encode assumed directional relationships between the genetic variants, the protein levels, the protein modifications, intermediate pathway activities, and the phenotypes so that counterfactual queries can be used to estimate outcome changes in response to hypothetically perturbed protein nodes or network edges. The instrumental variable designs use the genomic variants as the quasi-random perturbation to causally estimate the effect of proteins on the trait and reduce the confounding factor caused by environmental covariance. Network deconvolution decomposes both direct and indirect relationships in dense correlation matrices, which minimizes the domino effect of spurious relationships. Proteomic series based on time order can be used to achieve the dynamic causal discovery algorithms that use lagged dependencies to reveal cascades of regulation that can be triggered by certain stressors or nutritional conditions. Interventional scoring is used to assess the expected phenotypic effect of a specific modulation of the abundance of a particular protein, which is used to prioritize functional validation or selection index enhancement. Causal regularization built into predictive model training rewards avoidance of relying on features with no inferred causal torment, and improves generalization in the case of distributional shift, in which purely associative signals can be faulty. Through mediations, genomic effects are separated into the direct and proteome-mediated components, and the proportionality of trait variance passed through a particular biochemical pathway is quantified, a fact that helps in choosing multi-trait selection strategies that will not compromise the resilience or welfare-relevant molecular circuits (Mesinovic et al., 2025; Wu et al., 2025).

The operational needs that are satisfied by the use of automated feature selection, explainability, and model interpretation tooling include the need to give transparent, auditable, and biologically coherent model outputs to livestock proteomics. The sparse selection models are necessitated by the high dimensionality, collinearity, and partial observability, where there is a tradeoff between the sparsity and polygenic and network-diffuse signal preservation. Embedded selection is performed by inducing sparsity, which induces penalties on attention weight, attentional loads, or latent factor loads, and leads to small sets of predictors that can be maintained to support performance. The evolutionary and reinforced guided search strategies explore the subsets of features using multi-objective constraints that comprise predictive accuracy, stability across resampling, diversity of biological pathways, and cost of protein assay deployment. Post hoc explainability implements gradient-based attributions, perturbation analysis, and Shapley value approximation to quantify the contributions of individual proteins or pathways to sample-level predictions and concept activation methods to match latent dimensions and interpretable biological processes inferred about ontologies, such that the contribution of individual features can be abstracted. Counterfactual explanation generators propose small proteomic interventions that would transform the classes of predicted risks to give mechanistically plausible hypotheses to be observed through intervention or experiment. Prototype-based interpretation models project the samples into a latent space that is structured in terms of representative archetypes that have proteomic signatures that are analyzed to acquire insights into decision logic at the cluster level. Reliability overlays bring together estimations of uncertainty and importance attributions, and they focus on high characteristics of influence that are accompanied by a high level of epistemic uncertainty that should moderate confident action. Using visualization tools, network contexts are integrated such that the magnitude of differentiation expression, attribution weights, and estimation of causal effects can be actively learned on every node in a protein network, which promotes the synthesis of the overall interpretation of a visualization rather than the piecemeal reading of measurements (Liang et al., 2025; Madhukar, 2025).

Combining these methodological strands creates an ecosystem where raw spectral data go through deep representation learning, are placed in context in relational biological terms, given by cross-modality fusion, and are probed with causal plausibility and then simplified to exploitable as well as interpretable insights of clarity, which are in terms of transparent confidence. The key to livestock-specific deployment is constant monitoring of domain shift due to the changing management practices, new diseases, climate variability, and genetic enhancement trends modifying the underlying physiology. The adaptive model maintenance builds on the monitoring of embedding distributions, measures of calibration drift, and causal structure stability to invoke retraining or recalibration whilst protecting against the erosion of any previously validated knowledge. Ethical and governance issues overlap with the choice of method, where methods are favored that enable fair performance on a wide range of breeds and production systems and allow an audit trail of decisions when they come before regulatory or stakeholder review situations (Ghavi Hossein-Zadeh, 2026a). With the further development of foundation models and causal graph neural architectures, the methodological frontier is expected to evolve to integrative frameworks that will be able to simulate proteome-level responses in a multi-factor intervention situation, shedding light on tradeoffs between productivity, welfare, and environmental impact in a way never previously seen before with such granularity. The current environment already provides a solid scaffold; its wise implementation and continuous improvement will determine to what level proteomics, with the help of AI, will transform itself into a key pillar of sustainable, and ethically sensitive livestock production. A summary of AI methodological frameworks applied to proteomic analysis is presented in Table 6.4.

TABLE 6.4 A summary of AI methodological frameworks applied to proteomic analysis.

Framework	Description	Typical AI techniques	Key benefits of proteomic analysis	Major considerations/trade-offs
Classical machine learning	Models built on engineered proteomic features (e.g., selected protein abundances, ratios, and network metrics) for classification or regression tasks	Support vector machines (SVM), random forests, and gradient boosting machines (XGBoost and LightGBM)	• Relatively interpretable models • Works well when the sample size is moderate, and features are curated • Good baseline for biomarker discovery in proteomics	• May struggle with very high-dimensional proteomic datasets • Performance depends heavily on feature engineering and data quality • Less suited to learning hierarchical or complex patterns
Deep learning/ neural networks	Models that learn hierarchical representations from raw or minimally processed proteomic data (e.g., peptide spectra, protein networks, and embeddings)	Fully-connected neural networks, convolutional neural networks (CNNs), recurrent neural networks (RNNs), autoencoders, and variational autoencoders (VAEs)	• Can model complex non-linear relationships, such as proteome dynamics, PTMs, and interactions • Feature learning reduces the need for manual feature engineering • Enhanced performance in large-scale proteomic datasets	• Requires large training datasets to avoid overfitting • Often, "black box" interpretability is limited • Intensive computational resources and hyperparameter tuning needed
Graph and network-based AI	Frameworks that explicitly encode protein-protein interactions, pathways, or multi-omics networks, enabling relational and topological learning	Graph neural networks (GNNs), graph convolutional networks (GCNs), and network embedding models	• Captures relational information (e.g., protein networks, pathway context) in proteomics • Supports multi-omics integration and systems-level modeling • Improves prediction and interpretability when network structure is relevant	• Requires reliable network or interaction data (which may be sparse in livestock species) • Model complexity and training time increase • Interpretability of graph embeddings can still be non-trivial
Transfer learning and domain adaptation	Techniques that allow models trained on one dataset, species, or condition to be adapted for another, reducing the need for large labeled datasets in each context	Pre-trained neural networks, fine-tuning, few-shot learning, and domain adaptation algorithms	• Particularly valuable in livestock proteomics, where sample sizes for specific breeds or conditions may be limited • Enables reuse of existing models and reduces cost/time in new studies • Facilitates cross-breed or cross-species prediction	• Differences between domains (e.g., breeds, sample prep, environment) may reduce model transferability • Risk of negative transfer if domains are dissimilar • Requires careful validation in the target domain

Explainable AI (XAI) frameworks	Methods aimed at making AI models' outputs transparent and providing biologically meaningful insight from proteomic data	SHAP (SHapley additive ExPlanations), LIME (local interpretable model-agnostic explanations), attention mechanisms, and feature importance mapping	• Builds trust among domain experts (e.g., livestock scientists, breeders) • Connects learned model features back to proteins or pathways, facilitating biological interpretation • Helps identify actionable biomarkers or targets and supports regulatory/field adoption	• Often reduces model predictive performance if interpretability is prioritized • Additional computational or annotation overhead • Interpretations can still be misleading if not validated
Federated learning and privacy-preserving AI	Decentralized training of AI models across multiple institutions/farms without sharing raw data, preserving data privacy and breed-specific sensitivities	Federated neural networks, homomorphic encryption, and secure multi-party computation	• Enables multi-farm or multi-breed collaborative modeling without central data pooling • Supports inclusion of diverse livestock populations while protecting sensitive data (breed IP, farm privacy) • Enhances model generalizability and equity	• Increased complexity in system architecture, communication, and model updates • Heterogeneity of data across nodes may slow convergence • Requires strong governance and infrastructure support

6.6 Challenges in integrating AI with proteomic data

There are underlying issues in the integration of AI and livestock proteomic data, which lie in the inherent statistical geometry of data. Proteomic acquisition produces feature spaces where the dimensionality of these spaces is much greater than in usual samples, and the regimes formed are where variance estimates are ill-posed, covariance matrices are not well-conditioned, and spurious associations flourish. Sparsity is also caused by detection limits and by biological regulation, giving rise to zero inflation or low intensity distributions of proteins that are the subject of adaptive processes but are transiently expressed. They cause poor estimation of model parameters, overstate false discovery rates, and promote overfitting in the event that either deep architectures or highly flexible ensembles are used and no strict regularization is applied to them. Conventional dimensionality reducing methods will lose delicate but biologically meaningful variation, whilst sparsity induction methods can be overly aggressive, and selectively retain structural proteins which are abundant as opposed to situation-specific regulatory mediators. Fragments of data are created by stratification across breeds, management systems, or physiological states, which increases the limitations of statistical power. Synthetic data augmentation and transfer learning are solutions that mitigate but present the potential of increasing bias when the distributions used in augmentation do not represent biological diversity. This leaves the fundamental inferential stability thus far as unsound unless the experimental design, sample balancing, and the hierarchical modeling schemes are made to line up with the high-dimensional sparse topology of the data that it is (Lee et al., 2024; Haack and Kalogeropoulos, 2025).

On top of inherent statistical frailty is a widespread batch effect, instrument drift, and cross-lab harmonization impediments that complicate biologic interpretation and worsen model extrapolation. Systematic bias is introduced into matrices of intensities and fragment ion traces by subtle changes in chromatography performance, contamination of the ion source, calibration cycles, lots of reagents, and sample manipulation. These technical changes are frequently confounded with biological covariates in a pragmatic way of scheduling the runs, generating a form of partial confounding that compromises standard correction methods that make the assumption that the batch and phenotype are orthogonal. There is heterogeneity in facilities harmonization in the acquisition platforms, gradient patterns, spectral resolution, and data-dependent acquisition heuristics, which breaks down joint modeling potential. Retrospective batch correction based on global scaling or empirical Bayes shrinkage may propagate true differential signals in nonlinear portions of the intensity manifold, especially with post-translational modified peptides with greater detectability to instrument tuning. Longitudinal monitoring horizons produce an additional source of drift, further destabilizing the model calibration process and requiring constant adoption without catastrophic forgetting of previously learned structure. Without standardized reference material and with interleaved quality control designs planned to be machine learning ready, such intersite and time varying variabilities provide performance ceilings on reproducible AI performance (Liu et al., 2024; Tomo and Nakaki, 2025).

Diversity in post-translational modification and the gap in annotation offer one further complexity, defying the accuracy of AI-generated biological inference. The prevalence of species-specific modification in livestock species depends on diet, microbiome metabolites, environmental stressors, and breed-specific metabolic adaptations. However, there still are biased public spectral libraries and annotation repositories centered on a narrower catalog of canonical modifications described in the human and model organism context. Large-scale discrete molecular structures generate orphan features, which are discarded or misassigned, resulting in costly richness in features effective in regulatory circuits and customary constraints in regulatory circuiting of nonphosphorylation, glycosylation, acetylation, ubiquitination, oxidation, and novel noncanonical modifications. Combinatorial change states increase the theoretical search space, which raises the risk of false discovery and the computational cost of open search or a de novo pipeline. AI models that are trained on partially modified vocabularies air out prejudiced fragmentation a priori, lose sensitivity to underrepresented motifs, and introduce error during downstream network and pathway prediction. Lack of curated ground truth to modify site occupancy across physiological perturbations is a disadvantage in supervised learning to functionally predict site-specific occupancy, whereas limited structural context annotations reduce graph-based reasoning about allosteric and interaction-modifying effects of modifications (Kim et al., 2025a,b).

Fragmentary protein coverage and peptide redundancy work against it even more to destabilize integrative modeling. The proteomics processes tend to provide only partial coverage of the proteome, and coverage tends to be biased by large housekeeping proteins, whereas low-copy transcription factors, membrane receptors, and signaling adaptors do not have their presence detected. Missingness does not usually obey random ecology: remarkably often, it will be hierarchical with respect to physiochemical evidence, tissue matrix constitution, and timing of regulation, and it invalidates assumptions of naive imputation. Overlapping peptide sequences among paralogous proteins or isoforms bring about ambiguity in inferring proteins, which causes a lack of certain quantifications or a grouping of biologically different isoforms into a single aggregate. This is a weakness in the granularity of the association of molecular patterns with phenotypes that AI models can perform, and obscures isoform functional divergence affecting fertility, feed efficiency, immune responsiveness, or stress resilience. The probabilistic approach to solving ambiguity by using probabilistic protein assembly continues to spread uncertainty, but the resulting downstream predictive and causal models often ignore this probabilistic layer and convert provisional quantities into fixed ones, doubting their uncertainty and overselling specificity. Also, coverage gaps hinder network-level contextualization when important connectors or regulative hubs are not represented as quantitative, discontinuous inferred interaction topology and undermine graph model embeddings (Dayhoff et al., 2025).

A lack of standardization in the metadata for phenotypes and contextual variables creates semantic heterogeneity and decouples proteomic patterns from the actual drivers of environmental and physiological contexts. Unstructured or inconsistently coded examples of phenotypic annotations for diet composition and housing conditions, health interventions, reproductive status, stress exposures, and management practices are common across studies. There are cases of granularity differences, lack of a time stamp, and lack of a measurement procedure that, in turn, impair the process of alignment and normalization. Multi-cohort training on AI models, therefore, exposes latent confounding and label noise that propagates discriminative signal diffusion and causes distortion to inference. Temporal variability between the collection of samples and phenotype events that are being logged further obscures the dynamic relationship, and coarse categorical labels do not reflect quantitative progressions of environmental or treatment strength that are essential in finer modulation of proteomics. The lack of ontologies specific to livestock production situations restricts automated harmonization, which supports the existence of manual curation overheads that cannot be feasible at a large scale. Without organized metadata schemas and controlled vocabularies, interoperability of trained models between enterprises is limited, and purported association reproducibility is, at best, far-fetched (Zulueta-Coarasa et al., 2025).

Limited adoption of sophisticated AI practices is restricted by computational scalability, financial limitations, and infrastructure differences between research institutions and research production stakeholders. Longitudinal sampling combined with massive storage, memory, and compute requirements caused by high-resolution mass spectrometric datasets and multimodal integration are well above the abilities of smaller laboratories or geographic areas with limited resources. The large graph neural networks, foundation models, or uncertainty-aware ensembles need accelerated hardware and optimized data pipelines to be trained, but due to the cost sensitivities in the agricultural industries, investment cannot be sustained. Cloud solutions eliminate capital expenses yet create issues of data sovereignty, reliability of connectivity, and recurrent operational expenses, especially when the production systems are spread regionally. Sustainability requirements also overlap with energy efficiency and the environmental impact of compute-intensive training, and methodological sophistication conflicts with carbon footprint. Latency can be brought about by algorithmic complexity that cannot support health surveillance or welfare intervention with near-real-time decision support. Approximate inference or model compression methods can be scaled, which is an encouraging development, but compression can have negative impacts on calibration, interpretability, or rare event biomarkers, with tradeoffs that are yet to be systematically measured in livestock proteomics (Bianchini et al., 2025).

Generalization of work across breeds, production regimes, and environments has always been a challenging issue because of genotype-by-environment interactions and management-initiated physiological modulation. Embeddings trained on data heavily controlled by intensively managed systems are possible to be worse when used in large or mixed farming systems, where there is a change in the stressors and pathogen pressures and nutritional variability of the baseline proteomic distributions. Breed-specific allelic variation has an impact on protein isoform predominance, alteration patterns, and a baseline of expression, resulting in a covariate shift that can imprecisely calibrate discriminative thresholds or exaggerate false positive detection of a biomarker. Climatic variability and environmental gradients change the metabolic and stress response pathways in a manner that is not sufficiently represented in training corpora and result in brittle models that respond to heat stress, cold, or drought-induced nutritional flux. Domain adaptation strategies reduce a fraction of the discrepancies but may tend to over-smooth natural biological variation that is part of selection and management differentiation. The lack of representative sampling of the ecological and genetic diversity of livestock systems, therefore, constrains fairness, strength, and reliability in the long-term of AI-derived insights (Martínez-López et al., 2025).

Attempts to build up holistic system biology models are complicated by the integrative multi-omics alignment and time-synchronization restrictions. The disparate assay technologies yield streams of data that vary in terms of dynamic ranges, noise properties, sampling rates, and the lag time between introducing a biological perturbation and the formation of an observable signal. Proteomic and metabolomic responses can change on a shorter time scale than transcriptomic or phenotypic responses and need temporal alignment strategies that assume a latent time warp or phase shift. Subcomplete overlap of sample sets between omics layers creates a block missingness pattern, which compromises joint representation learning as well as biases correlation structures. Independent batch correction in each of the modalities can interfere with cross-modal covariance needed in integration, and naive concatenation rewards high-variance modalities. The absence of an agreed method to provide time stamps on interventions, feeding, health interventions, stress exposures is an obstacle to the modeling of causal lag structure between early proteomic perturbations and downstream phenotypes. A combination of missing data, non-concurrent sampling, and multimodal noise in a multi-task deep architecture compromises stability and interpretability and places onerous demands of validation to confirm that emergent latent factors are biologically relevant and not products of alignment assumptions (Kant et al., 2025).

Socio-technical barriers to predictive metrics include interpretability, trust, and adoption by domain experts, even though predictive metrics look good. Complex model systems that act as opaque black boxes are incompatible with practitioner notions of mechanistic plausibility and opaque justification of the decisions made in animal health, welfare, and breeding paths. The feature attribution techniques may be highly sensitive to small perturbations, leading to unsteady or inconsistent ranking of importance, undermining the belief in explanatory stories. Redundancy in attributions can be disguised by aggregation of attributions in correlated proteomic features, and this can result in an overinterpretation of particular proteins as uniquely causal. Off-target prediction of model-emphasized pathway versus physiological knowledge raises doubts, but consistency may be representative of confirmation bias when models are compelled to reflect existing knowledge instead of uncovering new processes. Data scientists, veterinarians, nutritionists, or breeding specialists can only communicate effectively through communication gaps that impede the process of model refinement based on domain critique. Ethical issues regarding possible abuse or overdependence on automated notifications without sufficient human control further increase the pace of integration, whereas the absence of educational equipment to visualize the non-concrete and causal/associational indicators hinders informed decision-making (Carles-Bou and Carmona, 2025).

The absence of regulatory standards and validation, and the shortage of benchmarking standards, imply that there are no universally agreed standards upon which an evidentiary sufficiency, in turn, should be decided upon before it can be employed in critical management circumstances. This is because there is no standardized reference dataset stratified by tissue, physiological state, and type of production system that can be used to make a fair comparison of algorithmic performance claims. Some studies will only rely on internal cross-validation, which may be susceptible to optimistic bias on leakage or inadequate stratification of time and environmental blocks, and may thus affect the level of validation. Minimum performance standards have not been widely agreed upon, and calibration standards or reasonable uncertainty tolerances have been disputed between the various areas of application selected, such as disease surveillance versus selection index augmentation. Rules and regulations have not set up a set of rules that rule out traceability on the changes of models, auditability of decision records, and post-deployment monitoring performance drift requirements. Without standardized reporting requirements to describe data preprocessing, batch correction, handling of missing values, hyperparameter setting, and an assessment of interpretability, reproducibility, and third-party trust, it has fewer limitations. The methodological innovation has also limited its priority because of the scarcity of benchmarks, as the research work may be optimized in small measures that are not correlated with the field utility or even ethical requirements (Anderson, 2025).

A combination of these issues is outlined in a landscape in which technical sophistication should be traded off with statistical rigor, biological verisimilitude, practicality of resources, and alignment with the stakeholders. The high-dimensional sparsity will have to be tackled by the concerted effort of the experimental design, the adaptive regularization, and the uncertainty propagation. Coherence of activities at the point of acquisition, creation of livestock-centric modification libraries, and rich metadata ontologies will help to increase signal consistency and contextual base. Disparities in benefit allocation can be minimized through scalable infrastructure and balanced model creation to encourage cross-system generalizability. Strong integrative models that are sensitive to time dynamics and modality-specific noise patterns will be necessary in authentic systems-level inference. Clearly interpretable systems that are verified against domain knowledge, combined with dynamic regulatory and benchmarking systems, can create the basis of trust to do this responsibly. Until such entwined shortcomings are mitigated in an organized manner, the transformative capability of AI that permeates livestock proteomics will always come in bounded sections, with spots of excellence being limited due to constraints in reproducibility, portability, and actionable certainty. Major challenges in AI-proteomics integration and potential mitigation strategies are presented in Table 6.5.

6.7 Validation, benchmarking, and lifecycle governance

Lifecycle governance of AI models in application to livestock proteomics is based on a bottom layer of standardized benchmarking assets and performance measurement suites that all make up a typical evidentiary grammar (see Fig. 6.1). Benchmark datasets need to capture multi-breed, multi-tissue, and multi-production system diversity, cover many acquisition platforms, preparation chemistries, videos of oxime nefecthes era, entail raw spectral files, curated peptide and protein identifications intoxicated with confidence markings, post-translational modification mappings, and standardized phenotype and contextual metadata coded using controlled vocabularies. Stratified partitioning plans are to be predefined

TABLE 6.5 Major challenges in AI–proteomics integration and potential mitigation strategies.

Challenge	Description	Impact on AI–proteomics integration	Mitigation strategies
Data heterogeneity and variability	Proteomic datasets vary substantially in sample preparation, instrumentation, species, tissue types, and experimental protocols	• Reduces model generalizability across breeds/species and labs • AI models may learn protocol artifacts rather than biological signals • Increases risk of batch effects and domain bias	• Standardize protocols and metadata annotation • Include batch correction and domain-adaptation methods • Use federated or multi-site training to capture diversity
High dimensionality with limited sample size (curse of dimensionality)	Proteomic data typically include thousands of protein features, but often relatively few samples	• AI models are prone to overfitting • Difficulty in distinguishing signal from noise • Reduced statistical power for trait prediction	• Employ dimensionality reduction (PCA, autoencoders), feature selection, and regularization • Use transfer learning from larger datasets or cross-species data • Increase sample size via collaborations or data sharing
Missing values, noise, and low-abundance proteins	Proteomic measurements often have missing values (drop-outs) and large dynamic ranges	• AI may misinterpret missingness patterns as a biological signal • Features tied to rare proteins may be ignored, reducing biological insight • Noise may degrade predictive performance	• Imputation strategies (k-NN, MICE), robust normalization • Exclude features/proteins with excessive missingness or low reliability • Use methods tailored for low-abundance signal detection
Interpretability and biological credibility of AI models	Deep learning and complex AI frameworks often act as "black boxes" with limited biological transparency	• Breeders and scientists may distrust model predictions • Harder to translate AI output into actionable biological insights or breeding decisions • Regulatory or adoption barriers for opaque models	• Incorporate explainable AI (XAI) frameworks (SHAP, LIME, attention) • Combine AI outputs with pathway/network analysis to link with biology • Validate model features with domain experts and experimental follow-up
Computational and infrastructure demands	Large proteomic datasets and advanced AI models demand high-performance computing, storage, and skilled personnel	• Resource constraints limit adoption by smaller research groups or farms • Slower model iteration and deployment in field settings • Potential for inequity between well-funded and under-resourced institutions	• Leverage cloud-based or shared infrastructure • Use scalable and efficient AI algorithms (lighter architectures, quantized models) • Provide training and capacity-building for end users

Continued

TABLE 6.5 Major challenges in AI–proteomics integration and potential mitigation strategies.—cont'd

Challenge	Description	Impact on AI–proteomics integration	Mitigation strategies
Insufficient species/breed-specific databases and annotation	Many livestock species lack well-annotated proteome reference databases, proteome atlases, or validated peptide libraries	• AI models may rely on proxies or annotations from other species, reducing accuracy • Discovery of breed-specific biomarkers is hampered • Errors in annotation propagate through AI training pipelines	• Develop and curate species- and breed-specific proteome databases • Use cross-species transfer learning but validate locally • Encourage community data sharing and standardization efforts
Data privacy, sharing limitations, and regulatory concerns	Data from farms, breeds, or commercial herds may be proprietary; genomic/proteomic data raise IP and privacy issues	• Fragmented datasets reduce training diversity and model robustness • Legal or ethical constraints may restrict centralized model training • Smaller sample pools limit generalization across populations	• Adopt federated learning or privacy-preserving AI techniques • Define clear data-sharing agreements and benefit-sharing mechanisms • Develop governance frameworks that balance openness and protection

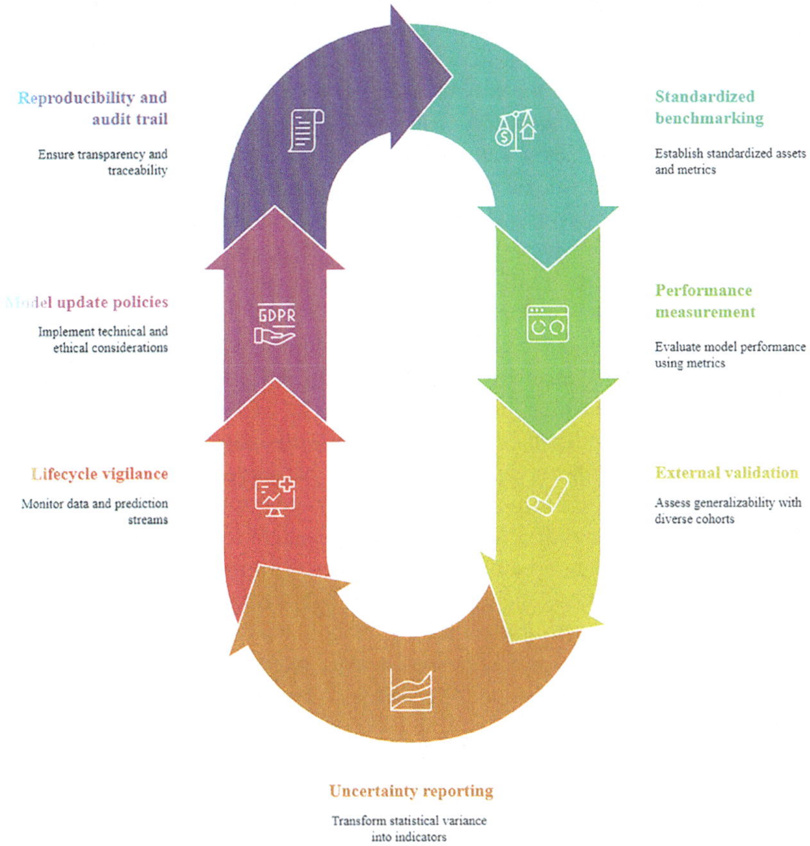

FIGURE 6.1 Lifecycle governance of AI models in livestock proteomics. This figure illustrates the continuous lifecycle governance framework for AI models in livestock proteomics, emphasizing transparency, accountability, and performance monitoring. It highlights key stages including standardized benchmarking, model validation, uncertainty reporting, lifecycle vigilance, and ethical model updates to ensure reliability, reproducibility, and long-term trust in AI-driven proteomic systems.

to avoid the unintentional leakage of near-duplicate spectra or animals close to that across training and evaluation partitions, and hold-out subsets are to be set aside to be used in community challenges. The metrics used to evaluate performance should go beyond the traditional performance in terms of identification accuracy or abundance correlation to include the performance in terms of calibration quality, resilience to induced missingness, and the ability to recover the structure of a network, and the ability to provide causal attribution. The composite scorecards that incorporate such dimensions encourage subtle comparisons, which will deter the over-maximization of a single headline statistic whilst harmonizing the methodology advancements with the practical deployment requirements (Rojahn and Grum, 2025).

External validation cohorts will always be necessary to determine generalizability, where models will have to face proteomic and phenotypic distributions that are not present in developmental corpora. Artificially diverse such cohorts can be characterized by geographic, climatic, nutritional, and managerial heterogeneity, which present changes in underlying metabolic phenotypes, stress response biomarkers, and immunological phenotypes. Temporal robustness analysis needs longitudinal analysis of model predictions and uncertainty quantifications of models followed through production cycles, seasons, or generations to show degradation patterns associated with the aging of instruments, the changing prevalence of pathogens, or genetic enhancement that changes proteomic baselines. This is because rolling origin evaluation frameworks, where models are frozen at specific time anchors and then prospectively applied to future windows, offer more realistic estimates as compared to retrospective random resampling. Scenario stressing by controlled perturbation of input feature sets, emergent modification motif introduction, or simulated loss of modality helps to better understand brittle failure modes that standard static validation would fail to detect. Semantic alignment of the external validation design to the application contexts desired in the end-use, whether early disease warnings, nutritional optimization, or selection decisions, guarantees the significance of performance claims to stakeholders (Gallitto et al., 2025).

Uncertainty reporting is a critical point of plausible implementation, which transforms the statistical variance into operationally relevant indicators. Diagnostic reliability, such as calibration curves, anticipated calibration error profiles stratified by breed or system of production, makes available systematic over- or underconfidence that may lead to false advice for threshold-based interventions. Distributional coverage measures are those that are used to measure the extent to which predictive intervals meet nominal containment at the level of intensities and densities of modification, and sharpness measures punish inappropriately diffuse estimates, which dissipate utility. Separating aleatoric and epistemic aspects helps to unify the meaning of uncertainty due to inherent biological variability or a fixable lack of data, and helps prioritize data obtaining. Decision impact audits superimpose uncertainty with the cost-benefit matrices of welfare, economic, and environmental externalities to assess the net expected value of the model-assisted actions compared to traditional protocols. Counterfactual risk reallocation studies evaluate the effect of recalibration or even a different thresholding on the false alert load of the various demographic layers, which enables fair governance and determines in which cases the conservative abstinence policies ought to override automated recommendations. Open sharing of uncertainty stories develops informed trust and matches expectations and restrictions (Hofman et al., 2025).

The lifecycle vigilance is being used continuously in monitoring structures by establishing automated data, prediction, and outcome stream monitoring of deployed models. Drift detectors are covariate shift, which is estimated by multivariate distributional distance statistics on spectral embeddings, abundance vectors; conditional shift, which is estimated by residual analyses of prediction errors over time; and label shift, which is estimated by ground truth phenotype happening with latency. The remedial thresholds based on statistical principles should be context-sensitive to agile stability. Canary models are simple baseline models that are run in conjunction with a complex architecture that serves as a sentinel, and the pattern of divergence is employed to provide early warning of pathologies of the dataset or instrumentation anomalies. Phase transitions of shadow deployments allow refreshing of candidate models with no apparent overhead, including the compilation of comparative performance and calibration information before the promotion. Governance policies are used to control the degree of automated adaptation that can be admitted, one of them being lightweight recalibration, that is, updating temperature scales, the other being complete representation retraining, which would need inspection and documentation of human work, and potential regulatory disclosure. Feedback on alert relevance and interpretability enhances monitoring in the end user, using qualitative indicators that cannot be quantified using quantitative metrics (Kim et al., 2025a,b).

The model update policies are some policies that are based on technical policies, ethical, regulatory, and economic considerations. Rather than individual improvement, version improvement should be determined by a known multi-metric gate, which must comprise accuracy, calibration, fairness, and interpretability, which must be made better. This is determined by backwards compatibility tests to establish that longitudinal measures due to models made after one another can be compared or that they have translation mappings where breeding value lines or health risk baselines can be extended. The processes of the depleted model sunset entail the need to keep archives, metadata capturing, and linking to historical decisions to have an audit. The level of risk in the application domain influences the time of update, as in the case of decision contexts with high consequences, a long-term shadow consideration and stakeholder discussion are required. High-risk vulnerabilities, such as new pathogen signatures, are reacted to using emergency update pathways, and post hoc validation cycles are used to ensure high-risk vulnerabilities were not introduced due to remediation. The incentive schemes that are aligned to the models that promote conservative, documentation-intensive updating in place of fast unrecorded iteration are promoted to gain systemic reliability (Gupta, 2025).

The mechanism of reproducibility, versioning, and audit trail constitutes the connective tissue of enforcing the transparency, traceability, and restorability of every analytical transformation of a proteomic AI pipeline. Declarative workflow specifications state the parameter choices to spectrum preprocessing, identification filtering, normalization, imputation, feature engineering, model training, and evaluation, and containerization states software requirements and hardware setup choices. Data lineage entries give model artifacts a hash of input data (immutable) and metadata schema versions used at the time an analytic state was derived by an analytic process, or an operation decision made. To make sure that unintentional or malicious changes have not been made, configuration manifests and cryptographic signatures of model binaries can be applied, and differential provenance comparison tools may be applied to determine differences between versions at sufficiently fine granularity to trace the sources of performance change. It is possible due to large-scale recording of inference time inputs and outputs, as well as the policies of retention that are sensitive to privacy and commercial needs, contributing to the retrospective analysis of incidents and regulatory inquiry. Publicly registering benchmark reports, descriptions of models with promised application, privacy, fair play audit, and uncertainty behavior and governance dashboard of monitoring metrics bring together rather than disempower one another. Within this hierarchical lifecycle model, the discovery and use decision support will be pushed in a disciplined, reliable, long-term companion to livestock proteomics using AI, which will be audit-friendly (Mason-Williams and Mason-Williams, 2025).

6.8 Future trends and innovations in AI for livestock proteomics

New directions in AI in livestock proteomics are moving toward constructing multi-omics foundation models, which internalize common latent biological grammars of transcriptomic, metabolomic, genomic, epigenomic, microbiomic, and environmental sensor inputs alongside proteomic intensity and fragmentation signatures. Such large capacity architectures take advantage of self-supervised goals that leverage co-occurrence structures, temporal consistency, contrastive cross-modality alignment, masked reconstruction of peptide fragments, and predictive modeling of post-translational modification contexts, accruing transferable representations that can be applied in traits prediction, health surveillance, nutritional optimization, and selection support. Transferability depends on the ability to separate modality-invariant physiological axes and context-specific perturbation signals in such a way that fine-tuning of the parameters by lightweight means can be used to adapt to a new breed, production system, or disease condition, without re-training. Recent progress in mixture of experts routing, hierarchical latent factorization, and sparsely activated transformers can enable scaling model expressivity without escalation of compute requirements, and curriculum strategies that interleave modalities gradually help reduce the dominance of representations represented by high-variance sources of data. The livestock domain sets its own specific needs for implementing environmental heterogeneity, seasonal feed changes, pasture and confinement dissimilarity, and changing microbial ecosystems, and these need to be encoded as conditioning inputs that influence proteomic interpretation. The foundation models will not be successful solely reliant on the architectural sophistication, but also based on federated data cooperatives that can collect enough varied pretraining corpora that will be governed by equitable benefit-sharing models that will encourage stakeholder involvement (Kant et al., 2025; Ukpoju-Ebonyi et al., 2025).

Parallel innovation occurs in real-time streaming proteomics and edge-oriented models of inference, which are intended to reduce the time interval between sample capture and actionable model outputs in manufacturing environments. Miniaturized separation technologies, fast digestion protocols, and high scan rate mass analyzers serve as input to continuous or near continuous spectral streams that may be processed by on-premise accelerators or low-power embedded systems deployed to the barns, milking parlors, or the mobile field units. This change requires architectural changes that focus on the incremental updates, sequence models with states, and the dynamic recalibration of uncertainties due to concept drift. Memory compression, recurrent state space model, and sketching-based feature extraction pipeline transformers are stream-compatible, allowing low-footprint inference that does not damage salient discriminative content constraints and is energy constrained. Temporal prioritization heuristics assume that the spectral signal should be given high fidelity processing versus a rough embedding, depending on the initial anomaly scoring. The edge deployment faces the issue of intermittent connectivity, where opportunistic synchronization with central repositories occurs to check the presence of weight updates; however, it adheres to resilience to unscheduled isolation. The increased level of security on device models lowers the movement of data and mitigates the privacy concerns and bandwidth expenses, but creates difficulties in the coordinated change and harmonized threshold configuration across geographically distributed nodes. Context-aware interpretation of animal proteomic variations is enhanced by real-time monitoring with environmental sensors to measure temperature, humidity, ammonia, and feed composition to promote predictive maintenance and animal health and welfare (Gao, 2025; Soundararajan, 2025).

Integrating structural proteomics based on predictive backbones that resemble AlphaFold increases the semantic richness of AI-mediated interpretation by positioning sequence-derived or empirically observed proteomic variation in three-dimensional conformational landscapes. With the addition of predicted or experimentally optimized structural ensembles, models can place the post-translational modifications, mutation-driven isoform differences, and proteolytic processing events in the context of solvent accessibility, domain architecture, interaction interfaces, and allosteric networks. Graph neural networks that use residue-level contact maps as input and overlay edge annotations representing hydrogen bonding, electrostatic potentials, or transient interaction propensities can learn structure-conditioned embeddings that go beyond functional discrimination with linear, peptide-centric features. In livestock species that are relatively sparsely covered by crystallographic data, transfer learning between human and model-organism structural manifolds needs to be counterbalanced by sequence divergence and adaptive residue replacements affecting stability and interaction kinetics. Propagation to downstream functional attribution of structural uncertainty quantification allows over-interpretation of low confidence regions to be mitigated. Comparison of structural processes between breeds demonstrates that there may be conformational effects of isolating variants, which can be used in selection to be sensitive to protein stability, enzymatic activity, or antigen presentation to immune systems (Molla and Meseret, 2025).

Synthetic spectra and data augmentation generative models will solve chronic data deficiency in undersampled tissues, physiological states, and low-abundance classes of regulatory proteins. Diffusion-generated generative models, variational sequence to spectrum translators, and adversarial models conditioned on peptide sequence, charge state, collision energy, and modification pattern learn forward mappings that are able to generate realistic fragment distributions of intensity and retention time profiles. Synthetic augmentation has the ability to equalize the distribution of classes in prediction tasks, tolerate the variability in instrument settings, and also facilitate the creation of a simulation of a potential experimental design before resource-consuming acquisitions. Nevertheless, mode collapse and microbiases that support the magnified chemistry often need to be guarded against fidelity measures mindful of subtle spectro-gradients in the fine-grained manners and calibration verifications that contrast downstream model actions not only with and without the augmentation but also en route to it. Conditional generation of modified peptides increases the training space of models that predict the location of modifications or the functional effects of those modifications, whereas optimization loops of sequence space to generate a desired spectral or functional property open options to rational peptide biomarker discovery. Governance models should clarify transparency requirements that characterize synthetic contributions to guarantee reproducibility and should not be confused in any way with the empirical evidence (Ghavi Hossein-Zadeh, 2026b; Marchesi et al., 2025).

Proteomic state dynamics of digital twin systems drive the predictive frontier at the union of mechanistic simulation with statistical inference and real-time data assimilation to discrete and distinct virtual flows of physiological trajectories, at individualized or cohort scale. These twins integrate models of metabolic fluxes network, signaling pathway logic, immune response cascades, and growth physiology modules, which are parameterized by genomic background, exposure to their environment, as well as longitudinal proteomic profiles. Hybrid models are based on the neural ordinary differential equations or latent stochastic differential systems, to dissipate the prohibited mechanisms that are not reflected in the mechanistic priors, and thus highly accurate short and medium-scale predictions of the threat of health events, manufacturing productivity, as well as adaptive actions are made. Bayesian filtering and smoothing algorithms, also known as particle filtering and variational message passing, are applied to combine a sequence of streaming observations of the proteomics system into state estimates and parameter posterior to support the evaluation of proactive intervention scenarios. Twin architectures cause causal isolation between modifiable and intrinsic causes of unwanted trends, which drive discriminative responses to management. Multi-scale coupling relates cellular-level perturbations of proteomics to both organismal performance measures and epidemiological risk measures of the population level to assist in strategic planning and resource distribution. The constant calibration on external standards would provide twin validity, the projection of uncertainty would be limited, and structural assumptions would be clear (Loginovsky et al., 2025).

The principles of sustainable computing and energy-aware training have become the primary design constraints as model complexity increases. Training and inference phase carbon accounting promotes the use of sparsity-inducing regularization, low-precision quantization, adaptive computation-time schemes, and modular architectures that are adaptive to the complexity of input-conditioned partial activations. Diversity-focused training of data curation minimizes redundant samples that do not provide incremental gain in representation of the data. Lifecycle assessment models compare tradeoffs between incremental performance and marginal energy cost, which would support evidence-based judgments regarding model scaling. Edge inference lowers the cumulative transmission power, whereas dynamic batching plans match processing with renewable energy on cloud backends. A sustainability and predictive fidelity algorithmic research in green hyperparameter optimization, early exit strategies based on confidence stabilization, and knowledge distillation pipelines halved to proteomic domain shift resilience can stabilize sustainability and predictive fidelity. Such efforts are in line with wider strategic agricultural sustainability requirements that tie chains of digital tools to the environmental stewardship goals (Dwivedi and Islam, 2025).

The potential breakthroughs in computationally intensive subroutines in large-scale proteome analysis and AI integration are provided by quantum and special hardware acceleration. Quantum-inspired optimization and new fault-tolerant quantum processors have the opportunity to deliver faster combinatorial search problems found in the de novo sequencing, modification pattern enumeration, and optimal transport alignment of multi-batch spectral embeddings. Expressive kernels based on variational quantum circuits entangled with classical deep models can provide high-dimensional proteomic manifold similarity assessment, but their application will require hardware maturation. Parallel to this, domain-specific accelerators, such as neuromorphic chips, neuromorphic and photonic, and specially configured field programmable gate arrays (FPGA) implementations of spectral preprocessing pipelines, result in lower latency and power. Sparse attention kernels are hardened kernels that can be scaled to sequence length and sample throughput independent of linear memory growth. Co-design techniques based on the correspondence of model architecture to both memory hierarchy and interconnect constraints provide end-to-end efficiency advantages that are vital to real-time applications. To establish equity of access to high-quality accelerators, it is necessary to have collaborative infrastructure sharing schemes that do not expand technological inequalities between resource-intensive research centers and scattered production actors (Fatunmbi, 2025).

It is expected that federated and privacy-preserving collaborative training ecosystems will reveal greater data diversity by considering proprietary, biosecurity, and ethical limitations. Variants of federated averaging that are designed to work with non-uniform distributions of proteomic data apply client-specific idiosyncrasy-driven drift and make use of customized layers of personalization and adaptively shaped weighting. The exposure risk can be reduced by secure aggregation, calculating noise in differential privacy, homomorphic encryption to compute gradients, and trusted execution environments, and utility is maintained. Hierarchical federation adds intermediate aggregation nodes based on the regional clusters, minimizing communication overhead and adapting to strata of climatic or environmental variations. Dividends of model performance that are incentive mechanisms to reward quality of contribution will encourage the inclusion of smaller farms to offset centralization tendencies. Accountable collaboration scaffolds are developed through governance charters that specify auditability, bias monitoring, and model card transparency. Combining with the foundation models, decentralized updates will provide more information to the common embedding spaces, but not replace representations of minority patterns important to robustness (Ayeelyan et al., 2024).

One such area that AI is progressing toward, in the context of strategic decision making, is adaptive causal simulation of intervention scenarios. They are based on causal discovery algorithms with variant learning to invariant risk minimization, counterfactual regression, and graphical structures in proteomic temporal panels to discover candidate mechanistic pathways between environmental manipulations, nutritional formulations, or health interventions and proteomic state transitions and downstream phenotypes. Counterfactual queries based on structural causal models with learned functions estimate the change in outcomes to recent management interventions, and are parameterized by confounding factors represented in multi-omics and sensor covariates. Agents of reinforcement learning located in causal simulators are able to assess sequential policy optimization, constrained by ethical and welfare guardrails coded as soft penalties or hard feasibility constraints. Uncertainty-aware causal effect estimation conveys the range of confidence on the intervention benefits that are predicted and discourages overcommitment to weak inferences. Coupling with digital twins provides consistency between personalized trajectory predictions and policy models, the adaptive control of which is increased through observations of feedback loops (Chen and Jing, 2025).

The inclusion of the carbon, welfare, and biodiversity measures in the optimization objectives will be a paradigm shift that will tie proteomic AI with holistic sustainability frameworks. The architectures of multi-objective learning achieve a balance between production efficiency, reduction of disease risk, intensity of greenhouse gases, animal welfare indicators, and ecological footprints proxies based on land use, pasture botanical diversity, or pollinator habitat measures. The application of scalarization strategies, Pareto front exploration, and constrained optimization formulations is used to ensure that the gains in one dimension are not used to cover regressions in others. Latent welfare biomarkers include proteomic signatures of physiological indicators of stress, inflammatory conditions, or metabolic stress, which can be used to predict poor conditions early and assess the side effects of interventions. In carbon-aware modeling, lifecycle emission factors associated with feed composition and manure management are used with proteomic indicators of feed conversion efficiency to simulate emission intensity trends at alternative conditions. Sensitive objectives of biodiversity take into account proteomic reactions to grazing plans affecting pasture ecosystem services, which co-optimize animal and environmental health. Publicity of trade-off surfaces makes stakeholders choose appropriate balance points in context in compliance with regulatory, market, and ethical drivers (Alevizos et al., 2025).

Together, these innovations paint one possible picture of the future where AI will be the foundation of a cyber-physical bioinformatic pipeline between molecular measurement, structural and functional interpretation, dynamic simulation, and sustainability-focused decision optimization in livestock systems. The achievement of this vision requires methodological progress to be brought to bear on governance, infrastructural equity, and participatory design that will see technological sophistication translated into resilient, ethical, and environmentally consonant practice. It will be important to continuously assess new risks, which might include: overdependence on artificial data, erosion of genetic diversity resilience by algorithmic homogenization, and sustainability returns by energy-intensive modeling, etc. Adaptive evolution of these structures will be maintained through interdisciplinary collaboration involving proteomics, quantitative genetics, systems physiology, computational modeling, and socio-economic expertise. Grounded in the principles of representation learning, causal reasoning, and open stewardship, the next epoch of AI, involving animal proteomics, implies further refinement of precision, vision, and sustainability in animal farming, supporting the multifaceted biological and ecological dependencies of its sustainability. Future innovations shaping AI-driven livestock proteomics are presented in Table 6.6.

6.9 Ethical considerations and societal implications

The aspects of ethics in the implementation of AI in the livestock proteomics lie at the base of the layer of data sovereignty and consent, since, despite being molecular in nature, proteomic signals are integrated into the socio-economic

TABLE 6.6 Future innovations shaping AI-driven livestock proteomics.

Innovation area	Description of innovation	Implications for livestock proteomics	Enabling technologies and methods
Single-cell and spatial proteomics with AI	High-resolution profiling of proteins at the level of individual cells or within tissue spatial context, integrated with AI for pattern recognition and inference	• Enables discovery of rare or tissue-specific proteome signatures in livestock (e.g., immune niches and muscle micro-architecture) • Supports linking micro-environmental changes to productivity, health, or welfare traits • Allows mapping of cell-type-specific responses to diet, stress, or disease in livestock	• Imaging-MS, micro-sampling, and laser-capture microdissection • AI methods for spatial pattern recognition (CNNs and graph-based models)
Foundation models and large-scale pre-trained proteome models	Development of large pre-trained AI models for proteomic data (e.g., embeddings of peptide spectra and proteome language models) that can be fine-tuned for livestock-specific tasks	• Reduces the need for large labeled datasets in every species or breed • Improves transferability of proteomic AI across breeds, conditions, and labs • Accelerates biomarker discovery and model deployment in livestock proteomics	• Transformer architectures pretrained on global proteomics/spectra databases • Transfer learning and few-shot learning frameworks
Real-time proteome monitoring and edge AI sensing	Deployable sensors and microfluidic or biosensor systems integrated with AI to monitor proteomic changes in near-real-time on the farm, with decision-support capabilities	• Enables early detection of physiological changes (e.g., stress and infection) through proteome shifts rather than only phenotypes • Supports precision management (feeding, health, and environment) based on proteomic signals • Bridges molecular measurement with operational farm systems	• Microfluidic proteomic assays, wearable sensors for biomarkers • Embedded AI, TinyML, and edge computing for on-site inference
Multi-modal multi-omics fusion with proteomics at the core	AI-driven frameworks that fuse proteomics data with genomics, transcriptomics, metabolomics, phenomics, microbiomics, and environment/context data to build holistic models	• Provides deeper mechanistic insight into complex livestock traits (productivity, disease, and adaptation) • Improves predictive accuracy of models by capturing multiple biological layers • Enables discovery of cross-omics biomarkers and actionable targets	• Graph neural networks, multimodal deep learning, and data-fusion algorithms • Integration pipelines, ontologies, and standardized metadata
Explainable and causal AI in proteomics	AI models that not only predict but also explain proteomic-phenotype links and infer causality (not just correlations) in livestock systems	• Builds trust among breeders, vets, and regulatory stakeholders by linking model output to biological mechanisms • Enables actionable intervention design (e.g., targeting specific proteins or pathways) • Helps to avoid over-reliance on "black box" predictions, supporting adoption	• SHAP, LIME, attention-based interpretability methods • Causal inference frameworks, counterfactual analysis, and longitudinal proteomics

Quantum/Next-gen compute for proteomic AI	Application of quantum computing, neuromorphic computing, or highly-accelerated hardware to solve complex proteomics-AI problems (e.g., ultra-large datasets and combinatorial PTM modeling) in livestock	• Allows modeling of ultra-high-dimensional proteomic spaces and interactions (e.g., PTMs and isoforms) beyond classical computational limits • Accelerates AI training and inference, enabling real-time or near-real-time proteome analytics on farms	• Quantum neural networks, neuromorphic hardware, and HPC with specialized accelerators • Federated/hybrid compute architectures and cloud-edge hybrid systems
Ethical, sustainable, and traceable proteomics with AI	Innovations that integrate proteomics and AI to support sustainable production, traceability, animal welfare, and ethical breeding in livestock systems	• Enables proteome-based markers of welfare or environmental footprint (sustainability traits) • Supports traceability of livestock product quality through proteomic signatures • Aligns proteomic AI innovation with ethical-societal imperatives in livestock industries	• Blockchain plus proteomic data signatures • AI frameworks designed for bias detection and fair modeling across breeds/species • Sustainability metrics embedded into proteomic-AI models

production ecologies and imply implicit information about the management practices, genetic assets, environmental conditions, and, in some cases, recognizable operational patterns that the owner may belong to. The work of authentic collection and secondary reuse of proteomic information needs assent structures that differentiate among original diagnostic applications, research-filled model development, business optimization, and derivative analytics to breeding or supply chain certification. Context-sensitive approval should be retractable, unambiguously defined, and tied to explicit expression of value streams releasing material or informational advantages to source stakeholders, even lesser actors whose small amounts of data would otherwise dilute their impact. The principles of data sovereignty also require transboundary transfer to focus on jurisdictional bioeconomic governance, practically physical concerns regarding the concentration of analytic advantage in a small group of technologically gifted actors, and apply access tiering that is able to maintain competitive viability and concurring scientific development. Equitable access positions questions of infrastructural disparity, as the subsidized empowerment of under-resourced production systems, capacity building in data custodianship, and the prevention of epistemic extraction, whereby the different proteomes support generalized models without making equal investments back to contributing communities. Ethical stewardship, therefore, enhances procedural justice, who will make what decision, will the decision benefit or harm, as well as distributive justice, how benefits and burdens are distributed, in the proteomic AI knowledge commons (Iseko, 2025; Taddese et al., 2025).

Systematic performance differences among breeds, lines, modes of production, or geographically differentiated health ecologies, induced by representational imbalances or latent confounders, result in algorithmic bias, whose downstream effects include selection pressure, intervention timing, and welfare triage. Breed-specific biochemical ground state, metabolic response, or variation in previous exposure history can be coded in proteomic features, and skewed in training corpora can challenge normalization to a single physiological state of being, the referent ground truth. In this type of fairness auditing, fairness goes beyond standard measures of parity to biologically motivated measures that evaluate calibration, symmetry of errors, and non-uniform dispersion of uncertainties separately across genetic strata, production intensities, and life cycle stages. Ethical questioning involves scrutiny of the implicit preference for high-yielding phenotypes in the payoff of optimization goals unrelated to resilience attributes in relation to the environmental robustness or disease tolerance, thus increasing monoculture susceptibilities. A feedback loop can also result in biased prediction performance, which can lead to preferential data acquisition that strengthens an initial skew. Active reweighting, goal-oriented data augmentation, and adaptive sampling techniques can correct the imbalance. Clear expression of tradeoffs on fairness compensations and residuals would allow the stakeholders to adapt acceptability levels instead of getting dichotomous guarantees, and continuous auditing would entrench fairness as a life cycle property and not a one-time certification (Agnes, 2025).

Transparency and interpretability are a relational construct that forms the basis of trust wherein the stakeholders want intelligible explanations of the proteomic risk stratifications, trait predictions, or intervention recommendations. The interpretation requirements of these different classes of stakeholders are diverse: molecular scientists can insist upon mechanistic attribution to pathways; producers may need simplified, actionable explanations; regulatory agents may require procedure certificates; and ethicists may accept all assignments to proclaimed purpose limits. Proteomics is a complex concept since high-dimensional patterns of peptide or post-translational modifications are not easily mapped to vernacular terms, and so abstraction layers that transform latent embeddings to pathway perturbation stories, structural impact summaries, or metabolic flux implications are needed. Ethical interpretability opposes oversimplification by reduction that hides uncertainty or contingent requirements, positive communication of the degree of certainty, and delimits the information adequacy thresholds. Artifacts of documentation of model scope, exclusion criteria, data provenance, and observed failure modes are part of long-lasting pools of trust, especially when they are supported by responsive grievance mechanisms to challenge the output perceived as inconsistent or maleficent. Most importantly, transparency should not deteriorate into performative disclosure; its ethical power is based upon providing significant agency, informed consent renewal, and participatory co-governance by means of model evolution (Nair et al., 2025).

Dual-use risk is a situation in which analytical powers used to advantage the welfare, disease surveillance, or productivity optimization are reused, whether deliberately or unintentionally, to support actions that are inconsistent with biosecurity, ecological integrity, or ethical breeding practices. The high-resolution proteomic signatures are able to deduce adaptation of a pathogen to antimicrobials or the appearance of antimicrobial resistance, or the effect of immunomodulators, which information, were it made public with no restrictions, would reduce barrier levels to desired manipulation or exploitation of vulnerabilities. Peptide or protein variant generative modeling, although potentially useful in biomarker design, is potentially overlapping with emergent synthesis approaches to hasten bioactive molecule manufacturing with a poorly understood ecological or health impact. Ethical governance thus brings in protocols of progressively graded dissemination, incorporating sorting filters of risk, delayed release of analytic detail of sensitivity, and compartmentalization of functionality of tools to match access to evidence of genuine intent of use. Biosecurity cognizant audit trails,

TABLE 6.7 Ethical challenges and societal impacts of AI in livestock proteomics.

Ethical/societal issue	Description	Implications for livestock proteomics and AI	Potential mitigation strategies
Animal welfare and instrumentalization	Use of AI and proteomic monitoring may shift animals from sentient beings to data objects, prioritizing production over well-being	• Proteomic surveillance may enable very fine-grained monitoring, but this may lead to over-optimization of animals for productivity at the expense of welfare • Intensive monitoring may reduce natural behaviors and the autonomy of animals	• Design AI systems with welfare metrics (not solely productivity) as key objectives • Include animal behavior scientists in system design to avoid reducing animals to "proteome data points" • Adopt ethical frameworks that balance productivity with the intrinsic value of animals
Data privacy, ownership, and farm stakeholder equity	The generation of proteomics and AI-derived data raises issues about who owns the data, who benefits, and whether smaller producers are disadvantaged	• Large farms or tech providers may monopolize proteomic/AI infrastructure, disadvantaging smaller farmers • Data from animals may be used for secondary commercial purposes without farmer or community consent	• Develop clear data governance frameworks where farmers and producers retain rights and share benefits • Promote open-access or cooperative models of proteomic data sharing • Ensure inclusive stakeholder participation in AI system design
Bias, fairness, and representativeness	AI models trained on proteomic data may reflect breed, species, or environment biases; certain breeds or smallholder systems may be underrepresented	• Models may perform poorly for underrepresented breeds, leading to inequitable benefits or misinformed decisions • "Speciesist" or "breedist" biases may emerge in decision systems	• Ensure training datasets include diverse breeds, species, and environments • Use fairness auditing and interpretability tools to identify bias in proteomic AI models • Apply domain adaptation or transfer learning to protect underrepresented groups
Transparency and explainability of AI decisions	Complex AI models driving proteomic interpretations may become "black boxes", limiting trust and accountability	• Breeders, veterinarians, and producers may not understand why certain proteomic-AI outputs are suggested • Lack of transparency may hinder adoption, and could conceal unintended biases or welfare risks	• Integrate explainable AI (XAI) tools to link proteomic features to biological or management decisions • Provide training and documentation for end-users on AI model logic and assumptions • Establish audit trails and human-in-loop decision structures

Continued

TABLE 6.7 Ethical challenges and societal impacts of AI in livestock proteomics.—cont'd

Ethical/societal issue	Description	Implications for livestock proteomics and AI	Potential mitigation strategies
Sustainability, resource equity, and environmental justice	AI-driven proteomics may accelerate intensification of livestock systems, which can negatively impact the environment or exacerbate inequities	• Increased productivity driven by AI-proteomics could lead to higher resource use, land intensification, or reduced biodiversity • Small-scale farmers may lack access to such technologies, widening inequality	• Align proteomic-AI tools with sustainable livestock practices (e.g., feed efficiency, methane reduction) • Frame technologies in inclusive models accessible to diverse farm sizes • Monitor the ecological footprint of proteomic-AI deployment and enforce ecological safeguards
Regulatory and ethical governance gaps	Emerging AI-proteomics systems are ahead of policy and regulatory standards, raising uncertainty about liability, animal rights, and data protection	• Lack of regulation may enable misuse of proteomic-AI systems, e.g., prioritizing profit over welfare or environmental standards • Producers may face unknown risks, and public trust may erode	• Develop industry-wide guidelines and codes of ethics specific to AI-proteomics in livestock • Engage multi-stakeholder governance (farmers, tech developers, ethicists, consumers) • Conduct ethical impact assessments and continuous monitoring of proteomic-AI deployments

query trend anomaly detection, and required red teaming exercises enhance resilience against misuse. The ethical pressure is in the need to strike a balance between the openness that drives collective innovation and precautionary restrictiveness of capabilities whose unregulated dispersion may increase systemic vulnerabilities (Trotsyuk et al., 2024).

Environmental responsibility redefines proteomic AI as an efficiency amplifier, but a negotiator of the multi-dimensional sustainability agreement. Proteomic signals of metabolic efficiency, stress physiology, immune activation, or adaptation may be leveraged in optimization loops, and functioning becomes unwittingly traded between short-term productivity gains and increased chronic stress, reduced genetic diversity, or externalized ecosystem degradation in case objective functions are poorly scoped. The cross-domain tradeoffs must be brought into the limelight in the process of ethical deliberation: the reduction of emissions intensity against the possible changes in nutrient excretion; the increased growth rate against the welfare indicators of the disrupted comfort; the increased feed ratio against the stability in times of climatic volatility. Proteomic evidence is also capable of both proving and disproving sustainability claims, and this puts a responsibility on the narrative framing of interventions. The consideration of intergenerational equity concepts means that there is no loss of adaptive capacity in a future climatic regime by selection pathways guided by molecular data. Ethical frameworks of negotiation are advantageous in that they have scenario modeling on the distribution of outcome in case of alternative weighting of environmental, welfare, and economic criteria, thus democratizing choice by stakeholders instead of supporting technocratic unilateralism (Freihat et al., 2025).

These ethical aspects are placed in the formal institutional, legal, and normative structures through governance systems, which translate principles of abstract values into concrete action. The approach of multi-layered governance between farm level protocols, industry consortia standards, national regulatory instruments, and transnational agreements creates consistency whilst creating contextual flexibility. Metadata schema, consent ontology, audit logging format, and fairness reporting template standardization enable interoperability and regulatory scalability. Proteomic AI pipeline certification regimes have the potential to make operational data stewardship, bias mitigation, interpretability provisioning, and sustainability integration thresholds, as well as periodically recertify proteomic AI pipelines as lifecycle drift becomes apparent and new normative expectations emerge. The alignment of policy requires alignment with animal welfare laws, genetic resource agreements, rules on privacy, and sustainability reporting, which minimizes the mixed compliance between jurisdictions and discourages ethical arbitrage between jurisdictions. Governance systems gain legitimacy and adaptability by including representatives of producers, animal welfare activists, environmental agencies, and technical specialists. Grievance adjudication can be administered by ethical oversight bodies that have an investigatory authority and transparency requirement, and provide binding directives to take corrective actions. Voluntary codes of conduct and model stewardship charters are soft law tools that are responsive to emergent problems that are not yet developed into hard law, and horizon scanning functions look to technological inflection points that create the need to formulate policy proactively (Barbudo Carrasco, 2025).

In these interrelated spheres, an example of ethical stewardship of AI in livestock proteomics represents a vibrant appeal between incentives to unlock innovations and incentives fed on distributive justice, precaution and constancy, and optimization efficiency and ecological middle and wholeness. Both the predictive precision that is enabled by the molecular granularity and the increased responsibility that it brings increase the risk of avoiding entrenching inequity, unintentionally eroding genetic resilience, or externalizing ecologically. It becomes imperative to have a reflexive ethic that constantly questions the implicit assumptions, power imbalance, and the inferred effects. Finally, long-term social acceptability will depend upon expressible compatibility between proteomic AI pathways with social values at large-scale on animal welfare, environmental responsibility, food system sustainability, and equitable distribution of benefits, thereby making sure that the high level of technological sophistication will be reflected in ethically sound group development. Ethical challenges and societal impacts of AI in livestock proteomics are presented in Table 6.7.

References

Adams, C., Bittremieux, W., 2025. A 2025 perspective on the role of machine learning for biomarker discovery in clinical proteomics. Expert Rev. Proteonomics 1–12. https://doi.org/10.1080/14789450.2025.2545828.

Agnes, K., 2025. Addressing bias in AI algorithms for health applications. IAA J. Biol. Sci. 13 (1), 37–43. https://doi.org/10.59298/iaajb/2025/1313743.

Alevizos, V., Gerolimos, N., Leligkou, E.A., Hompis, G., Priniotakis, G., Papakostas, G.A., 2025. Sustainable swarm intelligence: assessing carbon-aware optimization in high-performance AI systems. Technologies 13 (10). https://doi.org/10.3390/technologies13100477.

Anderson, D., 2025. Trustworthy multi-modal AI in healthcare: a comprehensive framework for bias detection, explanation, and mitigation. Artif. Intell. Quantum Computing Robot. Sci. Technol. J. 3 (1), 56–67. https://doi.org/10.64206/y8dc7h12.

Ayeelyan, J., Utomo, S., Rouniyar, A., Hsu, H.-C., Hsiung, P.-A., 2024. Federated learning design and functional models: survey. Artif. Intell. Rev. 58 (1). https://doi.org/10.1007/s10462-024-10969-y.

Azad, F., Bosnić, Z., Kukar, M., 2025. Meta-imputation balanced (MIB): an ensemble approach for handling missing data in biomedical machine learning. arXiv, Slovenia arXiv. https://doi.org/10.48550/arXiv.2509.03316.

Ballard, J.L., Dai, Z., Shen, L., Long, Q., 2025. JASMINE: a powerful representation learning method for enhanced analysis of incomplete multi-omics data. bioRxiv. https://doi.org/10.1101/2025.06.16.659949.

Barbudo Carrasco, L., 2025. Mitigating bias and advocating for data sovereignty: the role of metadata and paradata in ethical AI-driven information systems. J. Libr. Metadata 25 (3), 185–200. https://doi.org/10.1080/19386389.2025.2515766.

Benkirane, H., Vakalopoulou, M., Planchard, D., Adam, J., Olaussen, K., Michiels, S., Cournède, P.-H., Mendes, P., 2025. Multimodal CustOmics: a unified and interpretable multi-task deep learning framework for multimodal integrative data analysis in oncology. PLoS Comput. Biol. 21 (6), e1013012. https://doi.org/10.1371/journal.pcbi.1013012.

Bianchini, S., Müller, M., Pelletier, P., 2025. Drivers and barriers of AI adoption and use in scientific research. Technol. Forecast. Soc. Change 220, 124303. https://doi.org/10.1016/j.techfore.2025.124303.

Buchha, D., 2025. AI & ML for livestock welfare: analysing behaviour and stress patterns for better care. Int. J. Sci. Res. Eng. Manag. 09 (04), 1–9. https://doi.org/10.55041/ijsrem46227.

Carles-Bou, J.L., Carmona, E.J., 2025. Achieving faithful explainability in feedforward neural networks through accurately computed feature attribution. Neural Netw., 108277 https://doi.org/10.1016/j.neunet.2025.108277.

Chen, J.e., Jing, A., 2025. Recent advances in causal machine learning and dynamic policy learning. Wiley Interdiscip. Rev. Comput. Stat. 17 (4). https://doi.org/10.1002/wics.70050.

Dayhoff, G.W., II., Kortzak, D., Liu, R., Shen, M., Zhang, Z.-Y., Shen, J., 2025. Illuminating the druggable human proteome with an AI protein profiling platform. In: bioRxiv. Cold Spring Harbor Laboratory. https://doi.org/10.1101/2025.09.07.670677.

De Filippis, G.M., Amalfitano, D., Russo, C., Tommasino, C., Rinaldi, A.M., 2025. A systematic mapping study of semantic technologies in multi-omics data integration. J. Biomed. Inf. 165. https://doi.org/10.1016/j.jbi.2025.104809.

Durgun, M., 2024. Real-time milk quality control using multi-spectral sensing and edge computing: advancing on-site detection of milk components with XGBoost. Appl. Sci. 14 (23), 10916. https://doi.org/10.3390/app142310916.

Dwivedi, P., Islam, B., 2025. Balancing performance and energy efficiency: the method for sustainable deep learning. J. Supercomput. 81 (10). https://doi.org/10.1007/s11227-025-07610-2.

Eloff, K., Kalogeropoulos, K., Mabona, A., Morell, O., Catzel, R., Rivera-de-Torre, E., Berg Jespersen, J., Williams, W., van Beljouw, S.P.B., Skwark, M.J., Laustsen, A.H., Brouns, S.J.J., Ljungars, A., Schoof, E.M., Van Goey, J., auf dem Keller, U., Beguir, K., Lopez Carranza, N., Jenkins, T.P., 2025. InstaNovo enables diffusion-powered de novo peptide sequencing in large-scale proteomics experiments. Nat. Mach. Intell. 7 (4), 565–579. https://doi.org/10.1038/s42256-025-01019-5.

Enitan, D.S., Adejumo, E., Osakue, D.O., Eke, D.S., Akele, R., Edafetanure-Ibeh, O., Enitan, C., 2025. Integrating genomics and proteomics technologies in biological research: advantages, challenges, and prospects. Global South Health Horizons 1 (1), 12–35. https://doi.org/10.63950/gshh.2025.1.1.2.

Fatunmbi, T.O., 2025. Quantum computing and artificial intelligence: toward a new computational paradigm. World J. Adv. Res. Rev. 27 (1), 687–695. https://doi.org/10.30574/wjarr.2025.27.1.2498.

Freihat, L., Yang, Q., Wang, Q., Charki, A., Livina, V., Sun, X., 2025. The relation between artificial intelligence (AI), eco innovation (EI) and green innovation (GI) and effects on sustainability. MATEC Web of Conf. 413, 10004. https://doi.org/10.1051/matecconf/202541310004.

Gallitto, G., Englert, R., Kincses, B., Kotikalapudi, R., Li, J., Hoffschlag, K., Bingel, U., Spisak, T., 2025. External validation of machine learning models—registered models and adaptive sample splitting. GigaScience 14. https://doi.org/10.1093/gigascience/giaf036.

Gao, X., 2025. Toward real-time and efficient edge intelligence: advances and challenges in lightweight machine learning. Sci. Technol. Eng. Chem. Environ. Prot. 1 (4). https://doi.org/10.61173/bwn0ez36.

Ghavi Hossein-Zadeh, N., 2025. Artificial intelligence in veterinary and animal science: applications, challenges, and future prospects. Comput. Electron. Agric. 235, 110395. https://doi.org/10.1016/j.compag.2025.110395.

Ghavi Hossein-Zadeh, N., 2026a. Advancing climate-resilient livestock systems: Next-generation emission mitigation strategies and integrated technological innovations. Vet. Anim. Sci. 31, 100588. https://doi.org/10.1016/j.vas.2026.100588.

Ghavi Hossein-Zadeh, N., 2026b. Generative artificial intelligence in animal genomics for smart agriculture: Applications, challenges, and future prospects. Vet. Anim. Sci. 33, 100702. https://doi.org/10.1016/J.VAS.2026.100702.

Ghoneim, O., Dobias, P., Romain, O., 2025. Survey of neural network optimization methods for sustainable AI: from data preprocessing to hardware acceleration. Mach. Learn. Appl. 22, 100762. https://doi.org/10.1016/j.mlwa.2025.100762.

Grefsrud, A., Blaser, N., Buanes, T., 2025. Calibrated and uncertain? Evaluating uncertainty estimates in binary classification models. arXiv. https://doi.org/10.48550/arXiv.2508.11460.

Gupta, N.R., 2025. Regulatory policies on ethics evaluations for large-scale AI systems. Proc. AAAI/ACM Conf. AI Ethics Soc. 8 (3), 2878–2880. https://doi.org/10.1609/aies.v8i3.36778.

Haack, A.M., Kalogeropoulos, K., 2025. Data processing and analysis in positional proteomics. Proteomics. https://doi.org/10.1002/pmic.70069.

Happonen, L.J., Varjosalo, M., 2025. State-of-the-art and future directions in structural proteomics. Mol. Cell. Proteomics 24 (10), 101065. https://doi.org/10.1016/j.mcpro.2025.101065.

Hein, Z.M., Guruparan, D., Okunsai, B., Che Mohd Nassir, C.M.N., Ramli, M.D.C., Kumar, S., 2025. AI and machine learning in biology: from genes to proteins. Biology 14 (10). https://doi.org/10.3390/biology14101453.

Hofman, P., Sale, Y., Hüllermeier, E., 2025. Uncertainty quantification with proper scoring rules: adjusting measures to prediction tasks. arXiv. https://doi.org/10.48550/arXiv.2505.22538.

Horvath, P., Coscia, F., 2025. Spatial proteomics in translational and clinical research. Mol. Syst. Biol. 21 (6), 526–530. https://doi.org/10.1038/s44320-025-00101-9.

Hui, S., He, R., Li, H., Li, Y., Yuan, L., Zhou, H., Yu, R., Huang, P., 2025. Revealing dynamic transcriptomic and immune cell signatures underlying heterogeneous responses to influenza vaccination. Vaccine 64, 127777. https://doi.org/10.1016/j.vaccine.2025.127777.

Iseko, A., 2025. Diversity as ethical infrastructure: reimagining AI governance for justice and accountability. Int. J. Sci. Technol. Soc. 13 (5), 190–204. https://doi.org/10.11648/j.ijsts.20251305.13.

Jadav, D., Mandanaka, T., 2025. Artificial intelligence in mathematical modeling of complex systems: a comprehensive review of concepts, applications, and future directions. Int. J. Sci. Res. Eng. Manag. 09 (07), 1–9. https://doi.org/10.55041/ijsrem51328.

Jalbani, A., Weerawarna, R., Al-Zubaidi, K., 2025. Enhancing data provenance in AI with blockchain technology: a comprehensive quality model. CSI Trans. ICT 13 (2–3), 213–224. https://doi.org/10.1007/s40012-025-00419-7.

Jebet, A., Ma, X., He, L., Anthony, A.J., Kim, S., Mueller, E., McClain, C.J., Zhang, X., 2025. Optimization of data-dependent acquisition parameters in mass spectrometry for oligonucleotide identification. J. Am. Soc. Mass Spectrom. 36 (11), 2509–2519. https://doi.org/10.1021/jasms.5c00250.

Jitjumnong, J., Taweechaipaisankul, A., Lin, J.C., Wongchanla, S., Chuwatthanakhajorn, S., Lin, C.J., Khang, L.T.P., Linh, N.V., Sangsawad, P., Dinh-Hung, N., Tang, P.C., Moonmanee, T., 2025. An overview of advancements in proteomic approaches to enhance livestock production and aquaculture. Animals 15 (13). https://doi.org/10.3390/ani15131946.

Kancharla, S.S.R., 2025. Data quality and integration: the AI-driven evolution. J. Comput. Sci. Technol. Stud. 7 (7), 882–888. https://doi.org/10.32996/jcsts.2025.7.7.97.

Kant, S., Deepika, Roy, S., 2025. Integrative multi-omics and artificial intelligence: a new paradigm for systems biology. OMICS A J. Integr. Biol. https://doi.org/10.1177/15578100251392371.

Kim, G.Y.E., Corbin, C.K., Grolleau, F., Baiocchi, M., Chen, J.H., 2025a. Monitoring strategies for continuous evaluation of deployed clinical prediction models. J. Biomed. Inf. 168, 104854. https://doi.org/10.1016/j.jbi.2025.104854.

Kim, D.N., Yin, T., Zhang, T., Im, A.K., Cort, J.R., Rozum, J.C., Pollock, D., Qian, W.J., Feng, S., 2025b. Artificial intelligence transforming post-translational modification research. Bioengineering 12 (1). https://doi.org/10.3390/bioengineering12010026.

Kumar, A., Kumar, R.R., 2025. A comprehensive review of image denoising techniques: from traditional filters to deep learning. Int. J. Multidiscip. Res. 7 (4). https://doi.org/10.36948/ijfmr.2025.v07i04.46673.

Lee, K.H., Assassi, S., Mohan, C., Pedroza, C., 2024. Addressing statistical challenges in the analysis of proteomics data with extremely small sample size: a simulation study. BMC Genom. 25 (1). https://doi.org/10.1186/s12864-024-11018-2.

Li, J., Xiong, S., Shi, H., Cui, F., Zhang, Z., Wei, L., 2025. NeuroPred-AIMP: multimodal deep learning for neuropeptide prediction via protein language modeling and temporal convolutional networks. J. Chem. Inf. Model. 65 (9), 4740–4750. https://doi.org/10.1021/acs.jcim.5c00444.

Liang, S., Zhang, Y., Zheng, K., Bai, Y., 2025. FeatureX: an explainable feature selection for deep learning. Expert Syst. Appl. 282, 127675. https://doi.org/10.1016/j.eswa.2025.127675.

Liu, Y., Wei, Z., Wang, J., Zhu, X., Liu, R., Wang, X., Wang, S., Lai, X., 2024. RMComBat: a batch effect correction algorithm for repeated measurement sequencing data to prevent overcorrection. In: 2024 IEEE International Conference on Bioinformatics and Biomedicine (BIBM ..., 2024).

Loginovsky, O., Beliakov, A., Zvereva, E., 2025. The application of digital twin technology in information support systems. In: 2025 International Conference on Industrial Engineering, Applications and Manufacturing (ICIEAM), pp. 902–906. https://doi.org/10.1109/icieam65163.2025.11028358.

Lou, Y., Ma, Y., Xiang, L., Sun, J., 2025. Flexible modeling of left-truncated and interval-censored competing risks data with missing event types. Comput. Stat. Data Anal. 211, 108229. https://doi.org/10.1016/j.csda.2025.108229.

Ma, Y., Ren, Q., Hettinga, K., Fogliano, V., 2025. Leveraging foundation models and transfer learning for peptide transport prediction, molecular taste classification, and visual texture analysis. Innov. Food Sci. Emerg. Technol. 105, 104247. https://doi.org/10.1016/j.ifset.2025.104247.

Madhukar, E., 2025. Multi-level feature selection and transfer learning framework for scalable and explainable machine learning systems in real-time applications. J. Inf. Syst. Eng. Manag. 10 (46s), 1091–1101. https://doi.org/10.52783/jisem.v10i46s.9242.

Marchesi, R., Lazzaro, N., Leonardi, G., Rignanese, F., Bovo, S., Chierici, M., Jurman, G., 2025. Toward reliable synthetic omics: statistical distances for generative models evaluation. In: bioRxiv. Cold Spring Harbor Laboratory. https://doi.org/10.1101/2025.05.08.652855.

Martínez-López, R., Centurión, L.M., Pereira, W.E., 2025. Machine learning modelling for weighting and ranking of multiple variables related to genotype-environment interaction: innovative protocol proposal for selecting breeding cows in wetlands. J. Agric. Sci. 1–16. https://doi.org/10.1017/s0021859625100294.

Mason-Williams, I., Mason-Williams, G., 2025. Reproducibility: the new frontier in AI governance (version 1). arXiv. https://doi.org/10.48550/ARXIV.2510.11595.

Mathieu, T., Légaré, S., Nzekoue, A.F., Jauré, N., Lester, H., Dias, T., Kusters, R., 2025. Integrative multi-omics modeling for cultivated meat production, quality, and safety. Trends Food Sci. Technol. 166, 105364. https://doi.org/10.1016/j.tifs.2025.105364.

Mesinovic, M., Buhlan, M., Zhu, T., 2025. Causal graph neural networks for healthcare (version 2). arXiv. https://doi.org/10.48550/ARXIV.2511.02531.

Mo, Y., Shen, H.T., Zhu, X., 2025. Efficient self-supervised heterogeneous graph representation learning with reconstruction. Inf. Fusion 117, 102846. https://doi.org/10.1016/j.inffus.2024.102846.

Molla, A., Meseret, G., 2025. The role of artificial intelligence in protein structural and functional prediction: current status and future prospective. Innovation 6 (3), 130–138. https://doi.org/10.11648/j.innov.20250603.20.

Moozhippurath, B., Natarajan, J., 2025. Metabolomics pathway prediction using enhanced-graph convolutional networks with graph attention networks. Int. J. Online Biomed. Eng. (iJOE) 21 (10), 48–62. https://doi.org/10.3991/ijoe.v21i10.55539.

Mou, X., Du, H., Qiao, G., Li, J., 2025. Evaluation of imputation and imputation-free strategies for differential abundance analysis in metaproteomics data. Briefings Bioinf. 26 (2). https://doi.org/10.1093/bib/bbaf141.

Müller, J., Kuster, B., The, M., 2025. Computational approaches for pathway-centric analysis of protein post-translational modifications. Proteomics. https://doi.org/10.1002/pmic.70055.

Nair, R., Thomas, R., Sandra, M.V., Siji, K.B., 2025. Interpretable AI: enhancing transparency and fairness in decision-making. Int. J. Adv. Res. Sci. Commun. Technol. 438–443. https://doi.org/10.48175/ijarsct-23772.

Niloy, D.B., Hossen Jony, M.E., Ruhani, H.K., 2025. A self-adaptive data preprocessing pipeline for machine learning: an automated and dynamic approach. In: International Conference on Quantum Photonics, Artificial Intelligence, and Networking (QPAIN), pp. 1–6. https://doi.org/10.1109/qpain66474.2025.11171989.

Oh, V.K.S., Li, R.W., 2024. Wise roles and future visionary endeavors of current emperor: advancing dynamic methods for longitudinal microbiome meta-omics data in personalized and precision medicine. Adv. Sci. 11 (47). https://doi.org/10.1002/advs.202400458.

Palabiyik, A.A., Palabiyik, E., 2025. Proteomics in biomarker discovery: uncovering disease-specific interactome and post-translational modification networks. J. Protein Proteonomics 16 (3), 237–251. https://doi.org/10.1007/s42485-025-00190-y.

Phlairaharn, T., Shannon, A.E., Zeng, X., Truong, D.-J.J., Schoof, E.M., Ye, Z., Searle, B.C., 2025. Improving proteomic dynamic range with multiple accumulation precursor mass spectrometry. J. Proteome Res. 24 (10), 5116–5126. https://doi.org/10.1021/acs.jproteome.5c00469.

Prieto-Baños, S., Nevers, Y., Altenhoff, A., Warwick Vesztrocy, A., Dessimoz, C., Glover, N.M., 2025. Annotation matters: the effect of structural gene annotation on orthology inference. Bioinformatics 41 (7). https://doi.org/10.1093/bioinformatics/btaf365.

Rahman, Z.U., Asaari, M.S.M., Ibrahim, H., 2025. Advancing animal farming with deep learning: a systematic review. Comput. Electron. Agric. 237, 110674. https://doi.org/10.1016/j.compag.2025.110674.

Rojahn, M., Grum, M., 2025. Green AI: a systematic review and meta-analysis of its definitions, lifecycle models, hardware and measurement attempts (Version 1). arXiv. https://doi.org/10.48550/ARXIV.2511.07090.

Sanders, J., Wen, B., Rudnick, P.A., Johnson, R.S., Wu, C.C., Riffle, M., Oh, S., MacCoss, M.J., Noble, W.S., 2025. A transformer model for de novo sequencing of data-independent acquisition mass spectrometry data. Nat. Methods 22 (7), 1447–1453. https://doi.org/10.1038/s41592-025-02718-y.

Shao, Y., Yang, C., Ni, S., Pang, M., Liu, X., Kong, R., Chang, S., 2025. Applications and prospects of artificial intelligence in proteomics via mass spectrometry: a review. Curr. Protein Pept. Sci. 26. https://doi.org/10.2174/0113892037368295250520102121.

Singh, S., 2025. Artificial intelligence models for predicting fertility transitions: a systematic review. Asian J. Res. Comput. Sci. 18 (6), 122–129. https://doi.org/10.9734/ajrcos/2025/v18i6684.

Soundararajan, B., 2025. Developing new AI model compression techniques. Int. J. Sci. Res. Eng. Manag. 09 (03), 1–8. https://doi.org/10.55041/ijsrem43474.

Stoisser, J.L., Phillips, L., Misra, A., Lamb, T.A., Torr, P., Martell, M.B., Fauqueur, J., Märtens, K., 2025. Towards label-free biological reasoning synthetic dataset creation via uncertainty filtering (version 1). arXiv. https://doi.org/10.48550/ARXIV.2510.05871.

Taddese, A.A., Addis, A.C., Tam, B.T., 2025. Data stewardship and curation practices in AI-based genomics and automated microscopy image analysis for high-throughput screening studies: promoting robust and ethical AI applications. Hum. Genom. 19 (1). https://doi.org/10.1186/s40246-025-00716-x.

Tedeschi, L.O., Lopez, P.G., Menendez III, H.M., Seo, S., 2025. Advancing precision livestock farming: integrating artificial intelligence and emerging technologies for sustainable livestock management. Anim. Biosci. https://doi.org/10.5713/ab.25.0289.

Tomo, Y., Nakaki, R., 2025. iComBat: an incremental framework for batch effect correction in DNA methylation array data. Comput. Struct. Biotechnol. J. 27, 4121–4131. https://doi.org/10.1016/j.csbj.2025.09.014.

Trotsyuk, A.A., Waeiss, Q., Bhatia, R.T., Aponte, B.J., Heffernan, I.M.L., Madgavkar, D., Felder, R.M., Lehmann, L.S., Palmer, M.J., Greely, H., Wald, R., Goetz, L., Trengove, M., Vandersluis, R., Lin, H., Cho, M.K., Altman, R.B., Endy, D., Relman, D.A., Levi, M., Satz, D., Magnus, D., 2024. Toward a framework for risk mitigation of potential misuse of artificial intelligence in biomedical research. Nat. Mach. Intell. 6 (12), 1435–1442. https://doi.org/10.1038/s42256-024-00926-3.

Truong, T.F., Bepler, T., 2025. Understanding protein function with a multimodal retrieval-augmented foundation model (Version 1). arXiv. https://doi.org/10.48550/ARXIV.2508.04724.

Ukpoju-Ebonyi, O.M., Ebonyi, J.-M.A., Ogunjobi, T.T., Obasi, D.E., Ugwu, D.C., Anyanwu, G.O., Abolade, S.O., 2025. Integrating multi-omics and personalized medicine in one health: a systems biology approach to precision healthcare. Medinformatics. https://doi.org/10.47852/bonviewmedin52025831.

Wang, J., 2025. Multi-modal topology-aware graph neural network for robust chemical–protein interaction prediction. Int. J. Mol. Sci. 26 (17), 8666. https://doi.org/10.3390/ijms26178666.

Wang, Z., Zhu, G., Li, S., 2025a. Mapping knowledge landscapes and emerging trends in artificial intelligence for antimicrobial resistance: bibliometric and visualization analysis. Front. Med. 12. https://doi.org/10.3389/fmed.2025.1492709.

Wang, X., Zhang, Y., Zhou, Y., 2025b. Bidirectional probabilistic multi-graph learning and decomposition for multi-view clustering. IEEE Trans. Image Process. 34, 3609–3621. https://doi.org/10.1109/tip.2025.3574924.

Weng Nyam, P., Bishir, A., Mukhtar, U., Gali, A.M., Moses, N.Y., 2025. Advances in Bayesian approaches for stochastic process modeling and uncertainty quantification. Mikailalsys J. Math. Stat. 3 (3), 777–784. https://doi.org/10.58578/mjms.v3i3.7429.

Wu, A., Kuang, K., Xiong, R., Wu, F., 2025. Instrumental variables in causal inference and machine learning: a survey. ACM Comput. Surv. 57 (11), 1–36. https://doi.org/10.1145/3735969.

Xing, L., Wang, H., Yang, Y., Mei, B., 2025. An integrative multi-omics framework for genomic prediction in livestock using Bayesian deep neural networks. In: Frontiers in Epigenetics and Epigenomics.

Zeng, L., Liu, Y., Han, G., Yu, Z.-G., Liu, Y., 2025. ProGraphTrans: multimodal dynamic collaborative framework for protein representation learning. BMC Biol 23 (1). https://doi.org/10.1186/s12915-025-02432-3.

Zhang, S., Lai, C., Zhao, J., Wang, J., 2025. Big data and AI-powered modeling: a pathway to sustainable precision animal nutrition. Adv. Sci. 12 (41). https://doi.org/10.1002/advs.202507564.

Zhong, H., Chi, S., Magaña, A.A., Fordwour, O.B., Foster, L.J., 2025. Integrative omics and AI-driven systems biology: multilayer networks decoding *Apis mellifera* health and resilience. J. Proteome Res. 24 (11), 5305–5318. https://doi.org/10.1021/acs.jproteome.5c00294.

Zhu, R., Dong, Y., Guo, J., He, J., Chen, H., Wang, R., Ren, F., Raben, A., Martinez, J.A., 2025. From prediction of personalized metabolic responses to foods to computational nutrition: concepts of an emerging interdisciplinary field. Clin. Nutr. 55, 11–23. https://doi.org/10.1016/j.clnu.2025.10.009.

Zulueta-Coarasa, T., Jug, F., Mathur, A., Moore, J., Muñoz-Barrutia, A., Anita, L., Babalola, K., Bankhead, P., Gilloteaux, P., Gogoberidze, N., Jones, M.L., Kleywegt, G.J., Korir, P., Kreshuk, A., Küpcü Yoldaş, A., Marconato, L., Narayan, K., Norlin, N., Oezdemir, B., Hartley, M., 2025. MIFA: metadata, Incentives, Formats and Accessibility guidelines to improve the reuse of AI datasets for bioimage analysis. Nat. Methods. https://doi.org/10.1038/s41592-025-02835-8.

Chapter 7

Artificial intelligence integration in livestock metabolomics

The metabolome possesses a unique responsive niche in livestock biology in the transformation of genomic promise, proteomic execution, and cometabolic association of the microbes into the small molecule fluxes of instantaneous physiological priority, adaptive responses to stress, nutritional use, immune preparedness, reproductive reorganization, welfare condition, and ecological footprint expression. Its chemical diversity, rapid turnover, heterogeneous matrix, and context-dependent variability make it exceptionally sensitive, but also complicate standardized acquisition, annotation, longitudinal comparability, and causal interpretation. Artificial intelligence (AI) is not a phenomenon that works as an esthetic ancillary aspect of analysis but works as a framework of organization, capable of unlearning technical and biological variability, unraveling asynchronous readings of time, and aligning the patterns of metabolites with higher-order phenotypic and sustainability objectives.

This chapter presents AI-facilitated livestock metabolomics as an integrative inferential framework that exists above the disaggregated peak lists in semantically rich uncertainty-aware dynamically powered metabolic state specifications. It summarizes that the innovations in machine learning geometry, representation learning, graph- and pathway-conscious modeling, probabilistic reasoning, and causal structures may hasten metabolite detection, stabilize prediction on distributional shift, and expose a mechanistically interpretable targets with precision intervention, selection enhancement, and adaptive regulation. It bases its emphasis on the multimodal fusion between the metabolomic signatures and the genomic, transcriptomic, proteomic, microbiome, sensor-derived, and environmental streams to create the whole organism and system-level functions over time. It unveils the avenues of digital twins, foundation-style embeddings with biochemical ontologies, active production of augmentation in low-resource matrices, and active learning methodologies that rephrase the experimental design and sampling efficiency.

Model fairness, generalizability, and ethical legitimacy are perceived to be the main requirements of data engineering and governance. The chapter focuses on ontologies, provenance capture, quality control regimes, batch harmonization, privacy-preserving collaboration, federated analytics, and lifecycle stewardship, such as drift detection, recalibration, version lineage, and decommissioning. These ethical, legal, and societal factors are linked and address elements of transparency standards, the threat of algorithmic discrimination, ownership of data, dual-use sensitivity of metabolic pathway inference, and equitable sharing of advantages in a broad assortment of production environments. It is interspersed with sustainability in the discourse of metabolic indicators of resource efficiency, emission reduction, productivity in line with welfare, circular valorization, and carbon-conscious computational strategies.

Issues issues utilized to design the architecture and operations are problems of incomplete annotation, isomeric ambiguity, matrix-dependent variability of ionization, temporal variability, minor effect heterogeneity, integration friction with longer latency layers, and so forth. The chapter also explores emerging opportunities, including real-time sensing, spatial metabolomics, fluxomics, cross-species transferability, causal digital twins, and governance frameworks that integrate accountability and inclusivity. The chapter positions AI-driven livestock metabolomics to become a critical impetus to the resilient, adaptive, responsible future of combined omics intelligence in transforming livestock systems.

7.1 Introduction to AI-enabled livestock metabolomics

Metabolomics now represents a key integrative niche in livestock systems biology, intermediate between comparatively fixed genomic systems, more adaptable transcript and protein layers, and the emergent phenotypes modeled by health, productivity, welfare, and environmental performance. Since the metabolites are the proximate effectors and readouts of the biochemical pathway flux, endocrine regulation, immune regulation, nutrient partitioning, microbiome symbiosis, and negotiation of environmental stress, the metabolites reduce the regulatory complexity of the multiscale signatures to high information density. Unlike the sequence-based modalities, the metabolome is dynamic in nature and encodes the

Artificial Intelligence in Livestock Omics. https://doi.org/10.1016/B978-0-443-45663-3.00002-6

immediate adaptations in response to feeding, circadian cycles, developmental shifts, and microecological responses. This temporal immediacy provides a vulnerable insight into system status at the cost of longitudinal comparability, causal ascription, and translational standardization. To place metabolomics on a continuum of holistic multi-omics, conceptual and computational frameworks are thus needed that balance between different temporal resolutions, variance fields, the sparseness of different data sets, and representations of their semantic contexts within the molecular and phenotypic world. This reconciliation makes AI fundamental, with the ability to discover latent structures and cross-layer alignment that cannot be performed manually or through the standard statistical capabilities (Kim and Pathak, 2025; Wadood et al., 2025).

Unique features of the livestock metabolome are increasing the opportunity and complexity. Temporal lability indicates fast flux redirection and flux turnover of important intermediates and context-dependent flux redirection, which generate nonstationary distributions and microstates of phenotype. The circulating blood components; the secreted milk; the gastrointestinal contents; the fecal, urinary, salivary, and respiratory-derived condensates; and the tissue biopsies all have matrix diversity that places unique physicochemical limitations on the extraction efficiency, ionization characteristics, and repeatability of the analytical analysis between them. Variations in feed composition, housing, the variability of climatic stressors, physiological stage, and management intervention are further factors that influence compositional heterogeneity. In microbial co-metabolism (especially acute in ruminant fermentation communities, but also in monogastric hindgut and upper tract microbiota), the presence of an extended biochemical reaction space with interdigitating host and microbial pathways results in conjugated, reduced, oxidized, and otherwise transformed metabolites that complicate library-based identification paradigms. Scales of several orders of magnitude of isomeric complexity, adduct multiplicity in source fragmentation, and concentrations in principle increase analytical uncertainty, whereas all biological and technical changes interact through batch effects, instrument drift, and small sample handling differences. All these unique aspects require adaptive, uncertainty-conscious computational approaches that can include dynamic updating of baselines, calibration of matrices that are specific to each host, and integrative disentangling of host intrinsic and exogenous or microbial signal elements (Nunes et al., 2024; Wei et al., 2025).

The multiplicity of strategic value is many times over in resolving these complexities. Metabolomic signatures have sensitive surveillance indicators of the onset of physiological imbalance and thus offer an earlier and potentially less invasive intervention pathway than subsequent clinical phenotypes. The optimization of nutritional efficiency can be done based on metabolic fingerprints that indicate real-time nutrient levels of absorption, partitioning, and metabolic bottlenecks to support precision feeding and stewardship of resources. Metabolically encoded correlates of stress physiology, inflammatory tone, and oxidative balance have a role to play in welfare assessment, which provides objective supplements to behavioral and environmental assessments as well. Quality characteristics of products in relation to compositional integrity, sensory precursors, and stability of shelf life are metabolically supported, making avenues to active quality assurance in production pipelines. The requirements of environmental stewardship also overlap with metabolomics, including biochemical indicators of nitrogen and energy use efficiencies, enteric emission precursors, and metabolic stability in response to variations in climatic conditions, which justifies the multiobjective optimization framework, where productivity is largely independent of, but reduces ecological footprint. Integrating metabolomics into selection and management choices, therefore, offers systemic value creation in health, efficiency, welfare, quality, and sustainability scales in the case of highly mediated inference by AI (Zhao et al., 2025).

The drivers encouraging the use of AI in this field are related to the complexity, size, timeliness, and integration requirements that are beyond the classical analytical paradigms. High-dimensional untargeted profiles include a large number of partially redundant, co-linear, and sparsely annotated features with joint distributional geometries that are non-Euclidean and context-dependent. Conventional single-variable or low-dimensional multivariate methods have difficulties in higher-order interactions, nonlinear manifolds, conditional dependence, and hierarchical pathway constraints. Longitudinal sampling, multimatrix profiling, and growing herds present a scaling problem since they produce large, heterogeneously missing, and temporally discontinuous streams of data. Timeliness pressures arise when production systems demand near real-time state estimation to support adaptive management, when they require rapid preprocessing, drift correction, and inferential updating when subjected to streaming or edge constraints. The requirements of integration necessitate that metabolomic embeddings need to be aligned with genomic approximated breeding values, transcriptomic activation conditions, proteomic abundance patterns, microbiome functional profiles, sensor-acquired behavioral and environmental phenomena, and externally maintained pathway insights. Current AI learning frameworks offer representation learning designs, which learn silently and noisily sparse spectra; partially labeled spectral, probabilistic, and causal models; graph and sequence-aware architectures; and federated and privacy-conserving models that learn together without centralizing raw data (de Bodt et al., 2025; Hasan et al., 2025).

It is on this backdrop that the chapter establishes the scope and conceptual boundaries so as to establish analytical rigor and interpretive clarity. Its story is based on methodological and translational aspects of AI implementation in livestock metabolomics as opposed to a comprehensive listing of analytical chemistries or species-specific physiological detail. The emphasis is placed on untargeted and hybrid targeted/untargeted strategies as the main sources of high-dimensional complexity, and the targeted panels are mentioned in the context of validation and deployment. It focuses on integrative models, which connect metabolomic data with other layers of omics and phenomics, and on how representation learning, causal inference, uncertainty quantification, and pathway-aware graph modeling can be used to improve the quality of inferences and actionable precision. It provides consideration to data engineering basics that include standardization of sampling, quality assurance, batch harmonization, metadata ontologies, provenance capturing, and lifecycle management as they support the ideas of reproducibility, fairness, and ethical legitimacy. Ethical, legal, and societal implications are not seen as peripheral addenda but model design criteria that are crucial in shaping model development, implementation, and trust by the stakeholders. The constraints and open research opportunities, including incompleteness of annotation, isomer discrimination issues, cross-cohort generalizability, the physical binding of time with slower omics layers, and the scalability of larger models computationally, are clarified. The chapter provides a consistent framework by which both the present practice and future innovation paths can be determined by formulating organizing questions regarding how AI can best supplement metabolic identification, decouple technical and biological variability, become more predictive and causal, support multiobjective decision-making, and incorporate ethical and environmental responsibility.

The introduction constitutes a part of synthesizing such thematic strands in the sense that AI-enabled livestock metabolomics is a ground-breaking integrative science that will initiate more predictive, precise, and sustainable management paradigms. It proceeds with the argument that the achievement of this potential could only be possible by the corresponding development of analytical, computational, governance, and ethical systems in a way that methodological sophistication and success would be converted into meaningful, equitable, and strong developments in livestock systems.

7.2 Foundations of livestock metabolomics

The history of livestock metabolomics is interpreted by the fact that the diversity of chemical classes all encodes immediate biochemical states and long-term adaptive pathways. Central carbon metabolites, amino acid beds, nucleotide precursors, energy currency molecules, osmolytes, and redox couples of cofactors are primary metabolites, and they are the ones that support basal cellular functions and orchestrate the metabolic division of labor between maintenance, growth, reproduction, immune vigilance, and product synthesis. Secondary metabolites, despite being historically conceived as ancillary, do also participate in signaling regulation, oxidative signaling, host defense, and even interspecies communication, as well as contain phytochemically produced molecules, endogenously produced molecules, and stress-induced molecular patterns. Structural, energetic, and signaling roles of the lipid species lie along a diet and physiological phase as well as an inflammatory tone continuum extending the storage stores, emulsifying surfactants, eicosanoid-like mediators, and compositional remodeling. Volatile/semivolatile compounds resulting from fermentative/oxidative transformation of lipids and microbial/host enzyme reactants are a ready estimate of the metabolic flux in domains that can be sensed in the gases or headspace. Decarboxylation, transamination, polyamine synthesis, and microbial metabolism result in biogenic amines and nitrogenous small molecules that are modulators of both vascular and neuromodulatory and immunological functions and also indicate proteolytic and fermentative states. Microbial co-metabolites are conjugated, reduced, dehydroxylated, demethylated, and otherwise structurally remodelled entities produced through metabolic symbiosis between hosts and microbes. This extends the chemical alphabet beyond the enzymatic output encoded in the genome of the host organism, introducing an expansive chemical dark matter that challenges existing reference libraries and annotation pipelines. Every class will share characteristic polarity, volatility, ionization propensity, dynamic range, degradation dynamics, multiplexed analytical strategies, or prioritization schema that influence downstream AI processing patterns of feature sparsity distributions and class-specific noise behavior, as well as stratification of annotation confidence (Hailemariam and Dervishi, 2024; Park et al., 2025).

The biological matrices that define the segments of the metabolome that can be observed, and with what faithfulness, are the biological gateways to the metabolome. Circulant fluids mix systemic distribution and are sustained by compartmental boundaries, clearance ways, and quick buffering, which provides a composite, yet transiently smoothed, snapshot. Secreted matrices are specialized biosynthetic transport systems that not only entrap conditions of nutritional, immunological, and physiological conditions but also contain multiplex colloidal structures with consequences of extraction and ion suppression mechanisms. Gradients of pH, osmolarity, and enzyme activity, which differ to stabilize the molecules, trap end products of fermentation, secretions of the host, and transitional food in gastrointestinal luminal matrices. Excreta is a form of integrative processes of absorption efficacy and microbial biotransformation, nitrogen and mineral regulation, and detoxification pathways but is obscured by water balance and transit time variability. The least invasive access to volatile, semivolatile, and aerosol-linked metabolites is offered by oral and upper respiratory secretions and condensed breath phases, and thus, this offers more frequent sampling at the expense of introducing dilutional variance and contamination variance. Tissue biopsies give data on the level of cell-type-specific flux distributions and microenvironmental gradients at the tradeoff of invasiveness and time repeatability. The matrices incorporate matrix-dependent preanalytical factors, which encompass collection time, stabilization, incidence of freeze and thaw limits, effects of anticoagulant or preservation effect, and batch transportation logistics. These variables overlap the sensitivities of the analytical platforms to form manifolds of matrices with specific features, to which the AI systems must trade with domain adaptation, matrix-conscious normalization, and cross-matrix embedding schemes that minimize biologically meaningful differentials and minimize any spurious technical divergence (Eshawu et al., 2025; Tronel et al., 2025).

This structure of variability is a baseline of temporal and circadian dynamics. Feeding cycles cause oscillating changes in absorptive flux, hormonal changes, changes in the availability of microbial substrates, and changes in energy partitioning, which lead to pronounced postprandial patterns in carbohydrate, lipid, nitrogen, and short-chain fermentation domains. Circadian controls coordinate the expression of genes, the capacity to enzymatically perform particular tasks, and endocrine signaling, providing phase-dependent baselines, which react with light, feeding, and treated activity regimens. Metabolic reprioritization is found in physiological changes: development stages, initiation and termination of the productive cycle, reproductive inflection points, and changes in mobilization versus accretion ratio, oxidative versus reductive metabolic pathways, and anabolic versus catabolic hormone interactions. Homeorhetic setpoints are evolved through cumulative adaptive processes, which entail a memory, such as a pattern in metabolite covariance structures. Sample time irregularity exacerbates aliasing problems and makes naive comparisons with unsynchronized rhythmic/event-aligned baselines. In this way, the problem of temporal embedding, sequence modeling, interval-aware imputation, and disentanglement of intrinsic rhythms with exogenous perturbations are important elements of AI strategies that aim to arrive at robust inferences and predictions (Galasso et al., 2024).

Host microbiome metabolic interaction is one of the central axes of livestock metabolomic organization, particularly in fermentation-facilitated digestive communities but increasingly in monogastric ones. Microbial consortia expand catalytic repertoires, which carry out polysaccharide depolymerization, nitrogen salvage, reductive transformations, deconjugation, vitamin biosynthesis, amino acid products, and signaling molecule biosynthesis. Hosts secrete enzymes, bile-related molecules, and immunoglobulins, as well as antimicrobial peptides to regulate the microbial composition and activity and the microbial metabolites to regulate epithelial integrity, immunogenesis, endocrine regulation, and energy homeostasis. Transkingdom exchange gives rise to composite pools of metabolites, the attribution of which can be unclear in spectral data that do not have isotopic labeling or source tracing augmentation. Communal structure variations, phage interactions, protozoa interactions, and substrate availability determine cometabolite abundances and predict further variation beyond host genomic determinants. By combining metagenomic, metatranscriptomic, or predicted functional pathway profiles with measurements of the host metabolism, it is possible to disentangle host intrinsic metabolic changes and microbiome-mediated changes, allowing the construction of causal graphs, the mediation of these interactions, and the determination of modifiable microbial levers to optimize a given system. Latent factor models that include joint host microbial representations can project onto fewer dimensions and maintain biologically interpretable axes that are consistent with fermentative efficiency, nitrogen-utilizing, or barrier functioning (Lushchak, 2025; Shelat and Rawal, 2025).

Metabolic landscapes are further moderated by phenotypic and environmental drivers, which interact in highly complex web patterns. Genetic architecture determines enzyme isoform distributions, transporter capacities, receptor sensitivities, and regulatory network parameters and determines baseline metabolic predispositions on which the management and environmental conditions take effect. The developmental and environmental experiences are epigenetically tuned by developmental and environmental exposures to establish the responsiveness of gene expression that determines metabolic plasticity and resilience to perturbations. The physiological phenotypes that include the body composition, endocrine condition, immune tone, and behavioral tendencies precondition the division of resources, stress reactivity, and use of metabolic reserves. Temperature, humidity, altitude, photoperiod, stocking density, and microclimatic variability have direct impacts on thermoregulatory energetic rates, oxidative stress load, and feed consumption patterns, with indirect impacts in altering microbial fermentation processes and nutrient absorption curves. Substrate supply gradients, fermentation kinetics, and satiety signaling change in response to changes in nutritional composition, physical form, feeding frequency, and supplement inclusion and diffuse throughout lipid mobilization, amino acid catabolism, and excretion of nitrogen pathways. The modulators of metabolic stability are added by management practices, including housing design, social grouping, health interventions, and schedule consistency. The resulting variance structure is

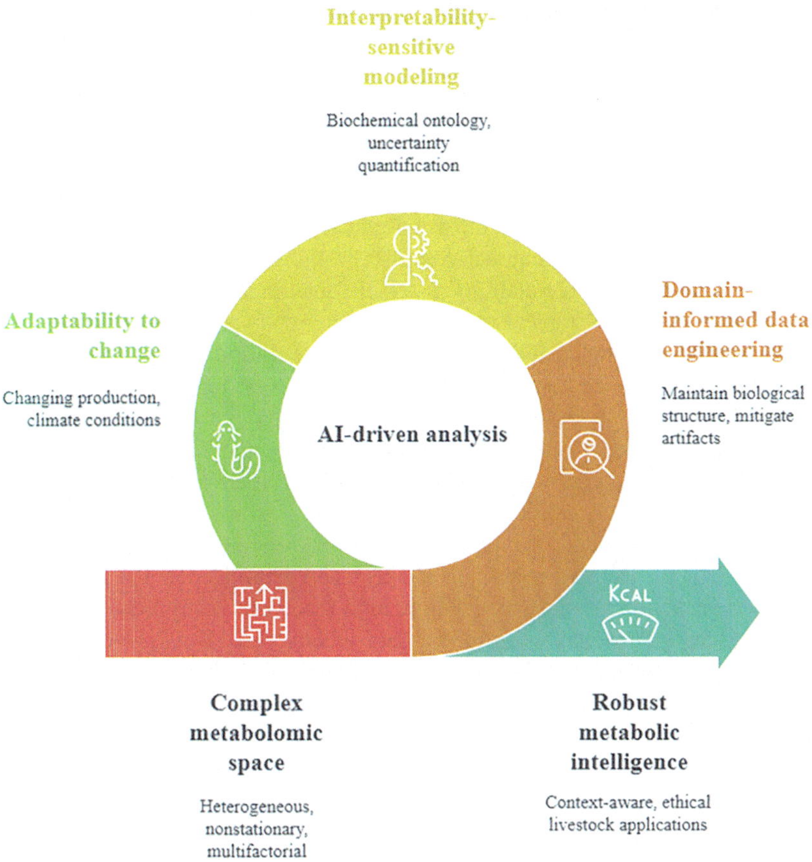

FIGURE 7.1 AI-powered livestock metabolomics. This figure illustrates a conceptual framework for how AI can empower livestock metabolomics by addressing the inherent complexity of metabolic data and producing robust, interpretable, and adaptable models for use in real-world production systems.

hierarchical, nonlinearly separable, and heteroscedastic, and needs modeling methods that are able to model interaction effects, context specificity, and possible causal directionality. Multilevel mixed-effects models, hierarchical Bayesian models, graph neural networks with relational representations, and invariant risk minimization models can help to separate the variance into genetic, management, environmental, and temporal levels. Pathway-calibrated feature attribution techniques help to discriminate between biologically meaningful signals and confounded associations (Al Zubaidi et al., 2025; Yasmin et al., 2025).

A combination of these underlying dimensions defines a metabolomic information space that is chemically heterogeneous, matrix-conditioned observable, temporally nonstationary, biochemically integrating across transkingdoms, and phenotypically and environmentally multifactorial. The AI systems that are designed to work in this space should be supported by domain-informed data engineering, which maintains the true biological structure and mitigates technical artifacts, and by modeling philosophies that favor interpretability that is sensitive to biochemical ontology, uncertainty quantification that is sensitive to annotation confidence, and the ability to adapt to changing production paradigms and climate conditions. The building of these foundations makes sure that future integration efforts based on predictive health surveillance, precision nutrition, welfare assessment, product quality optimization, and sustainability analytics are focused on robust, context-aware metabolic intelligence as opposed to fragile statistical associations and allows for much more robust and ethical application across various livestock systems (see Fig. 7.1).

7.3 Analytical platforms and data acquisition architectures

The biological reality and the abstraction of computation in livestock metabolomics are presented through analytical platforms, and the physicochemical principles of operation are the shape of the statistical characteristics of data fed into AI pipelines, the noise structure of data, and the semantic richness of data. The spectroscopy of nuclear magnetic resonance (NMR) fits uniquely in the category of being nondestructive, intrinsically quantitative, and highly reproducible, with little sample preparation-induced variability. However, its spectral output has a large amount of overlap of its peaks due to limited chemical shift dispersion in crowded blocks, enhanced by macromolecular background effects and baseline effects. The deconvolution of such composite signals has become an increasingly sophisticated art made possible by AI techniques that outperform traditional line shape fitting with probabilistic deep generative networks that learn the prior information about spin systems, coupling patterns, and chemical shift context, conditioned on the properties of matrices. Representation learning systems can separate overlapping resonances, combine component-specific distributions of concentration with measured uncertainty, and perform adaptive baseline subtraction that models slow-varying drift in free induction decay reconstructions. Moreover, multinucleus acquisitions and diffusion-edited schemes increase the available chemical subspaces, and AI-made joint inference across these orthogonal contrasts leads to increased discriminatory power without spectral curation by human rules. The comparative stability of NMR quantification makes it a useful anchor to harmonize cross-platform data and allows calibration transfer of more variable mass spectrometric characteristics and allows hybrid training regimes when high-confidence NMR-measured concentration labels guide semi-supervised classification models of lower annotable features in complementary data. However, reduced sensitivity in comparison to high-performance mass spectrometry places a detection limit determined by the dilution factors of matrices, and AI models should include censoring-conscious likelihood formulations to prevent bias in downstream integrative analyses between metabolite abundances and genomic or phenotypic predictors (Nunes et al., 2024; Mooz-hippurath and Natarajan, 2025).

Liquid chromatography-coupled mass spectrometry represents the broadest discovery engine in livestock metabolomics, encompassing both untargeted discovery methods, targeted quantitative assays, and data-independent acquisition methods like sequential window fragmentation and accurate mass instrumentation, high-resolution instruments including Orbitrap and time-of-flight instruments. Every modality develops a different data architecture. Untargeted profiling generates high-dimensional feature tables that contain retention time, mass to charge, and intensity attributes nested in run-level batch environments but has unfortunate stochastic ion suppression, retention drift, in-source fragmentation, and adduct or isotope redundancies. The technique of targeted assays focuses on a specific analytic panel with optimized transitions or inclusion lists that have greater quantitative accuracy at the cost of latitude in exploratory applications. Data-independent acquisition sums up multiplexed fragment ion data over large precursor isolation windows, and requires computational demultiplexing to separate cofragmenting species. AI solves such layered problems with built-in chromatographic shape priors combined with heteroscedastic noise models, retention time alignment methods that use differentiable warping transformations with landmark distribution constraints, and probabilistic grouping of features, bringing together isotopologs, adduct series, and in-source fragments using graph-based clustering that uses similarity kernels trained on annotated corpora. Using deep spectral unmixing networks learned with synthetic composite spectra in the setting of SWATH or similar data-independent schemes can help deconvolute sets of overlapping fragments, and attention-based networks train fragment intensity patterns to candidate molecular substructures to supplement the traditional library matching, in which livestock-specific spectral reference sets are dispersed. Although high-resolution mass accuracy enhances formula assignment, combinatorial isomer possibilities still proliferate. AI-augmented in silico fragmentation, integrated with retention prediction and ion mobility where available, yields composite scoring frameworks that explicitly quantify identification uncertainty. This enables downstream causal or predictive models to propagate confidence weights, rather than treating all features as equally reliable. Domain adaptation layers reduce cross-instrument and interlaboratory variability by formulating distributional alignment objectives on latent embeddings, reparametrized to impose adversarial or optimal transport constraints, without affecting biological gradients correlated with genetic or management variables and reducing analytical systematic error (Bosten et al., 2025; Kopecky et al., 2025).

Gas chromatography–mass spectrometry has been retained in volatile compounds and thermally derivatized small molecules, which convert polar primary metabolites into more hydrophobic derivatives, which have a better chromatographic separation and electron impact fragmentation reproducibility. Derivatization chemistry presents batch-specific variance of completeness of reactions, artifact generation, and index shifts in retention, which depend on column aging and affect complex error structures. Probabilistic calibration curves based on ambient laboratory conditions and reagent lot metadata are used to normalize retention indices with AI-enabled normalization, whereas using a transformer-based sequence to structure models on simulated spectra induced by quantum-informed fragmentation heuristics is used to disambiguate fragmentation patterns. The comparative consistency of electron impact fragmentation of instruments makes possible transfer learning that summarizes generalizable fragmentation semantics that help in annotating livestock matrices containing diet-derived xenobiotics or microbially transformed moieties not adequately represented in current collections. In volatile profiling, sequence-aware anomaly identification using time-stamped chromatographic baselines over volatile profiling variability due to dynamic headspace sampling introduces time fluctuations in concentration.

Incorporation of volatile signatures into less robust integrative metabolic response models would benefit more severely than sequence-aware anomaly detection (Ali, 2025; Sato and Ikeda, 2025).

Capillary electrophoresis mass spectrometry and other niche systems provide increased coverage of highly polar and charged, and low ionic metabolites that cannot be retained by reversed-phase chromatography or hydrophobic chromatography separation and provide high-efficiency separation of charge-to-size ratios with a given buffer system. The variation of electrophoretic mobility with changes in temperature, depletion of buffers, and variation in electroosmotic flow makes migration time alignment difficult, producing nonlinear distortions that are difficult to correct with simple warping. Physics-informed neural operators based on AI methods simulate electrophoresis transport dynamics conditional on real-time measurement of current profiles and internal standard migration, making them adaptively align themselves and correct drift. Narrow peak widths also lead to signal-to-noise issues that make denoising an interesting requirement using sparse coding or deep residual networks that are trained to conserve sharp edges of analysis and inhibit stochastic baseline contaminations. Other niche methods that include supercritical fluid chromatography or ion mobility separation characterize multidimensional data cubes in which mobility-matched collision cross-section provides structural discrimination. The fusion of the orthogonal dimensions based on AI builds more effective latent embeddings, which enhances the discriminative power of isomeric species that occur in biologically relevant pathways (Fang and Sun, 2025).

Direct analysis in real time and desorption electrospray ionization are ambient and rapid ionization techniques that show little to no sample preparation and may eventually lead to on-farm application, the result of which is spectra with lower chromatographic separation and, consequently, increased ion suppression and isobaric overlap. The lack of separation imparts high-resolution mass accuracy and patterns of differential ionization efficiency determined by matrix surface composition and environmental conditions onto the analytic specificity. AI replenishes with acquired deconvolution utilizing vast bodies of coionized spectra that have been annotated by paired conventional chromatographic measurements, and virtual separation by statistical disentangling of sets of covarying ions is achievable. The issue of calibration drift due to changes in temperature and humidity in field conditions is solved by continuous learning models, which update mass recalibration parameters with limited memory budgets and with sparse internal standard measurements. Compression and quantization models can be deployed on the resource-constrained edge device, and uncertainty estimation mitigates overconfident classification or quantification of spectra that are outside of the training distribution because of new preparative additives or an emerging change in the microbes. All the methods listed above provide access to fast metabolic screening as a component of decision support systems, but require a framework of governance to achieve the integrity of calibration and data provenance of decentralized acquisition (Liu, 2024; Wang, 2024).

Infrared, Raman, and near-infrared spectroscopies are examples of complementary rapid phenotyping modalities that detect vibrational signatures of aggregate chemical composition but not individual molecular species. They are cheap per sample, potentially high-throughput, and have convoluted overlapping band structures, scattering artifacts, baseline shifts, and sensitivity to physical sample characteristics, including moisture, particle size, and temperature. The conventional chemometric approaches are being supplemented with deep spectral encoders that train on hierarchical representations that are independent of nuisance variance but retain subtle compositional differentials that are associated with metabolic states, nutritional quality, or health measures. The approaches of data augmentation reproduce the realistic scattering perturbations and baseline changes, and improve the overall cross-instrument and cross-production setting generalization. Transfer learning aids intergenerational hardware recalibration and interlaboratory recalibration, which reduces performance loss related to instrument-specific response functions. This provides interpretability to couplings of vibrational spectroscopy embeddings with targeted metabolomic anchors through multitask learning to map abstract spectral latent dimensions back to biochemical axes with respect to reference panels that are measured simultaneously, thereby increasing their importance as coarse proxies to integrative components in multi-omics predictive ensembles (Liang et al., 2025).

In imaging mass spectrometry, a spatially resolved dimension is incorporated by ascertaining how metabolites are distributed in tissue sections, which offers an insight into the microenvironmental heterogeneity in metabolic flux, immune infiltration, vascularization, or pathological remodeling in terms of productivity, health, and welfare phenotype. Heteroscedastic pixel-level spectra generated by the matrix-assisted desorption or secondary ion-based techniques form high-dimensional spatial tensors, which are below the low ion counts, heteroscedastic noise that depends on local tissue chemistry, and heterogeneity of crystallization of the matrix. In this respect, AI is applying convolutional and graph neural networks that comprise spatial contiguity priors to amplify feature denoising, peak picking, and subdivision into functional microdomains. Multimodal data fusion that co-registers histological stains and immunofluorescence layers enables cross-modal attention mechanisms between structural morphologies and metabolic gradients to be inferred with greater precision on cellular compositions and microvascular influences on metabolite distributions. Spatial deconvolution tools deconstruct the combined pixel spectrum to create contributing cellular or subcellular source profiles when reference single-cell transcriptomic signatures have been provided, connecting metabolic phenotypes to programs of gene regulation and bridging the gap between metabolic phenotypes and underlying gene regulatory programs. The three-dimensional and time-resolved imaging requires scalable two-dimensional tensor factorization and distributed training methods that can make computations computationally viable, and ethical aspects of invasive sampling require a concrete foundation based on incremental predictive or mechanistic knowledge that cannot be achieved by bulk profiling (Jamzad et al., 2025; Seydoux et al., 2025).

Architectures of real-time sensing include electronic nose arrays, breath condensate analyzers, wearable biosensors, and sensor networks in the environment to produce continuous streams of metabolic and contextual data to redefine the paradigms of analysis from an episodic batch process to a streaming inference. These data streams are also nonstationary, diurnal, and concept drift due to changes in the dietary patterns or management intervention and have uneven sampling rates between sensor modalities. The online learning algorithms used to implement AI learn successively and keep past knowledge in replay buffers and adapt to emerging trends without catastrophic forgetting. Anomaly detection is based on probabilistic forecasting models that build predictive distributions on future sensor-generated metabolic signatures of deviations that indicate an impending health disturbance or environmental stress reactions. Multisensor fusion combines patterns of volatile organic compound measurements, heart rate variability, activity measurements, thermal imagery measurements, and ambient environmental measurements using hierarchical Bayesian or attention-based fusion models that assign modalities' weights in a dynamic manner based on estimates of reliability. Edge computing strategies prioritize the local processing of latency-sensitive inference tasks, while delegating the aggregation and deeper analytical routines to centralized infrastructure. This balances the need for bandwidth with privacy and data sovereignty obligations. The streaming paradigm also requires strong temporal alignment schemes and defective data processing in periods of transient sensor unavailability, as well as the prudent setting of alert thresholds to reduce false positives to a level that will not harm stakeholder trust (John et al., 2024; Ivanov et al., 2025).

Sampling, stabilization, storage, and chain of custody metadata standardization are requirements to ensure AI preparedness and are directly related to the generalizability, reproducibility, and equity of downstream models. Preanalytical variability, which consists of time relative to feeding or circadian phase, properties of devices and containers, choice of anticoagulants or preservatives, quenching and freezing delay, excursions in transport temperature, and freeze-thaw cycles, creates structured noise that can otherwise be misdiagnosed as a biological signal or create systematic bias between cohorts or between management systems. These variables are encoded in comprehensive metadata schemas based on interoperable ontologies and can be adjusted in covariates, cross-validated, stratified, and revealed as biased. Immutable audit trails are used to track provenance record processing actions, analytical run actions, instrument maintenance actions, and software versioning, which are used to retrospectively root cause analyze anomalies in batch output. The design of quality control incorporates pooled matrices, internal standards that are isotopically labeled, blank samples, and longitudinal performance monitors, where AI-powered monitoring control panels implement multivariate control charts and drift sensing algorithms. Data lifecycle regulations combine access control, needed consent where relevant, and retention regulations, which guarantee ethical adherence, as well as enable federated approaches to learning in which models can be collaboratively constructed without focusing on raw sensitive information. The use of harmonized file formats and standardised feature nomenclatures promotes consistent integration with genomics and other omics. This enables cross-species comparative analyses and meta-analytical synthesis, leveraging the breadth of the livestock production environment. Together, high standardization will change the heterogeneous raw analytical outputs into a semantically consistent substrate on which complicated AI architectures can be run with confounding reduced to a minimum to make insights that are robust across genetic backgrounds, management regimes, and geographical settings (Caufield et al., 2025; Hiniduma et al., 2025). Key metabolomic data types and acquisition methods in livestock are presented in Table 7.1.

7.4 Data engineering, preprocessing, and quality management

Translational metabolic coverage of livestock metabolomics is the extensive data engineering of heterogeneous raw analytical signals into machine-actionable representations that can be combined with genomic, transcriptomic, phenomic, and environmental layers to generate integrative models. The long-term result of the parameterization of the cascade of processes that convert raw detector outputs is on the downstream statistical structure. Data sets of chromatographic and electrophoretic techniques based on the peak-picking algorithms should be capable of distinguishing between real elution profiles and noisy stochastic variations with changes in overall peak width, asymmetry, and coelution. Adaptive probabilistic models and deep sequence architecture are becoming substitutes or complements to the classical local maxima and wavelet decompositions since they apply the physics of the chromatographic process (smoothness and uniformity of

TABLE 7.1 Key metabolomic data types and acquisition methods in livestock.

Data type	Acquisition method(s)	Description	Applications in livestock	Considerations or trade-offs	Species or breed example
Global (untargeted) metabolome	• NMR spectroscopy (e.g., ^{1}H-NMR) • LC-MS (liquid chromatography–mass spectrometry) • GC-MS (gas chromatography–mass spectrometry)	Provides a broad "snapshot" of many small-molecule metabolites without prior selection.	• Biomarker discovery across growth, health, or nutritional traits • Comparative metabolic profiling (e.g., between physiological states or breeds)	• NMR: highly reproducible, minimal sample prep, but less sensitive • LC-MS: very sensitive, high coverage, but more complex sample prep and variable quantitation • GC-MS: robust, good for volatile/derivatized metabolites but requires derivatization and is less suited to nonvolatile compounds	Cattle: NMR-based plasma metabolomics in dairy calves to detect infection biomarkers
Targeted (quantitative) metabolome	• LC-MS/MS (tandem MS) • Multiple reaction monitoring (MRM)	Focused measurement of a predefined set of metabolites, often with high sensitivity and quantitation	• Precise quantification of metabolites of known biological importance (e.g., amino acids, lipids) • Validation of candidate biomarkers for traits like feed efficiency or disease	• Limited to selected metabolites • Requires standards and often a more expensive method of development • May miss biologically relevant unknowns	Pigs: targeted metabolomics comparing heavy-pig breeds using the Biocrates panel
Ionic and highly polar metabolites	• Capillary Electrophoresis–MS (CE-MS)	Uses electrophoretic separation to resolve small, polar, or charged metabolites (e.g., amino acids, nucleotides)	• Profiling of ionic metabolites in biofluids (e.g., plasma, urine) in nutritional or health studies	• Requires specialized instrumentation • May have lower throughput compared to LC-MS • Sensitivity may vary by charge state and matrix	Sheep/goats: metabolomics in the milk of goats, sheep, and cows, a widely targeted approach across species
Metal ion/ elemental metabolome	• ICP-MS (inductively coupled plasma–MS)	Measures trace metal ions and elemental composition in biological samples	• Assessment of micronutrient status in livestock (e.g., rumen, blood) • Monitoring mineral nutrition and toxicity	• Requires digestion/ preparation of samples. • Not all metabolomics labs have ICP-MS • Challenging to integrate with nonelemental metabolome data	Cattle: review noted nearly no ICP-MS studies in livestock metabolomics, indicating a gap

Continued

TABLE 7.1 Key metabolomic data types and acquisition methods in livestock.—cont'd

Data type	Acquisition method(s)	Description	Applications in livestock	Considerations or trade-offs	Species or breed example
Flux metabolomics (fluxomics)	• Stable isotope labeling and MS (e.g., ^{13}C, ^{2}H)	Tracks metabolic fluxes and turnover by following isotopically labeled substrates through metabolic pathways	• Investigating metabolic efficiency, nutrient utilization, and metabolic bottlenecks in livestock • Assessing metabolic adaptation under dietary or environmental interventions	• High cost of isotopically labeled feed/substrates • Complex experimental setup, especially in large animals • Requires sophisticated modeling and computational analysis	Livestock: fluxomics data remain almost absent in livestock metabolomics studies, a major research gap
Spectroscopic/ fingerprint-based metabolomics	• NMR fingerprinting (1D NMR) • FTIR or IR spectroscopy	Rapid profiling based on spectral patterns rather than full metabolite identification	• High-throughput screening of biofluids (e.g., for quality control, authentication in meat or milk) • Potential for rapid screening in farm-scale applications	• Lower metabolite coverage/resolution • Identification can be ambiguous without further separation or MS • Requires good spectral libraries and chemometric models	Ruminants: comparative metabolite profiling in goat, sheep, and cow milk using targeted metabolomics includes spectral approaches

the components) constraint and enable flexible retention time drift on the sample boundaries. A compromise exists between spectral binning of NMR and high-resolution mass domain, where bin width selection is informed by information theoretic criteria, which is reinforced by a reinforcement learning agent due to the dynamic constraints or entropy-based bin width choice. The alignment processes correct the retention time, migration time, or chemical shift variation due to temperature variations, column aging, change of ionic strength, or magnetic field defects. Learn-time nonlinear warping, which is inferred using landmark features and a global smoothness penalty and multireference consensus alignment, eliminates reference bias in the case of multicohort consortia. Joint probabilistic shift modeling is applied in spectral alignment in NMR, and the correlated multiplet components are modeled to avoid artificial broadening that negatively affects quantitation. The cumulative goal is harmonization of features with cumulative goals, which has residual variance that is mostly biological rather than technical, which enhances the fidelity of latent structure discovered by unsupervised or supervised AI algorithms (Godon et al., 2025; Kant et al., 2025).

Noise reduction and baseline correction are joint minimization problems, with aggressive filters as an option that can attenuate low-abundance, but biologically important, metabolites, and conservative methods that result in heteroscedastic noise that inflates false discovery. Baseline wander is caused by variability in the current matrix ions, macromolecular background, or instrumental electronic drift. Current methods use the hybrid signal decomposition: empirical mode decomposition or variational mode extraction provides rough candidates in the form of baselines, which are tuned using convolutional neural networks trained to distinguish morphological features of edge baselines from genuine peaks. Denoising diffusion probabilistic models and sparse Bayesian learning models are used to denoise spectra by using learned priors on the shapes of peaks conditioned by local retention neighborhoods and instrument class, which preserve the structural integrity of partially overlapping peaks. Signal-to-noise optimization goes beyond denoising to encompass adaptive smoothing of extracted ion chromatograms, dynamically increasing integration boundary positions using stability of derivative crossings, and context-aware requantification of initial automated limits that cross abnormal baselines. Uncertainty measures based on the posterior variance of the Bayesian peaks models are used to inform downstream weighting schemes to avoid confident use of marginally discovered features in the multivariate predictive models involving the linkages between metabolite abundances and genetic markers or health outcomes (Tian et al., 2025).

Incomplete data saturates livestock metabolomic matrices by a variety of processes, in which incorrect classification is misleading in inference. Although the concept is straightforward, completely random absence is comparatively rare. More prevalent are patterns of missing at random driven by retention distortions or ion suppression specific to the batch, or linked to co-eluting high-abundance species. Missing not at random manifestations are also common, wherein low-abundance metabolites fall below the detection threshold, which is conditioned by physiological state or feeding status. Use of explicit censoring threshold modeling (instead of naive zero filling) can maintain the monotonic dose response interpretation, which is important to integrative quantitative trait locus mapping. Model-based imputation makes use of multivariate Gaussian or copula models and can be used in conjunction with variance-stabilizing transformations but cannot withstand compositional constraints and skewness. Generative adversarial networks, variational autoencoders, and diffusion-based imputers are trained on nonlinear manifolds that model higher-order metabolite covariance that is predictive of underlying pathway stoichiometry and host microbe co-metabolism (Ghavi Hossein-Zadeh, 2026a). Multiview approaches utilize jointly obtained orthogonal data, for example, by ion mobility, multiionization, and paired transcriptomic and microbiome measurements, to imprint imputations on cross-domain associations, minimizing the risk of hallucination. The context of time is added, whereby with longitudinal sampling, smoothing of state space becomes possible, extensions of the Kalman filter to nonlinear emission functions consistent with feeding cycle oscillations, and circadian organization. To evaluate it properly, one would require masking experiments stratified by abundance and detection frequency, and not use random masking that has the disadvantage of excessively favoring optimistic bias in the evaluation of the missing not-at-random (MNAR) dynamics (Rastogi et al., 2025; Wang et al., 2025).

Normalization and scaling adjust systematic differences in intensity and try to retain true biological changes in the file during cohorts that vary with diet formulation, physiological stage, or management systems. Simple total ion current or probabilistic quotient techniques can be inadequate in responding to heteroscedastic variance as well as global drifts that can be caused by matrix composition. Varied mappings of variance stabilization are generalized log or asinh mappings, which interrelate with scaling options like Pareto, unit variance, or robust median absolute deviation scaling to control the impact of high leverage characteristics in machine learning models. The mitigation of batch effect entails the separation of analytical drift and genuine age-associated biological evolution; the classic empirical Bayes models presuppose that the exchangeability is not broken in longitudinal livestock experiments with structured production cycles. Subtler correction is achieved using adaptive models that use latent temporal drift predictors, spline control of intensity differences anchored on recurrent pooled references, or domain adversarial neural networks that not only encourage batch-invariant latent embeddings but also respect label predictive gradients. Transfer component analysis and optimal transport alignment also bring the distributional geometry of cross-platform further into harmony and allow joint modeling of metabolomic data produced on Orbitrap and time-of-flight platforms without removing pathway-level biological variance. The autoregressive models are used to predict the instrument drift, which allows scheduling recalibration proactively and updating a normalization parameter dynamically (Jeanselme et al., 2025).

Quantitative traceability is based on internal standards and samples of shared quality control, providing anchor points on which calibration, drift detection, and partitioning of variances are determined. Isotopically labeled internal standards, which are chemically similar to classes of target analytes, offer local response factor correction, but resource limits in livestock scale studies require selectivity with the help of coverage optimization algorithms that base marginal variance reduction per standard. Temporal performance trajectories are produced by having pooled quality controls inserted at periodic intervals; AI-assisted drift detection uses multivariate control charts that are improved with change point detection algorithms based on Bayesian online updates or kernel cumulative sum statistics to indicate excursion of stochastic variability beyond the expected variability. Latent factor decomposition of QC series isolates instrument-specific trends not correlated with random noise, which feeds adaptive normalization layers, raising and lowering sample intensities in near-real time. Suspicious activity on internal standard intensity ratios marks the onset of anomalies in spray stability, column back press, or ion source contamination, which facilitates timely action that maintains longitudinal integrity, which is essential in predicting transitions in health (Aigensberger et al., 2025).

The issue of redundant entities due to isotopic variants and adducts, multimer complexes, and in-source fragments is solved by feature curation, which inflates the dimensionality of the naive inclusion, increases multiple testing penalties, and skews enrichment of the pathway. Graph-based clustering builds networks whose m/z differences are weighted nodes, which are matched by theoretical adduct spacings, retention time coelution similarity, and correlation between the sample cohort and the nodes. The nodes are identified by detecting communities, which are collapsed into representative nodes based on intensity stability or identification confidence. Probabilistic graphical models also decorrelate in source fragmentation; ions of a fragment are linked to precursor candidates with conditional dependency graphs that are trained on curated training sets with collision-induced fragmentation behavior. Redundancy pruning also handles near collinear features due to the splitting of peaks or poorly sparse deconvolution, with sparse regression and variance inflation diagnostics being used to make decisions. The generated curated set of features captures a more orthogonal basis and is consistent with biological pathways, as well as makes them more readable and less risky to overfit genomic association or predictive health models (Bernabé-Díaz et al., 2025).

The spectral library matching, in silico fragmentation prediction, retention prediction, and mobility prediction are merged into a pipeline known as annotation pipelines, with the final termination being the ontology mapping and resulting in tiered confidence assignments. Dietary phytochemical diversity, microbially transformed xenobiotics, and species-specific lipidome nuances not well represented in generalist libraries are among the livestock-specific challenges. To predict fragmentation in livestock molecules, transfer learning models predicting fragmentation in large chemical spaces are fine-tuned on small, high-quality tandem spectra to boost performance on undercharacterized subclasses. Predicting retention times using graph neural networks captures substructural interactions with the chromatographic phases. Meanwhile, collision cross-section prediction improves isomer discrimination. Bayesian evidence integration models are mass accuracy-based, isotope pattern fit-based, similarity-based fragmentation, deviation-based retention, and predicted based on physicochemical property congruence-derived independent likelihoods that are integrated downstream to produce posterior identification probabilities, which serve as uncertainty weights. Ontology mapping is used to match semantically defined hierarchical biochemical pathway vocabularies and controlled agricultural trait ontologies, annotated metabolites, allowing meta-analysis across studies as well as semantic querying. Confidence tiering separates unequivocal identifications to reference standards and putative class-level identifications to avoid conflation, which can give false causal interpretations when used in integrative multi-omics network construction (Moozhippurath and Natarajan, 2025).

The software engineering discipline is added to metabolomic analytics pipelines through data provenance, data versioning, and/or reproducible methodologies, such that the data that has been obtained after an analysis can be traced through various methodological correctness. The workflow languages are used to coordinate the individual tasks, that is, conversion, peak detection, alignment, normalization, and annotation, and these are internalized under containerized environments, and they map the dependency stacks of tools by pipelining and versioning of the tools to mitigate drift that can be attributed to library updates. These parameters, input checksums, random seeds, and computation environment fingerprints are stored in provenance metadata records that enable us to correct and recalculate results, and also conduct a differential analysis in case we want to change our methodology. Versioned data objects store serial transformation of raw files of vendors into processed feature matrices to allow auditing of information loss or change at each step and comparative regulatory or stakeholder viewing, where metabolomic results are consumed to make breeding or health intervention decisions. Interestingly, the validation tests are automated and embedded with continuous integration pipelines to measure the performance metrics, such as the highest recall, accuracy at alignment, or annotation coverage of the standardized reference datasets, and report regressions that were introduced by the changes in the code (Cheimarios, 2025).

The compliant metadata schema and ontology alignment that comply with FAIR (Findable, Accessible, Interoperable, Reusable) improve the fine-tuning of interoperability, discoverability, and reusability, which are the preconditions of federated analyses that merge geographically and genetically diverse cohorts of livestock. Minimal information standards assist in ensuring that the capture of parameters of the acquisition is uniform, and activities of sample handling and biological contextual parameters like species, breed, physiological stage, diet formulation, health status, and environmental exposures are ensured. The alignment of anatomical, developmental, nutritional, disease, and management ontology resources with other resources can be used to provide a semantic connection among these resources to allow integrative queries to be made between metabolomic patterns and genomic variants or phenotypic endpoints. Alternative identifiers, which include sample identifiers, instrument identifiers, analysis run identifiers, and sample metabolic identifiers, each make it easier to reference samples, instruments, and analytical runs in publications, in model training databanks, and in collaborative consortia. Machine-readable formats can automatically be absorbed into knowledge graphs, and AI-based reasoning can be used to draw relationships in the latent space or detect metadata inconsistencies. FAIR principles are used to govern access and licensing and establish unambiguous rights to use, which do not impose any limits on secondary analysis, but rather ethical and proprietary restrictions that dominate in commercial livestock production. Aligned metadata, which is comprehensive and ontology-based, can also enhance the explainability of AI models. This is achieved by providing contextual covariates that can be explicitly incorporated, thereby reducing the risk of latent confounders influencing predictive performance (Barret et al., 2025; Takahashi et al., 2025).

7.5 AI methodological landscape in livestock metabolomics

Livestock metabolomics-AI relationship is developed within a stratified methodology spectrum where classical machine learning will not be supplanted by deep and generative ones. Regularized regression models like ridge, lasso, elastic net, group sparse, and sparse partial least squares still provide intuitive variable selection in high-dimensional low-sample regimes of controlled nutritional or health challenge investigations. With these models, inductive bias determined by metabolic network modularity, but reducing the influence of spurious correlations due to technical covariation, is imposed in a manner that is sensitive to the encoding of structured penalties that are sensitive to biochemical groupings or pathway annotations. Kernel-based models, especially the support vector machines and kernel ridge regressors, map nonlinear correlations between metabolite concentrations and phenotypic endpoints onto the reconstruction of the kernel Hilbert space, which offers convenient general-purpose discrimination with no explicit feature construction but performs poorly with very strong multicollinearity and strong tail-risky noise, except with strong loss functions. Random forests, gradient-boosted decision trees, and extremely randomized trees are examples of ensemble methods that provide resistance to patterns of heterogeneous feature relevance, and when metadata, genomic markers, and metabolite intensities are pooled together under controlled scaling, they can accommodate mixed data modalities. They provide intrinsic contributions to the variable importance, but these need permutation-based debiasing to counteract inflation of correlated predictors, which are prevalent in lipid classes or amino acid pathways. Gaussian process regressions present probabilistic estimates of functions with calibrated error, but computational scaling is an issue because, as the size of a study grows via longitudinal follow-up, sparse inducing point approximations are made or structured interpolation of the kernel. When cross-validating in the livestock setting, there is a need to ensure that there is no information leakage between genetically related individuals or repeat measures, and this is achieved by taking into consideration the hierarchical splits, which ensure the soundness of the generalization argument. Hybrid pipelines are also based on classical methods, whereby their interpretable output structures, skeleton feature selection, or pseudo labels are produced, which are used to initiate semi-supervised refinement by data-greedy deep architectures. In this way, instead of being overshadowed, the traditional algorithms give a statistically principled baseline and a calibration point that other complex models are judged based on in terms of incremental gains (de Groot et al., 2025; Lee and Chen, 2025).

Deep learning methods have transformed spectral and chromatographic pattern recognition, in which hierarchical abstractions are learned directly in minimally processed tensors of mass spectral profiles, ion mobility-aligned feature maps, NMR-free induction decays, or combined multimodal inputs of retention time, mass-to-charge, and collision cross-section axes. Convolutional neural networks can use local translational invariance and feature reuse on shapes of peaks with isotopes and fragmentation motifs and identify both sharp and broad shoulders on elutions using dilated and multiscale kernels. By m/z matrix retention time convolution or spatial imaging mass spectrometry cube convolution, two-dimensional convolutions are used to impose continuity within contexts, which are useful in the deconvolution of overlapping or coeluting analytes. Transformers add self-attention, which is flexible in modeling long-range dependencies between sequences of entire chromatographic or spectral token streams, and allow the network to concentrate on clusters of informative ions while suppressing noise-dominated regions. Positional encoding schemes with continuous retention time or spectral order retain ordering information when there is no repetition. The hybrid convolutional neural network (CNN) transformer architectures exhibit localized feature extraction with global relational analysis to improve identification precision in complex matrices like rumen fluid, where localized signals of microbes and hosts pose challenges to an isolated pattern detector. A common challenge in livestock-specific metabolite annotation is data scarcity, which is solved by pretraining on large cross-species metabolomics corpora with subsequent fine-tuning using domain adaptive layer normalization, which adjusts internal statistics to species- and matrix-specific distributions. Stochastic masking of spectral segments, mixup between chromatographic traces, and intensity jitter are applied regularly to mimic realistic acquisition variations and to enhance resistance to instrument drift as well as sample preparation effects. The issue of interpretability also still exists; saliency mapping, integrated gradients, and attention weight visualization can help in making the decision of a model based on spectral areas, but attribution stability requires the use of ensembles to minimize the spuriousness of saliency. Deep models are also combined with calibration layers that convert raw outputs to well-calibrated probabilistic estimates that are needed when deploying risk-aware health surveillance in highly variable field conditions (Li et al., 2025; Yu, 2025).

The connective tissue of representation learning allows one to generalize, integrate multimodally, and reduce dimensionality without losing salient biochemical variance. Autoencoders effectively reduce high-dimensional matrices of metabolites into low-dimensional latent embeddings that reveal the leading coregulation axes, usually conserved cellular processes, including amino acid metabolism, lipid remodeling, or a single carbon flux. Variational autoencoders go further by imposing a probabilistic latent prior, which provides smooth manifolds for interpolation and generative sampling, in addition to quantifying uncertainty in occupancy of latent space to generate more factorized latent dimensions, which are more reminiscent of interpretable metabolic modules. Density estimators based on diffusion models and score-based generative processes can generate synthetic spectra under covariate conditioning, such as species, diet category, or physiological stage, and increase the representation of the minority classes in imbalanced classification problems such as early disease detection. Contrastive learning builds upon instance discrimination, temporal continuity, or cross-modality agreement goals to build embeddings in which samples that are biologically similar cluster in the face of batch noise. As an example, joint contrastive goals that match metabolomic profiles to comeasured transcriptomic or microbiome compositions identify latent axes that represent host microbe metabolic coadaptation. The techniques of masked signal modeling that involve the prediction of withheld spectral intensities promote the contextual encoding of structural dependencies similar to language modeling techniques in natural language processing applied to continuous analytical signals. Metric learning optimizes latent spaces to make distances similar to phenotypic gradients, and thus, with the help of metric learning, k-nearest neighbor retrieval of analog metabolic states across breeds or production systems is obtained. Embeddings generated with these paradigms can be used in downstream applications such as genomic association, where they work as aggregate quantitative traits; causal mediation analysis, where they are used as metabolic latent variables to predict phenotypes given genotypes; and anomaly detectors, where they forecast deviations to learned healthy distributions on a calibrated scale (Esser et al., 2025; Sheshanarayana and You, 2025).

Graph-based modeling builds on the built-in networked form of metabolism, which views metabolites, reactions, enzymes, genes, and phenotypic properties as nodes in a directed or undirected graph where biochemical, regulatory, or empirical associations link nodes into a single graph. Message passing in graph neural networks is operationalized in which the metabolite node embeddings are updated in an iterative process by learned transformations of neighboring node states as well as edge features and combines pathway topology and quantitative concentration data. The attention mechanisms in graph transformers are weights on the contribution of particular neighbors to the reaction, which learns context-dependent flux channeling, which can vary between physiological conditions such as lactation or growth. Integrated gene variants, annotated with genomic features as auxiliary nodes, can be used to propagate genetic perturbation signals across subnetworks of a metabolism to predict any downstream concentration changes and enable genotype-to-metabolite prediction and pathway-based genomic selection indices. Hypergraph formulations are generalizations of pairwise edges that represent multisubstrate multiproduct reaction stoichiometry in individual hyperedges and have principles of conservation that are more true to simple pairwise approximations. The addition of stoichiometric coefficients as edge attributes and indicators of thermodynamic feasibility brings about constraints that regularize the passing of messages in situations that are energetically implausible with respect to the inferences. Knowledge graphs that combine ontologies of anatomy, diet components, microbial taxa, and management interventions with metabolic entities give a wider semantic context; embedding algorithms, including translational distance models or tensor factorization, can add cross-domain relational semantics to metabolite representations, thus allowing the undocumented pathway links or diet metabolite interaction hypotheses. Causal pathway scoring combines observational changes in concentration based on network topology with diffusion kernels or network propagation and removes degree bias, which allows prioritization of points of intervention with the greatest distributed effect and minimum risk of off-target perturbation (Fan et al., 2025; Moozhippurath and Natarajan, 2025).

Multitasking and transfer learning solve fragmentation of data across species, breeds, biological matrices, and experimental designs by making use of shared structure and allowing domain specialization. In practice, joint models are trained to predict many phenotypes, including feed efficiency, susceptibility to disease, milk composition, etc., jointly with shared metabolomic embeddings and task-specific heads with task-specific sample sizes, where task-specific sample sizes are small; in this case, sharing overcomes overfitting. Weight factorization can be used to break down the weights in species-invariant cores and species-specific adapters that allow the parameterization to adapt to any underrepresented breed by only fine-tuning the weights, not retraining the entire model. Training in the domain adversarial objective introduces predictive target-invariant, batch, or matrix domain label, cross-matrix generalization to the plasma, milk, urine, and rumen fluid without sensitivity to biological variation. Meta-learning paradigms acquire an initialization state, which efficiently adapts to new farms or dietary regimes in a few gradient steps in response to the constant appearance of new management practices. Cross-species transfer makes use of a homology in metabolic pathways and the alignments between orthologous genes of an enzyme node anchor shared graph structures, and negative transfer can be avoided by careful calibration in the presence of divergence in a species-specific metabolic branching (such as rumen microbial fermentation products). To keep privacy and proprietary competitive interests intact, federated learning architectures coordinate the joint training of models on geographically dispersed farms or laboratories and aggregation algorithms that are resistant to nonindependent and identically distributed data distributions of clients; for example, FedProx or adaptive federated optimization damp out client drift where client environmental and genetic backgrounds differ systematically (Deng et al., 2025; Hirst et al., 2025).

Causal inference systems are equipped with predictive models to go beyond correlation patterns to mechanistic knowledge and to support intervention design. Structural causal models represent hypothesized directed acyclic graphs linking genomic variants to enzyme expression, metabolic intermediates, and phenotypic outcomes, while accounting for possible confounders such as diet composition or housing conditions. These graphs inform adjustment sets to estimate the effects in an unbiased manner and inform the selection of missing covariates, which will be targeted to be acquired when their omission would lead to backdoor bias. Although genetic variants are used as instrumental factors to establish directional causality between metabolites and traits under conditions of validity assessable through stress testing in heterogeneity metrics and pleiotropy robust estimators. Invariant risk minimization attempts representations in which there are predictive correlations across environment strata, which implicitly select out spurious correlations associated with management or batch artifacts, which can be used to ensure performance does not fail when models are applied to new production systems. Counterfactual modeling can simulate the change of metabolite trajectory under hypothetical dietary intervention or gene editing conditions by doing calculus interventions on structural equations combined with learned conditional distributions, allowing strategies that are predicted to change the predicted metabolic risk markers while adhering to constraints of feasibility. Causal discovery algorithms that use either conditional independence tests or gradient-based continuous optimization can find candidate edges on partially known metabolic regulatory networks, but their accuracy depends on a sufficient sample diversity and noise modeling; using prior pathway knowledge as structural regularizers increases accuracy. The uncertainty in the estimation of the causal effects, which are cascaded to the economic or welfare analysis of the decision-making, informs the risk-sensitive management policies instead of triggering binary interventions (Jiang et al., 2025).

The AI landscape is filled with probabilistic and Bayesian processes in order to measure uncertainty, combine multilevel hierarchical structures, and scale risk-dependent decisions. Hierarchical Bayesian models divide the variance into individual, breed, herd, temporal, and analytical elements, allowing condensing of sparse estimates of the stratum toward global priors and a more detailed interpretation of the sources of variability. Bayesian neural networks are weight distributions, which represent epistemic uncertainty, as it diminishes with new data, and likelihood expressions with terms representing heteroscedasticity, which represent aleatoric variability due to measurement error or biological stochasticity. Variational inference is faster in the estimation of the approximate posterior, but calibration procedures through simulation-based diagnostics or posterior predictive checks are still needed. Composite kernel Gaussian process models identify both additive and interaction effects of metabolites and environmental covariates and provide closed-form predictive distributions that provide early warning thresholds with explicit false alarm probability control. Predictive distributions are then superimposed with a set of utility functions based on economic, welfare, and antimicrobial stewardship to optimize intervention policies based on studies of the expected value of information that quantifies the marginal benefit of obtaining further metabolomic measurements before action. Bayesian model averaging overcomes structural model uncertainty in cases where there is a series of competing pathway-augmented architectures that have equal predictive adequacy and prevents overconfidence. Transmission of preprocessing uncertainty, for example, variance in retention time alignment or imputation posterior, by hierarchy models avoids coherent calibration and precludes spurious inflation of subsequent genomic association P-values or breeding value estimation (Hassan, 2025; Pasupuleti, 2025).

The methods of generative augmentation include the synthesis of plausible metabolomic data used to address class imbalance, increase robustness, and explore the limits of model sensitivity. Physics-informed simulators combine deterministic chromatographic retention models, adduct formation probabilities, and fragmentation trees with stochastic noise injection calibration to instrument-specific empirical distributions to generate synthetic training corpora that enhance rare metabolite class exposure to deep classifiers. Variational and diffusion generative models generate synthetic spectral fragments through conditionalization with latent descriptors, for example, carbon chain length, saturation degree, or pathway membership, and thus can be used to generate structural diversity on command with no manual acquisition of standards. The use of adversarial generation of chromatographic profiles that are subject to mass conservation and peak shape priors avoids unrealistic artifacts that may confuse learning systems. Conditional generation based on environmental covariates, such as ambient temperature or diet macro-composition, models forecasted metabolic changes under sampled conditions, which is used to provide information on the strength of domain extrapolation of the model. Nevertheless, evaluation procedures should be able to differentiate between fidelity (statistical resemblance to actual data) and utility (enhancement of downstream task execution) and avoid overfitting, in which generative models (probably unintentionally) reproduce private farm-specific patterns at the cost of confidentiality. Postprocessing of generated samples may be done using plausibility constraints based on stoichiometric flux balance analysis or thermodynamic feasibility filters to filter away those that violate the mass balance or redox neutrality and so maintain biochemical realism (Matyushin et al., 2024).

Active learning, adaptive sampling, and optimization of experimental design add feedback loops where the uncertainty of the model and information gain are expected to be used to control incremental data acquisition to maximize the inferential value per unit resource in cost-constrained livestock studies. The query strategies are used to sample candidate individuals into specific metabolomic profiling to maximize anticipated reduction in the prediction error or uncertainty in the predictions of causal effects, a cost-benefit between exploration with underrepresented genetic backgrounds and exploitation of high-leverage metabolic states that reflect new health changes. Bayesian optimization models optimize the parameters of nutritional interventions by repeatedly making suggestions on diet formulations and their predicted metabolic effect that traverse multiobjective tradeoffs between growth efficiency, proxies of methane emissions, and immune competence indicators. Longitudinal adaptive scheduling uses algorithms to adjust the sampling frequency dynamically in response to the real-time sensor-provided signals of variance or drift, increasing the intensity of measurement about the suspected transition points, such as the onset of subclinical disease, and reducing resources used when the measurement is in stable states. Multiarm bandit models distribute analysis ability across rival platforms and targeted/untargeted assays, depending on changing marginal information yield, such that costly high-resolution acquisitions are concentrated where they reduce the most decision-critical uncertainties. Power analysis is also incorporated in design optimization within hierarchical structures of variances that can be used to determine the sample sizes and the ratios of allocation between breeds or management groups to obtain prespecified credible interval widths to determine whether the estimated genetic effects of a metabolite are credible (Uribe-Guerra et al., 2024; Shobana, 2025).

Neuromorphic adaptations, edge, and on-device adaptations will take AI metabolomic analytics out of centralized laboratories and distributed farm settings, which pose low-latency inference, require resilience to occasional connectivity, and have energy-efficient operation. Pruning, low-rank factorization, weight clustering, and quantization are model compression techniques that minimize computational footprints of convolutional and transformer neural networks implemented on portable spectrometers or biosensor hubs, and knowledge distillation transfers representational capacity in large teacher networks onto lean student models with little performance loss. The pipelines of streaming inferences take incremental inputs of electronic nose arrays or wearable metabolite sensors and apply sliding window normalization and concept drift detectors, which require model recalibration when the distributional differences continue beyond adaptive confidence thresholds. Federated edge training also allows localized model refinement, which includes farm-specific environmental signatures, and communicates encrypted gradient updates at the central point, avoiding privacy loss and reducing the raw data bandwidth requirements. Neuromorphic computational paradigms based on spiking neural networks convert continuous sensor data into discrete time spike trains to achieve event-driven energy savings and provide asynchronous correspondence of streams of multiple sensors; surrogate gradient training schemes and temporal coding schemes train on the properties of the metabolic-related signal sparsity and burstiness. Model stability is testable, such as adversarial perturbation testing, which is used to simulate surges of sensor noise or spectrum truncation before being deployed to the field. Layers of security will be used to encrypt parameters of the models and provide attestation to deter interventions likely to alter automated decision support systems to influence health alerts or dietary modifications (Kanwar, 2025).

The changing methodological situation, therefore, is aimed at an integrative ecosystem in which classical statistical rigor, high hierarchical abstraction, probabilistic calibration, causal reasoning, and resource-aware deployment come together to open the translational potential of livestock metabolomics. These paradigms have to work synergistically: representation learning provides compact representations fed to causal graph models; generative reinforcement provides diversity that stabilizes transfer between breeds; Bayesian uncertainty quantification makes active learning the priority; edge-optimized architectures implement insights in real-time management processes. Issues such as the reduction of bias caused by unequal breed representation, interpretability and ethical transparency of stakeholders, data sovereignty in federated systems, or causal assertions in confusing confounds remain a problem. Dealing with these will involve standardization harmonized with flexibility, interdisciplinary codesign that grounds algorithmic innovation on domain constraints, and continual evaluation frameworks that not only measure predictive accuracy but also calibration, fairness, robustness, and economic utility. By converging in this way, AI approaches will gradually advance metabolomic data out of descriptive biochemical snapshots into dynamic and mechanistically based intelligent adaptation to sustainable productivity, farm animal welfare improvement, and robust adaptation to environmental and market pressures throughout the global livestock industry. AI techniques commonly used in livestock metabolomics are shown in Table 7.2.

7.6 Multi-omics and multimodal integration

Multi-omics and multimodal assembly of livestock metabolomics are aimed at lifting siloed layers of analysis and building coherent representations of mechanistic bases that cut across molecular variation and physiological performance, environmental exposure, and management intervention. This integrative project faces heterogeneity of scales of measurement, asynchronous sampling times, nonaligning noise patterns, and multifaceted hierarchically stratified populations (species, breeds, lineages, and production systems). AI techniques provide mathematical abstractions that can be uniformly instantiated, shared latent variables that can be embedded, graph embeddings, regularization, probabilistic data fusion, and differentiable simulators to bring these differences into harmony while maintaining lineage particular delicacy and pathway-level interpretability. The general goal is to transform sporadic evidence into predictive and causal inferences, which can boost selection accuracy, hone nutritional and health interventions, and anticipate system management in the volatile climatic and market conditions. The key to success lies in principled management of uncertainty spread, confounder control, disentangling of modality-specific batch effects, and ethical management of sensitive streams of the farm-derived data (Adam et al., 2025).

The genomic prediction through the use of metabolomics coupled with genomics is based on the idea that the intermediate biochemical phenotype acts as an intermediary that breaks down polygenic effects on complex traits to a certain extent. Multikernel and deep kernel models are systems that combine genomic relationship matrices with metabolomic similarity matrices, which enable the separation of additive genetic variance transmitted via metabolite-mediated pathways and that which would otherwise be in direct genetic effects, as not measured in measured metabolites. Sparse structural equation and Bayesian mediation models are used to quantify the indirect effects flowing through metabolite nodes, which can give mechanistic attribution that can be used in targeted selection strategies to focus on metabolic efficiency, resilience, or a smaller environmental footprint. Joint genomic metabolomic best linear unbiased prediction enhances predictive accuracy in small to medium training groups by the borrowing of strength between correlated metabolite phenotypes, but prevents redundancy by shrinkage priors that punish collinear sets of metabolites. Deep multimodal models embed the single-nucleotide polymorphism embeddings and tensors of metabolite intensities into an overlapping latent space, and the adversarial terms reduce breed or batch stratification. Transfer learning sprays latent genomic mappings in related breeds by preserving conserved pathway-aligned subspaces and recalibrating breed-specific modulators, which prevents negative transfer in case divergent allele frequency spectrums overlap with diet-specific metabolic baselines. The estimation of reliability encodes the posterior variance inflation on low-heritability metabolites to avoid the propagation of genomic values with high confidence. Metabolomic endophenotype incorporation controls selection indices to early life predictive benefits and reduce generation times without compromising strength due to explicit modeling of environmental-by-genotype interactions that are expressed in metabolic plasticity across changing feeding regimes (Montesinos-López et al., 2025).

Combination with proteomics, transcriptomics, and epigenomics further vertically extends the molecular inference axis that indicates regulatory cascades controlling metabolic flux. Hierarchical latent factor models consider epigenomic marks as upstream regulators of transcript abundance, which in turn regulate protein expression and enzymatic activity that affect the concentrations of metabolites. Structured sparsity is used to provide identifiability by imposing biologically

TABLE 7.2 AI techniques commonly used in livestock metabolomics.

AI technique	Typical applications in livestock metabolomics	Key advantages	Limitations or challenges	Examples or use cases
Support vector machine (SVM)	• Classification of metabolic profiles (e.g., healthy vs. diseased) • Biomarker discovery	• Robust to high-dimensional data • Effective in binary classification • Well-established and interpretable	• May not scale well to very large datasets • Requires hyperparameter tuning (kernel, regularization) • Limited in modeling highly nonlinear relationships	Used in metabolomics to discriminate metabolic states; SVM has been applied to livestock metabolomic data-sets for disease or condition stratification
Random forest and gradient boosting (XGBoost)	• Feature selection and ranking of discriminative metabolites • Regression models linking metabolite levels to traits (e.g., feed efficiency)	• Handles many features • Provides feature importance scores • Robust to overfitting when tuned properly	• Can be less interpretable when many trees are used • Requires careful hyperparameter tuning • Training time may be long for large datasets	Widely used in metabolomics for biomarker discovery and trait prediction, XGBoost often helps identify key metabolic markers
Principal component analysis (PCA) and partial least-squares discriminant analysis (PLS-DA)	• Dimensionality reduction • Grouping or classification of samples based on metabolic phenotypes	• Reduces data complexity • PCA unsupervised (exploration), PLS-DA supervised (discrimination)	• PCA may miss subtle but biologically meaningful features • PLS-DA can overfit without cross-validation • Interpretability depends on the variance explained	Common in metabolomics workflows; PCA helps explore global metabolic variation, and PLS-DA is used to classify groups (e.g., high vs. low production)
Neural networks and deep learning	• Learning from raw spectral (MS/NMR) data • Predicting metabolite identity or concentration • Reducing dimensionality via autoencoders	• Models complex, nonlinear relationships • Feature learning minimizes the need for manual engineering • Scales well with large datasets	• Data-hungry (needs large training sets) • Risk of overfitting • Often "black box," making biological interpretation difficult	Deep learning has been used to interpret LC-MS spectral data in metabolomics, reducing batch effects and improving signal reconstruction
Clustering and unsupervised learning	• Grouping animals or samples by metabolic phenotype • Detecting subpopulations and outliers	• No need for labeled data • Reveals hidden structure • Useful for hypothesis generation	• Cluster numbers can be arbitrary • May combine biologically distinct groups • Sensitive to data scaling	Used in metabolomics to explore sample stratification (e.g., by diet, breed, health status)
Explainable AI (XAI) (SHAP, LIME, and attention)	• Interpreting complex ML/DL model predictions • Linking metabolites to biological phenotypes transparently	• Highlights which metabolites drive model decisions • Builds trust with domain experts • Facilitates biological hypothesis generation	• Interpretation can be approximate • Additional computational cost • Might reduce predictive power if simplicity is prioritized	Emerging in metabolomics, XAI helps to explain which metabolites are most strongly associated with traits in predictive models

Reinforcement learning (RL)	• Optimization of experimental or sampling strategies (e.g., adaptive sampling timepoints) • Dynamic decision-making for metabolic interventions or feeding strategies based on metabolite feedback	• Learns optimal policies through trial-and-error • Adaptable to changing environments or livestock states • Useful where feedback loops exist	• Requires definition of a reward function • Sample inefficiency may need many iterations • Hard to validate in live animal studies	Although direct RL applications in livestock metabolomics are still emerging, RL has been used in genomic selection for resource allocation
Generative models and GANs	• Synthetic metabolomic data augmentation • Generating realistic metabolite spectra • Imputation of missing metabolite profiles	• Enhances dataset size via synthetic examples • Helps overcome data scarcity or class imbalance • Can model complex distributions of metabolite data	• Risk of mode collapse (generating limited diversity) • Training instability • Generated data may not fully reflect true biological variation	Used in metabolomics to augment data and correct batch effects; e.g., GANs have been used to simulate metabolomic spectra and remove interbatch variation
Variational autoencoders (VAEs)	• Learning latent representations of metabolomic profiles • Data denoising and dimensionality reduction • Generating new metabolite patterns	• Captures underlying structure in data • Probabilistic, supporting the generation of new samples • Useful for interpolation in latent space	• Latent space may not always map to biological meaning • Requires careful tuning of VAE architecture • Generated samples may be unrealistic if poorly trained	VAEs are increasingly used in metabolomics to embed high-dimensional metabolite data into a lower-dimensional latent space for downstream modeling
Transformer and large-language models (LLMs)	• Metabolite annotation from MS/MS spectra • Predicting retention times or chemical properties • Multimodal integration (e.g., metabolomics + genomics)	• Captures long-range dependencies in spectral or sequence data • Transfer learning potential • Scalable and flexible across tasks	• Requires large pretraining datasets • High computational demand • Risk of overfitting to a narrow domain unless fine-tuned	Transformer-based models have been applied to metabolite annotation from LC-MS spectra, and LLMs have been adapted to multimodal metabolomics tasks
Graph neural networks (GNNs)	• Modeling metabolite relationships (e.g., metabolic networks, pathway graphs) • Predicting metabolite stability, interactions, or function	• Captures relational and topological structure • Effective for integrating pathway information • Supports transfer across species or conditions	• Needs reliable graph structure (metabolite networks) • Scalability can be challenging • Interpretability of learned embeddings may be nontrivial	GNNs have been used in metabolomics to predict metabolic stability and network relationships; e.g., MetaboGNN uses graph learning for cross-species metabolic stability prediction

reasonable directional sparsity structures with the flexibility to allow feedback representation through residual covariance terms. Deep joint embedding models use modality-specific encoders with common bottlenecks that are trained on cross-reconstruction losses, and contrastive alignment is used to produce modality-agnostic latent coordinates that represent coordinated pathway activation states. Variational formulations represent latent spaces as composed of shared and private components, which contain modality-specific information like posttranslational modification or chromatin remodeling signatures but contain convergent metabolic control signals. Graph-constrained autoencoders are an extension of graph-constrained autoencoders, which consider adjacency between enzymes and metabolites, as well as regulatory edges between transcription factors and target genes, and which can pass messages that inject previous network structure, hence imposing less sample complexity requirements. Multiobjective optimization provides a compromise between reconstruction fidelity, pathway sparsity, and phenotype predictiveness, eliminating latent collapse, which prevents mechanistic interpretability. Temporal stratification introduces dynamic Bayesian networks and neural ordinary differential equations to capture lagged regulatory effects. For example, this could be used to model the process of epigenomic remodelling that precedes sustained metabolic shifts, with regularization achieved through kinetic priors on reaction turnover or transcriptional decay rates. Cross-platform batch harmonization combines mutual nearest neighbor alignment and domain adversarial loss that is used hierarchically to ensure the amplification of latent confounders across cascaded molecular layers is avoided (Fu et al., 2025; Habib et al., 2025).

Host microbiome–metabolome fusion is a fusion of metagenomic functional potential, metatranscriptomic expression, microbial metabolite production, and systemic metabolic regulation of the host. Functional inference transforms microbial genomes' catalogs of genes into enzymatic orthologies and computational reaction repertoires and forms bipartite networks between microbial functions and host-accessible malpractices. Joint probabilistic models divide observed metabolite variation into components of host genomic, host transcriptional, microbial composition, and microbial functional expression and measure the competitive versus synergistic effects on metabolic phenotypes. Graph neural networks pass through multikingdom networks, allowing cross-domain message exchange, which reveals fermentation pathways, cross-feeding interactions, and modulation of host immune regulation of microbial metabolism. Variational multiview embeddings reconstruct microbe taxonomic abundance, functional genes, and metabolite intensities along common latent ecological metabolic axes, which improve prediction of health or efficiency outcomes over independent modalities. Causal mediation models are used to test the hypotheses that microbial functional modules mediate the relationship between diet composition and host metabolite profile on the basis of instrumental variables that are based on feed formulation variance and adjustment of shared environment confounding. The models of generative cross-domain translation are developed: the models predict shifts in the metabolism of microorganisms in response to hypothetical changes in dietary fiber, based on which targeted prebiotic interventions are developed and designed to promote favorable metabolic remodeling. Uncertainty calibration provides a correction to the missing coverage of microbial reference genomes and variability of several strain levels, and thus amplifies ambiguity of annotations to downstream decision levels to avoid excessive functional assignments being given to tenuous and uncertain assignments (Dell'Olio et al., 2025; Kumar et al., 2026).

The situations of metabolic states are placed in behavioral, physiological, and environmental contexts by the fusion of phenomic and sensor data, which is gathered by continuous monitoring infrastructures. The wearable accelerometry, feeding station records, thermal-humidity indices, and welfare measures all provide high-frequency and irregularly sampled time series whose interaction with the episodic metabolomic snapshots requires temporally synchronous strategies. Representation learning uses temporal convolutional networks and recurrent variational encoders, generating latent dynamic embeddings of behavioral and environmental regimes that are close to sampling windows and concatenates them with metabolite latent vectors in predicting traits. Multiresolution architectures support disparate sampling rates by building pyramidal time feature hierarchies in time feature metabolomic draw timestamps. Attention mechanisms adaptively redistribute sensor modalities of the weight to the extent that they have a contemporaneously predictive role in the metabolism deviations, including paying attention to thermal data during heat load situations and deeming less predictive streams of activity by low variance. Fusion assists in early anomaly detection with sensor latent deviations, causing anticipatory samples to be metabolomic sampling requests under active learning policies, enhancing the temporal localization of the pathophysiological onset disease. Causal disentanglement isolates management-induced shifts in the feeding pattern modes of the organism and an intrinsic metabolic dysregulation through conditioning on learned representations of feeding microstructure. Federated sensor fusion with privacy-preserving aggregation does not store raw behavioral information in a central place, but passes encrypted gradient images that use metabolomic supervisory messages without violating the confidentiality of the farms (Ahmed and Gummeson, 2025).

Knowledge graph construction provides a converging semantic framework that connects entities of genes, transcripts, proteins, metabolites, reactions, pathways, microbial taxa, dietary constituents, environmental exposures, phenotypes, and interventions. Biochemical transformations, regulatory effects, substrate–product interactions, anatomical localization, functional hierarchy, diet ingredient compositional properties, and causal statements named in literature and domain experience are represented using ontology-aligned node types and relation predicates. Graph embedding algorithms generate a representation of the graph as a set of vectors that include multirelational context; translational distance, tensor factorization, and graph neural encoders propagate semantic and quantitative edge features and link prediction to generate an assumption of a missing pathway association or hypothesis of diet metabolite interaction. Symbolic structure is then based on empirical data by integrating quantitative edge weights that capture the strength of reactions, or association strength, and temporal edge tags that allow dynamic calculations to be made concerning condition-dependent activation of a pathway. Explainable AI is based on queryable knowledge graphs, which offer path-based explanations between predictive metabolites and the upstream genetic or environmental determinants, as well as contribute to trust and interpretability in decision support outputs. In downstream model training, constraint injection is used to regularize latent embeddings, and biologically implausible proximity is discouraged; coherent groupings of metabolically coregulated groups are encouraged by using graph structure (Zhang et al., 2025).

Digital twins' architectures combine mechanistic metabolic modeling and data-driven inference to generate personalized virtual states of livestock metabolisms able to perform simulations of scenarios and data optimization by optimizing interventions. The key elements consist of a structural backbone with stoichiometric reaction networks, compartmental nutrient flow models, and dynamic energy partition equations calibrated to species and physiological stage with machine-learning surrogate models that are used to approximate complex nonlinear kinetics or unmodeled outputs of fermentation by microbes. Data assimilation combines real-time or periodic metabolomic, sensor, and performance measurements with time-varying updates made to latent state estimates by ensemble Kalman filtering, particle filtering, or variational Bayesian smoothing of forecasting model predictions and observations with quantification of posterior state uncertainty. The parameter learning method uses hierarchical priors that allow population-level sharing and individual-specific adaptation, which allows twin personalization with sparse individual metabolomic time points to avoid overfitting. Scenario simulation is used to run virtual diet programs, management changes, or selective breeding manipulations by perturbing model inputs or structural parameters exogenously and running the resultant metabolic responses to obtain probabilistic outcome distributions that may be used to guide risk-conscious decision-making. Differentiable programming brings together mechanistic differential equation solvers and gradient-based optimization, which allows end-to-end training against observational loss functions and incorporates physical conservation constraints, which enhances extrapolation fidelity to novel intervention regimes. The feedback control loops use twin prediction to suggest an adaptive feeding or health monitoring schedule, and create a cyber-physical management loop that the twin gets more and more accurate as more data is acquired through a process of targeted data collection based on the information the prediction predicts (Métayer et al., 2025; Su et al., 2025).

The issue of mobile, microbial, physiological, and behavioral processes developing on different time constants and sampling cadences can be overcome by cross-scale temporal correspondence and dynamic data assimilation models. Multiscale state space models separately model the latent dynamics as fast metabolite turnover dynamics, intermediate microbial community adaptation dynamics, and slow genomic or epigenomic modulation dynamics, with the parameters of the transition operators being inferred together with cross-level coupling terms. Temporal alignment uses the dynamic time warping and monotonic spline warping of probabilistic kernels to time-warped asynchronous or discontinuous sample timestamps into a shared latent time coordinate frame without causal warping. Latent variable models with continuous time parameters based on neural stochastic differential equations can deal with unevenly spaced samples and allow interpolating missing metabolic states to produce dense sensor trajectories, which makes it easy to closely attribute rapid metabolic behaviors to momentary changes in the environment. The predictive uncertainty is taken into consideration in data assimilation cycles to organize adaptive sampling events when forecast divergence is out of threshold bounds, and the effective allocation of resources across cohorts is the objective. The mechanisms of missingness are clearly defined by the observation mechanisms that differentiate structural absence and latent state-driven censoring to maintain inferential validity to estimate the dynamic causal effects. Cross-scale coupling is also imposed by energy balance and mass conservation conditions that couple high-frequency feed intake-obtained nutrient inflow of high-frequency feeds to lower frequency integrated metabolite pool and performance indicators, creating cohesive multi-temporal consistency (Garcia-Oliva et al., 2025).

7.7 Application domains and translational pipelines

The field of AI-enhanced livestock metabolomics has applications spanning the whole gamut of early physiological perturbation signaling to strategic genetic enhancement and operational orchestration, and its advancement requires stringently designed translational pipelines that transform unproven analytical clues into validated and actionable information. These pipelines cover standardized sample acquisition, multi-platform pre-processing harmonization, metabolite identification with controlled confidence metrics, feature engineering or representation learning, multi-objective model training, internal and external validation under realistic population structures, assessment of assay feasibility for routine deployment, economic and risk utility appraisal, regulatory and ethical governance review, integration into farm management information systems, and post-deployment performance surveillance with continual learning loops. The enablers of cross-cutting are strong data provenance tracking, uncertainty quantification frameworks, which consolidate analytical as well as biological variance into downstream recommendations, interpretable attribution mechanisms that tie model outputs to biochemical mechanisms, privacy-preserving federated or edge architecture, and user-centered interface design, which integrates domain ontologies and context-sensitive decision narratives into interface design. All application domains are not necessarily only as successful based on the quality of predictive accuracy but also on a long chain of reliability, adaptation to environmental and genetic heterogeneity, and compatibility with sustainability, welfare, and stewardship requirements (Tedeschi et al., 2025).

Subclinical surveillance of metabolic disorders and prevention of disease through its early detection make use of the sensitivity of metabolite fluxes to the underlying pathological changes that can be seen before clinical symptoms manifest themselves. Longitudinal deviation modeling, which divides the individual metabolic streams into residual streams in relation to customized baselines learned through hierarchical Bayesian or autoencoder-induced latent reference states, is the main focus of modeling longitudinal deviation feature extraction pipelines. The anomaly detection algorithms combine probabilistic predictions of the state space or neural differential equation models with graph contextualization, where metabolites are embedded into pathway topologies, which allows prioritizing the deviations in accordance with the coherent biochemical units instead of stochastic anomalies. Multitask paradigms alternate across a range of different syndromic targets, sharing common pathophysiological axes, for example, oxidative stress, inflammation, or energy balance perturbation, thereby enhancing the strength of detection in sparse label conditions where no conclusive diagnostic validation can be made. The active learning strategies selectively examine more confirmatory tests on borderline metabolic risk scores to improve calibration of classifiers at optimal diagnostic resource consumption. Translational readiness requires the analytical stability evaluation of the batches, instruments, and sample storage conditions, and domain adaptation layers partially eliminate drift. Ethical deployment is based on the trade-off between high sensitivity of early warning and heavy load of false alarms that may cause unneeded interventions and combines the cost-sensitive utility functions with an operator feedback loop that retunes thresholds between the changing tolerance level and veterinary supervision capacity of the farm (Csorba et al., 2025; Roy et al., 2025).

The mathematical modeling of the balance between diet composition, digestive dynamics, microbial fermentation, absorptive efficiency, and systemic metabolic partitioning is the key to nutritional optimization and precision feeding based on metabolic fingerprinting. Based on metabolite profiles linked to amino acid balance, lipid mobilization, short-chain volatile products, nitrogen dynamics, and micronutrient status, AI systems can use the results to generate recommendations on adaptive ration adjustment. The agents of reinforcement learning, in conjunction with mechanistic energy and protein flow limits, model the alternative feed formulations and forecast metabolic reactions in the face of uncertainty, revising policies based on persistent integration of longitudinal metabolomic and performance feedback. The latent factor models separate genetic predisposition and ambient environmental modifiers with diet-induced variation and allow individual feeding programs to achieve metabolic homeostasis and pursue efficiency or product quality goals. Generative augmentation designs probable scenarios of metabolite responses in untested combinations of feed ingredients, which increases exploration space at a safe cost. The pipelines used to implement cloud-based optimization engines are combined with their edge-based quick assays or infrared surrogate sensors for calibration using core metabolomic markers, allowing near real-time ration adjustments. The transparent generation of rationale explains how certain metabolite changes suggest conditional nutrient deficits or excesses to justify the trust of nutritionists and help them to be adopted. Multiseason resilience is included in evaluation frameworks to determine whether optimized strategies maintain stability during climatic stress events or constraints in formulating supply chains (Abd El-Hack et al., 2025; Catussi et al., 2024).

The metabolic biomarkers used to construct predictive intermediates to determine feed efficiency and mitigate methane emission in the rumen are involved in the digestive fermentation patterns, hydrogen utilization, volatile fatty acid ratios, lipid oxidative status, and the recycling of nitrogen efficiency. Multimodal fusion integrates blood/milk metabolite phenotypes with ruminal/fecal microbial functional encodings that construct integrated phenomena that govern host microbial coadaptation of the energetic losses and greenhouse gas precursors. Multioutput regression assigns predictive variance to efficiency and emission endpoints, finding metabolite constellations that indicate the potential of efficient joint optimization or may signal potential trade-offs that need management priority. Causal mediation analysis identifies biomarkers as mechanistic mediators and incidental correlates, which helps to select targets of breeding or dietary modulation. Components of the digital twin model project downstream metabolomic variation and predicted emission reduction in response to proposed nutritional intervention or additive approaches by simulating flux modifications. Deployable biomarker panel selection focuses on analytical tractability and temporal stability, where surrogate sensor estimation is used in high-throughput estimation. Lifecycle economic analysis combines the effects of feed costs, emission intensity incentives, and potential productivity changes to determine the net benefit of biomarker-directed interventions and inculcates sustainability metrics together with profitability goals (Nunes et al., 2024).

Reproductive performance, fertility, and periparturient metabolic resilience applications are aimed at predicting energy balance problems, inflammatory cascades, and endocrine adjustments that determine conception success, embryonic maturation, and postpartum recovery. Contextualization of metabolic inflection points in the temporal format of metabolomic data with continuous sensor arrays (activity, temperature, and intake) and sequence modeling of lag structures where previous alterations in lipid mobilization or amino acid turnover were precedents of future reproductive behavior are the results of temporal alignment of those data sets. The joint model of fertility measures and health resilience endpoints through multitask learning uncovers common metabolic predictors of both phenotypes. Graph-based pathway enrichment determines coordinated steroidogenesis, one-carbon metabolism, and oxidative stress networks with their composite perturbation indices offering more stable prognostics than individual marks. Counterfactual simulation examines the possible advantages of timed nutritional supplementation or management intervention against the estimated openings of metabolic vulnerability to help provide proactive interventions instead of reactive ones. Deployment combines interpretive dashboards indicating the expected risk paths with operational leverages, and unpredictability ranges encourage risk-targeted monitoring intensity instead of decision-making. There is a genetic evaluation fairness that takes care to adjust parity and environmental confounding to incorporate metabolically mediated attributes of female longevity and reproductive persistency in integration into breeding decisions (Raj et al., 2025; Zhao et al., 2025).

The assessment of welfare and stress physiology with the help of metabolomics supplemented with the data on behavior and environment aims at quantifying the cumulative and acute stress loads, including exposure to heat, transport, disruption of social hierarchy, and handling practices. Representation learning combines stress-responsive patterns of metabolite groups associated with osmotic balance, oxidative damage mitigation, neuromodulator turnover, and immune activation with sensor-derived signals of behavioral agitation, thermoregulatory strain, or social interaction patterns. The purpose of using invariant risk minimization plans is to separate stress-causal signatures of confounding effects of diurnal or feeding routine variations and enhance overall facility generalization. Temporal anomaly detection differentiates both transient adaptive response and chronic maladaptation pathways with temporal windows of intervention. Recommendation engines are environmental change proposals, stocking density change proposals, or change in management practice ranked by the forecasted improvement of welfare as compared to resource expenditure and operations limits. Explainable models can be used to trace elevated risk scores to clear biochemical behavioral stories, which increases the level of stakeholder trust and compliance with welfare certification models. Monitoring methods also include fairness auditing to avert minority breed or subgroup circumstances of being systematically mischaracterized as a result of being underrepresented in training corpora (Buchha, 2025).

The optimization of product quality and safety incorporates metabolomic proxies of compositional properties, balance of flavor precursors, oxidative stability, resistance to microbial spoilage, and risk of contamination in the quality prediction and control of processes. Multistage models extrapolate preharvest metabolic conditions affected by genetics, diet, and environmental manipulation on downstream processing phenotypes and shelf-life patterns to bridge temporal and operational discontinuities using latent state persistence modeling. Generative modeling predicts postharvest metabolic evolution under new storage or processing conditions, which facilitates analysis of intervention scenarios in order to prolong retention of quality. Embedded reasoning knowledge graphs are computational networks connecting metabolite changes with sensory descriptors and physicochemical features, which are used to guide specific changes in feeding or processing parameters. Active learning gives high emphasis to further high-resolution analytical batch tests on batches whose quality prediction is ambiguous, and classifier boundaries are refined. Safety surveillance uses anomaly detection to ensure that indicative metabolic patterns of contamination or the beginning of spoilage are detected before standard endpoints to combine uncertainty calibration to reduce false alarm events that may interrupt operations when not required. Economic appraisal systems theorize incremental revenue or waste reduction attributed to any predictive intervention and serve as the foundation of return on investment discourses vital to industrial acceptance (Kanicheril Ambikalekshmi et al., 2025).

Antimicrobial stewardship is the use of metabolic signatures of infection status, host immune activation, pathogen metabolic interference, and therapeutic response dynamics in making wise choices of treatment and monitoring efficacy. Pretreatment stratification system models predict this probability of bacterial versus nonbacterial etiologies with multiomics fusion of metabolomics, proteomic acute phase reactants, and limited genomic markers to minimize unnecessary antimicrobial utilization. Monitoring treatment follows chronological metabolic normalization curves, and when deviation is observed, it may indicate failure of treatment or the development of resistance that may necessitate reassessment of the regimen. Mediated by drug effects, causes of changes in the metabolic patterns within the disease resolution confounders are isolated through causal inference to avoid misattribution that could lead to bias in the stewardship policies. Evidence-based guidelines are typically written in decision support modules, which combine model outputs with withdrawal period rules and risk of residue to produce actionable and compliant recommendations. Federated learning in veterinary networks enables the ecological validity of predictive models to be improved by making sense of sensitive health records, thereby making them less centralized. Stewardship performance indicators include antimicrobial usage intensity adjustment based on the severity of the case mix and based on baseline metabolic stratification to allow fair benchmarking of the herd or region (Kumar et al., 2025; Liu et al., 2025).

To optimize lifecycle impact measurement and mitigate objectives, environmental footprint attribution uses metabolomic proxies for the potential to excrete nitrogen and phosphorus, the presence of carbon- and methane-producing fermentation routes, and oxidative stress inefficiencies. Multioutput modeling predicts the intensity of emissions and the propensity of nutrient loss of inaccessible metabolite panels. It incorporates covariates of the environment, such as temperature and humidity, that affect metabolic allocation. Counterfactual studies compare how dietary reformulations, management change, or genetic rejection scenarios would affect projecting environmental indicators mediated by expected metabolic adaptation. Combination with remote sensing or facility area monitoring provides geographical information, which makes sustainability dashboards georeferenced. Propagation of uncertainty guarantees that the communication of footprint estimates conveys accurate intervals that are credible, taking into consideration metabolite assay variance, model structural uncertainty, and environmental stochasticity, and risk-wise regulatory reporting becomes possible. The correspondence to the sustainability certification and incentive schemes presupposes the disclosure of methodology, the ability to audit the provenance of data, and compliance with the standardized impact calculation protocols (Galmacci et al., 2025).

By incorporating into selection indices and breeding value optimization folklore, metabolomics increases its scope beyond mere phenotyping to a strategic genetic advantage by offering endophenotypes capable of being statistically more heritable or phenotypically measurable at an earlier stage of life. Multitrait genomic prediction models combine metabolite latent factors that are weighted economically, and the genetic correlation structure takes care of the fact that unintended correlated responses may occur otherwise. Structural equation modeling disaggregates the effects of genomes into pathways, which are mediated by a specific metabolic module, from effects that are not mediated by a metabolic module, narrowing the focus of precision selection on biologically relevant dimensions, including feed efficiency or resilience mediated by a metabolic pathway. Longitudinal metabolomic profiling provides prediction of lifetime performance at an early stage of life, which assists in minimizing generation periods and dynamic reranking as longitudinal data is collected. Fairness audits look at the performance of breed-specific predictions to prevent entrenchment of differences in the representation of the differences in the reference populations. Resource-efficient implementation investigates pooling designs or very sparse metabolite panels chosen by Bayesian experimental design, maximizing the expected incremental genetic gain per assay cost (Guo et al., 2025).

Upstream modeling is refined into operationally available systems to make precision intervention recommendations and decision support dashboards that coordinate data flow, inference, explanation, and tracking of actions in daily management processes. Some of the architectural layers are data ingestion pipelines with schema validation and anomaly filtering, feature stores with harmonized metabolomic and contextual variables, model orchestration services with ensemble inference with real-time uncertainty calibration, knowledge graph-enhanced explanation engines with pathway-grounded narratives, and user interfaces with prioritized alerts, predictive trajectories, and intervention simulations. Multiobjective optimization is recommended as advisory policies that strike a balance between productivity, welfare, environmental, and economic standards that have an overriding mechanism of audit of human decision errors to help in training feedback. Constant learning cycles reap results associated with implemented suggestions and update model parameters based on drift detection, retraining plans that avoid early overfitting to temporary perturbations. Governance modules introduce role-based access control, encryption, differential privacy on shared aggregates, and audit trails that are immutable ledgers recording the lineage of data and the version of the model used and its use to enable compliance and traceability. Human factors engineering guarantees the minimum of the cognitive load, the clarity of the actionability, and the compatibility with the farm management software ecosystems that already exist, whereas the participatory design with the producers and veterinarians will promote the feeling of trust and relevance (Murri, 2025).

All these areas of application emphasize that any successful translation of AI in livestock metabolomics would be dependent on the holistic fusion of methodological proficiency with high levels of validation, ethical responsibility, economic viability, and user focus and design. Recurring co-creation between scientists, technologists, producers, and

TABLE 7.3 Functional applications of AI in livestock metabolomics.

Functional application	AI role or function	Benefits for livestock metabolomics	Example or evidence in livestock or related fields
Biomarker discovery for health and disease	Pattern recognition, feature selection, and classification	• Enables identification of metabolites predictive of disease (e.g., mastitis, subclinical infections) • Supports noninvasive diagnostics via biofluid metabolite signatures	In dairy cows, ML (e.g., random forest) was used on fecal metabolites to detect subclinical mastitis, pinpointing key metabolic biomarkers
Prediction of production traits	Regression modeling and multi-omics integration	• Links metabolite profiles to economically important traits • Helps predict and select animals with optimal metabolic phenotypes	In cattle, metabolomics combined with AI has been reviewed for its potential to elucidate metabolic networks associated with traits like milk yield or residual feed intake
Metabolic pathway and network reconstruction	Clustering, network inference, and graph-based AI	• Reveals pathway perturbations underlying phenotypic states • Helps understand how metabolism reorganizes under stress, disease, or nutritional change	AI-based network modeling can infer pathway-level changes in metabolomic data, enabling deeper mechanistic insights
Early disease detection and health monitoring	Classification algorithms, anomaly detection	• Detects metabolome shifts before clinical symptoms appear • Supports proactive health management in herds	Application of AI to bovine biofluids via metabolomics may offer early-warning biomarkers for disease states. For instance, metabolomic profiling for mastitis has identified discriminative metabolites
Integration with multi-omics for systems biology	Machine learning, ensemble models, and feature fusion	• Combines metabolite data with genomics, proteomics, etc., for holistic prediction • Improves accuracy and interpretability of metabolic-trait models	In cattle, integration of metabolomics with other omics using AI has been proposed to understand the genetic architecture underlying economic traits
Quality control and sample classification	Unsupervised clustering and anomaly detection	• Classifies samples by metabolic phenotype (e.g., healthy vs. stressed) • Detects outliers, batch effects, or degraded samples via their metabolic fingerprint	AI-driven clustering can flag abnormal or low-quality metabolomic samples in livestock studies, improving downstream data reliability

policymakers, backed by open standards of interoperability of data and models, will have long-term innovations and ensure reliability. The ever-growing convergence of mechanistic understanding, probabilistic algorithms, and adaptable deployment structures would gradually change the metabolomic data into a strategic resource that will trigger healthier, more efficient, sustainable, and welfare-comparative livestock manufacturing frameworks. Functional applications of AI in livestock metabolomics are presented in Table 7.3.

7.8 Validation, benchmarking, and performance governance

The apparent spine that can reveal functionality improvement on whether AI integration in livestock metabolomics progresses beyond exploratory potential to a reliable working infrastructure includes validation, benchmarking, and performance governance. In the absence of rigorous assessment systems, perceived increases in predictive power are prone to the inclusion of overfitting to temporary cohort idiosyncrasy, latent confounding, or analysis artifacts in metabolite-identification pipelines. The multi-dimensional nature of livestock systems, which encompasses breed diversity, heterogeneous diets, climatic gradients, microbial ecologies, management practices and evolving market and regulatory pressures, makes them susceptible to dataset shift and undermines naive extrapolation. A paradigm of governance thus needs to incorporate the quantitative metrics selection, bio-interpretability, characterizing uncertainty, drift observation, engineering reproducibility, managing annotation confidence, and resourceful lifecycle control into a consistent, traceable fabric between statistical performance and physiological plausibility, economically valuable, ethically responsible, and goals toward long-term sustainability (Ghavi Hossein-Zadeh, 2025; Ghavi Hossein-Zadeh, 2026b; Rosati, 2025).

Standard performance measures give sufficient but inadequate pointers to model adequacy; their choice and interpretation should be aligned with biological meaning and decision context. The traditional measures of discrimination to index classification, continuous measures of errors in regression, and ranking measures of genomic prediction are complemented with the measures of calibration, stability, and fairness measures that are stratified by breeds, parity classes, production system, and environmental strata to reveal hidden performance inequities. The monetary value of the false interventions, welfare changes of the missed pathology, and the environmental cost of inefficient feed efficiency are combined in a multiobjective formulation and converted into expected utility surfaces, which are of use to the stakeholders. Hierarchical evaluation further breaks down variance explained at metabolite, pathway, and systems trait levels into more biochemical-based components and provides accountability where pathway-level predictive degradation correlates with a general loss of accuracy. Temporal consistency measures are used to measure change in predictive residual structure in successive production cycles, whereas net reclassification improvements are used to measure the incremental value added by new metabolomic panels beyond current genomic or phenomic baselines. The robustness measures over perturbed inputs and metabolite intensities or simulated analytical noise strengths to test resilience to variation in the laboratory, and for example, the sparseness or interpretability indicators assess correspondence to an underlying metabolic circuit and discourage the use of unsupported spectral artifacts. Combined, these layered views turn the metric dashboards into nontransparent numeric scoreboards to physiologically contextualized governance tools (Van Calster et al., 2025; Whittle et al., 2025).

Claims of generalizability are operationalized by external validation between independent cohorts, between time horizons, and across different geographic regions and are used to curb the inadvertent information leakage that is inherent to the associated pedigree structures or batch-linked sampling frames. Forward chaining evaluation is used to maintain chronological order, such that the training only has access to historically available data, replicating a realistic deployment where new metabolic responses to new feed additives or climatic extremes were not available during training. The geographic transfer experiment involves different agroecological zones and regimes of management interaction, showing the interaction between preexisting metabolic equilibria and environmental stressors that can be demanding domain-specific recalibration of environment–genotype interaction terms or addition of environment–genotype interaction terms. Domain adaptation diagnostics investigate analogous changes in latent representations and test whether embedding alignment techniques indeed bring about distributional matching or are just indicative of distributional cover-up. Trained out-of-distribution detection models, which forecast higher uncertainty on metabolic profiles that in training densities have small likelihood, can serve as a safety valve in deployment when incident disease syndromes, ingredient substitutions, or breeding innovations perturb the metabolic landscape beyond the history of training density estimators. Record of cohort provenance, inclusion and exclusion criteria, and sample handling comparability form the basis of interpretability of external performance differences and inform specific data augmentation actions (Jingyang, 2025).

Calibration is used to convert the predicted probabilities or continuous risk scores to empirically faithful incidence expectations that form the basis of rational choice of threshold and resource allocation. Reliability diagrams and related quantitative summaries like expected calibration deviations reveal systematic overconfidence, which would inflate premature intervention rates, or underconfidence, which may delay action that is required. Brier decomposition is used to break down the uncertainty into reliability, resolution, and uncertainty, and determine whether model refinement or a change in the sampling strategy should be the priority of improvement. To perform continuous trait predictions, probabilistic regression with predictive interval coverage measurement and quantile loss measurement will ensure that the bands of uncertainty cover realized values at nominal frequencies across the range of metabolic states, not just around the mean. Under exchangeability, conformal prediction structures provide distribution-free coverage whereby metabolite panel-wise sets of predictions generate widths used to inform dynamic decision levels modulated by opportunity cost and risk tolerance. Cost-sensitive calibration uses asymmetric penalties to reflect different welfare or economic consequences. Online recalibration routines adjust probability mappings as new validation data becomes available, preventing gradual misalignment in the event of slow environmental change. Governance requires periodical calibration audits that are timed with season changes, diets, and changes in genetic replacement, which are known to change the prevalence of base risk (Qian et al., 2025).

Drift detection protects the performance fidelity of predictive models by detecting changes in the input distributions, feature relations, and conditional target mappings before the quality of decisions is materially affected by the predictive degradation. Covariate drift occurs when the metabolite intensity distributions change as a result of alterations in diet composition, feed contaminants, heat stress, or changes in the laboratory protocol; they are detected through statistical divergence measures and cumulative sum surveillance, leading to root cause diagnostic analysis. Latent embedding feature drift is also applicable to the topology of latent embeddings that may be distorted with the addition of new metabolite adduct patterns or cometabolites in microbes; indices of representation stability measure geometric distortion. Concept drift is caused by changes in the biological dependence between metabolic condition and target phenotype, such as genetic selection that changes physiological set points or emergent microbial adaptations that change the fermentation behavior; performance monitoring of sentinel-labeled subsets separates error growth due to covariate versus concept drift. Windowed reference updating is needed to eliminate false positives in anomaly detection modules when there are new metabolic changes characterized by baselines at transitional physiological windows. Adaptive retraining pipelines use drift magnitude thresholds, cooling phases to prevent oscillatory overcompensation, and safeguard tests that ensure updates to the pipelines maintain the previously verified calibration and fairness properties. The logs of transparent drifts are used to feed retrospective audits and support meta-analysis of environmental and management predictive fragility drivers (Greco et al., 2025; Yang et al., 2026).

Open challenges and benchmark datasets contribute to reproducibility, comparability, and methodological progress. Common grounds of fair algorithmic comparison can be achieved by curated repositories of multispecies, multibreed, multiseason metabolomic profiles with standardized raw data, processed feature matrices, metadata ontologies, and explicit data partition suggestions. Things that limit p-hacking and post hoc analytic drift include blind evaluation schemes in which there are sequestered test labels, cryptographic hash-based submission locking, and preregistration of analysis plans. Reproducible audits must have containerized computational environments, input data checksums, and parameter settings; declare workflow specifications; and have provenance records between every model artifact and the input data checksums and parameter settings. It can be more widely participated in by having synthetic data generators that are calibrated to maintain covariance structures but obscure the identities of the individual farms without affecting confidentiality. Inflation of the leaderboard is being calibrated by including negative controls and randomized baseline competitors, and fairness diagnostics built into benchmark scoring motivate equitable performance of players at both demographic and environmental levels. Error taxonomy is condensed as community challenge reports and shows future research systematic failure modes (Barnard, 2025).

The fact that no single method can identify metabolites with uniform precision is dealt with by annotation confidence propagation, which finds that leaving out the heterogeneity causes undesirable predictive overconfidence and biased feature attribution. The probabilistic identification scores of spectral library matching, retention time prediction, isotopic pattern concordance, and in silico fragmentation modeling can be modeled into hierarchical priors or feature-level weights, and this weighs down the influence of tenuously assigned signals on the update of model parameters. Bayesian models assume unknown or ambiguous metabolites are latent categorical mixtures, the posterior uncertainty of which adds to predictive intervals, whereas the discriminative models use confidence-weighted loss modulation, which deters reliance on low-credibility features. The downstream pathway enrichment and feature importance analyses need to incorporate identification uncertainty in order to avoid spurring inference of mechanistic relations pegged to entities falsely assigned. Through joint modeling of the technical variance, batch correction residuals, and annotation probability, variance decomposition can be done to isolate the biological signal and analytical ambiguity and therefore be used to guide reacquisition in target analysis or confirmatory analysis resource allocation. Reporting of identification confidence distributions behind transparent panels deployed increases interpretative integrity for end users (Vitale et al., 2025).

Model lifecycle management makes processes institutionalized to guarantee performance governance over time after the first deployment. Maturity is a managed process with development, shadow testing, controlled deployment, full deployment, monitoring, adaptation, and decommissioning, where each stage, including development, is controlled by established quantitative and qualitative thresholds. Data lineage tracking includes the source of every observation used by the model to update, transformation transformations, quality control indicators, and consent dexterity of each observation included in the model updates, and model lineage includes data of model architecture, hyperparameters, training data snapshot identifiers, random seeds, and model results in an immutable registry allowing rollback in scenario of performance regression or emergent bias. Champion challenger models permit candidacy of updated models alongside current systems under live traffic simulating conditions before promotion choices based on statistically powered comparative studies. Training and retraining policies are characterized by continuous learning depending on the drift warning, calibration decay, or target phenotype taxonomy expanse, balancing responsiveness and stability. The proposed updates are reviewed by ethical oversight committees with respect to being updated with welfare, environmental, and equity implications, incorporating the feedback of the stakeholders and regulatory changes. The systems of retirement store decontextualized models through institutional memory, which enables longitudinal studies of methodological change. Federation deployment supports cryptographic aggregation, differential privacy, and secure inference enclaves on lifecycle tooling to ensure that the goals of governance can be reached across data ecosystem deployment (Dong, 2025).

7.9 Ethical, legal, and societal dimensions

The issues of AI integration in livestock metabolomics are framed by the ethical, legal, and societal aspects of technology integration, determining the legitimacy, acceptability, and sustainability of technology use in the long term and how molecular understanding can be translated into production, welfare, environmental, and economic outcomes. The ecosystem of any metabolic data comprises the acquisition of biospecimen, high-resolution mass spectral/spectroscopic characterization, context metadata of genetics, management, and environment, and derived feature or latent representations. Data sovereignty is a concept that holds that the authorities of originating producers, cooperatives, and, in certain jurisdictions, regional or indigenous custodial bodies maintain determinative control over subsequent use, cross-border transfer, and commercialization methods of metabolomic resources to their associated herds. Consent architectures need to no longer be the same, but to become dynamic, tiered, and revocable permissions represented in machine-readable policy frameworks, to allow individual consent to categories of research, algorithmic training, benchmarking consortia, or proprietary product development. Models based on smart contract-mediated benefits sharing can redistribute value criteria created by predictive tools or biomarker intellectual property in the form of proportional royalties, discounts on services, or credits on knowledge based on the volumes of contributed data, diversity, or strategic rarity as an incentive to include underrepresented breeds and agroecological settings and prevent data monopolization that may entrench asymmetrical competitive advantages (Barbudo Carrasco, 2025).

The privacy issue is that, when combined with genomic markers, geospatial environmental cues, or management records, the metabolic fingerprints can be reidentified as an individual operation or genetic line, which can be used to understand performance or health liability sensitive information useful in market bargaining or insuring an individual. In spite of the fact that livestock data does not possess the human personal health status dimension, confidentiality intersects with the trade secrets, breeding program strategy, and regional biosecurity posture. Continuous high-dimensional metabolite matrices can be trained on using different mechanisms of differential privacy, collaborative model training with secure multiparty computation, and federated learning architectures that do not store the raw spectra in local trust domains but maintain analytic usefulness. Strong deidentification would involve the study of linkage attacks in which rare metabolic phenotypes related to either niche feeding regimens or climatic extremes have a triangulating effect on farm identity. The traceability infrastructures required by food safety have a risk of further endangering privacy when they are implemented without separation between minimal compliance data and exploratory research characteristics. The legal frameworks should specify what the acceptable aggregation granularity, retention times, and cross-jurisdictional transfer protocols are between the animal health surveillance imperatives and proprietary interests (Zafar et al., 2025).

The fairness of algorithms requires fairness of breeds, composite genetic lines, production system intensities, and resource levels so that structural inequalities are not reinforced, in which data-rich industrial operations get disproportionately high optimization benefits and smallholder or extensive systems get poor recommendations because of covariate shifts. Stratified error decomposition, calibration parity evaluation, and assessment of disparate false alert burdens that may influence different workloads of veterinary or economic strain are needed in fairness auditing. Where there are inherent structural differences in the metabolic variance of baseline statuses because of genetic diversity or forage-based diets, in fairness interventions, care should be taken not to obscure the legitimate adaptive heterogeneity. Methods involve reweighting to balance training representation, learning domain-invariant representation that maintains causal metabolic pathophysiology and neutralizes confounding through management intensity, and site-modulating layers that can do site-specific calibration without global performance erosion. Fairness, remedial choices, and positioning of residual performance gaps through transparency underpin informed adoption of the stakeholders and help make decisions through the repetition of collaborative dataset enrichment to fill deficits ethically (Chakradeo et al., 2025).

Questions of dual use and biosecurity arise since deep metabolic pathway modeling has, in theory, the power to explain intervention leverage points that can be abused to harness growth-promoting effects, to avoid residue monitoring, or to obscure the emergent disease signals via metabolic masking. Besides, high-resistance pathway reconstructions with microbial and host fluxes can guide adversarial engineering of pathogen or microbiome perturbations that can weaken production efficiency. These should be governed to defined levels of access between generalizable biomarker indices of welfare or sustainability and proprietary or sensitive mechanistic models with vectors of misuse. Threat modeling incorporated in risk assessments can categorize the models based on the severity and probability of abuse and be used to direct the security measures, code escrow, controlled dissemination rights, and anomaly awareness of suspicious query patterns on deployed decision support systems. Ethical review that is harmonized with dual-use human control should also be applied to animal agriculture omics, and it is essential to ingrain the culture of precaution without preventing actual innovation (Pannu et al., 2025).

Transparency and interpretability form the basis of social license since predictive recommendations are made intelligible to the producers, veterinarians, auditors, and policy bodies. The influential latent dimensions should be mapped onto biochemically annotated metabolite sets, pathway nodes, and physiological processes and articulate narrative descriptions that should be consistent with domain ontologies and not reduced to one marker. Communication systems must be multiple layers of abstraction: succinct actionable explanations on frontline decision-making, more mechanistic development on experts, and higher-level system trends contextualization to governance boards. Risk gradations should not be misinterpreted by articulating uncertainty using probabilistic intervals in combination with semantic qualifiers. Transparency also implies the release of training data scope, exclusion criteria, known limitations, and update schedules, allowing the stakeholders to understand the conditions in which the extrapolation risk is relevant. User feedback during design cycles makes the proposed product/service more salient to the explanation and its cognitive ergonomics more attuned to user needs, so as to minimize automation bias or the use of black-box model power (Garouani et al., 2025).

The standards development models, the certification pathways, and accountability mechanisms are governance models that are based largely on the institutionalization of ethics and legal compliance. Interoperable data schema, metabolite identification confidence levels, validation reporting templates, and fairness auditing checklists can be promulgated by standardization bodies and can reduce the costs of transactions, and expectations can be aligned. The certification processes can check compliance with quality management systems, records of calibration and drift monitoring, secure development practices, and indicators of stakeholder consultation, resulting in trust marks that can distinguish responsibly engineered platforms. Accountability structures establish the allocation of liability for the occurrence of adverse events that can be traced to model error, inappropriate deployment, or a failure in the provenance of data, as an incentive to invest in a strong contractual understanding and risk avoidance. External assurance provided by independent audit capabilities, such as reproducibility verification and bias stress testing, allows convergence between jurisdictions on a regulatory basis. Adaptive governance envisages emergent societal values, for example, a stronger focus on climate management or antimicrobial stewardship, in which dynamic updating of the standards is anticipated and sunset provisions over outmoded practices are made (Cordeiro et al., 2026).

Equity and capacity building of low-resource production situations deal with the risk that metabolomics-driven decision support increases a digital divide. Such obstacles include inadequate cold chain logistics to preserve sample integrity, a lack of high-resolution instrumentation, infrequent connectivity, and a lack of analytical knowledge. Ethical integration approaches focus on the construction of tiered assay architectures such that core minimal panels or sensor-derived surrogate indices will give approximate but actionable metabolic information that is calibrated to resource conditions, with cloud-mediated improvement in case infrastructure allows. Local analytical competency and interpretative literacy are encouraged by open educational resources, modular training curricula, and regional centers of excellence. Moisture-sharing is a specific legislative spread of a fraction of acquired commercial value to offset hardware purchase, laboratory strengthening, and extension assistance in underrepresented territories, which are smartly rewarding inclusive information donations to enhance the robustness of global models. Interpretation of interfaces locally, appreciation of linguistic differences, and accommodation of cultural practices of husbandry lead to trust and contextual inspiration. Capacity building outcome measures that go beyond the level of model accuracy, like empowerment indices, resilience adoption, and knowledge transfer durability, incorporate equity goals into the performance evaluation (Ahmed et al., 2025).

All these ethical, legal, and societal strands highlight the fact that technical sophistication is not enough to ensure positive integration of AI-enabled metabolomics in livestock systems. The normative scaffoldings of institutional commitments to sovereignty, privacy, fairness, biosecurity vigilance, interpretability, accountable governance, and equitable capacity development are already part of the normative scaffoldings that molecular precision augurs for shared prosperity, stewarded welfare, and environmental sustainability instead of exacerbating existing inequalities or even creating new vulnerabilities. This adaptive alignment will require unceasing communication between scientists, producers, ethicists, regulators, and civil society as the technological, market, and sociocultural environments change. Ethical and societal considerations for AI in livestock metabolomics are presented in Table 7.4. In addition, challenges in integrating AI with metabolomics data in livestock systems are listed in Table 7.5.

TABLE 7.4 Ethical and societal considerations for AI in livestock metabolomics.

Ethical or societal issue	Description or concern	Specific implications in livestock metabolomics	Nuances or tensions
Animal welfare and instrumentalization	The risk that animals are reduced to sources of metabolic data rather than being viewed as sentient beings	• High-frequency metabolic sampling or monitoring (via AI-driven decisions) may lead to stress, overhandling, or invasive procedures • AI-informed selection may prioritize metabolic efficiency at the expense of welfare traits	• Technology may improve health (by early detection), but also risk "overoptimizing" animals without regard for broader well-being • There is a tension between using metabolomic data to improve animal welfare vs. using it purely for production gains
Transparency and explainability of AI	AI models (especially complex ones) may function as "black boxes," making their metabolomics-based predictions hard to interpret	• Stakeholders (farmers, veterinarians) may not trust decisions about metabolic interventions or selection if they can't understand how the AI reached its conclusions • Lack of interpretability can hinder acceptance of metabolomics-driven recommendations	• Explainability tools (e.g., SHAP, LIME) help but may not fully resolve concerns if stakeholders still distrust "automated" decisions • There is a trade-off: more interpretable models may be less powerful; more predictive models may be less transparent
Data ownership, privacy, and sovereignty	Questions about who owns the rich metabolic data generated and how it is used	• Metabolomic profiles may be proprietary assets (for breeding companies) or sensitive (linked to farm performance) • Without clear governance, smaller farms may lose control or access to their own metabolic data	• Farmers may fear "data colonization" by large agritech firms • Balancing open science (data sharing) with protection of farm-level intellectual property is challenging
Bias, fairness, and representativeness	Risk that AI models trained on metabolomics data perpetuate inequities or ignore underrepresented animals	• Metabolomics datasets may be skewed toward high-input or commercial breeds, underrepresenting smallholder, heritage, or local breeds • Models may perform poorly for underrepresented populations, leading to biased or inequitable outcomes	• Broader datasets can mitigate bias but are costly; generating metabolomic data for rare breeds is often underfunded • Domain-adaptation (transfer learning) may help, but biological differences can challenge model transfer
Loss of farmer autonomy and/or dependency	AI systems could make metabolic-based decisions that limit farmers' control or push them toward specific commercial platforms	• Farmers may become dependent on proprietary metabolomics-AI platforms to make decisions about nutrition, health, or breeding • Subscription models or "lock-in" may undermine farmer sovereignty	• Potential power imbalance: agritech firms may own the metabolic-AI tool, data, and business model • Ethical governance must address who controls the AI and who benefits

Validation, reliability, and "do no harm"	Ensuring AI-metabolomics tools are safe, accurate, and empirically validated before widespread deployment	• Poorly validated models may lead to incorrect metabolic interventions, harming animals • Real-world farm conditions may differ from controlled research settings, so model performance must be robust	• Ethical frameworks (akin to a "Hippocratic oath") should guide AI tool development to minimize risk • There is a need for regulatory or audit frameworks to enforce quality assurance and continuous monitoring • Accountability for AI errors must be clearly defined across stakeholders
Democratization and accessibility	Ensuring that AI-metabolomics benefits are accessible broadly, not just to large or wealthy operations	• Small-scale or resource-poor farmers risk exclusion from advanced metabolic AI technologies • Inequitable access may exacerbate existing disparities in livestock production	• Technical solutions like open-source platforms and federated learning can help • Ethical and policy efforts are needed to ensure inclusive deployment and avoid concentration of power • Trust-building and stakeholder participation (farmers and ethicists) are crucial
Speciesist bias and ethical consideration of animals	AI systems could perpetuate or worsen discrimination against certain species or breeds	• Models trained on data from certain species or production systems may implicitly devalue other animals • Ethical frameworks might omit or downplay species-specific welfare considerations	• AI fairness frameworks rarely include nonhuman considerations; "speciesist bias" is an emerging issue • Addressing this requires expanding fairness definitions and including animal-centric ethics
Sustainability and environmental externalities	The potential environmental impacts of intensified metabolic-AI optimization	• AI-optimized metabolic traits (e.g., faster growth, feed efficiency) may lead to environmental trade-offs (e.g., land use, emissions) • Focusing on productivity can sideline considerations of ecological resilience or welfare	• Ethical development needs to embed sustainability metrics, not just productivity • Governance frameworks should align metabolomic AI innovation with long-term environmental goals

TABLE 7.5 Challenges in integrating AI with metabolomics data in livestock systems.

Challenge	Description	Impact on AI-metabolomics integration	Underlying causes or nuances	Mitigation or best-practice strategies
Data quality (missing values and noise)	Metabolomics datasets often contain missing peaks, dropouts, and substantial technical noise	• Poor or biased model training • Reduced predictive accuracy • Potential misinterpretation of biological signal	• Low-abundance metabolites fall below detection limits (MNAR) • Instrumental variability or sample prep issues contribute to noise • Nonrandom missingness (e.g., missing because below LOD) complicates imputation	• Use advanced imputation methods tailored to metabolomics, such as Bayesian models (e.g., BayesMetab) that explicitly model missingness type • Utilize multiview or variational autoencoder methods to jointly model missingness and latent structure • Apply rigorous QC filtering and noise-reduction workflows (e.g., proper peak detection, outlier removal)
Batch effects and data heterogeneity	Systematic differences across samples due to instrument runs, lab protocols, or temporal drift	• AI may learn technical artifacts rather than true biological variation • Poor cross-study or cross-farm generalization	• Variability in sample handling, instrument calibration, and chromatographic conditions • Differences in labs or acquisition platforms causing nonbiological variance	• Use harmonization tools that correct batch effects without aggressive imputation (or minimal data loss). Tools like HarmonizR (originally for proteomics) demonstrate these principles • Employ normalized QC-based methods (e.g., internal standards, pooled QC samples) to monitor and correct for drift • Design experiments with randomized batch allocation when planning metabolomics sampling
High dimensionality or curse of dimensionality	More metabolite features than samples: a common issue in high-throughput metabolomics	• Overfitting risk in AI models • Difficulty in identifying truly predictive metabolites	• Many correlated metabolite features • Limited sample sizes for livestock metabolomics studies • Redundancy and multicollinearity among metabolites	• Dimensionality reduction (PCA, autoencoders) before modeling • Feature selection techniques (e.g., regularization, importance ranking) • Use of cross-validation, ensemble models, and regularization to avoid overfitting

Interpretability of AI models	Complex AI models (e.g., deep learning) behave like "black boxes"	• Biological insight is harder to extract • Lower trust and adoption by biologists or breeders	• Nonlinear relationships learned by deep networks obscure direct feature–output mapping • Few domain-specific interpretability tools for metabolomic features	• Employ explainable AI (XAI) methods (e.g., SHAP and LIME) to attribute model decisions to specific metabolites • Use interpretable architectures (e.g., sparse models) or hybrid models (interpretable and predictive) • Validate important features with domain experts or experimental follow-up
Incomplete metabolite annotation and identification	Many signals in untargeted metabolomics remain unidentified or ambiguous	• AI models may base predictions on "features" without known biological meaning • Harder to translate findings into actionable biology	• Limited spectral libraries, especially for livestock metabolites • Ion fragmentation, isomerism, and retention time ambiguity	• Use probabilistic or Bayesian matching frameworks (e.g., Bayesian matching of features) to manage uncertainty in annotation • Leverage multi-omics integration (e.g., genomics and transcriptomics) to help infer functional identity of metabolite features • Encourage data sharing and expansion of metabolite reference databases for livestock-specific metabolites
	High computational demand for training and deploying AI on metabolomics datasets	• Slow model iteration or retraining • Difficulty in on-farm or decentralized deployment	• Resource-intensive models (deep nets) on large, high-dimensional data • Limited compute infrastructure in field or farm settings	• Use resource-efficient modeling (lighter architectures, quantization) • Use cloud or shared high-performance infrastructure for training • Implement edge-oriented inference where feasible, or predeploy models for on-site use
Lack of standardized protocols or benchmarking	Variability in sample prep, normalization, and AI model evaluation across labs	• Poor reproducibility • Difficulty comparing models or transferring AI systems between studies	• No universal standards for metabolomics AI workflows • Differences in preprocessing (scaling, imputation) and evaluation metrics	• Develop common benchmarking datasets and protocols in the livestock metabolomics community

Continued

TABLE 7.5 Challenges in integrating AI with metabolomics data in livestock systems.—cont'd

Challenge	Description	Impact on AI-metabolomics integration	Underlying causes or nuances	Mitigation or best-practice strategies
				• Adopt FAIR principles for data (metadata and preprocessing) • Use cross-validation and external validation across independent cohorts to assess generalizability
Biological complexity and multi-omics integration	Integrating metabolomics with other omics layers introduces additional complexity	• Models may not capture cross-omics interactions • Integration can amplify missing data and noise	• Different modalities have different scales, missingness, and measurement error • Temporal mismatch between metabolite dynamics and other omics layers	• Use fusion models (e.g., multimodal deep learning and graph-based integration) that explicitly model combined omics interactions • Align sampling designs to capture temporal correspondence • Regularize integration to account for modality-specific noise and missingness
Species or breed representation and bias	Skewed dataset composition toward a few major breeds or farm types	• AI models may not generalize to underrepresented breeds or environments • Equity concerns in benefit distribution	• Limited metabolomics data from smallholder farms or rare breeds • Bias in training data toward commercial or well-studied populations	• Build diverse, representative datasets, including multiple breeds and production systems • Apply domain adaptation or transfer learning to handle underrepresented groups • Engage stakeholders (farmers and breeders) in study design to ensure inclusivity
Regulatory and ethical concerns	Data ownership, privacy, and responsible AI deployment	• Reluctance to share data • Slow uptake of AI tools due to governance uncertainty	• Proprietary nature of metabolomics data from commercial operations • Lack of clear regulatory frameworks for AI in livestock omics	• Develop data governance frameworks respecting farm and breed ownership • Employ privacy-preserving AI methods (federated learning) to allow collaborative model development without data sharing • Incorporate ethical impact assessments in AI-metabolomics projects

7.10 Computational infrastructure and deployment strategies

The deployment strategies and computational infrastructure of AI in livestock metabolomics have to balance the heterogeneous data acquisition environment in the farms, the complexity in laboratory analysis, and high privacy and sovereignty demands, as well as the demands of timely and interpretable outputs that impact nutritional formulation, health surveillance, genetic selection, and sustainability indices. A paradigm has appeared of cloud-edge hybrid, which can reconcile such requirements. On the edges, preliminary signal conditioning, low-latency anomaly screening, and adaptive sampling of the data are done using embedded controllers in milking parlors, feeding, or environmental sensors to give priority to biologically informative specimens to be profiled with laboratory metabolomics. These prestige feature extraction, spectral peak detection quality measures, or compressed forms of near real-time metabolite biosensor signals, compress bandwidth needs without compromising predictive salience. This cloud layer then coordinates intensive alignment, deconvolution, and compound identification, multi-omics data integration with breeding values of genomes or microbiome profiles, and advanced inference of a model under elasticity provisions to support changing loads of batches. Decision hooks, which are latency sensitive, like early warning scores, can be calculated in a stratified cascade by only executing more computationally expensive ensemble or causal inference modules when preliminary screening thresholds are surpassed. This multilevel solution addresses network failures that are prevalent in rural environments and continues to keep the network running essential processes, but synchronizes the provenance-rich transaction records when reconnected. Cross-region replication and infrastructure as code templating give resilience against local outages and help with compliance with the data localization law that regulates cross-border agrobiological data flows (Mahato and Neethirajan, 2025; Yu et al., 2025).

Federated and privacy-preserving analytics go one step further to provide architectural sophistication in the form of collaborative model enhancement, not based on centralized pooling of raw metabolomic spectra or sensitive herd contextual metadata. Within this paradigm, the update of parameters based on local gradient computations on matrices of site-specific metabolite features and coupled phenotypes is relayed in secure aggregation protocols in which the coordinator has no access to individual contributions to the aggregate. Differential privacy applies tuned noises to updates or outputs to trade off the risk of reconstruction of redeeming population metabolic patterns(s) corresponding to rare feeding routines it produces or tailored genetics line marks proprietary genetic ancestries but can retain utility to infer large populations. Adaptive clipping methods regulate gradient sensitivity with the heavy-tailed distribution of changes in metabolite intensities of various environmental stress events. Hierarchical federated topology enables regional aggregators to coordinate within agroecological clusters before global aggregation, resulting in statistical heterogeneity and overhead communication. Attestation procedures that ensure policy integrity and short-lived retention may be used to control integration of secure enclaves to perform occasionally centralized tasks, including anchor metabolite calibration or pathway library updating. The policy engines are audited to represent the allowable model goals and prohibit unintentional loss of confidentiality via side channel testing in order to ensure that privacy assurances are not compromised by auxiliary logging or performance measurement procedures (Varsani, 2025).

Reproducible, scalable pipeline design based on workflow orchestration and containerization forms the heart of metabolomics, where preprocessing steps, spectral centroiding, retention time alignment, isotopologue collapsing, batch normalization, scaling, missing values imputation, identification scoring, pathway mapping, and feature selection comprise a directed acyclic dependency graph, which is sensitive to parameter changes. Declarative workflow specification languages store these transformations in explicit versioning, resource requirements, and artifact hashing; permit deterministic replay of the transformations; and support selective recomputation in the event of a change in upstream parameters. Container images bundle algorithmic dependencies such as specialized libraries of high-resolution mass spectral processing, machine learning frameworks, and graphics processing unit (GPU) drivers, reducing configuration drift between development, staging, and production. Horizontal scaling methods mix metabolomic matrix sharding based on sample or feature dimension with microbatching of incoming data in a streaming process and optimize the normalization procedures to capture global distributional context by periodically synchronizing summary statistics. The provenance metadata, consisting of checksums of raw files, parameter manifests, and environment fingerprints on each derived dataset and model checkpoint, is what maintains interpretability, audit readiness, and scientific rigor (Xiao et al., 2026).

The practice of machine learning operations (MLOps) that is specific to metabolomics goes beyond the generic machine learning processes and integrates biochemical domain validation, multi-omics consistency assurances, and regulatory alignment into feedback loops of continuous delivery. Monitoring is not only traditional latency-based, throughput-based, and resource utilization-based, but also metabolic coverage stability-based, identification drift-based, level of residual distribution shifts at the pathway level, and indicators of fairness between breeds or production systems. The warning levels include a statistical control chart, which distinguishes random analytical variation and significant biological change. The workflows of governance have compulsory peer review of the model update pull requests, automated running of the benchmark suites, recalibration diagnostics, checking the reproducibility, and scanning license compliance of the third-party spectral libraries. Policy-based deployment gates combine ethical and legal limits, which check that the consent scope has been satisfied in any dataset addition or cross-regional data movement suggested by retraining the model. Artifact registries store lineage information between models and raw data flashpoints, feature engineering code hash, hyperparameter search entries, and validation documents and allow rolling back to an earlier state in the event post-deployment validation shows that performance has deteriorated or a systematic bias has been realized. The continuous integration pipelines load synthetic stress environments that simulate the loss of any one of the following scenarios: loss of batch correction, changed ionization efficiency, or abnormal metabolite spike profile. To be robust, it is tested before the promotion (Eken et al., 2025).

Specialized hardware acceleration can increase the computational capability of deep architectures over high-dimensional metabolomic profiles, pathway structure graph neural networks, fragmentation spectra interpretation transformer-based sequence models, or variational generative models. The GPUs are flexible to dense linear algebra and parallelized training tasks, whereas Tensor Processing Units (TPUs) or other matrix specialization architectures can execute large batch self-supervised representation learning when the scale of available data is high. Field-Programmable Gate Array (FPGA) accelerator provides flexible pipelines in deterministic low-latency inference in edge modules that do on-site screening to allow on-chip execution of quantization-aware models to minimize memory and power footprints. Although this increases upfront design cost, Application-Specific Integrated Circuits (ASICs) could prove to be warranted in standardized repetitive processes such as high-throughput spectral similarity scoring in large reference libraries. The cost-benefit analysis is forced to internalize not only the raw performance per monetary unit, but also the energy use, cooling structures, lifecycle emissions, purchasing lead times, programmability, and staff skill needs, as well as the risk of vendor lock-in in regard to long-term flexibility. Hybrid acceleration techniques are also able to partition workloads based on arithmetic intensity and communication patterns, which are coordinated by runtime schedulers that reassign tasks dynamically to match lab processing peaks during diurnal usage cycles (Silvano et al., 2025).

The energy efficiency profiling comes out as both an economic and sustainability requirement with the increasing computational demands of the advanced models and with the promises of the agricultural sector to reduce emissions. Granular telemetry collection is a measure of energy per pipeline stage, model training epoch, and inference, which are normalized based on predictive utility metrics and improvements in biological coverage. Some optimization strategies involve mixed precision training, operator fusion, locality of memory, tuning a batch size under hardware usage constraints, and an adaptive early termination based on convergence predictive heuristics. Pruning, quantization, knowledge distillation into lightweight student architectures, and exploitation of sparsity are modeling compression techniques that minimize the inference energy requirements, especially where required by edge deployment. Algorithms that have been designed to schedule both energy-intensive tasks to align with renewable energy availability or reduced carbon intensity in the grids, and carbon-aware orchestration can delay nonurgent retraining where marginal emissions are high. Centric refinement of data to improve signal-to-noise ratio through better preprocessing or feature curation brings accuracy improvements to scale smaller architectures to performance levels. Clearly reported energy intensity in conjunction with traditional metrics will create accountability and the procurement strategy oriented toward hardware with positive performance per watt trends (Ghoneim et al., 2025).

7.11 Emerging frontiers and research gaps

New horizons in AI-facilitated livestock metabolomics are characterized by a synthesis of sensory novelty, integrative multiscale modeling, and governance-conscious data ecosystems that attempt to go beyond descriptive profiling of now toward predictive, mechanistic, and intervention-based capabilities. Real-time metabolic sensing, ultra-high throughput, and real-time metabolic sensing are pivotal directions that no longer require episodic and laboratory batch analysis but continuous or quasicontinuous streams of data originating on miniaturized spectrometric, electrochemical, optical, or nanomechanical platforms integrated into production systems. The AI models will have to adapt to include streaming, variably sampled, heteroskedastic signals, make on-device drift correction, adaptive denoising, and context-sensitive feature extraction that conserves subtle temporal microdynamics, manifesting changes to incipient physiological transition. The literature still has research gaps in calibration harmonization across the distributed low-cost sensors with varying temperature, humidity, and biofouling pressure and in the development of uncertainty quantification to differentiate sensor failure from maladaptive metabolic perturbation without the need to resort to frequent destructive validation. It requires multiresolution temporal modeling architectures capable of effectively telescoping between subminute fluctuations and longitudinal seasonal patterns to avoid losing information or computational inefficiency, and self-calibration ensemble methods may be able to exploit redundancy to effectively impute missing intervals. Ethical aspects intertwine with the feeling of ubiquity, which requires principled approaches to proportionality, data redundancy, and informed consent on the part of the stakeholders, even in nonhuman surveillance sectors, so that the technosurveillance paradigm does not sideline trust or distort husbandry motivations (Khalid et al., 2025; Maulucci, 2025).

Fluxomics in production animals with the addition of stable isotope tracing with AI is expected to bring metabolomics beyond concentration-based snapshots to dynamic reconstructions of pathway throughput, allocation, and metabolic plasticity in response to nutritional, developmental, or other environmental stress. Infusion of the tracer with heavy isotopes or incorporation in the diet will provide a more complex distribution of mass isotopologues that require very stringent corrective measures to deconvolute results of high-purity tracer, natural abundance, and coincidental fragmentation profiles. Machine learning and the addition of compartmental modeling and differentiable programming methods can simultaneously estimate flux parameters and propagate spectral uncertainty, but existing frameworks typically make idealistic assumptions about steady states or compartment homogeneity between physiologically heterogeneous tissues and changing feeding states. Adaptive experimental design algorithms that are able to optimize the selection of tracers, scheduling of sampling, and depth of analysis through partial posterior information to minimize resource intensity and animal disturbance are required. Subscaling fluxomics to livestock would also impose logistical limitations on sample processing throughput and tracer cost, plus ethical implications of repeated invasive sampling, which can encourage surrogate modeling that can predict flux changes using minimally invasive proxy matrices or even combined sensing and limited tracer pulses. Cross-modality fusion with transcriptomic and proteomic measures of enzyme capacity has not been developed, especially in building mechanistic priors that are consistent with thermodynamic feasibility, yet allow for finding context-specific regulatory rewiring (Venkataraman and Huffmyer, 2025).

Spatial-temporal metabolic atlases are a more ambitious construct in which the tissue architecture and development chronology of three-dimensional tissues are superimposed with quantitative metabolite distributions and predicted reaction activities. These atlases would enable the explanation of microenvironmental gradients, the nutrient partitioning of cell types specifically, and the spatial coordination of the immune metabolic crosstalk, all of which apply to efficiency, resilience, and well-being. Spatial reconstruction based on AI techniques has to grapple with sparsity, ion suppression artifacts, and misregistration between imaging modalities and requires probabilistic image mass spectral fusion schemes to generate predictions of unobserved analytes and limit their predictions to known biochemical neighborhood graphs. Even more complexity is added by temporal continuity because repeated sampling can be destructive, so population-level statistical alignment and transfer learning are required to predict individual trajectories. Research gaps encompass standardized ontologies for anatomical and microanatomical location referencing on production species, domain adaptation strategies that have the potential to transfer spatial metabolic signatures between breeds or age groups, and ethical stewardship in the establishment of the right level of resolution that can have a reasonable balance between scientific understanding and economy of data. Cross-species transferability is a chance and a threat, which suggests the need to use meta-learning strategies to ensure that the dissimilarity between the opportunity presented by a well-annotated model organism and the diverse livestock lineages is not mistakenly used to infer learning results (Samarah et al., 2025).

Multi-omics that are grounded in metabolism foundation models will enable the reorganization of analytics, offering generative representations that are versatile and share a common latent structure of metabolomic, genomic, transcriptomic, proteomic, and microbiome modalities. These scale-capacity architectures, trained with self-guided tasks like masked reconstruction, contrastive alignment, or multimodal coembedding, can provide richer contextual embeddings to downstream tasks, which require less labeled data to forecast, detect anomalies, and discover cause-and-effect connections. However, the metabolic domain has special restrictions: the scale of quantitative intensity, the consistency of chemical formulas, the variability of adducts, and the stoichiometry of pathways must be adhered to so that learned representations do not challenge the plausibility of biochemistry. The gaps in current research are scalable tokenization schemes of spectral-based feature representation that maintain fine-grained fragmentation sensitivity without causing vocabulary explosion, schemes to encode hierarchical pathway knowledge and empirical covariation, and schemes to alleviate modality imbalance, whereby some omics layers are rich while others are sparse. Further, benchmarks of evaluation that are relevant to livestock production goals are still in their adolescent phases and hinder objective comparison of foundation model utility and fairness among breeds or production systems. Costs of pretraining energy and compute, and its sustainability, imply that future applications of pretraining must address issues of sustainability and potential inequity in access, which is why disseminating capabilities through distilled or parameter-efficient adapter models in a nonreplicative manner is important (Kant et al., 2025).

Explainable causal digital twins go beyond descriptive twins by considering mechanistic and structural causal models in continuous data assimilation to simulate intervention outcomes, including reformulation of diet, environmental modification, prophylactic health, or selective breeding strategies. To build such twins, dynamic systems modeling, causal graph inference, and real-time sensor fusion are necessary, and counterfactual reasoning in the face of uncertainty can be done. When there are latent confounders, feedbacks, and dynamic management responses that cause changes in exposure distributions, it becomes difficult to determine causal graph structure. The identifiability is limited by the data constraints of the rare stressors or events of the emergent disease and the ethical concerns of transparency regarding the limits of uncertainty to avoid premature overconfidence in the intervention prescriptions. The methods of Bayesian and invariant risk minimization can be used in order to separate the effects of robust causal signals and context-specific correlations, but demand strict validation procedures that embrace interventional or quasiexperimental evidence. The computational tractability can also be met by model reduction approaches, which would also conserve needed nonlinear dynamics and metabolite transport lags, especially when twins are being used at farm-edge conditions to provide timely decision support (Wang and Li, 2025).

Adaptive self-monitored annotation acceleration on new metabolites is a solution to the annotation bottleneck that has remained constant in untargeted profiling, with a large fraction of spectral features being unknown. Systems based on AI can use fragmentation tree structure, retention behavior, isotope patterns, and cross-sample coabundance networks to group unknowns, suggest candidate structures, and rank experimental validation. Few-shot extensions across compound classes can be achieved through meta-learning, and contrastive learning can be enhanced by synthetic spectra generation. Nonetheless, no exhaustive ground truth means that it is prone to propagating systematic misannotations, which biases downstream biological interpretation. Areas of research such as the quantification of structural hypotheses of strength in uncertainty, the combination of quantum chemical prediction with deep learning to enhance the faithfulness of fragmentation patterns, and the generation of anonymized spectral schedules that uphold the justice of information, and help move toward the annotation of digital collections, should also be explored. Active loops of learning that seek specific validation of high-influence unknown nodes in pathway inference graphs may be the best way to optimize resource allocation but need principled influence measures and ethical principles to guide sample prioritization (Vitale et al., 2025).

Strong approaches to low abundance and severely limited sample inference are also still sought after, especially in longitudinal research limited by welfare concerns or in infancy-age investigations in which sample volumes are limited. It is common practice in standard deep learning architectures to fail under small sample regimes, which has driven the development of Bayesian hierarchical models, physics-informed priors, and physics-inspired foundation embedding-based transfer learning, which attempts to add external knowledge to stabilize the estimation. Low-abundance signals interfere with detection sensitivity and create zero-inflated distributions that are prone to overimputation. False discovery propagation can be mitigated by the use of probabilistic censoring models that are parametrically different between biological and analytical dropouts and adaptive denoising autoencoders parametrically optimized on noise distributions. Cross-cohort harmonization can also further complicate inference of effects that interplay with true rareness. Standardized simulation frameworks are needed to realistically simulate noise, drift, and coelution effects. These frameworks should evaluate methodological resilience and fairness to ensure that performance improvements under sparse data conditions are not disproportionately benefiting already well-characterized breeds with additional modalities available for transfer (Bourel et al., 2025).

Measures of environmental and welfare impact attribution based on metabolomics are not sufficiently standardized, which prevents the interchangeability of research and incorporation into sustainability certification. The conversion of metabolic changes to measurable proxies of such traits as emissions intensity, nutrient use efficiency, or stress load requires causal attribution models, which correct for confounding management activity and genetic history. Composite indices that combine pathway-level perturbation scores with mechanistic emission models or welfare biomarkers that have been experimentally validated need to be sensitive and can be interpreted. The current use of ad hoc marker panels is at risk of overfitting to any particular situation and compromising generalization. Professionalization of this field can be achieved by the development of open metric ontologies, calibration procedures, and validation hierarchies, and the incorporation of uncertainty decomposition, which divides variance between biological, analytical, and model structural sources. Ethical implications occur when measures affect economic motivation; therefore, open administration, involvement of stakeholders, and periodic adjustment in the view of new science are essential so as to thwart perverse maximization or welfare trade-off (Salaudeen and Vuanghao, 2025).

Ethical instrumentation and real-time compliance monitoring software indicate an expanding boundary in which the sensors of the built environment, biochemical analytics, and policy engines could be united to implement requirements from the areas of medication withdrawal, feed additive consumption, environmental discharge, or welfare level. Compliance can be operationalized through AI, not by using retrospective auditing, but through continuous variable compliance, which identifies aberrations early enough and responds to them. Nonetheless, the introduction of compliance logic into the instrumentation creates the problem of ownership of data, vendor neutrality, and possibly punitive surveillance cultures. The principles of design, including privacy as an architectural feature, adjustable consent tiers, and rationales of alerts, are necessary to ensure legitimacy. The research gaps include the risk of differential access, under which resource-constrained operations may not be able to use compliance-enhancing technology and therefore may not be subjected to asymmetrical regulation. This could be reduced by open interfaces and interoperable standards, which would allow innovation and maintain a fair level of participation (Currie et al., 2025).

Enabling infrastructures on which many of the above innovations would not be able to mature are interdisciplinary consortia and the development of international data commons. The multifaceted nature of the metabolic, genomic, phenomenological, environmental, and managerial information requires a joint rule, which is both open and defensive of the ownership of legal business and cultural rights. Architectures of data commons should provide access control in a granular manner, provenance tracking, schema evolution, and embedded ethical licensing to ensure that the secondary use is in line with the expectations of the contributor. Consistency in metadata standards that include managed vocabularies of husbandry practices, feed formulations, health treatment, and contextual environmental factors is a precondition to powerful cross-study synthesis and machine readability. The accumulation of bias can be addressed by inclusive consortium models, where different production systems, geographic regions, and groups of stakeholders are given a decision-making representation, which can develop globally applicable models. Investing structures should encourage long-term maintenance, curation quality, and discoverability and interoperability tools instead of just focusing on the creation of new datasets. Expertise differences can be addressed through capacity-building programs that are integrated into the consortia so that there can be equal contributions and sharing of benefits. However, there are still tensions between competitive advantage and communal pooling, which require incentive alignment, recognition, citation credit systems, federated learning alternatives on sensitive subsets, and redistribution of value based on downstream commercialized products (Beyvers et al., 2025).

It is into this backdrop of future directions that a cohesive research gap exists regarding rigorous evaluation paradigms that capture the objectives of livestock production, which are multivariate and inclusive of productivity, welfare, environmental stewardship, resilience, and ethical integrity. Stratified datasets representing breeds, climates, and management intensities, with standardized tasks in prediction, causal inference, annotation, and fairness evaluation, would speed methodological iteration and reduce redundant or unproven assertions. A theory of integrative information value (the marginal decision improvement per unit analytical and computational cost) of metabolomic information might be used to prioritize between ultra-high-throughput sensing, fluxomics resolution, spatial mapping, and annotation depth and would therefore make resource allocation consistent with systemic rather than technological novelty. Finally, AI applications in the sphere of livestock metabolomics will be made available to pursue a reflective compromise between technological aspiration and ethical care, making adaptive governance and participatory partnership an essential part of infrastructure and algorithm engineering. By active development of these new frontiers, the discipline may evolve more into coherent and strong ecosystems of innovation, instead of isolated steps in innovation, to provide actionable, fair, and sustainable metabolic intelligence to animal agriculture on the planet.

7.12 Future perspectives and strategic roadmap

The future of AI-advanced livestock metabolomics is determined by a deliberate shift of limited pilot constructs to the adoption of new service ecosystems, where metabolic intelligence is integrated in day-to-day genetic enhancement, nutritional optimization, health management, and environmental accounting, as well as welfare custodianship. This shift needs to redefine metabolomics as a research-oriented add-on analytical tool to a controlled utility layer that gives breeding organizations, veterinarians, nutritionists, sustainability auditors, and policy actors access to validated indicators and decision guidelines that they can consume. Consolidation of disaggregated tooling into modular platforms with components that fit into common semantic schemes of samples, analytical parameters, quality metrics, and phenotypic situations must be a priority in strategic road mapping since the modules can be substituted at any time without involving the expensive reintegration costs. Blended value models with core data infrastructure being mutually maintained and differentiated analytics and advisory overlays creating innovation will ensure economic sustainability. Service level goals, including analytical turnaround time, interpretability levels, fairness among breeds and production systems, and ability to withstand laboratory or network outage, need to be institutionalized as part of risk management frameworks, which can foster stakeholder trust in the scaled adoption (Tedeschi et al., 2025).

The central principle of the scalable service ecology is that omics layers are harmonized in a way that the outputs of metabolomics are not a single silo signal but a part of a coherent, longitudinal phenotype intelligence continuum spanning all the way between the genome and exposome. The future ontological posteriors will require adaptive ontological mediation, which can harmonize genomic variation effects, transcript abundance practices, proteoform action conditions, microbiome functional attribute possibilities, and metabolic flux or concentration phenotypes to a comprehensive set of causal graphs susceptible to developmental changes and environmental regulation. Representation learning approaches should place biochemical and genetic constraints such that latent embeddings maintain stoichiometric feasibility, pathway topology, and gene-to-enzyme-to-metabolite interaction, while being flexible to form emergent cross-layer interaction. Harmonization involves statistical balancing of different granularities of sampling as well, where metabolomics is occasionally compared to continuous recordings of the environment or sensor measurements, and requires a temporal alignment model that is able to deduce latent physiological states in the state of genotype by environment using management-evaluation measures. The cross-platform calibration cohorts, multispecies reference atlases, and shared benchmark tasks to integrate prediction and causal attribution will be used to hasten the maturation of holistic phenotypic intelligence to provide more accurate genomic selection indices based on metabolic efficiency, robustness, and sustainability characteristics (Li and Lock, 2025; Montesinos-López et al., 2025).

The construction of resilient, adaptive, and ethically sound metabolic AI systems involves introducing the concepts of robustness, transparency, and accountability to the entire lifecycle of gathering and processing data from the retirement of the model. Resilience entails technical redundancy with cloud edge hybridization, automatic adaptation of the changing analysis instrument baselines, and graceful degradation in the face of partial loss of data without losing any of the minimum viable decision support. Adaptivity will be achieved by means of constant learning pipelines with drift identification that distinguishes between real biological evolution, like seasonal forage shift, and provides an analytic artifact to activate retraining with a governance-certified validation gate. Operationalization of the consent provenance, data minimization, and auditing of the fairness of algorithms alongside explainable portals connecting influential metabolites, pathways, and cross-omics characteristics with mechanistic accounts that domain professionals can understand, is necessary to provide ethical grounding. The level of cyber biosecurity should be taken to a higher level, and zero-trust architectures, cryptography, integrity sealing of raw and processed data, and detection of anomalies when adversarial models are manipulated by spectral inputs or model parameters should be implemented. Internalization of the sustainability goals should be done through carbon-conscious workload scheduling, energy intensity reporting, and optimization of model efficiency, which incorporates the environmental externalities in infrastructure planning (Kovac et al., 2025).

The development of strategic roadmaps should include the use of scenario planning in the context of compounding climate, regulatory, and market uncertainties that will put pressure on metabolomic AI value propositions. The shifting baseline of metabolite distributions posed by heat stress, water deficiency, forage composition change, and pathogen pressure also occurs due to climate variability, and this poses a risk of generalizing the model when corpora are historically biased during training. Regulatory evolution can introduce more stringent data ownership, animal welfare, and emission reporting and antimicrobial use restrictions, which require a flexible change in compliance and can assert the lineage of model outputs to underlying raw evidentiary layers. Market forces such as high prices of low-carbon or welfare-guaranteed products are motivators of credible metabolic indicators with the ability to audit purported characteristics. Analytic portfolio robustness can be assessed with scenario simulation frameworks that combine biophysical climate projection, policy pathway models, and economic elasticity of demand and with preemptive diversification of biomarkers, tailored region-specific calibration strategies, and contingency investment in alternative senses. Regular scenario drills should be institutionalized by the governance boards, and they have to evaluate preparedness, singleness of failure points, and reassess risk appetite (Kayusi et al., 2025).

A progressive future toward participatory, transparent, and sustainable harmonized metabolic innovation puts inclusive governance and cocreation of knowledge on an equal footing with the algorithmic novelty. Participatory systems will enable the manufacturers, veterinarians, and feed formulators, and local communities to determine the nature of priority traits, acceptable limits of data use, and benefit-sharing systems to reduce asymmetries in which value-making is centralized upstream and operations burdens extend down. The concept of transparency should not just be limited to explainability of models but should also be opened to include the publication of validation protocols, the decomposition of uncertainties, taxonomies of errors, and the lifecycle environmental footprint of computational tasks. Sustainability alignment entails developing metabolic indicators that provide incentive to systemic efficiency gains as opposed to the short-term optimization that may externalize resource or welfare costs. This involves multiobjective optimization models that trade off productivity, resilience, welfare, and emission intensity via clear trade-off surfaces that are relayed to the decision makers. The capacity building of education and ready tooling reduces the barriers to underresourced regions, creating globally representative datasets, which increase generalizability and fairness. The ethical foresight processes can scan the technology at the time of its emergence to identify any dual-use or unwanted welfare effects and place protection mechanisms in place before a technology is deployed at scale (Sarkar, 2025).

In general, the strategic roadmap is focused on planned development in the infrastructural, methodological, governance, and socio-economic vectors. The near-term milestones are focused on standardization of data/metadata ontologies, cross-omics integration benchmarks definition, implementation of lower layers of representation that are task-flexible, and standardized ethical and sustainability reporting. Mid-horizon goals aim to achieve ubiquitous deployment of resilient

TABLE 7.6 Future innovations in AI for livestock metabolomics.

Innovation area	Description	Potential impact on livestock metabolomics	Challenges or risks
Multi-omics integration and systems modeling	*AI frameworks that more deeply integrate* metabolomics with genomics, transcriptomics, proteomics, and phenomics to build mechanistic and predictive systems biology models	• *Enables more comprehensive* understanding of metabolic regulation and trait architecture in livestock • Improves trait prediction by combining metabolic data with genetic and environmental context • Supports data-driven breeding strategies (e.g., selecting for metabolic efficiency)	• *Requires very large, well-annotated* multi-omics datasets across breeds • Integrating different data modalities with varying scale, noise, and missingness is technically complex • Risk of overfitting if model complexity is not managed
Real-time metabolic monitoring and digital twins	AI-enabled sensors, edge computing, and digital twin models to monitor metabolite fluxes or surrogate markers in livestock in (near) real time	• Allows dynamic monitoring of animal metabolic states (e.g., stress, nutritional shifts) • Enables adaptive decision-making on the farm (feeding, health interventions) based on metabolic feedback • Minimizes need for frequent invasive sampling by using proxy/virtual models	• Development and deployment of metabolite-sensing hardware for the farm environment are challenging • Digital twin accuracy depends on model fidelity and data resolution • Infrastructure limitations on farms (connectivity, power) may hinder adoption
Advanced imputation and generative modeling	Use of generative AI (e.g., GANs) and variational autoencoders (VAEs) to impute missing metabolite data, simulate realistic metabolic profiles, or create synthetic metabolomics datasets	• Addresses common issues of missingness and low sample numbers in metabolomics • Generates synthetic data for undersampled breeds or conditions, improving AI model training • Helps in designing "what-if" scenarios (virtual experiments) for metabolic interventions	• Synthetic data may diverge from true biological distributions ("mode collapse" risk) • Validation of generated metabolite profiles is nontrivial • Ethical considerations around synthetic biological data generation
Conversational AI or knowledge-graph agents	AI agents (e.g., multiagent LLMs) that enable natural language querying of complex metabolomics knowledge graphs, integrating data interpretation and exploration	• Democratizes access to metabolomics knowledge for veterinarians, breeders, and scientists • Facilitates hypothesis generation and rapid exploration of metabolic pathways and relationships • Bridges domain knowledge with computational data, making metabolomics more accessible	• Building and maintaining high-quality, curated metabolomics knowledge graphs is resource-intensive • Ensuring accuracy and consistency of natural language to graph queries • Risk that nonexpert users misinterpret AI-generated responses

Continued

TABLE 7.6 Future innovations in AI for livestock metabolomics.—cont'd

Innovation area	Description	Potential impact on livestock metabolomics	Challenges or risks
Fluxomics-informed AI and metabolic flux prediction	AI methods that combine fluxomics (stable-isotope tracing) data with machine learning to infer metabolic fluxes, not just static metabolite concentrations	• Provides dynamic insight into metabolic pathway rates, enhancing understanding of metabolic efficiency, turnover, and bottlenecks in livestock • Supports the design of nutritional interventions to optimize fluxes for performance or health • Informs breeding decisions based on metabolic kinetics	• Fluxomics experiments (isotope tracing) are expensive, especially in large animals • Data complexity increases (time-series, labeling), making modeling difficult • Integration of flux data into AI models demands high-level computational expertise
Ethical, explainable, and sustainability-oriented AI	Development of AI models that are transparent, fair, and optimized for sustainability metrics (e.g., environmental footprint, animal welfare) rather than purely productivity	• Promotes trust and adoption by stakeholders (farmers, regulators) through explainable AI • Aligns metabolomics-AI innovation with sustainability goals (e.g., reducing methane and improving feed efficiency) • Ensures that AI-driven metabolic optimizations do not compromise animal welfare	• Trade-offs between model accuracy and interpretability • Defining and operationalizing sustainability metrics in AI models can be contentious • Requires multidisciplinary governance (ethics, animal science, and policy) to guide deployment

continuous learning systems, scaling of causal digital twins of intervention simulation, scaling fluxomic and spatial metabolomic systems to routine decision-making, and institutionalized participatory monitoring to achieve fair allocation of value. Long-horizon visions encompass metabolically federated commons of intelligence worldwide communicating via privacy-preserving protocols, providing real-time, causally stable, resource-efficient advice that has a positive effect on genetic and environmental development and animal welfare in a mutually supporting way. The effectiveness and consistency of such a direction will be based on strict compliance with reproducibility, readiness for interdisciplinary cooperation, and the ability to adhere to ethical care to ensure that the integration of AI in livestock metabolomics will meet its promise of transformation without harming social trust and environmental action. Future innovations in AI for livestock metabolomics are presented in Table 7.6.

References

Abd El-Hack, M.E., Allam, A.A., Aldhalmi, A.K., Kamal, M., Arif, M., Alawam, A.S., Rudayni, H.A., Taha, A.E., Swelum, A.A., Elolimy, A.A., Madkour, M., Ashour, E.A., 2025. Integrating metabolomics for precision nutrition in poultry: optimizing growth, feed efficiency, and health. Front. Vet. Sci. 12. https://doi.org/10.3389/fvets.2025.1594749.

Adam, E., Zanoaga, M.D., Rota, R., Cominetti, O., 2025. A comprehensive protocol and step-by-step guide for multi-omics integration in biological research. JoVE J. 2025 (222). https://doi.org/10.3791/66995.

Ahmed, A., Gummeson, J., 2025. Latent Sensor Fusion: Multimedia Learning of Physiological Signals for Resource-Constrained Devices (Version 1). arXiv. https://doi.org/10.48550/ARXIV.2507.14185.

Ahmed, Z., Safwan, A., Ahmad, A., 2025. Cold chain logistics: challenges and innovations in temperature-sensitive product distribution. ACADEMIA Int. J. Soc. Sci. 4 (3), 4093–4115. https://doi.org/10.63056/acad.004.03.0687.

Aigensberger, M., Bueschl, C., Metzler-Zebeli, B.U., Berthiller, F., Schwartz-Zimmermann, H.E., 2025. QuantyFey: an open-source tool for targeted LC-MS quantification with integrated drift correction. Anal. Chim. Acta 1375. https://doi.org/10.1016/j.aca.2025.344571.

Al Zubaidi, S.M., Nasar, M.I., Notebaart, R.A., Ralser, M., Alam, M.T., 2025. An enzyme activation network reveals extensive regulatory crosstalk between metabolic pathways. Mol. Syst. Biol. 21 (7), 870–888. https://doi.org/10.1038/s44320-025-00111-7.

Ali, H.H., 2025. Gas chromatography: principles, medicine and pharmaceutical, energy, fuel applications and environmental analysis. SAR J. Med. Biochem. 6 (04), 94. https://doi.org/10.36346/sarjmb.2025.v06i04.005.

Barbudo Carrasco, L., 2025. Mitigating bias and advocating for data sovereignty: the role of metadata and paradata in ethical AI-driven information systems. J. Libr. Metadata 25 (3), 185–200. https://doi.org/10.1080/19386389.2025.2515766.

Barnard, A.S., 2025. BenchMake: turn any scientific data set into a reproducible benchmark. Mach. Learn.: Sci. Technol. 6 (3). https://doi.org/10.1088/2632-2153/adf810.

Barret, N., Bernasconi, A., Bikbov, B., Pinoli, P., 2025. I-ETL: an interoperability-aware health (meta)data pipeline to enable federated analyses. BMC Med. Inf. Decis. Making 25 (1). https://doi.org/10.1186/s12911-025-03188-0.

Bernabé-Díaz, J.A., Franco, M., Vivo, J.M., Fernández-Breis, J.T., 2025. Optimizing clustering-based analytical methods with trimmed and sparse clustering. Comput. Biol. Med. 194. https://doi.org/10.1016/j.compbiomed.2025.110436.

Beyvers, S., Hochmuth, J., Brehm, L., Hansen, M., Goesmann, A., Förster, F., 2025. Towards FAIR and Federated Data Ecosystems for Interdisciplinary Research. arXiv, Germany. https://doi.org/10.48550/arXiv.2504.20298.

de Bodt, C., Diaz-Papkovich, A., Bleher, M., Bunte, K., Coupette, C., Damrich, S., Sanmartin, E.F., Hamprecht, F.A., Horvát, E.Á., Kohli, D., Krishnaswamy, S., Lee, J.A., Lelieveldt, B.P.F., McInnes, L., Nabney, I.T., Noichl, M., Poličar, P.G., Rieck, B., Wolf, G., Mishne, G., Kobak, D., 2025. Low-Dimensional Embeddings of High-Dimensional Data. arXiv, Belgium. https://doi.org/10.48550/arXiv.2508.15929.

Bosten, E., Chen, K., Hellings, M., Cabooter, D., 2025. Artificial intelligence for method development in liquid chromatography. TrAC, Trends Anal. Chem. 192, 118320. https://doi.org/10.1016/j.trac.2025.118320.

Bourel, B., Joly, A., Servajean, M., Bettinger, S., Sanabria-Fernández, J.A., Mouillot, D., 2025. From presence-only to abundance species distribution models using transfer learning. Ecol. Lett. 28 (7). https://doi.org/10.1111/ele.70177.

Buchha, D., 2025. AI & ML for livestock welfare: analysing behaviour and stress patterns for better care. Int. J. Sci. Res. Eng. Manag. 09 (04), 1–9. https://doi.org/10.55041/ijsrem46227.

Van Calster, B., Collins, G.S., Vickers, A.J., Wynants, L., Kerr, K.F., Barreñada, L., Varoquaux, G., Singh, K., Moons, K.G.M., Hernandez-boussard, T., Timmerman, D., Mclernon, D.J., Van Smeden, M., Steyerberg, E.W., 2024. Performance Evaluation of Predictive AI Models to Support Medical Decisions: Overview and Guidance (Version 1). arXiv. https://doi.org/10.48550/ARXIV.2412.10288.

Catussi, B.L.C., Lo Turco, E.G., Pereira, D.M., Teixeira, R.M.N., Castro, B.P., Massaia, I.F.D., 2024. Metabolomics: unveiling biological matrices in precision nutrition and health. Clin. Nutr. ESPEN 64, 314–323. https://doi.org/10.1016/j.clnesp.2024.10.148.

Caufield, J.H., Ghosh, S., Kong, S.W., Parker, J., Sheffield, N., Patel, B., Williams, A., Clark, T., Munoz-Torres, M.C., 2025. Standards in the Preparation of Biomedical Research Metadata: A Bridge2AI Perspective. arXiv, United States. https://doi.org/10.48550/arXiv.2509.10432.

Chakradeo, K., Huynh, I., Balaganeshan, S.B., Dollerup, O.L., Gade-Jørgensen, H., Laupstad, S.K., Malham, M., Nguyen, T.L., Hulman, A., Varga, T. V., 2025. Navigating fairness aspects of clinical prediction models. BMC Med. 23 (1). https://doi.org/10.1186/s12916-025-04340-3.

Cheimarios, N., 2025. Scientific software development in the AI era: reproducibility, MLOps, and applications in soft matter physics. In: Frontiers in Physics.

Cordeiro, C.M., Adomaitis, L., Huang, L., 2026. The AI-policy-governance nexus: how regulation and AI shift corporate governance toward stakeholders. Technol. Soc. 84, 103117. https://doi.org/10.1016/j.techsoc.2025.103117.

Csorba, S., Vribék, K., Farkas, M., Süth, M., Strang, O., Zentai, A., Farkas, Z., 2025. Data-driven early warning approach for antimicrobial resistance prediction–anomaly detection based on high-level indicators. Vet. Sci. 12 (10), 935. https://doi.org/10.3390/vetsci12100935.

Currie, W.L., Leimeister, J.M., Schlagwein, D., Willcocks, L., 2025. Rethinking technology regulation in the age of AI risks. J. Inf. Technol. 40 (3), 236–245. https://doi.org/10.1177/02683962251378815.

Dell'Olio, A., Biasioli, F., Fogliano, V., Rubert, J., 2025. Decoding microbial volatile signals in host–microbiome crosstalk. Trends Endocrinol. Metab. https://doi.org/10.1016/j.tem.2025.08.004.

Deng, X., Xiang, W., Lin, W., Zheng, Z., Yang, Y., 2025. Few-shot injection molding quality prediction method integrating deep transfer and multi-task learning. Int. J. Adv. Manuf. Technol. 141 (1–2), 609–632. https://doi.org/10.1007/s00170-025-16651-z.

Dong, S.C., 2025. Prudential Reliability of Large Language Models in Reinsurance: Governance, Assurance, and Capital Efficiency (Version 1). arXiv. https://doi.org/10.48550/ARXIV.2511.08082.

Eken, B., Pallewatta, S., Tran, N., Tosun, A., Babar, M.A., 2025. A multivocal review of MLOps practices, challenges and open issues. ACM Comput. Surv. 58 (2). https://doi.org/10.1145/3747346.

Eshawu, A.B., Gunda, T.J., Kumar, B., Matta, S., 2025. Advancements in sample preparation and extraction techniques for metabolomic analysis of animal fluids. Microchem. J. 218. https://doi.org/10.1016/j.microc.2025.115259.

Esser, P.M., Fleissner, M., Ghoshdastidar, D., 2025. Theoretical Foundations of Representation Learning Using Unlabeled Data: Statistics and Optimization. arXiv, Germany. https://doi.org/10.48550/arXiv.2509.18997.

Fan, W., Yonghong, H., Hou, X., Ding, C., Huang, D., Zheng, W., Dai, Z., 2025. Unraveling principles of thermodynamics for genome-scale metabolic networks using graph neural networks. Cell Syst. 16 (10), 101393. https://doi.org/10.1016/j.cels.2025.101393.

Fang, F., Sun, L., 2025. Quantitative native proteomics by capillary zone electrophoresis-mass spectrometry. Anal. Chem. https://doi.org/10.1021/acs.analchem.5c06099.

Fu, J., Yang, C., Triantafyllidis, C.P., 2025. Genotype-Phenotype Integration through Machine Learning and Personalized Gene Regulatory Networks for Cancer Metastasis Prediction (Version 1). arXiv. https://doi.org/10.48550/ARXIV.2510.23620.

Galasso, L., Castelli, L., Bruno, E., 2024. Editorial: roles of the circadian rhythms in metabolic disease and health. Metabolites 14 (11), 621. https://doi.org/10.3390/metabo14110621.

Galmacci, T., Fantaccione, L., Luzi, F., Verlezza, G., Ragni, G., Fantozzi, F., 2025. How accurate is your carbon footprint? Measuring the uncertainty with pedigree matrix in the design of an oxyfuel combustion gas turbine. J. Eng. Gas Turbines Power 147 (12). https://doi.org/10.1115/1.4069460.

Garcia-Oliva, L., Carrassi, A., Counillon, F., 2025. Exploring the influence of spatio-temporal scale differences in coupled data assimilation. Nonlinear Process Geophys. 32 (4), 439–456. https://doi.org/10.5194/npg-32-439-2025.

Garouani, M., Mothe, J., Barhrhouj, A., Aligon, J., 2025. Investigating the Duality of Interpretability and Explainability in Machine Learning.

Ghavi Hossein-Zadeh, N., 2026a. Generative artificial intelligence in animal genomics for smart agriculture: Applications, challenges, and future prospects. Vet. Anim. Sci. 33, 100702. https://doi.org/10.1016/J.VAS.2026.100702.

Ghavi Hossein-Zadeh, N., 2026b. Advancing climate-resilient livestock systems: Next-generation emission mitigation strategies and integrated technological innovations. Vet. Anim. Sci. 31, 100588. https://doi.org/10.1016/j.vas.2026.100588.

Ghoneim, O., Dobias, P., Romain, O., 2025. Survey of neural network optimization methods for sustainable AI: from data preprocessing to hardware acceleration. Mach. Learn. Appl. 22, 100762. https://doi.org/10.1016/j.mlwa.2025.100762.

Godon, T., Plante, P.L., Corbeil, J., Germain, P., Drouin, A., 2025. On selecting robust approaches for learning predictive biomarkers in metabolomics data sets. Anal. Chem. 97 (24), 12669–12678. https://doi.org/10.1021/acs.analchem.5c01049.

Greco, S., Vacchetti, B., Apiletti, D., Cerquitelli, T., 2025. Unsupervised concept drift detection from deep learning representations in real-time. IEEE Trans. Knowl. Data Eng. 37 (10), 6232–6245. https://doi.org/10.1109/tkde.2025.3593123.

de Groot, L.M., Twisk, J.W.R., Kok, A.A.L., Heymans, M.W., 2025. Incorporating longitudinal variability in prediction models: a comparison of machine learning and logistic regression in a cohort study with long follow-up. Ann. Epidemiol. 110, 51–65. https://doi.org/10.1016/j.annepidem.2025.07.060.

Guo, X., Sarup, P., Bay Nord, A., Henryon, M., Ostersen, T., Christensen, O.F., 2025. Metabolomic-genomic prediction realizes small increases in accuracy of estimated breeding values for daily gain in pigs. Genet. Sel. Evol. 57 (1). https://doi.org/10.1186/s12711-025-00972-4.

Habib, M.D.A., Alvi, R., Sarkar, S., Shuvo, M.D.A., Hasan, M., Azim, R., 2025. ADDomics: enhancing multiomics fusion with autoencoders and deep neural networks. In: IEEE Region 10 Symposium: Innovations in Technology: Shaping the Future, TENSYMP 2025. Institute of Electrical and Electronics Engineers Inc., Bangladesh https://doi.org/10.1109/TENSYMP63728.2025.11144973. http://ieeexplore.ieee.org/xpl/mostRecentIssue.jsp?punumber=11144927.

Hailemariam, D., Dervishi, E., 2024. Editorial: diving deeper with metabolomics into animal physiology. Front. Mol. Biosci. 11. https://doi.org/10.3389/fmolb.2024.1503412.

Hassan, M.M., 2025. Bayesian Kolmogorov-Arnold networks: uncertainty-aware interpretable modeling through probabilistic spline decomposition. Phys. Stat. Mech. Appl. 680, 131041. https://doi.org/10.1016/j.physa.2025.131041.

Hiniduma, K., Byna, S., Luca Bez, J., 2025. Data readiness for AI: a 360-degree survey. ACM Comput. Surv. 57 (9), 1–39. https://doi.org/10.1145/3722214.

Hirst, D.P., Térézol, M., Cantini, L., Villoutreix, P., Vignes, M., Baudot, A., 2025. MOTL: enhancing multi-omics matrix factorization with transfer learning. Genome Biol. 26 (1). https://doi.org/10.1186/s13059-025-03675-7.

Hossein-Zadeh, N.G., 2025. Artificial intelligence in veterinary and animal science: applications, challenges, and future prospects. Comput. Electron. Agric. 235, 110395. https://doi.org/10.1016/j.compag.2025.110395.
Ivanov, S., Wilk-Jakubowski, J.Ł., Ciopiński, L., Pawlik, Ł., Wilk-Jakubowski, G., Mihalev, G., 2025. Modern trends in the application of electronic nose systems: a review. Appl. Sci. 15 (19), 10776. https://doi.org/10.3390/app151910776.
Jamzad, A., Warren, J., Syeda, A., Kaufmann, M., Iaboni, N., Nicol, C.J.B., Rudan, J.F., Ren, K.Y.M., Hurlbut, D., Varma, S., Fichtinger, G., Mousavi, P., 2025. MassVision: an open-source end-to-end platform for AI-driven mass spectrometry imaging analysis. Anal. Chem. 97 (39), 21588–21597. https://doi.org/10.1021/acs.analchem.5c04018.
Jeanselme, V., Palma, M., Barrett, J.K., 2025. Assessing the Impact of Variance Heterogeneity and Misspecification in Mixed-Effects Location-Scale Models. arXiv, United Kingdom. https://doi.org/10.48550/arXiv.2505.18038.
Jiang, H., Wang, Y., Feng, K., Yin, C., Chang, Y., Sun, H., 2025. Biological Regulatory Network Inference through Circular Causal Structure Learning (Version 1). arXiv. https://doi.org/10.48550/ARXIV.2511.02332.
Jingyang, C., 2025. Construction of the theoretical framework for the generalization ability of machine learning models in the big data environment. Am. J. Big Data 6 (3), 28–38. https://doi.org/10.71465/ajbd2366.
John, A., Cardiff, B., John, D., 2024. A Review on Multisensor Data Fusion for Wearable Health Monitoring (Version 1). arXiv. https://doi.org/10.48550/ARXIV.2412.05895.
Kanicheril Ambikalekshmi, A., Pushparaj, P., Cherian, E., Rajan, R., Mohan, L., 2025. Advancing food safety through artificial intelligence: a detailed review of monitoring and detection technologies. Food Safety Health. https://doi.org/10.1002/fsh3.70050.
Kant, S., Deepika, Roy, S., 2025. Integrative multi-omics and artificial intelligence: a new paradigm for systems biology. OMICS A J. Integr. Biol. https://doi.org/10.1177/15578100251392371.
Kanwar, G., 2025. Edge AI revolution: compressed neural networks enabling real-time mobile processing. Eur. Modern Stud. J. 9 (5), 114–128. https://doi.org/10.59573/emsj.9(5).2025.11.
Kayusi, F., Chavula, P., Lungu, G., Mambwe, H., 2025. AI-Driven Climate Modeling: Validation and Uncertainty Mapping – Methodologies and Challenges, vol. 3. LatIA, p. 332. https://doi.org/10.62486/latia2025332.
Khalid, Z., Chen, Y., Liu, X., Noureen, B., Chen, Y., Wang, M., Ma, Y., Du, L., Wu, C., 2025. Recent advances and unaddressed challenges in biomimetic olfactory- and taste-based biosensors: moving towards integrated, AI-powered, and market-ready sensing systems. Sensors. https://doi.org/10.3390/s25227000.
Kim, J.M., Pathak, R.K., 2025. Editorial: vetinformatics: an insight for decoding livestock systems through in silico biology, volume II. Front. Vet. Sci. 12. https://doi.org/10.3389/fvets.2025.1672786.
Kopecky, L., Sands, C.J., Gómez-Romero, M., Misra, S., Want, E.J., Ebbels, T.M.D., 2025. MAMSI: integration of multiassay liquid chromatography-mass spectrometry metabolomics data using multiview machine learning. Anal. Chem. https://doi.org/10.1021/acs.analchem.5c01327.
Kovac, F., Neumaier, S., Pahi, T., Priebe, T., Rodrigues, R., Christodoulou, D., Cordy, M., Kubler, S., Kordia, A., Pitsiladis, G., Soldatos, J., Zervoudakis, P., 2025. Towards a Framework for Supporting the Ethical and Regulatory Certification of AI Systems. arXiv, Austria. https://doi.org/10.48550/arXiv.2510.00084.
Kumar, M., Ballamoole, K.K., Shetty, V.A., Rao, R.S.P., Gollapalli, P., 2025. Perspective on integrated multi-omics approaches and constraint-based modeling to explore metabolic functionality on the evolution of bacterial antibiotic resistance. Microb. Pathog. 208. https://doi.org/10.1016/j.micpath.2025.107999.
Kumar, A., Xu, C., Dakal, T.C., 2026. Microbiome based precision medicine through integrated multiomics and machine learning. Microbiol. Res. 303, 128384. https://doi.org/10.1016/j.micres.2025.128384.
Lee, H., Chen, S., 2025. Systematic bias of machine learning regression models and correction. IEEE Trans. Pattern Anal. Mach. Intell. 47 (6), 4974–4983. https://doi.org/10.1109/tpami.2025.3552368.
Li, G., Lock, E.F., 2025. Integrative analysis of multimodal omics data. Annu. Rev. Stat. Appl. https://doi.org/10.1146/annurev-statistics-042424-113016.
Li, H., Yi, C., Liu, J., Zhang, Z., Dong, Y., 2025. ESSTformer: a CNN-transformer hybrid with decoupled spatial spectral transformers for hyperspectral image super-resolution. Appl. Sci. 15 (21), 11738. https://doi.org/10.3390/app152111738.
Liang, J., Ling, J., Xu, L., Zhu, X., 2025. A dataset of Raman and infrared spectra as an extension to the ChEMBL. Sci. Data 12 (1). https://doi.org/10.1038/s41597-025-05289-x.
Liu, Y., 2024. Enhancing drug analysis through advanced mass spectrometry techniques and quantitative modeling. Theor. Nat. Sci. 37 (1), 180–185. https://doi.org/10.54254/2753-8818/37/20240180.
Liu, C., Rosen, E.A., Stohs, E.J., Imlay, H., Nigo, M., Gottesdiener, L.S., So, M., Tverdek, F., Dadwal, S., Gudiol, C., Satlin, M.J., Seo, S.K., Trubiano, J.A., Banerjee, R., Hanson, K.E., Abbo, L.M., 2025. Tackling antimicrobial resistance in people who are immunocompromised: leveraging diagnostic and antimicrobial stewardship. Lancet Infect. Dis. https://doi.org/10.1016/S1473-3099(25)00311-1.
Lushchak, V.I., 2025. Symphony of digestion: coordinated host–microbiome enzymatic interplay in gut ecosystem. Biomolecules 15 (8). https://doi.org/10.3390/biom15081151.
Mahato, S., Neethirajan, S., 2025. Dairy DigiD: an edge-cloud framework for real-time cattle biometrics and health classification. AI 6 (9), 196. https://doi.org/10.3390/ai6090196.
Mahmudul Hasan, Md, Fariha, N., Uddin Mahmood, T., Rahman, A., 2025. AI-enhanced data engineering for high-performance big-data processing and advanced analytics optimization. J. Comput. Sci. Technol. Stud. 7 (8), 724–732. https://doi.org/10.32996/jcsts.2025.7.8.84.

Matyushin, D.D., Burov, I.A., Sholokhova, A.Y., 2024. Uncertainty quantification and flagging of unreliable predictions in predicting mass spectrometry-related properties of small molecules using machine learning. Int. J. Mol. Sci. 25 (23). https://doi.org/10.3390/ijms252313077.

Maulucci, G., 2025. Rethinking metabolic imaging: from static snapshots to metabolic intelligence. Biophysica 5 (3), 42. https://doi.org/10.3390/biophysica5030042.

Métayer, C., Ballesta, A., Martinelli, J., 2025. Data-Driven Discovery of Digital Twins in Biomedical Research. arXiv, France. https://doi.org/10.48550/arXiv.2508.21484.

Montesinos-López, O.A., Montesinos-López, A., Mosqueda-González, B.A., Delgado-Enciso, I., Chavira-Flores, M., Crossa, J., Dreisigacker, S., Sun, J., Ortiz, R., 2025. Genomic prediction powered by multi-omics data. Front. Genet. 16. https://doi.org/10.3389/fgene.2025.1636438.

Moozhippurath, B., Natarajan, J., 2025. Metabolomics pathway prediction using enhanced-graph convolutional networks with graph attention networks. Int. J. Online Biomed. Eng. (iJOE) 21 (10), 48–62. https://doi.org/10.3991/ijoe.v21i10.55539.

Murri, S., 2025. Data engineering for intelligent systems and generative AI: architectures, pipelines, and strategy. In: 2025 1st International Conference on Secure IoT. Assured and Trusted Computing (SATC), pp. 1–5. https://doi.org/10.1109/satc65530.2025.11137298.

Nunes, A.T., Faleiros, C.A., Poleti, M.D., Novais, F.J., López-Hernández, Y., Mandal, R., Wishart, D.S., Fukumasu, H., 2024. Unraveling ruminant feed efficiency through metabolomics: a systematic review. Metabolites 14 (12). https://doi.org/10.3390/metabo14120675.

Pannu, J., Bloomfield, D., MacKnight, R., Hanke, M.S., Zhu, A., Gomes, G., Cicero, A., Inglesby, T.V., 2025. Dual-use capabilities of concern of biological AI models. PLoS Comput. Biol. 21 (5). https://doi.org/10.1371/journal.pcbi.1012975.

Park, J.D., Lee, S.R., Dhennezel, C., Taylor, N., Dame, A., Kadoki, M., Pishchany, G., Graham, D.B., Xavier, R.J., Seyedsayamdost, M.R., 2025. Elucidating the role of Campylobacter concisus–derived indole metabolites in gut inflammation and immune modulation. Proc. Natl. Acad. Sci. U. S. A. 122 (34). https://doi.org/10.1073/pnas.2514071122.

Pasupuleti, M.K., 2025. Stochastic computation for AI: Bayesian inference, uncertainty, and optimization. Nat. Educ. Serv. https://doi.org/10.62311/nesx/rriv325.

Qian, S.S., Jaffe, S., Gionfriddo, E., Wang, H., Richardson, C.J., Godage, N.H., 2025. Rethinking calibration as a statistical estimation problem to improve measurement accuracy. Anal. Chim. Acta 1372. https://doi.org/10.1016/j.aca.2025.344395.

Raj, A., Mahar, K., Kumar, P., Chandra Goli, R., 2025. Metabolic alterations during peripartum period and its effect on fertility in bovines. Int. J. Bio-resour. Stress Manag. 16 (Oct, 10), 01–08. https://doi.org/10.23910/1.2025.6397.

Rastogi, T., Jonsson, D., Karlström, A., 2025. Population Synthesis Using Incomplete Information (Version 1). arXiv. https://doi.org/10.48550/ARXIV.2510.00859.

Rosati, A., 2025. Invited paper Artificial intelligence and the future of animal husbandry: rethinking knowledge, practice, and governance. Arch. Latinoam. Prod. Anim. 33 (Suppl. 1), 567–572. https://doi.org/10.53588/alpa.330509.

Roy, P., Boker, A., Mili, L., 2025. Beyond Marginals: Learning Joint Spatio-Temporal Patterns for Multivariate Anomaly Detection. arXiv, United States. https://doi.org/10.48550/arXiv.2509.15033.

Salaudeen, A.O., Vuanghao, L., 2025. Metabolomics approach in environmental studies: methodologies, application and challenges. Crit. Rev. Anal. Chem. https://doi.org/10.1080/10408347.2025.2521734.

Samarah, L.Z., Zheng, C., Xing, X., Lee, W.D., Afriat, A., Chitra, U., MacArthur, M.R., Lu, W., Jankowski, C.S.R., Ma, C., Hunter, C.J., Neinast, M., Weilandt, D.R., Raphael, B.J., Rabinowitz, J.D., 2025. Spatial metabolic gradients in the liver and small intestine. Nature. https://doi.org/10.1038/s41586-025-09616-5.

Sarkar, A., 2025. Rethinking AI inclusion: a systems thinking framework for policy, education, and participation. AI Ethics 5 (6), 6297–6311. https://doi.org/10.1007/s43681-025-00810-9.

Sato, M., Ikeda, K., 2025. Untargeted analysis of hydrophilic metabolites using enhanced LC-MS separation with a pentafluoro phenylpropyl-functionalized column and prediction-based MS/MS spectrum annotation. Metabolomics 21 (4). https://doi.org/10.1007/s11306-025-02272-w.

Seydoux, C., Bryan, M., Barnes, J.P., Jouneau, P.H., 2025. Efficient denoising of mass spectrometry images by PCA-assisted self-supervised deep learning. ChemRxiv. https://doi.org/10.26434/chemrxiv-2025-wcm8l.

Shelat, P., Rawal, R.M., 2025. Role of metabolomics in uncovering the mechanistic link between microbiome modulation and host metabolism. J. Microbiol. Biotechnol. Food Sci. 15 (2). https://doi.org/10.55251/jmbfs.12612.

Sheshanarayana, R., You, F., 2025. Molecular representation learning: cross-domain foundations and future Frontiers. Digital Discovery 4 (9), 2298–2335. https://doi.org/10.1039/d5dd00170f.

Shobana, D., 2025. AI-Powered Biosensors for Real-Time Physiological Monitoring and Response. Rademics Research Institute, pp. 373–403. https://doi.org/10.71443/9789349552975-13.

Silvano, C., Ielmini, D., Ferrandi, F., Fiorin, L., Curzel, S., Benini, L., Conti, F., Garofalo, A., Zambelli, C., Calore, E., Schifano, S., Palesi, M., Ascia, G., Patti, D., Petra, N., De Caro, D., Lavagno, L., Urso, T., Cardellini, V., Perri, S., 2025. A survey on deep learning hardware accelerators for heterogeneous HPC platforms. ACM Comput. Surv. 57 (11), 1–39. https://doi.org/10.1145/3729215.

Su, C., Wang, P., Foo, N., Ho, D., 2025. Optimizing metabolic health with digital twins. Npj Aging 11 (1). https://doi.org/10.1038/s41514-025-00211-6.

Takahashi, M., Frøslev, T.G., Paupério, J., Thalinger, B., Klymus, K., Helbing, C.C., Villacorta-Rath, C., Silliman, K., Thompson, L.R., Jungbluth, S.P., Yong, S.Y., Formel, S., Jenkins, G., Laporte, M., Deagle, B., Rajbhandari, S., Jeppesen, T.S., Bissett, A., Jerde, C., Berry, O., 2025. A metadata checklist and data formatting guidelines to make eDNA FAIR (findable, accessible, interoperable, and reusable). Environ. DNA 7 (3). https://doi.org/10.1002/edn.3.70100.

Tedeschi, L.O., Lopez, P.G., Menendez III, H.M., Seo, S., 2025. Advancing precision livestock farming: integrating artificial intelligence and emerging technologies for sustainable livestock management. Anim. Biosci. https://doi.org/10.5713/ab.25.0289.

Tian, Z., Wu, J., Zhang, Z., Dai, Y., Zhang, W., Wang, S., 2025. Study on noise reduction method for bridge temperature signal using adaptive parameter selection and improved wavelet threshold function. Measurement 253, 117683. https://doi.org/10.1016/j.measurement.2025.117683.

Tronel, A., Roger-Margueritat, M., Plazy, C., Cunin, V., Mohanty, I., Dorrestein, P.C., Soranzo, T., Le Gouellec, A., 2025. Untargeted and semi-targeted metabolomics approach for profiling small intestinal and fecal metabolome using high-resolution mass spectrometry. Metabolomics 21 (4). https://doi.org/10.1007/s11306-025-02288-2.

Uribe-Guerra, G.D., Múnera-Ramírez, D.A., Arias-Londoño, J.D., 2024. Swine Diet Design Using Multi-Objective Regionalized Bayesian Optimization. arXiv, Colombia. https://doi.org/10.48550/arXiv.2409.12919.

Varsani, T., 2025. Evaluation of Differential Privacy Mechanisms on Federated Learning (Version 1). arXiv. https://doi.org/10.48550/ARXIV.2510.09691.

Venkataraman, Y.R., Huffmyer, A.S., 2025. Interrogating metabolic plasticity in marine organisms: a framework for best practices using metabolomic and lipidomic approaches. Integr. Comp. Biol. 65 (4), 1166–1184. https://doi.org/10.1093/icb/icaf138.

Vitale, G.A., Xia, S.-N., Dührkop, K., Zare Shahneh, M.R., Brötz-Oesterhelt, H., Mast, Y., Brungs, C., Böcker, S., Schmid, R., Wang, M., Hughes, C.C., Petras, D., 2025. Enhancing tandem mass spectrometry-based metabolite annotation with online chemical labeling. Nat. Commun. 16 (1). https://doi.org/10.1038/s41467-025-61240-z.

Wadood, A.A., Bordbar, F., Zhang, X., 2025. Integrating omics approaches in livestock biotechnology: innovations in production and reproductive efficiency. Front. Anim. Sci. 6. https://doi.org/10.3389/fanim.2025.1551244.

Wang, Y., 2024. Recent advances in the application of direct analysis in real time-mass spectrometry (DART-MS) in food analysis. Food Res. Int. 188, 114488. https://doi.org/10.1016/j.foodres.2024.114488.

Wang, J., Li, L., 2025. Innovative application of knowledge graph-driven causal inference in digital twin of chronic disease progression. Theor. Nat. Sci. 130 (1), 195–200. https://doi.org/10.54254/2753-8818/2025.25630.

Wang, W., Zhang, B., Liu, Z., Zhu, P., 2025. Dynamic multiple GAN for missing data imputation with awareness of sample distribution and attribute characteristics. Clust. Comput. 28 (11). https://doi.org/10.1007/s10586-025-05375-0.

Wei, S., Liu, Y., Zhang, H., Wang, Y., Jin, M., 2025. 37 The dynamic interplay between microbial consortia and arabinoxylan reveals the remodeling of metabolism by swine gut microbiome variation. J. Anim. Sci. 103 (Suppl. 3), 182–185. https://doi.org/10.1093/jas/skaf300.222.

Whittle, R., Ensor, J., Archer, L., Collins, G.S., Dhiman, P., Denniston, A., Alderman, J., Legha, A., van Smeden, M., Moons, K.G., Cazier, J.-B., Riley, R.D., Snell, K.I.E., 2025. Extended sample size calculations for evaluation of prediction models using a threshold for classification. BMC Med. Res. Methodol. 25 (1). https://doi.org/10.1186/s12874-025-02592-4.

Xiao, Q., Yang, J., Xie, J., Liu, Q., Huang, H., Tao, Y., Ding, B., 2026. Impact of Preprocessing and data analysis strategies on metabolite annotation in biological samples: a mass spectrometry-based metabolomics study using feature-based molecular networking. J. Chromatogr. B 1268, 124828. https://doi.org/10.1016/j.jchromb.2025.124828.

Yang, S., Han, M., Zhu, S., Yang, W., Dai, Z., Li, J., Ding, J., 2026. A survey of processing methods for different types of concept drift. Data Knowl. Eng. 161. https://doi.org/10.1016/j.datak.2025.102525.

Yasmin, A., Mittal, G., Yadamuri, R., Goud, S., Jeniffer, D.J.T., Bhuyan, K.B., 2025. Genetic adaptation of species to environmental change: insights from molecular biology. Int. J. Environ. Sci. 780–784. https://doi.org/10.64252/55fqjt69.

Yu, M., 2025. Application of deep learning to automatic modulation recognition. Appl. Comput. Eng. 175 (1), 18–29. https://doi.org/10.54254/2755-2721/2025.ast24988.

Yu, P., Teng, F., Zhu, W., Shen, C., Chen, Z., Song, J., 2025. Cloud–edge–device collaborative computing in smart agriculture: architectures, applications, and future perspectives. Front. Plant Sci. 16. https://doi.org/10.3389/fpls.2025.1668545.

Zafar, O., González, R.S., Namazi, M., Morales, A., Ayday, E., 2025. Empowering Digital Agriculture: A Privacy-Preserving Framework for Data Sharing and Collaborative Research (Version 1). arXiv. https://doi.org/10.48550/ARXIV.2506.20872.

Zhang, J., Qi, S., Dong, Y., Yuan, L., Shen, T., Fu, W., Wei, B., Zhu, H., Liu, J., 2025. GKG-LLM: a unified framework for generalized knowledge graph construction. In: Information Fusion.

Zhao, X., Tsermoula, P., Khakimov, B., 2025. Applications of metabolomics in cow health assessment. Metabolomics 21 (6). https://doi.org/10.1007/s11306-025-02364-7.

Chapter 8

Artificial intelligence integration in livestock nutrigenomics

The integration of artificial intelligence (AI) in the context of livestock nutrigenomics represents the intersection of biological understanding, computational intelligence, and strategic change in the fundamental understanding and application of dietary interventions. The goal of nutrigenomics is to unravel the mechanisms by which nutritional inputs can affect the health, productivity, resilience, and resource efficiency in the presence of the genome and the regulatory layers. The inclusion of AI makes this pursuit move from descriptive interpretation to predictive, adaptive, and decision-oriented frameworks that can deal with biological complexity. This chapter, in its essence, discusses how intelligent systems can map dynamic relationships between genomic variation, epigenetic modulation, transcriptomic responses, metabolic states, microbiome interplay, physiological phenotypes, and environmental contexts to inform both individualized and system-wide sustainable nutrition.

One of the central storylines in the chapter is that the development of ration formulation moves toward responsive nutritional ecosystems. With the advent of AI, continuous learning processes can be realized where multimodal streams of data are curated, harmonized, and fused into models that predict instead of respond to outcomes. Such a shift transforms the role of nutrigenomics as a research-focused science into a functional component of precision livestock management. It also reimagines the usefulness of biological markers, based on prospective use as a guide to intervene at the right time, refine a formulation, and enhance resilience during a period of climatic or management variability.

Such integration depends on a knowledge architecture based on solid data governance, semantic interoperability, model transparency, and biological validation. Along with the increasing dimensionality and depth of datasets, AI aids in both the untangling of causal signals and confounding noise and provides a framework within which hypothesis generation and hypothesis refinement can be performed. Hybrid modeling methods that combine mechanistic nutritional science with data-driven inference are becoming a basis for reliable deployment, by balancing empirical performance with interpretability and practical utility. Powerful interactions within this paradigm highlight a transition of models not being a black box, but a mutually beneficial tool that supplements human knowledge.

There are ethical, economic, and ecological factors that permeate the integration landscape. Fair access to the best nutrigenomic instruments, protection of genetic diversity, reduction of environmental impact, and alignment with welfare priorities are some of the guardrails that are necessary in responsible innovation. The chapter highlights the necessity of open benchmarking, lifecycle responsibility of models, and adapting frameworks that may change with the changing regulatory, societal, and climatic demands. The strategic foresight turns out to be as important as technical aptitude and determines investment in infrastructure, human capital development, and cooperation with multi-stakeholders.

Finally, this chapter sets AI in livestock nutrigenomics as the driver of a more resilient, efficient, and ethically sound paradigm of production. It expresses the theoretical principles, methodological directions, integration issues, and trajectories into the future, which are all aimed at shaping the maturation of this discipline. With this in perspective, what appears to be defined as nutrigenomics is restated not merely as a field of molecular research activity, but as an integrated decision-support framework in sustainable animal farming.

8.1 Introduction to AI in nutrigenomics

The field of AI in livestock nutrigenomics is a result of the intersection between three knowledge domains that were formerly closely related but separated by space and time: nutrition science, with its emphasis on the quantitative

Artificial Intelligence in Livestock Omics. https://doi.org/10.1016/B978-0-443-45663-3.00014-2

description of dietary intake and physiological needs; genomics and the layers of molecular profiling that demystify the architecture and regulation of biological characteristics; and algorithmic intelligence that strips away latent structure and actionable inference to high-dimensional, heterogeneous, and time-changing data. Nutrigenomics has moved beyond listing the nutrient-responsive genes to include the coordination of transcriptional, epigenetic, metabolic, microbiome-mediated, and endocrine response mechanisms, which together define phenotypic performance and adaptive capacity. AI offers the methodological bedrock to navigate this multidimensional environment, showing the conditional dependencies, emergent interactions, and context-contingent response surfaces that more traditional pipelines of analysis find hard to describe. The resulting integrative paradigm re-contextualizes nutrition as a non-dynamic provisioning activity to active modulation of dynamical systems of biological networks in reference to specified production, welfare, and sustainability goals (Zhang et al., 2025).

The justification for the implementation of AI in this field is rooted in the complexity, magnitude, time dynamics, and multi-objective nature of livestock nutritional biology by its very nature. This is complicated by the nonlinear interaction of genotype by diet and environment, feedback manifested in host and microbial metabolism, and regulation at hierarchical levels beginning with the accessibility of chromatin and ending with phenotypic expression and post-translational modification. Scale is a manifestation of increased omics resolution, sparse longitudinal streams of sensors, and an increased repository of formulated diet compositions and performance history. Digestion kinetics, circadian regulation of metabolism, developmentally sensitive nutritional windows, and adaptive nutritional remodeling in response to climatic or management disturbances are all temporal in nature. There must be multi-objective optimization between growth or production and immune competence, reproductive efficiency, resource utilization, emission mitigation, robustness under stress, and economic viability. The AI approaches have a unique advantage in their ability to model the nonlinear response surfaces, propagate uncertainty, rank the most influential variables, integrate asynchronous time series, and introduce intervention policies to balance opposing goals and constraints across changing environmental priors. Predictive accuracy is a single dimension in this framing, and interpretability, adaptability, scalability, and fit to ethical and sustainability requirements are equally important performance indicators (Nawaz et al., 2025; Zhang et al., 2025).

It is essential to clarify the scope and the boundary conditions to place this section in the context of parallel chapters on omics, which are on breeding, health management, welfare monitoring, microbiome ecology, environmental sensing, and system-level digital twins. In this case, we are dealing with the two-way interface between dietary inputs and the two more layers of genomic and post-genomic regulation, as well as their impact on functional phenotypes. Although microbiome analytics, immune modulation, or reproductive tuning are recognized as mediators, being mediated by, or mediating, nutritional signaling, much of the focus is on the integrative modeling of nutrient-driven molecular cascades and resultant phenotypic modulation. Differences are made based on genomic selection models that view long-term frequency changes of alleles instead of short-to medium-term changes in expression phenotype and metabolic phenotype, and disease surveillance models that view pathogen dynamics instead of host nutrition determinants of nutritional resilience. The mention of edge sensing architectures and more general supply chain optimization is limited to the extent that it provides or demands the outputs of models of nutrigenomics. Similarly, the economics of pure feed formulation and logistical optimization cannot be in scope without being enhanced with omics-derived functional response modeling. This limitation provides conceptual consistency as well as the ability to interoperate across chapter boundaries with shared data standards, ontologies, and validation standards (Farzand et al., 2025; Yang et al., 2025).

Strategy relevance cuts across interlaced realms of productivity, health, wellbeing, sustainability, and economic resilience. The benefits of productivity shift from generic ration enhancement to a precision modulation of metabolic efficiency, nutrient partitioning, and adaptive resilience in changing environmental or managerial situations. The health is addressed by predictive host defense, and suppression of subclinical dysregulation is countered by aligning the nutrient-caused immunometabolic states with genomic risk profiles. Welfare involves nutritional support for behavioral stability, stress reduction, as well as prevention of metabolic distress states with a focus on early identification of maladaptive responses at molecular or physiological levels, before an apparent phenotypic deterioration. The aspect of sustainability comes in with optimization of nutrient utilization efficiency, minimization of excretory losses, regulation of enteric and manure-related emission pathways, and better lifecycle resource intensity indicators. Economic resilience is the resultant ability to forecast and flex dietary plans to fluctuating input expenses, climate-associated shocks, altering regulatory limitations, and changing consumer demands, as far as ethical and environmental management is concerned. These strategic levers are supported with AI due to the continuous learning cycles, simulation of scenarios, and uncertainty-aware decision support, being able to quantify trade-offs and opportunity gradients in heterogeneous production systems (Džermeikaitė et al., 2025).

The weaknesses of traditional paradigms of analysis highlight the disruptive nature of AI. Linear mixed models, univariate association studies, and isolated pathway analyses tend to make assumptions of additivity, stationarity, and small orders of interaction, and therefore fail to capture the potential existence of nonlinear synergy, emergent thresholds, and context-specific regulatory rewiring. The static formulations do not take into account time-phase relationships and adaptive feedback, limiting the proactive intervention ability. Manual feature selection creates subjective bias and can overlook combinatorial signatures of weak individual but strong collective influence. The standard statistical correction schemes are biased to overinflate the false negatives of high-dimensional sparse effect landscapes, which restricts the discovery of subtle but operationally significant nutrient genotype contingencies. In addition, compartmentalized pipelines subdivide the data provenance, hinder the reproducibility, and enhance batch effects between laboratories and platforms. To overcome these shortcomings, AI helps in supporting multi-view representation and learning to establish a hierarchical attention to disentangle scale-specific signal, probabilistic graphical, or causal structure inference to go beyond correlation, sequence, and temporal convolution to learn dynamic transitions, two-tier optimization, or reinforcement to derive policy-level advice in line with multi-criteria objective architectures. Variations on hybrid mechanistic data-driven paradigms are more resistant to black box issues and incorporate rigorously established concepts of nutritional physiology into flexible learning frameworks (Androulakis et al., 2025; Aydın et al., 2025).

In this chapter, the development of a systematic movement toward the initial concepts to integrative frameworks, translational implementation, governance, and future directions is presented. This chapter poses questions such as what data collection methods can make models more robust, what models are most appropriate to strike a balance between performance and interpretability in multilevel biological systems, how uncertainty should be quantified to make better choices of risk-aware intervention, and what paradigms of validation can be argued to provide validity between experimental and commercial tasks. It examines how knowledge graphs, semantic ontological as well as standardized metadata schema, catalyze interoperability between cross-platform and model transferability. It looks at the ethical and equity aspects of access, algorithmic favoritism among breeds, systems of production, or regions, and the duty of stewardship of genomic and phenotypic data. It considers infrastructural enablers to include federated learning, data sovereignty, privacy limitations, edge-to-cloud orchestration decision loops with latency requirements, and performance drift with recalibration lifecycle monitoring.

Instead of listing, guiding questions run through it, inquiring how integrative models may unravel causally relevant modulation and epiphenomenal correlation, how adaptive algorithms may be reliable in climatic volatility and changing dietary ingredient scenery, how interpretability systems can balance operational transparency with intellectual property security, how multi-objective optimization may exemplify Pareto efficient nutritional strategies balancing productivity and environmental footprints, how data governance may be equitable value distribution and trust among stakeholders and how capacity building programs can align workforce competencies and emerging methodologies. These questions place an agenda that is forward leaning, where AI is not treated as an adjunctive addition to the analysis but a form of organization intelligence, where discovery, validation and deployment and continuous improvement cycles along the nutrigenomic innovation pipeline are re-defined.

When combining these items, the introduction develops a conceptual and methodological background that will be developed in later parts, which will expound on the data infrastructures, algorithmic methods, areas of integration cases, validation plans, implementation issues, ethical standards, and future innovation directions. The general thesis is that an ethically regulated, but rigorously applied, AI can facilitate the shift in nutritional evaluation of the past to a more adaptive and resilient, anticipatory, and nutrigenomic stewardship that balances the biological potential with sustainability and humane production necessities.

8.2 Overview of livestock nutrigenomics

The central premise of livestock nutrigenomics lies in the fact that signals of nutrient derivatives are not just substrates to be metabolically burned or structurally packaged, but dynamic regulators of gene expression, cellular differentiation, and organism change. The gene-nutrient interactions include direct ligand-mediated regulation of transcription factors, cofactor-dependent enzymatic flux regulation, substrate provision effects on epigenetic methyl donor pools, as well as feedback via metabolite-sensing nodes that re-tune expression networks. This metabolic regulation is transformed into a interplay of part-whole allocation over anabolic and catabolic pathways responsive to immediate nutrient abundance as well as longer-term adaptively plastic regulation coded in regulatory forms. Adaptive plasticity is the ability of physiological systems to reparameterize response parameters and allocation priorities at stages of development, levels of production, and environmental demands, utilizing temporary and semi-permanent molecular marks to bring the changing resource environments into agreement with homeostatic demands and performance requirements. Nutrigenomics, therefore, places diet as an information-dense input that influences phenotypic pathways by multilayer molecular mediation, as opposed to an energetic bottom line (Gundersen and Anas, 2025). Main nutrigenomic mechanisms in livestock are shown in Table 8.1.

The regulated lattice at the center of which are the canonical and connected nutrient-sensing networks coordinating energetic charge, amino acid adequacy, lipid-derived messages, oxidative conditioning, and inflammatory intensity. The mTOR axis serves as an integrator of amino acid cues, growth factor signaling, and energy conditions to control the initiation of translation, autophagic flux, and biosynthetic capacity, and therefore the accretion and metabolic priority of tissues. AMPK detects the energy charge in the cell and coordinates the compensatory changes to support catabolism, open the cell to substrates and mitochondrial biogenesis, and adjust anabolic restraint. Retracing signal boards onto lipid oxidation and adipogenesis transcriptional modulations mediated by PPAR mediate lipid ligand landscapes into inflammatory compensation results and link these functions to energy scenario and immune setting. Antioxidant and cytoprotective gene expression coordinated by NRF2 to electrophilic and oxidative signals is associated with the dietary intake of phytochemicals and endogenous redox responses and is linked to resiliency to metabolic stress. These hubs are inflammatory crosstalked on by metabolically signaling cytokines and danger-associated molecular patterns that can reorganize substrate allocation, mitochondrial activity, and appetite. These pathways never exist as single pathways; rather, they form conditional networks in which they are subject to reciprocal feedback, competitive cofactor requirements, shared transcriptional coactivators, and convergent posttranslational modification control nodes, all of which contribute to form the system-level phenotype arising in a particular dietary situation (Smiles et al., 2024).

The epigenetic modulation is a critical dimension by which nutritional cues have the capacity to convert short-term environmental cues into a set of altered regulatory set points with a variable duration of persistence. DNA methylation trends are responsive to the availability of methyl donors, one-carbon metabolic flux, and oxidative stress, and thus encode nutrient provision into promoter or enhancer accessibility changes, which determine transcript abundances. Another layer is the histone post-translational marks, with acetylation indicating the availability of acetyl CoA and the more general energy state, methylation indicating the level of S-adenosylmethionine pools, and acylation variants representing a variety of intermediates. These marks affect chromatin compaction, measurement of transcription factor binding, and communication of enhancer promoters and regulate transcriptional permissiveness between sets of metabolism, stress response, and development genes. Non-coding RNAs, such as microRNAs and long non-coding transcripts, regulate post-transcriptional stability, translation efficiency, regulatory protein decoying, and chromatin remodeling, with these extended effects of nutritional regulation, therefore, extending beyond the primary events of transcription initiation. These layers of epigenetics, combined, offer a response calibration mechanism to nutritional experience, noise attenuation, and multi-gene program coordination, which underlie metabolic flexibility, immune readiness, and reproductive allocation programs. They have partial reversibility and dependence on the context, which gives them not only the possibility of targeted modulation but also complexity in the process of deciphering directionality and persistence of the causes of life over its stages (Del Vecchio, 2025; Ibragić et al., 2025).

The microbiome provides an interface of metabolism and signaling that has the capability to convert primary dietary substrates to bioactive metabolites that mediate host transcriptional, epigenetic, and endocrine environments. Microbial consortia mediate fermentation, deconjugation, hydrogen management, nitrogen salvage, and biotransformation of bioactive compounds to generate the products, short-chain metabolites, secondary lipid derivatives, amino acid analogs, redox active molecules, which serve as host receptor ligands, energy sources, and epigenetic substrate precursors. The microbial community structure is influenced by host genotype and immune monitoring, and the types of diets and feeding behaviors give the selective pressures that, at times, control the abundance of functional genes in the microbiome. This mutual interaction has an effect on nutrient absorption potential, intestinal obstinacy, inflammatory facilitation, and metabolic signaling gradient, entrenching microbiome-mediated diversity into the larger nutrigenomic response ecosystem. Interference by temporal instability, heterogeneity in strain level, and ecological feedback loops makes it very challenging, but when microbial functional-profile data, coupled with host omics, are processed, multi-layered regulatory loops can be confirmed, and the data can be better decomposed by inferential control by AI (Zeng et al., 2025).

Further levels of differentiation in the dynamics of nutrigenomics are added by species-level physiological architecture and production stage-specific priorities. Compartmentalized fermentation in the rumen species preconditions nutrient profiles at the small intestine, which allows the nutrient signaling locus to be shifted to microbial fermentation products and hepatocentric coordination pathways. There is a stronger direct relationship between dietary formulation and post-absorptive metabolite kinetics in monogastric species, with a strong focus on enterocyte nutrient sensing and pancreatic endocrine adjustment. The gastrointestinal transit dynamics, thermoregulatory factors, and unique lipid and nitrogen

TABLE 8.1 Main nutrigenomic mechanisms in livestock.

Mechanism	Definition/Description	Biological pathway(s) involved	Examples of nutrients or genes affected	Role of AI in analyzing or enhancing this mechanism
DNA methylation	The addition of methyl groups (–CH_3) to cytosine residues in DNA (often at CpG sites) modulates gene expression without changing the underlying DNA sequence	One-carbon metabolism; DNA methyltransferases (DNMT1, DNMT3A, and DNMT3B)	• Methyl donors such as methionine, folate, choline, betaine, and vitamins B12/B6 influence methylation • DNMT genes (e.g., DNMT1 and DNMT3A) in ruminants are nutritionally regulated	• AI (e.g., machine learning) can predict methylation states from nutrient intake data • AI-driven "epigenome-wide association studies" (EWAS) can model the relationship between diet, methylation, and phenotype. • Deep learning can help identify methylation biomarkers predictive of production traits
Histone modifications	Post-translational modifications (e.g., acetylation, methylation, and phosphorylation) of histone proteins that affect chromatin structure and gene accessibility	Chromatin remodeling; histone acetylases/deacetylases; histone methyltransferases/demethylases	• Nutrients such as methionine influence histone methylation in livestock • Transcription-factor genes like PPARα may be regulated via diet-induced histone marks in cattle	• AI models can use multi-omics data (nutrient and histone mark) to predict epigenetic states • Feature-selection algorithms can identify which histone marks are most responsive to dietary interventions • Explainable AI (XAI) can help interpret how histone-modification patterns influence gene expression in response to nutrients
Non-coding RNAs (ncRNAs)	Regulatory RNA molecules (e.g., microRNAs and lncRNAs) that modulate gene expression post-transcriptionally without encoding proteins	Post-transcription regulation; RNA-mediated gene silencing; chromatin interaction via lncRNAs	• Macronutrients and bioactive compounds can alter the expression of miRNAs and lncRNAs • Non-coding RNAs target metabolic-regulation genes (e.g., in adipogenesis or immune function) in livestock	• AI (e.g., clustering or neural networks) can identify ncRNA signatures that respond to dietary changes • Predictive models can link nutrient-regulated ncRNAs with metabolic phenotypes • Integration with AI-driven network inference can reconstruct regulatory networks involving ncRNAs
RNA methylation	Covalent modifications (e.g., N^6-methyladenosine, m6A) on RNA molecules that influence RNA stability, splicing, translation, and decay	RNA processing pathways; "writers" and "erasers" of RNA methylation; one-carbon metabolism for SAM production	• Micronutrients (e.g., folate, vitamin B12) that impact SAM availability influence RNA methylation • RNA methylation can regulate transcripts related to metabolism and development	• AI models can infer RNA methylation patterns from dietary inputs and transcriptomic data • Machine learning can be used to predict how changes in nutrient supply alter the epitranscriptome

Continued

TABLE 8.1 Main nutrigenomic mechanisms in livestock.—cont'd

Mechanism	Definition/Description	Biological pathway(s) involved	Examples of nutrients or genes affected	Role of AI in analyzing or enhancing this mechanism
Genomic imprinting	Parent-of-origin–specific gene expression controlled by epigenetic marks (such as DNA methylation) established during gametogenesis	Imprinting control regions; DNA methylation; histone marks; chromatin structure	• Imprinted genes in livestock (e.g., growth, development-related) may respond to maternal nutrition	• AI can model how prenatal dietary interventions influence imprinting patterns. • Predictive models could incorporate imprinting status to forecast long-term trait outcomes
Transgenerational/ Epigenetic memory	Epigenetic marks (methylation and histone modifications) that persist across generations, transmitting information about ancestral nutrition	Epigenetic inheritance; chromatin remodeling; germline methylation maintenance	• Nutrition during gestation can induce epigenetic changes that persist in offspring and even grand-offspring • One-carbon nutrients (e.g., B-vitamins) influence methylation patterns across generations	• AI-driven longitudinal and prediction models can forecast transgenerational trait effects from maternal diet • Multi-generational machine-learning models can help disentangle inherited epigenetic variation driven by nutrition

processing patterns with diverse effects on nutrigenomic response windows are unique to poultry systems. Traditional production stage shifts, including growth acceleration, reproductive cycling, lactational demands, and finishing or post-peak transitions, change hormonal milieus, immune prioritization, and rates of tissue remodeling, reweighting pathway sensitivities, and adaptive plasticity levers. Such species and stage disparities require stratified modeling methods that accommodate divergent baseline metabolic schemes, microbial organization, and genetic historical context, in addition to permitting transfer learning in contexts of dominance of conserved regulatory motifs (Belanche et al., 2025).

Examples of the trait domains that are regulated by nutrigenomic modulation include the efficiency of nutrient uptake and use, immune competence tuning, strength of the reproductive axis, product quality traits, and the environmental emission profile. The synchronization of modulating appetite regulation, digestive enzyme capacity, transporter expression, mitochondrial coupling, and microbial synergism leads to the development of feed efficiency. The functions of immunometabolic reprogramming, antioxidant defense capacity, barrier integrity, and inflammatory resolution dynamics are integrated during immune functioning that is determined by nutrient-derived signaling and energetic partitioning rules of the genomic regulatory nature. Reproductive outcomes reveal that nutrients have an effect on the support of gametogenesis, endocrine stability, optimization of the uterine or embryonic environment, and epigenetic programming events throughout developmental checkpoints. Nutrient-mediated biosynthetic pathways that dictate the composition of tissues, structural protein crosslinking, lipid deposition patterns, and bioactive compounds incorporation condition product quality attributes. Nutrient-driven microbial fermentation profiles, nitrogen and sulfur metabolism, and the overall metabolic efficiency in the system all regulate emission intensities and broader environmental externalities by means of excretory endpoints. Nutrigenomics, therefore, offers leverage sites at the boundaries between which numerous domains of traits may be developed by modulation of convergent regulatory nodes and allocation circuits in concert (Deng et al., 2025).

Nutrigenomics inquiry is based on data modalities that bridge a continuum of compositional, molecular, physiological, behavioral, and environmental levels. The data of feed composition record proximate and functional traits and ingredient diversity that are due to effects of origin, processing, and season, and which can be used as explanatory vectors of downstream biological diversity. Genomic variant data sets encode population diversity in both the coding and regulatory sequences that affect nutrient receptor affinity, transporter efficiency, metabolic enzyme kinetics, and regulatory motif occupancy. Transcriptomic maps define the state of the situation in gene expression, whereas proteomic strata provide information on the process of filtering of post-transcriptional regulation, assembly, and post-translational modifications. Proximal signatures in metabolomics are a combination of proximal regulation and systemic metabolic state, which is a causal inference between the upstream regulation and phenotypic outcomes. Microbiome functional and taxonomic signatures provide exogenous metabolic scale and signaling abilities, whereas phenomic streams are used to quantify growth patterns, production rates, behavior, disease indicators, and reproductive rates. Sensor and environmental feeds supply high-resolution time data on thermal load, humidity, locomotion, rumination of feeding patterns, and facility-level variables to allow them to find molecular snapshots that are dependent on dynamic contextual drivers. Regularization of these modalities produces an integrative framework of disentangling causal structure and correlative entanglement (Alem, 2025).

The problems associated with this multidimensional landscape are that they are highly heterogeneous in terms of data sources, platforms, and temporal sampling structures, which result in batch effects and alignment problems. Mismatches between temporal latency are apparent both in situations where molecular shifts precede phenotypic divergence or in situations where microbial community restructuring lags dietary reformulation, making the situation difficult to calibrate and predict. Context-dependence influences environment interactions between genotype and diet, and it scales up effects and makes environmentally static models fragile to regime changes. Another obstacle is sparse mechanistic annotation, and a significant proportion of differentially responsive transcripts, metabolites, or microbiome functional components do not have a definite pathway incorporation, decreasing the interpretability and transferability of predictive signatures. Furthermore, physiological states are partially observable, which adds the latent variable confounding possibility, and longitudinal continuity is limited in a commercial setting, limiting the power to model dynamics of adaptation and resilience. There are also ethical and logistical limits on invasive sampling to make some tissues or life stages sparse, which strengthens the need to use proxy biomarkers, the mechanistic specificity of which can be incomplete (Khan et al., 2025).

The integrative conceptual models attempt to close the gap between diet composition and emergent phenotypic outcomes via molecular cascades to represent hierarchical regulation, feedback, and allocation logic. Ingestion events are transformed into luminal physicochemical conditions, microbial metabolic transformation, and enterocyte nutrient sensing to trigger conceptualization, which unfolds via portal circulation mediating the delivery of metabolites to the liver and integrating endocrine axes of liver-receptive systemic energy balance signaling with peripheral tissue responses. Nutrient-sensing pathways within tissues synchronize transcriptional programs that are dependent on epigenetic context to regulate the concentration of enzymes, the density of transporters, and the biogenesis of organelles, which establish metabolic flux distributions. The outputs of metabolites are fed back into transcriptional and signaling nodes and also have an effect on the availability of substrates of the microbiome via excreted intermediates, and therefore, multi-scale loops. In phenotypic outputs at the organism level, the resolved allocation is manifested in the areas of growth, reproduction, immune surveillance, thermoregulation, and adaptation to stress, and each of these areas is regulated by regulatory prioritization algorithms through emergent layered molecular interactions. To simulate an intervention scenario, resilience to perturbation, competitive resource allocation, and context-contingent rewiring, integrative models should encode nonlinearity, time lags, saturation thresholds, competitive resource allocation, and context-contingent rewiring (Segev et al., 2025).

AI serves as an enabling layer of abstraction that has the ability to balance high-dimensional heterogeneity, align asynchronous temporal signals, and uncover latent structure that projects diet-induced molecular perturbation to phenotypic trajectories in a condition of uncertainty. Representation learning methods transform sparse, noisy, and partially labelled features into semantic-enriched representations, which are easy to generate mechanistic hypotheses that predict with fidelity. Graph-oriented models are in a position to internalize pathway topology, metabolism network stoichiometry, and microbiome-host interaction edges, which aid in generalizing and deciphering compared to purely opaque function approximators. The temporal modeling structures support latency structures and state dependence, and delineate transient variations and lasting trajectory changes, which have long-standing phenotypic consequences. Causal discovery and counterfactual inference methodologies go beyond association, investigating the directional influence and intervention efficacy, and confounding is considered. Multi-objective optimization interfaces combine modeled surfaces of outcomes and constraints that represent welfare, environmental, and economic limits and give actionable nutritional strategy guidance in accordance with the stakeholder priorities. In such a conceptual matrix, reliable and powerful data management, standardized ontologies, and calibration procedures form the foundation of reliability, fairness, and scalability (Wei et al., 2024).

The changing assembly of such components makes livestock nutrigenomics a key channel in which AI can translate the molecular understanding of adaptive nutritional stewardship. Through the formulation of fundamental principles, the description of regulatory circuits, the exposure of epigenetic and microbiome-mediated regulation, the recognition of species and stage-specific organization, the mapping of influence domains of traits, a catalog of data modalities, the inherent challenges and formalization of integrative conceptual models, the field provides a systematic basis for future methodological innovation and translational application. The general story supports the idea that diet is a planned informational carrier whose optimization needs systems-level modeling that goes beyond and across the reductionist fragmentation, with AI offering the computing framework to make complexity manageable at the same time, maintaining biological sensitivity and ethical fit. Major nutritional factors studied in livestock nutrigenomics are presented in Table 8.2.

8.3 Applications of AI in nutrigenomics research

The AI-based applications in the nutrigenomics of livestock are converging into a functional ecosystem that interprets the streams of molecular, physiological, behavioral, and environmental data into adaptive nutritional responses. Its goal is to maintain a balance between the biological potential, resource limitations, and sustainability imperatives by constantly reconciling the data caused by learning loops, which map the diet composition and feeding dynamics to multiscale responses. It is a predictive modeling and causal inference, uncertainty-based optimization, and simulation-based augmented decision support, which is based on biologically informed representations that are aware of metabolic, immunological, reproductive, and ecological constraints. The resulting storyline runs through a range of areas of focus in which AI enhances accuracy, anticipatory governance, and multi-objective equilibrium, and systematically introduces mechanistic interpretability and ethical custodianship (Zhang et al., 2025).

Precision feed formulation has developed out of static ration balancing to dynamic ration optimization that incorporates actual or nearly actual time indicators of consumption, growth patterns, environmental burden, ingredient quality variation, and biological state indicators. AI controllers find combinations of these signals and refresh marginal response surfaces to include energy, amino acid, mineral, and functional additive, and make use of reinforcement learning or Bayesian optimization models with nutrient requirement prior and safety constraints. Representation learning is a compression-based method that uses the high-dimensional ingredient composition, spectral, and processing characteristics to latent vectors that enhance transferability across batches. Probabilistic models spread uncertainty of ingredient variability and measurement error to risk-adjusted formulation decisions. Multi-scale integration enables digital control of feeding curves in reaction to early non-concurrence between expected and observed phenotypes, and thus stabilizes performance and eliminates over-formulation and waste. Hybrid mechanistic data-driven models use nutrient partitioning

TABLE 8.2 Major nutritional factors studied in livestock nutrigenomics.

Nutritional factor	Role in nutrigenomics/One-carbon metabolism	Biological mechanisms affected	Examples in livestock or model systems	Relevance for AI analysis/Predictive modeling
Methionine	Essential amino acid; a primary methyl donor via S-adenosyl-methionine (SAM)	• Transmethylation reactions • Protein synthesis and methylation of DNA, RNA, and histones	• In neonatal piglets, dietary methyl donors (including methionine) influence whole-body protein synthesis • In cattle, methionine cycle flux is critical during gestation for epigenetic programming	• AI models can integrate dietary methionine intake with metabolomic and epigenomic data to predict methylation states and growth traits • Feature-selection algorithms can identify methionine's relative contribution to epigenetic markers or production outcomes
Folate (vitamin B_9)	Key cofactor in the one-carbon (folate) cycle; contributes one-carbon units for methylation	• Provides methyl groups via tetrahydrofolate (THF) forms • Supports remethylation of homocysteine to methionine via methionine synthase	• Folate supplementation or restriction in ruminants influences hepatic one-carbon metabolites and expression of methionine-cycle genes • Folate deficiency may impair SAM production, reducing methylation capacity	• AI can be used to model how fluctuations in dietary folate affect metabolic fluxes in one-carbon metabolism • Predictive models may identify livestock individuals particularly sensitive to folate supply for epigenetic programming
Vitamin B_{12}	Essential cofactor for methionine synthase in remethylation of homocysteine	• Supports conversion of homocysteine to methionine • Supports maintenance of DNA methylation	• In beef heifers, one-carbon metabolite (OCM) supplementation (including B_{12}) significantly increases fetal and maternal OCM intermediates • Correlations between B_{12}, folate, and methionine-cycle gene expression have been documented in fetal liver	• AI-driven association analyses can link B_{12} status to epigenetic outcomes and production phenotypes • Machine learning models may identify optimal B_{12} levels for epigenetic stability across breeds

Continued

TABLE 8.2 Major nutritional factors studied in livestock nutrigenomics.—cont'd

Nutritional factor	Role in nutrigenomics/One-carbon metabolism	Biological mechanisms affected	Examples in livestock or model systems	Relevance for AI analysis/ Predictive modeling
Choline/Betaine	Methyl donor via betaine, and precursor to phospholipids; contributes to remethylation	• Betaine-homocysteine methyltransferase (BHMT) pathway • Phosphatidylcholine synthesis and lipid metabolism	• In dairy cows, choline acts as a methyl donor and supports methionine regeneration, affecting liver metabolism • Choline-derived betaine supports remethylation and influences epigenetic methylation through SAM supply	• AI models can predict how choline supplementation affects methylation and production traits (e.g., milk yield) • Feature importance methods can reveal how choline versus methionine contributes to methylation variability. • Deep learning could model non-linear interactions between choline, betaine, and other methyl donors
Vitamin B_6 (Pyridoxine)	Cofactor for enzymes in one-carbon metabolism (e.g., serine-to-glycine conversion)	• Supports serine hydroxymethyltransferase (SHMT) activity • Influences folate cycle and nucleotide biosynthesis	• As reviewed in epigenetic studies, B_6 is required for one-carbon flux via serine/glycine interconversion	• AI-based pathway models (e.g., flux balance modeling) can estimate how B_6 levels modulate one-carbon flux • Comparative modeling can identify when B_6 becomes a limiting nutrient for methyl group supply
Other bioactive nutrients (e.g., short-chain fatty acids (SCFAs) and trace minerals)	non-traditional methyl donors or epigenetic modulators	• SCFAs like butyrate modulate histone acetylation (inhibiting HDACs). • Trace minerals such as zinc or selenium influence epigenetic enzymes (e.g., histone deacetylases)	• In ruminants, SCFAs derived from fiber fermentation might influence epigenetic regulation via acetyl-CoA supply • Mineral nutrients may regulate chromatin-modifying enzymes with downstream effects on gene expression	• AI (e.g., multi-omics integration) could connect rumen fermentation profiles, SCFA levels, and epigenetic states • Predictive models might evaluate how dietary fiber and mineral supplementation influence long-term epigenetic outcomes

equations in adaptive residual learners, improving biological realism while learning context-specific nonlinearities. The emergent system facilitates individualized feeding in technologically empowered contexts as well as cohort-level optimization in the face of infrastructure limitations and bridges the gap with economic, welfare, and environmental performance feedback (He et al., 2025).

Feed efficiency and residual feed intake predictive models are based on genome-wide variation, transcript abundance markers, metabolite assays, microbial functional profiles, and phenomic time series to separate biological factors of differential nutrient use. Graph-based architectures combine metabolic pathway topology and gene-level regulatory interactions to become more faithful to feature aggregation, and integrate correlated features using multitask learning to utilize similar latent structure. Causal discovery techniques question the directionality between microbiome-mediated fermentation changes, proxies of mitochondrial functioning, and realized efficiency measures, shifting the need to predict with associativity to intervention advice. Transient perturbations caused by heat stress or health challenges are identified as temporal encoders because persistent individual efficiency signatures are viewed as time-independent. Uncertainty quantification can be used to inform selective phenotyping in order to sample more omics areas where predictive entropy is still large. These models drive the selection indices, real-time feeding adjustments, and dashboards for benchmarking, which enable genetic advancement and optimization of short-horizon management. The process of interpretability is used to find nutrient-responsive gene networks that can be manipulated to further reduce residual feed intake without compromising the robustness or the welfare (Liu, 2025).

AI approaches to the modulation of immune competence and disease resistance are promoted with AI frameworks that combine immunotranscriptomic data, cytokine sets, acute phase proteins, metabolomic inflammation proxies, and dietary bioactive compound characteristics. Nutrient, receptor, signaling cascade, and effector functions are connected in knowledge graphs, which are used to predict immunometabolic set points with lower disease susceptibility. Counterfactual simulation is a proposed test of the effect of hypothetical changes in the nutrient profile and its simulated effect on immune preparedness, which allows proactive immune modulation of the diet in high-risk seasons. The multi-omics integration separates the overlapping signatures of subclinical inflammation and adaptive remodeling, which enables nutritionists to prevent misclassification and over-supplementation. Mechanistic regularization prevents the model hypotheses from known immunological principles, reducing spuriousness in the correlations. Probabilistic risk stratification is constantly enhanced by continuous learning pipelines updating based on new data on the spread of an outbreak or challenge, completing a process of adaptive feedback resulting in possible adjustments in nutrient allocation to enhance immune fortification without reducing growth and reproduction (Thoniyot and Balakrishnan, 2026).

Temporal anomaly detection across ingestion patterns, rumination or activity sensor streams, spectral milk or excreta biomarkers, and blood or saliva metabolic panels is useful in the early detection and prevention of metabolic and digestive disorders. Examples of sequence models that can be used to detect preclinical drift patterns before overt manifestations of disorder include multivariate change point detection as an indicator of coordinated disturbances in metabolic and microbial signals that may be indicators of imminent dysbiosis or organ dysfunction. One way to enhance the specificity of preventive recommendations is to use causal structure learning to isolate primary triggering shifts, such as a sudden change in fermentable carbohydrate load, and secondary downstream changes. The reinforcement learning agents model intervention timing and magnitude, and strike a balance between corrective adjustments of dietary fiber, buffering ability, or additive incorporation and expenditure, and possible palatability impacts. Privacy-preserving federated learning updates model changes in an environment where the sensitive health data are not centralized, enhancing generalization and robustness of model updates across management systems. Interpretability layers bring visible, actionable clusters of biomarkers, which makes them more approachable to practitioners and speeds up their adoption in the field (Roy et al., 2025).

Reproductive performance optimization through nutritional programming takes advantage of developmental and periconceptional plasticity to introduce micronutrient supply, energy balance, and specific bioactive compounds that leave an epigenetic and endocrine signature of fertility, embryonic development, and offspring vigor. The AI models combine longitudinal body condition, metabolic hormone profile, uterine environment metabolomics, composition of follicular fluid, and diet formulation metadata with the prediction of reproductive outcomes and the possibility to modify nutritional levers. Temporal alignment algorithms make synchronous the asynchronous sampling between estrous cycles, gestation phases, and lactation transitions to build phase-specific predictive embeddings. Causal mediation analysis is the quantification of the percentage of nutritional effects that are mediated causally by insulin-like, steroidogenic, or oxidative balance pathways, with mechanistic targets taking priority. Optimization algorithms of intervention propose customized supplementation plans, using probabilistic outcome uplift estimates that could be applied for risk-stratified resource allocation. Predictive granularity is enhanced by the incorporation of epigenomic markers in gametes or early embryos, but models impose ethical protection of the use of this data or decision-making concerning reproductive issues (Marinello et al., 2025).

Integrative models of dietary techniques to reduce methane, nitrogen, and phosphorus emissions combine feed composition, the abundance of rumen or gut microbial functional gene abundances, fermentative end products, measures of nitrogen partitioning, and metadata of manure management. Systems identification techniques generate emission response surfaces at different diet structural carbohydrate, lipid, and additive levels of inclusion, and multi-objective optimization expresses trade spaces between the emission intensity, production output, and the cost of feed. Graph-based causal inference separates the direct dietary control of methanogenic archaea and the indirect change through hydrogen sink processes or fiber degradation rates. Uncertainty propagation deals with the variability in emission measurement technology to provide directions for conservative or adaptive interventions. Reinforcement learning models sequential dieting changes during climatic and market shocks, and keep the emission targets without damaging the resilience. Transfer learning makes cross-species adaptation of emission mitigation models possible by matching conserved metabolic modules without interfering with species-specific fermentation ecologies (Xue et al., 2025).

To inform on diet personalization and modeling of probiotic or prebiotic effectiveness, microbiome-informed research has to separate host genetic background, baseline microbial ecology, and features of nutrient matrices that precondition intervention responsiveness. Variational representation learning reduces the strain level compositional data and functional gene catalogs into stability and resilience embeddings, which predict the magnitude of responses. Causal uplift modeling is used to calculate the estimated benefit probability of a specific candidate prebiotic fiber or microbial consortium to an individual, which can then be used to deploy it to an individual, in contrast to blindly supplementing. Dynamic Bayesian networks combining feedback in diet changes, microbial metabolite changes, epithelial signaling, and feed intake behavior explain the sequence ordering and delay parameters, which are important in scheduling interventions. The concept of simulation in virtual cohorts examines trajectories of complete community assembly under variant dietary scaffolds that inform the formulation that encourages positive guild growth and pathogen suppression. Federated microbiome modeling overcomes data sparsity and privacy challenges across regions and improves the global robustness of the model, but allows for biogeographic specificity (Nisa et al., 2025).

AI structures for assessing nutraceuticals and functional additives combine cheminformatic descriptors, in vitro bioactivity analyses, transcriptomic perturbation profiles, metabolomic changes, and in vivo performance results. The method infers the mechanism of action by using contrastive learning between additive-treated and control multi-omics profiles to isolate patterns of pathway modulation. Knowledge graphs represent interactions between molecular targets, signaling pathways, and phenotypic outcomes, and permit knowledge graph link prediction algorithms to identify potential mechanistic relationships between new compounds. Dose-response modeling is a technique that uses Gaussian process regression or neural spline architectures to learn nonlinear efficacy and saturation dynamics and uncertainty quantification. Meta learning is used to estimate the extrapolation of mechanisms and efficacy estimates to new production environments with sparse field data much faster. Safety signal detection incorporates the use of subclinical biomarker deviations and behavioral predictors in anomaly scoring systems, providing an opportunity to predict the risk of adverse interactions during existing diet components (Bhattarai et al., 2024).

The improvement of product quality and compositional traits, including milk, meat, and egg characteristics, capitalizes on predictive models connecting diet-derived fatty acid profiles, amino acid balance, provisioning of minerals and antioxidants, and inclusion of bioactive compounds to the final product texture, flavor precursor reservoirs, shelf life stability, and health-associated nutrient fractions. Multimodal fusion models are a combination of near-infrared or spectral imaging data, proteomic or lipidomic composition, and sensory panel embeddings to narrow the prediction of indices of consumer-relevant quality. To define the ultimate product characteristics, causal mediation models separate the effects of diet on muscle fiber metabolic phenotype, lipid deposition dynamics, or casein synthesis signaling. Optimization algorithms achieve a trade-off between the quality improvement with production efficiency and environmental cost by considering the Pareto efficient formulation sets. Domain adaptation methods can be used to transfer learned quality determinants among breeds or different production systems that have divergent baseline composition profiles. Interpretability achieves actionability, that is, it identifies the nutrient classes or adjustment of feeding patterns that result in the most marginal effects on the desired compositional results (Wang et al., 2025).

The models of latent state estimation that combine cortisol or analogous stress hormone proxies, acute phase proteins, oxidative balance indicators, feeding microstructure, locomotion or posture sensor streams, and ambient environmental load are used for welfare and stress reduction by metabolic and endocrine nutritional markers. These models deduce psychophysiological conditions that can be improved through special nutritional buffering using certain amino acids, electrolytes, antioxidants, or adaptogenic substances. Counterfactual policy assessment is based on the comparison of predicted welfare paths in different supplementation regimes, and interventions are designed to avoid being unintentionally stressful on metabolic capacity or social competition. Constant checking of variance, not merely mean levels, can be used to identify dysregulated adaptability, which is an indicator of chronic stress. To avoid the strategies that conceal the symptomology rather than the underlying burden, ethical restrictions are integrated in optimization goals to exclude strategies that hide their true intention without enhancing the actual welfare (Khaldy and Gheraíbia, 2025).

It relies on AI to valorize alternative feed resources and enable circular bioeconomy integration, based on the heterogeneous by-products, insect, algal, or novel fermentation-derived ingredients, which possess variability of composition and anti-nutritional effects disrupting traditional formulation paradigms. Spectroscopic and imaging measurements are supplemented by sparse chemical measurements that provide self-supervised models to predict the digestible nutrient fractions, presence of functional secondary metabolites, and possible toxin concentrations. Scenario simulation evaluates the volatility in supply availability, processing energy requirements, and the effect of substitution on the current ration structure and the emission footprint. The information on life cycle inventory is included in the multi-objective optimization of circularity gains with performance risk and cost. Uncertainty mindful recommendation systems suggest the use of ramped-based inclusion strategies on new ingredients, which is in tandem with monitoring protocols, which result in reassessment in case of behavior contrary to the anticipated biological reaction. Transferable latent ingredient embeddings enable the quick acquisition of novel characterized resources into formulation knowledge bases (Vatsanidou et al., 2025).

Multi-objective sustainability optimization is a step further than individual efficiency or emission indicators towards concomitant productivity, greenhouse gas intensity, nitrogen and phosphorus excretion, water footprint, land use displacement, antimicrobial reliance proxies, welfare indices, and economic margins. The evolutionary multi-objective algorithms or constrained Bayesian optimization present different Pareto optimal nutritional strategies that provide stakeholders with transparent trade-offs. Preference eliciting interfaces enable policy, market, and ethical priorities to be dynamically weighted with changes in policy, market, and ethical priorities. The robust optimization takes into account scenarios of climatic and supply chain perturbation to have recommended strategies that perform acceptably in case of variance. There is constraint handling, which includes regulatory nutrient discharge limits, welfare minimums, and this makes compliance early in solution generation. Interpretable trade space topology visualization reveals the nutrient changes or ingredient replacements that create a frontier movement, which can be used to create informed bargaining between competing goals (Sharma et al., 2025).

Genomic selection with nutrient responding biomarkers and gene diet interaction models expands classical estimations of breeding value by combining the context-specific expression quantitative trait locus, epigenetic marks, metabolite quantitative trait locus, and microbiome interaction coefficients that change the genetic merit achieved under different nutritional conditions. Hierarchical Bayesian models or deep kernel models are models that use genotype by diet covariance structure, allowing the prediction of different genetic rankings between possible feeding strategies. Selection indices can be dynamically adjusted to meet strategic changes toward emission reduction or product quality focus, but protect against the loss of strong qualities of robustness. The prioritization of causal variants is facilitated by the usage of chromatin accessibility and transcription factor occupancy data under different dietary conditions, which helps in the refinement of functional interpretation. Ethical supervision is used to ensure fair access to the advanced selection tools in different regions to avoid further development of inequality in performance due to data asymmetry (Ghosh et al., 2025).

The concept of digital twins of personalized or community-based adaptive nutritional control realizes virtual models of animals or cohorts that take into account continuous sensor data, periodic omics measurements, health history, environmental conditions, and feeding history to model state dynamics under alternative dietary conditions. Hybrid mechanistic machine learning systems are physiologically interpretable and enable idiosyncratic variation. State estimation filters compare predictions and observations in real time, and the latent variables, for example, the maintenance energy requirements of the dynamic maintenance, the functional capacity of the gut microbiota, or the immunity preparedness, are updated. In silico stress testing of intervention policies on stochastic perturbations is applied before field implementation, eliminating risk. Twins transfer learning accelerates new entrant calibration using population-level priors. Governance procedures provide privacy of data, computer security, and transparency in the justification of automated decisions, which promotes trust in algorithm-based nutritional modifications (Ebadi and Selamoglu, 2025).

Supply chain resiliency applications utilize scenario-based AI to predict ingredient volatility, climatic perturbations, regulatory changes, and geopolitical disruption, which affect nutritional strategy feasibility. Generative models predict future price, availability, and quality trajectories that are plausible given macroeconomic and weather conditions. Stochastic optimization determines hedging formulation strategies, inventory buffering, which ensure nutritional adequacy and performance in unfavorable conditions with the minimum variance in costs. Causal impact analysis calculates the anticipated deviation of performance against the base in every situation, which guides contingency planning. Coupling with emission and welfare objective models is necessary to rule out the situation where the resilience strategies push the risk to the environmental or ethical aspect. Active learning focuses data gathering on new areas of uncertainty demands, for example, new ingredient supply sources or climate shift baselines, keeping adaptive systems relevant (Lin et al., 2025).

Themes of convergence can be found across these interdependent fields of application: there is a clear need to ensure that data governance is robust enough to achieve integrity and ethical stewardship, there is a need to ensure that mechanistic augmentation is done to bring about stability in generalization and interpretability, there is a need to ensure that uncertainty quantification is the center of attention to risk-aware decision support, and that federated, privacy respecting collaboration is the key strategic element in democratizing innovation. Instead of replacing domain expertise, AI increases its bandwidth, converting the complexity of nutrigenomics to operational, adaptive, and multi-objective nutritional management strategies. The result is a gradual shift from a fixed input system to adaptive optimization of biological, economic, and ecological pathways, all guided by transparent and responsible decision-making structures. Functional applications of AI in livestock nutrigenomics research are presented in Table 8.3.

8.4 Data mining and machine learning techniques in nutrigenomics

Within the framework of livestock nutrigenomics, data mining and machine learning technologies form an integrated methodological continuum that begins with disciplined data acquisition pipelines and progresses to representation learning, predictive and causal modeling, and adaptive and lifecycle control of deployed systems. The nutrigenomic multi-faceted nature, which includes different characterizations of feed ingredients, multi-layered host omics, dynamic microbiome functional statuses, longitudinal phenomics, environmental telemetry, and economic/sustainability indicators, requires strict preprocessing structures that maintain the preservation of analytical soundness and do not lose the small biological signal. Raw streams of data are fed into staging layers where they match the schema, ontology mapping, and timestamping against divergent measure granularities. Modality-specific normalization still reduces scaling effects by applying variance-stabilizing normalizations to counts of sequencing, probabilistic quotient, or robust scaling to intensities of metabolites, ratio centering to spectral structures, and temperature-humidity index calibration to environmental variables. Microbiome profiles, proportions of ingredients included, and vectors of nutrient allocation require compositional models of data; log ratio transforms such as centered, isometric log ratio transforms, normalized simplex geometry, and sparse zero inflation are corrected by Bayesian multiplicative replacement or Bayesian model-based zero classification. Effects induced by batch sequencing, analysis platforms, or feed laboratory instruments are solved by empirical Bayes harmonization moderated by maintenance of biology-consistent covariation. Imputation makes use of the conditional distributions trained with denoising autoencoders, low-rank matrix completion, or hierarchical Bayesian data augmentation, and propagates imputation uncertainty into eventual downstream modeling, as opposed to silent replacement. Temporal interpolation is physiologically plausible with respect to kinetics and shuns leakage by the future through window-constrained smoothing. Anomaly detection is used to identify quality control layers that measure improbable nutrient assays or transcriptional outliers, which could be the result of poor sample handling. Reproducibility is based on provenance metadata, and cryptographic hashing and semantic enrichment through controlled vocabularies are essential to guarantee machine-actionable interoperability between repositories (Mia et al., 2025; Uhler and Zhang, 2025; Zhang et al., 2025).

With high-dimensional, sparse, and noisy inputs on a stabilized data substrate, feature engineering and representation learning convert the inputs to structured embeddings that reflect latent biological and functional correlations. The representation of feed ingredients is derived from crude proximate analysis to multi-channel embeddings of amino acid spectra, fiber physicochemical fractions, secondary metabolite fingerprints, processing-induced structural changes, and digestibility coefficients. In these embeddings, information is learned through contrastive frameworks that match analytical profiles of ingredients to observed post-absorptive metabolite phenotypes or performance behaviors, thereby distilling functional, salient dimensions. The representations of nutrient levels have hierarchical ontologies between chemical classes, metabolic functions, and regulatory interactions, allowing relational embeddings that reflect patterns of substitutability and synergy between amino acids, volatile fatty acids, minerals, lipid moieties, and bioactives. A graph-neural encoder defines pathway embeddings on top of the metabolic and signaling network topology, completed with dynamic edge weights representative of context-specific flux or activation states. K-mer informed transformer or haplotype graph representation in genetic variant representation presents the linkage structure and regulatory context of genetic variants, but incorporates chromatin accessibility and transcription factor motif occupancy scores to rank variants with potential nutrigenic modulation effect. The cross-modality co-embedding is used to feed higher-latent learners' genotype, microbiome functionality, and phenotypic tracks that can learn relationships through cross-modality transfer and be able to predict together datasets that were previously siloed. Sparse feature selection is also a viable direction that is becoming more frequent in latent space, with disentangling goals promoting the orthogonality of biological axes like energy metabolism, immune modulation, and reproductive signaling (Chiregi et al., 2025; Hu et al., 2025).

TABLE 8.3 Functional applications of AI in livestock nutrigenomics research.

Application domain	AI role or function	Benefit for nutrigenomics	Considerations or challenges
Precision diet design	Use predictive models (e.g., supervised learning and ensemble models) to tailor diets based on animal genotype, epigenomic state, and nutritional requirements	• Enables formulation of diets that optimize gene expression, epigenetic marks, and metabolic efficiency • Improves feed efficiency and production performance by aligning nutrient supply with genetic background	• Requires large integrative datasets (genotype, diet, transcriptome, or epigenome) for accurate training • Risk of overfitting if diet-genotype interactions are complex and under-sampled • Practical implementation cost: "precision diets" may be expensive or difficult for farms to deliver
Biomarker discovery for nutrient response	Feature-selection ML techniques (e.g., Random Forest and LASSO) or deep learning to identify nutrient-responsive genes, transcripts, or epigenetic marks	• Identifies molecular signatures (e.g., methylation sites, ncRNA, and transcripts) that predict how animals respond to specific nutrients • Facilitates the development of biomarkers for nutritional status or responsiveness	• Biomarker candidates may not generalize across breeds or species • Validation in independent cohorts is essential • Interpretability: It's critical to understand whether a marker is causally related to nutrient response or is correlative
Longitudinal nutritional intervention optimization	Reinforcement Learning (RL) or adaptive models to adjust diets over time in response to measured molecular responses	• Supports adaptive feeding strategies that evolve with an animal's changing developmental or physiological state (e.g., gestation and growth) • Potential to maximize beneficial epigenetic programming while minimizing detrimental effects	• Definition of reward function is challenging (what molecular and or phenotypic endpoint should be maximized?) • Requires repeated sampling (e.g., for methylation and gene expression), not always feasible • Ethical considerations in dynamically changing diet strategies in live animals
Multi-omics mechanistic modeling	AI or hybrid models (multi-layer networks and graph-based models) integrating nutritional data with genomics, transcriptomics, and epigenomics	• Captures the complex interactions between diet, gene regulation, and epigenetic modifications • Enables predictive models of production traits that incorporate molecular-level nutritional effects • Helps generate mechanistic hypotheses about how nutrients influence gene networks	• Data heterogeneity: different omics layers may have vastly different scales and missingness • Higher model complexity can make interpretation difficult • Requires substantial computational resources and domain expertise

Continued

TABLE 8.3 Functional applications of AI in livestock nutrigenomics research.—cont'd

Application domain	AI role or function	Benefit for nutrigenomics	Considerations or challenges
Regulatory and explainable models for nutrigenomics	Explainable AI (XAI) methods (SHAP, LIME, and attention) to make AI decisions transparent	• Enables researchers and nutritionists to understand which nutrients or genes are driving AI predictions • Supports trust and adoption by revealing the biological rationale behind AI-driven diet recommendations • Facilitates hypothesis-driven follow-up experiments (e.g., test the effect of nutrient X on gene Y)	• Interpretability methods may simplify or distort complex relationships • Trade-off: more explainable models might have lower predictive accuracy • Additional computational cost and complexity
Population-level nutrigenomic optimization	Federated learning or collaborative AI frameworks combining data from multiple farms or breeds	• Enables the development of models that generalize across different farm environments and genetic backgrounds • Encourages data sharing while preserving privacy and proprietary constraints • Supports scaled-up precision nutrition strategies that benefit diverse production systems	• Federated architectures are complex to implement and coordinate • Variability in data quality, protocols, and sampling among participants may hinder model training • Governance, consent, and data ownership issues must be carefully managed

The supervised learning paradigms include regression of continuous phenotypes, such as average daily gain, milk component concentrations, intensity of methane emission, or plasma metabolite concentration; classification of categorical health states, levels of risk of rumen dysbiosis, or likelihood of reproductive success; survival modeling to predict when a metabolic disorder will occur or days open metrics; and multitask prediction where correlated endpoints can be represented using shared representation. Network biology priors are enforced in regularization schemes through sparse grouping with respect to pathway or gene ontology categories to guard against potential overfitting and to make results easier to understand. Ensemble learning is a synthesis of neural architecture-based gradient boosted trees that takes advantage of the modeling of nonlinear interactions and strong resistance to modest sample regimes inherent to high-cost omics phenotyping. Calibration methods modify probabilistic outputs using isotonic regression or Bayesian recalibration to reduce overconfidence, which is damaging to risk-sensitive nutritional decision-making. Hierarchical supervised models can take a nested hierarchy: animal within pen within farm within region, extend to partial pooling, and make the local specificity remain intact at the still-estimates at those data-scarce strata. Cost-sensitive learning is a type of learning that deals with imbalanced classification that occurs, such as the prediction of rare digestive disorders, giving more weight to false negatives in response to welfare and other economic consequences. Temporal supervision adds a sequence to prediction alignment, in which input history at the horizon targets forecasts, which makes use of encoder-decoder or temporal convolutional architectures to capture lag structure between the shift in diets and the phenotypic expression of this shift (Fan et al., 2025).

Self-supervised and unsupervised learning methods are efficient and fast methods of extracting structure that do not need labeled data, which is expensive and incomplete in experimental settings. Clustering can erroneously identify animal nutritypes that have convergent metabolic and microbiome responses to dietary interventions, which can be stratified using nutrition strategies. Clustering schemes based on density and hierarchy replenish topology-preserving manifold learning to continuity in adaptation paths, which are nonlinear. Unsupervised contrastive learning explicitly aims to build positive pairs by foregoing unnecessary augmentation that goes beyond the bounds of analytical errors of an organism, such as noise addition to omics profiles or masking subsections of omics profiles and training decoders to respect these invariances, making use of empirical robustness to pseudotime strength. Masked modeling in transcriptomics or metabolomics trains models to make features that have been withheld, and it learns to make useful predictions at the downstream stage of fine-tuning on limited tasks that are labeled, in addition to allowing imputational robustness. Multiview co-training simultaneously captures microbiome, host transcriptome, and feed composition and finds coordinated latent axes of variation that would be missed in univariate or pairwise correlation analyses. Biological enrichment tests are used to contextualize clustering validity indices to prevent algorithmically forced but biologically spurious clustering (Choi et al., 2025; Sharon, 2025).

Deep learning models are able to offer agnostic role approximation and inductive biases that are more inductive to nutrigenomic structure. Transformers modify sequence attention algorithms to longitudinal phenomic data, allowing irregular sampling and long-range interdependences of cumulative energy balance effects on subsequent reproductive performance to be captured. Graph neural networks are used on metabolic reaction networks, gene regulatory graphs, the bipartite structure of host-microbe interactions, and supply chain dependency graphs, and propagate contextual information by passing messages that indicate biochemical constraints or ecological interaction strengths. Variational autoencoders discover orthogonal biological processes in the form of disentangled latent factors, enabling the simulation and counterfactual exploration of dietary intervention as well as the measurement of reconstruction uncertainty. Hybrid sequence graph models combine dynamic behaviors in the form of metabolite profiles with variable network topology, also enhancing the predictor accuracy of adaptive responses in which the structure of pathways limits the execution of feasible dynamics. Mediating mechanisms of attention used across modalities provide emphasis on context-contingent salience, such as dynamically prioritizing microbiome functional pathways as compared to host epigenetic modifications in the prediction of immune resilience during thermal stress versus pathogen encounter scenarios. Extrapolative robustness is increased by regularization through biologically-based restrictions, for example, by conservation of mass (or monotonicity of some dose-response curves) of a dose (Fontanesi et al., 2025).

Knowledge graphs are graphs that combine structured objects that are nutrients, feed ingredients, metabolite intermediates, genes, transcripts, proteins, epigenetic marks, microbial taxa and functions, phenotypic traits, environmental exposures, and additive compounds, connected by typed relations that can be metabolic conversion, regulatory activation or inhibition, transport, dependency on a cofactor, gene-diet interaction, or causal association. Graph completion algorithms infer missing edges, making predictions of new interactions between foods or other possible mechanistic mediators of the relationship between a diet and a phenotype. Heterogeneous graph embedding produces representations in the form of vectors that are used to make downstream supervised tasks and path ranking algorithms to reveal mechanistic stories where dietary interventions model phenotypic outcomes in the form of multihop relational chains. Ontology alignment resolves overlapping controlled vocabularies of feed chemistry, gene functions, and microbiome categories, causing a decrease in the semantic fragmentation. Temporal layering supports relation modulation at the production stage or environmental regimes so that time-aware reasoning can occur. Provenance tagging also provides inference traceability, which facilitates audit and scientific validation (Liu et al., 2025).

Causal inference and structural modeling aim to separate directional effect and confounded relationship, which is a crucial criterion for reliable nutritional advice. Structural equation models are a formalization of hypothesized relationships between diet composition by way of intervening biomarkers to performance outcomes that produce the ability to partition total effects into direct and mediated parts. In the instrumental variable designs, exogenous variation, like the substitution of ingredients by supply chains without the correlation between the latent animal health conditions, is used to obtain the unbiased causal effects. Graphical causal discovery-based algorithms determine partial relationships of ancestry between variables of omics and phenotype, making use of conditional independence tests and score-based search with domain knowledge constraints to eliminate implausible edges. The invariant risk minimization method finds predictor elements that are predictable across environments, in isolation of causal signals that are context-specific correlations that degrade generalization. Counterfactual estimators estimate the difference in outcomes in the situation of alternative nutrient provisioning policies, which justify the individual selection of the feeding intervention. The sensitivity analysis measures how much the information on the causes of an outcome has been inferred to be robust to unmeasured confounding, used in confidence calibration. Causal graphs coupled with reinforcement learning can be used to improve safety by limiting exploration in interventions that are prone to cause adverse causal pathways (Ahmad et al., 2026).

Reinforcement learning and adaptive control models have considered feeding strategy optimization as a sequence of decisions over time in a state of uncertainty, where the states represent dynamic animal metabolic status, load on the environment, microbial functional capacity, and recent feeding history. Policy networks, which suggest ration adjustments or additive inclusiveness, obtain reward messages that are conditioned by multi-horizon production productivity, proxies of welfare, emission intensity, and risk penalties. Model-based variants use predictive models, which are informed by mechanistic nutrient partitioning equations with learned residual dynamics, allowing planning through rollouts that consider long-term consequences of short-term nutritional actions. Causal model-based constraints or hard-coded physiological constraints are safety layers used to ensure that exploration does not take place at the expense of health. Off-policy evaluation builds on recorded historical feeding decisions and performances to predict future policy performance without taking the risk of immediate deployment. Multi-agent formulations can be used to model group feeding cases in which social hierarchy and bunk space competition expose the effects of interaction, which organizes strategies such that they reduce intake dispersion without affecting aggregate efficiency. Constant learning tackles nonstationarity caused by seasonal variations, life stage dynamics, or changing ingredient structure space (Duan et al., 2025).

Multi-objective and evolutionary optimization methods traverse intricate trade spaces in which diets need to meet productivity, cost, environmental emissions reduction, nutrient excretion reduction, animal health risk reduction, and quality improvement of the product set of goals simultaneously. The evolutionary algorithms maintain a wide range of candidate diet populations, and they use crossover, mutation, and shedding of Pareto frontier solutions, which disclose trade-off curvature. The surrogate modeling through Gaussian processes or neural surrogates can be used to hasten the evaluation in cases where the mechanistic simulation or high-fidelity prediction model is computationally complex. Preference articulation maps the priorities of the stakeholders to scalarization weights or aspiration levels, which would allow exploration of the frontier with a focus. Interactive visualization in the negotiation of competing objectives aids in highlighting marginal gains and opportunity costs with respect to changes along the frontier. Constraint handling guarantees adherence to inclusion limits and nutrient discharge thresholds for particular feed ingredients or additives. Dynamic optimization is solved when the prices of ingredients or policy environments change, where existing knowledge of previously learned surrogate structures are employed to compute efficiency (Shao et al., 2025).

Nutrigenomic analytics relies on probabilistic models and Bayesian inference because it explicitly models uncertainty in both parameters, latent states, and predictions. The hierarchical Bayesian models are used to model multi-level structure in such a way that population-wide priors can regulate farm- or cohort-specific estimates, while posteriors are updated as new data is added. The latent variable models are used to infer the unobservable constructs, such as metabolic flexibility, immune preparedness, or rumen functional efficiency, which is a combination of heterogeneous biomarkers. Bayesian nonparametric processes have the advantage of adapting the complexity with the richness of the data without arbitrary truncation in tasks of clustering or mixture models. Model misspecification is diagnosed using posterior predictive checks, which are then refined before being put into practice. Decision-theoretic models integrate predictive distributions with utility functions to maximize actions of feeding under uncertainty between the expected improvement of performance and risk. Variational techniques or Hamiltonian Monte Carlo make use of approximate inference that is scalable and has high fidelity. Probabilistic programming increases the level of reproducibility and transparency through the explicit declaration of model structure (Shah et al., 2024).

Transfer learning and domain adaptation can be used to practice cross-species, system, and geographic generalization by learning the representations anew but correcting the differences in the distributions in feature space and conditional relationships. In methods of feature alignment, maximum mean discrepancy is reduced, or domain predictive information is adversarially eliminated at latent layers, and this encourages invariance to extraneous contextual variations and maintains biologically viable variability. The parameter freezing and fine-tuning approaches calibrate pre-trained models with small-scale local data, speeding up deployment in under-resourced localities. The orthologous sets of genes, metabolic modules, and similar functional groups in the microbiome are aligned by cross-species transfer, and the emphasis in attention is skewed to emphasize species-specific divergence that needs to be adapted to. The sequences of curriculum adaptation transform intermediate domains to close distributional gaps progressively to enhance stability. Domain adaptation uncertainty estimation detects extrapolative areas of transferred models and initiates target data collection or risk-averse decision-making (Park et al., 2024).

Meta-learning and few-shot adaptation are used in quickly updating novel nutrient interventions or infrequent feed additives, in which large-scale supervised training is impossible. Gradient-based meta-learners learn to update rules to quickly specialize to new tasks using few labeled examples, whereas metric-based approaches learn an embedding space to which small sets of prototypes are used as definitions of classification or regression targets. The curation of task distribution means that the meta-training has variability in the patterns of nutrient interactions, which increases the level of adaptation reliability. The Bayesian meta-learning combines prior uncertainty about task-specific parameters, which results in calibrated adaptation paths. Multimodal meta-frameworks permit the concurrent combination of chemical descriptors and in vitro bioassay signatures and early in vivo response markers to extrapolate efficacy or optimum dosing. Robustness tests protect against meta overfitting, in which superficial similarities between training tasks can be used to erroneously adapt to entirely new intervention mechanisms (Zhang et al., 2023).

Active learning efficiently chooses experimental diets, animals, or sampling times whose analysis would minimally decrease the uncertainty in the model or maximally improve the quality of decisions. Query strategies trade off exploration of underrepresented regions of nutrient composition and exploitation based on boundary conditions of emission or health risk limits. Bayesian experimental design computes the information gain that is expected to be obtained through the use of entropy reduction of predictive distributions of target characteristics. Multi-objective active selection combines the cost of sample acquisition, the welfare of experimental deviation from optimal diets, and logistic constraints, and maximizes net utility. Temporal active sampling picks biomarker measurement windows whose marginal contribution to the state estimation fidelity is expected to be the largest, which minimizes the invasive sampling frequency and maintains predictive strength. The pool-based models have a connection to formulation generators to suggest candidate diets and constraint-aware filtering that ensures safety and regulatory constraints are met before implementation on trials. The use of feedback loops introduces actualized outcomes, constantly adjusting the focus of acquisition (Isenberg et al., 2024).

Federated and privacy-preserving analytics provide cross-enterprise operation without centralizing raw data, covering proprietary and regulatory issues. Federated learning is model training in a decentralized manner, in which the updates of the parameters, instead of data, are centralized; the adaptive weighting considers the heterogeneity of the local data quality or quantity. A secure aggregation protocol and the use of differential privacy perturbations prevent the risk of reconstruction of sensitive records using shared gradients. Global models are also tailored to local distribution idiosyncrasies by personalization and fine-tuning or a mixture of expert gating. Cross-silo validation is an approach that assures equal performance and fairness of various production systems and detects latent bias created by skewed participation. Homomorphic encryption and secure multiparty computation provide the ability to apply collaborative inference to more complicated models, even though only with computational trade-offs mitigated by model compression and sparsity induction. Levels of data use, the data retention policies, and the data accountability channels are determined by governance structures (Legler et al., 2024).

Deploying AI on edges and on the farm, laboratory-quality analytics can be applied in resource-hardened environments where latency, power usage, connectivity intermittency, and hardware variability have been considered to pose design-level constraints. Pruning, quantization, knowledge distillation, and low-rank factorization of the model compression can be used to achieve lower memory and computational requirements without compromising predictive faithfulness to acceptable tolerance levels. The streaming inference pipelines deal with the asynchronous arrival of sensors and poor data quality, but they use minimalistic anomaly filtration before calling on the more resource-intensive cloud adjudication of uncertain sensor results. Adaptive sampling jobs vary the sensor data rate based on the stability of an estimated latent state, maintaining battery lifetime in wearable computers. Intermittency connectivity resistance encompasses buffering, opportunistic synchronization, and decentralized fallback policies to switch back to conservative feeding heuristics when the model becomes stale. The monitoring agents keep track of the performance drift on device models, which obstructs incremental retraining or rollback. Hardening of security deals with adversarial manipulation attacks, such as falsified streams of intake sensors that may lead to misinformed automated feeders (Sander et al., 2025).

Digital twins implementation combines mechanistic physiological modeling, including the kinetics of digestion, rumen fermentation, hepatic metabolism, endocrine control, and tissue accretion dynamics, with data-driven update models that incorporate real-time observations to correct model state and model parameters. State observers use variants of the Kalman or particle filters to balance the simulation results with the sensor streams, with the neural correction terms added as variations of the latent variables, such as dynamic maintenance energy or microbial community functional potential. The structured parameter learning has incorporated species-specific physiology based on parameters, which allows the application of controlled elasticity, enabling the system to adapt to personal variation while respecting the basic conservation rules. Counterfactual sequences of diets, additive interventions, and environment-stress exposures, and scenario simulation in twins use projective multi-horizon outcomes with uncertainty envelopes to make comparisons. Reinforcement learning coupling results in the optimization of policies in silico before deployment and minimizes the cost of trial and error. Inter-twins communication trades anonymized latent dynamics to speed up population-level patterns-based calibration, respecting privacy (Hunter et al., 2024).

Explainable and interpretable AI brings transparency to the complex model pipelines, making them easier to trust, comply with regulations, and generate hypotheses. Attribution methods break down predictions into contributory feature effects, modified with structured embeddings by learning predictions of contribution decades later, and devolved back to original biological objects using decoder pathways. Pathway-centric stories combine gene-level attributions to metabolic or signaling pathway influence scores, which allow the comparison with known physiology and the finding of new regulatory circuits. Counterfactual explanations assume low levels of nutrient content or biomarker changes, which would reorient anticipated risk categories and act as a basis of action-related responses. Concept activation vectors measure the presence and effect of the higher biological concept of oxidative stress or amino acid imbalance in latent representations. Causally-aware explanations differentiate between associative feature prominence and effectual leverage, using interventional feature perturbation based on causal graphs. Multiscale interpretive layers, including ingredients, nutrients, molecular, and trait domains, are displayed through visualization interfaces to the domain experts (Kasmi et al., 2024).

Model lifecycle establishes a drift detection process, recalibration process, versioning, monitoring, and audit trail to ensure its reliability over time. Drift detection observes input feature distributions, latent embedding manifolds, and error residual distributions of statistically significant deviation of divergence of both domains with training baselines, including domain shift thresholds parameterized by biological plausibility ranges. The activated recalibration can be incremental, retraining with enlarged datasets, or revision of the architecture in cases where the magnitude of shifts is larger than the capacity to adapt. The version control catalogs are structured to capture the lineage, snapshots of training data, hyperparameter settings, and evaluation metrics, allowing reliable revisiting and comparative assessment of the results. Audit trails record the inference requests, inputs, outputs, and decision rationales to aid post-hoc inquiry of negative results or unexplained deterioration of performance. Policy engines are used to force old models to retire following planned transition intervals. Human-in-the-loop control adds a domain professional signature on significant model changes that affect vital welfare or safety domains (Merkow et al., 2024).

Adversarial resilience and robustness address vulnerabilities to distributional shift vulnerability, sensor failure vulnerability, data corruption vulnerability, or intentional manipulation vulnerability. Stress testing subjects models to synthetic perturbations of plausible environmental extremes, anomalies in the composition of ingredients, or noise increases in assays of biomarkers, and tests graceful degradation characteristics. Ensemble diversity minimizes the covariate shift correlated risk of failure. Adversarial training defends models against gradient-based perturbations in continuous feature space, and domain-specific defenses identify unlikely joint feature representations. Cases of inference detection of out-of-distribution that do not have analogs in training data are diverted to a human review or conservative fallback policy. Single-stream compromise can be researched by means of redundancy in sensor modalities and corroborative signal cross-validation, such as predicted energy balance cross-checking of intake, activity, and body condition proxies. The verification of constrained decision components at the formal level guarantees that there is strict adherence to important safety limits within the range of inputs (Panpatil and Gaikwad, 2025).

The carbon- and energy-aware model design incorporates sustainability aspects in the computational architecture decisions, with the understanding that the large-scale training and inference of models have an environmental footprint. Efficient architectures implement sparsity-causing regularization, mixed-precision computation, and dynamic depth execution, in which only salient layers are used on individual inputs, minimizing energy used. Scheduling allows non-urgent retraining activities to be scheduled in periods of renewable energy supply or when the grid is not at full load. Model selection is a sophisticated method that uses an explicit objective of energy per inference instead of predictive accuracy, which motivates Pareto optimal trade-offs that do not invite unnecessarily complicated models. Knowledge distillation cascades facilitate the creation of hierarchical levels of service where simple models can be used to make routine, low-risk choices, and only more complex models are used where the decision under consideration is ambiguous or impactful. The emissions of carbon accounting frameworks are assigned to model lifecycle phases and used to inform the governance policies that require periodic reporting of the footprint and continuous improvement objectives. The marginal value of adding more samples versus training energy costs is taken into account in dataset curation, which focuses on the efficiency of using the maximum volume of data to obtain the best possible amount of information. AI and computational methods applied in livestock nutrigenomics are shown in Table 8.4.

8.5 Ethical considerations of AI in livestock nutrigenomics

The overlaps of biological complexity, heterogeneous stakeholder interests, transboundary regulatory regimes, and rapidly expanding technological capabilities are some of the factors that raise ethical concerns in AI in livestock nutrigenomics. Decisions concerning nutrient distribution, genetic selection bias, microbiome regulation, and sustainability trade-offs are now mediated by data and model-driven systems. This mediation also reinstates the old agency relationships between producers, veterinarians, nutritionists, genetic companies, regulators, consumers, and the larger ecological commons. Ethical reflection should therefore be able to state ideal guardrails and operational governance mechanisms that would not respond post-hoc to harm but rather would predict systemic externalities. The coherent ethical system combines the promises of animal welfare, environmental stewardship, distributive justice, epistemic transparency, privacy, and responsible innovation, and acknowledges the tensions between these imperatives and the need to balance them in context-specific ways encouraged by deliberative processes (Rosati, 2025).

Data governance is the substance of ethical integrity since the quality of nutrigenomic intelligence and equity is constrained by the integrity of datasets that are the originators of information. The issues of ownership and sovereignty are compounded by multi-tier data generation, where genomic panels, sensor telemetry, feed formulation records, environmental measures, and microbiome profiles can be co-produced between farms, laboratories, compilers of service providers, and intermediaries of platforms. Claims against unilateral ownership tend to obscure the equities of relations of co-creation of data and dependency of derived value on the aggregate scale within the network. Ethical governance goes beyond proprietary to stewardship forms of acknowledging valid interests of primary producers in retaining control over strategic performance information and pooled data architectures that provide a collective good in terms of higher model generalizability. Compliance fragmentation that arises due to jurisdictional diversity in data protection law, agricultural policy, and bioresource sovereignty creates an ethical obligation on the implementer that obligates them to take into account the local mandates on localization, consent scope, and secondary use limits, and indigenous custodianship principles regarding biological information. Contractual structures must be non-coercive and asymmetric, in which smaller operations grant broad rights to access analytics in exchange for access to analytics by entrenching a structure of standardized, understandable licensing with revocation mechanisms, different levels of access, and shared audit capacities. Governance charters define the domains of use that can be done, the retention limits, and the de-identification standards that would reflect the privacy and scientific reproducibility objectives (Mendonca et al., 2025; Tattersall et al., 2025).

The concept of privacy and confidentiality is further expanded to include not only traditional personal data questions but also the performance indicators of the farm at the strategic level and its exclusive genetic stocks and microbiome or metabolomic signatures that might expose some weaknesses in competition. Although individual identity is not salient, a combination of geospatial, temporal, and trait patterns makes leakage risk likely to occur. Ethical design proponents advocate for data minimization, proportionality in collection, pseudonymization or anonymization in line with re-identification threat modeling, and separation of raw identifiers and analytic workspaces achieved through technical and organizational means. Lateral movement of sensitive genomic assets is limited by cryptographic hashing, access privileges of various levels, and secure enclaves. The requirement of confidentiality will remain in the form of model outputs; aggregated benchmarking dashboards should not suffer any inferential leakage, allowing one to reverse engineer the data of a specific participant. The disclosure of residual privacy risk and mitigation constraints is respectful of stakeholder autonomy, as it allows them to make informed decisions when there is uncertainty (Gajavalli, 2025).

To handle structural biases that may be present in the development of models in cases where corpora may be trained to disfavor certain breeds, species forms, production systems, or geographic ecologies, algorithmic fairness can be used. Inequality in sensor use, acquisition of genomic testing, and composition of latent diversity based on introductory factors can generate biased latent representations that induce predictive loss or overemphasize uncertainty in disproportionately represented marginalized groups, and performance disparities contribute to a forecasting effect and discourage equitable

TABLE 8.4 AI and computational methods applied in livestock nutrigenomics.

Method	Typical nutrigenomic application	Advantages or strengths	Limitations or challenges	Examples and use cases
Supervised machine learning	• Predicting nutrient-driven gene expression changes or epigenetic marks (e.g., methylation) • Classifying animals by response to nutritional interventions (responder vs. non-responder)	• High predictive performance when labeled data is available • Feature importance allows insight into key genes or nutrients • Proven, mature algorithms	• Needs sufficiently large, high-quality labeled datasets • Risk of overfitting if feature space (omics and diet) is large • Limited ability to model very complex interactions	Random Forest trained on diet and methylation data to predict which animals show epigenetic response; SVM classifying transcriptomic response to methionine supplementation
Unsupervised machine learning and clustering	• Discovering subpopulations of animals with similar nutrigenomic profiles • Detecting latent patterns in molecular data in response to diet	• No need for outcome labels • Reveals natural structure and heterogeneity • Useful for hypothesis generation	• Clusters may not align with biologically meaningful groups • Sensitive to scaling and distance metrics • Does not inherently predict outcomes	Clustering of methylation or gene-expression data to identify "diet-sensitive" versus "diet-stable" groups among livestock
Deep learning and neural networks	• Modeling complex, non-linear relationships between dietary inputs and multi-omics responses • Dimensionality reduction (autoencoders) • Learning from raw sequence or epigenetic data	• Captures high-order interactions among nutrients, genes, and epigenetic marks • Reduces need for manual feature engineering • Scalable to very high-dimensional data	• Requires large training datasets • High computational cost • Often difficult to interpret ("black-box")	Autoencoders used to compress methylome and transcriptome into latent features, neural nets to predict gene expression response to choline/folate
Reinforcement learning (RL)	• Optimizing dietary strategies over time to maximize favorable genomic or epigenomic outcomes • Adaptive, individualized nutrition protocols	• Learns a policy from dynamic feedback (e.g., omics read-outs) • Can adapt interventions based on changing biological state • Suitable for long-term, iterative feeding or supplementation experiments	• Defining an appropriate reward function (e.g., epigenetic stability) is non-trivial • Requires many interventions and measurements, costly in farm settings • Ethical or practical constraints in live-animal RL	Simulated RL agents optimizing feed composition to maintain the desired methylation signature across growth phases
Multi-omics integration and ensemble models	• Combining diet data with genomics, transcriptomics, and epigenomics to predict traits or health outcomes • Capturing the interplay of nutrition, gene expression, and epigenetic regulation	• Holistic modeling of nutrient–gene–epigenome interactions • Often yields better predictive accuracy • Supports mechanistic insight through cross-layer feature interactions	• Integrating heterogeneous datasets is technically challenging (scale and missing data) • Requires careful normalization and alignment • Danger of over-complex models with overfitting	Hybrid models combining gradient boosting on nutritional features, neural nets on expression data, and linear models on methylation data to predict growth or disease resilience

Explainable AI	• Interpreting which nutrients drive changes in gene expression or epigenetic marks • Identifying key molecular drivers of dietary responses	• Increases transparency and interpretability • Helps generate testable biological hypotheses • Facilitates trust among nutritionists and geneticists	• Post-hoc explanations may oversimplify complex relationships • Extra computational overhead • Interpretability does not guarantee causality	SHAP used to rank nutrients (e.g., methionine and choline) by their importance in driving predicted methylation changes; attention mechanisms in deep models highlight key dietary-omics interactions
Graph neural networks (GNNs)	• Modeling nutrient–gene–regulatory networks (e.g., gene co-expression and metabolic networks) • Predicting how perturbations (dietary changes) propagate through biological networks	• Naturally handles graph-structured biological data (e.g., gene regulatory networks) • Captures relationships between genes, metabolites, and regulatory elements • Can predict cascade effects of nutritional interventions	• Requires well-defined network structures (interactions) • Graph construction (which nodes, edges) may be non-trivial • GNNs can be computationally intensive for very large graphs	While specific studies in livestock nutrigenomics are limited, GNNs have been used in biological network modeling (e.g., metabolic or co-expression networks) to infer gene function and regulatory influence (GNNs are well-suited for modeling nutrient–gene networks per general biological AI frameworks)
Transformer and DNA large-language models (LLMs)	• Modeling sequence-based regulatory elements or gene regions to predict how nutrients may influence regulatory motifs • Capturing long-range dependencies in DNA or epigenomic sequence relevant to nutrient-gene interactions	• Self-attention captures long-range sequence relationships • Pre-training on large genomic datasets allows transfer to downstream nutrigenomic tasks • Highly flexible and scalable	• High computational cost and memory requirements • Requires very large pre-training datasets • Interpretability of transformer models remains challenging	DNA large-language models (DNA-LLMs) treat genomic sequences as "language" to learn regulatory features; such models could be fine-tuned to predict nutrient-responsive regulatory elements

diffusion of technology. The ethical framework must be just diagnostic, sensitive to agronomic situations as opposed to blind importation of generic parity indicators. Fairness can be assessed through calibration parity, symmetry of error distributions, and robust quantification of uncertainty that is stratified between breeds, systems, or regions, and this allows for assessment. The remediation can be re-weighting, domain adaptation, synthetic data augmentation based on physiological constraints instead of unconstrained generative fabrication, and participatory investments in data collection that increase diversity. Equity involves downstream decision policies that ought to utilize fairness-aware maximization, wherein marginal expected benefits of already advantaged groups are exchanged with uplift of underserved groups within welfare and sustainability constraints (Garwe et al., 2025).

Legitimacy and informed consent throughout the spectrum of stakeholders are maintained through transparency and explainability. AI systems traverse the web of opaque biochemical interdependencies in nutrigenomics, but their intelligent use requires articulated descriptions of their provenance, input data provenance, feature processing strategy, and performance validation. This limits the use of AI systems in nutrigenomics. Uncertainty should be explicitly quantified, and extrapolated zones of certified performance reveal extrapolation risk, which is explicitly identifiable through intelligent use. Documentation regimes define standard model cards and data sheets, which represent the contextual assumptions, labeling protocols, missingness treatment, and bias reduction steps. Explainability should be provided at multiple levels, with different interpretative narratives supporting actionable rationales in terms of ration changes at the local level, pathway-centric aggregations to explain mechanistic plausibility, and systemic summaries to explain trade-off rationales in emissions, cost, and welfare complexity. Informed consent does not just mean single-event acceptance of the continued dialog about new model functionality, secondary uses of derivative embeddings, and inferences about federated learning updates. Consent granularity is a selective opt-out that allows participation in more global precompetitive knowledge creation. Human agency is maintained by a clear separation of decision support and automated authorization, which prevents the abdication of professional judgment as a result of the bias of algorithmic authority (Farooqui et al., 2025).

The ethics of nutrition warns against optimization that is too narrow, unintentionally causing higher productivity measures and a metabolic burden, immune suppression, frustration of behavior, or a greater risk of injury. Ends-oriented machine learning objective functions can cross over physiological resilience limits or confound early signs of distress by implementing feedback methods that postpone but increase later health outcomes. Integrating welfare sentinel variables and hard constraints into optimization formulations imposes non-maleficence. Utilitarian drift is countered by multi-objective balancing, which does not consider welfare a soft penalty but a coequal measure of success. Continuous post-deployment surveillance of variance and minor deviations in biomarker supplements assessment of mean outcomes in search of emergent welfare erosion. Regulation procedures must include red lines in which algorithmic suggestions that cause closeness to known risk conditions are brought to human examination, regardless of anticipated short-term performance gain. Ethical theories also question the use of intensification, which is facilitated by accurate nutrition to compress patterns of rest or social interaction, and the redesigning of reward structures so as to internalize holistic welfare (Day et al., 2025; He et al., 2025).

The concept of environmental sustainability refers to nutrigenomic ethics in two ways: the environmental impact of feeding methods and the computational environmental impact of AI infrastructure itself. Although optimized diets may decrease emissions intensity and nutrient excretion, rebound effects may arise when efficiency gains are used to promote production, but there is no overall decrease in the footprint. The scope of ethical analysis requires the presence of broad system boundaries to include upstream ingredient farming, the risk of land use change, and downstream waste management externalities. Allowing large-scale model training, hyperparameter searches, and regular retraining cycles (at the same time) also uses energy and embodied hardware resources and can effectively wash away net sustainability benefits. The inflection points of the marginal accuracy gains applied to model lifecycle evaluation are identified by carbon accounting. The design of lean models, hardware efficiency alignment, renewable balanced scheduling, and hierarchical cascades inference reduces the digital environmental load. Energy reporting enables responsibility and innovation in low-footprint designs that are responsive to the agricultural setting instead of being imported based on resource-rich territories (Kim et al., 2025).

Socio-economic inclusivity touches on the danger that AI-facilitated nutrigenomics would increase digital divides between resource-intensive activities and smaller or resource-constrained manufacturers. Such factors as sensor acquisition price, connectivity constraints, differences in data literacy, and platform subscription strategies that harvest rising returns and concentrate the market are barriers. The ethical strategy promotes tiered service architectures under graceful degradation in case of low bandwidth conditions, offline-capable edge analytics, open reference implementations of core precompetitive algorithms, and collaborative data trusts to give contributors greater bargaining power. Training and available documentation, and participatory co-design lead to capacity building in forming local interpretive competence, against the dependence on opaque external intermediaries. Elimination of exclusion risk is achieved through pricing models that involve the use of sliding scales, or benefit sharing based on derived value, instead of extractive uniform charges. Resistance to monopolistic data lock-in is through interoperability standards, portable data export formats, and anticompetitive practice surveillance so that innovation and choice are not lost (Kruhlov and Dvorak, 2025).

Intellectual property, benefit sharing, and open science balancing face a dilemma of either rewarding innovation or avoiding enclosure of underlying farm data based on the overall contributions of data. The value streams can be privatized through exclusive rights over derivative models that are trained on pooled genomic and nutrigenomic data, with data originators receiving little back. Ethical models distinguish between precompetitive levels, including standardized ontologies, core physiological models, and base embeddings that are openly licensed, and application-specific customizations that are limitedly protected. Revenue sharing schemes are a system whereby a part of the commercial model licensing proceeds is shared amongst data cooperatives to support further data quality improvement and fair access programs. The explicit attribution requirements identify the work of data stewardship. The maintenance of epistemic integrity can be achieved through the licensing that guarantees rights to research, reproduction, and independent audits. Protections against the exploitation of traditional feed knowledge or regionally adapted landrace trait knowledge would need prior informed consent in a culturally sensitive manner and need to be negotiated on a framework of benefits in line with principles of fairness (Lvovs et al., 2025).

Regulatory compliance has an overlap with ethical responsibility in that normative standards are converted into enforceable requirements that regulate feed additive claims, safety testing, labeling, data protection, and animal health regulation. The systems of AI affecting additive inclusion should respect hierarchies of evidence that are set up to allow authorization and not extrapolated claims that are not substantiated by any mechanistic or clinical validation. The processing of the data is consistent with the privacy laws due to the limitation of the purpose, the formulation of a lawful basis, and the facilitation of data subject rights. The labeling transparency in which AI is used to customize nutrients allows building consumer trust and overcoming unnecessary techno-optimism. Cross-jurisdiction heterogeneity comes with the responsibility of sustaining adaptive compliance mapping as well as avoiding regulatory arbitrage, where deployments are made on lax frameworks to evade the safeguards. The positive effects of harmonization initiatives include ethical advocacy (which puts a premium on minimum protection irrespective of geographical coverage), as it realizes the interdependence of supply chains and externalities on the environment globally (Jenko et al., 2025).

The possibility of misuse will include situations when AI-enhanced nutrigenomic data is reused in a manner that is against welfare, ecological harmony, or societal standards. When models maximize on physiological extremes that cause a loss of adaptive resilience or, on the other extreme, on emergent pathogen susceptibility, there is a loss of performance enhancement boundaries. There are ecological disruption issues whereby strategies to modify microbiomes or synthetic feed organisms escape confinement to modify the ecology of soil or water microbes. Dual-use risk is also applicable to the wicked use of model outputs to design feed with undesired characteristics or to avoid any regulatory restrictions on restricted additives. Threat modeling, sensitive algorithmic capability access tiering, anomaly detection of patterns of abnormal intervention, and red teaming exercises to explore system defenses are some of the ethical resilience strategies. Disclosure policies deal with reporting vulnerabilities and reducing the latency of exploitation. Developing a precautionary and vigilant response culture reverses the normalization of boundary-pushing optimization, which gradually erodes ethical foundations (Pannu et al., 2025).

The automation trends do not eliminate human control and responsibility. Deliberate demarcation of roles is an assurance that negative results are returned to discrete governance nodes and not treated as algorithmic black art. Reviewing high-impact model deployments, risk assessment evaluation, and authorization to progress through predefined maturity gates is done by multidisciplinary ethics committees. Achieving auditability is ensured by thorough data transformation, model parameters, inference inputs, decision rationale, and override events logging. Ethical breaches or near misses have codified incident response procedures that include detection, triage, mitigation, disclosure, and learning procedures. Ongoing professional growth is based on competence in the interpretation of AI results, the range of uncertainty, and ethical considerations. Accountability systems combine penalties for negligent misuse and rewards for ethical innovation, strengthening normative commitment rather than mere compliance (Verma et al., 2025).

There is a greater need to scrutinize the ethical considerations of microbiome engineering and synthetic feed ingredient interventions since they may involve irreversible horizontal gene transfer pathways and uncertainty of ecological cascades. A shift in microbial community resilience threshold may also occur by engineering the fermentation profiles or strengthening nutrient release with engineering consortia and unwanted pathogen facilitation or resistance gene transfer. Synthetic nutrient analogs or bioactive carriers can build up or convert in ecosystems following excretion. Precautionary performance incorporates ecological network marketing and containment validation. The benefit-risk analyses include welfare uplift, emission containment, and dependency trade-offs, including decreased dietary diversity flexibility. Not only are immediate farm operators involved in consent discourse, but also extended stakeholders of environmental spillover. Metadata governance is a mechanism to ensure that granular microbiome datasets that are utilized to train engineering design models are governed based on principles of privacy and fairness, not commodified without comparable ecological stewardship obligations (Bakkeren et al., 2025).

In these intersecting fields, AI with regard to livestock nutrigenomics is not fixed codification, but it is an iterative praxis in response to advancing scientific understanding, the changing value of society, and feedback regarding the environment. Anticipatory ethics foresight activities, scenario planning, and participatory deliberation embedded in ethics would help to widen the scope of the limited technical risk frames. Ethical performance metrics and key performance indicators supplement the conventional measure of accuracy or efficiency and combine factors such as welfare stability measures, fairness disparity attenuation, carbon intensity of computation, inclusivity progression, and transparency satisfaction measurements. Finally, an ethically-informed nutrigenomic AI ecosystem acknowledges interdependence: animal physiological integrity has sustainable productivity; environmental health conditions have long-term resource viability; equitable access has innovation legitimacy; and transparent and accountable governance has gained trust that allows further collaboration of data. With internalized ethical vectors in the design, deployment, and oversight cycles, stakeholders can use AI to promote nutrigenomic knowledge and ethical responsibility toward animals, humans, and planetary systems. Challenges and limitations in AI-driven livestock nutrigenomics are presented in Table 8.5.

8.6 Future trends and innovations in AI for livestock nutrigenomics

The design spaces of the applications of AI in nutrigenomics of livestock all point toward an architectural transformation from episodic, retrospective analytics to continuously learning, cyber-physical ecosystems, which sense, assume, and act in uncertain situations within the tightly controlled space of uncertainty. Real-time adaptive feeding represents a future advancement that integrates various technologies to monitor and optimize animal nutrition. This includes analyzing dense physiological data on ingestion behavior, using accelerometers, rumen bolus telemetry, and micro-dynamics of pH and temperature. It also involves infrared thermography, acoustic monitoring of fermentation, and wearable microneedles for sampling metal metabolism. Additionally, near-infrared spectral analysis bypasses traditional methods like milk or excreta testing, while Bayesian state estimators help link unpredictable observations to individual digestive kinetics and nutrient absorption profiles. The reinforcement learning schedules earlier trained in high-fidelity simulation and guarded by constraint satisfaction layers that model metabolic safety limits are updated multiple times a day to exploit short-term opportunities of augmented amino acid uptake efficiency or to preempt the risk of developing acidosis before symptoms manifest themselves. Latency-aware edge inference gives the ability to tolerate intermittent connections, and hierarchical uncertainty decomposition gives the ability to decouple sensor artifact chance ambiguity and actual physiological perturbation to remove oscillatory overcompensation. The process where batch-formulated diets are converted to dynamically titrated nutrient provisioning changes the definition of efficiency ceilings and necessitates new assessment instruments where the stability indices, maintenance of metabolic resilience, and dynamic welfare indicators are treated jointly as opposed to a one-dimensional expression of each of the independent variables supplied by way of feed conversion (Ghavi Hossein-Zadeh, 2025; Tedeschi et al., 2025).

In the meantime, multimodal foundation models would have the capability to retrain representational baselines by pretraining at scale on heterogeneous collections of genomic SNP and structural variation panels, transcriptome time series, metatranscriptomic and metagenomic microbiome assemblies, untargeted metabolomics, feed chemical fingerprints, ingredient provenance chains, environmental telemetry, and continuous phenomics. The tokenization methods map various modalities with semantic embedding spaces with genes, microbial enzymes, nutrient moieties, spectral peaks, and sensor-derived temporal motifs placed co-contextually by masked reconstruction, contrastive alignment, and cross-attention fusion. These models use compositional and regulatory regulations, which enable rapid low-level fine-tuning downstream on emergent properties, obtained or arbitrary feed additives, with a small labeled set of exemplars. Architectural innovations adopt sparsity, which leads to a mixture of experts routing to the scale parameter space and topologically constrained positional encodings to respect chromatin conformation loop adjacency and metabolic pathway topology in genomes. Long-horizon temporal transformers have irregular sampling alignment layers to maintain coherence across life stages. The responsible deployment expects domain optimization of adaptation to prevent erroneous extrapolation in the case of encountering a different breed of genomic structure or region-specific feedstuff chemical peculiarities not sampled in pre-training corpora (Ghalkha et al., 2025).

Generative AI is not limited to text or molecular design but can maximize diet with sustainability, multifactor, cost, health risk, and supply volatility objectives. Variational generators are augmented with diffusion models and reinforcement learning to produce ingredient inclusion vectors that adhere to the linear constraint of nutrients, the nonlinear

TABLE 8.5 Challenges and limitations in AI-driven livestock nutrigenomics.

Challenge	Description	Impact on nutrigenomics or AI integration	Underlying causes or nuances	Potential mitigation strategies
High cost and infrastructure barriers	The financial burden associated with generating large, high-quality omics datasets (genomics, epigenomics, and transcriptomics), and running AI models	• Limited adoption by smaller farms or research groups • Slower generation of training data for AI, restricting model generalizability	• High cost of sequencing, sample preparation, and multi-omics assays • Need for computational hardware, data storage, and bioinformatics expertise • Limited funding or infrastructure in developing regions	• Promote collaborative consortia to share omics data and computing resources • Use cloud-based or shared AI infrastructures • Explore cost-effective sampling strategies and lower-cost omics technologies
Data complexity, heterogeneity, and quality	Nutrigenomic datasets are often highly heterogeneous (nutritional, genetic, and epigenetic) and noisy, making integration and modeling difficult	• AI models may learn artifacts rather than true nutrient–gene interactions • Poor model robustness across different populations or diets	• Differences in sample collection, sequencing depth, batch effects • Biological variability in response to diet • Sparse or missing data in longitudinal studies	• Implement rigorous data quality control and normalization pipelines • Use AI methods robust to missingness (e.g., generative models and imputation) • Standardize protocols and metadata across studies
Limited biological interpretability	Many AI-driven models (e.g., deep learning) are "black boxes" and do not easily reveal biological mechanisms	• Difficulty in deriving actionable nutritional strategies • Lack of trust or adoption by nutritionists, geneticists, and farmers	• Complex non-linear models obscure feature–outcome relationships • Nutrigenomics involves multi-layered mechanisms (e.g., epigenetics) that are not straightforward to interpret	• Combine AI with explainability methods (e.g., SHAP and LIME) • Use hybrid models that balance interpretability and predictive performance • Train domain experts (nutritionists and geneticists) in AI interpretation
Insufficiently representative data and bias	Nutrigenomic datasets may not represent all livestock species, breeds, or production systems	• Models trained on narrow datasets may not generalize broadly • Risk of biased recommendations favoring well-studied breeds	• Most nutrigenomics research focuses on common or high-yield breeds • Production system variation (feed and environment) is often not captured	• Actively collect data across diverse breeds, species, and management systems • Use transfer learning or domain adaptation to adapt models • Promote open data sharing to broaden datasets

Continued

TABLE 8.5 Challenges and limitations in AI-driven livestock nutrigenomics.—cont'd

Challenge	Description	Impact on nutrigenomics or AI integration	Underlying causes or nuances	Potential mitigation strategies
Ethical, regulatory, and data governance challenges	Applying AI to nutrigenomics raises ethical issues (data ownership, consent, and misuse) and lacks clear regulation	• Resistance to adoption; risk of misuse or inequitable benefit distribution • Potential lack of governance over data sharing or AI-driven dietary advice	• Ambiguity around who owns molecular and nutritional data (farmers and companies) • Few regulatory frameworks for AI-based nutritional-genomic interventions in livestock • Ethical concerns about "manipulating" epigenetics for production gains	• Establish data governance policies that respect farm ownership and benefit sharing • Develop regulatory guidelines for nutrigenomics and AI (involvement of ethicists and policymakers) • Promote transparency and stakeholder engagement (farmers, consumers, and scientists)
Temporal and causal complexity	Nutrigenomic responses to diet are dynamic, context-dependent, and may involve long time frames (e.g., developmental programming)	• Static models may poorly capture long-term or transient effects • Difficulty in establishing causality (do nutrients change epigenetics, and it change the understudy trait?)	• Epigenetic and transcriptional changes may lag behind dietary changes • Longitudinal multi-omics data are expensive and logistically challenging to collect • Nutrient–gene relationships may be context-dependent (developmental stage and stress)	• Use longitudinal sampling and modeling (time-series AI) • Apply causal inference methods (e.g., counterfactual modeling) • Design experiments (e.g., multi-generational) to validate model predictions

constraint of surfaces of digestibility alteration, and the discrete inclusion constraint, and simultaneously reduce latent welfare and emission surrogate scores as conditional foundation model heads predict (Ghavi Hossein-Zadeh, 2026). Structured comparative/similarity ingredient substitution engines are trained to incorporate compositional, functional, anti-nutritional factor, and processing tolerance characteristics, depending on the geopolitical shock, climatic extremes, and pathogen-related trade barriers in the supply chain that lead them to require reformulation contingency. Counterfactual scenario-generating studies the nutritional adequacy margins to establish a strong, viable space that is resilient to stochastic changes in ingredient quality to inform policies on procurement hedging actions. The generative uncertainty quantification is a significant aspect: swarms of stochastic generators with conformance discriminators find low plausibility formulations, and constraint satisfaction auditors assure safety thresholds of minerals, specific fatty acid isomers, or mycotoxin dilution policy (Necula and Rieder, 2025).

Recent developments in advanced causal discovery algorithms will narrow mechanistic interpretability through the induction of directed nutrient-gene-phenotype pathway influence structures in the face of dense confounding and feedback interaction networks. The hybrid frameworks combine the principles of invariant risk minimization with constraint-based and score-based graphical search, using prior knowledge graphs representing known enzymatic transformations, transcription factor binding, and microbiome-mediated transformations of metabolites. The lack of interventional data is also overcome by pseudo interventions based on natural experiments, such as sudden climatic changes or ration changes caused by supply and handled as a quasi-instrumental perturbation. Continuous time causal modeling uses the formalism of neural stochastic differential equations to encode lagged dynamics of nutrient pulse propagation through transcriptomic oscillations and eventually alters the rate of accruals in the phenotypic basis. Determined causal subgraphs drive mechanistic simulation co-calibration, which boosts digital twin faithfulness and allows focused intervention planning, which perturbs intermediate biomarkers as opposed to inducing bulk nutrient supply changes. Strict sensitivity analysis measures remnants of ambiguity to trigger ethical restraint against overweight mechanistic assertions (Tao et al., 2025).

A combination of single-cell and spatial nutrigenomics with systemic metabolic flux modeling promises a surge of resolution that puts macroscopic results of performance within the microenvironmental heterogeneity of cells. The expression of transporters, mitochondrial biogenesis, and inflammatory signaling gradients to nutrient composition are revealed by spatial transcriptomics, single-nucleus chromatin accessibility, and high-throughput metabolite imaging in intestinal epithelium, liver lobular zones, muscle fiber types, and rumen papillae. Multiscale AI models are used, which have neighborhoods of cells annotated with omics signatures represented as nodes, and diffusion/paracrine/vascular coupling represented as edges. Individualized fluxome predictions obtained through coupling to genome-scale metabolic models parameterized by flux balance augmented neural regressors are consistent with stoichiometric constraints and adjust to experimental levels of regulatory state. The cross-scale attention processes convert cellular adaptation in the nutrient fitness of an entire organism into a predictive diet that capitalizes on the metabolic plasticity of a specific individual or limits the adverse effects of localized hot spots of predicted pathology. Cost and data sparsity encourage the scheduling of active acquisition, in which model epistemic uncertainty determines regions of the tissue that should be sampled more in longitudinal studies (Sabit et al., 2025).

The revitalized paradigm of co-design will be utilized in microbiome-directed nutritional engineering, in which the diet composition is co-optimized with microbiome modulators, for example, phages, precision probiotics, oligosaccharide prebiotics, or bacteriocin-producing consortia. Ecological interaction networks recreated on AI platforms cross-feed metabolically, exclude each other, and experience the threat of horizontal genetic transfer and phage host range dynamics on a shifting landscape of fiber and protein degradability. The multi-objective model targets the convergence of fermentation volatile fatty acids, efficiency of nitrogen use, reduction of methane, and control and balance of pathogens stably and flexibly during switching of feed. The explorations can be expedited by the embeddings of the latent ecological states, which are optimized by the contrastive prediction of the future community state in the presence of the candidate interventions. Safety governance takes proactive designs to prevent unintentional collateral oppression of beneficial keystone taxa and the probability of mobilization of antibiotic resistance genes and ecological dumping. The outputs of synthetic biology, that is, engineered probiotic strains, are combined with the outputs of nutrient matrix modification, scaled by in vivo deployment, that is, with engineered probiotic strains (in a defined consortium in vitro fermentation system using the systems of iterative design-build-test) (Duhan et al., 2025).

Synthetic biology interfaces will also develop with engineered enzymes to rumen-bypass degradative pathways that are countermeasured with precision encapsulation matrices to release achievable quantities of desired amino acids or micronutrients in symbiotic-sister breadth and rationally designed bioactives to regulate nutrient-sensing arcades like mTOR or AMPK to efficiently deposit lean tissue without augmenting the risk of metabolic disorder. In AI-driven protein engineering, structure-aware diffusion models or transformers, conditioned by rumen pH, redox, and protease environments, are employed to enhance catalytic resistance. To predict the interactions of the matrix and microbiome, formulation optimizers employ pharmacokinetic and release profile predictions, which rely on mechanistic capsule dissolution models using neural residuals. Risk assessment is a combination of predictive immunogenicity screening, probability of horizontal gene element containment, and post-excretion environmental degradation kinetics in ensuring biosafety. The synthetic component lineage is captured in the traceability infrastructure and is utilized to facilitate the incorporation of post-market surveillance in conjunction with real-world performance and welfare data to make improvements in a continuous manner (Jiang et al., 2025).

Multi-agent and game-theoretic optimization extends nutrition decision models beyond the individual farm, and interdependent objectives are modeled between the feed mills, producers, processors, retailers, and environmental regulatory bodies. The agents whose utility functions are divergent minimize costs, comply with emission reduction, maintain consistency in the quality of their products, achieve animal welfare certification, and interact through market pricing signals, contract design, and shared sustainability credits. Differentiable auction or contract reinforcement learning studies incentive schemes that bring the local ration formulation behaviors in line with the globally optimal emission and resource utilization outcomes. Mechanism design limits strategy spaces to discourage free riding or withholding of data that may negatively affect the performance of collective intelligence models. In federated multi-agent simulations, distributional effects of scale class and geographic area of policy proposals, like nutrient excretion levies or feed additive eligibility to carbon credits, are generated by counterfactuals. The evolutionary game dynamic analysis of stability is used to test the ability to resist perturbations of collaborative equilibria, such as the abrupt shortage of an ingredient (Mai et al., 2025).

The quantum-informed and neuromorphic computing directions emerge as a reaction to the problem of combinatorial explosion in the context of multi-dimensional optimization of the diet with stochastic constraints. Tensor network and quantum annealing heuristic factorization are scalable to substantially accelerate the exploration of high-order interaction landscapes with nonlinear factors of digestibility and conditional additive efficacy with variables of temporal nutrient delivery scheduling. Hybrid algorithms send subproblems, like binary inclusion selection with multi-residue non-overlap constraints, to quantum-inspired solvers, while continuous inclusion rate refinement is conducted on a classical gradient architecture. Neuromorphic spiking designs provide ultra-low-power inference on on-farm devices learning anomalies on high rates of intake or rumination signals, which conserves power and can be densely deployed in remote applications. Ethical assessment requires strict performance at classical levels to support environmental benefit assertions and prevent myopic techno uptake, which results in the misappropriation of resources in favor of less practical optimization alternatives (Wei et al., 2025).

Green AI has evolved into rhetorical hopes, which are converted into operational selection criteria in which model format, model training, and model inference deployment are evaluated as per the lifecycle sustainability criterion, which is integrated into procurement and research funding decisions. Marginal utility analysis of parameter scaling and carbon intensity per percentage point of predictive improvement generates go/no-go thresholds for retraining a large model, and embodied hardware impact analysis does so as well. Adaptive training pipelines utilize the techniques of early exit accuracy convergence detection, cascades of knowledge distillations, dynamic precision scaling, and data curriculum scheduling to select high information gradient batches early so as to truncate unnecessary epochs. This annotation of energy provenance should be a model registry, which allows it to favorably reuse the existing foundation models rather than pretrain the ones that are not beneficial to a model. The use of edge filtering reduces the computation load of cloud computing by pre-set aggregation of sensor data in lossy compression that is selectively tuned to retain nutritionally significant frequencies. Industry regulations require clear reporting on energy and water footprints, promoting innovation in collective efficiency (Malik, 2025).

The federated global consortia linked with interoperable ontologies would be in a position to alleviate the inadequacy of data in regions and offer robust cross-environmental generalization of unconcentrated proprietary data. Feed ingredient taxonomies are reconciled to ontologies of functional genes, markers of metabolic pathways, welfare scoring schemes, and expression of emission factors, and semantic friction is minimized as much as possible; model parameters may be exchanged with limited propagation of errors of misalignment. Federated optimization is a method that adjusts the aggregation weight through the assistance of cross-site validation residual and Shapley value contribution estimates to fairly assign marginal performance advantages. Privacy-enhancing technologies, including secure aggregation, a calibration of differential privacy noise that guarantees the signal of rare diseases to be viable, and homomorphic encrypted inference of sensitive computations that include genotyping, form the basis of trust. The systems of rotational stewardship, unfettered auditing, and equitable sharing of intellectual property are formed by the structures of government to ensure the interests of different classes of economies (Papadakis et al., 2024).

Examples of such participatory data ecosystems include increased contributions to scale in terms of contributors, whereby there is a citizen science-style feedstock, where the contributions are based on low-cost handheld spectral

TABLE 8.6 Future directions and innovations in AI-enabled livestock nutrigenomics.

Future direction or innovation	Description or concept	Expected impact on nutrigenomics	Challenges and considerations
Behavior-driven precision nutrition	Use of AI systems (e.g., reinforcement learning and computer vision or sensor data) that adapt diet in real-time based on observed animal behavior (feeding, movement, and stress)	• More responsive and individualized feeding strategies that reflect real metabolic or motivational states • Improved welfare and nutritional efficiency by aligning nutrient supply to actual needs • Reduction of over- or under-nutrition and resource waste	• Requires robust behavior-to-nutrition mapping (how behavior patterns translate to metabolic needs) • High demand for wearable devices, cameras, and farm infrastructure • Ethical and animal-welfare implications of continuous behavior monitoring
Sustainable nutrigenomics via multi-objective optimization	AI optimization frameworks that balance productivity, sustainability, and epigenetic health, e.g., multi-objective optimization integrating environmental metrics, dietary inputs, and genetic or epigenetic outputs	• Ability to create diets that maximize production while minimizing environmental footprint (e.g., nitrogen and phosphorus emissions) • Supports long-term strategies for sustainable breeding and nutrition that account for epigenetic programming • Encourages resource-efficient feeding aligned with "green livestock" goals	• Defining and weighting multiple objective functions (production versus sustainability versus health) can be complex • Trade-offs may be controversial (e.g., minimizing emissions might conflict with maximal growth) • Requires policies and incentives to support the adoption of sustainable AI-nutrigenomic strategies
Federated and privacy-preserving AI models	Collaborative learning frameworks (e.g., federated learning) where nutrigenomic models are trained across farms or organizations without sharing sensitive raw data	• Wider, more diverse datasets build better-generalizing models while preserving data privacy • Enables small farms or producers to contribute to and benefit from shared AI-nutrition models • Fosters equitable access to advanced nutrigenomic tools	• Requires technical coordination across farms with different data systems • Data heterogeneity in diet, genetic background, and measurement protocols may hinder model convergence • Governance frameworks needed to manage data ownership, consent, and benefit sharing
Transformer and large-language models (LLMs) for nutrigenomic knowledge representations	Use of transformer-based models (e.g., genomic LLMs) to learn regulatory grammar, long-range interactions, epigenetic motif prediction, or nutrient–gene regulatory relationships	• Capture long-range dependencies in regulatory sequences, enabling better prediction of nutrient-responsive regulatory elements • Support transfer learning: pre-train on general genomic data, then fine-tune for nutrigenomic tasks • Enable novel hypothesis generation about nutrient-gene interactions at the regulatory and epigenetic levels	• Requires large pretraining datasets (genome sequences and regulatory annotations) • Models are computationally expensive and may overfit without careful fine-tuning • Interpreting transformer models (attention weights) in a biological context remains non-trivial

Continued

TABLE 8.6 Future directions and innovations in AI-enabled livestock nutrigenomics.—cont'd

Future direction or innovation	Description or concept	Expected impact on nutrigenomics	Challenges and considerations
Graph neural networks (GNNs) for nutrient-gene regulatory networks	Application of GNNs to model networks of genes, metabolites, and regulatory elements (e.g., epigenomic graphs) responding to nutrition	• Can model nutrient effects on entire regulatory/metabolic networks rather than just features • Predict how perturbing one nutrient or gene propagates through the network • Supports systems-level understanding of nutrient-driven epigenetic changes	• Requires robust and high-quality graph construction (defining nodes and edges) • Graph size and complexity may lead to high computational burdens • Validation of predictions (network perturbations) is experimentally demanding
Longitudinal and transgenerational nutrigenomic modeling	AI-driven time-series and multi-generational models that capture how dietary interventions influence epigenetics, gene expression, and phenotype over life stages and across generations	• Helps predict long-term and inherited effects of nutrition (e.g., epigenetic memory and imprinting) • Supports breeding strategies that integrate nutritional programming • May inform maternal or developmental nutrition designs that benefit future generations	• Requires multi-omics longitudinal datasets (across lifespan and across generations), which are rare and expensive • Modeling transgenerational inheritance is complex and may require new causal AI approaches • Ethical and regulatory issues around programmed epigenetic inheritance in animals

scanners, smartphone-captured morphological images, and co-designed assay kit development to be applicable in resource-constrained environments. The crowdsourced information regarding ingredient variability can be further used to provide insights into exposure to region-specific forages and agro-industrial byproducts to strengthen the foundation model, thereby strengthening substitution resilience and reducing reliance on imported concentrates. Weighted contributions to the models of historical accuracy are used as a probabilistic model of truth inference for quality assurance. Metadata of contextual sensor calibration and consensus of redundancy are also considered. Even the privileged purchase into cooperative purchasing pools or access to advanced analytics modules with incentive systems could be tokenized with access credit to an advanced analytics module or other incentive schemes to make the incentives consistent with data quality improvement. The design of inclusivity deals with language localization, offline caching, and transparent feedback loops, and this is why the perceived value is reciprocal, since the provided data enabled the alterations in the model (Mupangwa et al., 2025).

The simulation-intervention cycle of digital twins has evolved to a closed loop, where twins do not just receive streams of sensors and omics but also take action to generate experimental design suggestions that optimally solve high-uncertainty state dimensions given maximum expected utility. Hierarchical ensembles of mechanistic digestion-fermentation models, dynamic energy partition equations, and neural residual modules compute the uncertainty propagation by Monte Carlo propagation or polynomial chaos expansion through risk-aware stochastic outcome distribution conditioned on candidate feeding paths. Active learning controllers optimize a set of minimal interventions, pulse dosing of a given amino acid mix, and phage insertion, which are speculated to most effectively contract anterior uncertainty about the response surface of the herbiculation of the methane emissions or amino acid throbbing constraint. The ethical limitations imposed by the governance on the surplus workload of testing on individual animals reassigns exploration and findings on populations, yet does not influence welfare floors. The meta-analysis research studies show structural parameter changes, which point to the emergence of new metabolism adjustments or population-wide issues (Puniya, 2025).

The edge-to-cloud orchestration is more semantically sensitive, no longer focused on data logistics but on ontology-driven message transfer in which sensor packets are marked with machine-interpretable nutritional context tags and triggered into the appropriate transform pipelines. Middleware decreases the destabilizing inference of the downstream, permitting the addition of new sensor modalities or assays. Linearization of the calibration models can be applied on devices with incremental learning to examine NIR ingredient or bolus pH changes with self-supervised goals pegged on cloud-verified periodic ground truth to reduce the latency and bandwidth requirements. The adaptive compression schemes use information-theoretic salience scores to sparsify the dynamics of anomaly or transition phases and heavily characterize steady-state phases. Trust models ensure the integrity of the edge models and firmware, nullifying the security threats of the supply chain that can corrupt the ration recommendations (Jakku et al., 2025).

This corresponds with the convergence of welfare measures, emission accounting, and the circular bioeconomy analytics through staging nutrigenomic AI as a platform of a nexus of holistic sustainability orchestration and no longer as a productivity enhancer. Unified modeling layers are synthetic, so they can internalize the externalities by making multi-objective ration changes to include real-time enteric methane estimate curves, nitrogen excretion partition predictions, thermal stress indicators, locomotion-derived behavior welfare proxies, and feed ingredient life cycle assessment data. Circularity optimization will aim at optimizing regional co-product streams to incorporate streams with low embedded carbon intensity and minimize the effect of anti-nutritional and palatability-limiting factors through proposed enzyme preprocessing innovations based on synthetic biology modules. Together with generative diet search, dynamic shadow pricing of emission and welfare effects exhibits opportunity cost surfaces, of which strategic investment can be undertaken in additive innovation or infrastructure upgrade, such as precision cooling. Open dashboarding presents uncertainty bands of sustainability key performance indicators and prevents over-interpretation of temporary gains and other performance improvement trends by developing verifiable metrics. Future directions and innovations in AI-enabled livestock nutrigenomics are indicated in Table 8.6.

References

Ahmad, S., Shah, K., Debbouche, A., 2026. Structural equation modeling for causal effect estimation with machine learning. J. Comput. Appl. Math. 475, 117020. https://doi.org/10.1016/j.cam.2025.117020.

Alem, D., 2025. Innovations in livestock nutrition: precision feeding, microbiome modulation, and sustainable resources for future productivity a review. Int. J. Eng. Manag. 9 (1), 24–29. https://doi.org/10.11648/j.ijem.20250901.13.

Androulakis, I.P., Cucurull-Sanchez, L., Kondic, A., Mehta, K., Pichardo, C., Pryor, M., Renardy, M., 2025. The dawn of a new era: can machine learning and large language models reshape QSP modeling? J. Pharmacokinet. Pharmacodyn. 52 (4). https://doi.org/10.1007/s10928-025-09984-5.

Aydın, S.K., Ali, R.H., Faiz, S., Ali Khan, T., 2025. An integrated AI framework for personalized nutrition using machine learning and natural language processing for dietary recommendations. Appl. Sci. 15 (17), 9283. https://doi.org/10.3390/app15179283.

Bakkeren, E., Piskovsky, V., Foster, K.R., 2025. Metabolic ecology of microbiomes: nutrient competition, host benefits, and community engineering. Cell Host Microbe 33 (6), 790–807. https://doi.org/10.1016/j.chom.2025.05.013.

Belanche, A., Belzecki, G., Hernandez-Sanabria, E., Martínez-Fernandez, G., Ramos-Morales, E., 2025. Gabriel de la Fuente, editorial: unravelling the unknown of the rumen microbiome: implications for animal health, productivity, and beyond. Front. Microbiol. 16. https://doi.org/10.3389/fmicb.2025.1720795.

Bhattarai, P., Rai, S., Koirala, P., Prakash Nirmal, N., 2024. Computational Approach and Its Application in the Nutraceutical Industry. Springer Science and Business Media LLC, pp. 449–468. https://doi.org/10.1007/978-1-0716-3601-5_18.

Chiregi, M., Mazinani, M., Mitra, M., 2025. A new feature selection method using deep learning and graph representation in high-dimensional datasets. Knowl. Base Syst. 329, 114338. https://doi.org/10.1016/j.knosys.2025.114338.

Choi, J.J., Kalafut, N.C., Gruenloh, T., Engelman, C.D., Lu, T., Wang, D., 2025. COSIME: cooperative multi-view integration and scalable and interpretable model explainer. https://doi.org/10.1101/2025.01.11.632570.

Day, J., Ben Haddou, M., Kylling, R., Vasdal, G., van de Weerd, H., 2025. Optimising the selection of welfare indicators in farm animals. Front. Vet. Sci. 12. https://doi.org/10.3389/fvets.2025.1661470.

Del Vecchio, D., 2025. Epigenetic memory: the role of the crosstalk between histone modifications and DNA methylation. Comput. Struct. Biotechnol. J. 27, 4019–4025. https://doi.org/10.1016/j.csbj.2025.08.034.

Deng, Z.C., Liu, M., Cao, K.X., Khalil, M.M., Guan, L.L., Sun, L.H., 2025. Gut microbiome and postbiotics: bridging the dietary nutrition and feed efficiency in food-producing animals. Sci. China Life Sci. https://doi.org/10.1007/s11427-025-3063-5.

Duan, W., Yu, Y., He, J., Shi, Y., 2025. Adaptive Context length Optimization with Low-Frequency Truncation for Multi-Agent Reinforcement Learning (Version 1). https://doi.org/10.48550/ARXIV.2510.26389.

Duhan, P., Gupta, B., Ahmed, M., Bansal, P., 2025. Gut microbiome engineering with probiotics: current trends and future directions. Discov. Appl. Sci. 7 (11). https://doi.org/10.1007/s42452-025-07622-w.

Džermeikaitė, K., Šidlauskaitė, M., Antanaitis, R., Anskienė, L., 2025. Enhancing genomic selection in dairy cattle through artificial intelligence: integrating advanced phenotyping and predictive models to advance health, climate resilience, and sustainability. Dairy 6 (5), 50. https://doi.org/10.3390/dairy6050050.

Ebadi, A.G., Selamoglu, Z., 2025. Personalized nutrition in the genomics and digital age: current issues, future directions, and evidence. Wah Acad. J. Health Nutr. 1 (1), 33–37. https://doi.org/10.63954/sns00e14.

Fan, Z., Yu, Z., Yang, K., Chen, W., Liu, X., Li, G., Yang, X., Chen, C.L.P., 2025. Diverse models, United goal: a comprehensive survey of ensemble learning. CAAI Trans. Intell. Technol. 10 (4), 959–982. https://doi.org/10.1049/cit2.70030.

Farooqui, Y., Shekokar, K., Bhatt, P., Macwan, K., 2025. Explainable fuzzy logic in AI: enhancing transparency and trust in deep learning models. J. Mach. Comput. 5 (4), 2241–2253. https://doi.org/10.53759/7669/jmc202505174.

Farzand, A., Rohin, M.A.K., Awan, S.J., Ahmad, A.M.R., Akram, H., Saleem, T., Imran, M.M., 2025. Nutrigenomics of obesity: integrating genomics, epigenetics, and diet–microbiome interactions for precision nutrition. Life 15 (11), 1658. https://doi.org/10.3390/life15111658.

Fontanesi, M., Micheli, A., Podda, M., Tortorella, D., 2025. Bridging XAI and spectral analysis to investigate the inductive biases of deep graph networks. Mach. Learn. 114 (11). https://doi.org/10.1007/s10994-025-06884-0.

Gajavalli, S.H., 2025. Human-centered interactive anonymization for privacy-preserving machine learning: a case for human-guided k-anonymity. https://doi.org/10.48550/arXiv.2507.04104.

Garwe, E.C., Arpaci, S., Moss, A.J., Bailey, K., Suchith, A., Oyewole, O.B., Thondhlana, J., 2025. Ethical crossroads in agricultural data: perspectives on trust, equity and sustainability. Outlook Agric. 54 (4), 323–331. https://doi.org/10.1177/00307270251391436.

Ghalkha, A., Tian, Z., Issaid, C.B., Bennis, M., 2025. SheafAlign: a sheaf-theoretic framework for decentralized multimodal alignment. https://doi.org/10.48550/arXiv.2510.20540.

Ghavi Hossein-Zadeh, N., 2026. Generative artificial intelligence in animal genomics for smart agriculture: Applications, challenges, and future prospects. Vet. Anim. Sci. 33, 100702. https://doi.org/10.1016/j.vas.2026.100702.

Ghosh, P., Bhattacharya, A., Pati, D., 2025. On quantification of borrowing of information in hierarchical Bayesian models. https://doi.org/10.48550/arXiv.2509.17301.

Gundersen, K., Anas, M., 2025. Developmental programming and postnatal modulations of muscle development in ruminants. Biology 14 (8), 929. https://doi.org/10.3390/biology14080929.

He, K., Chen, C., Chen, S., Chen, B., Zhang, A., Chen, P., Wang, Z., Wu, Z., 2025. Reinforcement learning for multi-objective optimization: a review. Arch. Comput. Methods Eng. https://doi.org/10.1007/s11831-025-10389-3.

Hossein-Zadeh, N.G., 2025. Artificial intelligence in veterinary and animal science: applications, challenges, and future prospects. Comput. Electron. Agric. 235, 110395. https://doi.org/10.1016/j.compag.2025.110395.

Hu, Z., Wang, J., Mańdziuk, J., Ren, Z., Pal, N.R., 2025. Unsupervised feature selection for high-order embedding learning and sparse learning. IEEE Trans. Cybern. 55 (5), 2355–2368. https://doi.org/10.1109/tcyb.2025.3546658.

Hunter, P., de Bono, B., Brooks, D.H., Christie, R., Hussan, J.R., Ma, L., Nickerson, D., 2024. The physiome project and digital twins. IEEE Rev. Biomed. Eng. 1–15. https://doi.org/10.1109/rbme.2024.3490455.

Ibragić, S., Dahija, S., Karalija, E., 2025. The good, the bad, and the epigenetic: stress-induced metabolite regulation and transgenerational effects. Epigenomes 9 (2), 10. https://doi.org/10.3390/epigenomes9020010.

Isenberg, N.M., Mertins, S.D., Yoon, B.J., Reyes, K.G., Urban, N.M., 2024. Identifying Bayesian optimal experiments for uncertain biochemical pathway models. Sci. Rep. 14 (1). https://doi.org/10.1038/s41598-024-65196-w.

Jakku, P.C., Chippagiri, S., Uppalapati, N.S.R., Dasi, U., 2025. Multi-cloud semantic orchestration of AI-driven pipelines for privacy-first LLM inference. In: International Conference on Computing Technologies and Data Communication, ICCTDC 2025. Institute of Electrical and Electronics Engineers Inc. https://doi.org/10.1109/ICCTDC64446.2025.11158733

Jenko, S., Papadopoulou, E., Kumar, V., Overman, S., Krepelkova, K., Wilson, J., Dunbar, E., Spice, C., Exarchos, T., 2025. Artificial intelligence in healthcare: how to develop and implement Safe, ethical and trustworthy AI systems. AI 6 (6), 116. https://doi.org/10.3390/ai6060116.

Jiang, X., Ji, J., Yang, Q., Dou, Y., Li, Y., Yang, X., Liu, C., Dou, S., Dong, L., 2025. Key technologies of synthetic biology in industrial microbiology. Microorganisms 13 (10). https://doi.org/10.3390/microorganisms13102343.

Kasmi, G., Brunetto, A., Fel, T., Parekh, J., 2024. One wave to explain them all: a unifying perspective on post-hoc explainability. https://doi.org/10.48550/arxiv.2410.01482.

Khaldy, M.A., Gheraibia, Y., 2025. Ethical considerations in the development and deployment of large language models. In: 2025 1st International Conference on Computational Intelligence Approaches and Applications (ICCIAA), pp. 1–6. https://doi.org/10.1109/icciaa65327.2025.11013722.

Khan, M., Arslanturk, S., Draghici, S., 2025. A comprehensive review of spatial transcriptomics data alignment and integration. Nucleic Acids Res. 53 (12). https://doi.org/10.1093/nar/gkaf536.

Kim, M.-K., Yoo, T.-A., Chung, J.-B., 2025. Toward Sustainable Generative AI: A Scoping Review of Carbon Footprint and Environmental Impacts Across Training and Inference Stages (Version 1). https://doi.org/10.48550/ARXIV.2511.17179.

Kruhlov, V., Dvorak, J., 2025. Social inclusivity in the smart city governance: overcoming the digital divide. Sustainability 17 (13), 5735. https://doi.org/10.3390/su17135735.

Legler, T., Hegiste, V., Anwar, A., Ruskowski, M., 2024. Addressing Heterogeneity in Federated Learning: Challenges and Solutions for a Shared Production Environment. https://doi.org/10.48550/arXiv.2408.09556.

Liu, R., Wen, Y., Sun, T., Wu, Y., Huang, P., Yu, Z., Li, S., 2025. Semantic-Condition Tuning: Fusing Graph Context with large language models for Knowledge Graph Completion. https://doi.org/10.48550/arXiv.2510.08966.

Lin, K.Y., Wu, S.Y., Matsuno, K., 2025. Generative AI-driven resilience in supply chain management: UNISONE framework for disruption modelling and capacity optimisation. Taiwan Int. J. Prod. Res. https://doi.org/10.1080/00207543.2025.2574412.

Liu, G.E., 2025. Exploring cattle structural variation in the era of long reads, pangenome graphs, and near-complete assemblies. J. Anim. Sci. Biotechnol. 16 (1). https://doi.org/10.1186/s40104-025-01294-7.

Lvovs, D., Creason, A.L., Levine, S.S., Noble, M., Mahurkar, A., White, O., Fertig, E.J., 2025. Balancing ethical data sharing and open science for reproducible research in biomedical data science. Cell Rep. Med. 6 (4). https://doi.org/10.1016/j.xcrm.2025.102080.

Mai, W., Liwang, M., Yi, X., Xia, X., Hosseinalipour, S., Wang, X., 2025. Game-Theoretic Safe Multi-Agent Motion Planning with Reachability Analysis for Dynamic and Uncertain Environments (Extended Version) (Version 1). https://doi.org/10.48550/ARXIV.2511.12160.

Malik, F., 2025. Carbon-conscious intelligence: life cycle assessment and green standards for generative AI. Int. J. Innovat. Sci. Res. Technol. 1030–1037. https://doi.org/10.38124/ijisrt/25aug585.

Marinello, D., Reschini, M., Di Stefano, G., Carullo, G., Casalechi, M., Tarantini, L., Albetti, B., Bollati, V., Viganò, P., Somigliana, E., Piani, L.L., 2025. Epigenetic age and fertility timeline: testing an epigenetic clock to forecast in vitro fertilization success rate. Reprod. Biol. Endocrinol. 23 (1). https://doi.org/10.1186/s12958-025-01429-5.

Mendonca, F., DiMarzo, G., Abdennadher, N., 2025. Data Cooperatives: Democratic Models for Ethical Data Stewardship (Version 1). https://doi.org/10.48550/ARXIV.2504.10058.

Merkow, J., Dorfner, F.J., Yang, X., Ersoy, A., Dasegowda, G., Kalra, M., Lungren, M.P., Bridge, C.P., Tarapov, I., 2024. Scalable Drift Monitoring in Medical Imaging AI. https://doi.org/10.48550/arXiv.2410.13174.

Mia, N., Sarker, T., Halim, M.A., Alam, A.M.M.N., Ali, M.S., Rahman, M.M., Hashem, M.A., 2025. Machine learning overview and its application in the livestock industry. Meat Res. 5 (1). https://doi.org/10.55002/mr.5.1.109.

Mupangwa, W., Maharaj, U., Taylor, J., Dickens, C., Choruma, D., Pike, T., Dirwai, T.L., Nyawira, S., Geck, M.S., Braithwaite, I., 2025. Developing a conceptual framework for a citizen science-based agroecology self-assessment (ASA) tool. In: Frontiers in Agronomy. https://doi.org/10.3389/fagro.2025.1646470.

Nawaz, U., Zaheer, M.Z., Khan, F.S., Cholakkal, H., Khan, S., Anwer, R.M., 2025. AI in Agriculture: A Survey of Deep Learning Techniques for Crops, Fisheries and Livestock. https://doi.org/10.48550/arXiv.2507.22101.

Necula, S.C., Rieder, E., 2025. Generative and adaptive AI for sustainable supply chain design. J. Theor. Appl. Electron. Commer. Res. 20 (3). https://doi.org/10.3390/jtaer20030240.

Nisa, P., Kirthi, A.V., Sinha, P., 2025. Microbiome-based approaches to personalized nutrition: from gut health to disease prevention. Folia Microbiol. 70 (5), 961–978. https://doi.org/10.1007/s12223-025-01337-6.

Pannu, J., Bloomfield, D., MacKnight, R., Hanke, M.S., Zhu, A., Gomes, G., Cicero, A., Inglesby, T.V., 2025. Dual-use capabilities of concern of biological AI models. PLoS Comput. Biol. 21 (5). https://doi.org/10.1371/journal.pcbi.1012975.

Panpatil, A.Y., Gaikwad, V.P., 2025. A method to study the robustness of ML models against adversarial attacks. Int. J. Environ. Sci. 707–716. https://doi.org/10.64252/dpk19t90.

Papadakis, N., Bouloukakis, G., Magoutist, K., 2024. CCDUIT: a software overlay for cross-federation collaboration between data spaces. In: Proceedings—IEEE 21st International Conference on Software Architecture Companion, ICSA-C 2024. Institute of Electrical and Electronics Engineers Inc., pp. 143–150. https://doi.org/10.1109/ICSA-C63560.2024.00031

Park, Y., Muttray, N.P., Hauschild, A.C., 2024. Species-Agnostic transfer learning for cross-species transcriptomics data integration without gene orthology. Briefings Bioinf. 25 (2). https://doi.org/10.1093/bib/bbae004.

Puniya, B.L., 2025. Artificial-intelligence-driven innovations in mechanistic computational modeling and digital twins for biomedical applications. J. Mol. Biol. 437 (17), 169181. https://doi.org/10.1016/j.jmb.2025.169181.

Rosati, A., 2025. Invited paper artificial intelligence and the future of animal husbandry: rethinking knowledge, practice, and governance. Arch. Latinoam. Prod. Anim. 33 (Suppl. 1), 567–572. https://doi.org/10.53588/alpa.330509.

Roy, P., Boker, A., Mili, L., 2025. Beyond marginals: learning joint spatio-temporal patterns for multivariate anomaly detection. https://doi.org/10.48550/arXiv.2509.15033.

Sabit, H., Arneth, B., Pawlik, T.M., Abdel-Ghany, S., Ghazy, A., Abdelazeem, R.M., Alqosaibi, A., Al-Dhuayan, I.S., Almulhim, J., Alrabiah, N.A., Hashash, A., 2025. Leveraging single-cell multi-omics to decode tumor microenvironment diversity and therapeutic resistance. Pharmaceuticals 18 (1). https://doi.org/10.3390/ph18010075.

Sander, J., Cohen, A., Dasari, V.R., Venable, B., Jalaian, B., 2025. ON accelerating edge AI: optimizing resource-constrained environments. https://doi.org/10.48550/arXiv.2501.15014.

Segev, T., Barak, D., Zahavi, L., Godneva, A., Rein, M., Krongauz, D., Rossman, H., Weinberger, A., Segal, E., 2025. Diet shapes the gut microbiome: cross-sectional and longitudinal insights from the Human Phenotype Project. In: medRxiv. Cold Spring Harbor Laboratory. https://doi.org/10.1101/2025.10.12.25337559.

Shah, B.A., Astudillo, R., Frazier, P.I., 2024. Multi-attribute optimization under preference uncertainty. In: Proceedings—Winter Simulation Conference. Institute of Electrical and Electronics Engineers Inc., pp. 3241–3252. https://doi.org/10.1109/WSC63780.2024.10838798

Shao, S., Tian, Y., Zhang, Y., Yang, S., Zhang, P., He, C., Zhang, X., Jin, Y., 2025. Evolutionary computation for sparse multi-objective optimization: a survey. ACM Comput. Surv. 57 (11), 1–35. https://doi.org/10.1145/3734865.

Sharma, S., Anirudh, K.M., Bakshi, B.R., Sandeep, S., Kodamana, H., Ramteke, M., 2025. Multi-objective optimization for sustainable agricultural strategies considering future climatic scenarios: a study in the context of India. Comput. Chem. Eng. 203, 109306. https://doi.org/10.1016/j.compchemeng.2025.109306.

Sharon, T.J., 2025. Self-supervised learning for small data environments. Int. J. Sci. Technol. Eng. 13 (7), 1875–1880. https://doi.org/10.22214/ijraset.2025.73282.

Smiles, W.J., Ovens, A.J., Kemp, B.E., Galic, S., Petersen, J., Oakhill, J.S., 2024. New developments in AMPK and mTORC1 cross-talk. Essays Biochem. 68 (3), 321–336. https://doi.org/10.1042/EBC20240007.

Tao, Z., Gao, Y., Sun, Y., Tu, Q., Gao, F., Chen, L., Wang, W., Chen, H., 2025. Active differentiable structure learning for clinical causal discovery. Knowl. Base Syst. 327, 114145. https://doi.org/10.1016/j.knosys.2025.114145.

Tattersall, E., Cardinal-McTeague, W., Myers-Smith, I., Jenkins, D.A., Burton, A.C., 2025. Affirming Indigenous data sovereignty in collaborative wildlife conservation in the era of open data. People Nat. 7 (11), 2659–2677. https://doi.org/10.1002/pan3.70161.

Tedeschi, L.O., Guarnido Lopez, P., Menendez III, H.M., Seo, S., 2025. Advancing precision livestock farming: integrating artificial intelligence and emerging technologies for sustainable livestock management. Anim. Biosci. https://doi.org/10.5713/ab.25.0289.

Thoniyot, S., Balakrishnan, V., 2026. Integrating multi-omics data for personalized nutrition using knowledge graphs and graph neural networks: a comprehensive review. Comput. Biol. Chem. 120 (Pt 1), 108772. https://doi.org/10.1016/j.compbiolchem.2025.108772.

Uhler, C., Zhang, J., 2025. Causal structure and representation learning with biomedical applications. https://doi.org/10.48550/arXiv.2511.04790.

Vatsanidou, A., Konstantinidi, S., Bonos, E., Skoufos, I., 2025. Comparative life cycle assessment of animal feed formulations containing conventional and insect-based protein sources. AgriEngineering 7 (9), 275. https://doi.org/10.3390/agriengineering7090275.

Verma, H., Padh, K., Thelisson, E., 2025. Can AI be auditable? (Version 3). https://doi.org/10.48550/ARXIV.2509.00575.

Wang, J., Yan, Y., Peng, X., Gao, X., Luo, Q., Luo, Z., Wang, K., Liu, X., 2025. Linking lipidomics to meat quality: a review on texture and flavor in livestock and poultry. Food Chem. 492 (Pt 1), 145402. https://doi.org/10.1016/j.foodchem.2025.145402.

Wei, J., Li, Z., Zhuo, L., Fu, X., Wang, M., Li, K., Chen, C., 2024. Enhancing drug-food interaction prediction with precision representations through multilevel self-supervised learning. Comput. Biol. Med. 171, 108104. https://doi.org/10.1016/j.compbiomed.2024.108104.

Wei, J., Dektor, A., Shen, C., Wen, Z., Yang, C., 2025. Numerical optimization for tensor disentanglement. https://doi.org/10.48550/arXiv.2508.19409.

Xue, B., Yan, T., Chen, X., Baral, K.R., Dabiri, A., Cristobal-Carballo, O., Smith, L., Stergiadis, S., Theodoridou, K., 2025. Integrated rumen-animal-manure analysis of dairy emission mitigation by feeding apple pomace and hempseed cake. Environ. Technol. Innovat. 40, 104508. https://doi.org/10.1016/j.eti.2025.104508.

Yang, Q., Cai, Y., Guan, Y., Wang, Z., Guo, S., Qiu, S., Zhang, A., 2025. Metabolic phenotypes: molecular bridges between health homeostasis and disease imbalance. Comput. Struct. Biotechnol. J. 27, 4710–4719. https://doi.org/10.1016/j.csbj.2025.10.057.

Zeng, L., Qian, Y., Cui, X., Zhao, J., Ning, Z., Cha, J., Wang, K., Ge, C., Jia, J., Dou, T., Chen, H., Liu, L., Bao, Z., Jian, Z., 2025. Immunomodulatory role of gut microbial metabolites: mechanistic insights and therapeutic frontiers. Front. Microbiol. 16. https://doi.org/10.3389/fmicb.2025.1675065.

Zhang, Y., Wang, C., Shi, Q., Feng, Y., Chen, C.-Y., 2023. Adversarial gradient-based meta learning with metric-based test. Knowl. Base Syst. 263, 110312. https://doi.org/10.1016/j.knosys.2023.110312.

Zhang, S., Lai, C., Zhao, J., Wang, J., 2025. Big data and AI-powered modeling: a pathway to sustainable precision animal nutrition. Adv. Sci. 12 (41). https://doi.org/10.1002/advs.202507564.

Chapter 9

Artificial intelligence integration in livestock metagenomics

Integration of artificial intelligence (AI) into livestock metagenomics is proving to be a pivotal development, providing a data-driven ability to query, predict, and adjust the microbial community that characterizes health, production, nutrient efficiency, product quality, and environmental effects. Metagenomics is transforming the poorly characterized biological component into a measurable, dynamically changing component of production systems, but the sheer diversity, functional redundancy, horizontal gene flow, and temporal instability of microbial consortia are complicating attempts to make mechanistic inferences and provide actionable decision support. The represented AI technologies can break up high-dimensional abundance and functional expression of genes, recapitulate latent interaction networks, and characterize the state of microbial communities in host-relevant metabolic, immunological, and emission states. Integrative models can transcend existing commonplace taxonomic snapshots to a predictive ecological intelligence that informs the time and manner of intervention, fine-tuning feeding behaviors, selective bioadditive deployment, and microbiome-directed selective breeding or management transition by a multimodal combination of metagenomic, metatranscriptomic, metabolomic, host omics, sensor-displayed physiological signals, and environmental circumstances.

The path toward operational microbiome stewardship must evolve the descriptive cataloging paradigm into predictive and ultimately prescriptive models, which should be supported by explicit uncertainty quantification and biological plausibility constraints. These can be resolved with ecological mechanistic priors, learning deep representations, and using causal structure discovery, reinforcement, or constrained optimization to navigate a counterfactual intervention space and mitigate the dangers of overfitting and spurious correlation. Semantic interoperability, microbial function ontologies, reference genomes used to support their repeated reproducibility and cross-system transferability, and federated learning paradigms based on the requirement to cooperatively improve model quality without central storage of all data in a single location are working in support of regional biosecurity requirements. Farm anomaly detection, fermentation stability monitoring, and edge-to-cloud adaptive additive dosing will increasingly be based on edge-to-cloud orchestration, reducing feedback loops between model understanding and managerial response. Digital twins models, where a microbial metabolic unit is instantiated in host energetic and immunometabolic simulators, further permit the investigation of the consequences of scenarios, give priority to sampling to reduce epistemic imprecision, and organize ethically constrained adaptive experimentation.

The technological course of action cannot be isolated from ethical, ecological, and sustainability considerations. The data sovereignty, equitable access to advanced analytical technologies across livestock production systems and geographical areas, biased algorithms in the display of breeds or management systems, and biosafety problems related to engineered interventions of microbes or phage-based modulation should be addressed by the governance systems. Computational pipeline lifecycle assessment encourages energy-efficient model design and reuse in line with innovation and environmental stewardship. These measures of emissions, monitoring of antimicrobial resistance genes, welfare, and the evaluation of circular bioeconomies are balanced so that microbiome optimization has an overall sustainability agenda rather than individual productivity objectives. In order to build trust, interpretive literacy, and adaptive governance, it is essential to build capacity, a real performing benchmark, model interfaces that are fine-tuned, and a participatory data ecosystem among producers, veterinarians, nutritionists, and policymakers. The chapter introduces AI-augmented metagenomics in the light of an integrated computational decision-support layer that converts the complexity of microorganisms into a strategic resource and a lever of resilient and sustainable livestock production that is ethically oriented.

9.1 Introduction to AI in livestock metagenomics

The integration of AI into livestock metagenomics is due to the rapid convergence of progress in culture-independent microbial ecology and the maturation of computational paradigms that are able to derive latent structure, predict

Artificial Intelligence in Livestock Omics. https://doi.org/10.1016/B978-0-443-45663-3.00007-5

dynamic transitions, and suggest intervention strategies in the face of uncertainty. The science of microbiomes has moved beyond the listing of the presence of community members and has realized that assemblages of microbes are distributed metabolic and signaling systems in which the emergent properties of these systems precondition nutrient transformation, immune calibration, developmental trajectories, and emission pathways. However, the extreme diversity defines this complexity, with redundancy in functions, contingent interaction networks with context, the possibility of horizontal gene exchange, and plasticity in time, all of which are driven by host physiology and environmental change. AI provides representational and inferential machines capable of resolving high-dimensional sparsity, compositional constraints, nonlinear dependencies, multiscale temporal links, as well as cross-domain heterogeneity, and in doing so convert raw sequence or functional annotation matrices into biologically and operationally meaningful state descriptors and decision levers (Rizzi et al., 2025).

The strategic value of integrating AI in metagenomic processes cuts across collateral goals of health maintenance, productivity growth, welfare improvement, sustainability improvement, and economic sustainability. It has been found that microbiomes can mediate modulation of barrier, inflammatory, and metabolic functionality, nutrient reclamation, and creation or dampening of environmentally impactful metabolites. Anticipatory knowledge of microbial functional changes can be used to make proactive nutritional or management changes to stabilize physiological homeostasis, limit reactive responses, and match resource use with ecological stewardship. The intersection of welfare considerations is the ability of microbial ecosystems to affect the neuroendocrine stress axis, priorities in energetic allocation, and subclinical dysregulation courses. Sustainability aspects are involved through microbial roles in greenhouse gas precursors, nitrogen and sulfur cycling, and the valorization of residual substrates in a circular bioeconomy. As a result, anchoring metagenomics becomes an integrative layer of intelligence that coordinates multiobjective trade-off navigation, with forecasts and quantification of uncertainty driving balanced intervention design that is determined in line with regulatory, societal, and market expectations (Rizzi et al., 2025; Shah et al., 2025).

It is necessary to explain the scope and definitions to define the sphere of analysis. In this regard, metagenomics refers to shotgun sequencing-based characterization of the total genomic contents of microbial communities, which provides access to both taxonomic and functional gene and accessory element landscapes simultaneously. It contrasts with metabarcoding methods, which use the intensive amplification of conserved marker sequences to deduce taxonomic structure with restricted functionality resolution and frequently restricted phylogenetic sensitivity. The concept of meta-omics is used to denote a broader spectrum of integrative continuum profiling, expressed functional potential, realized enzymatic presence, and terminal biochemical products, as well as viromics and mobilome description, where applicable. The chapter is focused on AI algorithms that consume metagenomic raw data and its consistent association with complementary host and environmental modalities to provide actionable ecological intelligence. Host-centric genomic selection pipelines, narrow pathogen surveillance analytics, and feed supply chain logistics are not only intersected where there exists material interaction between their data streams or decision outputs and microbiome-informed modeling. Similarly, single-marker diversity surveys, which are so narrow as to lack functional integration, are beyond the operational focus unless used as a methodological baseline of legacy (Ghozlane et al., 2025).

The shift from descriptive taxonomies to functional, predictive, and finally, causal intelligence is one of the key thematic changes. Early community profiling focused on relative abundance tables and indices of diversity, which, although informative on coarse ecological stratification, are poor measures of metabolic flux capacity, resilience properties, or responsiveness to interventions. Inference-based layers of functional annotation are based on gene family, pathway, and metabolic module inference and bring about a change to a more mechanistic interpretability, but with the limitation of database incompleteness, horizontal gene transfer making orthology mapping challenging, and variable genome reconstruction fidelity. AI enhances the progress even more than capturing static annotation, through the process of learning latent embeddings in which covariation patterns across genes, operons, plasmids, and environmental host contextual factors represent emergent functional modules not entirely enumerated in reference ontologies. Temporal sequence modeling compares transitions in community state with transitions in host physiological states and environmental disturbances, and identifies dynamical predictors of an impending condition of dysbiosis or an optimum adaptation period. The various versions of causal discovery that include invariant risk principles, constraint-based graphical inference, and counterfactual simulation combine a prior ecological understanding with observational heterogeneity to prioritize putative directional effects among microbial guilds, functional pathways, and host outcome proxies. The predictive and causal structure is then transformed into the best intervention strategy, with safety, welfare, and sustainability constraints, by reinforcement and constrained policy learning layers, shifting the axis of analysis away from what currently happens, toward what is likely to happen and can be prudently implemented to achieve desirable trajectories (Cai et al., 2026; Wang et al., 2025).

AI techniques must compete with the unique statistical and biological characteristics of metagenomic data. The constant sum constraints of compositionality do produce spurious correlations, unless dealt with by suitable log ratio transformations or generative models, which can handle the uncertainty of absolute abundance. The consequence of ecological rarity and the limitations of the sampling depth is represented by sparsity and zero inflation, which requires probabilistic architectures that can differentiate between structural absence and stochastic undersampling. The level of heterogeneity in strain-level sequences adds much subtle but functionally significant variation in sequence that is sometimes hidden in the aggregate feature tables, and deep representation learning and graph-based clustering of k-mer sequences or gene patterns of cooccurrence can rediscover subpopulation signals underlying metabolic specialization or resistance dominance. The dynamics of horizontal gene transfers and mobile elements pose a challenge to the current assumptions of a static network, and provide an impetus to adopt dynamic graph models that are able to update edge weights and potential adjacency structure as new associations form. Multiscale temporal irregularity arises due to asynchronous sampling of molecular, physiological, and environmental layers; sequence-to-sequence and continuous time modeling is a reconciliation of these differences, which maintains a causal temporal order. New measures of uncertainty are essential to both the ethical and practical reliability, and Bayesian neural processes, ensemble variance decomposition, and counterfactual consistency checks are used to identify epistemic gaps that may be addressed by targeted sampling versus aleatoric variability of ecological stochasticity (Kublashvili, 2025; Ma et al., 2025; Sharma et al., 2025).

The infrastructure of data governance, semantic interoperability, and reproducibility is enabling substrates of credible integration of AI. The minimization of semantic drift is achieved using ontology alignment of semantic annotations of functional genes, semantic identifiers of metabolic pathways, semantic identifiers of antimicrobial resistance, and environmental descriptors. Connecting work across studies assists with model transfer across studies. These can be used to reduce biosecurity and intellectual property interests and propagate ecological representation by learning parameters collaboratively with datasets that are geographically or institutionally dispersed using federated learning and privacy-preserving cryptographic aggregation. Named provenance and version-controlled preprocessing pipelines, and standardized metadata schemes are used to combat batch effects and version-controlled undocumented analytical divergence, which would otherwise destroy generalizability. Green computational practices that focus on effective model designs, effective hyperparameter optimization, and pretrained models reuse focus on methodological progress and sustainable resource management (Shu, 2025).

The questions within this framing reflective of the analytical journey are as follows: how can the analogy of latent representations retain biologically significant variability and reduce spurious association; how can dynamic modeling differentiate reversible fluctuation and irreversible inflection of trajectories; how can causal inference be able to distinguish between direct microbial functional influence and host mediated or environment confounding; how can strategies oriented toward maximizing productivity, welfare and mitigating emissions be maintained without extrapolating too far; how can risk-taking be enabled by relevant communication and how can understanding the literature serve. The chapter is organized based on ideas and information architecture to representational and causal modeling, predictive to prescriptive shift, governance and ethical custodianship, translational implementation in decision systems and forward-thinking innovation. AI, as practical in metagenomics, is thus not to be viewed as an analytical booster, but as a change agent, an adaptive, decision-aware intelligence surface that reconceptualizes microbial ecosystems as actively steerable agents of resilient and sustainable livestock production.

9.2 Overview of livestock metagenomics

The field of livestock metagenomics involves the description of multifaceted microbial communities living in a variety of anatomical and ecological environments that, together, affect organism physiology, productivity, well-being, and ecological imprint (see Fig. 9.1). Niche-specific consortia comprise distinct physicochemical gradients, substrate availabilities, redox states, host secretions, and immunological filtering processes to form niche-specific consortia between housed or grazed animals and the agroecosystems surrounding them in the interconnected soil-manure interface. The complex hierarchical organization of fermented polysaccharide breakdown, biohydrogen regulation, and volatile fatty acid production, and downstream gut components represents downstream gut processes under anaerobic conditions, as well as host-derived substrate exploitation, bile and mucus-adapted guilds, and immune modulation. The skin-associated communities regulate the keratinized matrices, moisture dynamics, ultraviolet radiation, as well as the antimicrobial peptides, and this leads to the integrity of the barrier and inflammatory tuning. The microbiota of the respiratory tract inhabits aerated and regulated mucociliary processes where there are unique ecological filters of particle interception, gaseous exchange surfaces, and immune surveillance. The reproductive tract is a cyclically changing environment with physicochemical fluctuations that are hormonally regulated, with oscillating mucosal conditions impacting the

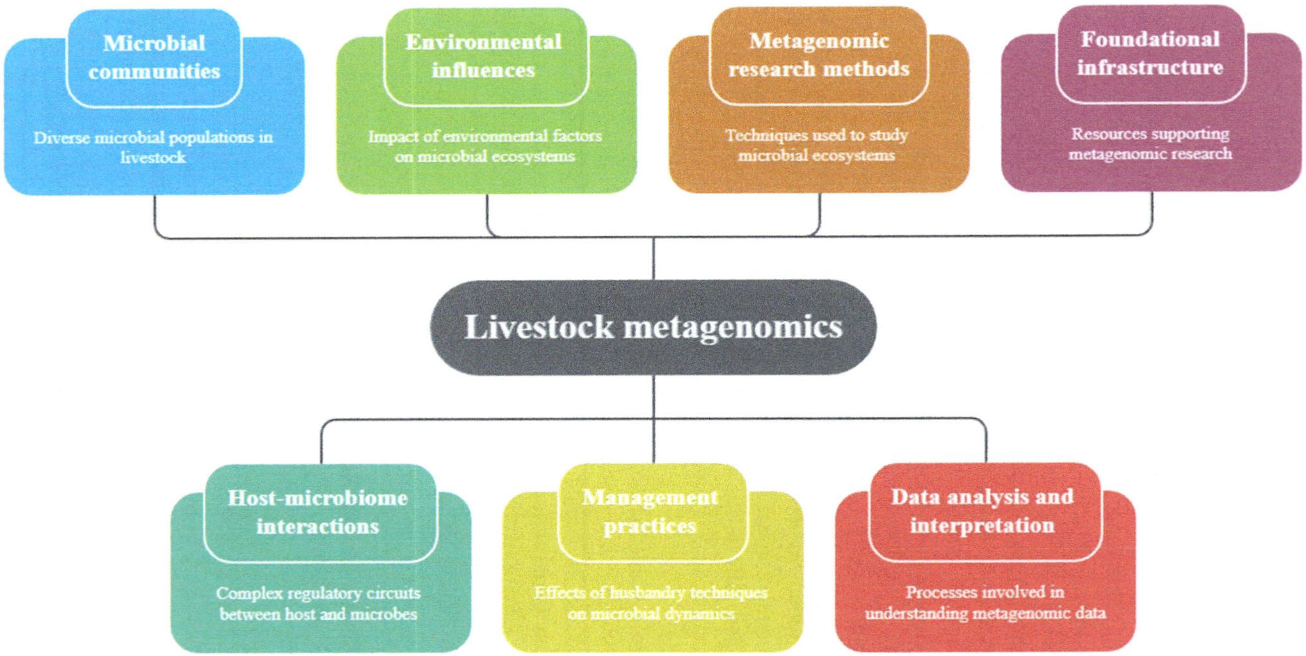

FIGURE 9.1 **An overview of livestock metagenomics.** The figure conceptualizes livestock metagenomics as an integrative discipline that links diverse microbial communities associated with animals to environmental conditions, management practices, and analytical methods, and shows how these feed into understanding host–microbiome interactions, on-farm management, and data-driven interpretation for livestock health, productivity, and sustainability.

colonization and tolerance of microbes. The soil-manure continuum is a dynamic reservoir of biogeochemical reactors in which excreted microbiota, residual feed, antimicrobial compounds, and environmental microbes interact and result in nutrient mineralization, gaseous emissions, the propagation of resistance genes, and possibly the reseeding of animal-related ecological niches. Knowledge of cross-compartmental connectivity, pathways of microbial dispersal, and selective bottlenecks is crucial to an integrated system-level understanding (Yamada and Palm, 2025; Liang et al., 2026).

The community assembly in these niches is the result of interaction between deterministic and stochastic processes, in line with the ecological theory, but within the context of husbandry. Diet is the primary source of energy and nitrogen substrates, regulates the rate of passage, alters the pH microenvironment, and migrates fermentative electron flow, which together provide considerable selective pressure on the prevalence of metabolic guilds and expression of functional genes. The genetics of a host affect the mucosal secretions, immune receptor repertoires, the epithelial glycan landscapes, individual motility patterns, and physiological throughput rates, which establish hereditary microhabitats that support some consortia and functional capabilities. The management practices of housing density, hygiene practices, feeding practices, additive use, stress alleviation, and biosecurity practices alter transmission patterns, environmental microbial pools, and opportunistic bloom patterns. The rest of the environmental parameters, such as temperature changes, humidity, oxygen partial pressure with altitude, photoperiod, and soil physicochemical conditions, further limit or promote the movements of certain taxa and mobile elements along the animal-environment interface. Selective sweeps, collateral perturbation, and recovery paths with long-term structural and functional effects are brought about by therapeutic and prophylactic interventions that include antimicrobial compounds, antiparasitic agents, disinfectants, acidifiers, and immunomodulatory inputs. The resultant community structure is a combination of priority effects, resource partitioning, ecological drift, dispersal limitation, and selective filtering, which are controlled by host developmental stage and cumulative exposure history (Week et al., 2025; Yang et al., 2025).

The metagenomic research employs a range of data modalities that vary in their resolution, breadth, and functionality. This approach is cost-effective but is limited by biases in the primer sets, limited phylogenetic coverage, and low density of inferences of direct functionality. Shotgun whole-genome sequencing provides full-genome content, allowing for simultaneous performance of taxonomic assignment, family of genes, pathway reconstructions, identification of mobile genetic elements, and analysis of strain-level variants depending on coverage depth and assembly complexity. Auxiliary viromic surveys complement the knowledge of phage and eukaryotic viral communities, which govern the population dynamics of bacteria, horizontal gene transfer, and host immunology. Metatranscriptomic data clarify expressed functional potential and patterns of temporal regulation, and indicate context-specific activation of metabolic modules that are not present in the genome. Community-level metabolomics measures the quantity of small-molecule products, intermediates, and signaling molecules that combine microbial activity with host metabolic fluxes. The establishment of connections between these layers of data and standardized records of nutrient intake, physiological sensor streams, immune biomarkers, and environmental monitoring creates multimodal data to facilitate ecological modeling in its entirety. Intermodality harmonization requires strict metadata acquisition, time-scale matching, as well as normalization plans that recognize various dynamic and detection limits, and batch designs (Cunningham-Oakes et al., 2025; Durand et al., 2025).

Functional annotation is an operative linking point to attend to raw sequence reads to mechanistic interpretability. This is then followed by binning, which produces metagenome-assembled genomes, which are approximate draft genomic reconstructions of associated populations that can be used to do comparative genomics, make inferences about metabolic capacity, and population genetics among uncultivated lineages. Extensive gene catalogs are unified nonredundant open reading frame collections in cohorts providing reference sets to abundance measurement, diversity monitoring, and pangenomic examination of core and accessory functional repertoires. Annotation pipelines signal gene sequences to orthologs, enzyme commission families, metabolic scenarios, families of transporters, resistance establishments, virulence-related structures, and signatures of mobile elements. Pathway and enzyme profiling reconstitutes the possible metabolic pathways involved in the breakdown of carbohydrates, fermentation, transformation of nitrogen and sulfur, lipid biohydrogenation, synthesis of vitamins, processing of xenobiotics, and redox regulation. The level of faith in annotation is balanced by the completeness of the database, divergence of the sequences, structural variation in domain architecture, and the high occurrence of hypothetical proteins having no known functionality. New machine learning-aided protein structure prediction and embedding with based function inference are also used to cover orphan genes, with network-based cooccurrence and genomic context studies providing predictions of operon-level functional interactions. The mapping of gene-level signals into pathway activity indices, as well as combining with host phenotypic or environmental data, helps to integrate the two, but becomes vulnerable when one assumes about stoichiometry, redundancy, and situational regulation of a pathway (Moitra and Larrouy-Maumus, 2025).

Host-microbiome interaction axes are immune, metabolic, endocrine, and neural circuits that have bidirectional regulation. Immune interfaces include pattern recognition receptors, modulation of antimicrobial peptides, immunoglobulin secretion, cytokine signaling cascades, and tolerance versus activation thresholds, which mediate the composition of the microbial community and its adaptive response to perturbations. Microbial products and structural elements regulate immune maturation, effector cell differentiation, and interactions in inflammatory resolution, thereby affecting systemic resilience. The focus of metabolic interactions is on dietary and endogenous substrate microbial conversion to short-chain compounds, gases, amino acid derivatives, lipid intermediates, vitamins, and bioactive conjugates that control host energy utilization, epithelial structure, and metabolic signaling, directing the use of glucose, lipids, and nitrogen. The endocrine axes are influenced by microbial biotransformation or production of hormone-like molecules, changes in receptor ligand availability, changes in enteroendocrine cell activity, and stress-related hormonal changes, which reciprocally affect motility, secretory profile, and mucosal barrier properties, influencing microbial niches. Neural interaction is realized through neuroactive metabolites, vagal signals, immune-mediated neuromodulation, and enterochromaffin cell interactions, and feeding behavior, stress sensitivity, and gastrointestinal transit patterns are feedback-regulated to reciprocally restructure microbial habitats. Such coupled axes can work on a system network where perturbation spreads over a molecular, cellular, and organ scale domain that understands the need to have integrative modeling that reflects distributed causality (Ouyang et al., 2025; Wu et al., 2025).

The effects of the microbiome extend to feed efficiency, emissions, health, product quality characteristics, and perturbation-resilience. The polarization of nutrient utilization requires the synchronized breakdown of polysaccharides, fermentative expenditure, distributions of the electron sinks, recycling of nitrogen, and repression of wasteful or counterproductive metabolic pathways, functions controlled by community make-up, genomic promise, expression of enzymes, and metabolic cross-feeding webs. Emission profiles are based on microbial pathways yielding gaseous products, nitrogenous volatiles, and downstream conversion of waste substrates in manure matrices, and the changes in the magnitude and partitioning in fluxes are changes at the community level. Colonization resistance to opportunistic agents, preservation of epithelial barrier, tone regulation of inflammatory reactions, and toxin production are some of how health outcomes are being molded as a result of supporting preventative over reactive paradigms. Metabolic intermediates, the processes of biohydrogenation, proteolytic and lipolytic activities, vitamin biosynthesis, production of flavor precursors or structural modifiers are met through both preharvest and postharvest inoculum reservoirs in product quality characteristics in animal derivatives. Resilience is the ability to withstand dietary, environmental, or infectious perturbations and stabilize the functional output and recover quickly due to diverse properties, functional redundancy, keystone interaction networks, and adaptive regulatory plasticity. Separating causal microbial influences by correlated host or environmental confounders is an analytical accomplishment towards translating associations into sound interventions (Deng et al., 2025).

Livestock microbial ecosystems exhibit complex temporal dynamics and successional pathways, both during initial colonization, maturity, and in response to episodic disturbances. The sequential establishment of pioneer taxa and functional guilds, which are limited by maternal, environmental, and dietary inocula, immune maturation, and the physicochemical evolution of anatomical niches, is known as developmental succession. The phases of weaning, restructuring of diets, reproductive cycles, environmental transitions, and management manipulations present resource reallocation, luminal parameter shifts, and variations in exposure that reorganize the community structure and functioning. Transient or permanent dysbiosis, determined by changes in diversity, functional imbalance, metabolic inefficiencies, and possible opportunistic expansions, is triggered by perturbations that comprise acute stressors, infectious challenges, therapeutic remedies, and environmental extremes. The trajectories of recovery are determined by the remaining functional redundancy, the dispersal of refugia, the compensatory strategies of the host, and the foreign inputs of management, which are directed at ecological stabilization. Dense longitudinal designs, along with analytical models that support irregular time intervals and nonstationary dynamics, are crucial. The understanding of early warning, hysteresis, and the occurrence of alternative stable states is used to inform the strategic timing of interventions against trajectory correction before the degradation of functional states becomes entrenched (Lubin et al., 2025).

The process of analyzing microbial ecosystems involves collecting samples, preserving them, extracting nucleic acids, preparing a library, sequencing, quality control, host sequence depletion, assembling or mapping the sequencing library, taxonomic classification, functional annotation, quantification, statistical analysis, and integrative modeling. The preanalytical steps focus on standardized sampling procedures that reduce the handling-induced compositional changes and provide fully traceable metadata. Methods of extraction involve a trade-off between wide lysis performance and nucleic acid integrity, and eliminating nucleic acid inhibitors. Furthermore, variability of protocols is a source of cross-study heterogeneity. Low-quality reads, contaminants, and chimeric sequences are eliminated by quality control filters, and host read subtraction decreases the computational cost and privacy concerns of host genomic material. The strategies employed in assembling can be either co-assembly or single-sample assembly with successive refinement, and reference-based mapping is efficient in well-characterized genes but can fail to capture novel sequences. Taxonomic and functional classification plans are based on curated reference databases in which prokaryotes, archaea, fungi, viruses, and plasmids are integrated and ontologies that help in uniting naming rules and taxonomic connections. Semantic interoperability is promoted by ontological representations of functional annotation, metabolism, enzyme nomenclature, drug resistance determinants, and ecology related to organisms, which can be integrated across different research and analysis systems. Containerized pipelines, workflow description languages, version-controlled configuration, and standardized reporting of parameters support workflow reproducibility, which is necessary to achieve transparency and allow comparative validation (Klimina et al., 2025; Sobolev et al., 2025).

Foundational infrastructure, reference databases, and ontologies are inherently biased in nature, with imbalanced sampling of geographies and production systems, as well as among host species and in environmental settings, which results in uneven annotation depth and may misclassify divergent sequences. Representativeness is improved through continuous curation, the addition of new metagenome-assembled genomes, and the relationships in hierarchies. Ontologies provide controlled vocabularies to anatomical locations, environmental characteristics, management practices, illnesses, ailments, and nutritional constituents that assist in the strict encoding of metadata and machine-readable combinations. Ontological harmonization helps to avoid semantic fragmentation and enables cross-domain reasoning, which is necessary in integrative applications of AI that are based on consistent semantic grounding to facilitate extrapolation to support applications outside narrowly maintained datasets (Chorlton, 2024).

Although progress has been made, some restrictions reduce the interpretation and translational reliability. The difficulty in strain resolution still exists because genomic similarity between adjacent groups makes it hard to genetically distinguish them, recombination, or disproportional coverage makes it difficult to identify a functional condition of microdiversity such as metabolic specialization, resistance determinants, or stress tolerance. Horizontal gene transfer alters accessory genomes, disrupts congruence of phylogenetic signal to functional profiles, and forms mosaic genomic structures, which undermine assembly continuity and binning accuracy. Mobile genetic elements that contain adaptively moving functions can cross ecological barriers, making it difficult to trace their origin and evaluate the risk. Sparsely represented functional assignments of uncharacterized genes hinder the possibility of reconstructions of entire pathways and suppress the ability of the organism to harbor previously unknown metabolic potential. Computational prediction is ahead of experimental validation, and culture-dependent follow-up is resource-intensive and biased in favor of taxa that can easily be cultured. The effects of compositional data constraints are spurious associations when not modeled correctly, and zero inflation of actual absence, as well as detection limits, which affect the tools of differential abundance and cooccurrence. The confounders of batch effects caused by variable extraction kits, sequencing platforms, and bioinformatic parameterization can be so overwhelming that they can mask real biological signals when not corrected. Temporal sparsity constrains dynamic inference, as well as decreases the capacity to differentiate causal precedence and synchronous correlation. The invasive sampling of some niches may be ethically and logistically constrained, which then biases the learning of more accessible compartments. All these restrictions encourage further theoretical innovation, further reference curation, integrative multi-omics frameworks, superior statistical and machine learning methods based on compositional and sparse structures, and cross-system standardization efforts to increase robustness, reproducibility, and solvability of the livestock metagenomics field (Ghiotto et al., 2025). Common metagenomics techniques and data types in livestock breeding are presented in Table 9.1.

9.3 Role of AI in data analysis and interpretation

AI will be very helpful during the first stages of livestock metagenomic data analysis and redefine preprocessing, denoising, and compositional processing as bona fide, information-preserving transformations rather than arbitrary filtering. The raw reads are provided in a form that includes sequencing errors, chimeric constructs, excess host fragments, lab contaminants, index-hopping artifacts, and batch-specific signatures that have been introduced by extraction chemistries or library preparation reagents. Increased sensitivity through AI-enhanced quality pipelines, ensemble classifiers that have been trained on the k-mer spectra, base quality trajectories, and positional error likelihoods, distinguishes true biological variation from artifactual noise with increased sensitivity than hard thresholding. Probabilistic denoising models can be explicitly split between sampling dispersion and zero inflation, between hierarchical Bayesian or variational count formulations, structural absence, and depth-limited nondetection. Learned log ratio embeddings that respect subcomposition disjunction and exploit phylogenetically inspired equilibrium, overcoming relative count normalization, leading to compositional constraints, are exploited. Deep normalization models apply covariate adjustment layers, which identify biological signal, batch, and instrument effects, borrowing statistical strength to distinguish biological signals from technical variation. Adaptive processes of distinctive feature conservation under predictive importance implied assist in preventing the untimely loss of low-prevalence strains or gene families with uneven contributions of functionality or epidemiological importance. Through these innovations in preprocessing the heteroskedastic, compositional, sparse count

TABLE 9.1 Common metagenomics techniques and data types in livestock breeding.

Technique/approach	Brief description	Typical data types produced	Common bioinformatics outputs and file or data formats	Notable strengths and limitations
Amplicon/marker-gene sequencing	PCR amplification of phylogenetic marker genes (e.g., 16S rRNA for bacteria/archaea; 18S or ITS for eukaryotes/fungi), then high-throughput sequencing of the amplicons	- Raw sequence reads (FASTQ) - Sequence variants (after processing), ASVs (amplicon sequence variants) or OTUs (operational taxonomic units) - Taxonomic profiles (relative or absolute abundance) - Diversity measures (α-diversity and β-diversity)	- FASTQ files for raw reads - OTU/ASV tables (e.g., as TSV or CSV count tables) - BIOM format files (common microbiome community table format)—taxonomy mapping files (linking OTU/ASV IDs to taxonomic classification) and optional phylogenetic trees	Strengths: • Cost-effective and scalable for many samples • Well-established pipelines and reference databases Limitations: • Limited taxonomic resolution, often to genus level; strain-level differences are missed • No direct functional (metabolic) information: only who is present, not what functions they carry • Bias due to primer choice or differential amplification of taxa
Shotgun metagenomic sequencing (whole-community DNA sequencing)	Random sequencing of all DNA from the sample, capturing genomes of bacteria, archaea, viruses, fungi, and other microbes, both cultured and uncultured	- Raw sequence reads (FASTQ)—taxonomic classification at species/strain level where reference exists - Gene catalogs: lists of genes present in the community - Functional gene annotation (metabolic pathways, resistance genes, virulence factors, etc.) - Metagenome - Assembled genomes (MAGs)/bins (for reconstructing genomes of individual community members)	- FASTQ files for raw reads - Gene-family abundance tables (e.g., counts or normalized abundances per sample) are often in tabular formats (TSV/CSV) or specialized formats. - Pathway abundance/functional profiling tables, e.g., metabolic pathway abundance per sample (e.g., through HUMAnN or other functional annotation tools) - Metagenome-assembled genome (MAG) files/genome bins (FASTA or contigs files), plus annotation files (e.g., GFF/GTF) - Taxonomy and metadata tables for community composition (similar structure as OTU tables) when taxonomic classification is applied to reads or bins	Strengths: • High taxonomic resolution, including strain-level detection (when references exist) • Provides functional potential of the community (metabolic pathways, antibiotic resistance, virulence, etc.) • Detects all domains of life, bacteria, archaea, fungi, viruses, and uncultured taxa Limitations/challenges: • More expensive and computationally intensive than amplicon sequencing • Assembly and binning require high sequencing depth and quality, difficult in highly diverse or uneven communities • Annotation depends on existing databases; novel taxa may remain uncharacterized or mis-annotated

Metatranscriptomics (community RNA sequencing)	Sequencing of total RNA (after rRNA depletion or mRNA enrichment) from the microbial community, capturing genes that are being actively transcribed, reflecting functional activity rather than mere presence	- Raw RNA-seq reads (FASTQ) - Expression profiles: counts/normalized transcript abundances per gene or gene-family - Taxonomy-resolved expression (which taxa are active and what genes they express) - Functional pathway expression profiles (which metabolic pathways are active)	- FASTQ files for raw RNA reads—gene/transcript expression tables (counts, RPKM/TPM normalized) in tabular formats (TSV/CSV) - Pathway expression/functional activity tables (per sample) indicating pathway-level transcript abundances - Metadata and sample annotation files (conditions, diet, time, and host info) matching expression data	Strengths: • Captures active microbial functions, telling us not just who is present, but what they're doing under given conditions (diet and stress) • Allows dynamic studies (e.g., how the microbiome responds to feed changes, antibiotics, and environment) Limitations: • RNA is less stable than DNA; sample collection and storage are more challenging • Gene expression is highly dynamic; a snapshot may not represent long-term activity • Requires matched metadata (diet, time, and environment) for correct interpretation • Higher cost and more complex library prep than DNA sequencing
Metagenome-assembled genomes (MAGs)/Genome binning and assembly	From shotgun metagenomic reads, assembly and binning to reconstruct genomes (or genome fragments) of individual microbial community members, including previously uncultured taxa	- Genome bin contigs/scaffolds (FASTA), annotated genome sequences (with genes, operons, and regulatory elements), gene annotation files (e.g., GFF/GTF), strain-level taxonomic assignments and phylogenetic placement - Functional annotation per genome (metabolic pathways, resistance, virulence)	- FASTA files for genome bins/assembled contigs - GFF/GTF/feature annotation files listing genes and features per genome bin - Genome completeness/quality reports (e.g., from tools like BUSCO) for evaluating bin quality - Functional annotation tables per bin: gene presence/absence, pathway assignments, metabolic potential tables	Strengths: • Allows reconstruction of genomes of uncultured or novel microbes, expanding microbial reference databases, especially important in livestock microbiomes, where many microbes remain uncharacterized • Strain-level resolution, critical when strain differences influence function (e.g., pathogenic versus. commensal strains, resistance) • Supports detailed functional and evolutionary analysis (pathways, gene content variation, mobile elements) Limitations/challenges: • Requires high sequencing depth and quality, especially

Continued

TABLE 9.1 Common metagenomics techniques and data types in livestock breeding.—cont'd

Technique/approach	Brief description	Typical data types produced	Common bioinformatics outputs and file or data formats	Notable strengths and limitations
				for complex or diverse communities • Assembly and binning may yield fragmented or incomplete genomes; mobile elements (plasmids, phages) may be missed or mis-binned • Annotation depends on existing databases; novel genes may remain unannotated or mischaracterized
Targeted/functional-gene/resistome/virulome amplicon sequencing	PCR amplification and sequencing of specific functional genes (e.g., antibiotic-resistance genes, virulence factors, and fiber-degradation enzymes) across samples, rather than universal marker genes	- Raw sequence reads (FASTQ) - Amplicon-based gene variant tables (counts per sample) - Relative or absolute abundance profiles of targeted genes across samples	- FASTQ files for raw reads - Gene-variant abundance tables (TSV/CSV) indicating counts or normalized abundance per sample and gene target - Metadata tables (sample info, treatment, environment) to interpret abundance changes	Strengths: • Cost-effective way to monitor specific functional aspects (e.g., resistome, pathogenic potential, and fiber-degradation) across many samples • Enables sensitive detection of rare but important genes (e.g., low-abundance resistance genes or virulence factors) Limitations: • Only captures what is targeted, no broader community context • PCR bias and primer specificity may skew results • Does not provide information on whether the genes are expressed (i.e., functionality), only presence/abundance
Metagenomic/metatranscriptomic long-read sequencing	Use of long-read sequencing technologies (e.g., PacBio and Oxford Nanopore) applied to community DNA or RNA, producing longer contiguous reads to improve assembly, detect structural variation, plasmids,	- Long DNA or cDNA reads (FASTQ/FAST5 depending on platform) - Improved assemblies/genome bins (long contigs)	- FASTQ or platform-specific raw read files (e.g., FAST5 for Nanopore) - Improved contig/scaffold FASTA files postassembly	Strengths: • Better assemblies, improved genome completeness, better resolution of repeats, plasmids, phages, mobile genetic elements (important for horizontal gene transfer,

	phages, full-length genes/operons, and better resolve complex genomic regions	- More complete gene sequences (full-length genes, operons, and mobile elements) - Detection of structural variants, plasmids, viral genomes, and repeat regions	- Annotation files (GFF/GTF) for genes/features, including mobile elements - Functional/gene-family/pathway abundance tables (TSV/CSV) similar to shotgun workflows but with better genome completeness	antibiotic resistance, and virulence) • Enables strain-level resolution and accurate functional annotation (full-length genes and operons) Limitations/challenges: • Generally lower throughput and higher per-base error rates (though these are improving) • Higher cost per sample compared with short-read; less accessible for large sample sets • Bioinformatics pipelines for long-read metagenomics are still being improved and require expertise

matrix, the combined preprocessing of the count server into a latent abundance representation is jointly trained to perform fair comparative analysis, longitudinal trajectories modeling, and integrative fusion with host and environmental modalities (Oppliger et al., 2024; Alonso-Ramos et al., 2025).

Representation learning offers a single representation layer that provides a constant, dense representation of high-dimensional, sparse taxonomic and functional profiles that preserve hierarchical form, cofunctional association, and ecological setting. Trained sequence-level encoders based on masked k-mer prediction schemes provide embeddings that encode codon usage bias, structural motif likelihood, and adaptive signal of selective pressures that can then be clustered and used to obtain improved functional recovery of open reading frames of unstructured proteins. Graph-aware neural networks on phylogenetic trees, as used to compute taxonomic embeddings, encode patterns of cofluctuation of abundance and score ecological exchangeability. Conversely, learned paired metagenomic and metabolomic samples are learned under contrastive learning frameworks, where applications map microbial feature representations to representations of functional outcomes manifolds by maximizing agreement of biologically coordinated views. Such nonlinear gradients as diet-induced fermentative shift or progressive developmental succession can be learned using diffusion map hybrids and variational autoencoders, and can similarly interpolate between the poorly sampled states. Multiresolution embedding structures are attention models that dynamically rank importance on either families of genes, plasmid-encoded traits, or mobile element signatures in the existence of a predictive salience to a target phenotype without ignoring the context of the wider community. These learned representations give succinct representations to predictive and causal models, enhance computational efficiency, and disclose latent ecological axes that can be observed only partially in the crude proportional tables (Xie et al., 2025; Wang et al., 2026).

Graph and network-based modeling utilizes the relational structure of existing microbial ecosystems and points of interaction to host physiology to describe interaction scaffolds that extend beyond the marginal cooccurrence. Putative direct ecological interrelations at the cost of transitive artifacts are estimated through the estimation of conditional dependence on the sparsity of inverse covariance with Bayesian structure learning or with score-based graphical search. Graph neural networks may also be positively (but not necessarily locally) regressive of ordinary feature information between edges, such as neural embeddings, to (but not only to) the entire structure, such as modular fermentative guilds or hub-mediated vitamin-exchange complexes. Response repertoires of genome annotation are constructed into metabolic interaction graphs, by which possible cross-feeding, syntrophic electron transfer, or competitive substrate overlap and edge weight priors, which are either expression or metabolite evidence, are used to refine constraint-based flux approximations. Host-microbe bipartite diagrams have been applied in associating microbial activity or taxon with host transcriptomic pathways, immune signatures, or metabolic flux signatures to facilitate the identification of interface bottlenecks and mediate taxa. Temporal extensions model edge weights stochastically, in which case interaction rewiring can be found when there is a diet shift or during perturbation therapy. Multiplex graph models are a combination of layers in taxonomic codynamics, gene exchange by mobile elements, metabolic complementation, and host immune coupling into a coherent learning model, and AI models can isolate convergent signals in both emergent modalities of interaction. The uncertainty-aware graph inference resists sparse cosampling over interpretation as well as provides confidence bounds on centrality or existence of edges, useful in making concrete intervention targeting (Cao et al., 2025; Gholamzadeh et al., 2025).

Multi-omics of metagenomes, host genomes, transcriptomes, proteomes, metabolomes, and phenomes require AI structures that allow heterogeneity when it comes to scale, noise structure, missingness patterns, and time warping. Joint latent variable models break down cross-modality data into shared and modality-specific data, and thus can isolate the effects of core ecological axes, which have an effect on both the state of the microbes and the host, and idiosyncratic measurement noise. Variational multiview autoencoders represent the data layer coordinates in correlated latent coordinates, and variational multiview autoencoders by minimizing reconstruction fidelity and maximizing cross-modality coherence, adversarial distribution alignment, or contrastive mutual information losses are applied. Graph augmented fusion models map each of the modalities to parallel manifolds anchored on ontological semantic connectivity between genes, pathways, metabolites, and phenotypic traits, and preserve the biological interpretability. Missing modality imputation applies conditional generative models to complete the missing transcriptomic or metabolomic measurements on the metagenomic and environmental environment to get more analytical inclusivity in the sparsely-sampled cohorts. Temporal integration applies attention-based sequence fusion to the asynchronous sampling that is fixed by learning alignment weights that rely on the latent dynamical phase. The introduction of host genomic variation in polygenic environmentally modulated versus genetically-mediated variations in functional shifts in the microbes can be dissected using polygenic environmentally modulated versus genetically-mediated outcomes, which will form the basis of personalized management approaches. These integrative models, besides enhancing predictive performance, also show the mediating mechanism of the genotype-microbial functionality chain and phenotypic expression (Han et al., 2025; Park et al., 2025).

Supervised AI algorithms optimized to the specifics of metagenomic feature spaces are used to predictively model operationally critical phenotypes, including feed efficiency, disease susceptibility, emission intensities, the quality of products, and resilience. Ensemble gradient boosting and kernel methods have also been found to have a high baseline, but they tend to be overtaken by deep architectures, which take advantage of hierarchical taxonomic structure and learned functional embeddings to better generalize. Multitask learning models take advantage of similar microbial determinants between related targets, and are jointly predictive of efficiency and emission outputs, and separate specific and common latent variables. Time-aware prediction makes use of recurrent or temporal convolutional models to utilize longitudinal patterns of microbiome dynamics as predictors of an imminent deviation in health or efficiency. This implementation of loss functions as a cost-sensitive incorporation is to balance the optimization of models with economic and welfare interests by giving false negatives representing pathogen bloom prediction a higher weight than the marginal efficiency estimation errors. Pathway groupings or phylogenetic branchings can be used to construct regularization strategies that avoid overfitting and do not reduce interpretability. Domain adaptation methods rectify differences in distributions between laboratory testing data of experimental trials and commercial farm implementation settings by aligning feature representations on the basis of adversarial minimization of disparities. The calibration evaluation is crucial in that probabilistic predictions are based on the real likelihood of events taking place, so that they can be incorporated in risk-based decision models to determine when and how intensively to intervene (Dond, 2025; Peng et al., 2025).

Causal inference brings in rigor to enable the ability to determine microbial characteristics, the manipulation of which is anticipated to induce desirable host reactions within realistic management situations. Structure learning algorithms are algorithms that build candidate directed acyclic graphs between microbial taxa, functional pathways, environmental inputs, and phenotypes using invariance across different intervention strata, production stages, or geographical clusters to filter out spurious edges that are context-specific. Counterfactual outcome modeling is a doubly robust estimator that uses targeted regularization to measure the anticipated change of phenotype in the case that microbial characteristics are hypothetically modulated, with sets of high-dimensional confounding variables (host genetics and management variables) being allowed. Mediation analysis separates total diet-induced phenotype effects into direct host pathways and indirect microbial functional contributions, and provides leverage points on which nutritional strategy improvements can be made. Instrumental variable designs make use of exogenous variation that has been added by a controlled feed additive or a change in management in order to provide stronger causal attribution in situations where unmeasured confounding is suspected. Intervention prioritization systems integrate estimated magnitudes of causal effects, uncertainty ranges, feasibility limitations, as well as multiobjective trade-off scores into ranked action portfolios to support evidence-based distribution of experimental validation resources. The value of information analysis determines the microbial features in which further targeted sampling would best reduce the critical uncertainty in the decision, and thus causal discovery and adaptive data acquisition are aligned (Kitov, 2025; Han et al., 2025; Mohan, 2025).

Deep sequencing embeddings, as a way of improving strain level resolution, deal with the real-world issue of functionally salient microdiversity often being found at the aggregation of the species level. In the presence of recombination and repetitive elements, assembly graph-aware neural encoders analyze contig adjacency and coverage gradients to make inferences on strain-specific haplotypes. Transformer-based models trained on long k-mer sequences or compressed minimizer streams are trained on context-sensitive embeddings that distinguish very subtle patterns of polymorphism, structural variant signatures, or accessory genes in the genome, without having to rebuild the entire genome. Pangenome graph embeddings are core and accessory gene neighborhoods, which are used to identify clusters of strains as related to variation in traits, antimicrobial resistance determinants, or metabolic specialization. Bayesian strain deconvolution integrates read alignment inconsistencies, copy number variations, and existing strain frequency distributions to generate posterior abundance estimates. These estimates, accompanied by credible intervals, are valuable for downstream risk assessment following deconvolution. The association of strain-resolved phenotypes or causal graphs with phenotypic prediction maps or causal graphs is used to refine the granularity of interventions, allowing deleterious variants to be selectively suppressed but conserving adaptive conspecific functions (Kazantseva et al., 2024).

The gap between functional potential and the actual activity of the real world necessitates integrative models for transforming the inert genomic inventory into dynamic expression and metabolite production structures. Coupled metagenome-metatranscriptome analysis is a method that employs hierarchical generative models to break down measured expression into simple genomic capacity, regulatory modulation, and stochastic transcriptional noise, all dependent on environmental covariates. Attention-based cross-layer transformers learn to embed the abundance of genes into the pattern of gene expression conditional on nutrient availability and host inflammatory markers, learning regulatory nodes that are disproportionately sensitive to perturbations. Multioutput Gaussian process or neural ordinary differential equation (ODE) models that solve the equations of time-dependent metabolite concentrations as functions of predicted enzymatic flux proxies with stoichiometric priors are used to perform integration with metabolomic profiles. The differences between the potential and the actual activity are evidence of the latent regulation, restriction of the substrates, or the activity of competition, and serve to guide specific intervention measures to disinhibit the latent benefit pathways or to restrain the wasteful production of metabolites. The metaproteomic evidence incorporation balances the accuracy of translation efficiency estimates and supports a higher fidelity of flux inferences, and unsupervised clustering of matched multilayered profiles characterizes functional states that define the discrete ecological operating regimes (Campbell and Baldridge, 2024).

Generative AI has the capability to learn probability distributions over feasible microbiome composition and simulate synthetic consortia by learning probability distributions over feasible microbiomes and their functional outputs. Variational community generators encode observed abundance and functional panels into latent ecological coordinates that can be sampled to produce novel but plausible configurations to explore the compositional landscapes that have not yet been experimentally measured, but which are postulated to give good metabolic profiles. Diffusion models train iteratively, remove noise, and initialize abundances on conclusions with conditions that promote compositional closure, taxonomic plausibility, and pathway coverage limits to be able to perturb existing communities towards desired functional changes. The embedding of constraints includes the thermodynamic feasibility, stoichiometric balance, and ecological stability requirements in order to punish the unrealistic assemblage that has a high probability of collapse. Evaluation of synthetic communities by multiobjective scoring functions based on achieving efficiency increase, emissions reduction, and pathogen suppression can be done by coupling generative models to multiobjective scoring functions with metabolic network simulators or learned phenotype predictors. These are used to speed up the hypothesis generation of consortium-based interventions before validation is done in wet labs, which is resource-intensive, and uncertainty is cautiously factored in to prevent the tendency to overreliance on extrapolations beyond validated ecological envelopes (Yu et al., 2025; Ghavi Hossein-Zadeh, 2026a).

Reinforcement learning structures the adaptive modulation of the microbiome by treating changes in diet composition, dosing behaviors, environmental parameter adjustments, and management interventions as actions within a dynamic state space. This space includes existing microbial functions, host physiological responses, and environmental influences. Model-based strategies acquire transition dynamics between conditioned states of action to successive microbial and host behavior, allowing the planning strategies to simulate rollout of policies and measure rewards accumulation over long horizons across multiobjective measures such as feed conversion, welfare measures, emission levels, and risk penalties. The constrained reinforcement learning instills safety conditions against short-term inducing dysbiosis or other welfare-damaging policies. Introducing action complexity in curriculum learning is done in stages, with limited dietary changes being implemented before incorporating combinatorial additive strategies, which enhance policy stability. Offline reinforcement learning uses historical farm data and observational data to correct policy bias in the behavior policy by weighting by importance or using a conservative value estimate, eliminating the need to rely on expensive prospective testing. The use of causal graphs in the evaluation of policies makes the evaluation more interpretable and minimizes spurious exploitation of confounded associations (Shimpi, 2025).

Active learning methods can optimize sampling and experimental design by sequencing additional animals, time points, anatomical locations, or omics layers through the selective learning of a new sample or time points to reduce predictive or causal uncertainty with budget limitations. The acquisition functions are based on Bayesian predictive variance, mutual information, expected model change, or counterfactual disagreement regions of the ecological state space that are not well represented in the existing training data. Active learning schedules involve temporal schedules of active learning, which devote denser sampling to periods of expected transition, such as diet shifts, and reduce uncertainty in dynamic parameter estimation. Multiobjective acquisition also takes into account the logistical feasibility and immediate improvement of the model, trading high information and invasive sampling off against minimal impact alternatives. Adaptive laboratory workflows combine real-time bioinformatic preprocessing and uncertainty assessment with later sampling decisions in one longitudinal study, leading to more efficient and economical statistics and less redundancy (Moran and Morato, 2025).

An ecologically diverse and generalizable training of federated and privacy-preserving cross-farm model training is considered, taking into account data governance, proprietary management practices, and biosecurity implications and concerns. Federated averaging explicitly combines the locally trained model updates of distributed farm nodes without centrally communicating the raw data, supporting heterogeneity in sample size, sequencing depth, and husbandry settings. Each node can personalize shared backbone representations to local idiosyncrasies and thus enhance the local predictive performance without introducing global convergence problems. The danger of reconstructing sensitive information based on gradient updates is reduced by secure aggregation protocols and differential privacy, and the imbalance caused by large contributors is solved by adaptive weighting. Domain adaptation models of the federated framework correct systematic shifts in distribution, matching the latent representations in common semantic spaces. Strong aggregation helps to filter outliers and deviant updates and achieves integrity when the system could have been misconfigured, or an adversary may have been active. These facilities promote the joint development of AI models that help the livestock industry to spread innovation faster (Li et al., 2025).

Explainable AI builds trust, helps to achieve regulatory compliance, and facilitates scientific discovery by explaining how complex models transform inputs from the microbiome into their predictions or suggested interventions. The methods of feature attribution can be used to decompose model outputs in terms of the contributions of taxa, genes, pathways, or metabolite proxies, which provide quantitative descriptions of the relationship between mechanistic factors and outcomes. Pathway-centric explanations aggregate fine-grained feature attributions into biologically interpretable modules, and thereby simplifying interpretation for end users. The emphasis on representation of a series of architectures or graph interactions demonstrates subnetworks of interaction or time windows that result in the predictions, such as early divergence in fermentative balance before alterations in efficiency. Counterfactual generation methods propose minimal microbiome modifications, which would change a risk category depending on what target modulation policies would be implemented and whether they thereby revealing the limits of biologically plausible interventions in cases where unrealistic changes are required. Concept bottleneck models directly incorporate specified biological concepts as an intermediate predictor that injects a mapping between the internal representations and information about the domain. Human-centered assessment defines the clarity of explanations, completeness, and reliability of explanations between different stakeholder groups, that is, nutritionists, veterinarians, sustainability auditors, etc., and refines the communicative efficacy by means of repetition (İlgen et al., 2025).

Uncertainty quantification forms the basis of reliability in deployment to distinguish between confident model advice and guesses. Both the epistemic and aleatoric uncertainty of the ecological variability are the inferential and predictive posterior distributions of Bayesian neural networks, deep ensembles, and stochastic variational approximations. Conformal prediction surrounds arbitrary predictive models using distribution-free calibration layers of arbitrary predictive models, resulting in set-valued or prediction interval set-valued outputs with known error rates on mild exchangeability assumptions and risk-sensitive decision levels. Reliability scoring is a composite of the calibration measures, local density models in latent representation space, and out-of-distribution detector models, which imply that the new states of the microbiome are out of the validated training manifolds. The uncertainty is included in decision-theoretic models that practice expected utility maximization and that recommend conservative intervention strategies when the risk of undesirable unintended changes is higher than the potential marginal benefits. Free communication of uncertainty levels and explanations reduces overconfidence, which can lead to the proper use of scientific evidence and congruence of stakeholder expectations and constraints on scientific evidence (Salem et al., 2025).

The integration of digital twins brings all the above aspects of AI to dynamic and personalized computational counterparts of host-microbiome ecosystems that internalize new information to enhance predictive and prescriptive capacity. Mechanistic submodels of digestion, fermentation, immune activation, and metabolic flux are integrated with data-driven modules to explain the remaining variation and not modeled interactions to form scenario evaluation in the instance of hypothetical dietary changes, environmental stressors, or intervention regimes. The feed intake and periodic multi-omics snapshots are fed into the system sensor data are assimilated into state-estimation algorithms of the latent system to retune the forecast of efficiency, emission pathways, and health hazards. Multiscale interaction of ruminal fermentative dynamics and system circulation of metabolites and energy allocation to behavior provides a combined understanding of trade-offs. Relative to other intervention policy choices, simulation of candidate intervention policies by reinforcement learning rules, and subject to causal effect estimates, leads to comparative outcome distributions, which are used to implement policies in the real world, where limited experimentation periods are used. The provenance of the updated models is verified through a series of governance checks. Discrepancies between the forecasted and actual outcomes are logged to enable continuous implementation of ethical constraints. These constraints do not permit recommendations for intervention in cases where there is insufficient safety evidence. The digital twins continues to be converted into an adaptive decision support resource by serial coevolution with the physical system to gain the ability to proactively control microbial ecosystems as controllable knobs to sustainable livestock production (Chen et al., 2025; Ye et al., 2025).

The resultant development of these AI methods, termed progressive, transforms the livestock metagenomics analysis into a dynamic, predictive, causal, and actionable intelligence system of the traditional descriptive profiling. State-of-the-art, powerful preprocessing ensures validity; representation and graph learning learn latent ecological structures; multimodal fusion contextualizes the microbial function within citizen and environmental networks; predictive and causal modeling is given intuitive advantages; generative, reinforcement, and active learning extend the design and optimization boundary; federated infrastructures democratize innovation; explainability and uncertainty quantification push responsible adoption; synthesis in digital twins operationalizes continued adaptive control. Future challenges include balancing methodological complexity with practical field deployment, the practical implementation, and equal access to the scale of production, and rigorous validation to prevent premature translational excitement. Their solution will establish AI as a framework that will enable it to attain some degree of accuracy in the management of the host-associated microbial ecosystem to meet the productivity, welfare, and sustainability objectives. AI and/or computational methods applied to livestock metagenomics data are presented in Table 9.2. In addition, functional applications of AI in livestock metagenomics are listed in Table 9.3.

9.4 Challenges and limitations of AI integration in metagenomics

There are natural challenges to the application of AI to livestock metagenomics, which are exerted by heterogeneity, sparsity, and batch effects in underlying data streams. On any single analytical project, in breeds, physiological stages, anatomical niches, diets, climates, and management regimes, a representative sample can be gathered and exhibit a divergent distribution of abundance and dispersion parameters using unique microbial selective filters. Laboratory layers of variability are introduced, further varying the data by encompassing extraction chemistries, bead beating intensity, enzymatic cleanup, library preparation kits, indexing technologies, and sequencing platform peculiarities that have an overall effect to create batch-specific signatures with actual biological signal. Excessively high frequencies of zeros in count matrices, as well as structural dropouts where rare but ecologically significant taxa or genes are below the threshold of statistical inference to covariance, time-dependence, or intervention effects, are both manifested by a sparse structure. Traditional imputation or prevailing naive filtering is prone to losses of weak but true associations, and naive pooling of nonhomogeneous cohorts is an exaggeration of false discoveries by violating downstream model distributions. Insofar as AI architectures that have been trained on aggregated data contain no specific hierarchical variance decomposition or adversarial batch harmonization, they may accidentally learn features of a batch that are discriminative and not predictive, breaking down when used in new production systems. Controllable layers are accompanied by principled latent variable modeling, which distinguishes between technical and contextual variation, at the cost of more computation, and faces identifiability problems when incompletely copied within a stratum, and residual uncertainty remains an important consideration during downstream analysis and model interpretation (Hui et al., 2024; Buchha, 2025).

The inferential weakness of metagenomic abundance measurements is further exacerbated by compositionality, where any increase in one component may mechanically force other components to decrease in spite of the underlying absolute biomass processes. The default correlation statistics with raw or library size-normalized counts generate artificially generated negative correlation and dense network effects, which enhance AI systems to learn artificially antagonistic or alliance. Log ratio transformations mitigate some of the issues at the expense of paying special attention to the choice of reference frame/balance that is fixed between heterogeneous environments; the inappropriate choice of denominators or data that stimulates pivot taxa that are undergoing ecological instability induces bias and overestimation of the record. Compositional vectors fed to deep learning models that do not understand anything about the closure effects may encode representations that are confounded by closure effects, in which some taxa with an absolute but variable relative abundance are allowed to achieve a higher discriminative capacity than other members of the community that develop in isolation. Furthermore, when compositional phenotypes of hosts or environmental factors are modeled together with noncompositional phenotypes of hosts or environment, the attribution of causality to compositional artifacts can occur unless absolute load or biomass proxies or internal standards are added. Attempts to estimate latent absolute abundances using statistical deconvolution are also obstructed by identifiability constraints when there are no external scale anchors, which constrains the validity of downstream causal statements that follow a process of inferred growth or suppression. Powerful handling is therefore an open methodological boundary, particularly when there is inclusion of multi-omics layers that contaminate the compositional and unrestrained measurement space (Ma et al., 2025; Tang et al., 2025; Sharma et al., 2025).

The lack of depth in functional annotation and the relentless absence of sequence-derived features hinder the conversion of these features into biologically interpretable constructs that can be subjected to hypothesis-directed intervention. Metagenome-assembled assemblies isolated from livestock niches contain large proportions of open reading frames with no orthologs in curated databases, and even in annotated sequences, functional specificity is often conserved, with many different enzymatic activities or regulatory functions collated within high-level families. Embedding schemes based on AI, which cluster unlabeled proteins either by sequence or structure similarity, are more comprehensive at the cost of spreading misannotations in case of training on biased or mislabeled reference datasets. The transfer of functional

TABLE 9.2 AI and/or computational methods applied to livestock metagenomics data.

Method/computational approach	Typical application(s) in metagenomics	Key advantages or strengths	Limitations and or challenges
Supervised machine learning	Classification or regression tasks, e.g., predicting host phenotype, production trait, disease susceptibility, or health status from microbial community profiles; biomarker discovery from taxonomic or functional profiles	- Often good performance when training data are labeled - Can handle high-dimensional microbiome/metagenome feature sets (many taxa or gene-family abundances) - Feature-importance outputs help identify taxa or functions associated with outcomes (e.g., production traits and health) - Relatively mature and widely available algorithms (many software packages)	- May overfit if the sample size is small relative to the feature number (curse of dimensionality) - Performance depends heavily on the quality and representativeness of training data (metadata, diet, environment, host factors) - Interpretability can be limited: complex interactions and nonlinear relationships may not be fully explained by feature-importance alone
Unsupervised learning/clustering/ dimensionality reduction methods	Exploring structure in metagenomic data; identifying natural clusters of microbiome composition; stratifying samples by community type; reducing dimensionality for downstream analysis or visualization	- Does not require labeled outcome data, useful when exploring new datasets - Helps to reveal latent patterns, subcommunities, or variation in microbiome structure across livestock groups (e.g., diets, farms, and breeds) - Useful for hypothesis generation (e.g., clusters linked to feed type, environment, and management)	- Choice of clustering parameters (distance metric, number of clusters) can strongly influence results, not always biologically meaningful - Sensitive to data preprocessing (normalization, compositional data issues) - Clusters may not map clearly to phenotypes; further validation is usually needed
Deep learning/neural networks	Complex classification or regression tasks; feature learning; taxonomic classification; functional prediction; representation learning (embeddings) of metagenomic profiles	- Capable of modeling nonlinear, high-dimensional relationships common in microbiome/metagenome data (e.g., interactions among taxa or nonlinear associations with traits) - Feature learning reduces the need for manual feature engineering; it may uncover latent patterns or representations not obvious to standard ML - Deep-based taxonomic classifiers (for reads) can perform well even with noisy or long-read data	- Requires large, high-quality datasets for effective training (often a limiting factor in livestock studies) - Risk of overfitting, especially given the sparsity and compositional nature of microbiome data - Often, black box interpretability is challenging; difficult to trace which taxa/ functions drive predictions - High computational resources needed
Graph- or phylogeny-aware neural networks	Classification of metagenomic/microbiome samples, integrating information about relationships among taxa (phylogeny), or structural relationships in microbial networks; improved taxonomic classification and community-based prediction	- Incorporates phylogenetic/relational structure of microbial taxa or features, which may improve classification accuracy and biological relevance - Better captures dependencies among taxa (e.g., cooccurrence and phylogenetic	- Requires construction of a reliable phylogenetic/network structure, which can be challenging if taxa are poorly characterized or references are incomplete - Computationally more complex than traditional ML

Continued

TABLE 9.2 AI and/or computational methods applied to livestock metagenomics data.—cont'd

Method/computational approach	Typical application(s) in metagenomics	Key advantages or strengths	Limitations and or challenges
		similarity) rather than treating features as independent - Potentially more robust to noise or sparsity since relationship-based patterns are used	- For many livestock-associated microbiomes, phylogenetic information may be incomplete or uneven, limiting applicability
Generative models and data augmentation	Augment sparse or small metagenomic datasets to improve model training; simulate microbiome profiles under hypothetical conditions (e.g., diet change or environment), enable "what-if" analyses	- Helps overcome sample-size limitations, generating synthetic but plausible data to expand training sets - Allows exploration of hypothetical scenarios (diet shifts and farm conditions) when experiments are difficult or impractical - Improves robustness and generalizability of downstream predictive models when real data are limited	- Synthetic data may not perfectly reflect biological reality, with the potential for generating unrealistic or biased samples - Risk of "overfitting" to synthetic patterns; models may learn artifacts rather than true biological variation - Validation and biological verification of generated data are required before applying conclusions
Multi-omics and integrative modeling	Predicting complex traits (e.g., feed efficiency, health, and disease resilience) by integrating microbiome data with host omics, environmental variables, and management practices; modeling interactions among microbial functions, host genetics, and diet	- Provides a more holistic view of livestock production systems, microbiome effects do not work in isolation but interact with the host and environment - Helps in understanding causality and interactions (e.g., how diet , microbiome , and host genotype influence health or productivity) - Can improve predictive power and personalization (e.g., tailored feeding based on microbiome and genotype)	- Requires extensive, high-quality multilayer data (microbiome, host genomics, environment, and diet), often logistically and financially demanding - Integrative models can become very complex; interpretation and validation are challenging - Data harmonization, normalization, and dealing with missing data across layers are major hurdles
Explainable AI (XAI) and feature-importance methods	Making AI predictions interpretable; identifying taxa, gene-families, or functions that most influence model output; supporting biological insight and hypothesis generation	- Enhances transparency and trust in AI-based predictions, important for adoption in livestock science and management - Helps bridge computational findings with biological understanding (which microbes/functions drive trait associations) - Supports hypothesis generation for functional studies, experimental validation, and targeted interventions (e.g., probiotics and diet changes)	- Post-hoc interpretability may oversimplify complex relationships; it may not reflect causal effects - Interpretations can vary depending on model architecture and feature representation - Additional computational overhead; for complex deep models, interpretability remains challenging

TABLE 9.3 Functional applications of AI in livestock metagenomics.

Functional application	AI techniques used	Target outcomes	Data types and sources	Challenges and limitations
Prediction of feed efficiency/nutrient utilization	Supervised ML (e.g., ensemble methods, random forest, and gradient boosting), feature-engineering	Predict which animals will utilize feed more efficiently; identify microbiome biomarkers linked to feed-conversion efficiency	Rumen or gut microbiome shotgun or 16S/metagenomic sequencing data and feed intake/production records (e.g., residual feed intake and milk yield)	• Microbiome composition is variable and may not capture temporal dynamics • High dimensionality and limited sample size may cause overfitting. • Correlation does not imply causation; biomarker associations need biological validation
Prediction of methane emissions/environmental impact	Predictive models (ML/possibly deep learning) based on microbial community data	Estimate which animals are likely to produce lower methane emissions, enabling selection for lower-carbon-footprint livestock or targeted feed additive interventions	Deep metagenomic sequencing of rumen microbiomes, plus methane emissions measurements (gas collection or proxies)	• Methane emission is influenced by many factors beyond microbiome (diet, management, and host genetics) • Functional potential inferred from metagenome may not reflect actual microbial activity • Need for large, well-characterized datasets for robust predictions
Disease risk prediction and health monitoring	Supervised ML/deep learning; classification algorithms on metagenomic features (taxonomic or functional profiles)	Predict susceptibility to infections, gut dysbiosis, or health disorders; early warning of disease outbreaks	Microbiome (gut, respiratory, skin, etc.) metagenomic or 16S data , health/disease records, and metadata (age, diet, and environment)	• Microbiome-disease associations are often weak or inconsistent; high variability across individuals/farms • Confounding factors (environment, diet, and management) • Need for longitudinal data to capture the temporal dynamics of microbiome and disease development
Optimization of feed additives/microbiome-based interventions	ML/AI predictive modeling to forecast efficacy of feed additives (e.g., probiotics and methane-suppressing agents) before application; model-assisted selection of interventions	Identify which additives or dietary changes are likely to be effective in a given herd or animal based on their microbiome, thus optimizing intervention strategies and reducing trial-and-error costs	Rumen/gut metagenome data before intervention, metadata on diet/additive use, subsequent production or emissions outcomes	• Need large datasets to train reliable models • Microbiome complexity may limit predictability • Variation across farms, diets, and environments may reduce generalizability • Ethical/regulatory considerations around microbiome manipulation

Continued

TABLE 9.3 Functional applications of AI in livestock metagenomics.—cont'd

Functional application	AI techniques used	Target outcomes	Data types and sources	Challenges and limitations
Microbiome-based phenotyping for genetic selection	Integrative models combining host genomic data, metagenomic features, and production traits through ML/AI	Use microbiome signatures as phenotypic markers to complement traditional breeding values, to select animals with favorable microbiome–host interactions for productivity or robustness	Host genotype data (SNPs or genomic profiles), microbial metagenome profiles, performance/productivity records	• Complexity of host–microbiome interactions • High data requirements (genome , microbiome , and phenotype) • Risk that microbiome is influenced by environment more than genetics, may reduce heritability or stability of microbiome-based traits
Microbiome-driven nutrition/diet planning (precision nutrition)	ML/unsupervised clustering, predictive modeling, and possibly reinforcement-learning frameworks	Design diets tailored to the microbiome of a herd or animal to maximize health, nutrient absorption, and reduce waste or emissions	Gut/rumen microbiome sequencing data, diet composition, feed intake, production, or health outcomes	• Need comprehensive metadata (diet, feed composition, and environment) • Microbiome responds to diet; models must account for temporal and environmental variation • Risk of overfitting and limited generalizability across contexts/herds
Functional profiling and discovery of microbial biomarkers (e.g., for metabolism, resistance, virulence, etc.)	Deep learning, unsupervised feature learning, gene-family/pathway abundance modeling, network analysis	Identify microbial genes or pathways associated with health, productivity, resistance to pathogens or environmental stress; discover novel microbial biomarkers or bioindicators	Shotgun metagenomic data (functional gene catalogs), annotated microbial gene/pathway data, host metadata	• Functional annotation depends on database completeness; many microbes remain uncharacterized • Gene presence does not guarantee expression or functional activity • Need follow-up validation (transcriptomics, proteomics, and metabolomics) to confirm predicted functions
Rapid surveillance and early warning systems (pathogen/resistome monitoring)	Deep learning/classification models on metagenomic reads (taxonomic, virulence, and antimicrobial resistance (AMR) gene detection), anomaly detection algorithms	Early detection of pathogens, AMR genes, zoonotic risk, monitoring herd health, and biosecurity status	Metagenomic sequencing of fecal, environmental, water, or air samples; resistome/virulome profiling data	• High diversity and low abundance of pathogens/AMR genes can challenge detection sensitivity • Risk of false positives/negatives depending on sequencing depth and database completeness • Data privacy, data-sharing, and biosecurity regulatory concerns

information to more distant homologs remains probabilistic, and there may be determinants of activity that are captured by more specialized language models but not by the general-purpose language models, resulting in overconfidence in potential pathway completeness measurements. These uncertainties are propagated by building metabolic network models, predicting flux potentials, connecting microbial functions to host phenotypes, and missing or incorrectly classified reactions disfigure inferred bottlenecks and spurious additive or probiotic design. Reconciliation of multi-omics that aims to bridge the lack of knowledge in potential and activity generated by multi-omics is compromised when transcripts or metabolites map to obscure collections of genes, which dilute the signal and reduce the statistical strength of potential differential expression or difference in abundance analysis based on functional collections. This can result in AI systems giving more attention to well-characterized, historically examined pathways and less attention to innovative new metabolic ways of adapting to particular diets or stressors, which leads to the emergence of a kind of epistemic inertia (Shaw et al., 2025; Spicer et al., 2025).

The resolution of strain levels creates ambiguity based on high genomic similarity between coexisting variants, widespread horizontal gene transfer, structural rearrangements, and modular accessory genome turnover. Repeated assemblies disperse when there are near-identical paralogs, creating chimeric bins that merge closely related strains or divide individual pieces into multiple artifacts, making it difficult to enumerate abundance and bin variants. Horizontal gene transfer separates the presence of functional genes and vertical inheritance in a phylogeny, which challenges taxon-centric predictive conceptions where traits are expected to be constant in linkage. AI techniques that infer strain embeddings by looking at k-mer or graph neighborhoods may implicate mobile element-derived variation in core genomic divergence and inflate apparent ecological divergence. Recombination causes blurring of lineages and creates difficulties in clustering algorithms based on tree-like evolution, as well as creates mosaics that cannot be categorized simply based on categories that epidemiological tracking or targeted modulation requires. In addition, in cost-constrained livestock surveillance, limited depth sequencing restricts further coverage that can be critically used to construct a haplotype, forcing models to extrapolate with broad uncertainty that may not always be clearly reported. The interaction of these influences reduces causal attribution fidelity in attributing particular strain variants to emissions, efficiency, or pathogenic risk, leading to possibilities of misdirected intervention that does not discriminate between beneficial and deleterious microdiversity (Mishra and Lercher, 2025; Qiu et al., 2025).

Unregularized time and irregular sampling rates impair the ability to simulate the dynamics of microbiome-host interactions, train predictive antecedents of adverse events, or deconfound causation. Experimental farm conditions, animal welfare, and resource allocation restrict collection cadence, causing nonuniform intervals punctuated by intervals of gaps, which occur at management or environmentally relevant high biological events. Temporal models of AI trained on regularly sampled domains fail to deal with informative but random patterns of missingness, since informative missingness patterns are frequent in scarce sampled underlying domains when the sample is being perturbed. The strategy of interpolation or imputation can create illusory smooth changes where, in fact, the underlying dynamics are a sudden change of regimes, which will overstate the early warning indicators and underestimate volatility. When developmental landmarks or dietary shifts appear at uneven chronological ages, multianimal time series will become problematic and phase-shifted, which will lose statistical power to identify common trajectories. Latent time warping strategies add complexity and identifiability issues to the model, especially in the case of sparse observations per subject. Devoid of intensive, positionally timed sampling, reinforcement learning, or control-oriented AI modules do not have finely spaced feedback upon which strong policy assessment may be formed, limiting adaptive administration hopes. Therefore, temporal sparsity is an intrinsic structural limitation that requires novel experimental frameworks to be cooptimized with analysis tools (Hsieh et al., 2025).

Narrow data, including generalizability constraints, overspecification, and domain shift, are associated with models of AI trained on small datasets, encoding context-dependent noise, administration peculiarities, or location-specific topography-dependent microbial structures, which then react unpredictably when presented with a new production system, new breed, new-feed regime, or new climatic conditions. High-capacity architectures are capable of remembering spurious multivariate signatures, which predict well in training farms, but are not causal, and do not predict well on deployment. Distributional shift is not only found as marginal developments of abundance but also as changes in covariance structure, network of interactions, and pattern of functional redundancy, and invalidates assumptions that are implicit in learned representations. The existence of transfer learning and domain adaptation technologies alleviates certain discrepancies, but the lack of labeled outcome data in target domains, differences in sequencing depth regimes, and changes in unmeasured confounders such as stress levels or microclimate interactions make them ineffective. Reported metrics of performance that have not undergone rigorous cross-domain validation or leave-site-out cross-validation give an illusion of robustness and promote premature operationalization. Constant learning methods can lead to catastrophic forgetting of updating models with new domain data unless highly regularized, whereas naive aggregation of heterogeneous data can also cause performance to be worse in original domains by washing out the domain-specific signal needed to make some niche predictions. The issue of specialization and generalization is still unsolved (Qi et al., 2025).

Interpretability and transparency contradictions increase with increasing model complexity to model nonlinear, hierarchical, multimodal dependencies. Deep ensemble networks, graph attention layers, and diffusion-based generative modules are better predictors but still do not have internal representations that are easily biologically mapped. The methods of post-hoc attribution provide estimates; however, they are prone to instability with perturbation of inputs or clusters of related features and give inconsistent mechanistic stories. Making models simple to linear or shallow interactions reduces the explanatory power, but also results in a loss of interactions, which may be combinatoric microbial interactions or regulatory nonlinearity that forms the basis of phenotypic variation. The stakeholders, such as veterinarians, nutritionists, and regulators, need feature importance rankings, plausible cause-effect relationships, and counterfactual plausibility tests to be convinced of the suggested interventions. Excessive focus on black box performance encourages the development of risky (undetected) failure modes, whereas overinterpretability can early eliminate creative architectures. Building interpretable models inherently, and flexible to the structure of the microbiome compositions, is still an immature methodology design space (Arya, 2025).

The expansion of multi-omics, strain-resolved, longitudinal datasets is limited by practical and sustainability-oriented constraints, which are computational scalability, storage demand, and energy intensity. Large language models or the training of graph neural networks on large pangenome assemblies using metagenomic k-mer corpora demand large amounts of high-performance computing resources, which smaller research groups or producers may lack, contributing to inequities. Recurrent hyperparameter optimization and generative ensemble increase attention to carbon footprint, which is incompatible with sustainability goals that are at the center of livestock sector change. Raw read retention, the existence of multiple versions of annotations, and model checkpoints that are needed to ensure reproducibility increase storage costs, and data transfer bandwidth constraints make federated collaboration difficult. Strategies in approximation, such as sketching, subsampling, and quantization, reduce the demands of resources at the cost of casting doubt on the fidelity of downstream causal inference or the detection of rare features. Cloud-based solutions present peripheral issues on data sovereignty, access cost, and continuity risk of the service. The efficient algorithm design sensitive to the distribution of biological signals and the adaptive compression is a current requirement (Ma et al., 2025).

The pooling of datasets that would otherwise improve the model's robustness and fairness is complicated by data governance, ownership, and confidentiality. Although microbiome profiles at the farm level that are correlated with production measures are seen as possible sources of competitive advantage, there is little interest in sharing them outside of well-managed consortia. The human-centered privacy paradigms have to be modified to safeguard the vulnerable business data, genetic intellectual property in relation to the elite breeding sources, and management secrets of trade detectable through microbial biomarkers. In part, federated learning helps to combat direct sharing, but the residual leakage threats remain through gradient inversion or model update inference assaults. Elucidating legal ownership of multi-stakeholder-trained models, creating mechanisms of sharing benefits, and legalizing acceptable secondary uses are preconditions to scalable collaboration. Without a solid governance structure, data silos will persist, and AI models will have too many chances of overfitting to the local or management conditions (Ghavi Hossein-Zadeh, 2025; Shu, 2025).

Algorithms' bias and representational injustices arise when corpora are trained biased against a breed or geographic area or production system, or age category, or minority management practices, resulting in model performance differences and may further promote unfair distribution of resources. Climate, type of forage, genetic background, and local antimicrobial stewardship patterns produce significant variation in the composition of microbiomes; inability to sample this variation leads to latent embeddings that are specific to the majority of conditions. Phenotypic labels such as efficiency or disease status can be the source of systemic biases when methods of measurement are different or diagnostic practices vary across locations and transfer label shift to model evaluation. This is because the compositional data constraints and multicollinearity complicate bias detection by making it difficult to decompose simple performance. Mitigation methods such as reweighting, adversarial debiasing, or fairness-constrained optimization are associated with a trade-off between the overall accuracy and can even be inconsistent with the pattern of biological variance. Open disclosure of demographic and contextual coverage, and inclusion of fairness measures suitable to ecological prediction activities, is still rare, but a commitment to good progress (Panarese et al., 2025).

It still has integration friction with the rest of the omics and phenomics layers due to all four of the issues of mismatched temporal granularity, different noise distribution, variable sample availability, and inconsistent ontological annotation. Host transcriptomes can be sampled at discrete stress events, microbiome at a diet change schedule, and metabolomics snapshots at the time when staff can access logs and create incompatible matrices, which is a challenge to joint modeling. Missingness may not be random, and it is common to have cost or invasiveness leading to less collection in certain subgroups, which is a biasing of integrative latent representations. The gaps in ontology between microbial functional annotations and host pathway vocabularies do not support the semantic matching required by graph fusion or multimodal transformers, whereas irregular definitions of phenotypic traits do not support cross-study harmonization. Models of AI that operate in synchronized multimodal data cannot survive the reality of asynchronous sparsity caused by commercial farm operations, which can either result in aggressive imputation with possible hallucination or the removal of informative samples that are not complete, which reduces the biological understanding (Liu et al., 2025).

A lack of validation and testing, as well as a lack of standardized benchmarking systems, weakens comparative evaluation of AI methods and exaggerates published performance assertions. There is heterogeneity in preprocessing pipelines, versions of the taxonomic reference, functional annotation databases, and measures of evaluation, which hinders reproducibility and fair comparison among models. Microbiome public benchmark datasets used in human microbiome studies do not well represent niches unique to livestock, and researchers are left to collect custom validation sets without a consensus in the community. Access control usually limits external validation even when using independent farms, climatic regions, or even production periods, and holdout approaches used internally to one cohort overfit generalization due to maintaining the underlying batch structure. Causal inference, efficacy of interventions, or actual strain composition place ground truth paucity on synthetic data simulation, and its assumptions have the potential to be stacked against certain algorithm families. It cannot be used by regulators and in practice, as stakeholders lack transparency in reporting confidence intervals, calibration metrics, and negative control results, which are important measures of robustness and failure risk (Palmer et al., 2025).

As soon as AI outputs start entering feed formulation or additive implementations, or microbial modulation interventions, biosecurity risks such as adversarial manipulation will arise. Hackers might also use model-driven management routines to inject contaminants or data distortions that bias inputs to microbial surveillance, leading to an inappropriate response or obscuring emerging pathogen threats. The poisoning of data in the stages of training the models can introduce certain implicit biases in favor of certain commercial goods or in favor of competitors. Proprietary microbial indicators or management strategies might be exposed to competitive positioning by model extraction and inversion attacks on deployed cloud-based decision support platforms. There are strong defenses such as adversarial training, anomaly detection, and secure enclaves that have computational overhead and sufficiently impair baseline accuracy unless well-tuned. Biosecurity governance should then move beyond the containment of pathogens to include cyberbiological integrity of AI-enhanced microbiome management pipelines (Improta, 2025).

Scientific challenges are overlaid with regulatory and ethical issues, namely in relation to the assertions around AI-directed probiotic or microbial consortia interventions, suggestions of the implications of prescribing the genetic selection strategy based on the mediation of traits by microbiomes, and the environmental externalities of aggressive microbial manipulation. Regulatory authorities might fail to have predefined assessment guidelines for AI-inferred functional forecasts, which will overlook cultured assays, leaving the commercialization routes unclear and postponing advantageous innovation. Ethical issues can occur when the optimization goals emphasize efficiency of production or emission reduction in a manner that might unintentionally harm animal welfare because such trade-offs are not adequately brought to light through any of the model interpretability layers. Smallholder or resource-limited environments present the risk of exploitation of data collected through such environments without fair distribution of benefits when AI models built through such inputs produce proprietary products not accessible to the sources of data. Also, the environmental stewardship requirements provide a reason why ecological fringe effects during large-scale modulation of manure microbiomes or phage-mediated biocontrol treatments, as suggested by generative models, would be cautiously evaluated. Creating open and participatory institution systems that combine ethical audit, postimplementation surveillance, and adaptive supervision systems becomes part of socially responsible AI integration. Challenges and limitations in AI-driven livestock metagenomics are shown in Table 9.4.

9.5 Future perspectives and innovations in AI for metagenomics

Future AI in livestock metagenomics will be determined by the emergence of multimodal foundation models that concurrently encode genomic, transcriptomic, epigenomic, and metabolomic data on hosts and microbes in shared representation spaces. Such models will likely utilize massive self-supervised goals that simultaneously conceal nucleic acid markers, anticipate structural or functional labels, recreate metabolic flux prices, and correlate the patterns of host genetic variability with the variations of microbial pathways, producing embeddings having inherent cross-kingdom coadaptation. Implementing unsupervised pattern discoveries on top of existing biological ontologies could break down long-existing barriers of analytic segregation between host quantitative genetics and microbiome ecology by directly estimating context-dependent microbiome interaction effects on productivity, resilience, and emissions. Transfer learning would enable generalized embedding fine-tuning to breed specific, life stage-specific, or environment-specific predictive tasks by

TABLE 9.4 Challenges and limitations in AI-driven livestock metagenomics.

Challenge	Description	Impact on AI-metagenomics studies	Underlying causes or nuances
Data compositionality and sparsity	Metagenomic data represent relative abundances (not absolute), often with many zeros and large variation in sequencing depth, making standard statistical assumptions problematic	• Risk of spurious correlations or misleading associations when applying ML/AI on raw or inadequately normalized data • Reduced reliability of predictive models; limited generalizability across datasets	• The compositional nature, increasing one taxon automatically reduces the relative proportion of others • Nonbiological zeros (due to detection limits and sampling depth) complicate the interpretation of absence • High dimensionality: many taxa or gene-features, but limited sample size increases sparsity
High dimensionality/curse of dimensionality	Microbiome/metagenomic datasets typically have far more features (taxa, gene families, and pathways) than samples, especially problematic for livestock studies with limited sample numbers	• Overfitting risk in ML/AI models (particularly complex models), leading to poor external validity • Difficulty in feature selection; many potentially correlated or redundant features complicate model building	• Many variables (taxa and genes) compared with a few samples • Redundancy and correlation among features (e.g., cooccurring taxa or coregulated genes), which complicates feature selection or dimensionality reduction • Imbalanced datasets (common core taxa versus rare taxa), rare taxa often underrepresented, reducing statistical power to detect associations
Heterogeneity, variability, and lack of standardization	Differences in sampling protocols, DNA extraction, sequencing platforms, bioinformatics pipelines, metadata annotation, etc., across farms, studies, and labs lead to data heterogeneity and technical variability	• Limits comparability and reproducibility across studies • Reduces transferability of AI models trained on one dataset to other farms, breeds, or environmental conditions • Increases risk of batch effects, confounding, and misleading predictions	• Diverse farm conditions: diet, environment, housing, and management differ widely, contributing to biological heterogeneity • Variation in lab protocols: sample collection, storage, DNA extraction, library prep, and sequencing depth • Lack of consensus on metadata standards for livestock metagenomics; missing or inconsistent host/management data

Incomplete reference databases and poor annotation	Many microbes in livestock systems, especially in the gut, rumen, and environment, are uncharacterized; reference databases are biased toward well-studied taxa, limiting taxonomic and functional assignment accuracy	• Taxonomic assignment errors or inability to assign many reads may lead to the loss of information or misclassification • Functional inference (e.g., metabolic pathways and resistance genes) may be incomplete or inaccurate, reducing the utility of models aiming at function prediction	• Novel or rare taxa not yet in reference databases • High microbial diversity and rapid evolution (especially in gut or environmental microbiomes) • Viral metagenomics faces the dark matter problem: a high percentage of viral sequences remain unclassified due to a lack of reference genomes
Large data volume and computational burden	Shotgun metagenomic datasets are very large and complex, requiring high-performance computing, storage, and sophisticated bioinformatics analysis pipelines	• Resource constraints may limit the number of samples or depth of sequencing, reducing statistical power • Barrier to adoption in resource-limited settings (small farms, labs without high-end infrastructure) • Long processing times and possible delays in analysis impede scaling to large cohort studies	• Rapid growth in sequencing capacity increases data volume, but bioinformatics throughput and standardization lag • Need for specialized skills (bioinformatics and computational biology), which may be scarce in many livestock research groups • Costs associated not only with sequencing but also with computing infrastructure and data storage are often underestimated
Biological and temporal variability	Microbiome composition and function can vary over time (age, diet, and season), across individuals, and under different environmental or management conditions	• Difficulty in building stable predictive models, what holds for one-time point or condition may fail under another • Cross-sectional studies may miss dynamic changes; models may oversimplify complex temporal dynamics • Requires longitudinal sampling and metadata, costly and logistically challenging	• Host factors (age, breed, and health status), diet, environment, stress, medication, and season, all contribute to microbiome variation • The microbiome exhibits rapid responsiveness to temporal fluctuations; snapshot sampling may not reflect longer-term state • Lack of longitudinal datasets in livestock metagenomics due to cost and sampling burden
Model overfitting, poor generalizability, and lack of interpretability	Especially when applying complex ML/DL methods, models may perform well on training data but fail on new datasets; also, the black-box nature of many models limits biological interpretability	• Predictions may not hold when applied to other farms, breeds, or environmental contexts • Reduced trust by practitioners (farmers and veterinarians) if the biological basis is unclear • Hinders mechanistic insights, reduces the value of AI beyond black-box prediction	• Large feature space versus limited samples results in overfitting risk • Complex nonlinear models are difficult to interpret biologically • Lack of standard validation practices: some studies omit independent test sets or external validation, leading to inflated performance claims

Continued

TABLE 9.4 Challenges and limitations in AI-driven livestock metagenomics.—cont'd

Challenge	Description	Impact on AI-metagenomics studies	Underlying causes or nuances
Integration of multimodal and metadata variables	Effective AI-driven metagenomics often requires combining microbiome data with host genomics, diet, environmental, and management metadata, but these data types are heterogeneous, often incomplete, or poorly standardized	• Incomplete or inconsistent metadata reduces the ability to build robust, generalizable models • Limits capacity to interpret microbiome–host–environment interactions; may mask confounding effects, leading to spurious associations	• Differences in how metadata are recorded across farms or studies (diet composition, feeding regime, housing, and health records) • Lack of standards for metadata annotation for livestock metagenomic studies • Missing data, low granularity, or lack of longitudinal metadata undermines integrative modeling
Ethical, governance, and resource equity challenges	The deployment of AI-metagenomics in livestock production systems raises concerns regarding data ownership, access to technology, unequal resource distribution (rich versus. small farms), and ethical use of microbiome data (biosecurity, dual-use, privacy)	• Unequal adoption across farms, potentially widening technological disparities • Risk of misuse or unintended consequences (e.g., overreliance on AI predictions without biological validation) • Potential resistance from stakeholders due to a lack of transparency or trust in AI-driven recommendations	• High cost and technical expertise required for metagenomics and AI pipelines create barriers for small or resource-limited farms • Lack of regulatory frameworks or guidelines for the use of microbiome data in livestock production • Data privacy, consent, and biosecurity concerns, especially if microbiome data is used for disease surveillance or antimicrobial resistance monitoring

using limited labeled data, relieving current limitations of small phenotyped cohorts. Nevertheless, the effort to achieve this vision will require strict control of balanced and ethically considered training corpora covering the various production systems to prevent entrenching representational bias, and the development of parameter-efficient adaptation methods so that resource-limited stakeholders can make use of model capacity without repeating resource-prohibitive computational costs (Fu et al., 2025).

At the same time, analytics provided by real-time sequencing at the network edge are expected to revolutionize the microbiome monitoring paradigm of diagnostic methods applied in the past to retrospective monitoring, adopting a form of continuous situational awareness. Portable nanopore platforms coupled with streaming AI pipelines can update incrementally as the signal is produced, including base, error correction, taxonomic classification, antimicrobial resistance, and functional potential, indicating dysbiotic switches or emergent pathogen signals capable of changing into outcome phenotypes that are adverse to health. The reduction of latency without exceeding bandwidth constraints in remote or infrastructure-constrained farm contexts will be done by lightweight on-device inference with distilled foundation model fragments or sparsity-aware architectures. Dynamic remote control of read depth in adaptive sampling algorithms based on interim classification uncertainty could significantly increase successful resolution at a cost that is not too high. Multimodal anomaly detection would be further supported by the addition of edge-resident sensor feeds that capture temperature, humidity, feeding patterns, and behavioral clues with regard to an animal, in addition to state representations. The safeguards should be robust against signal drift, pore fouling, and adversarial perturbation, entrenching signal quality conditioning and probabilistic confidence monitoring to indicate output evidence that will require centralized reanalysis. The emerging cyberphysical layer preconditions the arrangement of proactive intervention closely linked with fermentation processes and environmental emissions patterns (Mahmood et al., 2025; Yan, 2025).

Hybrid mechanistic and data-driven ecosystem simulators are likely to occupy a middle ground between knowledge-based rumen fermentation models and flexible machine learning strategies. These coupled models will incorporate stoichiometric reaction constraints, thermodynamic feasibility, microbial guild interaction matrices, and host physiological compartmental flows as differentiable programmatic models that may be optimized by gradient-based methods to empirical multi-omics time series. Neural components will be trained on residual dynamics caused by regulatory changes, metabolic plasticity to strain levels, and unmodeled cross-feeding interactions, and mechanistic cores will maintain extrapolative credibility during diet or climate perturbations that are out of observed historical distributions. Hybrid models can be sensitized to focus on the measurement allocation to reaction groups or interaction coefficients with high posterior uncertainty and significant impact on policy outputs of interest, for example, the intensity of methane emissions or the efficiency of nitrogen use. With a higher data density, Bayesian calibration layers will become unlimited in terms of propagating parameter uncertainty into downstream management advice and will help in making risk-aware decisions. The identifiability issues, which high-dimensional mechanistic augmentation is prone to, will have to be overcome by having structured sparsity priors and causal regularizers that penalize implausible shortcuts to ensure that black box modules do not subsume interpretable mechanistic contributions (Balsa-Canto et al., 2025; Boussange et al., 2025).

The current state of generative community design is set to move past exploratory sampling of plausible taxonomic assemblages to function-focused synthetic consortia engineering. Future generation generative models will encode explicit metabolic flux objectives, redox balance constraints, cross-feeding compatibility scores, and host mucosal immune tolerance pointers directly into their objective functions, and sample communities that have been optimized to achieve multiobjective application envelopes of efficiency, emission reduction, welfare, and pathogen suppression. Microbiome configurations resilient to expected management regimes based on conditional generation conditioned with respect to particular diet formulations or environmental stressors will be custom-crafted. Active feedback loops between in silico generation and microfluidic coculture screening, iteratively boosted with convergent convergence on robust candidate consortia, would support faster convergence to robust candidate consortia. Improvement of interpretability by decomposing the predicted functional benefits of each of the proposed communities into pathway-level contribution maps will help build biological trust and enable regulatory appraisal. The postdeployment ecological stability will require the implementation of disturbance response simulations into generative evaluation loops that discourage designs that are based on tenuous mutualisms that are prone to breakdown under typical variability in feed or environmental conditions (Mkilima, 2025; Ma et al., 2025).

Further developments in causal discovery to be applied in interventional constraints that are common in livestock systems will probably transform the prioritization of candidate microbiome modulation approaches. New algorithms will take advantage of the heterogeneity across staggered or quasi-experimental management modifications, instrumental variation created by developed feed additives, and naturally-caused experiments signaled by climatic change to tighten causal graphs connecting microbial taxa, gene modules, host immune phenotypes, metabolic intermediates, and production results. By including principles of invariance between geographical layers and time scales, this will assist in differentiating the structural edges, and context-specific correlations and counterfactual limits of risk will be quantified by the amount of unintended consequences of manipulations of specific microbial levers. The use of constrained optimization models can incorporate feasibility measures that reflect the costs, welfare effects, regulatory allowability, and environmental impact to deliver ranked portfolios of interventions in accordance with the multistakeholder priorities. Closeness between the result of causal inference and reinforcement learning policy assessment will diminish spurious utilization of confounded predictors, and control strategies based on the structural validation of the pathways, as opposed to superficial associations. Nonetheless, the key to a strong performance is a clear statement of sets of assumptions and systematic sensitivity analysis with respect to latent confounding, which supports the necessity of standardized reporting norms (Fan et al., 2025; Khelfaoui et al., 2025).

The integration of microbiomes at the single-cell level will open up the layers of functional heterogeneity that are invisible in ruminal, intestinal, and respiratory niches. Combining high-resolution imaging mass spectrometry, spatial transcriptomics, and in situ sequencing of single-cell genomic and transcriptomic profiles of both the microenvironment and biofilm structures, as well as immunological synapses, will provide datasets of microenvironmental gradients, biofilm architecture, and immunological synapse configurations. The AI models will be trained on joint spatial embeddings that are indicative of diffusion-constrained metabolite accessibility, localized immune effector location, and microcolony lineage variation, and predict the spread of microscale rearrangements to macroscale fermentation efficiency or pathogen colonization resistance. Graph neural networks in the form of contact-mediated interactions with one continuous coordinate encoding can be trained to separate contact-mediated interactions with diffusible signaling, and generative spatial simulators can be trained to test intervention effects on community stratification and metabolite plume geometry. Spatial priors in the deconvolution of strains will improve the disambiguation of highly similar variants that occupy microniches, though functionally segregate them. The ethical and practical implications of the invasion of the sampled tissue and animal welfare will promote the creation of minimally invasive imaging modalities and in silico reconstruction techniques that do not rely on broad tissue sectioning (Zhu, 2025).

The development of microbiome, metabolite, and emission-targeted digital twins will continue previous host-centric constructs into fully coupled biogeochemical optimization platforms. These twins will continuously process metagenomic and metatranscriptomic data, volatile-compound sensor inputs, greenhouse gas emissions, rumen pH sensor data, and feed intake information to refine probabilistic predictions of emission intensities, nitrogen-loss pathways, energy distribution, and productivity metrics. Multiscale integration will fill the gap between intracellular level enzymatic control, community metabolic network flux distribution, and whole animal nutrient use to provide scenario testing of diet reformulations, microbiome-targeted additives, and selective breeding plans to achieve combined efficiency and sustainability benefits. Optimization modules may provide minimization efforts both inside Pareto frontiers on scaling back emissions or product quality limitations, which offer transparent trade-off surfaces to stakeholders. Uncertainty dashboards will indicate situations where a small amount of data will cause a difference in emission forecasts and will instruct specific sampling campaigns. These digital twins might eventually form a core tool for checking adherence to environmental reporting systems and, more importantly, recreating long-term cumulative effects of small-scale microbiome interventions in the context of the projected climate scenarios (Bagabaldo and Hackl, 2025; Han et al., 2025; Nathany, 2025).

The investigation of quantum-inspired and neuromorphic algorithms can offer revolutionary performance for high-dimensional inference problems that are inherent to metagenomic AI. The resource-efficient approximations provided by quantum annealing analogs and tensor network factorizations may be used to solve combinatorial optimization problems, for example, strain haplotype assembly, sparsifying the interaction graph in a microbial community, or group assignment of reads across overlapping pangenomes. Neuromorphic event-driven architectures that simulate the behavior of spiking neural networks could be effective at sparsity, irregular time resolution sequences of sequencing signal streams, or irregular sensor events with less energy at the expense of persistent on-farm inference, by operating on lower energy budgets. Hybrid pipelines may offload computationally expensive subroutines, such as large-scale graph embedding fresh to quantum accelerators, but these remain in classical differentiable training loops, which refine them to empirical loss functions. To achieve material benefit, strict benchmarking is necessary to separate the serious algorithmic improvements realized by the algorithm against hype instigated speculative accommodation, and create biologically inspired encoding frameworks to translate microbial sequence and functional information into structures that support fundamentally novel computational primitives (Iovane, 2025).

Green AI approaches will be unavoidable when model sizes, data scope, and complexity of simulation grow. The emission of life cycle related to microbiome analytic pipelines can be substantially lowered by carbon-aware schedulers that coordinate training and inference to events of low grid intensity or attend to renewable-powered data centers. Algorithms to address these issues, such as mixed precision arithmetic, adaptively stopping based on confidence convergence criteria, low rank factorization of weight matrices, and pruning of neural architectures dynamically based on compositional sparsity, will reduce computation costs without much performance degradation. The optimization of data centricity based on high-value, nonredundant, and diverse subsets of training has the potential to be more effective than brute force scaling, which is consistent with environmental stewardship and methodological rigor. Open disclosure of energy use, hardware design, and trade-offs, and optimization of performance and predictive performance will promote responsibility and motivate social norms of efficiency. This sustainability spirit is echoed in area-level emissions cut targets and corroborates consistency between analytical infrastructure decisions and production practice change (Malik, 2025).

A worldwide federated microbiome knowledge graph infrastructure has the potential to offer a substrate for interoperable discovery with respect to data sovereignty. This kind of graph would have cooccurring taxonomic objects, families of genes, metabolism reactions, environmental objects, host genotypes, phenotypic traits, intervention histories, and causal statements in a provenance-rich ontology framework. Federated update protocols would allow distributed nodes to provide localized embeddings, associations, and contextual metadata without necessarily providing raw sensitive datasets, and schema alignment services to address hierarchical inconsistency and annotation drift by release. AI agents would be able to use this graph to find latent multihop relationships between dietary components and emission-relevant metabolic modules mediated by particular microbial guild shifts controlled by host genetic variants, and thereby trigger hypothesis generation. Confidence weighting of edges using evidence type, breadth of replication, and methodology strength metrics would help block the propagation of lowly supported statements. To ensure the integrity and inclusiveness of governance structures, community adjudication of conflicting annotations and contribution incentives, and access tiers should be codified into a governance structure (Pudavar et al., 2025).

Individualized and adaptive microbiome modulation platforms will tend to coalesce around a closed-loop system that incorporates individual animal digital twins predictions, real-time microbiome state embeddings, feed bunk intake analytics, and welfare sensor signatures into the system to provide dynamically optimized interventions. They will be predictive phenotyping coupled to causal policy assessment to provide predictive additive dosing, precision or controlled supplementation, release encapsulated microbial inoculants responsive to microscale changes in fermentation balance, immune activation, or stress biomarkers. By transferring policy priors and protecting against overfitting to transient early life microbiome volatility, cross-animal meta-learning will enhance adaptation to new environments. Welfare prioritization constraints and transparency modules should be embedded in ethical safeguards that will transform the results of algorithms into a comprehensible biological rationale that is understandable by caretakers. Microbiome-mediated trait heritability estimates based on foundation model embeddings can be used upstream to selectively breed to enhance synergistically optimizing host genomic potential and microbiome stability to alleviate dependence on ongoing exogenous modulation (Greathouse and Choudhury, 2025).

Incorporating ethical by design governance and participatory data ecosystems into AI innovation cycles will be the key to continuing social license. Co-creation models, which involve producers, veterinarians, genomicists, environmental stakeholders, and consumer representatives in setting acceptable optimization goals, discourage prioritization of limited productivity measures. Fairness auditing, welfare impact assessment, and environmental footprint estimation can be part of algorithmic accountability pipelines that ensure continuous integration testing, where model changes that trigger unfavorable changes in the above metrics are automatically quarantined. The access permissions, benefit-sharing arrangements, and dispute resolution avenues operated by data trusts or cooperative stewardship organizations can provide contributors with agency over secondary use, facilitating the generation of pooled insights. Designs of consent capture systems that are optimized to apply in a business livestock environment need to be both practical and transparent in regard to downstream trajectories of analysis. Integrating the explainability artifacts reports and uncertainty quantification into the participatory dashboards supports informed decision-making and sets the expectation for management (Mushkani et al., 2025).

Reliable operational deployment will be based on secure pipeline designs that are based on the principles of zero trust and full provenance tracing. All the data entry points, model objects, intermediate processing, and inference queries can be signed cryptographically and recorded in tamper-evident ledgers, which allow end-to-end analytical lineage traceability and fast forensic recovery in case of anomalies. Fine-grained access control based on least privilege assumptions will limit the lateral movement opportunities to malicious actors, and constant behavioral analytics will be used to detect deviations from normal usage patterns across edge devices, sequencing instruments, and cloud orchestration layers. Collaborative learning and confidentiality will be balanced with privacy-preserving computing methods such as secure enclaves, homomorphic encryption to apply selective feature aggregation, and injecting noise in the gradient interfaces during differentiation of privacy. Simulated data poisoning, evasion, and model extraction attacks will be utilized in adversarial robustness testing and will be integrated as part of quality assurance tests before promotion to production. Balancing the

TABLE 9.5 Future directions and potential innovations in AI-driven livestock metagenomics.

Future direction or innovation	Description	Potential benefits	Challenges or limitations	Examples or potential applications
Deep-learning-driven functional and unknown-genome annotation	Use of advanced deep-learning models (e.g., neural networks, embedding models) to annotate genes, proteins, metabolic pathways, including from yet-uncharacterized microbial genomes (novel or uncultured taxa), bypassing reliance on existing reference databases	• Greater discovery of novel microbial functions in livestock microbiomes, beyond what reference-based annotation can reveal • Improved understanding of microbial contributions to nutrition, digestion, disease resistance, and environmental resilience • Enhanced ability to predict functional potential from metagenomes even in poorly characterized communities	• Deep models require large training datasets with high-quality annotated examples • Risk of false positives or mis-annotation if models overgeneralize • Interpretability remains difficult; linking predicted functions to real biological activity can be uncertain	Application of DL-based tools to reconstruct metabolic capabilities of rumen or gut microbiomes in cattle, sheep, or pigs, identifying novel enzyme systems or metabolic pathways relevant to feed efficiency or methane mitigation
Real-time and field-level metagenomic monitoring with AI	Integration of rapid sequencing (e.g., portable sequencers and long-read tech) with AI pipelines for on-farm or field-level microbiome monitoring, e.g., pathogen surveillance, microbiome stability check, and antibiotic-resistance monitoring	• Timely detection of disease outbreaks, zoonotic pathogens, or microbial imbalances • Ability to monitor resistome, microbiome shifts under changing diets or environments • Supports proactive management for health, welfare, and biosecurity in livestock operations	• Requires robust, user-friendly AI workflows that can run outside research labs (farm setting), technical and computational challenges • Need to handle noisy data (environmental contamination or host DNA), low abundances, and sequencing errors • Regulatory, privacy, and biosecurity considerations for microbiome data from farms	On-farm metagenomic screening of fecal or environmental samples to detect enteric pathogens, monitor antimicrobial resistance genes, or assess microbiome shifts following feed changes or environmental stress
Multi-omics and AI integration (host-microbiome-environment models)	Combining livestock metagenomics data with other omics layers (metabolomics, proteomics), host genomic data, diet, and environmental metadata, analyzed with AI (e.g., multimodal models, network models) to understand complex interactions	• Holistic understanding of how diet, host genetics, microbiome, and environment jointly influence health, productivity, and disease susceptibility • Potential to develop precision livestock management: diet or environmental intervention	• Requires collection of large, harmonized, high-quality datasets across multiple data types, logistically and financially intensive • Data integration challenges: heterogeneity, missing data, and normalization across omics layers	Integrated studies using microbiome data, host genotype, feed data, and metabolome to predict feed conversion efficiency in cattle, or disease resilience in poultry, enabling targeted breeding and intervention strategies

		tailored to microbiome and host genotype • More reliable predictions of complex traits (feed efficiency, disease resistance, and methane emission) than using single data types	• Complexity may reduce interpretability and increase computational cost	
Use of generative models and synthetic metagenomes for data augmentation	Application of generative AI (e.g., variational autoencoders and GANs) to create synthetic metagenomic profiles, to augment sparse datasets, simulate hypothetical scenarios (diet/environment changes), or explore rare microbial-community states	• Overcomes limitations of small sample sizes common in livestock metagenomic studies; improves training of predictive models • Enables what-if simulations, e.g., the prediction of microbiome response to a new feed, environmental change, or additive before real-world trial • Helps explore rare microbial states (e.g., rare but important pathogens or resistome events)	• Synthetic data may not faithfully represent real-world biological complexity; the risk of models learning artifacts rather than biology • Validation of synthetic-data-based predictions is difficult; it may lead to misleading conclusions • Ethical/regulatory concerns if synthetic data drives management decisions without validation	Using synthetic metagenomes to simulate the effect of dietary changes on gut microbial communities, to prescreen feed additives or management practices for impact on microbiome before field trials.
Long-read and single-cell metagenomics combined with AI for strain-level resolution.	Leveraging advances in long-read sequencing (e.g., Nanopore and PacBio) and single-cell metagenomics, combined with AI-based assembly, binning, and annotation, enables accurate strain-level taxonomy and functional resolution, detection of mobile elements (plasmids and phages), and resistomes	• Better detection of strain-level variation, horizontal gene transfer, resistance genes, and virulence factors, essential for disease monitoring, resistome surveillance, and microbiome function • More complete genomes (MAGs), better functional annotation, detection of structural variations, and mobile genetic elements • Improved ecological understanding of microbial dynamics, and more precise interventions (e.g., targeted probiotics, phage therapy, and resistome control)	• High cost and technical complexity; lower throughput than short-read metagenomics • Bioinformatics workflows for long-read or single-cell metagenomics remain challenging and require expertise • Data quality and noise (e.g., error rates in long reads), assembly/binning challenges in complex communities	Application in livestock virome and resistome surveillance: reconstructing complete plasmids/phages from gut or environmental samples to monitor antimicrobial resistance, mobile genetic elements, or novel viruses in farm ecosystems.
Explainable AI and causal inference models for microbiome-trait associations	Development of AI models that not only predict associations (e.g., between microbiome	• Provides actionable insights rather than black-box predictions, useful for	• Causal inference from observational metagenome data is inherently	Using attention-based deep models or causal-network models to link microbiome

Continued

TABLE 9.5 Future directions and potential innovations in AI-driven livestock metagenomics.—cont'd

Future direction or innovation	Description	Potential benefits	Challenges or limitations	Examples or potential applications
	composition and production traits) but also allow interpretation and inference of causal relationships, e.g., through attention models, causal ML, and interpretable networks	designing interventions (feed and management), breeding programs, and regulatory decisions • Helps identify key microbial taxa or functions causally linked to desired traits (e.g., feed efficiency and disease resilience) • Builds trust among stakeholders (farmers, vets, and regulators) by making mechanisms transparent	challenging (confounders, longitudinal variability, and causality versus correlation) • Requires high-quality, well-annotated metadata (diet, environment, and host genetics) and possibly longitudinal or interventional data • Interpretability methods may simplify complex interactions, potentially misleading if overinterpreted	features to methane emission reduction, feed conversion efficiency, or disease resistance, offering mechanistic hypotheses for experimental validation
Real-world deployment (farm-level decision support systems (DSS) integrating metagenomics and AI)	Development of practical decision-support tools for farmers, integrating microbiome data (gut and environment), AI predictions, and actionable recommendations (feed formulation, additive use, hygiene, and biosecurity)	• Helps translate research findings into actionable farm management strategies, improving productivity, health, welfare, and sustainability • Enables precision livestock management tailored to microbiome and environment • Can support large-scale adoption of microbiome-guided practices across farms of different sizes	• Requires standardization of sampling, sequencing, data processing, and metadata recording across farms • Infrastructure, cost, and technical capacity may be limiting, especially for small or resource-limited farms • Regulatory, data governance, privacy, and biosecurity concerns around the use of microbiome data in production decisions • Necessity for robust validation of AI recommendations under diverse farm conditions	Pilot systems that recommend feed additive regimes or hygiene protocols based on periodic metagenomic monitoring of herd gut- or environmental-microbiome, aiming to reduce disease risk, antibiotic use, or emissions

cybersecurity measures with biosecurity strategies will guarantee the integrity of the microbiome management system (Hossain et al., 2025).

Performance, robustness, and reproducibility benchmarking consortia will become standardized to enhance methodological transparency and accelerate cumulative advances. Multifarm longitudinal cohort sets, diet shift challenge sets, emission-labeled sets, and perturbation-annotated experimental trials that will be shared will enable head-to-head comparison of preprocessing pipelines, representation learning strategies, causal inference frameworks, and policy optimization modules under controlled conditions. Benchmarks will move beyond predictive accuracy, perhaps to encompass calibration quality, breed and region fairness, computation efficiency, energy intensity, high interpretability quality, and resistance to corruption by synthetic noise or change of distribution. Containerized reference implementations of defined dependency snapshots will contribute to verifiable replication and leaderboards under disclosure requirements, prevent cherry-picked reporting, which fosters positive competitive pressure. The overfitting to fixed test sets and the lack of innovative capacity will be avoided using dynamic benchmarks that periodically add previously unavailable edge cases or new data modalities to spur generalizable innovation (Mehta, 2025).

The merging of metagenomics and precision nutrition, welfare metrics, and the accounting of emissions will be condensed into integrated decision support ecosystems to support the holistic stewardship of sustainability. Microbiome-derived fermentation efficiency predictions, host genotype metabolizable energy partition predictions, and projected greenhouse gas flux responses will be assimilated in nutritional formulation engines to propose diet changes according to the emission intensity reduction targets and maintain welfare indicators, including behavioral stress proxies and ruminal acidosis risk scores. The divergence of the predicted and observed results will be measured in feedback loops, which will initiate the model refinement cycle and disclose the systematic biases that are likely to be the outcome of the changing microbial adaptation or climate conditions. Cross-domain optimization may give rise to trade-offs in which marginal emission reductions would come at disproportionate costs in terms of welfare or productivity, and enable deliberation of policy based on evidence. The integration of standardized emission accounting specifications into analytic processes simplifies the reporting on regulations and increases the plausibility of sustainability reports that are distributed to the supply chain partners and consumers. The microbiome will then become an operating lever that is hybridized with nutritional and environmental control, rather than a descriptive outer layer (Nsabiyeze et al., 2025; Ghavi Hossein-Zadeh, 2026b). Future directions and potential innovations in AI-driven livestock metagenomics are presented in Table 9.5.

References

Alonso-Ramos, J.J., Aguilera-Martos, I., Herrera-Poyatos, A., Herrera, F., 2025. Denograd: deep gradient denoising framework for enhancing the performance of interpretable AI models. arXiv. https://doi.org/10.48550/arXiv.2511.10161.

Arya, J., 2025. Optimizing ML model interpretability in high-stakes decision-making: balancing transparency and performance. J. Sustain. Solut. 2 (3), 22–27. https://doi.org/10.36676/j.sust.sol.v2.i3.72.

Bagabaldo, A.R., Hackl, J., 2025. Digital twins for intelligent intersections: a literature review. arXiv. https://doi.org/10.48550/arXiv.2510.05374.

Balsa-Canto, E., Campo-Manzanares, N., Moimenta, A.R., Roudaut, G., Troitiño-Jordedo, D., 2025. Quantifying and managing uncertainty in systems biology: mechanistic and data-driven models. Curr. Opin. Syst. Biol. 42, 100557. https://doi.org/10.1016/j.coisb.2025.100557.

Boussange, V., Vilimelis Aceituno, P., Schäfer, F., Pellissier, L., 2025. A calibration framework to improve mechanistic forecasts with hybrid dynamic models. Methods Ecol. Evol. 16 (12), 2992–3007. https://doi.org/10.1111/2041-210x.70190.

Buchha, D., 2025. AI & ML for livestock welfare: analysing behaviour and stress patterns for better care. Int. J. Sci. Res. Eng. Manag. 09 (04), 1–9. https://doi.org/10.55041/ijsrem46227.

Cai, J., Xu, C., Yang, H., Chen, X., Zheng, A., Jing, H., Xun, Y., Zhao, X., 2026. A Novel Framework for temporal knowledge graph reasoning with complex causal relations. Expert Syst. Appl. 299, 130089. https://doi.org/10.1016/j.eswa.2025.130089.

Campbell, D.E., Baldridge, M.T., 2024. A new piece of the microbiota pie: mining 'omics for DNA inversion states. Cell Host Microbe 32 (3), 293–295. https://doi.org/10.1016/j.chom.2024.02.009.

Cao, W., Tam Mai, N., Liu, W., 2025. Adaptive knowledge assessment via symmetric hierarchical Bayesian neural networks with graph symmetry-aware concept dependencies. Symmetry 17 (8), 1332. https://doi.org/10.3390/sym17081332.

Chen, S., Turanoglu Bekar, E., Bokrantz, J., Skoogh, A., 2025. AI-enhanced digital twins in maintenance: systematic review, industrial challenges, and bridging research–practice gaps. J. Manuf. Syst. 82, 678–699. https://doi.org/10.1016/j.jmsy.2025.07.006.

Chorlton, S.D., 2024. Ten common issues with reference sequence databases and how to mitigate them. Front. Bioinform. 4. https://doi.org/10.3389/fbinf.2024.1278228.

Cunningham-Oakes, E., Perez-Sepulveda, B.M., Li, Y., Hinton, J.C.D., Nelson, C.A., McIntyre, K.M., Wardeh, M., Haldenby, S., Gregory, R., Iturriza-Gómara, M., Hertz-Fowler, C., O'Brien, S.J., Cunliffe, N.A., Darby, A.C., Yonova, I., Shinneman, S., O'Sullivan, S., Hriskova, M., Ferreira, F., de Lusigna, S., Davie, R., Hughes, D.A., Bhatt, R., Wood, B., Vivancos, R., Smith, G.E., Shah, A., Schinaia, N., Pegorie, M., McCann, C.M., Lighton, L., Lamden, K.H., Elliot, A.J., Cleary, P., McEwan, A., Dodgson, A.R., Wilson, L., Orr, D., Gray, K., Cheesbrough, J., Bolton, L., Bennett, M., Owen, V., Corless, C., Anson, J., Rowlingson, B.S., Hale, A.C., Diggle, P.J., Dodd, K., Winstanley, C., Williams, D., Thiele, T.,

Stanistreet, D., Snape, D.A., Ruddock, H.K., Reaves, D., Radford, A.D., Pulman, J.A., Orton, L., Jones, T., Jones, L., Jackson, K., Hyseni, L., Howarth, D., Hall, N., Durie, A.E., Deja, E., Dawson, S., Clough, H.E., Christley, R.M., Bolton, F.J., 2025. Enhancing infectious intestinal disease diagnosis through metagenomic and metatranscriptomic sequencing of 1000 human diarrhoeal samples. Genome Med. 17 (1). https://doi.org/10.1186/s13073-025-01478-w.

Deng, Z.C., Liu, M., Cao, K.X., Khalil, M.M., Guan, L.L., Sun, L.H., 2025. Gut microbiome and postbiotics: bridging the dietary nutrition and feed efficiency in food-producing animals. Sci. China Life Sci. https://doi.org/10.1007/s11427-025-3063-5.

Dond, V.T., 2025. Supervised learning approaches for robust predictive modelling in data science. Int. J. Res. Appl. Sci. Eng. Technol. 13 (8), 1237–1243. https://doi.org/10.22214/ijraset.2025.73756.

Durand, K., Ogier, J.C., Nam, K., 2025. The evaluation of shotgun sequencing and rpoB metabarcoding for taxonomic profiling of bacterial communities. BMC Microbiol. 25 (1). https://doi.org/10.1186/s12866-025-04149-3.

Fan, Y., Zhang, J., Cai, K., Yang, J., Wang, K., 2025. Causal invariance and counterfactual learning driven cooperative game for multi-label classification (version 1). arXiv. https://doi.org/10.48550/ARXIV.2512.00812.

Fu, J., Ge, X., Xin, X., Karatzoglou, A., Arapakis, I., Zheng, K., Ni, Y., Jose, J.M., 2025. Efficient and effective adaptation of multimodal foundation models in sequential recommendation. IEEE Trans. Knowl. Data Eng. 37 (12), 7076–7089. https://doi.org/10.1109/TKDE.2025.3608071.

Ghavi Hossein-Zadeh, N., 2025. Artificial intelligence in veterinary and animal science: applications, challenges, and future prospects. Comput. Electron. Agric. 235, 110395. https://doi.org/10.1016/j.compag.2025.110395.

Ghavi Hossein-Zadeh, N., 2026a. Generative artificial intelligence in animal genomics for smart agriculture: Applications, challenges, and future prospects. Vet. Anim. Sci. 33, 100702. https://doi.org/10.1016/J.VAS.2026.100702.

Ghavi Hossein-Zadeh, N., 2026b. Advancing climate-resilient livestock systems: Next-generation emission mitigation strategies and integrated technological innovations. Vet. Anim. Sci. 31, 100588. https://doi.org/10.1016/j.vas.2026.100588.

Ghiotto, G., Xirostylidou, A., Gaspari, M., Kougias, P.G., Campanaro, S., Treu, L., 2025. Exploring genetic adaptation and microbial dynamics in engineered anaerobic ecosystems via strain-level metagenomics. Cell Genom. 5 (9), 100949. https://doi.org/10.1016/j.xgen.2025.100949.

Gholamzadeh, E., Singla, K., Scherf, N., 2025. Predicting microbial interactions using graph neural networks. arXiv. https://doi.org/10.48550/arXiv.2511.02038.

Ghozlane, A., Thirion, F., Plaza Oñate, F., Gauthier, F., Le Chatelier, E., Annamalé, A., Almeida, M., Ehrlich, S.D., Pons, N., 2025. Accurate profiling of microbial communities for shotgun metagenomic sequencing with Meteor2. Microbiome 13 (1). https://doi.org/10.1186/s40168-025-02249-w.

Greathouse, K.L., Choudhury, A., 2025. Precision nutrition and the gut microbiome: harnessing AI to revolutionize cancer prevention and therapy. Cell Host Microbe 33 (6), 766–776. https://doi.org/10.1016/j.chom.2025.05.011.

Han, S., Yu, S., Shi, M., Harada, M., Ge, J., Lin, J., Prehn, C., Petrera, A., Li, Y., Sam, F., Matullo, G., Adamski, J., Suhre, K., Gieger, C., Hauck, S.M., Herder, C., Roden, M., Casale, F.P., Cai, N., Peters, A., Wang-Sattler, R., 2025. LEOPARD: missing view completion for multi-timepoint omics data via representation disentanglement and temporal knowledge transfer. Nat. Commun. 16 (1). https://doi.org/10.1038/s41467-025-58314-3.

Hossain, A.A., Pk, M.K., Amsaad, F., Ahner, D., 2025. Secure and privacy-preserving AI: a zero trust architecture for federated machine learning. In: 2025 1st International Conference on Secure IoT, Assured and Trusted Computing (SATC), pp. 1–7. https://doi.org/10.1109/satc65530.2025.11137026.

Hsieh, Y.-H., Chien, T.-J., Huang, C.-K., Sun, S.-H., Lin, C., 2025. MedFuse: multiplicative embedding fusion for irregular clinical time series (version 1). arXiv. https://doi.org/10.48550/ARXIV.2511.09247.

Hui, H.W.H., Kong, W., Goh, W.W.B., 2024. Thinking points for effective batch correction on biomedical data. Briefings Bioinf. 25 (6). https://doi.org/10.1093/bib/bbae515.

İlgen, B., Dubey, A., Hattab, G., 2025. PHAX: a structured argumentation framework for user-centered explainable AI in public health and biomedical sciences. arXiv. https://doi.org/10.48550/arXiv.2507.22009.

Improta, C., 2025. Detecting stealthy data poisoning attacks in AI code generators (version 1). arXiv. https://doi.org/10.48550/ARXIV.2508.21636.

Iovane, G., 2025. Quantum-inspired algorithms and perspectives for optimization. Electronics 14 (14), 2839. https://doi.org/10.3390/electronics14142839.

Kazantseva, E., Donmez, A., Frolova, M., Pop, M., Kolmogorov, M., 2024. Strainy: phasing and assembly of strain haplotypes from long-read metagenome sequencing. Nat. Methods 21 (11), 2034–2043. https://doi.org/10.1038/s41592-024-02424-1.

Khelfaoui, I., Wang, W., Meskher, H., Shehata, A.I., El Basuini, M.F., Abouelenein, M.F., Degha, H.E., Alhoshy, M., Teiba, I.I., Mahmoud, S.S., 2025. Beyond just correlation: causal machine learning for the microbiome, from prediction to health policy with econometric tools. Front. Microbiol. 16. https://doi.org/10.3389/fmicb.2025.1691503.

Kitov, I., 2025. Combining causal analysis and machine learning to predict the effects of interventions. Am. J. Eng. Technol. 07 (04), 134–140. https://doi.org/10.37547/tajet/volume07issue04-18.

Klimina, K., Zoruk, P., Morozov, M.D., Veselovsky, V., Strokach, A., Babenko, V., 2025. Impact of DNA extraction techniques and sequencing approaches on microbial community profiling accuracy. Front. Microbiomes. https://doi.org/10.3389/frmbi.2025.1688681.

Kublashvili, S., 2025. Probabilistic neuro-symbolic reasoning for sparse historical data: a framework integrating Bayesian inference, causal models, and game-theoretic allocation (version 1). arXiv. https://doi.org/10.48550/ARXIV.2512.01723.

Li, Z., Sinha, A., Li, Y., Chard, K., Kim, K., Madduri, R., 2025. Experiences building enterprise-level privacy-preserving federated learning to power AI for science (version 1). arXiv. https://doi.org/10.48550/ARXIV.2511.08998.

Liang, Y., Yu, J., Yao, Z., Sun, Y., Feng, J., Shen, R., Luo, J., Zhao, L., 2026. Decoding microbial interactions: interaction networks and regulatory strategies for medium-chain fatty acid biosynthesis through anaerobic chain elongation. Biotechnol. Adv. 86, 108735. https://doi.org/10.1016/j.biotechadv.2025.108735.

Liu, B., Liu, Y., Xu, S., Wu, Q., Wu, D., Zhan, L., Liao, Y., Mai, Y., Zheng, M., Wang, S., Chen, Y., Huang, Z., Luo, X., Xie, Z., Pan, X., Yu, G., Xie, L., 2025. EasyMultiProfiler: an efficient multi-omics data integration and analysis workflow for microbiome research. Sci. China Life Sci. https://doi.org/10.1007/s11427-025-3035-0.

Lubin, J.B., Planet, P.J., Silverman, M.A., 2025. Microbial succession at weaning is guided by microbial metabolism of host glycans. Microbiome 13 (1). https://doi.org/10.1186/s40168-025-02193-9.

Ma, W., Gao, L., Gong, Q., Wang, Z., 2025. Modeling the environmental footprint and sustainability of high-performance computing. Theor. Nat. Sci. 104 (1), 8–17. https://doi.org/10.54254/2753-8818/2025.ch24595.

Mahmood, K., Khan, S., Abdelhaq, M., Hassan, M.U., Uddin, M., Alsaqour, R., Awan, K.A., Alsoufi, M.A., 2025. Adaptive resource aware and privacy preserving federated edge learning framework for real time internet of medical things applications. Sci. Rep. 15 (1). https://doi.org/10.1038/s41598-025-23398-w.

Malik, F., 2025. Carbon-conscious intelligence: life cycle assessment and green standards for generative AI. Int. J. Innov. Sci. Res. Technol. 1030–1037. https://doi.org/10.38124/ijisrt/25aug585.

Mehta, S., 2025. When are learning biases equivalent? A unifying framework for fairness, robustness, and distribution shift (version 1). arXiv. https://doi.org/10.48550/ARXIV.2511.07485.

Mishra, S., Lercher, M.J., 2025. Horizontal gene transfer inference: gene presence–absence outperforms gene trees. Mol. Biol. Evol. 42 (7). https://doi.org/10.1093/molbev/msaf166.

Mkilima, T., 2025. Synthetic biology approaches for restoring gut microbial balance and engineering disease-specific microbiome therapeutics. Microb. Pathog. 207, 107931. https://doi.org/10.1016/j.micpath.2025.107931.

Mohan, A., 2025. Causal inference in data science: a framework for attribution systems. Eur. J. Comp. Sci. Inf. Technol. 13 (36), 107–113. https://doi.org/10.37745/ejcsit.2013/vol13n36107113.

Moitra, T., Larrouy-Maumus, G., 2025. Integrated approaches for discovery and functional annotation of proteins of unknown function. Trends Biochem. Sci. https://doi.org/10.1016/j.tibs.2025.11.001.

Moran, J.A., Morato, P.G., 2025. Balancing the exploration-exploitation trade-off in active learning for surrogate model-based reliability analysis via multi-objective optimization. arXiv. https://doi.org/10.48550/arXiv.2508.18170.

Mushkani, R., Berard, H., Ammar, T., Chatonnier, C., Koseki, S., 2025. Co-producing AI: toward an augmented, participatory lifecycle. arXiv. https://doi.org/10.48550/arXiv.2508.00138.

Nathany, D., 2025. Digital twins for resilient and sustainable supply chains. Int. J. Multidiscip. Res. 7 (1). https://doi.org/10.36948/ijfmr.2025.v07i01.40768.

Nsabiyeze, A., Zhang, M., Li, J., Zhao, Q., Zhang, X., 2025. Precision livestock farming for climate-resilient livestock management: a review of real-time monitoring and decision support systems. J. Clean. Prod. 524, 146454. https://doi.org/10.1016/j.jclepro.2025.146454.

Oppliger, J., Denner, M.M., Küspert, J., Frison, R., Wang, Q., Morawietz, A., Ivashko, O., Dippel, A.C., Zimmermann, M.v., Biało, I., Martinelli, L., Fauqué, B., Choi, J., Garcia-Fernandez, M., Zhou, K.J., Christensen, N.B., Kurosawa, T., Momono, N., Oda, M., Natterer, F.D., Fischer, M.H., Neupert, T., Chang, J., 2024. Weak signal extraction enabled by deep neural network denoising of diffraction data. Nat. Mach. Intell. 6 (2), 180–186. https://doi.org/10.1038/s42256-024-00790-1.

Ouyang, H., Yang, Y., Zhang, X., Cui, Y., Zhang, Y., 2025. Microbial orchestration of neuroimmune crosstalk: from homeostasis to disease. Front. Immunol. 16. https://doi.org/10.3389/fimmu.2025.1679286.

Palmer, S.N., Mishra, A., Gan, S., Liu, D., Koh, A.Y., Zhan, X., 2025. Identifying optimal machine learning approaches for human gut microbiome (shotgun metagenomics) and metabolomics integration with stable feature selection. bioRxiv. https://doi.org/10.1101/2025.06.21.660858.

Panarese, P., Grasso, M.M., Solinas, C., 2025. Algorithmic bias, fairness, and inclusivity: a multilevel framework for justice-oriented AI. AI Soc. https://doi.org/10.1007/s00146-025-02451-2.

Park, S., Lee, K., Yim, S., Hwang, D., Kim, D., Lee, S., Dunn, A., Gatti, D., Chesler, E., O'Connell, K., Kim, K., 2025. Robust multi-omics integration from incomplete modalities significantly improves prediction of Alzheimer's disease. arXiv. https://doi.org/10.48550/arXiv.2509.20842.

Peng, X., Wei, Y., Zhou, X., 2025. Enhancing pathogen identification through AI-assisted metagenomic sequencing. Front. Microbiol. 16. https://doi.org/10.3389/fmicb.2025.1634194.

Pudavar, A.E., Baksi, K.D., Pokhrel, V., Kuntal, B.K., 2025. Microbiome knowledge graph as a tool to understand bacteria-host associations. Arch. Microbiol. 207 (9). https://doi.org/10.1007/s00203-025-04413-0.

Qi, Y., Liu, D., Zhang, J., Zhang, Y., 2025. A shift reduction domain generalization network for hyperspectral image cross-domain classification. IEEE Trans. Geosci. Rem. Sens. 63, 1–16. https://doi.org/10.1109/tgrs.2025.3599172.

Qiu, J., Tao, H., Li, H., Liu, X., Liu, R., Nawaz, M.N., Wang, X., Ma, L., 2025. The environmental adaptation of acidophilic archaea: promotion of horizontal gene transfer by genomic islands. BMC Genom. 26 (1). https://doi.org/10.1186/s12864-025-11875-5.

Rizzi, S., Saroglia, G., Kalemi, V., Rimoldi, S., Terova, G., 2025. Artificial intelligence in microbiome research and beyond: connecting human health, animal husbandry, and aquaculture. Appl. Sci. 15 (17), 9781. https://doi.org/10.3390/app15179781.

Salem, Z., Sahl, S., Ibrahim, M.H.K., 2025. From prediction to trust: enhancing deep learning models for insurance fraud detection through uncertainty quantification. PeerJ Comput. Sci. 11, e3365. https://doi.org/10.7717/peerj-cs.3365.

Shah, M., Wever, M., Espig, M., 2025. A framework for assessing the potential of artificial intelligence in the circular bioeconomy. Sustainability 17 (8), 3535. https://doi.org/10.3390/su17083535.

Sharma, S., Narahari, H.P., Raman, K., 2025. Harnessing machine learning for metagenomic data analysis: trends and applications. mSystems 10 (11). https://doi.org/10.1128/msystems.01642-24.

Shaw, R., Love, S.D., McWhite, C.D., 2025. Evaluating pretrained protein language model embeddings as proxies for functional similarity. J. Mol. Evol. https://doi.org/10.1007/s00239-025-10282-4.

Shimpi, P., 2025. Reinforcement learning in real life applications. Int. J. Sci. Res. Eng. Manag. 09 (01), 1–9. https://doi.org/10.55041/ijsrem40881.

Shu, W., 2025. Legal logic of AI data governance based on federated learning: institutional evolution from privacy protection to rights distribution. Int. J. Ethical AI Appl. 1 (4), 7–12. https://doi.org/10.64229/efs37007.

Sobolev, A., Sibiryakina, D., Chevokina, E., Slonova, D., Yurikova, D., Kozlova, S., Trofimova, A., Zubarev, V., Kiselev, A., Konovalova, O., Sutormin, D., Isaev, A., 2025. Benchmarking cost-effective DNA extraction kits for diverse metagenomic samples. Int. J. Mol. Sci. 26 (23), 11616. https://doi.org/10.3390/ijms262311616.

Spicer, R., Raychawdhary, N., Danwada, S., Udomprasert, P., Seals, C., Bhattacharya, S., 2025. Evaluating the significance of embedding-based protein sequence alignment with clustering and double dynamic programming for remote homology. Sci. Rep. 15 (1). https://doi.org/10.1038/s41598-025-23319-x.

Tang, G., Carr, A.V., Perez, C., Sarmiento, K.R., Levy, L., Lampe, J.W., Diener, C., Gibbons, S.M., 2025. Metagenomic estimation of absolute bacterial biomass in the mammalian gut through host-derived read normalization. mSystems 10 (8), 01–19. https://doi.org/10.1128/msystems.00984-25.

Wang, L., Qi, G., Shi, Y., Ma, Y., Gao, J., 2025. A systematic longitudinal study of microbiome: integrating temporal-spatial dimensions with causal and deep learning models. BMC Genom. 26 (1). https://doi.org/10.1186/s12864-025-12282-6.

Wang, J., Deng, H., Kan, S., Xu, J., 2026. Multi-level contrastive learning with graph convolutional network for multi-view clustering. Expert Syst. Appl. 302, 130573. https://doi.org/10.1016/j.eswa.2025.130573.

Week, B., Russell, S.L., Schulenburg, H., Bohannan, B.J.M., Bruijning, M., 2025. Applying evolutionary theory to understand host–microbiome evolution. Nat. Ecol. Evol. 9 (10), 1769–1780. https://doi.org/10.1038/s41559-025-02846-w.

Wu, G., Wang, R., Wang, Y., Sun, S., Chen, J., Qi, Z., 2025. Crosstalk among gut microbiota, microbial metabolites, and inflammatory cytokines: current understanding and future directions. Foods 14 (22), 3836. https://doi.org/10.3390/foods14223836.

Xie, Y., Jia, J., Wen, C., Li, D., Li, M., 2025. Multi-topology contrastive graph representation learning. Sci. China Inf. Sci. 69 (2). https://doi.org/10.1007/s11432-024-4467-3.

Yamada, T., Palm, N.W., 2025. A host-centric view of the microbiota metabolome. Immunity 58 (12), 2940–2956. https://doi.org/10.1016/j.immuni.2025.11.006.

Yan, M., 2025. Federated edge intelligence: a collaborative learning framework for multi-object detection on mobile platforms. Internet Technol. Lett. 8 (6). https://doi.org/10.1002/itl2.70145.

Yang, C., Snelson, M., El-Osta, A., Marques, F.Z., 2025. Parental diet and offspring health: a role for the gut microbiome via epigenetics. Nat. Rev. Gastroenterol. Hepatol. 22 (11), 755–772. https://doi.org/10.1038/s41575-025-01106-3.

Ye, X., Sahana, G., Lund, M.S., Cai, Z., 2025. The crosstalk between host and rumen microbiome in cattle: insights from multi-omics approaches and genome-wide association studies. World J. Microbiol. Biotechnol. 41 (8). https://doi.org/10.1007/s11274-025-04504-6.

Yu, Y., Wang, J., Faizan, A.S., Cheng, H., Liu, A., Liu, Y., Tian, S., Liang, J., Zhu, L., Zhang, G., 2025. Enhancing sourdough fermentation with AI and multi-omics: from natural diversity to synthetic microbial community. Trends Food Sci. Technol. 165, 105233. https://doi.org/10.1016/j.tifs.2025.105233.

Zhu, B., 2025. Emerging insights into the principles and advantages of single-cell and spatial transcriptomics. Acad. J. Sci. Tech. 15 (3), 50–54. https://doi.org/10.54097/gf62sx56.

Chapter 10

Artificial intelligence integration in livestock epigenomics

Artificial intelligence (AI) is transforming livestock epigenomics from a descriptive discipline into a predictive and intervention-aware science. Epigenomics involves DNA patterns of methylation, histone modification, chromatin accessibility, three-dimensional architecture of the genome, noncoding RNA regulation, and positioning of nucleosomes, which constitute a complex of modulation and coordination of transcriptional programs and metabolic, immune, and developmental pathways in response to context. These layers are plastic and temporal compared to the static genomic variation, tissue and cell-type-specific, and responsive to nutritional, climatic, microbial, and stress-related signaling that typify the heterogeneous livestock production ecosystems. This plasticity has provided both opportunity and complexity: opportunity in terms of biomarkers and probably steerable regulatory states underlying resilience, efficiency, welfare, and reduced environmental footprint, and complexity via ubiquitous confounding, cell mixture effects, measurement noise, and impermanence of some marks.

AI representation capabilities can be used to balance the multi-scale, multi-modality, and sparsely sampled properties of epigenomic streams of data, and learn latent structures that orchestrate sequence context, chromatin topology, and regulatory mark co-dependencies. The self-supervised and foundation-style architectures are purported to use large unlabeled corpora to extract transferable embeddings across species, across breeds, developmental stages, and environmental gradients, and to address data scarcity in annotated trait-linked cohorts. Graph-based and geometric deep learning techniques have the capacity to capture spatial chromatin interaction networks, regulatory element promoter enhancer connectivity, and continuous time and temporal models can separate stable regulatory remodeling of cells from acute changes. A combination of epigenomic layers with genomics, transcriptomics, proteomics, metabolomics, microbiome profiles, physiological sensor information, and environmental telemetry grounds the regulatory state in the context of the whole organism and production system related to the prediction of emergent phenotypes, as well as leverage points helping to predict the value of precision nutrition, breeding prioritization, and adaptive management.

Causal discovery models and counterfactual simulation would be needed to either differentiate genetic and environmental effects mediated by epigenetic variation or merely label upstream drivers. Known chromatin dynamics, enzyme kinetics, and transcriptional feedback motifs can be encoded within hybrid mechanistic data-motivated models, and machine learning components can be used to learn the residual context specificity. Generative modeling and probabilistic programming make interpretation further, into the analysis of scenarios, the assessment of possible control implications of management changes, or selection under conditions of uncertainty. Dimensionality and sparsity issues are exacerbated in single-cell and spatially resolved assays, but open prospects can be identified in defining cell lineage-specific regulatory plasticity and microenvironmental heterogeneity to require scalable imputation, denoising, and multi-resolution fusion methods based on biological plausibility.

Technical innovation is connected with governance and ethical considerations, including equitable representation of different breeds and production systems, transparency in feature attribution based on which decision recommendations are made, privacy-preserving utilization of sensitive longitudinal data, responsible reflection of novel modalities of epigenome editing, and energy-efficient computational practices in line with the sustainability imperative. Trustworthy deployment has its support in the standardization of metadata, ontological harmonization of regulatory feature definitions, rigorous benchmarking, and infrastructures of reproducibility. This chapter places AI not just as a teaching speeding up factor but as an organization's intelligent layer between molecular regulation and the agribusiness system, being stewards. It charts out conceptual underpinnings, analytical progress, integration, governance architectures, translational pathways, and future innovation vistas that are necessitated to transform the complexity of the epigenome into ethics-based, sustainability, and precision-oriented decision intelligence.

Artificial Intelligence in Livestock Omics. https://doi.org/10.1016/B978-0-443-45663-3.00005-1

10.1 Introduction to AI in livestock epigenomics

The merging of AI and livestock epigenomics is an indicator of a convergence between dynamic regulatory biology and the complex goals of contemporary production systems, which aim to achieve concomitant benefits in productivity, resilience, welfare, product quality, and environmental stewardship. Epigenetic regulation facilitates the conditional translation of genomic potential to realized phenotype by conditionally modulating chromatin state, DNA methylation landscapes, patterns of histone modification, higher-order genome architecture, noncoding RNA-mediated regulation, and the positioning of nucleosomes and other layers of structural and biochemical control. These are inherently plastic regulatory processes, cell type sensitive, time sensitive, and responsive to nutritional, microbial, climatic, developmental, and management-related stimuli that define heterogeneous husbandry environments. The representational, inferential, and optimization capabilities provided by AI are well adapted to the high-dimensional, sparse, multimodal, and irregular time series that occur once these epigenomic layers are profiled across tissues, developmental stages, physiological conditions, and environmental gradients. An attained synthesis of analysis places the epigenomic variation not only in the role of a documented correlate of the phenotype but in the informational role of an interface that can be manipulated and, in certain circumstances, manipulated to interface genotype, the environment, and management intervention (Sindhu et al., 2024; Zhu et al., 2025).

The motivational strategies of promoting AI have opened up epigenomic analysis across intersecting fields. The goal of productivity is determined by the identification that regulatory plasticity affects metabolic allocation, nutrient use efficiency, development pathway fine-tuning, and physiological homeostasis strength. Resilience priorities put more weight on the anticipatory determination of the regulatory remodeling that relates to environmental stressor adaptation, microbial ecological variation, or immune challenge, with the view of predicting transition points prior to performance deterioration becoming apparent. The regulation influences inflammatory tone, stress axis modulation, barrier integrity, and neuroendocrine signaling, which are potential sources of biomarkers and decision support substrates that go beyond the superficial level of observational assessment. The sustainability imperatives will include the regulatory control of pathways associated with emissions, efficiency of resource conversion, and circular bioeconomy valorization of residual substrates, which is the alignment of epigenomic intelligence with environmental responsibility frameworks. Precision breeding methods are also progressively attempting to combine regulatory state heritability, genotype by epigenome interaction, and stability of beneficial regulatory states, and complement traditional selection indices with epigenetically informed predictors without losing awareness of the reversibility of some of these marks over time and contingency to the environment. These motivations are orchestrated by AI, which relates disparate signals into consistent latent state descriptors that assist in multi-objective trade-off navigation with uncertainty constraints (Alsayaydeh et al., 2025; Seebacher and Little, 2025).

It is necessary to define the scope and delimitations to avoid conceptual confusion. Epigenomics, in this regard, refers to the overall, genome-wide description of chemical and structural changes or modulatory characteristics of chromatin accessibility, transcriptional potential, and regulatory element interaction topology without any change in the underlying DNA sequence. It is contrasted with genetics, which deals with variation of sequences that are inherited, and with transcriptomics, which measures actual RNA expression outputs that are the consequence of regulatory control layers, but which in turn are the result of network dynamics. It is also in contrast to proteomics and metabolomics, which are indicators of, respectively, enzyme abundance and biochemical flux endpoints that are affected by transcriptional and post-transcriptional regulation. Although integration amongst these modalities is analytically beneficial, conceptual differentiation ensures that modeling assumptions on causality, time sequence, and stability are proper. AI architectures should acknowledge the fact that some regulatory marks are context-dependent in their transience, others confer semi-stable developmental memory, and some express composite signals of cell composition as opposed to intrinsic per-cell state. The data preprocessing, normalization, latent representation, and causal inference design are informed by the distinction between cell intrinsic regulation and mixture confounding, between primary and secondary mediating mechanisms, and between data preprocessing and causal inference design (Dibaeinia et al., 2025; Wu et al., 2025a).

The shift from cataloging regulatory marks descriptively to predictive and actionable intelligence is a main theme. The initial paradigms of analysis focused on the presence or intensity of differential marks across conditions, without a systematic combination of temporal dynamics, chromatin architecture, multi-layered regulatory-interaction dependencies, or downstream phenotypic relationships in the face of intervention conditions. AI accelerates the time-consuming comparison of statistics and the maturation of latent vectors to the process where the co-variation of marks, genomic background, sequence pattern, and three-dimensional contact likelihoods and environmental annotation are condensed into vectors to preserve the mechanistic interpretability of enhancer promoter communication, insulator activity, and coordinate module interaction. Temporal and continuous time modeling separates long-lasting regulatory remodeling and oscillatory or stochastic variation, which allows predicting an imminent shift in the transcriptional regime. Causal discovery models, which include invariance principles and counterfactual inferences, attempt to distinguish between mediating regulatory interactions and epiphenomenal relationships that are caused by a common upstream factor like genotype or environmental perturbation. Hybrid mechanistic machine learning methods incorporate biochemical plausibility constraints and chromatin physics priors in the differentiable architectures to prevent overfitting and limit exposure to spurious shortcut learning. Predictive and causal structure can then be converted into intervention policy advice by optimization and decision layers, either to inform adaptive management, direct nutritional modulation, or selection prioritization, within the constraints of articulated welfare, safety, and sustainability constraints (see Fig. 10.1) (Hegde et al., 2025; Tejada-Lapuerta et al., 2025).

In this chapter, a linear analytical approach is developed, which connects the basic ideas to translational implementation. The exposition is guided by the following questions; how can the regulation biological signal be preserved by the latent representations and confounding by the cell mixture and compositional artifacts mitigated; how can the dynamic models be used to decouple reversible environmental responsiveness and stable developmental reprogramming; how can causal inference could also be used to distinguish between mediation and correlation in the context of interwoven genetic, microbial, nutritional, and climatic effects and how can equitable access, reproducibility, transparency and frugally computationally forms be instantiated at the structure of the chapter starts with the conceptual grounding and data architecture, advances through the idea of representation learning, dynamic and causal modeling, integrative multi-omics fusion, translational decision support, governance and ethical stewardship and forward innovation perspective including single cell and spatial integration, foundation models, and adaptive personalization. AI is therefore being placed in all these areas not as computational augmentation but as an integrated computational framework that aligns the plasticity and complexity of epigenetic regulation with the pragmatic exigency and ethical responsibilities of modern livestock production systems, which makes a possible transition to proactively directed, resilient, and sustainable regulatory phenotype management based on scientifically sound, transparent, and context-aware inferences.

10.2 Overview of epigenomics in livestock

Epigenomics in livestock refers to the interactions of biochemical and structural regulation that determine the accessibility, three-dimensional organization, and utilization of the genome without changing its nucleotide sequence, which, in turn, determine transcriptional potential and mediate translation of the genetic background and environmental context into observable phenotypes of economic, welfare, and sustainability interest. These core mechanisms include covalent DNA methylation concentrated around cytosine, a large repertoire of post-translational histone modifications deposited or erased by groups of enzymes that are context-dependent, chromatin accessibility modulation by nucleosome remodeling and higher-order fiber compaction, and a variety of non-coding RNA classes that direct chromatin modifiers, promote transcript stability, or scaffold regulatory complexes. DNA methylation in livestock represents vertebrate paradigms of regulatory region marking, gene body marking, and repetitive element marking, which may affect transcription initiation, exon alternative use, and transposable element silencing, and is involved in imprinting and developmental lineage choices. Combinatorial signatures of active promoters, poised enhancers, distal regulatory priming or facultative repression, and

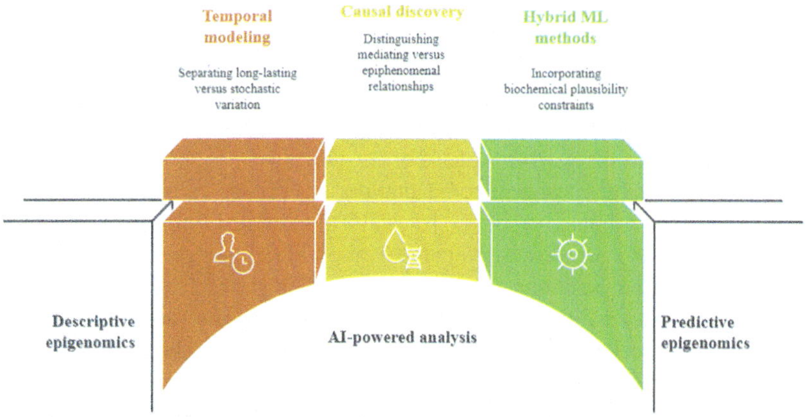

FIGURE 10.1 AI-driven epigenomic manipulation for livestock improvement. This conceptual framework illustrates how AI enables a transition from descriptive epigenomics, focused on cataloging regulatory marks, to predictive epigenomics that supports intervention-aware decision-making in livestock systems. Temporal modeling distinguishes stable regulatory programming from stochastic variation, causal discovery separates mediating epigenetic mechanisms from epiphenomenal associations, and hybrid machine learning methods integrate biological plausibility constraints into data-driven models. In general, these AI capabilities form an integrative analytical layer that links dynamic epigenomic regulation with phenotypic outcomes, supporting precision breeding, adaptive management, and ethically grounded livestock improvement.

histone tail acetylation, methylation, phosphorylation, ubiquitylation, and others are all marked by various combinations of histone tail modifications, which incorporate metabolic state information connecting nutrient availability to chromatin state. The accessibility of chromatin indicates the interplay of nucleoside occupancy, pioneer binding of transcription factors, ATP-dependent remodeler activity, and higher-order architectural constraints, which make accessible regions preferred targets of regulatory protein engagement. Non-coding RNAs, which span long intergenic transcripts, antisense species, enhancer RNAs, small interfering species, and microRNA precursors, are implicated in both cis and trans regulatory loops, and regulate DNA methylation establishment as well as transcript abundance and translation in cell-type-specific ways (Sindhu et al., 2024; George et al., 2025).

The three-dimensional genome structure provides further expansion of the linear regulatory vocabulary by juxtaposing regulatory components (promoters, enhancers, insulators, and silencers) spatially via looping, compartmental segregation, and connectivity of domain structure, which builds the probability of contact landscapes that determine transcriptional bursting and regulatory resilience. The structural dynamics of architectural proteins and chromatin extrusion define the boundaries of insulated neighborhoods, limiting enhancer promiscuity and limiting the propagation of repressive neighborhoods, and compartmental segregation between broadly active and repressed genomic neighborhoods. In livestock species, these structural rules are found to apply to the regulation of traits, as observed when non-coding genetic variants, epigenetic remodeling events, and environmental perturbations tend to have a phenotypic effect by increasing or decreasing or redirecting long-range regulatory contacts instead of changing proximal promoter sequences. Therefore, three-dimensional topology is essential to integrative reflection, both to understand distal epigenetic signatures and to define predictive models based on mechanistic faithfulness that combine sequence affinity motifs, chromatin condition, and spatial vicinity to create estimators of coherent control (Ambati and Choudhary, 2025).

It has been demonstrated that developmental programming brings about a temporal stratification, with epigenetic landscapes formed, remodeled, and reconfigured during embryogenesis, gametogenesis, imprint establishment, neonatal adaptation, and late development. Methylation erasure and reinstatement waves, reorganization of chromatin accessibility, histone mark redistribution, and sequential insulation of domain architecture are together matrices of defining lineages and organogenesis and mediate maternal, paternal, and early environmental signals. An example of phenomena induced by imprinting is the imprinting specificity of methylation maintenance that confers allelic bias in gene expression, with possible effects on the development of growth and resource allocation pathways. Developmental windows postnatally are characterized by increased plasticity, whereby nutritional composition, microbial infection, thermal exposures, as well as handling practices can tune the stabilization of regulatory networks and metabolic partitioning, immune preparedness, stress axis calibration, and transmit early life history cues in persistent phenotypic trajectories with epigenetic memory processes that may remain semi-stable but remain resistant to further remodeling under sustained stimuli (Edwards-Lee et al., 2025).

Nutritional, environmental, and management factors mediate the epigenomic regulation by providing metabolites as cofactors of methyltransferases, demethylases, acetyltransferases, and deacetylases; by oxidative and inflammatory conditions that distort the activity of chromatin-modifying enzymes; and by circadian, photoperiodic, thermal, housing density, and handling factors that alter endocrine and autonomic signaling with subsequent epigenetic effects. The interplay between one-carbon metabolism and acetyl coenzyme supply and macro and micronutrient equilibrium, as well as feed additive regimes and feeding regimes, influences the capacity to donate methyl groups and histone acetylation processes. Epigenetic enzyme fidelity can be altered by environmental agents or mycotoxins or triggered to respond to DNA damage, remodeling the chromatin landscapes. The action of housing and management of stress gradients through glucocorticoid actions is capable of triggering chromatin modification in neural, immune, and metabolic tissues, restructuring the responsiveness thresholds and adaptive capacity. The net effect of these multifactorial effects is the generation of context-dependent epigenomic signatures, which may reflect as well as potentially mediate phenotypic adaptations and make the task of causal attribution more complex, whilst providing levers of precision management based on regulatory state surveillance (Bernasocchi and Mostoslavsky, 2024; Perkasa et al., 2025).

Stress, disease, and plastic adaptive mechanisms also involve epigenetic adaptation, as immune cells, barrier organs, and metabolic organs reorganize regulatory networks to coordinate responses, re-establish tolerance limits, and distribute energy resources. The signaling cascades of inflammation trigger transcriptional activation of defense genes by recruiting chromatin remodelers to enhancer repertoires and to poised positions for secondary challenge acceleration. Pathogen-associated effectors may take over host epigenetic processes, re-directing transcriptional products, and host defense can include epigenetic silencing of invasion-promoting receptors or metabolic rerouting genes. Sustained exposures to chronic stress can entrench maladaptive regulatory phenotypes, which undermine welfare, productivity, and disease resistance by lasting rearrangement of gene networks that regulate neuroendocrine feedback or immune homeostasis. On the other hand, employing epigenetic training phenomena to prime protective responses, controlled low-level stressors, or vaccination strategies, the two-way nature of epigenetic plasticity in terms of vulnerability or resilience can be seen (Caldwell and Li, 2024).

Traits of economic significance, including growth dynamics, feed efficiency, reproductive performance, fertility, lactation characteristics, carcass structure, immune capacity, and product quality, are affected by epigenetic input that alters gene networks that mediate nutrient homeostasis, hormonal regulation, gametogenic capability, mammary differentiation, muscle fiber specification, adipocyte metabolism, diversification of immune repertoire, and oxidative stability of products. Epigenetic marks may provide environmentally responsive regulatory states between fixed genomic variation and organized trait expression, intermediate molecular phenotypes between fixed genomic variation and complex trait expression. The epigenetic integrity of germ cells and early embryos in the context of fertility determines developmental competency and offspring vigor, and chromatin accessibility to metabolic regulators and epigenetic evidence of mitochondrial biogenesis in the context of growth and efficiency associations to energetic partitioning. The quality features of the products can capture post-transcriptional regulatory cascades, which are triggered by chromatin-mediated transcriptional priming of lipid metabolism, protein turnover, and antioxidant pathways. To transform such associations into breeding or management decisions, a highly conservative disentangling of cause and consequence and correlation is necessary, keeping in mind the temporal stability and heritable and transient nature of observed patterns of epigenomic variation (Seem et al., 2024; Sindhu et al., 2024).

Host genetics, epigenetics, and gene expression interact in a multilayered causal lattice in which sequence variants have an effect either by direct coding changes, altered binding motifs of transcription factor binding sites, alteration of insulator elements, or by altering the recruitment site of chromatin remodelers or by acting on non-coding RNA loci. The quantitative trait loci of expression, the quantitative trait loci of methylation, the quantitative trait loci of histone modification, the quantitative trait loci of chromatin accessibility, and the allele-specific binding or methylation patterns are all mapping the points of genetic control that define regulatory landscapes that determine transcriptional outputs. Mediation analyses aim to assess the statistical and mechanistic transmission of genetic variants to either expression or phenotype by epigenetic variation, which can be applied to identify regulatory intermediates that can be modulated by management. On the other hand, the induced epigenetic changes by the environment can adjust the penetrance of genetic predispositions, which adds to the complexity of genotype-by-environment interaction. The data required to understand the directional flow is between genotype, epigenetic state, and expression. Joint modeling frameworks are required that include linkage disequilibrium, co-regulatory network topology, cell type composition, and time-dependent sampling to alleviate confounding and increase causal resolution (Stefansson et al., 2024; Aguirre et al., 2025).

Further complexity is brought by tissue specificity and cell type heterogeneity due to the fact that epigenomic landscapes are exquisitely biased to cellular identity, metabolic need, developmental lineage history, and microenvironment. Bulk tissue profiling mixes signals between different cell populations, which might obscure cell-restricted regulatory events, or may produce apparent different patterns due to compositional changes caused by compositional remodeling instead of inherent remodeling. This heterogeneity is superimposed with temporal dynamics, with cyclical physiological dynamics, seasonal changes, transitions in reproductive stages, and developmental changes, providing time-varying regulatory reconfiguration. To measure these dynamics, one will need a careful sampling design, a longitudinal cohort structure, as well as statistical deconvolution approaches that have the potential to disaggregate cell-intrinsic change and cellular admixture variation. Spatially resolved and single-cell data can be refined, but must address sparsity and noise issues. Useful imputation, aggregation, and cross-modality alignment techniques are essential to derive the patterns of interest that are biologically relevant and cannot create artifacts (Wang et al., 2024).

Regulatory architecture can be provided through complementary windows that are presented by standard experimental assays that form the basis of livestock epigenomics. Whole-genome bisulfite sequencing offers base-resolution methylation maps at the cost of high sequencing coverage requirements and conversion-related biases, which must be firmly checked. The reduced representation of bisulfite sequencing is cost-effective by enriching CpG-dense regions at the expense of overall homogeneity of the genome, and may also favor promoter-rich environments. Chromatin immunoprecipitation sequencing is a method of profiling histone modifications or transcription factor binding based on antibodies being able to enrich by antibody-specific binding, which requires appropriate quantities of input material. The transposase-accessible chromatin sequencing assays utilize open chromatin capture through integration events to give high-resolution accessibility footprints at a relatively low input cost. New enzyme tethering technology, including CUT and Tag and similar technology, scales up signal to noise and lowers input requirements to profile histone modifications, and makes many more limited or valuable tissues of livestock possible. Hi-C chromatin conformation assays using adaptations to derivative deflections and protein-centric conformation protocols are used to map frequencies of interactions between target promoters and distal regulatory elements and domain structures. Single-cell solutions to both accessibility and methylomics profiling offer accessibility and methylome resolution on single cells, and nascent single-cell multimodal solutions seek to resolve accessibility, methylations, and transcript production concomitantly. Both assays have quantitative, technical, and batch bias that must be corrected by normalization, batch effects, and integrative modeling that is sensitive to the quality of livestock-specific genome assembly, repetitive genome content, and structural variation (Cao et al., 2025; Tian et al., 2026).

The information repositories, annotation sources, and ontologies provide the background information needed to interpret the livestock epigenomic data sets, as well as to allow cross-study synthesis. Raw and processed sequencing data are available in public archives under an expanding range of tissue types and developmental stages and in a variety of environmental conditions and are accompanied by additional efforts at providing functional annotation to define reference sets of regulatory elements, chromatin states, and transcript isoform repertoires. Genome browsers and integrative portals facilitate visualization and integrative query layers between variations of sequences, regulatory marks, and gene models. Ontological schemas put harmonic relationships among tissue, cell type, stage of development, experimental condition, and trait metadata to help in semantic interoperability, discovery of machine-readable datasets, and reporting normalizations. Cross-species comparative annotation programs are based on the evolutionary conservation and synteny to extrapolate regulatory annotations of well-studied model organisms to livestock genomes, to guide candidate regulatory element prioritization, and also to point to lineage-specific innovations potentially involved in species-specific production traits. Bringing together standards on metadata harmonization forms the basis of reproducibility and enables AI models to consume heterogeneous data with reduced confounding due to terminological inconsistency (Meng et al., 2025; Zhou et al., 2025).

Some limitations moderate the existing scene, which affect the study design, interpretation, and translation in another group. The limitations of resources and costs limit sample sizes, the depth of time, and the concurrence of multidimensional profiling and replication, which may constrain statistical power and representativeness of the population. Poor coverage and signal drop-off are a problem for single-cell and low-input assay reliability, requiring imputation methods that may add synthetic structure without rigorously verified methods. The lack of functional annotation of non-coding regions, enhancers, silencers, insulators, and non-coding RNA loci in livestock species makes assigning regulatory functions with confidence difficult, whereas cross-species projection results in false classification of lineage-specific elements (Bi and Weng, 2025; Cao et al., 2025). Any technical variability across library preparation, antibody specificity, enzyme batch behavior, sequencing platform heterogeneity, and bioinformatics pipeline heterogeneity can cause batch effects and biases, which in turn confound biological inference without a sufficient model. Cellular heterogeneity generally confuses all bulk analyses, and deconvolution models rely on reference atlases available to date only in a small fraction of tissues and physiological conditions. Temporal dynamics are typically not well sampled, which obscures short-lived yet functional critical regulatory changes. Causal inferences are still complicated by pervasive confounding, reverse causation, and scarcity of longitudinal interventional data. Functional equivalence of varying combinations of histone modifications can be a complicated interpretation in context-dependent or bi-directional effects of modifying co-regulatory pleiotropic enzymes. The heritability and stability of epigenomic modifications induced by the environment across generations are not fully determined, and this casts doubt on assumptions of transgenerational stability. Sustainability issues are encountered with the computational requirements and use of energy in large-scale integrative modeling, and data sharing is restricted by privacy, proprietary breeding programs, and inequalities in the infrastructural capacity of different regions, which hinder equal access (Kelly et al., 2024; Ohishi and Au Yeung, 2025). All these restrictions underscore the need to develop methods, standardize, and organize consortia and integrative AI systems that explicitly consider the uncertainty, bias, and incompleteness, and attempt to bring the complexity of epigenomics into functional biological understanding and decision intelligence to livestock production systems. Main epigenomic mechanisms and functional roles in livestock are presented in Table 10.1.

10.3 Role of AI in epigenomic data analysis

AI is increasingly playing a central role in the analysis of livestock epigenomic data by providing algorithmic approaches to convert heterogeneous, noisy, sparsely sampled, and partially annotated measurements of molecular quantities into consistent regulatory inferential beliefs in directions consistent with the goals of complex trait improvement, welfare assurance, and sustainability. The unique limitations of livestock research, such as relatively smaller publicly available cohorts, breed stratification, changes in management regimes, and compromises to experimental depth imposed by costs, increase the applicability of adaptive, data-efficient, and uncertainty-aware machine learning paradigms. The shift to translational decision support, causal discovery, predictive modeling, and representation learning by means of layered integration through raw signal processing, representation learning, causal discovery, and predictive modeling makes AI accessible to the shift in the scholarly literature between descriptive catalogs of chromatin marks and

TABLE 10.1 Main epigenomic mechanisms and functional roles in livestock.

Epigenomic mechanism	Core molecular or structural basis	Primary regulatory functions	Functional roles in livestock biology	AI-enabled analytical and modeling roles
DNA methylation	• Covalent methylation of cytosine residues (primarily CpG) • Enzymatically written, maintained, and erased • Semi-stable yet environmentally responsive	• Transcriptional repression or modulation • Genomic stability via transposon silencing • Allele-specific and developmental regulation	• Developmental programming of growth and metabolism • Fertility, embryonic viability, and imprint control • Stress and immune responsiveness	• Longitudinal AI models separating stable versus plastic methylation • Causal mediation between genotype, environment, and phenotype • Prediction of regulatory memory and resilience
Histone modifications	• Post-translational histone tail modifications (acetylation, methylation, etc.) • Combinatorial and context-dependent patterns	• Activation, repression, or poising of regulatory regions • Integration of metabolic state with transcription	• Muscle growth, carcass composition, and lactation • Immune activation and inflammatory regulation	• Deep learning chromatin-state segmentation • Multitask modeling across tissues and conditions • Explainable AI for pathway-level interpretation
Chromatin accessibility	• Nucleosome displacement and remodeling • Pioneer factor–driven chromatin opening	• Defines regulatory element availability • Enables rapid transcriptional response	• Acute adaptation to nutrition, stress, and disease • Immune cell activation and tissue remodeling	• AI-assisted peak calling and denoising • Temporal modeling of regulatory responsiveness • Sparse-data imputation frameworks
Three-dimensional genome architecture	• Chromatin loops, compartments, and TADs • Architectural protein-mediated spatial folding	• Long-range enhancer–promoter communication • Insulation of regulatory neighborhoods	• Interpretation of distal non-coding variation • Coordination of complex production traits	• Graph and geometric deep learning • AI-based enhancer–gene linking • Spatially informed causal modeling
Non-coding RNA regulation	• lncRNAs, miRNAs, eRNAs, and small RNAs • Cis- and trans-acting regulatory roles	• Recruitment of chromatin modifiers • Post-transcriptional gene regulation • Network stabilization	• Fertility, immune competence, metabolic regulation • Stress and inflammatory fine-tuning	• Network inference and regulatory graph learning • Multilayer integration with epigenome and transcriptome
Nucleosome positioning	• Ordered nucleosome spacing along DNA • Sequence- and remodeler-dependent organization	• Modulation of transcription factor binding • Control of transcriptional noise	• Developmental stability and tissue differentiation • Stress-induced regulatory variability	• High-resolution chromatin reconstruction • Modeling transcriptional robustness and variability

Continued

TABLE 10.1 Main epigenomic mechanisms and functional roles in livestock.—cont'd

Epigenomic mechanism	Core molecular or structural basis	Primary regulatory functions	Functional roles in livestock biology	AI-enabled analytical and modeling roles
Genomic imprinting	• Parent-of-origin-specific DNA methylation marks • Maintained across development	• Monoallelic gene expression control • Dosage regulation during development	• Fetal growth, placental function, fertility • Resource allocation and offspring vigor	• AI detection of allele-specific methylation patterns • Causal inference for parent-of-origin effects • Integration into breeding and selection models
Chromatin remodeling	• ATP-dependent chromatin remodeling complexes • Dynamic nucleosome repositioning	• Rapid alteration of chromatin accessibility • Regulatory flexibility under stimuli	• Adaptive responses to nutrition, stress, and infection • Tissue remodeling and immune readiness	• Dynamic AI models capturing regulatory transitions • Integration of accessibility and histone data
Transgenerational epigenetic inheritance	• Partial persistence of epigenetic marks across generations • Incomplete erasure during germline reprogramming	• Transmission of regulatory memory • Modulation of offspring phenotypes	• Intergenerational effects of nutrition and stress • Long-term resilience or vulnerability patterns	• Long-horizon AI modeling across generations • Uncertainty-aware inference of inheritance stability
Developmental epigenetic programming	• Epigenomic reconfiguration during critical windows • Interaction of nutrition, environment, and genome	• Establishment of long-term regulatory trajectories • Canalization of phenotype	• Early-life nutrition effects on growth, immunity, and reproduction • Lifelong performance shaping	• Temporal and causal AI models identifying critical windows • Policy optimization for early-life interventions
Cell-type–specific epigenomic regulation	• Distinct epigenetic landscapes across cell types • Masked in bulk-tissue measurements	• Cell identity maintenance • Context-specific gene regulation	• Immune heterogeneity, tissue function, disease resistance	• Single-cell and deconvolution AI frameworks • Graph-based cell-state modeling
Epigenetic plasticity and reversibility	• Context-dependent gain or loss of regulatory marks • Interaction with environmental stimuli	• Adaptive versus maladaptive regulatory shifts • Phenotypic flexibility	• Climate resilience, nutritional adaptation, and welfare	• Reinforcement learning and counterfactual simulation • Prediction of reversible versus fixed states

functionally interpretable, temporally resolved, and mechanistically grounded regulatory intelligence (Moran and Aragam, 2025).

Automated quality control is the root protection against the spread of technical artifacts into the inference of downstream data. Traditional read depth, duplication rate, conversion efficiency, or fragment size distribution threshold heuristics may be enhanced by supervised and self-supervised models that are trained to know normative multi-variable associations among sequencing statistics, and signal anomalous libraries containing degraded, contaminated, or procedurally unconventional libraries. Denoising autoencoders and diffusion-based generative architectures can be used to recover latent clean signal representations of corrupted representations, reducing batch-specific spectral noise and maintaining biological variance. Convolutional models, when used on genome-wide coverage profiles, distinguish real peaks from systemic bias of GC composition, mappability difference, or biases in enzyme cut sites. The detection of outliers using a model based on the strong estimators of covariance and deep density scoring detects the samples that are not within the manifold structure characterized by high-quality references, which aids in automated quarantine and remediation through iterative methods. These pipelines reduce feedback loops in data production in consortia and supply quality scores that can be weighted down to a set of calibrated scores instead of binary inclusion-exclusion choices, and to extract as much information as possible with limited resources (Chen, 2025; Zhuang et al., 2025).

Improvement of peak calling and signal segmentation by deep learning overcomes the shortcomings of rule-based enrichment detection, which requires arbitrary window sizes and parametric background assumptions. Neural sequence signal hybrid models may be fed raw coverage tracks, nucleotide sequence, and predicted nucleosome occupancy and generate probabilistic occupancy maps defining promoters, enhancers, insulators, and repressed domains, and refining the delineation of boundaries. Attention mechanisms allow context-aware resolution of ambiguous areas where overlapping regulatory programs interact, and multitask formulations learn mark-specific and shared latent factors together, and based on this knowledge are able to generalize to marks profiled at a lower depth. Weak supervision helps alleviate the paucity of exhaustive manual annotations due to the patterns of cross-mark co-localization and orthology projected labels, and avoids depending on manually curated training sets. Calibration layers encode raw logits to well-calibrated posterior probabilities that allow for a principled choice of threshold based on downstream false discovery cost functions, which are useful in trait association studies. Such adaptive segmentation models encourage batch and breed consistency in livestock contexts, where the noise is increased by both antibody specificity variability and variable sample heterogeneity in the fields, leading to tissue heterogeneity (Dukić, 2025).

Imputation strategies allow for the compensation of missing marks, shallow coverage, and incomplete tissue by condition matrices that are associated with cost-constrained experimental designs. Matrix completion, tensor factorization, variational inference, and graph neural networks use a common structure between marks, genomic loci, cell types, and developmental stages to make predictions on unobserved signals, as well as to estimate epistemic and aleatoric uncertainty. The sequence-informed embeddings combine motif grammar and evolutionary conservation, to which they give priors in low data regimes, and cross-species transfer regularizes latent spaces. Diffusion and autoregressive generative models predict plausible high-resolution profiles conditioned on sparse observations, which are used to augment downstream supervised tasks with augmented training corpora. Of particular importance, livestock applications have to protect against the overconfident hallucination of control characteristics, which may lead to causal prioritization; hence, imputation outputs have voxel or locus-specific credibility intervals and abstention releases that shelter inferences in out-of-distribution situations, which preserve interpretive integrity (Skok Gibbs et al., 2024).

Multi-omics integration provides an opportunity to integrate genome, epigenome, transcriptome, proteome, metabolome, microbiome, and phenomic layers into a joint representation, including hierarchical information flow (see Table 10.2). Cross-modality alignment uses canonical correlation neural networks, contrastive embedding, and manifold alignment to transform feature spaces that are heterogeneous into feature shared sameness latent coordinates and to learn modality-specific noise. Graph-based models depict genes, regulatory factors, metabolites, proteins, and traits as nodes connected by experimentally designed or computationally determined edges to facilitate message passing designs to spread evidence and embed node representations. Bayesian and probabilistic graphical models have hierarchies, which encode the conditional dependencies and allow explicit propagation of uncertainty between layers, which enables the ability to reason about causal relationships. Designing integrations should consider the temporal resolution differences as well as sampling latency differences, such as between fast-reacting chromatin modifications and slower proteomic changes, by adding dynamic alignment or time warping mechanisms. The issue of livestock-specific annotation sparsity is addressed by semi-supervised training, which exploits partially labeled nodes, with distant supervision based on orthologous networks, which enhances the robustness of trait predictive signatures based on composite molecular encodings (Dou et al., 2025).

TABLE 10.2 Multi-omics data integration frameworks in livestock epigenomics and AI.

Integration framework	Omics layers integrated	Core computational strategy	Biological and functional objectives	AI-specific advantages and use cases
Early (feature-level) integration	• Genomics (variants, breeding values) • Epigenomics (methylation, histone modifications, and accessibility) • Transcriptomics	• Direct concatenation of normalized features • Dimensionality reduction or regularization applied post-merging	• Capture combined molecular signatures influencing phenotype • Identify correlated regulatory patterns across layers	• Simplicity and computational efficiency • Suitable for smaller datasets common in livestock studies • Enables classical ML and interpretable models
Intermediate (representation-level) integration	• Genomics • Epigenomics • Transcriptomics • Proteomics/metabolomics	• Separate modality-specific encoders • Shared latent space via deep learning or factor models	• Learn biologically meaningful latent regulatory states • Reduce noise and modality-specific bias	• Robust to missing data across omics layers • Enables transfer learning and cross-breed generalization • Supports explainable latent factor analysis
Late (decision-level) integration	• Any combination of omics and phenomics • Environmental and sensor data	• Independent models per modality • Ensemble or weighted decision fusion	• Combine complementary predictive signals • Preserve modality-specific interpretability	• Flexible inclusion or exclusion of data streams • Improved robustness under heterogeneous data availability • Facilitates uncertainty-aware ensemble prediction
Graph-based multi-omics integration	• Genomics • Epigenomics • Transcriptomics • Proteomics • Metabolomics	• Construction of biological graphs (genes, enhancers, and metabolites) • Graph neural networks or probabilistic graphical models	• Model regulatory topology and interaction networks • Link distal epigenetic marks to target genes	• Captures non-linear and long-range dependencies • Aligns with 3D genome architecture and regulatory biology • Enables causal pathway discovery
Hierarchical/Bayesian integration	• Genomics • Epigenomics • Transcriptomics • Phenotypes	• Probabilistic hierarchical models • Explicit uncertainty propagation across layers	• Disentangle direct, indirect, and mediated effects • Quantify confidence in regulatory inferences	• Well-suited for sparse and heterogeneous livestock data • Supports causal inference and policy evaluation • Transparent uncertainty quantification
Temporal multi-omics integration	• Epigenomics (dynamic marks) • Transcriptomics • Metabolomics • Longitudinal phenotypes	• Time-series and state-space models • Neural ODEs, transformers, or Gaussian processes	• Capture regulatory dynamics across development and stress • Distinguish transient from stable regulatory changes	• Enables forecasting of phenotypic trajectories • Identifies critical windows for intervention • Supports adaptive management strategies

Causal multi-omics integration	• Genomics (instrumental variables) • Epigenomics • Transcriptomics • Environmental data	• Structural causal models • Mediation analysis and counterfactual simulation	• Identify epigenetic mechanisms mediating genotype-phenotype links • Separate causation from correlation	• Essential for intervention-aware decision intelligence • Reduces spurious associations • Guides precision nutrition and breeding
Cross-species / transfer-learning integration	• Livestock omics • Model organism or human reference data	• Domain adaptation and transfer learning • Orthology-aware representation alignment	• Leverage conserved regulatory grammar • Compensate for limited livestock annotations	• Improves performance in low-resource species or tissues • Accelerates functional annotation • Enhances generalizability
Digital twins-oriented integration	• Genomics • Epigenomics • Transcriptomics • Microbiome • Sensor and environmental data	• Hybrid mechanistic–AI models • Continuous state updating and simulation	• Simulate individualized regulatory and phenotypic trajectories • Evaluate intervention scenarios	• Enables precision livestock management • Supports counterfactual reasoning under uncertainty • Integrates welfare, productivity, and sustainability goals
Federated/Privacy-preserving integration	• Distributed multi-omics datasets across institutions	• Federated learning and secure aggregation • Privacy-preserving model updates	• Enable collaboration without raw data sharing • Preserve data ownership and confidentiality	• Expands training data without violating governance constraints • Improves model robustness and equity • Supports scalable multi-center AI deployment

The domain adaptation and cross-species knowledge transfer help to alleviate sample scarcity by taking advantage of synteny-conserved regulatory syntax and orthology. Transfer learning pipelines pre-train a deep sequence model to operate on deep sequences on large data sets of phylogenetically related organisms and then fine-tune on epigenomes of livestock of interest, thus acquiring motif recognition as well as context-dependent dependency modeling. Adversarial domain adaptation discriminators promote species-invariant latent representations but retain task-relevant structure, which reduces covariate shift-induced performance degradation. The meta-learning models maximize starting states to allow quick adaptation to new breeds, tissues, or environmental conditions with a small number of gradient steps, which is a problem of heterogeneity in management systems. Graph alignment techniques are added to ortholog mapping, and they take into account conserved network neighborhoods and not one-to-one functional projections, which improves the projection fidelity of functions. Domain generalization methods use environment embedding to enhance strong predictive tasks on unobserved regimes of management, which is a key feature that is required to deploy predictors into geographically varied production settings (Liu et al., 2026).

Regulatory variation is then operationalized by predictive modeling of complex traits with epigenetic features to explain phenotypic risk stratification or predict performance. Direct chromatin mark intensities aggregated over biologically specified regions, latent factors inferred using variational encoders, graph embeddings of network positionality, and interaction terms of enhancers-promoters coupling are used in feature construction. Ensemble techniques, which are a combination of gradient boosted decision trees, sparse linear models, and neural architectures, are used to achieve a trade-off between interpretability, the ability to capture nonlinearity, and overfitting, particularly in light samples. Methods of regularization, such as group lasso and hierarchical shrinkage, only limit the complexity of a model, but maintain sets of co-regulated regions. The combination of genomic breeding values and epigenetic predictors into multitask structures separates additive sequence actions of context-dependent regulatory modulation, which may furthermore lead to superiority of the selection indices when epigenetic variation is partially stable or is mediating the genotype-by-environment interaction. Performance assessment concentrates on cross-evaluation plans stratified by management cluster or breed to promote generalization, and calibration assessment is required to support probabilistically significant risk products on determination limits (Yu et al., 2025).

Causal inference of nutrient environment epigenome phenotype pathways requires modeling paradigms that go beyond correlation to determine mediating regulatory processes. Structural causal models represent hypothesized directional interactions between dietary factors, metabolic intermediates, epigenetic states, transcriptional outputs, and phenotypes, which can be used to query the possible effects of interventions through counterfactual queries. Instrumental variable designs employ genetically anchored quantitative trait loci of methylation or histone modification that can be used as a type of quasi-random perturbation to determine the mediation of genetic effects by epigenetic intermediates as well as confounding. The invariance-based discovery takes advantage of the fact that true causal mechanisms do not change between strata of the environment and blocks spurious association interactions caused by unobserved confounders. Propensity modeling and outcome regression are combined in targeted maximum likelihood estimation and doubly robust learning to result in unbiased estimates of the mediation effect under mild model misspecification. The isolation of directionality in longitudinal nutriepigenomic data is achieved using temporal causal discovery algorithms based on differences in differences abstractions and Granger causal arguments, strengthened by sparsity-inducing penalties. Validity requires intensive sensitivity tests that measure susceptibility to unmeasured confounding and consistency with biological plausibility limits obtained using the knowledge of enzymatic and metabolic pathways (Faraji and Metz, 2025).

Epigenetic trajectory temporal modeling captures active regulatory changes in development, physiological change, stress exposure, and recovery. Recurrent networks, time convolutional networks, transformers using relative positional encodings, neural versions of the ordinary differential equations of motion, and sequence models are all sequence models that learn latent dynamic states that control the evolution of observed marks. Change point detection algorithms detect sudden regulatory regime changes, which are a sign of key developmental windows, or oncoming dysregulation resulting from stress, such that preemptive changes in management can be made. State space models combine the partially observed linear or nonlinear dynamics with the decomposition of the noise in observations to generate the smooth trajectory estimates and predict the future states under the given environmental conditions. Multiresolution decomposition can work with nested periodicities in the form of circadian modulation over developmental trends. Gaussian process interpolation with covariance capturing kernels between developmental and environmental time series data counteracts the sparse time series sampling common to livestock research and is then fed into deep hybrid models. Temporal modeling can be used to augment causal inference by refining the temporal precedence, and it can optimally plan sampling design by using active learning based on time points that are most likely to yield the highest information gain (Tahir et al., 2024).

The AI techniques used to predict three-dimensional chromatin interactions and link enhancers to promoters simultaneously using sequence context, linear mark distributions, and structural priors. Siamese networks are used to encode loci sequences and epigenetic profiles, and then predict interaction scores by masking chromatin contacts before graph-based inference and chromatin is modeled as a dynamic graph with edges indicating a predicted or observed contact and node features are updated through message passing. A regularization of predictions to biophysically accessible topologies by the incorporation of polymer physics minimizes spurious long-range contact inference. Multiscale architectures facilitate the integration of fine-grained loop detection with the larger compartmental structure involving hierarchical pooling. Enhancer target gene association makes use of attention mechanisms assign contribution weights to candidate enhancers in a topological domain to promoter activity predictions, aiding in prioritizing effect sizes of variants in distal regulatory states. Cross-assay training that simultaneously take into account accessibility, histone modifications, and methylation enhances specificity since enhancers of interest by interaction tend to have characteristic combinatorial patterns. Dependable interaction maps enhance trait mapping by directing the attribution of non-coding genetic variations and epigenetic changes to probable effector genes (Wall et al., 2024).

The high percentage of livestock genomic space that has no experimentally validated functions is solved by functional annotation of non-coding regions, where AI-mediated prioritization is done. Multitask objective-trained sequence-to-function models on chromatin marks, accessibility, transcription factor binding, and conservation are then used to classify candidate enhancers, silencers, insulators, and promoters based on their embeddings. Zero-shot learning makes use of semantic attribute vectors, based on ontology-coded functional descriptors, to learn the roles of regions lacking explicitly labeled instances. Patterns of regulatory coactivation are modeled probabilistically using probabilistic topic models to identify regulatory modules within a particular physiological context, and add to the granularity of the annotation. Generative perturbation models are in silico sequence edits that evaluate the effect of the predicted influence on regulatory output, allowing scoring of variants. Predicted three-dimensional contacts coupled with gene expression quantitative trait associations enhance attributing functional links. The uncertainty in models, agreement among the ensemble of models, and distance to the training data manifold are combined in annotation confidence scoring to rank experimental validation and discourage the propagation of poor labels (Ma et al., 2025).

This use of machine learning to isolate a few, strong feature sets whose measurement can produce actionable predictive information is the basis of epigenetic biomarker discovery for early disease detection and performance prognostics. Biomarker robustness depends on sampling stability, constitutes feature importance evaluation through subsampling, and Bayesian sparsity priors that overcome overfitting and accentuate the identification of loci that remain consistently informative across cohorts. Multi-view feature selection algorithms combine complementary multi-view information of the methylation, accessibility, and histone landscapes, which identify convergent regulatory information and discourage redundant inclusion. Prognostic risk models that use time-varying epigenetic covariates can be used to estimate prognostic risk in longitudinal health or productivity outcomes. Domain adaptation is used to create transferable sets of biomarkers that have identical predictive calibration between breeds and in the management environment. At-scale deployment requires simplifying the assays; thus, model compression methods can reduce a large signature to specific panels with no difference. Translational feasibility is achieved through rigorous external validation, decision curve analysis, and cost-sensitive evaluation (Benčik et al., 2025).

With digital twins integration, there is a dynamic epigenetic layer of computational representations of individual animals synthesizing physiological, genomic, microbiome, environmental, and management streams of data to predict responses and optimize interventions. In such twins, mechanistically informed modules of differential equations of metabolic fluxes and endocrine regulation interact with data-informed epigenetic state evolution modules that continuously update gene-regulatory modules. The parameter identification combines real-time sensor and periodic molecular sampling, where Bayesian filtering is used to correct latent states and propagate the uncertainty. Counterfactual simulation studies the scenarios of nutritional or environmental intervention and estimates the trajectory changes in epigenetic mediation of growth, immune preparedness, or reproductive timing. Adaptive experimental design algorithms discover the presence of individual molecular measures on which the expected value of information is the greatest and apply this to reduce decision-critical uncertainty to inform sampling schedules that utilize welfare and cost limits. The epigenetic component is therefore a memory substrate that can be modified to facilitate individualized and anticipatory management in accordance with the multi-objective optimization based on the metrics of productivity, welfare, and sustainability (Karin et al., 2023; Hansen et al., 2024).

Quantification of uncertainty and reliability scoring is spread across all layers of analysis to avoid overconfident use of models that are trained on small and heterogeneous data. Different epistemic and aleatoric uncertainty is divided among Bayesian neural networks, deep ensembles, and evidential regression schemes, and used for selective prediction and abstention strategies. Calibration methods bring about the adjustment of probability outputs in order to include empirical frequencies, whereas conformal prediction generates prediction sets that are guaranteed to be covered by exchangeability criteria. Out-of-distribution detection based on density estimation in latent spaces, energy-based scoring, and reconstruction error profiling detects inputs unsupported by model training, at which model accuracy declines. Uncertainty propagation during data imputation and multi-omics integration should be explicitly incorporated into downstream causal inference and risk assessment. The transmission of uncertainty in easy but rigorous forms facilitates the trust of stakeholders and ethically based decision-making, especially where welfare implications are involved (Gillis et al., 2025).

Explainability frameworks render AI-derived epigenomic insights biologically interpretable in regulatory mechanisms, which allows biological refinement and accountability to the stakeholders. Such techniques as gradient-based saliency, integrated path, Shapley value approximations, and perturbation analysis are feature attribution methods that locate loci, marks, and interaction edges that have a significant effect on predictions. The concept bottleneck models directly use biologically relevant intermediate variables, like enhancer activation scores or domain boundary integrity, in order to make mechanistic arguments and generate specific hypotheses. Attention-weight visualization in transformer architectures gives insights into long-range dependencies used to predict traits, but causal graph extraction aligns learned conditional dependencies with established biological pathway knowledge to point at new edges to be validated. Distilled model representations can find compact sets of rules that can be used to approximate neural decision surfaces and fill the gap between opaque statistical associations and human-readable regulatory logic, using symbolic regression. The complexity is offset by a commitment to explainability, which allows for analyzing the risks of bias, promotes reproducibility, and facilitates compliance with regulations in data governance situations (Medapati, 2025). High-throughput epigenomic technologies and data outputs in livestock breeding are listed in Table 10.3.

10.4 Machine learning techniques for epigenomic insights

The epigenomics of livestock with machine learning algorithms are a stratified range of methodological strategies that range in complexity between the classical statistical learning, to the sophisticated deep generative and adaptive paradigms, which in turn introduce certain inductive biases and trade-offs to their operation and, in combination, offer regulatory information subject to the limitations of heterogeneous sampling, small samples relative to human consortia, stratification by breed, and selective exposures to environments. These methods are strategically chosen and combined to determine the fidelity with which latent regulatory frameworks are restored, functional non-coding parts ranked, causal pathways disentangled, and predictive models integrated into breeding and management decisions that align with the targets of welfare, productivity, and sustainability. A balanced approach to uncertainty, minimizing the chances of overconfidence when faced with significantly sparsely annotated or cross-study heterogeneous situations, with a degree of computational scaling and openness that can be tested biologically through iterating the research, has been identified as a common methodological requirement (Ghavi Hossein-Zadeh, 2025; Zhu et al., 2025).

The relevance of classical supervised learning is enduring because regularized regression, support vector machines, and ensemble tree models can offer insight into the data, high resistance to low sample and high feature regimes, and principled approaches for inducing feature sparsity. The domain knowledge is manifested as penalization schemes such as elastic net or structured group penalties, which exploit CpG clusters (co-localized histone) or co-localized enhancer promoter pairs that stabilize the prediction of the coefficients of multicollinearity inherent in the genomic proximity and coordinated regulatory efforts. The idea behind support vector machines is that the maximization of margin may be applied to classify regulatory states or phenotypic classes using the composite epigenetic signatures; kernel engineering can make use of sequence similarity, chromatin mark correlation structure, or 3D distance proxies. Nonlinear interactions between marks and sequence-derived features, such as conditional dependencies where, for example, a discriminatory role of methylation changes may interact with concomitant chromatin states, are described by ensemble gradient boosting and random forests. Probabilistic outputs are calibrated with calibration layers or isotonic regression, such that risk stratification can be done in selection indices. These models were unable to scale to large datasets relative to deep architectures, but their sample efficiency and feature importance estimators offer initial helpful information and benchmarks on the direction to further progress in method development (Briscik et al., 2024; Lourenço et al., 2024).

The ground truth labels are not required to address the structure discovery task, and denoising and representation learning tasks, which are the foundational ones, rely on the assistance of unsupervised clustering, dimensionality reduction, and latent factor models. Clustering techniques can be used to differentiate epigenomic profiles across tissues, developmental stages, or regimes of exposure to a stress, but livestock heterogeneity and batch confounding necessitate the combination of batch-aware similarity measures and graph-based community detection techniques that have the capacity to distinguish between biological communities and technical covariation. Manifold learning, typically with nonlinear dimensionality reduction, can provide visual representation and hypothesis generation, but requires the incorporation of neighborhood-preservation constraints and density adjustments in order to avert artificial lineage hypotheses. In both matrix factorization and probabilistic latent variable models, multi-mark matrices are decomposed into

TABLE 10.3 High-throughput epigenomic technologies and data outputs in livestock breeding.

Technology/assay	Epigenomic layer profiled	Primary data outputs	Key applications in livestock breeding	AI-relevant analytical opportunities and challenges
Whole-genome bisulfite sequencing (WGBS)	• DNA methylation at single-base resolution • Genome-wide CpG and non-CpG contexts	• Base-level methylation percentages • Differentially methylated regions (DMRs) • Allele-specific methylation patterns	• Developmental programming and imprinting analysis • Fertility, growth, and stress-resilience marker discovery • Regulatory mediation of genotype–phenotype effects	• High-dimensional continuous data, ideal for deep learning • Enables causal mediation and longitudinal modeling • Challenges: high cost, coverage variability, and batch effects
Reduced representation bisulfite sequencing (RRBS)	• DNA methylation is enriched at CpG-dense regions • Promoters and regulatory hotspots	• Regional methylation profiles • Targeted DMRs	• Cost-efficient methylation profiling in large cohorts • Biomarker discovery for breeding programs	• Suited for feature selection and sparse modeling • Easier integration with genomic selection pipelines • Challenge: non-random genomic coverage
Chromatin immunoprecipitation sequencing (ChIP-seq)	• Histone modifications or transcription factor binding • Mark-specific chromatin states	• Enrichment peaks for histone modifications • Chromatin state annotations	• Muscle differentiation and lactation regulation • Immune and inflammatory pathway mapping	• Deep learning–based peak calling and segmentation • Multitask learning across marks and tissues • Challenges: antibody specificity and signal-to-noise variability
CUT&Tag/CUT&RUN	• Histone modifications or DNA-binding proteins • Targeted chromatin profiling	• High signal-to-noise enrichment maps • Low-input chromatin profiles	• Epigenomic profiling of scarce or valuable tissues • Improved regulatory annotation in breeding populations	• Enables AI modeling under low-input conditions • Reduces technical noise for representation learning • Challenge: protocol standardization across labs
ATAC-seq	• Chromatin accessibility • Open regulatory regions	• Accessibility peaks • Transcription factor footprinting	• Rapid assessment of regulatory responsiveness • Stress, nutrition, and immune adaptation studies	• AI-assisted denoising and peak detection • Integration with histone and methylation data • Challenges: cell composition confounding
DNase-seq	• Chromatin accessibility (endonuclease hypersensitivity)	• DNase hypersensitive sites • Regulatory region maps	• Identification of active regulatory elements • Functional annotation of non-coding regions	• Sequence–signal hybrid deep learning models • High-resolution regulatory grammar learning • Declining use due to higher input requirements

Continued

TABLE 10.3 High-throughput epigenomic technologies and data outputs in livestock breeding.—cont'd

Technology/assay	Epigenomic layer profiled	Primary data outputs	Key applications in livestock breeding	AI-relevant analytical opportunities and challenges
Hi-C and chromatin conformation capture	• Three-dimensional genome architecture • Chromatin interactions and domains	• Contact matrices • Loops, compartments, and TADs	• Linking distal regulatory elements to target genes • Interpretation of non-coding variants in breeding	• Graph and geometric deep learning applications • Enhancer–gene linking via AI • Challenges: data sparsity and scale
Single-cell ATAC-seq	• Cell-type–specific chromatin accessibility	• Sparse cell-level accessibility matrices • Cell-state embeddings	• Resolving tissue heterogeneity • Immune and developmental cell-state mapping	• Self-supervised and graph-based AI models • Deconvolution reference construction • Challenges: sparsity and imputation uncertainty
Single-cell DNA methylome sequencing	• Cell-specific methylation landscapes	• Sparse single-cell methylation profiles • Cell lineage–specific methylation patterns	• Developmental trajectory analysis • Early-life programming effects	• Advanced imputation and generative modeling • Integration with spatial and bulk data • High computational complexity
Spatial epigenomic technologies	• Epigenetic marks with spatial context	• Spatially resolved regulatory maps • Tissue microenvironment data	• Linking epigenetic states to tissue structure • Disease and stress microenvironment analysis	• Spatial transformers and diffusion models • High interpretive value with high complexity
Multi-modal single-cell assays	• Accessibility, methylation, and/or transcriptomics	• Joint multi-layer cell profiles	• Direct inference of regulatory causality • High-resolution cell-state modeling	• Ideal for multimodal AI foundation models • Challenges: cost, sparsity, and data integration

shared and mark-specific components, in such a way that latent axes can be interpreted as epigenetic programs associated with metabolic adaptation, immune priming, or reproductive maturation. Bayesian formulations address uncertainty during factor loading, which is beneficial during ensuing association tests with phenotypes and controlling false discoveries made by overfitting to noise. The smoothness priors of the temporal factor extensions reveal smoothly varying regulatory modules. These self-trained representations offer small embeddings that facilitate downstream supervised processes, allow a smaller learning cost, and enable anomaly detection by considering samples with irregular factor values that can be explored through experimentation (Simpson et al., 2024; Xiropotamos et al., 2024).

Convolutional neural networks and transformer-based deep sequence models increase their representation capacity by learning off-the-shelf hierarchical motif languages and long-range relationships on top of genomic sequence, along with aligned epigenomic signals. Convolutional layers include local sequence motifs and positional combinations that predict the accessible chromatin or humanistic version placement, but dilated convolutions broaden the receptive fields without growing parameters. Transformer self-attention models instantiate distributed dependency structures, for example, co-regulated groups of motifs or the syntax of enhancers at kilobase to mega scales, which is useful for defining super enhancers and understanding insulator boundaries. Attention is tuned by genome-specific positional encodings based on the weighting of attention in topologically constrained space using relative distance. Multi-input designs are designed to take in raw sequence, methylation density vectors, and normalized histone mark tracks and learn joint embeddings that improve regulatory element classification in multi-input designs relative to sequence alone methods. Masked k-mer or next segment order pretraining prediction offers the basis of linguistic-like knowledge of genomic syntax, and requires less data to be fine-tuned on species-specific tasks. Synthetic data augmentation using in silico motif shuffling and reverse-complement consistency regularization can be used to reduce overfitting and biases in species with limited annotation. To extend the current results of biological annotation, model interpretability techniques disaggregate learned filters and attention matrices and recreate de novo motif catalogs and cooperative motif assemblies (Cheng et al., 2024; Fan et al., 2025).

Graph neural networks provide an expressive paradigm to improve the association of enhancer genes and simulate chromatin interaction networks, modeling regulatory topology as graphs with nodes denoting genomic loci or regulatory elements or gene promoters and edges denoting physical contacts or co-accessibility or predicted functional interactions. The message passing layers sequentially update node embeddings using a combination of neighbor information with learned attention scores, which may include the interaction confidence, distance decay, or mark concordance priors. Formulations in heterogeneous graphs take into consideration specific types of nodes and edges, with enhancer-promoter, promoter-promoter, and enhancer-enhancer links being distinguished. The temporal extensions of graphs are used to trace dynamic rewiring in response to developmental or stressful stimuli, measuring persistence of edges and new hub formation. The prediction of edges is advantageous to negative sampling techniques that are aware of genomic distance patterns to prevent trivial learning of proximity priors. Hybridization with sequence encoders allows access to integration with node initialization at the motif architecture scale, as well as graph propagation at the contextualization of local regulatory syntax in higher-order chromatin neighborhoods. Graph regularization promotes smooth syntenic block embeddings, which promotes cross-species alignment. These architectures improve the accuracy of the association of distal variants and perturbations of epigenetics with target gene expression and trait associations (Tenekeci and Tekir, 2024; Zhang et al., 2024).

Variational autoencoders and generative adversarial networks make significant contributions to easily obtain probable epigenetic mark profiles, scale small datasets, and explore latent perturbation landscapes. Variational encoders encode high-dimensional conjunctions of marks into latent distributions, the disentangled dimensions of which may be biological axes (developmental maturity or metabolic state) as well as controlled perturbation, allowing in silico hypothesis generation about the co-variation of marks and compensatory dynamics. Conditional variants create sparsely filled mark matrices and are trained to create conditioned label sets on either sequence conditioned on or tissue identity conditioned on or environmental covariates, and are used to perform multitask predictive modeling. The adversarial networks enhance synthetic realism by training the generators on the adversarial training method to generate mark tracks that are indistinguishable from the empirical distributions, with the discriminators providing gradient information that enhances boundary resolution. Complex multimodal covariances of marks are additionally enhanced using hybrid diffusion to generate processes that enhance fidelity. Ensuring biological plausibility requires the implementation of such constraints as non-negativity, count conservation with sequencing depth, and a monotonic relationship among hierarchical mark states. Synthetic data should be used downstream in a way that uses domain discriminators to ensure that augmentation does not create distributional shifts that adversely affect model calibration (de Weerd et al., 2024; Wu et al., 2025b).

Multi-task learning is a concept that combines the prediction of several marks, tissues, stages of development, or species into common architectures, which capitalize on shared common regulatory properties and yet specialize toward tasks. A common backbone encoders are trained to acquire generalized motif grammar or chromatin context embeddings or temporal dynamics, whereas task-adaptable output heads adjust to idiosyncratic ones like tissue-specific variance of methylation. When there is limited data to perform a particular task, parameter sharing alleviates the problem of overfitting, and task routing or soft parameter decoupling alleviates negative transfer when tasks become separated. The hierarchical multi-task structures exhibit similarities to biological lineage trees, limiting related tissues from sharing intermediate representations. The loss balancing strategies redistribute the contributions of the gradients based on the uncertainty in the tasks or convergence processes to prevent the dominance of many tasks. Cross-species multi-task experiments take advantage of ortholog mapping to find relationships between tasks across genomes to make the representation more robust. Effective generalization of low-resource tasks, reduced variation in prediction errors, and greater co-feature consistency of attribution of features across subsets of tasks measure success (Sandnes et al., 2024; Lin et al., 2025).

The issues of meta-learning and few-shot adaptation can be applied to situations where a quick model adaptation is needed for rare tissues, breeds, or environmental conditions with minimal labeled epigenomic profiles. Optimization-based meta-learning models are trained to incorporate inductive priors that can be quickly adapted by generating a few gradient steps to new contexts, which, in effect, encode an initialization close to a broad regulatory optimum. The metric-based meta-learning builds the similarity embeddings according to which classification or segmentation in new domains is done by nearest neighbor decisions based on learned distance metrics, which are robust to domain shift. Bayesian meta-learning layers spread the uncertainty on the adaptation parameters, and they generate a posterior predictive distribution that captures the confidence in adaptation. Episodic training is a simulation of deployment, where pseudo tasks are sampled in the available tissues and marks, which improves strength. It trains meta-learners to handle variability in the real world by data augmentation that is specific to agricultural variability, including simulated partial coverage, temperature stress caused mark shifts, or injection of breed-specific polymorphism. Meta-learned models save time to actionable knowledge as field conditions require real-time or near-real-time understanding of emergent states of epigenetics (Jiang et al., 2025; Rossi and El-Sayed, 2025).

Contrastive and self-supervised learning methods initiate the value of unlabeled raw signal, k-mer embeddings, and multi-mark co-occurrence patterns by creating pretext tasks that instruct models to produce invariant and discriminative representations that can be applied to downstream supervised tasks. Contrastive models relate the augmentation of the same genomic locus or multi-mark slice and repel embeddings of different loci, encouraging embeddings that pick up regulatory identity but reject noise artifacts. Augmentations of sequences are reverse complement, motif dropout, and simulated sequencing noise; augmentations of signals to simulated noise, visible coverage jitter, shifting of peaks within biologically relevant windows, and masking of marks. To model the language, masked signal modeling tasks are similar, except that mark intensities are being modeled with a mask on them, a task that involves learning contextual dependency modeling. Cross-modality contrast brings together sequence and epigenetic cues by aligning embeddings and promotes the existence of multimodal coherence. The representations acquired using these paradigms improve performance on enhancer classification, predicting effects of variants, and methylation imputation tasks, and lead to a decrease in label dependence (Qiang et al., 2025).

Bayesian and probabilistic hierarchical models enable the accommodation of heterogeneity between breeds, herds, management systems, and laboratory batches through an explicit model based on latent grouping structures and parameter uncertainty. Hierarchical priors learn task- or group-specific parameters toward global distributions, opting for adaptation or noise suppression. Mixture models are used to deconvolute composite epigenetic signals from bulk tissue data, which removes latent subpopulations that may vary by cell type composition in the tissue. Gaussian process priors on genomic coordinates or developmental time model smoothness, irregularly sample, and can perform posterior interpolation with credible intervals. Bayesian variable selection provides the posterior inclusion probability of being part of the model of potential regulatory features, enhances the transparency of feature prioritization, and orders the experimental validation. Variational Bayes and integrated nested Laplace approximations are applied for fast inferences of large-scale genomic matrices. Probabilistic programming languages enable the compositional assembly of models, which makes it possible to replace any likelihood component with a count-based, zero-inflated, or overdispersed mark distribution. The outcome of posterior predictive checks is used to assess model adequacy, and sensitivity analysis is used to measure the effect of prior specification in sparse regimes (Ping et al., 2026).

Causal discovery algorithms aim to infer directed dependency networks between guidance factors in the environment, diet, epigenomic modifications, transcript modules, and phenotypes. The constraint-based methods examine conditional independences in the presence of genomic confounding on the basis of a partial correlation or kernel conditional independence statistic, and orient the edges on the basis of assumptions about acyclicity. Score-based search involves penalized likelihood criteria on candidate graph structures, along with heuristic search or differentiable continuous relaxations to sample combinatorial graph space. Invariant risk minimization models are models that find groupings of features whose predictive links to results are consistent across environmental strata, and are core to causal procedures that are unaffected by changes in distribution. Orientation is enhanced by the integration of temporal ordering, which makes use of the power of precedence of environmental or nutritional interventions over epigenetic changes. In cases where Mendelian randomization instruments are available, exogenous variation is injected to sharpen directional inference. Algorithms are tested by causal sufficiency violations and latent confounders; thus, latent variables are discovered, latent uncertainty is quantified, and domain knowledge constraints are enforced by known metabolic or chromatin remodeling pathways to make the solution space constrained (Jin et al., 2025).

Reinforcement learning proposes sequential decision optimization in an adaptive sampling strategy and intervention scheduling in epigenomic monitoring initiatives and digital twins. State representations provided to agents include existing epigenetic summary features, environmental sensor data, physiological indicators, and uncertainty measures, and choose actions including ordering particular assays, adjusting nutritional inputs, or housing conditions. These reward functions combine predictive information gain, welfare effect, cost penalty, and long-term productivity effect, and inform policies that must trade between exploration of uncertain regulatory states and exploitation of known beneficial interventions. Model-based reinforcement learning uses learned transition dynamics to evolve epigenetic states to plan interventions in the future, whereas model-free methods are those that explicitly approximate the value or policy functions. Importance sampling or doubly robust estimators in off-policy evaluation are both safe, as the candidate policies are assessed against historical logs before being deployed. Ethical considerations are added to the algorithmic optimization through safety restrictions and risk-sensitive objectives, which guarantee the evasion of interventions that can prompt adverse stress reactions (Kallenberg et al., 2025).

Federated and privacy-protecting machine learning designs allow collaboration between multiple institutions of breeding companies, research centers, and geographical jurisdictions without aggregated raw data. Federated optimization transports the updates in model parameters, but not epigenetic or phenotypic data, whereas the mechanisms of secure aggregation and differential privacy mechanisms do not allow the reconstruction of sensitive local datasets. Domain heterogeneity, such as disagreement in assay protocol and population structure between nodes, is solved by customized federated layers that adapt global representations to local peculiarities. Communication-efficient update compression minimizes bandwidth and energy expenses, which are essential in large-scale collaborations that are sustainable. Privacy-preserving computation of gradient contributions of tasks with finer-grained interactions of features is done using secure multiparty computation and homomorphic encryption. Client attacks, such as strong aggregation rebates, poisoning, or drift, which can degrade a model, are mitigated. Based on such frameworks, prediction models of higher caliber are democratized, and benefits are shared without infringing on the proprietary limitations (Kumar et al., 2025).

Active learning methods select experiments or assays that are likely to provide the greatest improvement to the model, given a budget. Uncertainty sampling, expected gradient length, information gain in latent factor spaces, or the expected reduction of variance of the expected causal effect is used to rank candidate loci, marks, tissues, or animals to be profiled by query criteria. The diversity constraints are introduced in the batch selection algorithms to prevent redundancy between queries. Multi-fidelity active learning takes into account the fact that the cost of full genome-wide assays and targeted panels is different, and the tiered sampling pipelines are optimized. The breeding programs that are closed-loop integrated allow the model of regulatory annotations and variant effects to be refined iteratively, leading to rapid convergence of the predictive ability of the model with fewer overall assays. Opportunity cost is quantified in deferred sampling to make decisions where a lack of logistical capacity to do the assays inhibits the execution of the assays (Shen et al., 2025).

The components of multi-omics data fusion architectures are genomics, epigenomics, transcriptomics, proteomics, metabolomics, microbiome, and phenomics, which are jointly embedded, projected to graph multilayer models, and factorized. Joint embedding networks align modality-specific encoders within a joint latent space, which achieves cross-modality correspondence with either contrastive or adversarial losses specific to partial sample coverage overlap. Tensor decomposition methods break down sample × feature × modality tensors into factor matrices, the interaction of which shows coordinated multi-layer regulatory modules, which can fill in missing data using low rank assumptions. Mechanisms of cross-modality attention, which indicate important epigenetic cues regulating transcript abundance or metabolic flux, can be interpreted mechanistically. The integration structures have scale normalization and modality-specific noise modeling that avoids high-variance layers dominating. Learned fusion representations can be transferred to trait prediction and causal mediation tasks, which is more powerful to identify multi-layer dependencies that are not obvious in single modality studies, particularly when epigenetic modifications mediate the functional effects of genetic variation on phenotypic outcomes (Baião et al., 2025).

Model compression, distillation, and edge deployment can handle real-world constraints of deploying complex epigenomic models to an on-farm decision support system or a portable sequencing machine. Knowledge distillation can be used to extract predictive capacity of large deep architectures with softened probability targets and feature map alignment into small student models, and maintain accuracy, incurring less inference latency and less energy. The quantization and pruning methods remove unnecessary parameters and lower numerical accuracy with graceful accuracy degradation, and can be used to conduct real-time inference on a resource-constrained device. Architecture search, which is livestock-constrained, yields compact networks with dimensionality of the epigenetic features and sequence length. Strategies of edge deployment combine intermittent capability of connectivity, local inference caching, and asynchronous synchronization to central models, and fallback heuristics in case of model unavailability. Energy efficacy and thermal monitoring help to sustain digital infrastructure in agro-ecosystems (Gao, 2025).

Explainable AI systems, such as those represented by graphs, provide model internal logic into ideas like biologically interpretable narratives of feature attribution, counterfactual edits to the genome, and pathway-oriented explanations. Possessing loci, marks, and interaction contributions to predictions, the attribution methods also hierarchically map fine-grained attributions onto regulatory modules and pathway ontologies. Counterfactual generation models approximate the changes in minimal epigenetic events needed to alter a forecasted outcome category, shedding light on feasible intervention targets and separating changes that are both possible and modifications that cannot be made, thereby causing biochemical conflicts. Causal explanation superimposes attribution and directional dependence to prevent the possibility of confusing correlational attributes with manipulable levers. The narrative generation systems can produce attribution outputs and present them in the form of structured textual justifications with references to categories of regulatory elements, putative transcription factor motifs, and interaction neighborhoods, which can help communicate with multiple disciplines. Transparency constructs trust, allows the validation process to take less time, and enables an iterative process of refinement in which invalid rationales manifest data bias or model misalignment (Ali, 2025).

Adversarial resilience and robustness are crucial protection mechanisms, since epigenomic classifiers can face issues regarding distribution shift, noisy measurements, or maliciously perturbed inputs. Strong training methods include noise injection that is reflective of real-assay variation, interval-bound propagation that is used to certify model prediction stability in the face of limited perturbations, and the development of adversarial example generation under structured inputs to genomic data by gradient-based attacks. Invariant representation learning improves domain shift robustness because it disables flexible biological characteristics and context-specific artifacts. Outlier-robust loss functions are less sensitive to mislabeled samples or contaminated samples. The methods of certification offer testable assurances of the largest perturbations that do not change predictions, as part of reliability tests of high-consequence choices. Aggregate predictions that result from ensembles with diversity that encourages perturbations are more resilient (Angioni et al., 2025).

Lifecycle management encompasses the lifecycle model initialization, model validation, model deployment, model monitoring, drift detection, model recalibration, model versioning, model audit trails, sustained model reliability, and model governance compliance. Statistical drift detection monitors track distributions, embedding shifts, decay of calibration, and error profiles with time, and generate retraining or adaptation to thresholds that are violated. Replay buffers or parameter isolation are utilized to enforce continual learning structures to incorporate new information and reduce the devastating forgetting of previously acquired regulatory signatures. Online recalibration is an approach employed in maintaining the same calibration methods of probabilistic outputs in the face of changing prevalence or sampling biases. The extensive metadata logging includes model version, provenance of training data, and hyperparameters, as well as metrics of evaluation, which allows reproducibility and the ability to forensically investigate an anomaly in behavior. Governance policies are the policies that outline the access controls, the approval of the update workflow, and ethical use restrictions that encompass accountability in the operational pipelines. Lifecycle transparency will help stakeholders in gaining confidence and computational innovation in line with the regulatory requirements and societal expectations in terms of animal welfare and data stewardship (Nogare et al., 2024). AI and machine learning techniques for epigenomic data analytics in livestock are presented in Table 10.4.

10.5 Challenges in implementing AI in epigenomic studies

The introduction of AI in the field of livestock epigenomics faces a set of interconnected limitations that distinguish it as a less developed human biomedical model and make the methodology, infrastructure, and ethical considerations highly stringent. First and foremost is the constant conflict between the high cost of assays and the necessity of having statistically empowered sample sizes. Whole genome bisulfite sequencing, whole genome histone modification maps, chromatin accessibility mapping, and three-dimensional conformation experiments are all costly and impractical on a large scale involving multiple breeds, development stages, management treatments, and environmental manipulation scenarios. Programs focused on breeding typically allocate more resources to the genotyping and phenotyping pipelines, which offer

TABLE 10.4 AI and machine learning techniques for epigenomic data analytics in livestock.

AI/ML technique category	Representative methods	Primary analytical functions	Epigenomic data types addressed	Strengths, limitations, and livestock-specific considerations
Classical supervised machine learning	• Linear and regularized regression (LASSO and Elastic Net) • Support vector machines • Random forests and gradient boosting	• Trait prediction from epigenomic features • Feature selection and importance ranking • Baseline benchmarking	• DNA methylation profiles • Histone mark intensities • Accessibility peak summaries	• Strengths: interpretable, sample-efficient, robust in small cohorts • Limitations: limited capacity for complex non-linear interactions • Well-suited for routine breeding analytics
Unsupervised and semi-supervised learning	• Clustering (k-means and hierarchical) • Matrix and tensor factorization • Probabilistic latent variable models	• Discovery of regulatory programs and epigenomic states • Dimensionality reduction and denoising • Pattern discovery without labeled traits	• Multi-mark epigenomic matrices • Longitudinal and multi-tissue data	• Strengths: reveals latent structure under sparse labels • Limitations: interpretability depends on post hoc analysis • Useful for hypothesis generation
Deep learning (convolutional and transformer models)	• CNNs for sequence-signal modeling • Transformers with long-range attention • Hybrid sequence-epigenome architectures	• Learning regulatory grammar and combinatorial patterns • Enhancer, promoter, and boundary detection • Variant and regulatory effect prediction	• Genomic sequence • ChIP-seq, ATAC-seq, and methylation tracks	• Strengths: captures non-linear, long-range dependencies • Limitations: requires large training datasets and higher computational cost • Requires careful regularization in livestock datasets
Graph-based and geometric learning	• Graph neural networks (GNNs) • Heterogeneous and dynamic graphs • Network embedding methods	• Modeling regulatory topology and chromatin interactions • Enhancer-gene linking • Pathway and network inference	• 3D genome architecture (Hi-C) • Multi-omics regulatory networks	• Strengths: biologically intuitive representation of regulation • Limitations: graph construction uncertainty and scalability • Highly aligned with epigenomic biology
Generative and imputation models	• Variational autoencoders (VAEs) • Diffusion and GAN-based models • Matrix completion methods	• Denoising and reconstruction of sparse data • Imputation of missing marks or tissues • Data augmentation	• Sparse epigenomic assays • Single-cell and low-input data	• Strengths: maximizes the value of costly assays • Limitations: risk of hallucinated features • Requires uncertainty-aware deployment
Temporal and dynamic modeling	• Recurrent neural networks • Temporal transformers • Neural ODEs and state-space models	• Modeling epigenomic trajectories over time • Detection of regulatory transitions • Forecasting phenotypic outcomes	• Longitudinal epigenomic profiles • Developmental and stress-response data	• Strengths: captures dynamic regulatory plasticity • Limitations: sparse time points in livestock studies • Essential for developmental programming analysis

Continued

TABLE 10.4 AI and machine learning techniques for epigenomic data analytics in livestock.—cont'd

AI/ML technique category	Representative methods	Primary analytical functions	Epigenomic data types addressed	Strengths, limitations, and livestock-specific considerations
Causal inference and mediation learning	• Structural causal models • Mediation analysis frameworks • Instrumental variable approaches	• Distinguishing causation from correlation • Identifying regulatory mediators of traits • Counterfactual reasoning	• Genomics-epigenomics-phenotype data • Nutritional and environmental covariates	• Strengths: intervention-aware inference • Limitations: strong assumptions, confounding sensitivity • Critical for ethical and reliable decision-making
Reinforcement learning and optimization	• Model-based and model-free RL • Policy optimization algorithms	• Adaptive intervention and sampling strategies • Optimization of management decisions	• Integrated epigenomic, phenotypic, and sensor data	• Strengths: supports precision livestock management • Limitations: reward specification and safety constraints • Emerging application area
Explainable AI (XAI)	• Feature attribution (SHAP and gradients) • Concept bottleneck models • Counterfactual explanations	• Interpretation of model predictions • Linking AI outputs to biological mechanisms	• All epigenomic modalities	• Strengths: transparency and trustworthiness • Limitations: explanation variance across methods • Essential for adoption in breeding programs
Federated and privacy-preserving learning	• Federated optimization • Secure aggregation • Differential privacy	• Collaborative model training without data sharing • Governance-compliant analytics	• Distributed multi-omics datasets	• Strengths: enables large-scale collaboration • Limitations: communication overhead, heterogeneity • Supports equitable AI deployment

short-term selection dividends, but leave deep epigenomic profiling to tighter cohorts that may not reflect the population variation. This results in a data regime where p≫n imbalance occurs, leading to extreme overfitting when naive predictive models are used, feature attribution can be unstable, and uncertainty quantification is highly unreliable. Efforts to reduce costs by pooling, reduced representation, or targeted panel strategies present additional statistical complications since the sampling design determines the inductive bias of downstream AI models. Despite the ability of transfer learning, multitask schemes, and generative augmentation to alleviate data paucity, they are not able to completely replace the diverse, well-balanced empirical observations when the underlying biological signal shows context-dependent modulation in nutritional state, climatic stressors, and microbiome alterations. Investment models and investments in joint-stock consortia have lagged in capturing the compounded valuation of common epigenomic initial maps in livestock compared to our genomic reference investment, resulting in an additional bootstrapping period whereby AI method sophistication exceeds the raw data base (Ohishi and Au Yeung, 2025; Zaccaria et al., 2025).

The heterogeneity of tissues and cell types is also another issue that complicates signal attribution and model interpretability, since bulk tissue epigenomic data combines isolated cellular subpopulations that have different regulatory states. Tissues of livestock production (immune defense, metabolic adaptation, reproduction, or muscle accretion) can typically experience active compositional changes in response to age, diet, exposure to pathogens, and physiological condition. AI models trained on composite marks can falsely give predictive power to loci, the apparent variance of which is due to varying cell mixture proportion, but not to cell intrinsic regulatory reprogramming. The classical normalization methods cannot be used when there is a correlation between mixing proportions and phenotypes or covariates in the environment, so that they produce confounding relationships that would mislead the causal prioritization. Reference-based correction requires reference methylomes or chromatin profiles of purified cell types, which are often unavailable across different breeds or developmental stages, and therefore limit the applicability of reference-based correction. Sometimes, self-supervised representation learning can identify mixture axes and regulatory program axes, but disentangling is rarely complete, and the interpretability of latent factors is weak. This heterogeneity is carried over to downstream tasks like enhancer gene mapping, variant effect prediction, and trait association fine mapping, where failure to solve cellular context compromises biological plausibility. More uniform tissues should be sampled strategically, histological metadata should be included, and more effort should be invested in single-nucleus or single-cell modalities, which are, however, limited by logistical, ethical, and cost challenges peculiar to agricultural research (Liu et al., 2025; Zhu et al., 2025).

The accuracy of estimation and bias are compounded by scant coverage and incomplete data, significantly affecting the reduced representation. Non-random missingness patterns of enrichment-based chromatin assays, selective sequencing of CpG dense regions, and partial depth trade-offs save money, but not random patterns of missingness, which several machine learning algorithms assume. Model-based imputation strategies using matrix completion, tensor factorization, or deep generative models can be used to estimate missing signal under structured low rank or manifold conditions; but in a biological setting where the variation between genes and regions being regulated is confounded with coverage variability, imputation can be used to falsely discover active regulation in under-sampled genomic neighborhoods. Locus-level imputations have confidence intervals that are rarely propagated in downstream feature engineering pipelines, resulting in an unknown and unacknowledged accumulation of uncertainty. The sequence-informed priors are somewhat regularized by the density of motifs or conservation, but also face the danger of strengthening conservation bias that excludes species-specific regulation innovations that are essential in species adaptation to breeding. Upon sparse sampling, differential signal calling is plagued with inflated dispersion estimates, which reduces the statistical power. Theoretically optimal but computationally expensive in practice, joint models of coverage generation mechanisms and latent regulatory intensities make parameter identification under limited replication challenging. As a result, AI models that have been trained using hybrid datasets that combine full coverage and sparse assays can overcome complex changes in domains that necessitate cautious calibration and explicit missingness encoding (Pereira et al., 2024; Fan, 2025).

Systemic biases in AI models can be caused by batch effects, protocol variability, and platform drift, which AI models can unintentionally amplify. Spatial variations in library preparation chemistry, antibody lots, sequencing later generations, bisulfite conversion efficiency, or bioinformatic preprocessing pipelines introduce stratified biases in the form of overall intensity differences, distortions in coverage distributions, or mark-specific noise patterns. These artifacts may be consistent with the study design variables of breed, farm, or temporal cohort, mixing technical and biological variance. High-capacity supervised learning models might treat batch-linked covariates as artifacts and yield overly high internal validation results, but do not generalize to a new batch of data. Traditional means of correction assume linear batch contribution or shared control characteristics, which are often not met in the case of covarying environmental and technical variables. New domain adaptation and adversarial disentanglement approaches will extract invariant representations but may deemphasize valid biological variations should the signal structure have a similar structure as those of batch trajectories. There is an additional layer of complexity with multi-site federated learning because asynchronous heterogeneous local preprocessing updates can incorporate latent drift, which is hard to detect on a central node. It may be necessary to introduce routine inclusion of calibration standards, open reporting of processing metadata, and the adoption of audit trails that can be used to introduce harmonization retrospectively to maintain model reliability (Lara-Gutierrez et al., 2025; Li and Zhu, 2025).

Transfer learning and the interpretation of functional genomic data are limited by gaps in cross-species annotation, and the lack of reference maps is a significant limitation. Regulatory elements have not been cataloged in many species of livestock, at both tissue and developmental window levels, which restricts the use of gold standard labels in the definition of enhancer, promoter, insulator, and chromatin states. Better-annotated species used as orthology sources can cause undesirable misalignment in genomic rearrangements, lineage-specific expansions, or species-specific adaptive regulatory innovations. The human or model organism pre-trained AI models can potentially be biased in the representation of motif or chromatin grammar distributions, including a complete or inaccurate reflection of livestock regulatory architectures, and therefore give systematic biases in the inference of subsequent enhancers or scoring of variant effects. The limited comparative epigenomic datasets make it difficult to deconvolve conserved and divergent regulatory logic in any given study, and it is hard to estimate the amount of uncertainty that can be attributed to the extrapolation of evolution. Graph alignment, as well as domain adversarial training, reduces cross-species covariate shift partially, but is incapable of generating biological subtleties unobserved in the training data. In the absence of enriched annotation, explainability tools can be used to project the attribution of predictions to poorly-characterized sequence patterns in a biologically validated cycle. Organized efforts to create multi-tissue, multi-mark atlases, including underrepresented breeds and production systems, would make a significant positive impact on AI model grounding and make it less reliant on cross-species heuristics (Ma et al., 2025; Zhou et al., 2025).

The epigenetic response has temporal sampling limitations and latency that produce a discordance between measurement time points and dynamic regulatory mechanisms that mediate phenotypic adaptation. Pragmatic constraints of repeatedly sampling useful animals, animal welfare issues generate sparsely spaced temporal series that are insufficient to extensively model chromatin reconfiguration, remodeling of methylation, and downstream transcriptional effects. Epigenetic mark establishment and environmental stimuli complicate the development of a causal temporal order when observational windows are coarse, and there is a temporal lag between environmental stimuli, epigenetic mark establishment, and phenotypic manifestation. The sequence models and the Gaussian process interpolators can smooth trajectories but are likely to introduce unwarranted continuity, which masks punctuated regulatory transitions. Reinforcement learning models that aim to maximize sampling schedules have partial observability and delayed feedback when the reward value of an intermediate epigenetic perturbation can only be observed in physiological measures that are revealed much later. Active learning techniques that suggest active temporal sampling are limited by the logistics of husbandry and cost constraints. As a result, time sparsity beats uncertainty in the inferred regulatory kinetics, which is applied to the predictions of intervention timing in nutritional supplementation or other stress dissipation. The regularization of temporal inference by incorporating mechanistic priors concerning known developmental milestones and metabolically seasoned periodic shifts can be useful, though there is a risk of entrenching vestigial assumptions that conceal new genotype-by-environment interactions (Olmeda et al., 2025; Xiong and Zhu, 2025).

Separating causation from correlation in regulatory signal relationships with polygenic traits is a major methodological issue. The high-dimensional epigenomic matrices exhibit many statistical relationships with phenotypes that can be a result of reverse causation, shared environmental confounding, or cellular compositional changes, as opposed to causal mediation in genotype-to-phenotype pathways. Causal discovery algorithms rely on testing conditional independence and score functions, whose assumptions of the absence of latent confounding appear to be commonly violated in unmeasured management variables, microbial exposures, or metabolic states. The instrumental variable methods that utilize genetic variants require robust methods of identifying variants that affect epigenetic phenotypes without pleiotropic pathways that affect phenotypes directly, which is a non-trivial requirement in polygenic structures. Temporal ordering is beneficial for orientation, but it is often undermined by the sparsity of sampling. The frameworks of AI-driven mediation analysis that include representation learning could reduce confounding but introduce collider bias when taking steps inadvertently that condition on common effects. The over-reliance on simple observational inference increases the risks of biasing functional tests and breeding choices. Combining both evidence of controlled perturbations (where possible) and mechanistic constraints that reflect biochemical kinetics, chromatin remodeling energetics up to transcription factor occupancy dynamics offer partial protective measures at the price of increased model complexity and increased thresholds in how much data they demand (Cai et al., 2025; Wang et al., 2025).

Computational and conceptual barriers to integration between epigenomic data and other genetic, transcriptomic, proteomic, metabolomic, microbiome, and phenotypic datasets are imposed by integration friction. This friction arises due to differences in scale, the patterns of sparsity, the noise distributions, the alignment of the temporal information, and the coverage of the sample, which make direct concatenation or joint modeling especially difficult. As an example, genetic variation is non-specifically static, methylation and histone modifications are non-specifically moderately dynamic, transcripts are rapidly fluctuating, and phenotypes may be cumulatively integrating over long durations. The AI designs will need to balance between asynchronous sampling and missing modality samples, and avoid causing the appearance of alignment artifacts that will disguise themselves as regulation. Canonical correlation and contrastive embedding teams can be applied to align modalities in shared latent spaces, but they tend to assume that the covariance structure of every modality is homogeneous and will fail when a modality has context-specific information that none of the other modalities share. Tensor factorization enforces low rank structure, which can result in simplistic multi-layer regulatory interactions, whereas graph-based fusion has to make arbitrating edge confidence decisions between heterogeneous sources of evidence with varying reliability distributions. The inability to weight modalities adequately has the power to smother the weak epigenetic mediation signals that lie beneath the large genetic additive effects or environmental variance. Layered propagation of uncertainty that is grounded in principled uncertainty propagation is further challenged by incompatible statistical-based assumptions in modality-specific preprocessing pipelines, such that downstream decision support is coherent (Schumann et al., 2025; Baião et al., 2025).

Generalization and overfitting in the presence of domain shift remain timeless operational risks because models trained on limited groups are now exposed to novel breeds, regimes, management, climatic stresses, feed diets, disease demands, or technological infrastructure. Predictive calibration can be eroded by covariate shift on the frequencies of sequence variants, the baseline distribution of methylation, or mark noises. Conventional cross-validation can provide optimistic forecasts when folds are related by the latent structure of some common farm practice or sires. Without task relevance measures to ground domain adaptation methods that reduce divergence of feature distribution, some biologically relevant differentiation may be spurred out unintentionally. Meta-learning enhances quick adaptability but can still be based on assumptions that new domains are covered by the convex hull of training variability. Drift detection monitoring systems do not have much ground truth, using unsupervised discrepancy measures whose levels are not biologically interpretable. Another form of overfitting can be observed in the process of spurious motif attribution, in which deep sequence models can store redundancy-specific motifs in the dataset but not general combinatorial grammar. The regularization, data augmentation, and ensembling methods minimize the variance but increase the computational load and complexity of interpretability. Hierarchical population structure and environmental stratification can be explicitly modeled to strengthen generalization, but this puts special strain on the parameters of rare strata (Lu and Wang, 2025; Xia et al., 2025).

The trade-offs of interpretability versus complexity of deep models remain controversial with respect to the level of tolerable opaqueness of decision pipelines in implemented choices of breeding or intervention strategies. High-capacity architectures provide higher predictive utility in enhancer detection, mark imputation, or phenotype risk scoring, but provide distributed representations at an internal state that is not easily mapped onto biology. The feature attribution approaches add a level of methodological variance where divergent explanations are based on the choice of baselines, gradient saturation, or treatment of the interaction, which may compromise the confidence of the stakeholders. High-order interaction capture, which is important to capture traits well, can be lost to simplification by distillation into linear or small-tree surrogates. Application of biologically organized inductive bias, for example, layers or groups of chromatin states or families of transcription factors, balances the problem of opacity but restricts flexibility and can represent old ontologies. Technical saliency is not limited to epistemic humility: conveying residual uncertainty, model boundary conditions, and unmodeled confounders to the practitioners requires narrative structures and methodological support structures, only given secondary attention in method development cycles (Shuvra et al., 2024).

Computational scalability and energy requirements increase with the growth of multi-tissue, multi-mark, and multi-species datasets, particularly when including high-resolution, three-dimensional chromatin contact matrices and long-read accessibility data. This is particularly challenging for small tokenizations of genomes or dense chromatin interaction graphs using deep graph neural networks, which require training transformer architectures with very large memory footprints, leading to very long training times and high costs for institutional infrastructure. The associated sustainability issues are further exacerbated as typical cloud-based acceleration is subject to restrictions on data sovereignty and privacy. Other approximation techniques, such as sparse attention, locality-sensitive hashing, subgraph sampling, mixed-precision arithmetic, and model pruning, provide partial relief; however, these demand stringent accuracy audits to ensure that biological fidelity is preserved. The need for resources makes quick iterative cycles essential to responsive model refinement in response to new assay batches, and this increases the latency between the delivery of data and the consumption of valuable information. The challenges of evaluating the trade-offs of environmental accounting in large-scale epigenomic AI pipelines in sustainability-aware agricultural ecosystems are compounded by the underreporting of energy accounting (Xu and Shen, 2025).

The obstacles to integrated AI implementation are data governance, ownership, and interoperability restrictions that are both organizational and legal. Raw epigenomic data and even derived embeddings have little to no willingness to share due to proprietary breeding datasets, clauses of contractual confidentiality, and competitive dynamics limiting the potential of federated learning. The standards of divergent metadata prevent automatic harmonization and inconsistent ontological labeling of tissues, development stages, or environmental variables, also preventing cross-study integration and cross-study model portability. The regulatory frameworks for animal data privacy are not as well codified as human regimes of data protection, and this creates uncertainty, contributing to risk aversion due to the fear of risks. There are no established mechanisms for the fair sharing of benefits based on AI models trained on a pool of multi-stakeholder data, and this raises the question of asymmetrical value capture. The interoperability has a detrimental effect on proprietary file formats, inconsistent versioning of preprocessing pipelines, and a lack of universally accepted and standardized minimal information reporting standards adapted to livestock epigenomics. These frictions in governance extend to duplication of effort, hinder benchmarking, and hinder the diffusion of innovation (Taddese et al., 2025).

The bottlenecks of validation are present since experimental perturbation and functional assays required to validate AI-predicted regulatory mechanisms are labor-intensive, costly, and sometimes limited by ethical considerations. Editing regulatory contents of candidates, regulating chromatin status, or performing controlled nutritional interventions in large animals implies longer generation rates, increased costs of facilities, and more stringent welfare controls than on model organisms. This results in the fact that many or all AI-obtained hypotheses have not been validated but may be indirectly corroborated, which undermines confidence in translational statements. The low throughput in validation leads to publication bias in favor of confirmatory findings, which hides negative findings that may refine model priors. It becomes difficult to prioritize the allocation of resources: which of the predicted enhancers, causal methylation mediators, or variant regulatory effects are to be validated requires strict triage frameworks that take into consideration predicted effect size, uncertainty measures, conservation data, and economic utility. In the absence of systematic feedback of functional validation to model retraining, iterative improvement becomes progressively slower, and error modes remain uncorrected (Ribeiro-Dos-Santos & Maurano, 2025).

The normative considerations arising from the ethical limits of possible epigenetic manipulation come into conflict with technical capabilities. As AI models learn putatively causal regulatory loci that could be edited or epigenetically modified to promote productivity, disease resistance, or environmental robustness, there are concerns about what is acceptable manipulation of adaptive plasticity, unanticipated and multi-generational effects, and conserving genetic diversity. The ability to manipulate epigenetic states may conceal early signals of stress, which would be detrimental to welfare checks. The perception of AI-guided interventions as a threat to natural adaptive processes and regulatory scrutiny can become more intense, posing a threat to market acceptance. Ethical frameworks should be able to differentiate between the analytical use of epigenetic data to enhance better management strategies and active manipulation strategies that alter regulatory landscapes. Transparency must exist regarding limitations of the model in use, the reversibility of the intervention, and off-target effect monitoring. The ethical limits embedded in the role of reinforcement learning rewards or optimization goals may anticipate the avoidance of pursuing strategies that provide short-term benefits at the cost of long-term resilience (Li et al., 2025).

The lack of reproducibility and benchmarking worsens the objective evaluation of methodological progress and prevents the cumulative construction of knowledge. The lack of standardized reference datasets with pre-determined training, validation, and blind test partitions based on livestock species enables unwanted data leakage and a lack of calibrated performance reporting. Preprocessing choices, such as peak calling stringency, normalization, or alignment, can vary significantly in terms of performance, and especially these aspects of pipelines are not reliably reported, although they can vary more than improvements do. The absence of consensus evaluation metrics for activities such as enhancer prediction with class imbalance or cross-breed generalization encourages selective reporting of metrics that obscure trade-offs between precision, recall, calibration, and robustness. The scarcity of open-source implementations that have a specification of environment reproducibility reduces independent verification and slows the adoption of a methodology. The challenges of benchmarking, similar to those faced by other subfields of omics, are still in their infancy and are complicated by heterogeneity in lawful permission to redistribute uncooked animal information. By implementing audit trails that record the provenance of code, parameter settings, random seeds, and data snapshot identifiers, it may be possible to advance the culture of reproducibility, and institutional incentives that favor novelty argue in favor of non-traditional approaches. Without systematic benchmarking, there is a risk of resource allocation being made based on the latest and greatest architectures, rather than one proven to be empirically valid and appropriate in the context it is used in (Brooks et al., 2024).

The solution to these multifaceted challenges is the convergence of strategic investment in data generation, governance structures that facilitate secure collaboration, methodological innovation that is sensitive to the unique limitations of livestock, ethical deliberation that involves integration of socio-economic settings, and cultural transformations toward transparency and reproducibility. To address the issues of cost barriers, the disentangling of heterogeneity, rigorously modeling uncertainty with sparsity, harmonizing protocols, suppressing batch effects, creating species-specific regulatory atlases, improving functional validation throughput, codifying ethical limits, and institutionalizing benchmarking norms, AI in livestock epigenomics will realize its transformative potential only by addressing the cost barrier in cooperative consortia, becoming disentangled from heterogeneity, refined sampling, and deconvolution, rigorously modeling uncertainty with sparsity, and harmonizing protocols. These interdependencies should be resolved as a cumulative process to create reliable, fair, and biologically founded AI systems that enhance livestock health, productivity, and sustainability without violating the complexity of the epigenetic control of living production ecosystems (Zhu et al., 2025). Major challenges in AI-driven epigenomic analysis in livestock are shown in Table 10.5.

10.6 Future perspectives and innovations in AI for livestock epigenomics

Future developments of AI will transform the field of livestock epigenomics, a backward-looking analytical field, into a predictive and intervention-based ecosystem, which constantly incorporates multi-scale biological signals and environmental context. The new paradigm focuses on convergence: molecular granularity and real-time phenotyping, mechanistic causality and generative optimization, distributed collaboration harmonized using shared ontologies, and ethically directed translational pathways. This transformation will be achieved by coordinating methodological advances that can internalize complexity without having adverse impacts on interpretability, operational sustainability, or social license. The geography of the future is set to move past fragmentary modeling of individual tracks toward unified computational systems, training teachable regulatory abstractions at the level of genome-scale architecture, individual cell differentiation, adaptation to changing environmental conditions, and overall animal behavior under changing agroecological conditions (Eckhardt et al., 2025).

Single-cell and spatially resolved epigenomics are two major pillars of this evolutionary advancement, as they break down the ambiguity created by averaging bulk tissues and allow AI systems to localize regulatory plasticity by defined cellular neighborhoods and microenvironmental gradients. Multi-resolution graphs, with nodes representing cells and spatial tiles, and edges representing lineage paths, localization to niche, and mutual signatures of regulatory programs, will be constructed using integrated analysis pipelines that will take a single nucleus chromatin accessibility, base-level methylation distributions, histone modification microdomains, and co-registered spatial transcript or protein abundance maps. Adaptive priors based on AI-driven deconvolution of residual bulk datasets will make use of these high-fidelity atlases, which will serve as dynamically updated embeddings of cell type references as cellular taxonomy is expanded by new breeds, developmental stages, or pathophysiological states. Bulk mark probabilistic decoding of latent mixtures into cell-resolved marks will scale calibrated uncertainty between subsequent causal mediation computations and intervention priorities with quantified confidence levels. Spatially informed transformers and diffusion models will learn morphogenetic gradients and inflammatory microfoci, allowing for theorizing about microenvironment-induced epigenetic remodeling, which could be used to target nutritional or other management interventions to regulate local immune or metabolic resilience (Preissl et al., 2023; Ohishi and Au Yeung, 2025).

Multimodal foundation models provide a scaffolding layer covering a broad range of data through the visualization of self-supervised representations that learn to match genomic sequences, multi-mark epigenomic data, patterns of transcript abundance, proteo-metabolic summaries, obtained images derived from structural signals, and phenotypic performance gradients in a shared latent space. Such models will learn generalized regulatory grammar, including motif syntax, co-occurrence of chromatin states, three-dimensional proximity heuristics, and dynamic response motifs to environmental perturbations, through masked reconstruction, cross-modality attention, contrastive alignment, and retrieval-augmented conditioning. The parameter-efficient adaptation systems will allow for quick fine-tuning of species-specific or breed-specific idiosyncrasies while retaining the broadly-conserved principles. The representation of spaces with disentangled axes of cell identity, developmental maturation, stress load, and nutritional status will be utilized to rapidly annotate zero-shot instances of newly sequenced tissues and will permit causal inference modules to factor out genetic variation, intermediate regulatory transitions, and phenotypic consequences. The architectural design of governance consciousness will encompass privacy-preserving representation restrictions that will permit federated pretraining across international cohorts whose raw data cannot exit sovereign territories (Khan et al., 2025).

The generative modeling will increasingly become active as it involves proactively designing nutritional and environmental intervention regimes to drive epigenetic states toward desirable phenotypic equilibria, without jeopardizing welfare or diversity. The conditional generative adversarial networks, variational diffusion processes, and policy-guided transformers will collaboratively investigate possible perturbation vectors for diet composition changes, micro-mineral

TABLE 10.5 Major challenges in AI-driven epigenomic analysis in livestock.

Challenge category	Underlying causes	Impacts on epigenomic analysis	Consequences for AI modeling	Mitigation strategies
Limited sample size and high assay cost	• High cost of genome-wide epigenomic assays • Resource prioritization toward genotyping and phenotyping • Limited replication across breeds and environments	• Reduced statistical power • Incomplete coverage of biological variability • Overrepresentation of narrow cohorts	• High risk of overfitting • Unstable feature attribution • Inflated predictive confidence	• Transfer learning and multitask models • Cross-study integration and consortia • Uncertainty-aware modeling
Tissue and cell-type heterogeneity	• Bulk tissue sampling • Dynamic shifts in cell composition • Limited livestock single-cell reference atlases	• Confounding of regulatory signals • Misinterpretation of epigenomic variation	• Learning of spurious correlations • Reduced biological interpretability	• Deconvolution and single-cell integration • Cell-aware latent representations • Strategic tissue sampling
Sparse coverage and missing data	• Reduced-representation assays • Variable sequencing depth • Non-random missingness	• Biased estimation of regulatory states • Loss of genome-wide context	• Imputation uncertainty • Risk of hallucinated features	• Generative and probabilistic imputation • Explicit modeling of missingness • Confidence propagation
Batch effects and technical variability	• Differences in protocols, reagents, and platforms • Lab- and time-specific effects	• Technical signals masquerading as biology • Reduced cross-study comparability	• Shortcut learning by AI models • Poor generalization across cohorts	• Harmonization and metadata standards • Domain adaptation techniques • Audit trails and calibration controls
Incomplete functional annotation	• Limited regulatory atlases in livestock species • Reliance on cross-species projection	• Ambiguity in assigning regulatory roles • Reduced biological validation capacity	• Weak explainability of model outputs • Annotation bias	• Orthology-aware transfer learning • AI-driven functional annotation • Expansion of livestock epigenomic maps
Temporal sparsity and dynamic complexity	• Limited longitudinal sampling • Welfare and logistical constraints	• Inability to resolve regulatory kinetics • Missed transient regulatory events	• Poor temporal causality inference • Uncertain forecasting	• Temporal models with mechanistic priors • Active learning for sampling design • State-space modeling
Causality versus correlation	• Observational study designs • Unmeasured environmental confounders	• Misidentification of regulatory drivers • Reverse causation risk	• Invalid intervention recommendations • Ethical and economic risk	• Structural causal models • Mediation analysis • Counterfactual simulation

Multi-omics integration complexity	• Different scales, noise profiles, and timing across omics • Partial modality availability	• Loss of regulatory signal coherence • Misaligned biological interpretation	• Model instability • Dominance of certain modalities	• Hierarchical and graph-based fusion • Modality-aware uncertainty weighting
Generalization and domain shift	• Breed, environment, and management heterogeneity • Climate and nutritional variability	• Reduced reproducibility across contexts	• Calibration drift • Unreliable deployment performance	• Domain generalization and meta-learning • Continuous monitoring and recalibration
Interpretability-performance trade-off	• High complexity of deep learning models • Distributed representations	• Difficulty linking predictions to biology	• Reduced trust and adoption	• Explainable AI frameworks • Concept bottleneck models • Mechanistic constraints
Computational scalability and energy use	• High dimensionality of epigenomic data • Large graph and sequence models	• Long training cycles • Sustainability concerns	• Limited accessibility of advanced models	• Model compression and pruning • Efficient architectures • Energy-aware design
Data governance and ownership	• Proprietary breeding datasets • Legal and contractual constraints	• Fragmented data landscape • Reduced collaboration	• Smaller effective training datasets	• Federated learning • Privacy-preserving AI • Standardized governance frameworks
Validation bottlenecks	• Costly and slow functional assays • Ethical constraints in livestock experimentation	• Limited confirmation of AI-derived hypotheses	• Persistence of uncorrected errors	• Prioritization via uncertainty and effect size • Iterative model-experiment feedback
Ethical and societal constraints	• Welfare considerations • Concerns over epigenetic manipulation	• Limits on intervention strategies	• Constraints on optimization objectives	• Ethical governance frameworks • Transparency and accountability
Reproducibility and bench-marking gaps	• Lack of standardized datasets and metrics • Inconsistent reporting	• Difficulty comparing methods	• Inflated or non-transferable performance claims	• Shared benchmarks • Reproducibility infrastructures • Open evaluation protocols

supplementation plans, photoperiod adjustment plans, thermal stress buffer plans, and enrichment practices. Metabolic flux limits, feed resources, and welfare criteria will be coded as integrated feasibility constraints to make the synthetic recommendations feasible and ethical. Overlays in reinforcement learning will be tested on long-horizon reward structures with productivity, climate variability resistance, intensity of emission, and disease challenge resilience, training generative priors with counterfactual simulation feedback based on causal digital twins. The loop of this generative intervention will rely on a strict degree of uncertainty quantification in order to avoid undue extrapolation outside of validated regimes of physiology (You et al., 2025).

Real-time convergence sensing will further reduce feedback latency by directing continuous feeds of wearable accelerometers, rumination microphones, thermal imaging, local environmental sensors, metabolite biosensors, as well as potentially micro-sampling devices to online epigenetic state inference engines. By linking rapid physiological proxy fluctuations with slower changing latent regulatory trajectory approximations through sequential Bayesian filters and neural state space models, it will be possible to anticipate imminent epigenetic changes before overt performance decline or health incidents. Predictive uncertainty will dynamically guide through adaptive logic of resource allocation, leading to increased resolution omics sampling or targeted single-cell profiling under circumstances of critical levels of predictive uncertainty, minimizing assay budgets. Continuous learning structures will be used to distinguish normal circadian or seasonal cycles of oscillation from abnormal drift, which will avoid the proliferation of spurious alarms. Privacy-preserving streaming analytics (along with edge compression) integration will be done to ensure that on-farm inference is not compromised by intermittent connectivity, and synchronization protocols will be used to consolidate local updates into global models without compromising data sovereignty (Wolfrath et al., 2024).

Causal digital twins are expected to become an operational framework for scenario simulation and intervention optimization, which integrates structural causal graphs, mechanistic metabolic and endocrinological submodels, biophysical growth kernels, and AI-learned mappings between regulatory perturbations and trait distributions. These hypothetical models of individual animals, herds, or genotype-characterized cohorts will integrate longitudinal multi-omics snapshots and sensor data to keep calibrated posterior beliefs regarding latent regulatory and physiological states. The counterfactual reasoning engines will extrapolate alternative actions by management, nutrition, or genomic selection decisions to predict changes in the epigenetic marks, transcriptional modules, and phenotypic endpoints, taking into account uncertainty. Multi-objective optimization that balances productivity, welfare resilience, environmental footprint, and economic viability will be included in policy evaluation layers. Causal learning loops of refinement will compare the results of simulated outcomes with the achievements of the realized trajectories to modify causal structure and parameterization based on Bayesian model criticism, and as a result, increase the reliability of the recommendations and their interpretability (Luo et al., 2025).

New technologies in long-read and nanopore sequencing will overlap with on-farm edge inference to allow real-time streaming analytics that identify methylation signatures, structural variants, and phase haplotypes. Adaptive basecalling, methylation calling, and on-the-fly compression neural decoding models that run on specialized accelerators will be used to reduce bandwidth and energy consumption by operating at downstream AI model sensitivities. The emergent epigenetic transitions indicative of acute stresses or pathogenic invasion will be detected by sliding window anomaly detectors, and targeted deeper sequencing or management changes will be initiated. Auto-piloting of cloud-orchestrated retraining and edge-distilled inference modules will be compatible with maintaining consistency between the fast-changing foundation models and resource-limited deployment endpoints, and safe attestation of model integrity against hijacking. Methylome attribution to allele-specific epigenetic modulation of complex trait variation will be enhanced by integration of long-read haplotype-resolved methylomes into predictors of regulatory variations (Ghavi Hossein-Zadeh, 2024; Tian et al., 2026).

Three-dimensional genome topology will be incorporated into predictive systems, enhancing the mapping of distal regulatory factors to signals of gene expression modulation and trait association. Graph neural networks will incorporate chromatin contact maps, loop domains, compartmentalization patterns, and insulation boundary dynamics, and the sequence of linear genome and epigenetic marks will generate context-aware regulatory influence scores. The mechanisms that are spatially aware of attention will be loaded with interactions that are weighted based on dynamic occupancy probabilities, based on the stage of development, metabolic condition, or environmental stressors. Topological reconfiguration temporal modeling will clarify structural mediation channels by which a perturbation of the environment rewires enhancer-promoter communication, enabling digital twins and generative intervention systems to factor in structural leverage points. Causal loci will be fine-mapped, and selection indexes will be built with increased mechanistic resolution by jointly embedding variant positions and contact-mediated regulatory neighborhoods (Wall et al., 2024).

Quantum-inspired and neuromorphic algorithms indicate an avenue for accelerating intricate regulatory system search and energy-saving inference. Quantum annealing formulations can be used to approximate the optimization of combinatorial motif interactions and a set of enhancer-promoter assignments, and can be further used to factorize massive joint probability landscapes describing multi-mark dependencies. Neuromorphic spiking networks can support low-latency anomaly detection and ongoing sensor fusion reminiscent of event-driven biological computations with reduced energy consumption compared to dense synchronous computing. Hybrid pipelines will redirect the coarse neuromorphic filtering output to high-fidelity conventional deep learning modules, saving predictive performance while offloading routine pattern recognition. The evaluation systems will require being able to test whether approximations offered by these new substrates do not alter biologically relevant uncertainty structures or cause systematic bias in the task of causal inference (Goyal et al., 2024).

Green AI promises to bring carbon-efficient pipeline coordination to a fundamental design requirement. Scheduler systems will match large-scale training jobs to times of reduced grid carbon intensity, trade accuracy with energy reductions using elastic early stopping policies as incremental improvements in performance decrease below the sustainability constraints, and tend to choose architecture variants whose accuracy-energy Pareto frontier is optimized. Lifecycle accounting will not only be used on training energy, but also on data acquisition, storage, model distribution, and edge inference, and will be used in a cost-benefit analysis, which will include environmental externalities. Compression of models, mixed precision arithmetic, skipping of layers during dynamic execution, and adaptive resolution encoding are expected to be considered not only based on latency but also on the overall resource utilization. Dashboards of transparency will be created to measure the energy usage relative to any legitimate biological understanding, which will create accountability and promote the uptake of best practices in consortia federated (Rojahn and Grum, 2025).

Federated international epigenome initiatives will evolve into organized consortia that are united by intersectional ontologies of breed taxonomy, hierarchies of developmental phases, tissue morphology, husbandry programs, nutrition preparations, environmental exposures, and lexicons of phenotypic traits. The alignment of semantic layers will facilitate metadata traversal for interoperability, automated compilation of cohorts, and strong cross-site stratification correction. Secure multi-party computation, differential privacy, and homomorphic encryption primitives will guarantee the input of gradient updates or representation statistics without the need to reveal proprietary raw signal, and provenance tracking will assign value to data contributors to share in the benefits. Supportive governance charters will specify model update voting, standards of curation, and dispute resolution, to make them adaptable to changing regulatory environments and societal demands. Community feedback loops will be used to incorporate ontological evolution processes that will reconcile the findings of an experiment with legacy classifications to ensure semantic currency (Singh et al., 2025).

The adaptive sampling and autonomous experimental design loops will further integrate AI with laboratory and farm operations by reducing the gap between computation-based generation of hypotheses and their empirical validation. The active learning engines will prioritize animals, tissues, time points, and marks for profiling, which are most likely to yield the most expected information within the limits of welfare and logistics. Prioritized assays can be performed in robotized sample handling and microfluidic preparatory platforms with reduced human latency, whereas simulation-augmented Bayesian optimization will balance between the discovery of undercharacterized biological niches and the utilization of known information-rich areas. Ongoing incorporation of newly generated data will recalibrate any uncertainty landscapes, and further proposals of sampling will be improved. This flexibility hastens the convergence toward comprehensive regulatory coverage, together with reducing redundant spending toward fair inclusion of different breeds and administration conditions (Liu, 2025).

Epigenetic editability measurement and ethical guardrails modeling will be invaluable as the technical capability of targeted chromatin modulation and locus-specific methylation editing becomes available. The success probability of edits, their durability, reversibility, or off-target collateral, and the possibility of transgenerational persistence of candidates across regulatory loci will be estimated using predictive models. Welfare impact scoring, biodiversity preservation priorities, societal acceptance sensitivity, and regulatory compliance constraints will be incorporated into ethical decision layers that will be used to filter the intervention portfolios before their consideration by the experiment. Unintended consequences that will be investigated via counterfactual risk scenario simulation will include disruption of adaptive stress response capacity or selection in favor of narrower genetic resilience (Roth et al., 2024).

Integration with the microbiome, metabolome, and proteome will make it possible to have holistic phenotype control approaches capable of modeling host-microbial co-regulation loops. Multi-compartment latent variable models will separate causal interactions between microbial metabolite synthesis, host chromatin remodeling, nutrient uptake routes, immune tolerance tuning, and performance phenotypes. The microbial taxa or metabolic flux nodes that have a disproportionate effect on epigenetic plasticity will be recognized by cross-domain attention modules and used to direct precision prebiotic or probiotic interventions in conjunction with nutritional regimens. The alterations in the host epigenetic state that mediate microbiome compositional stability or resilience will be quantified with the help of reciprocal feedback modeling and used to schedule sustainable interventions without causing ecological overshoot. The proteomic post-translational modification landscapes will further be integrated to provide finer details of regulatory cascade mapping of transcription factor activation to chromatin remodeling to the execution of metabolic pathways (Chetty and Blekhman, 2024). This will be based on standardized robustness and reproducibility benchmarking ecosystems, which provide public leaderboards, perturbation stress test suites, and blind evaluation repositories of tasks such as enhancer identification, methylation imputation, cross-breed generalization, causal mediator discovery, and sensor-fused epigenetic state prediction. The resilience will be tested with synthetic forms of biologically consistent perturbations, such as a distortion of coverage, a batch shift, the absence of a modality, and noise from an adversarial sequence. Common metric frameworks will also report on discrimination, calibration, the sharpness of uncertainty, fairness between breeds or sexes, and energy efficiency, which will discourage metric gaming. The longitudinal monitoring of methodological developments and regression identification will be made possible by the use of containerized reference pipelines and version-locked datasets (Guo et al., 2025).

The socio-economic impact modeling will reveal AI-based-derived epigenetic insights into system-level analysis of breeding and management strategies, giving consideration to adoption heterogeneity, market signals, regulatory changes, labor dynamics, and externalities. The role of diffusion of epigenetic-informed decision support tools at different scales of the farm will be investigated by way of agent-based simulations and system dynamics models, which will project productivity gains, changes in input efficiency, outcomes of emissions, welfare, and economic returns under different incentive regimes. Equity studies will evaluate benefit distribution between large integrated systems and smallholder systems in order to guide policy formulation to prevent making inequalities worse. Sensitivity of risk to climate volatility, feed price shock, or disease outbreak will be measured in risk analysis modules, and this will shed light on the dividends of resilience that can be attributed to epigenetic-mediated adaptive management. Public perception reaction to epigenetic intervention of various strengths will be assessed in ethical scenario modeling to determine the tactics of communication and transparent accountability systems (Alotibi, 2025). Future directions and technological innovations in AI-enabled livestock epigenomics are presented in Table 10.6.

10.7 Ethical considerations in AI use in research for livestock epigenomics

Ethical thinking in the application of AI in livestock epigenomics should align with the developing and growing capacity to predict and possibly manipulate molecular control landscapes associated with productivity, health, and strength. In contrast to the traditional genetic selection process, which is based on relatively fixed inherited variation, epigenomic profiling provides layers of plastic, environmentally sensitive regulation, the interpretation and manipulation of which challenge the notion of stewardship responsibilities, intergenerational responsibilities, the distribution of benefits, and the lawful limits of biological interventions. The ethical space in this case is plural in nature; it incorporates animal science, information jurisdiction, innovation economics, sustainability demands, biosecurity vigilance, and participatory justice. It takes active design principles to navigate such complex terrain, rather than pursuing crisis management of the situation after it occurs, by integrating normative limitations into technical structures, collective governance, and patterns of communication before path dependencies become solid (Sajjanapu, 2025).

The ownership of data, consent, and equal access on scales from one production system to another is a fundamental issue since epigenomic data collection not only captures molecular states but also codifies the historical presence of environmental management, nutritional regimes, health interventions, and microecological exposure. Ownership rights may include producers, breeding organizations, research institutions, technology sellers, and occasionally public funders, all of whom claim a legitimate interest in the derivative value generated by means of AI models. Livestock data stewardship presents the problem of proxy decision makers, as opposed to the human biomedical environment, where individualized consent is used, with their incentives potentially conflicting with the wider societal interests in the conservation of genetic diversity or new disease surveillance as a public good. The risk of smaller operations and under-resourced production systems being pushed out of the market arises when data aggregation platforms take disproportionate rents or charge access fees that are prohibitively expensive and consolidate the informational leverage of already powerful players. Ethical governance must thus also seek to encourage tiered access schemes, a scheme of reciprocity that provides practical insights to data donors, and provisions where a given prediction scheme is not unilaterally enclosed by a community-generated model under which contrived community-generated molecular variation is implicitly encoded. Further complicating data appropriation are cultural heritage inherent to specifically adapted genetic resources, whereby benefit-sharing arrangements are required to address collective custodianship and minimal bio-informational erosion of local autonomy (Lawson et al., 2025).

The epistemic basis of trust and accountable action is transparency, explainability, and displays of the unionized AI products to the stakeholders. Models of epigenomics can incorporate high-dimensional signals that yield multifaceted forecasts concerning trait risk, intervention timing, or putative regulatory targets. When such outputs are put into practice

TABLE 10.6 Future directions and technological innovations in AI-enabled livestock epigenomics.

Innovation domain	Emerging technologies or approaches	Scientific and biological advances enabled	Implications for livestock breeding and management	AI-specific opportunities and requirements
Single-cell and spatial epigenomics	• Single-cell ATAC-seq and methylome assays • Spatially resolved epigenomic profiling • Multi-modal single-cell technologies	• Resolution of cell-type-specific regulatory programs • Mapping of microenvironment-driven epigenetic remodeling • Elimination of bulk tissue confounding	• Precision targeting of immune, metabolic, and reproductive cell states • Improved disease resistance and welfare monitoring	• Graph- and transformer-based spatial modeling • Advanced imputation and uncertainty quantification • Scalable integration of sparse and high-dimensional data
Foundation and self-supervised models	• Large-scale sequence-epigenome foundation models • Masked modeling and contrastive learning frameworks	• Learning universal regulatory grammar across species and tissues • Transferable representations for low-resource breeds	• Faster functional annotation in understudied species • Reduced dependence on costly labeled datasets	• Parameter-efficient fine-tuning • Cross-species domain adaptation • Robust out-of-distribution detection
Multi-modal and multi-scale integration	• Unified modeling of genomics, epigenomics, transcriptomics, metabolomics, microbiome, and sensors • Hierarchical and graph-based fusion	• Coherent modeling of genotype-regulatory-phenotype flow • Identification of multi-layer regulatory bottlenecks	• Holistic breeding value estimation • Improved prediction of complex traits and resilience	• Modality-aware latent representations • Explicit uncertainty propagation across layers
Causal and counterfactual modeling	• Structural causal models • Invariance-based learning • Counterfactual simulation engines	• Distinction between causal regulators and correlates • Mechanistic understanding of regulatory interventions	• Safer and more reliable precision nutrition and management strategies • Reduced risk of unintended phenotypic effects	• Integration of biological priors • Sensitivity analysis and causal validation pipelines
Generative and optimization frameworks	• Generative diffusion and adversarial models • Reinforcement learning for intervention design	• Exploration of feasible regulatory state transitions • Simulation of intervention outcomes under uncertainty	• Optimization of feeding, housing, and stress-mitigation strategies • Balancing productivity, welfare, and sustainability	• Reward functions incorporating welfare and ethics • Safe exploration constraints
Digital twins systems	• Individualized computational livestock models • Continuous state updating from omics and sensors	• Dynamic simulation of regulatory and phenotypic trajectories • Anticipatory detection of adverse states	• Personalized management and early-warning systems • Improved lifetime performance prediction	• Hybrid mechanistic–AI architectures • Real-time Bayesian updating and forecasting
Real-time and edge AI deployment	• Lightweight models for on-farm deployment • Integration with wearable and environmental sensors	• Near–real-time inference of regulatory stress or adaptation	• Rapid response to environmental or health challenges • Reduced reliance on centralized computation	• Model compression and distillation • Energy-efficient inference

Continued

TABLE 10.6 Future directions and technological innovations in AI-enabled livestock epigenomics.—cont'd

Innovation domain	Emerging technologies or approaches	Scientific and biological advances enabled	Implications for livestock breeding and management	AI-specific opportunities and requirements
Federated and collaborative AI	• Federated learning across institutions and regions • Privacy-preserving analytics	• Larger effective training populations • Reduced bias from localized datasets	• Equitable access to advanced AI tools • Enhanced robustness across production systems	• Secure aggregation protocols • Personalization under heterogeneity
Ethically constrained AI and governance	• Embedded welfare and sustainability constraints • Transparent and explainable AI pipelines	• Responsible interpretation of epigenetic plasticity • Avoidance of maladaptive regulatory manipulation	• Public trust and regulatory acceptance • Alignment with sustainability goals	• Explainability and auditability by design • Governance-aware optimization
Standardization and reproducibility infrastructure	• Harmonized metadata standards and ontologies • Benchmark datasets and reproducible pipelines	• Improved cross-study comparability • Accelerated cumulative knowledge building	• Faster translation of research into practice	• Automated provenance tracking • Continuous benchmarking and validation
Active learning and adaptive experimentation	• AI-guided assay selection • Information-gain-driven sampling	• Efficient exploration of regulatory space • Identification of critical developmental windows	• Reduced experimental cost • Faster optimization cycles	• Uncertainty-driven query strategies • Closed-loop learning systems

in breeding or management, and the nature of the underlying rationale is not articulated intelligibly, these stakeholders may be disempowered to critically cast doubt or challenge, or intelligently adopt. Explainability is not about providing an algorithmic score of saliency to contextual explanations to understand and explain the origins of data, sources of uncertainty, the latent assumptions, and where the extrapolation is not reliable. Ethical practice involves creating multiple-layered communication modes adapted to the various degrees of expertise without the process of simplification degenerating into false reductionism. Furthermore, transparency obligations include reporting of possible conflicts of interest, data training development, and subpopulation performance differentials, which will allow making an informed judgment on whether the application of a model will improve structural injustices (Iyer et al., 2025).

To prevent welfare trade-offs in performance-driven optimization, it is essential to exercise caution when aligning objective functions with overall measures of animal well-being. AI systems set up to optimize production may unintentionally prioritize settings of their epigenotypes related to physiological stress response at the expense of comfort or even behavioral expression unless the training labels of the underlying training data portray holistic indicators of welfare. Embedded reinforcing loops have the danger of hastening the distance between measurable productivity and latent welfare decline, particularly in the case where epigenetic biomarkers of allostatic load or chronic stress are poorly modeled or understood as healthy robustness. The multi-objective optimization frameworks of ethical integration, which internalize the validated welfare biomarkers, penalty terms that discourage exploiting the physiological reserve capacity, and constant monitoring of the emergent unintended consequences in the form of altered epigenetic aging patterns or immune dysregulation indicators, are required. The welfare scientists should be incorporated in the development of the model on a participatory basis, so that the surrogate targets do not overshadow the main ethical duties of humane treatment (Foris et al., 2025).

The limitations of epigenetic manipulation and unintentional multigenerational effects present unique issues since controlled remodeling of regulatory marks could have consequences on more remote phenotypes, potentially redefining adaptive plasticity, or even giving rise to feedback with developmental programming in future generations. Interventions to stabilize or redirect the state of epigenetics should consider imperfect comprehension of pleiotropic networks, interdependence with microbiome co-regulation, and contingency by context, reversible. Ethical boundary setting is characterized by careful consideration of persistence, the ability to predict unpredictable gene-environment interaction changes in conditions of climate variability, and possible constriction of phenotypic variance needed for future resilience (Ghavi Hossein-Zadeh, 2026). The level of intervention intensity should be categorized, analytical observation should not be confused with active modulation, longitudinal surveillance should be mandatory after the deployment, and the explicit justification of the interventions whose benefits will be collected not by the improvement of welfare or sustainability, but by economic endpoints. Intergenerational stewardship views the future adaptive potential as a fundamental evaluative dimension (Banushi et al., 2025).

The intellectual property versus open science balance is a strategic conflict that defines the contributions to justice in innovations and the equal distribution of benefits. Algorithms, edited datasets, or risk scores created by using epigenetics can be proprietary to encourage investment, but will disrupt interoperability and limit validation. Excessive exclusivity can slow the identification of bias, error, or misuse, whereas a complete absence of restrictive controls can encourage dual-use risks. An intermediate form of ethical calibration is associated with hybrid models of core data standards, benchmarking resources, and underlying representation layers that are freely available under licenses requiring attribution and responsible usage, but on which application-specific derivative products can maintain partial protection under the condition of non-discrimination and open disclosure of performance. Commitments to benefit sharing should recover a share of commercial value for contributors of collective data and re-invest in the public infrastructures that increase the baseline capacity of production systems that are underrepresented (Lvovs et al., 2025).

Systemic identification and management of performance differences that may exacerbate structural disparities or diminish genetic diversity are needed to achieve fairness among breeds, environments, and management systems in algorithmic fairness. Training datasets tend to be overrepresentative in high-intensity production settings, and as a result, the latent domain shift bias occurs when models are used on other agroecological settings. Epigenomic variations shaped by local environmental pressures may be poorly understood, leading to lower predictive accuracy and inadequately calibrated intervention recommendations for underprivileged populations. Model development based on ethics includes parity analysis, domain adaptation validation, and location-specific data augmentation of under-sampled strata, and does not unnecessarily superficially equalize across true biologically supported differences. The test of fairness should go beyond individual aggregate measures, including the welfare consequences, false positive intervention cost, and genetic diversity loss in case the selection indices are skewed toward epigenetic patterns that are dominant in dominant production regimes (Mayuravaani et al., 2024).

The dual-use potential is seen in biosecurity and misuse risks, in which epigenomic knowledge may be used to control the manipulation of regulatory systems by adversaries to affect animal health, undermine tracing, or create traits with ecological spillover. Models that can be used to identify regulatory nodes that can regulate immune responses or stress endurance have the potential, when abused, to contribute to the spread of vulnerabilities. Mitigation is done through ethical controls, risk classification of sensitive functional annotation output, unusual query patterns that are detected through anomaly, and red teaming to reveal the exploitation pathways. Coordinated disclosure policies with a balance between warning dissemination and containment reduce the unintended amplification of the threat vectors. The inclusion of security checkpoints in model releases makes vigilance institutionalized and does not excessively suppress legitimate research (Pannu et al., 2025).

Computational pipeline sustainability borders on environmental sustainability by acknowledging that the environmental footprint of training models using large resources and of large data storage has indirect effects, which can be inconsistent with the generally more commendable promises of sustainable agriculture. The energy consumption, the effects of the hardware life cycle, and the opportunity costs of the computational resources allocation should be incorporated into the ethical assessment of AI strategies. Recurrence of components that are pre-trained, carbon conception scheduling, model effectiveness norms, pruning approaches, adaptive inaccuracy methodologies, and reuse assistant cut overall footprint per biological insight. Open disclosure of energy measures puts into perspective trade-offs between marginal incremental predictability and sustainable costs that allow stakeholders to make informed choices of what will be an acceptable level of computational cost in relation to planetary boundaries (Sharma, 2024).

Ethical principles are operationalized in the form of governance structures, auditability, and accountability mechanisms. Multilevel governance is a combination of institutional review, a cross-stakeholder advisory panel, and an automated compliance check, ensuring that use of data, model development, and intervention implementation follow a set of normative standards. Auditability demands irrevocable recording of data provenance, preprocessing transformation, model change, hyperparameter settings, and access history, to allow past decision history to be reconstructed. Accountability delineates the allocation of responsibility in case of undesirable consequences in order to differentiate between model developers, data custodians, deploying practitioners, and oversight agencies. Adaptive governance enables the process of continuous improvement through a series of iterations as more knowledge is built up on model impact, without strict codification, which could fall behind technological change. The provision of dispute resolution procedures helps parties who have been negatively impacted by model recommendations to be redressed (Chen et al., 2025).

Ethical considerations of predictive aging and selection forces have to weigh the possible benefits of predicting longevity and predicting health, or the ability to optimize productive lifespan, against the risk of reducing the phenotypic variability that is acceptable or increasing selection on epigenetic signatures whose etiological bases are not fully comprehended. Predictive epigenetic clocks are likely to unwittingly discriminate against animals with non-convergent developmental histories, hastening culling or biased breeding focus toward molecular phenotypes that are associated with short-term productivity as opposed to robustness. Ethical deployment models support multifaceted appraisal that incorporates welfare, flexibility, and diversity protection measurements, and need to reveal uncertainties to avoid premature study indulgence on biomarkers. Surveillance of correlated changes in behavioral characteristics or stress indicators ensures that selection is not based on unwanted welfare (Newediuk et al., 2025).

The strategy of inclusivity and digital divide reduction addresses the differences in the ability to create, comprehend, and use epigenomic AI products at geographic, production, and socioeconomic levels. As a matter of course, technological sophistication can become centralized in resource-rich networks, and small or distant producers may no longer have access to decision support or representation in training corpora. The focus of the ethical strategy is on capacity building, providing subsidized infrastructure to support participants who are under-resourced, interoperable low-bandwidth tool variants, multilingual interfaces, and joint training programs that promote local analytical independence. The participatory design processes seek feedback at all stages, which effectively customizes the requirements of systems to operate on realities in operations outside of high-input contexts. The use of metrics to monitor adoption equity and outcomes distribution makes adaptive alterations to the policy (Kettani et al., 2025).

Proper expression of probabilistic risk and uncertainty forms the basis of honest decision-making to avoid the misuse of predictive results to give out certainties that are not deterministic. The nature of epigenomic AI models is such that the models exist under various layers of epistemic uncertainty and aleatory uncertainty due to the sparsity of the sampling, noise in measurements, model approximation, and confounding that is not explicitly modeled. Ethical communication is not too simplistic but rather tends to provide scaled probability ranges, confidence bands, and contingencies in scenarios while explaining leftover uncertainty. The visualization and narrative framing techniques should not contain such cognitive biases as exaggerated perception of accuracy or low-probability, high-impact risks. Statistical literacy training

TABLE 10.7 Ethical considerations in AI use in research for livestock epigenomics.

Ethical dimension	Key ethical concerns	Risks if unaddressed	Implications for epigenomic research and AI models	Ethical safeguards and governance approaches
Animal welfare and well-being	• Potential for AI-driven optimization to prioritize productivity over welfare • Use of invasive sampling for epigenomic profiling • Risk of masking early stress signals through regulatory manipulation	• Compromised physical and psychological welfare • Undetected chronic stress or disease states • Ethical and societal backlash	• Biased reward functions favoring output metrics • Misinterpretation of epigenetic plasticity as purely exploitable	• Welfare-aware objective functions in AI models • Non-invasive or minimally invasive sampling strategies • Continuous welfare monitoring integration
Responsible use of epigenetic plasticity	• Ethical boundaries of intentionally modifying epigenetic states • Long-term and multigenerational uncertainty	• Unintended phenotypic consequences • Reduced adaptive resilience	• Overconfident causal inference • Inadequate modeling of long-term effects	• Precautionary modeling assumptions • Long-horizon uncertainty quantification • Explicit limits on intervention recommendations
Transparency and explainability	• Opaque AI decision-making processes • Difficulty interpreting deep learning outputs	• Loss of trust among researchers, breeders, and regulators • Inability to audit decisions	• Reduced adoption of AI tools • Misalignment between model logic and biology	• Explainable AI frameworks (XAI) • Mechanism-aware model design • Clear documentation and reporting standards
Bias, fairness, and representation	• Overrepresentation of specific breeds, systems, or regions • Underrepresentation of low-resource production systems	• Systematic bias in predictions and recommendations • Widening inequities	• Poor generalization across diverse contexts • Reinforcement of existing disparities	• Inclusive data collection strategies • Bias auditing and fairness evaluation • Domain generalization methods
Data ownership and privacy	• Proprietary breeding data concerns • Ambiguity in ownership of derived AI models	• Restricted collaboration • Misuse or unauthorized exploitation of data	• Fragmented datasets • Reduced training data diversity	• Federated and privacy-preserving learning • Clear data governance agreements • Transparent benefit-sharing frameworks
Accountability and responsibility	• Ambiguity in responsibility for AI-driven decisions • Automated recommendations without human oversight	• Ethical and legal liability disputes • Misguided management or breeding actions	• Overreliance on automated inference • Reduced critical oversight	• Human-in-the-loop decision frameworks • Clear accountability assignment • Audit trails and decision logging

Continued

TABLE 10.7 Ethical considerations in AI use in research for livestock epigenomics.—cont'd

Ethical dimension	Key ethical concerns	Risks if unaddressed	Implications for epigenomic research and AI models	Ethical safeguards and governance approaches
Environmental sustainability	• Energy-intensive AI computation • Resource-heavy experimental pipelines	• Increased carbon footprint • Misalignment with sustainability goals	• Trade-offs between model complexity and ecological cost	• Energy-efficient model design • Sustainable computing practices • Lifecycle assessment of AI pipelines
Reproducibility and scientific integrity	• Non-transparent preprocessing and modeling pipelines • Lack of standardized benchmarks	• Irreproducible findings • Erosion of scientific credibility	• Inflated performance claims • Difficulty validating AI-derived insights	• Open methods and code sharing • Reproducibility infrastructures • Benchmark datasets
Cross-generational and long-term effects	• Potential transgenerational epigenetic consequences • Limited empirical validation	• Accumulation of unforeseen effects across generations	• Overly short-sighted optimization objectives	• Longitudinal and multigenerational modeling • Conservative policy recommendations
Societal acceptance and public trust	• Perception of excessive technological intervention • Concerns about engineering livestock biology	• Public resistance and market rejection	• Regulatory and adoption barriers	• Transparent communication strategies • Stakeholder engagement • Alignment with societal values
Regulatory compliance and oversight	• Evolving legal frameworks for AI and animal research • Lack of epigenomics-specific regulations	• Legal uncertainty and compliance risk	• Hesitation in deployment and innovation	• Proactive engagement with regulators • Adaptive governance models
Equitable access to AI innovations	• Concentration of AI capabilities in well-funded institutions	• Technological divide across regions and producers	• Unequal distribution of benefits	• Federated learning and open tools • Capacity-building initiatives

of the stakeholders and the creation of institutional cultures in which recognition of uncertainty is rewarded, and uncertainty is not punished, alleviate the pressure to overstate reliability to gain a competitive edge (Jiménez et al., 2025).

These ethical dimensions can be integrated into a consistent practice architecture only when the change happens from reactive compliance to proactive ethical engineering. The structural reinforcement of ethical conduct through embedding fairness diagnostics, welfare protection, sustainability measures, security checks, uncertainty measurement, and explainability layers in centralized model lifecycle management platforms generates structural reinforcement of ethical behavior. Cross-disciplinary teamwork can be used to ensure that technical design decisions are able to reflect domain-specific normative knowledge and that iterative stakeholder involvement is used to bring system evolution and emergent values and social expectations into alignment with each other. Ethically growing AI in livestock epigenomics will depend on bringing innovation pathways and a whole concept of value together that will combine animal welfare, ecological integrity, social justice, economic viability, and intergenerational resilience. This kind of fit leads to both ethical legitimacy and long-run resilience of the scientific and production ecosystems, which rely on the trust of the people and ecological stability (Tripathi, 2025). Ethical considerations in AI use in research for livestock epigenomics are shown in Table 10.7.

References

Aguirre, M., Spence, J.P., Sella, G., Pritchard, J.K., 2025. Regulatory Network Topology and the Genetic Architecture of Gene Expression. bioRxiv, United States. https://doi.org/10.1101/2025.08.12.669924.

Ali, A.M.O., 2025. Explainability in AI: interpretable models for data science. Int. J. Res. Appl. Sci. Eng. Technol. 13 (2), 766–771. https://doi.org/10.22214/ijraset.2025.66968.

Alotibi, Y.S., 2025. A socio-technical agent-based simulation model for predicting smart agriculture adoption dynamics. Sci. Rep. 15 (1). https://doi.org/10.1038/s41598-025-27523-7.

Alsayaydeh, D.A.J., Narayanamurthy, V., Futane, A.S., Alsayaydeh, J.A.J., Najib, S.B.M., 2025. Advances of artificial intelligence-enabled epigenetics. Health Technol. 15 (4), 663–676. https://doi.org/10.1007/s12553-025-00998-0.

Ambati, M., Choudhary, A., 2025. Topologically associating domains in 3D chromatin organization. https://doi.org/10.31219/osf.io/hrjd3_v1.

Angioni, D., Demetrio, L., Pintor, M., Oneto, L., Anguita, D., Biggio, B., Roli, F., 2025. Robustness-congruent adversarial training for secure machine learning model updates. IEEE Trans. Pattern Anal. Mach. Intell. 47 (9), 7457–7469. https://doi.org/10.1109/tpami.2025.3573237.

Baião, A.R., Cai, Z., Poulos, R.C., Robinson, P.J., Reddel, R.R., Zhong, Q., Vinga, S., Gonçalves, E., 2025. A technical review of multi-omics data integration methods: from classical statistical to deep generative approaches. Briefings Bioinf. 26 (4). https://doi.org/10.1093/bib/bbaf355.

Banushi, B., Collova, J., Milroy, H., Echoes, E., 2025. Bridging nature, nurture, and healing across generations. Int. J. Mol. Sci. 26 (7), 3075. https://doi.org/10.3390/ijms26073075.

Benčik, I., Saftić Martinović, L., Mladenić, T., Ostojić, S., Pavlić, S.D., 2025. From DNA methylation and histone modifications to non-coding RNAs: evaluating tools for epigenetic research. Appl. Sci. 15 (18), 9940. https://doi.org/10.3390/app15189940.

Bernasocchi, T., Mostoslavsky, R., 2024. Subcellular one carbon metabolism in cancer, aging and epigenetics. Front. Epigenet. Epigenom. 2. https://doi.org/10.3389/freae.2024.1451971.

Bi, H., Weng, X., 2025. Single-cell epigenomics and proteomics methods integrated in multiomics. Fundam. Res. 5 (5), 1988–2002. https://doi.org/10.1016/j.fmre.2023.11.014.

Briscik, M., Tazza, G., Vidács, L., Dillies, M.A., Déjean, S., 2024. Supervised multiple kernel learning approaches for multi-omics data integration. BioData Min. 17 (1). https://doi.org/10.1186/s13040-024-00406-9.

Brooks, T.G., Lahens, N.F., Mrčela, A., Grant, G.R., 2024. Challenges and best practices in omics benchmarking. Nat. Rev. Genet. 25 (5), 326–339. https://doi.org/10.1038/s41576-023-00679-6.

Cai, R., Huang, X., Chen, W., Li, Z., Hao, Z., 2025. Temporal latent variable structural causal model for causal discovery under external interferences. Neurocomputing 640, 130281. https://doi.org/10.1016/j.neucom.2025.130281.

Caldwell, B.A., Li, L., 2024. Epigenetic regulation of innate immune dynamics during inflammation. J. Leukoc. Biol. 115 (4), 589–606. https://doi.org/10.1093/jleuko/qiae026.

Cao, J., Zhou, X., Liu, J., Liu, D., Liu, G., 2025. Single-cell multi-omics sequencing technologies and their applications in livestock and poultry research. J. Agric. Food Chem. https://doi.org/10.1021/acs.jafc.5c13040.

Chen, Y.C., 2025. A frequentist statistical introduction to variational inference, autoencoders, and diffusion models. arXiv. https://doi.org/10.48550/arXiv.2510.18777.

Chen, Z., Wang, Y., Zhao, X., 2025. Responsible generative AI: governance challenges and solutions in enterprise data clouds. J. Comput. Electron. Inf. Manag. 18 (3), 59–65. https://doi.org/10.54097/02teq773.

Cheng, L., Tong, Y., Khalitov, R., Yang, Z., 2024. Self-supervised learning for DNA sequences with circular dilated convolutional networks. Neural Netw. 171, 466–473. https://doi.org/10.1016/j.neunet.2023.12.002.

Chetty, A., Blekhman, R., 2024. Multi-omic approaches for host-microbiome data integration. Gut Microbes 16 (1). https://doi.org/10.1080/19490976.2023.2297860.

de Weerd, H.A., Guala, D., Gustafsson, M., Synnergren, J., Tegnér, J., Lubovac-Pilav, Z., Magnusson, R., 2024. Latent space arithmetic on data embeddings from healthy multi-tissue human RNA-seq decodes disease modules. Patterns 5 (11), 101093. https://doi.org/10.1016/j.patter.2024.101093.

Dibaeinia, P., Ojha, A., Sinha, S., 2025. Interpretable AI for inference of causal molecular relationships from omics data. Sci. Adv. 11 (7). https://doi.org/10.1126/sciadv.adk0837.

Dou, Y., Mirzaei, G., Cowen, L., 2025. MO-GCAN: multi-omics integration based on graph convolutional and attention networks. Bioinformatics 41 (8). https://doi.org/10.1093/bioinformatics/btaf405.

Dukić, D., 2025. Improving transfer learning for sequence labeling tasks by adapting pre-trained neural language models. arXiv. https://doi.org/10.48550/arXiv.2510.20033.

Eckhardt, R., Arablouei, R., Ingham, A., McCosker, K., Bernhardt, H., 2025. Livestock behaviour forecasting via generative artificial intelligence. Smart Agric. Technol. 11, 100987. https://doi.org/10.1016/j.atech.2025.100987.

Edwards-Lee, C.A., Jarred, E.G., Western, P.S., 2025. Coordinated regulation of chromatin modifiers reflects organised epigenetic programming in mouse oocytes. Epigenetics Chromatin 18 (1). https://doi.org/10.1186/s13072-025-00583-9.

Fan, J., 2025. An interdisciplinary and cross-task review on missing data imputation. arXiv. https://doi.org/10.48550/arXiv.2511.01196.

Fan, Q., Zhao, S., Zurada, J.M., Huang, T., Qin, X., Zhang, R., 2025. Bidirectional multiscale efficient dilated convolutional recurrent neural network improved by swarm intelligence optimization. IEEE Transact. Neural Networks Learn. Syst. 36 (12), 20009–20023. https://doi.org/10.1109/TNNLS.2025.3596244.

Faraji, J., Metz, G.A.S., 2025. Environmental Epigenetics: New Horizons in Redefining Biological and Health Outcomes. https://doi.org/10.31234/osf.io/sz9xv_v1.

Foris, B., Sheng, K., Dürnberger, C., Oczak, M., Rault, J.L., 2025. AI for One Welfare: the role of animal welfare scientists in developing valid and ethical AI-based welfare assessment tools. Front. Vet. Sci. 12. https://doi.org/10.3389/fvets.2025.1645901.

Gao, X., 2025. Toward real-time and efficient edge intelligence: advances and challenges in lightweight machine learning. Sci. Technol. Eng. Chem. Environ. Prot. 1 (4). https://doi.org/10.61173/bwn0ez36.

George, K., Neininger, K., Schmitz, A.E., Walter, J., Helms, V., 2025. Structural DNMT-nucleosome contacts are related to DNA methylation patterns. Epigenetics Chromatin 18 (1). https://doi.org/10.1186/s13072-025-00626-1.

Ghavi Hossein-Zadeh, N., 2024. An overview of recent technological developments in bovine genomics. Vet. Anim. Sci. 25, 100382. https://doi.org/10.1016/j.vas.2024.100382.

Ghavi Hossein-Zadeh, N., 2025. Artificial intelligence in veterinary and animal science: applications, challenges, and future prospects. Comput. Electron. Agric. 235, 110395. https://doi.org/10.1016/j.compag.2025.110395.

Ghavi Hossein-Zadeh, N., 2026. Advancing climate-resilient livestock systems: next-generation emission mitigation strategies and integrated technological innovations. Vet. Anim. Sci. 31, 100588. https://doi.org/10.1016/j.vas.2026.100588.

Gillis, H.M., Xu, I., Trappenberg, T., 2025. Uncertainty estimation using variance-gated distributions. arXiv. https://doi.org/10.48550/arXiv.2509.08846.

Goyal, R., Kumar, K., Sharma, V., Bhutia, R., Jain, A., Kumar, M., 2024. Quantum-inspired optimization algorithms for scalable machine learning in edge computing. In: Proceedings - 4th International Conference on Technological Advancements in Computational Sciences, ICTACS 2024. Institute of Electrical and Electronics Engineers Inc, India, pp. 1888–1892. https://doi.org/10.1109/ICTACS62700.2024.10840586.

Guo, X., Chen, Q., Zhang, Y., Zhang, Y., Liu, Y., Duan, S., Ma, Y., Ni, P., Wang, J., He, B., Ren, L., Ma, R., Hou, W., Yu, Y., Li, B., Qiu, F., Sun, Y., Zhang, Z., Xu, W., Fang, X., Li, J., Shi, L., Zhang, R., Zheng, Y., Dong, L., 2025. Methylation reference datasets from quartet DNA materials for benchmarking epigenome sequencing. Nat. Commun. 16 (1). https://doi.org/10.1038/s41467-025-64250-z.

Hansen, J., Jain, A.R., Nenov, P., Robinson, P.N., Iyengar, R., 2024. From transcriptomics to digital twins of organ function. Front. Cell Dev. Biol. 12. https://doi.org/10.3389/fcell.2024.1240384.

Hegde, A., Nguyen, T., Cheng, J., 2025. Machine learning methods for gene regulatory network inference. arXiv. https://doi.org/10.48550/arXiv.2504.12610.

Iyer, V., Manshad, M., Brannon, D., 2025. A value-based approach to AI ethics: accountability, transparency, explainability, and usability. Mercados y Negocios 26 (54), 3–12. https://doi.org/10.32870/myn.vi54.7815.

Jiang, W., Liu, G., He, Di, He, K., 2025. Boosting meta-training with base class information for robust few-shot learning. Eng. Appl. Artif. Intell. 152, 110780. https://doi.org/10.1016/j.engappai.2025.110780.

Jiménez, S., Jürgens, M., Waegeman, W., 2025. Why machine learning models fail to fully capture epistemic uncertainty. arXiv. https://doi.org/10.48550/arXiv.2505.23506.

Jin, W., Lang, L., Spence, A.B., Rubin, L.H., Xu, Y., 2025. Bayesian causal discovery with cycles and latent confounders. arXiv. https://doi.org/10.48550/arXiv.2511.12333.

Kallenberg, M., Baja, H., Ilić, M., Tomčić, A., Tošić, M., Athanasiadis, I., 2025. Interoperable agricultural digital twins with reinforcement learning intelligence. Smart Agric. Technol. 12, 101412. https://doi.org/10.1016/j.atech.2025.101412.

Karin, O., Miska, E.A., Simons, B.D., 2023. Epigenetic inheritance of gene silencing is maintained by a self-tuning mechanism based on resource competition. Cell Syst. 14 (1), 24. https://doi.org/10.1016/j.cels.2022.12.003.

Kelly, K., Scherer, M., Braun, M.M., Lutsik, P., Plass, C., 2024. EpiCHAOS: a metric to quantify epigenomic heterogeneity in single-cell data. Genome Biol. 25 (1). https://doi.org/10.1186/s13059-024-03446-w.

Kettani, D., Moulin, B., Barbosa, L.S., 2025. Digital divide (2.0): the shadow of AI technology. Perspect. Global Dev. Technol. 24 (3–4), 484–513. https://doi.org/10.1163/15691497-12341720.

Khan, S.A., Martínez-de-Morentin, X., Alsabbagh, A.R., Maillo, A., Lagani, V., Gomez-Cabrero, D., Lehmann, R., Tegner, J., 2025. Multimodal foundation transformer models for multiscale genomics. Nat. Methods. https://doi.org/10.1038/s41592-025-02918-6.

Kumar, A., Chaurasia, M., Singh, R., Kumar, A., Chaurasia, M., Singh, R., 2025. Privacy-preserving federated learning: challenges, techniques, and prospects for distributed AI. Int. J. Multidiscip. Res. 7 (5). https://doi.org/10.36948/ijfmr.2025.v07i05.57558.

Lara-Gutierrez, A., Fernandez-Gago, C., Onieva, J.A., 2025. A framework for drift detection and adaptation in AI-driven anomaly and threat detection systems. Int. J. Inf. Secur. 24 (5). https://doi.org/10.1007/s10207-025-01118-9.

Lawson, C., Englezos, E., Rourke, M., 2025. Access and benefit sharing biological materials for machines: artificial intelligence, machine learning and deep learning. Plants People Planet 7 (5), 1485–1497. https://doi.org/10.1002/ppp3.70007.

Li, Z., Khan, W.U., Bai, G., Dong, C., Wang, J., Zhang, Y., Wang, C., Zhang, H., Wang, W., Luo, M., Chen, F., 2025. From code to life: the AI-driven revolution in genome editing. Adv. Sci. 12 (30). https://doi.org/10.1002/advs.202417029.

Li, Z., Zhu, C., 2025. Auditing cognitive drift in AI-driven recommendation: a responsible AI methods protocol with a health case demonstration. Front. Neurosci. 19. https://doi.org/10.3389/fnins.2025.1697053.

Lin, X., Yu, Q., Cao, Y., Zou, L., Zhou, C., Wu, J., Li, C., Zhang, P., Pan, S., Lin, X., Yu, Q., Cao, Y., Zou, L., Zhou, C., Wu, J., Li, C., Zhang, P., Pan, S., 2025. Generative causality-driven network for graph multi-task learning. IEEE Trans. Pattern Anal. Mach. Intell. 1–16. https://doi.org/10.1109/tpami.2025.3610096.

Liu, J., Zhou, Z., Chen, Y., Islam, M.T., Xing, L., 2025. Unveiling tissue heterogeneity through genomic interaction-encoded image representation of RNA-sequencing data. Am. J. Hum. Genet. 112 (11), 2739–2750. https://doi.org/10.1016/j.ajhg.2025.08.021.

Liu, S., Peng, T., Yang, N., Bi, H., Liu, L., 2026. Rethinking structural semantics for unsupervised graph domain adaptation. Knowl. Base Syst. 332, 114878. https://doi.org/10.1016/j.knosys.2025.114878.

Liu, Y., 2025. (Invited) Gated active learning: integrating prior knowledge and expert insights in autonomous experiments. ECS Meet. Abstr. MA2025–01 (44), 2359. https://doi.org/10.1149/ma2025-01442359mtgabs.

Lourenço, V., Ogutu, J., Rodrigues, R.P., Posekany, A., Piepho, H.P., 2024. Genomic prediction using machine learning: a comparison of the performance of regularized regression, ensemble, instance-based and deep learning methods on synthetic and empirical data. BMC Genom. 25 (1). https://doi.org/10.1186/s12864-023-09933-x.

Lu, W., Wang, J., 2025. Self-ensemble post learning for noisy domain generalization (version 1). arXiv. https://doi.org/10.48550/ARXIV.2512.10818.

Luo, S., Qian, J., Geng, Y., Zhou, Q., Lin, Q., 2025. Toward efficient digital twin simulation: a causal representation learning approach. Knowl. Base Syst. 329, 114442. https://doi.org/10.1016/j.knosys.2025.114442.

Lvovs, D., Creason, A.L., Levine, S.S., Noble, M., Mahurkar, A., White, O., Fertig, E.J., 2025. Balancing ethical data sharing and open science for reproducible research in biomedical data science. Cell Rep. Med. 6 (4). https://doi.org/10.1016/j.xcrm.2025.102080.

Ma, R., Kuang, R., Zhang, J., Sun, J., Xu, Y., Zhou, X., Han, Z., Hu, M., Wang, D., Fu, Y., Zhang, Y., Li, X., Zhu, M., Zhao, S., Xiang, T., Shi, M., Zhao, Y., 2025. Annotation and assessment of functional variants in livestock through epigenomic data. J. Genet. Genomics 52 (12), 1588–1599. https://doi.org/10.1016/j.jgg.2025.03.013.

Mayuravaani, M., Ramanan, A., Perera, M., Asanka Senanayake, D., Vidanaarachchi, R., 2024. Insights into artificial intelligence bias: implications for agriculture. Digit. Soc. 3 (3). https://doi.org/10.1007/s44206-024-00142-x.

Medapati, S., 2025. Explainability in artificial intelligence: giving a method to the madness. J. Inform. Systems Eng. Manag. 10 (59s), 605–613. https://doi.org/10.52783/jisem.v10i59s.12916.

Meng, Z., Chu, M., Yang, H., Zhang, S., Wang, Q., Chen, J., Ren, C., Pan, Z., Zhang, Z., 2025. Regulatory element map of sheep reproductive tissues: functional annotation of tissue-specific strong active enhancers. Front. Vet. Sci. 12. https://doi.org/10.3389/fvets.2025.1564148.

Moran, G., Aragam, B., 2025. Towards interpretable deep generative models via causal representation learning. arXiv. https://doi.org/10.48550/arXiv.2504.11609.

Newediuk, L., Richardson, E.S., Bohart, A.M., Roberto-Charron, A., Garroway, C.J., Jones, M.J., 2025. Designing epigenetic clocks for wildlife research. Mol. Ecol. Resour. 25 (7). https://doi.org/10.1111/1755-0998.14120.

Nogare, D., Silveira, I.F., Cabral, P.P., Hauy, R.J., Neves, V., Nogare, D., Silveira, I.F., Cabral, P.P., Hauy, R.J., Neves, V., 2024. Machine Learning Model: perspectives for quality, observability, risk and continuous monitoring. Anais Do XXI Congresso Latino-Americano de Software Livre e Tecnologias Abertas (Latinoware 2024) 181–187. https://doi.org/10.5753/latinoware.2024.245679.

Ohishi, H., Au Yeung, W.K., 2025. Recent advances in methodologies of epigenomics. Epigenomics. https://doi.org/10.1080/17501911.2025.2571391.

Olmeda, F., Gupta, M., Bektas, O., Rulands, S., 2025. Spatio-temporal patterns of active epigenetic turnover. arXiv. https://doi.org/10.48550/arXiv.2505.13216.

Pannu, J., Bloomfield, D., MacKnight, R., Hanke, M.S., Zhu, A., Gomes, G., Cicero, A., Inglesby, T.V., 2025. Dual-use capabilities of concern of biological AI models. PLoS Comput. Biol. 21 (5). https://doi.org/10.1371/journal.pcbi.1012975.

Pereira, R.C., Abreu, P.H., Rodrigues, P.P., Figueiredo, M.A.T., 2024. Imputation of data missing not at random: artificial generation and benchmark analysis. Expert Syst. Appl. 249. https://doi.org/10.1016/j.eswa.2024.123654.

Perkasa, T.A.B., Puspasari, A., Priskila, L., Tarigan, N.P.I., Widyawati, D., Neldi, V., 2025. Exploration the multifaceted role of vitamins in epigenetic regulation: insights into DNA methylation and histone modification. Jambi Medical Journal: Jurnal Kedokteran dan Kesehatan 13 (1), 76–90. https://doi.org/10.22437/jmj.v13i1.42536.

Ping, M., Yan, W.-J., Papadimitriou, C., 2026. Variational inference for hierarchical Bayesian learning framework for model updating with non-stationary prediction errors. Reliab. Eng. Syst. Saf. 268, 111944. https://doi.org/10.1016/j.ress.2025.111944.

Preissl, S., Gaulton, K.J., Ren, B., 2023. Characterizing cis-regulatory elements using single-cell epigenomics. Nat. Rev. Genet. 24 (1), 21–43. https://doi.org/10.1038/s41576-022-00509-1.

Qiang, W., Song, Z., Gu, Z., Li, J., Zheng, C., Sun, F., Xiong, H., 2025. On the generalization and causal explanation in self-supervised learning. Int. J. Comput. Vis. 133 (4), 1727–1754. https://doi.org/10.1007/s11263-024-02263-9.

Ribeiro-Dos-Santos, A.M., Maurano, M.T., 2025. Iterative improvement of deep learning models using synthetic regulatory genomics. Genome Res. 35 (11), 2539–2549. https://doi.org/10.1101/gr.280540.125.

Rojahn, M., Grum, M., 2025. Green AI: a systematic review and meta-analysis of its definitions, lifecycle models, hardware and measurement attempts (Version 4). arXiv. https://doi.org/10.48550/ARXIV.2511.07090.

Rossi, M., El-Sayed, A., 2025. Meta-learning driven few-shot diagnostics: addressing rare disease classification in medical AI. Int. J. Adv. Artif. Intell. Res. 2 (05), 7–14. https://doi.org/10.55640/ijaair-v02i05-02.

Roth, G.V., Gengaro, I.R., Qi, L.S., 2024. Precision epigenetic editing: technological advances, enduring challenges, and therapeutic applications. Cell Chem. Biol. 31 (8), 1422–1446. https://doi.org/10.1016/j.chembiol.2024.07.007.

Sajjanapu, M., 2025. Ethical AI and sustainable digital governance: an interdisciplinary framework. Eur. Mod. Stud. J. 9 (4), 804–814. https://doi.org/10.59573/emsj.9(4).2025.76.

Sandnes, A.T., Grimstad, B., Kolbjørnsen, O., 2024. Multi-task neural networks by learned contextual inputs. Neural Netw. 179, 106528. https://doi.org/10.1016/j.neunet.2024.106528.

Schumann, Y., Gocke, A., Neumann, J.E., 2025. Computational methods for data integration and imputation of missing values in omics datasets. Proteomics 25 (1–2). https://doi.org/10.1002/pmic.202400100.

Seebacher, F., Little, A.G., 2025. Mechanisms underlying phenotypic plasticity in response to environmental change. Biol. Rev. https://doi.org/10.1111/brv.70100.

Seem, K., Kaur, S., Kumar, S., Mohapatra, T., 2024. Epigenome editing for targeted DNA (de)methylation: a new perspective in modulating gene expression. Crit. Rev. Biochem. Mol. Biol. 59 (1–2), 69–98. https://doi.org/10.1080/10409238.2024.2320659.

Sharma, A., 2024. Green AI: minimizing environmental cost of AI model training and deployment. Adhyayan: A J. Manag. Sci. 14 (02), 28–30. https://doi.org/10.21567/adhyayan.v14i2.06.

Shen, Y., Kudla, G., Oyarzún, D.A., 2025. Optimization of regulatory DNA with active learning. Comput. Struct. Biotechnol. J. 27, 4384–4392. https://doi.org/10.1016/j.csbj.2025.09.033.

Shuvra, M.K., Gony, Md N., Fatema, K., 2024. Explainability vs. Performance: bridging the trade-off in deep learning models. Int. J. Adv. Res. Comput. Sci. Technol. 07 (05). https://doi.org/10.15662/ijarcst.2024.0705006.

Simpson, C., Tabatsky, E., Rahil, Z., Eddins, D.J., Tkachev, S., Georgescauld, F., Papalegis, D., Culka, M., Levy, T., Gregoretti, I., Chernyshev, A., Koeppen, H., Walther, G., Ghosn, E.E.B., Orlova, D., 2024. Lifting the curse from high dimensional data: automated projection pursuit clustering for the variety of biological data modalities. bioRxiv. https://doi.org/10.1101/2024.04.18.589981.

Sindhu, P., Magotra, A., Sindhu, V., Chaudhary, P., 2024. Unravelling the impact of epigenetic mechanisms on offspring growth, production, reproduction and disease susceptibility. Zygote 32 (3), 190–206. https://doi.org/10.1017/s0967199424000224.

Singh, J.P., Aqsa, A., Ghani, I., Sonani, R., Govindarajan, V., 2025. Privacy-aware hierarchical federated learning in healthcare: integrating differential privacy and secure multi-party computation. Future Internet 17 (8). https://doi.org/10.3390/fi17080345.

Skok Gibbs, C., Mahmood, O., Bonneau, R., Cho, K., 2024. PMF-GRN: a variational inference approach to single-cell gene regulatory network inference using probabilistic matrix factorization. Genome Biol. 25 (1). https://doi.org/10.1186/s13059-024-03226-6.

Stefansson, O.A., Sigurpalsdottir, B.D., Rognvaldsson, S., Halldorsson, G.H., Juliusson, K., Sveinbjornsson, G., Gunnarsson, B., Beyter, D., Jonsson, H., Gudjonsson, S.A., Olafsdottir, T.A., Saevarsdottir, S., Magnusson, M.K., Lund, S.H., Tragante, V., Oddsson, A., Hardarson, M.T., Eggertsson, H.P., Gudmundsson, R.L., Sverrisson, S., Frigge, M.L., Zink, F., Holm, H., Stefansson, H., Rafnar, T., Jonsdottir, I., Sulem, P., Helgason, A., Gudbjartsson, D.F., Halldorsson, B.V., Thorsteinsdottir, U., Stefansson, K., 2024. The correlation between CpG methylation and gene expression is driven by sequence variants. Nat. Genet. 56 (8), 1624–1631. https://doi.org/10.1038/s41588-024-01851-2.

Taddese, A.A., Addis, A.C., Tam, B.T., 2025. Data stewardship and curation practices in AI-based genomics and automated microscopy image analysis for high-throughput screening studies: promoting robust and ethical AI applications. Hum. Genom. 19 (1). https://doi.org/10.1186/s40246-025-00716-x.

Tahir, M., Norouzi, M., Khan, S.S., Davie, J.R., Yamanaka, S., Ahmed, A., 2024. Artificial intelligence and deep learning algorithms for epigenetic sequence analysis: a review for epigeneticists and AI experts. Comput. Biol. Med. 183, 109302. https://doi.org/10.1016/j.compbiomed.2024.109302.

Tejada-Lapuerta, A., Bertin, P., Bauer, S., Aliee, H., Bengio, Y., Theis, F.J., 2025. Causal machine learning for single-cell genomics. Nat. Genet. 57 (4), 797–808. https://doi.org/10.1038/s41588-025-02124-2.

Tenekeci, S., Tekir, S., 2024. Identifying promoter and enhancer sequences by graph convolutional networks. Comput. Biol. Chem. 110, 108040. https://doi.org/10.1016/j.compbiolchem.2024.108040.

Tian, Y., McDonnell, S.K., Wu, L., Larson, N.B., Wang, L., 2026. Fine mapping regulatory variants by characterizing native CpG methylation with nanopore long-read sequencing. Hum. Genet. Genomics Adv. 7 (1). https://doi.org/10.1016/j.xhgg.2025.100532.

Tripathi, G., 2025. AI ethics and responsible AI development: navigating the rapid evolution of artificial intelligence. World J. Adv. Res. Rev. 26 (3), 145–152. https://doi.org/10.30574/wjarr.2025.26.3.2126.

Wall, B.P.G., Nguyen, M., Chuck Harrell, J., Dozmorov, M.G., 2024. Machine and Deep Learning Methods for Predicting 3D Genome Organization. Springer Science and Business Media LLC, pp. 357–400. https://doi.org/10.1007/978-1-0716-4136-1_22.

Wang, C., Lin, Y., Li, S., Guan, J., 2024. Deconvolution from bulk gene expression by leveraging sample-wise and gene-wise similarities and single-cell RNA-seq data. BMC Genom. 25 (1). https://doi.org/10.1186/s12864-024-10728-x.

Wang, X., Liu, J., Hu, S.S., Liu, Z., Lu, H., Liu, L., 2025. HILAMA: high-dimensional multi-omics mediation analysis with latent confounding. BMC Med. Res. Methodol. 25 (1). https://doi.org/10.1186/s12874-025-02686-z.

Wolfrath, J., Achanta, A., Chandra, A., 2024. Leveraging multi-modal data for efficient edge inference serving. In: Proceedings - 2024 IEEE 24th International Symposium on Cluster, Cloud and Internet Computing, CCGrid 2024. Institute of Electrical and Electronics Engineers Inc., United States, pp. 408–418. https://doi.org/10.1109/CCGrid59990.2024.00053.

Wu, A.P., Singh, R., Walsh, C.A., Berger, B., 2025a. Unveiling causal regulatory mechanisms through cell-state parallax. Nat. Commun. 16 (1). https://doi.org/10.1038/s41467-025-61337-5.

Wu, Y., Liu, J., Xiao, Y., Zhang, S., Li, L., 2025b. CoupleVAE: coupled variational autoencoders for predicting perturbational single-cell RNA sequencing data. Briefings Bioinf. 26 (2). https://doi.org/10.1093/bib/bbaf126.

Xia, Z., Tan, Z., Zhang, B., 2025. Overcoming domain shift in violence detection with contrastive consistency learning. Big Data Cogn. Comput. 9 (11), 286. https://doi.org/10.3390/bdcc9110286.

Xiong, J., Zhu, B., 2025. Epigenetic preparation of future gene induction kinetics. Annu. Rev. Genet. 59 (1), 25–49. https://doi.org/10.1146/annurev-genet-012825-093148.

Xiropotamos, P., Papageorgiou, F., Manousaki, H., Sinnis, C., Antonatos, C., Vasilopoulos, Y., Georgios, K., 2024. Georgakilas, aPEAch: automated pipeline for end-to-end analysis of epigenomic and transcriptomic data. Biology 13 (7), 492. https://doi.org/10.3390/biology13070492.

Xu, K., Shen, L., 2025. Predicting 3D chromatin interactions using transformer-enhanced deep learning models. bioRxiv. https://doi.org/10.1101/2025.04.10.647995.

You, J., Yao, A., Ma, Z., Feng, S., Guo, M., 2025. Synergistic application of deep generative models and reinforcement learning in automated decision-making. In: 4th International Symposium on Computer Applications and Information Technology, ISCAIT 2025. Institute of Electrical and Electronics Engineers Inc., China, pp. 5–9. https://doi.org/10.1109/ISCAIT64916.2025.11010749. http://ieeexplore.ieee.org/xpl/mostRecentIssue.jsp?punumber=11010171.

Yu, T., Zhang, H., Chen, S., Gao, S., Liu, Z., Wang, J., Crossa, J., Montesinos-López, O.A., Hearne, S., Li, H., 2025. EXGEP: a framework for predicting genotype-by-environment interactions using ensembles of explainable machine-learning models. Briefings Bioinf. 26 (4). https://doi.org/10.1093/bib/bbaf414.

Zaccaria, E., van der Valk, E., Kar, S.K., Rebel, J.M.J., Schokker, D., 2025. Mini review: studying epigenomic alterations can shed light on coping and adaptive abilities during heat stress in monogastric livestock. Front. Genet. 16. https://doi.org/10.3389/fgene.2025.1561804.

Zhang, T., Zhao, X., Sun, H., Gao, B., Liu, X., 2024. GATv2EPI: predicting enhancer–promoter interactions with a dynamic graph attention network. Genes 15 (12), 1511. https://doi.org/10.3390/genes15121511.

Zhou, H., Clark, E., Guan, D., Lagarrigue, S., Fang, L., Cheng, H., Tuggle, C.K., Kapoor, M., Wang, Y., Giuffra, E., Egidy, G., 2025. Comparative genomics and epigenomics of transcriptional regulation. Ann. Rev. Anim. Biosci. 13 (1), 73–98. https://doi.org/10.1146/annurev-animal-111523-102217.

Zhu, L., Yang, X., Xian, Y., Jiang, W., Pu, X., Wang, S., Cheng, J., Niu, L., Zhao, Y., Chen, L., Zhou, X., Wang, Y., Gan, M., Zhu, L., Shen, L., 2025. Empowering precision livestock farming: artificial intelligence applications in animal genomic breeding and multi-dimensional phenotypic measurement. Smart Agric. Technol. 12, 101655. https://doi.org/10.1016/j.atech.2025.101655.

Zhuang, B., Gallet, A., Smyl, D., 2025. Inverse structural design with generative and probabilistic autoencoders and diffusion models. Eng. Appl. Artif. Intell. 161, 112143. https://doi.org/10.1016/j.engappai.2025.112143.

Chapter 11

Artificial intelligence integration in livestock reproductive omics

The most significant factor in determining the sustainability, economic viability, and genetic development in livestock production systems is reproductive efficiency, which influences outcomes in breeding, health management, resource utilization, and animal welfare. Biological mechanisms of reproduction are complex, encompassing both molecular control of reproduction, cellular differentiation, tissue remodeling, endocrine control, immunological tolerance, interactions with microbes, and environmentally responsive physiological adjustments. These processes occur in a dynamic form across time, starting with gametogenesis and fertilization up to embryonic development, implantation, gestation, and parturition, which are further disrupted by management practices, nutritional status, climatic variability, and disease pressure. This, in turn, implies that reproductive performance cannot be best comprehended or optimized using single measures or fixed analytical frameworks, but rather with integrative approaches that can capture multilayered interactions and time dependency.

The development of high-throughput omics technologies has allowed the study of reproductive biology to go beyond gene-centric views and encompass the regulation, metabolic, and signaling space in its entirety. Genomic, epigenomic, transcriptomic, proteomic, metabolomic, microbiomic, and immunological data currently offer a level of resolution into the determinants of fertility, developmental competence, and reproductive resilience. The implications of these molecular layers are becoming more and more complemented by imaging data, physiological and behavioral measurements derived from biosensors and fine-grained environmental monitoring, which have dense and heterogeneous data ecosystems. As much as these advances potentially provide the ability to transform, they might also reveal serious shortcomings of analytical approaches because traditional statistical approaches tend to have difficulties with the heterogeneity of data, sparsity, nonlinear relationships, and the absence of regularly timed sampling of such reproductive systems.

AI has become a convergent paradigm of analysis that can solve these issues by enabling the systematic combination, representation, and inference of a variety of reproductive data modalities. Instead of acting as a monistic tool, AI represents an extensive constellation of computational procedures that induce latent structure in intricate information, model dynamic procedures, deduce causal relationships, and assist in uncertainty-aware prediction and decision-making. In livestock reproductive omics, the features will support the shift from the descriptive characterization of the molecular states to the predictive modeling of the reproductive pathways and the prescriptive recommendation of management and breeding interventions. AI then redefines reproductive omics as a dynamic intelligence that provides an interface between molecular regulation and phenotypic consequences in a contextually sensitive and prospective fashion.

Another conceptual and ethical dimension is presented by the introduction of AI into reproductive omics. Informed decisions based on reproductive analytics have significant implications for the welfare of animals, genetic diversity, future adaptation, and the trust that society has in technology-mediated agriculture. The use of algorithmic predictions can affect decisions made by insemination, embryo selection, culling, breeding allocation, and the intensity of intervention, and transparency, fairness, communication of uncertainty, and accountability should be carefully considered. Besides, the growing ability to detect regulatory and epigenetic factors of reproductive characteristics raises the question of the limit of acceptable intervention, the impact on the next generation, as well as the compromise between productivity and biological integrity. Such considerations reveal why it is essential to incorporate the principles of governance, ethical reflection, and sustainability into the analytical design, rather than considering them as external constraints.

In this chapter, the author describes the integration of AI in livestock reproductive omics by integrating the scientific background of biological principles, analytical procedures, practical implementation scenarios, and regulatory and governance systems logically. It frames reproductive omics as a multiscale and dynamically changing field that requires integrative intelligence capable of reconciling molecular insight and practical decision-making. Through the analysis of the transformations of AI in data incorporation, predictive modeling, causal inference, and prescriptive optimization, the chapter aims to clarify the opportunities and obligations related to the implementation of advanced computational systems

Artificial Intelligence in Livestock Omics. https://doi.org/10.1016/B978-0-443-45663-3.00003-8

in the area of reproductive management. In achieving this, it will not only introduce AI as a technological supplement but as an organizational change in the way in which reproductive biology is modeled, perceived, and ethically acted upon in modern livestock systems.

11.1 Introduction to AI in livestock reproductive omics

The development of AI as a paradigm of analytic change in livestock reproductive omics is one of the outcomes of the increasing realization that reproductive performance is regulated by multilayered biological systems that can no longer be effectively modeled in reductionist or holistic terms. The process of reproduction involves a combination of molecular regulation, cellular differentiation, tissue remodeling, endocrine signaling, immune modulation, metabolic allocation, and environmental responsiveness in tightly coordinated processes that dynamically occur over time. These are nonlinear interactions between these processes and are influenced by biological potential and external conditions of management, which result in inherently context-dependent reproductive outcomes. AI has the computational power to model, combine, and query this complexity in a manner that is not limited by the power of traditional analytical methods (Mansour, 2025).

The traditional study of reproductive biology has used physiological indicators and genetic correlations that are focused on elucidating the variation in fertility-related phenotypes. Although these methods have played a significant role in the study of reproductive biology and enhancement of breeding performance, they tend to consider regulatory strata as distinct or suppose linear associations between predictors and performance (Hassani and Ghavi Hossein-Zadeh, 2025). These assumptions are now increasingly being called into question by the evidence that it is the coordinated regulation of genomic, epigenomic, transcriptomic, proteomic, metabolomic, and microbiome layers that underlie reproduction phenotypes, and which all respond dynamically to environmental and management factors. AI rethinks the analysis of reproduction as it allows these layers to be modeled simultaneously, so that fertility can be conceived as an emergent property of the system instead of an addition of independent effects (Sengupta et al., 2024).

In this context, AI acts not only as a collection of predictive instruments but as an integrative intelligence layer that encodes heterogeneous data on reproductive phenomena into consistent forms of biological condition. Machine learning architectures can learn latent structures that can represent regulatory coordination between the molecular and physiological world, thus converting high-dimensional data into reproductive signatures that are understandable. Such representations make it possible to study reproductive processes holistically, retaining the relationships between regulatory layers and minimizing noise and redundancy. This has been instrumental in the reproductive omics field, where most of the signals of interest are often minor, fleeting, and hidden in complicated biological contexts (Moustakli et al., 2025).

One of the main characteristics of reproductive systems is the fact that they are well-organized in time. Reproductive processes are organized with cycles, developmental behavioral windows, and adaptive responses existing within compounding periods. There are sensitive transitions, regulated by time and order, in gametogenesis, fertilization, embryonic development, implantation, and gestation. With AI, it is possible to represent these temporal dependencies in an explicit form, even with irregularly sampled data or under practical considerations. Constrained by implementing reproductive metrics into presumed paths of progress, AI assists in segregating brief fluctuations of regulation and consistent developmental reprogramming, thus increasing the operation of biology and predictive accuracy (More et al., 2025).

The other major role of AI in addressing the issue of causality in reproductive omics is that it is essential. The reproductive datasets are usually observational, which would represent the management practices and environmental exposures that cannot be complete. Associations between molecular signals and reproductive outcomes might thus be due to indirect associations, common confounders, or downstream effects, but not direct biological determinants. Causally based AI models enable reproductive regulation to be probed more carefully in support of differentiating causal mediating processes and epiphenomenal cues. This difference is vital to responsible use because associations that allow the use of interventions grounded in noncausal relationships may be inefficient or have unanticipated effects (Wu and Xie, 2024).

Another key reason why AI should be incorporated in reproductive omics is predictive capability. The reproduction of livestock is defined by a high level of economic and welfare interest, and any delays in the detection of reproductive dysfunction may lead to significant losses (Ghavi Hossein-Zadeh, 2013). AI facilitates proactive modeling that predicts the future reproductive circumstances with reference to existing and past data, aiding in making more informed and better choices. Notably, this type of prediction should be provided with a clear demonstration of uncertainty since the results of reproduction are affected by stochastic uncertainty and incomplete information. Probabilistic inference is being used more to provide estimates of the confidence in prediction, as AI techniques are designed to provide a quantitative understanding of likely results of predictions so that decision-makers can balance risk and expected results (Giordano, 2024; Mansour, 2025).

In addition to prediction, AI is used to employ prescriptive and adaptive reproductive management by combining inference with decision optimization. The use of AI helps to investigate alternative solutions to the desired results by assessing the possible management or breeding approaches based on learned models of reproductive dynamics. This ability brings about a change in the form of reactive to proactive reproductive control, whereby interventions are based on future projections, as opposed to retrospective evaluation. Prescriptive applications, however, need to be applied with caution so that the goal of optimization is indicative of biological constraints, welfare, and long-term sustainability, and not a low-performance benchmark (Giordano, 2024; Yin et al., 2024).

The use of AI in livestock reproductive omics also requires the redefinition of the way of knowledge creation and validation. The conventional experimental assumptions tend to focus on the manipulated interventions and the hypothesis testing approach that is linear, whereas AI can be viewed as the pattern discovery, the approach of representation learning, and the process of further refinement. These two methodologies are not in conflict with each other, and their combination must be methodologically sound and transparent. The AI models should be tested against the biological knowledge and put through sensitivity analysis to test their resilience in different settings. Being used in the right way, AI may also serve as a discovery driver, helping to develop hypotheses and create experiments, as well as speed up the synthesis of intricate reproductive data (Lac et al., 2024; Wadood et al., 2025).

The introduction of AI in reproductive omics cannot be discussed without considering ethical issues. Not only are reproductive decisions one of the factors that affect the animal's personality, but also the population structure, genetic diversity, and intergenerational consequences. AI enhances decision-making abilities, and consequently, it increases the level of ethical responsibility. Optimizing reproductive performance models should be crafted with specific consideration to the welfare, fairness, transparency, and accountability. The approach to AI as a neutral tool masks its normative implications, but rather, it should be viewed as a socio-technical system in which designs make value and priority representations (Tsonis and Khlifa, 2025).

11.2 Overview of reproductive omics in livestock

Reproductive omics in livestock is the holistic systems perspective of fertility, developmental competence, and reproductive resilience via the integrated molecular and physiological characterization (see Fig. 11.1). The process of reproduction is not controlled by one pathway or regulatory layer, but it is a result of the combined action at the genomic, regulatory, metabolic, immunological, and environmental levels. Reproductive omics aims to understand this coordination through increasing analysis of single markers to systematically represent the reproductive state of sexes, tissues, and lifespan. The paradigm represents the change in the understanding of reproduction as a discrete biological process toward the concept of a dynamic, multiscale process that is part of the larger system of physiology and ecology (Wadood et al., 2025).

Reproductive omics has its roots in genomic variation that determines the inherited capability of reproductive performance. Genetic architecture has an impact on gametogenesis, endocrine responsiveness, tissue development, and disposition to reproductive dysfunction. Nonetheless, genomic data can only give a definitive and partial description of the reproductive capacity since it fails to describe the dynamics of the expression of genetic potential in various situations. The crucial element of reproductive outcomes is the regulatory mechanisms that determine the pattern of gene expression according to the developmental cue, physiological state, and environmental exposure. This identification encourages the incorporation of other omics layers that interpose the genetic information into the reproductive phenotype (Khan et al., 2024; Sengupta et al., 2024).

Epigenomic regulation plays a central role in the reproductive omics, especially due to the comprehensive chromatin reprogramming that drives the process of reproduction. Large-scale remodeling of patterns of DNA methylation, histone modifications, and chromatin accessibility orchestrate gametogenesis and early embryonic development, setting up patterns of regulatory activity to direct further development. These epigenomic conditions are vulnerable to both internal and external factors and can affect reproductive success, embryonic survival, and development trajectories. In the reproductive omics, the data of epigenomics provide an understanding of regulatory plasticity and biological memory, which connects short-term environmental cues to long-term reproductive effects (Burton and Torres-Padilla, 2025).

Additional articulation of reproductive regulation comes in the form of transcriptomic and noncoding RNA expression, which captures context-specific programs of gene expression in the reproductive tissues and developmental stages. Transcriptional dynamics are the consequence of the combined product of genomic and epigenomic regulations, which convert regulatory potential into functional signaling. Noncoding RNAs also provide further complexity of control through the regulation of transcription, posttranscriptional modification, chromatin organization, and cell-cell communication. Such RNA-mediated processes are important in reproductive biology, specifically when a high level of temporal

FIGURE 11.1 AI-enabled multiscale integration of reproductive omics and physiological data in livestock. This figure illustrates how AI integrates multiple layers of reproductive omics data, including genomic, epigenomic, transcriptomic, proteomic, metabolomic, and microbiomic information, into a unified analytical framework. Each data layer captures distinct but complementary aspects of reproductive regulation, ranging from inherited genetic potential to dynamic metabolic and microbial influences. The upward convergence of these data streams emphasizes the role of AI in synthesizing heterogeneous biological signals into coherent, system-level representations that support predictive, interpretative, and decision-oriented reproductive analysis.

and spatial regulation of gene expression is necessary to coordinate cellular differentiation, tissue remodeling, and immune tolerance (Anjum et al., 2025).

The layers of proteomic and posttranslational modifications expand the reproductive omics to the functional one by defining the protein abundance, activity, localization, and interaction. Proteins are the main performers of reproductive processes, which facilitate hormonal signaling, cellular communication, and structural organization. Regulatory complexity is further added by posttranslational changes in the dynamic regulation of protein activity by metabolic and signaling signals. In reproductive omics, proteomic data fill the gap between molecular regulation and physiological outcome, and provide an understanding of the process by which regulatory cues are acted upon at the cellular and tissue levels (Peña et al., 2024; Werner et al., 2024).

Metabolomic and lipidomic data will add a different angle to the data in that they will be reflective of the energetic and biochemical context in which reproductive processes exist. Production places excessive demands of a metabolic nature that have to be met against other competing physiological needs. Metabolic responses will build systemic and regional resource distribution, redox condition, and biosynthetic potential, each of which affects both reproductive nativity and maturation endowment. These endocrine and immune interactions are also bidirectional and point to the interconnections of reproductive regulation (Anandaram et al., 2024).

The reproductive microbiome is another aspect of reproductive omics, which notes the impact of host-attached microbial communities on reproductive environments. Microbial flora and activity have an effect of modulating immunity, maintaining the integrity of the barrier, and taking part in the exchange of metabolites in reproductive tissues. Such interactions are involved in reproductive health and resilience, and a more general conceptualization of reproduction as an ecosystem-scale process with both host and microbial actors. The addition of microbiome information to reproductive omics further highlights the need to have integrative analytical methods that can be used to describe cross-domain interactions (Lingasamy et al., 2024).

Reproductive omics is not limited to molecular profiling, but is applied to physiological, behavioral, and environmental aspects, which puts the molecular states in the context of lived biological situations. Physiological signals provide systemic explanations of reproductive regulation, and behavioral indicators provide information on reproductive readiness and adaptation. Nutrition, climate, and management practices are environmental factors that have a continuous effect on reproductive biology, both in the regulation of molecules and the manifestation of physiological functions. The addition of these dimensions takes reproductive omics to become indeed a more integrative systems science, in which fertility is a combination of biological potency and environmental condition (Sengupta et al., 2024; Fatima et al., 2025).

One of the issues that renders reproductive omics challenging is the diversity and time fluctuations of reproductive processes. Reproductive regulation varies among tissues, sexes, developmental stages, and individuals, and gives rise to patterns that cannot be adequately measured by bulk or by static measurements. There is also further complexity brought about by temporal dynamics because reproductive states change across cycles and developmental windows, which are typically irregularly sampled. Reproductive omics thus needs to be designed to support partial observability, asynchronous data acquisition, and interpretation, which is context-dependent, and thus needs analytical structures that allow preservation of biological detail while making generalizations (Liu et al., 2024).

The importance of integrating various omics layers in the field of reproductive biology can be explained by the constraints of classical analytical tools that analyze data modalities separately or that are based on the premise of linear relationships. Reproductive systems have feedback, threshold effects, and nonlinear dependencies that are a challenge to reductionist approaches. Reproductive omics, therefore, requires an analytical approach that can reflect hierarchical organization, temporal development, and cross-modal interaction. The need sets the initial groundwork for the rationale of AI integration, as further described in the chapter.

In general, reproductive omics provides the biological basis on which AI works to produce predictive and prescriptive knowledge. Reproductive omics changes the paradigm of livestock reproduction by identifying its complex determinants of fertility and developmental competence, making it a dynamic system that can be modeled, viewed, and led responsibly. It is against this backdrop that this overview positions reproductive omics both as a scientific model and a viable baseline for AI-enabled reproductive management, which then preconditions the further analysis of the topic of data integration, predictive modeling, and ethical governance in the succeeding sections. Major omics layers contributing to livestock reproductive biology and their functional relevance are shown in Table 11.1. In addition, key reproductive processes and biological states captured through reproductive omics are presented in Table 11.2.

11.3 Applications of AI in livestock reproductive omics

AI will allow a paradigm shift in the methods of analyzing, interpreting, and applying reproductive omics information to useful biological and managerial information in livestock systems. The features of reproductive biology include layered regulation, time dynamics, and high sensitivity to both environmental and physiological contexts, which challenge the paradigms of traditional analysis. AI is attempting to solve these problems by offering adaptable computational models that have the capacity to discover intricate patterns, incorporate heterogeneous information, and make inferences about latent reproductive conditions that cannot be observed directly. Consequently, AI in reproductive omics is not limited to efficiency improvements to transform the conceptual and operational basis of reproductive analysis (Kılıç et al., 2024).

One of the most significant uses of AI is the integration of a variety of data streams of reproductive data to representations of reproductive state. Reproductive omics encompasses genomic information, epigenomic information, transcriptomic information, proteomic information, metabolomic information, microbiomic information, physiological information, behavioral information, and environmental information, each with distinct scales, noise profiles, and ranges of temporal variability. The AI models have the special capability of resolving these disparities by learning hierarchical representations that maintain the structure that is modality-specific but allows interaction across modalities. Reproductive status can be concluded as a coordinated system state, not as an aggregate of independent indicators, through such representations, and thus, further insight into regulatory coherence and dysfunction is obtained (Nayana et al., 2025).

AI also allows the reproductive processes to be modeled as dynamic rather than mechanical states. Reproductive systems are evolving on a cyclic, developmental, and adaptive time scale, but there is commonly sparse and intermittent data collection in the livestock setting. Reproductive measurements can be latent in the temporal modeling structures of AI to support the explanation of change in terms of phase and progression. The ability facilitates the differentiation between temporary regulatory changes and enduring reprogramming, which would be very critical in learning about

TABLE 11.1 Major omics layers contributing to livestock reproductive biology and their functional relevance.

Omics layer	Biological scope in reproduction	Key regulatory functions	Reproductive processes informed	Relevance for AI-based modeling
Genomics	• Inherited DNA sequence variation • Structural and regulatory genetic architecture	• Establishes baseline reproductive potential • Influences gametogenesis, endocrine sensitivity, and tissue development	• Fertility predisposition • Gamete quality and viability • Developmental competence	• Provides stable priors for modeling • Enables population-level stratification • Anchors dynamic regulatory layers
Epigenomics	• DNA methylation patterns • Histone modifications • Chromatin accessibility and 3D organization	• Regulates gene expression plasticity • Mediates developmental and environmental effects • Encodes biological memory	• Gametogenesis reprogramming • Embryonic and placental development • Long-term reproductive adaptation	• Captures dynamic regulatory states • Supports temporal and causal inference • Links environment to phenotype
Transcriptomics	• mRNA expression profiles • Noncoding RNA landscapes	• Translates regulatory potential into active programs • Coordinates tissue- and stage-specific responses	• Oocyte and sperm maturation • Embryo-maternal signaling • Hormonal and immune regulation	• Reflects integrated regulatory output • Enables state classification and trajectory inference
Proteomics	• Protein abundance and localization • Interaction networks	• Executes reproductive signaling pathways • Mediates cellular communication and structure	• Hormonal signaling cascades • Cellular differentiation • Tissue remodeling	• Bridges molecular regulation and phenotype • Improves biological interpretability of models
Posttranslational modifications	• Phosphorylation, acetylation, and ubiquitination • Dynamic protein modification states	• Fine-tunes protein activity and stability • Enables rapid regulatory response	• Cell-cycle regulation • Signal transduction in reproduction	• Captures rapid regulatory dynamics • Enhances temporal resolution of AI models
Metabolomics	• Small-molecule metabolites • Energy and redox state indicators	• Reflects energetic and biochemical constraints • Integrates metabolic and hormonal signaling	• Reproductive readiness • Embryo support • Stress and resilience assessment	• Provides functional phenotype proxies • Supports predictive and prescriptive modeling
Lipidomics	• Lipid species and membrane composition • Steroid precursors	• Supports membrane dynamics and signaling • Regulates hormone synthesis	• Gamete membrane integrity • Steroidogenesis • Embryonic development	• Adds biochemical specificity • Enhances metabolic context in AI inference
Microbiomics	• Host-associated microbial communities • Functional microbial profiles	• Modulates immune tolerance • Influences local metabolic environment	• Reproductive tract health • Infection susceptibility • Inflammatory regulation	• Enables ecosystem-level modeling • Captures host-microbe interactions

Immunomics	• Immune cell signaling and cytokine profiles • Inflammatory mediators	• Maintains immune tolerance • Balances defense and reproduction	• Embryo implantation • Pregnancy maintenance	• Supports causal modeling of immune-reproductive trade-offs
Endocrinomics	• Hormonal profiles and rhythms • Feedback regulation	• Coordinates systemic reproductive timing • Integrates metabolic and reproductive cues	• Estrous cycling • Ovulation and gestation	• Enables temporal alignment • Improves trajectory-based predictions
Physiomics	• Physiological indicators • Behavioral and sensor-derived signals	• Reflects whole-organism integration • Captures real-time reproductive status	• Estrus detection • Reproductive stress responses	• Enables real-time inference • Supports digital reproductive twins
Exposomics/ Environomics	• Nutritional, climatic, and management factors • External stressors	• Shapes regulatory and physiological responses • Modulates reproductive resilience	• Seasonal fertility variation • Adaptation to production conditions	• Contextualizes biological data • Enables scenario simulation and optimization

TABLE 11.2 Key reproductive processes and biological states captured through reproductive omics.

Reproductive process or state	Biological definition and scope	Major omics layers involved	Regulatory and functional characteristics	Relevance for AI-driven reproductive inference
Gametogenesis	• Formation and maturation of male and female gametes • Establishment of developmental competence	• Genomics • Epigenomics • Transcriptomics	• Extensive chromatin remodeling • Tight temporal regulation of gene expression • Sensitivity to metabolic and environmental cues	• Enables early detection of compromised fertility • Supports modeling of developmental potential
Gamete quality and viability	• Functional integrity of oocytes and spermatozoa • Capacity to support fertilization	• Transcriptomics • Proteomics • Metabolomics	• Coordination of energy metabolism and structural integrity • Regulation of oxidative balance	• Improves prediction of fertilization success • Supports quality stratification
Fertilization competence	• Ability of gametes to successfully fuse and initiate development	• Proteomics • Lipidomics • Epigenomics	• Membrane dynamics and signaling compatibility • Epigenetic readiness for reprogramming	• Enables assessment of compatibility and success probability
Epigenetic reprogramming	• Resetting and re-establishment of regulatory states postfertilization	• Epigenomics • Noncoding RNA profiles	• Global DNA methylation remodeling • Histone modification turnover	• Critical for modeling long-term developmental outcomes
Early embryonic development	• Preimplantation cellular division and lineage specification	• Transcriptomics • Epigenomics • Metabolomics	• Rapid transcriptional shifts • Metabolic reprogramming	• Supports trajectory inference and viability prediction
Embryo-maternal communication	• Molecular signaling between embryo and maternal tissues	• Transcriptomics • Proteomics • Immunomics	• Immune modulation and signaling coordination • Endocrine cross-talk	• Enables detection of implantation readiness
Implantation	• Attachment and invasion of the embryo into the uterine tissue	• Epigenomics • Immunomics • Proteomics	• Immune tolerance induction • Tissue remodeling	• Supports causal inference of implantation failure
Placental development	• Formation and maturation of placental structures	• Epigenomics • Transcriptomics • Metabolomics	• Nutrient transport regulation • Vascular and endocrine development	• Improves prediction of gestational success
Gestational maintenance	• Sustained pregnancy progression	• Endocrinomics • Immunomics • Physiomics	• Hormonal stability • Immune equilibrium	• Enables monitoring of pregnancy stability
Fetal growth and programming	• Developmental shaping of offspring phenotype	• Epigenomics • Metabolomics • Exposomics	• Nutrient allocation and epigenetic imprinting • Sensitivity to maternal environment	• Supports modeling of long-term offspring outcomes
Estrous cycle regulation	• Cyclical reproductive hormonal regulation	• Endocrinomics • Transcriptomics • Physiomics	• Feedback-controlled hormonal oscillations • Phase-dependent gene expression	• Enables precise cycle phase inference

Ovulation	• Release of mature oocyte	• Proteomics • Endocrinomics	• Acute signaling events • Tissue rupture and repair	• Supports timing optimization and prediction
Lactational-reproductive interaction	• Interaction between lactation and reproductive recovery	• Metabolomics • Endocrinomics	• Resource allocation trade-offs • Hormonal suppression or recovery	• Enables modeling of reproductive recovery trajectories
Reproductive stress and resilience	• Capacity to withstand environmental and physiological stressors	• Epigenomics • Metabolomics • Physiomics	• Adaptive regulatory plasticity • Stress-response integration	• Supports resilience assessment and intervention planning
Reproductive senescence	• Age-related decline in reproductive function	• Genomics • Epigenomics • Transcriptomics	• Accumulation of regulatory drift • Reduced cellular responsiveness	• Enables lifecycle-aware reproductive modeling

reproductive preparedness, recovery, and long-term resilience to different environmental conditions (Tedeschi et al., 2025).

The other valuable use of AI in reproductive omics is the discovery and description of regulatory pathways behind fertility and developmental competence. The common methods of analysis used traditionally are often based on correlation-related associations that can confound causative drivers with downstream outcomes or common confounders. AI techniques, which use causal reasoning, allow deeper interrogation of reproductive control through dependence modeling between molecular layers, physiological signals, and outcomes. This ability increases the biological interpretability and minimizes the chance of taking action on spurious associations, especially in situations in which interventions can be irreversible and/or long-term in effect (Hegde et al., 2025).

AI is also quite important in predictive modeling of reproductive outcomes. With the interpolation of both past and current reproductive omics data, machine learning models can predict future reproductive conditions and possible disturbances before their occurrence in an overt manner. Such predictive ability aids anticipatory reproductive management and results in an earlier recognition of risk and a greater timely change in approaches. However, in reproductive situations, prediction will have to be paired with a clear expression of uncertainty, since biological variability and incomplete measurability will present irreducible ambiguity. This need is being fulfilled more by AI frameworks that generate probabilistic rather than deterministic predictions, rather than deterministic predictions, which allow informed risk assessment (Moustakli et al., 2025).

In addition to prediction, AI can assist with prescriptive analytics, which is used to make decisions based on the comparison of other possible approaches to reproductive management. AI can be used to investigate potential futures within known biological limits by creating the mechanisms of response of reproductive trajectories to various management or breeding interventions by simulation. This prescriptive ability makes reproductive omics more like a decision-support system to balance between competing goals like reproductive efficiency, animal welfare, genetic diversity, and sustainability. Notably, the prescriptive applications are based on the models that incorporate biological reality and ethical constraints, such that the optimization is not more than physiological tolerance, and the integrity of the system in the long term is not violated (Hynniewta et al., 2025).

AI is also a leading cause of personalization and stratification in reproductive management. There is great heterogeneity in the reproductive response of the livestock populations based on genetic background, developmental history, exposure to the environment, and conditions of management. They are also able to learn common reproductive patterns and, at the same time, address individual or subgroup deviations, allowing reproductive insights to be customized instead of applied in the same manner. This ability allows for a more accurate adjustment of reproductive policy to biology and less inefficiency and unintended consequences of generalized policies (Emsen et al., 2025).

At the population level, AI allows for the exploration of the reproductive trends of different production systems to facilitate knowledge transfer and accommodate the differences in context. Through both general regulatory structure modulation and environment-specific regulation modeling, the AI makes generalization possible without obliterating meaningful biological variation. Such a balance is of great significance in animal systems with heterogeneity in climate, management systems, and genetic endowment. The AI serves as a connection between localized reproductive knowledge and system-wide knowledge (Garcia et al., 2025).

Experimental and analytical processes in reproductive omics are also transformed through AI to allow adaptation, learning, and hypothesis refinement. As models ingest new data, they are able to identify where they are uncertain or where the laws are ambiguous, and direct their specific data gathering and experiment prioritization. This feedback mechanism of knowledge generation increases the speed of knowledge generation by devoting resources to the most informative elements of reproductive systems. In this regard, AI is involved in the active scientific process, and reproductive omics is turned into an analytical enterprise that changes and develops continuously (Nayana et al., 2025).

The use of AI in reproductive omics also brings up the issue of validation and interpretability, which cannot be envisaged outside of its utility. The models of behavior and results of reproductive choice require trust in the model behavior and outputs because the choice has serious biological and ethical consequences. AI systems should consequently be considered not only based on predictive accuracy, but also based on biological realism and cross-contextual strength. The mechanisms of interpretability linking model outputs with the underlying reproductive regulation are necessary to build trust and allow responsible oversight (Pradhan, 2024).

In general, AI reproductive omics applications in livestock are an exemplar of changing the paradigm of reproductive biology and transforming descriptive analysis to dynamic, integrative, and decision-focused modeling. AI provides the analytical framework, which is needed to convert reproductive omics into action and ethically sound insight, and can be applied to identify the multimodal integration of information, temporal inference, causal reasoning, prediction, and prescriptive optimization. The applications can be seen as an essential interface between biological knowledge and actual reproductive control, and pave the way to further exploration of the data integration approaches, methodology issues, and governance questions in the following sections. Applications of AI across stages of livestock reproduction are listed in Table 11.3.

11.4 Data integration and analysis techniques

The core methodological approach in AI-enabled livestock reproductive omics consists of data integration and analysis methods because reproductive biology is distributed by nature across molecular, cellular, physiological, and environmental levels. These dimensions all describe different, but interdependent, parts of the reproductive regulation, and none of them can be used alone to explain reproductive outcomes. The main issue that data integration aims to solve is thus never merely the aggregation of various datasets, but rather the development of coherent analysis representations that are biologically meaningful, and allow inference to be made in heterogeneous and often incomplete data (Lu et al., 2024).

The data of reproductive omics is distinguished by a significant range of variability in magnitude, resolution, and dynamics. Genomic variation offers comparatively constant data throughout the lifespan of an animal; epigenomic, transcriptomic, proteomic, and metabolomic layers are relatively receptive in relation to the phase of development, physiology, and exposure to the environment. An additional variation is seen in microbiome composition and functionality, which is a product of host regulation and external factors. Physiological, behavioral, and environmental data add further complexity levels and are usually sampled in real time or semireal time and are controlled by other noise structures. These differences, instead of being flattened or distorted by the hierarchical organization of reproductive systems, have to be reconciled by effective data integration (Sengupta et al., 2024; Wadood and Zhang, 2024).

AI-based integration methods cope with this problem through learning structured representations that represent the relationships within and between data modalities. Instead of assuming that all inputs are interchangeable features, high-end models do not eliminate modality-specific properties but do not restrain information to interact at common latent levels. This is consistent with the biological fact that reproductive regulation is achieved by nonidentical but coordinated mechanisms that run the processes at both molecular and physiological levels. With this structure, integrated representations allow reproductive states to be determined as emergent behavior of interacting systems instead of being a product of data aggregation (Kan et al., 2025).

The temporal location is a key issue in reproductive data integration because the reproductive processes occur in cyclical and developmental time that may not be monitored at the same time. Sample data taken at varying stages can be representative of different developmental stages of reproductive evolution, and naive integration is then misleading. In models of AI, a time model is included to introduce the concept of asynchronous observations into the process of inferred reproductive trajectories, which enables the interpretation of the molecular and physiological measurements with respect to the latent reproductive stage instead of actual time. This contextualization over time is what is necessary in differentiating regulatory signals, which could be progression, adaptation, or disruption (Moustakli et al., 2025).

The other data integration dimension that is considered critical is the reconciliation of individual-level and population-level information. Reproductive outcomes are a combination of inherent biological variability and external factors that are present due to management practices and environmental factors. AI facilitates hierarchical modeling approaches, which are effective at expressing shared reproductive organization within populations and enable individual-specific dispensations to be modeled. By allowing the generalization of reproductive insight, such methods allow erasing biologically meaningful heterogeneity, and they enhance the interpretability and applicability of the techniques across different livestock systems (Tedeschi et al., 2025).

Unfinished data and disparate sampling density are the bane of omics in reproduction. Some of the measured molecular layers can be sparsely measured by cost or capability constraints, and nonmeasurement can frequently be related to biological or management reasons, rather than being random. The analysis methods based on AI solve this problem by modeling the latent states probabilistically, signifying that unobserved information can be estimated, and the uncertainty can be clearly expressed. The strategy does not fall into the trap of deterministic imputation and can be more responsible in making inferences in the presence of realistic data constraints (Hussein et al., 2024).

The combination of environmental and management data is especially relevant in the area of reproductive omics because reproductive biology is very sensitive to the external environment. Reproductive regulation and outcomes are subject to continuous effects of nutritional status, climatic variability, housing, and health interventions. The external drivers can be modeled using AI along with the internal biological states, and through these methods, reproductive phenotypes are explained as contextually relevant manifestations of biological potential. This collaborative modeling is necessary to convert reproductive omics understanding into practical advice in an actual production setup (Fernandez et al., 2025).

TABLE 11.3 Applications of AI across stages of livestock reproduction.

Reproductive stage	Primary biological focus	AI-enabled analytical applications	Data and omics layers integrated	Value and impact for reproductive management
Pregametic development	• Establishment of reproductive potential • Early germline regulation	• Identification of latent fertility risk factors • Stratification of reproductive capacity	• Genomics • Epigenomics • Transcriptomics	• Enables early-life reproductive assessment • Supports long-term breeding strategy planning
Gametogenesis	• Formation and maturation of gametes • Regulatory reprogramming	• Modeling of developmental competence trajectories • Detection of regulatory disruption	• Epigenomics • Transcriptomics • Metabolomics	• Improves prediction of gamete quality • Reduces uncertainty in reproductive readiness
Gamete quality assessment	• Functional integrity of sperm and oocytes • Energetic and structural status	• Multivariate quality scoring • Pattern recognition of suboptimal states	• Proteomics • Metabolomics • Lipidomics	• Enhances selection reliability • Supports precision reproductive decision-making
Estrous cycle monitoring	• Hormonal oscillation and timing • Physiological synchronization	• Cycle phase inference • Detection of atypical hormonal dynamics	• Endocrinomics • Physiomics • Transcriptomics	• Enables accurate timing of reproductive interventions • Reduces missed or mistimed opportunities
Ovulation prediction	• Acute follicular rupture events • Hormone-triggered transitions	• Short-horizon temporal forecasting • Phase transition detection	• Endocrinomics • Proteomics	• Supports optimal synchronization • Improves fertilization success probability
Fertilization	• Gamete interaction and fusion • Initiation of embryonic development	• Compatibility inference • Risk assessment of fertilization failure	• Proteomics • Epigenomics • Lipidomics	• Reduces fertilization inefficiency • Improves outcome predictability
Epigenetic reprogramming	• Regulatory resetting after fertilization • Establishment of developmental trajectories	• Identification of stable versus stochastic regulatory changes • Causal modeling of developmental potential	• Epigenomics • Noncoding RNA profiles	• Improves long-term outcome prediction • Supports developmental risk assessment
Early embryonic development	• Cell division and lineage specification • Metabolic reorganization	• Trajectory modeling of embryo viability • Detection of developmental deviation	• Transcriptomics • Metabolomics • Epigenomics	• Enables early viability assessment • Reduces downstream reproductive loss
Embryo-maternal signaling	• Bidirectional molecular communication • Immune and endocrine coordination	• Signal integration and compatibility inference • Detection of communication failure	• Transcriptomics • Immunomics • Proteomics	• Improves implantation success prediction • Supports intervention timing
Implantation	• Tissue remodeling and immune tolerance • Structural integration	• Causal inference of implantation success or failure • Risk stratification	• Epigenomics • Immunomics • Proteomics	• Reduces unexplained implantation loss • Supports targeted management

Gestational progression	• Maintenance of pregnancy • Systemic physiological balance	• Longitudinal monitoring of pregnancy stability • Early detection of gestational risk	• Endocrinomics • Physiomics • Metabolomics	• Enhances pregnancy management • Improves welfare outcomes
Fetal development and programming	• Growth and developmental regulation • Long-term phenotype shaping	• Prediction of developmental trajectory outcomes • Assessment of environmental influence	• Epigenomics • Metabolomics • Exposomics	• Supports intergenerational planning • Improves offspring performance prediction
Postpartum recovery	• Restoration of reproductive function • Resource reallocation	• Modeling of recovery trajectories • Identification of delayed return to fertility	• Endocrinomics • Metabolomics • Physiomics	• Optimizes breeding interval planning • Reduces reproductive downtime
Reproductive aging	• Progressive decline in reproductive capacity • Regulatory drift	• Lifecycle-aware fertility modeling • Risk assessment of senescence	• Genomics • Epigenomics • Transcriptomics	• Enables age-informed breeding strategies • Supports population sustainability
Reproductive stress and adaptation	• Response to environmental and physiological stressors • Regulatory resilience	• Detection of adaptive versus maladaptive responses • Scenario simulation	• Epigenomics • Metabolomics • Physiomics	• Improves resilience-oriented management • Supports welfare-aware intervention

Another problem that adds complexity to the reproductive data integration is technical variability and batch effects. The systematic variation caused by differences in experimental protocols, platforms, and analytical pipelines can be a confounding factor to biological interpretation. Such effects can be detected and removed using AI techniques that learn to distinguish technical artifacts from biological signals, but this process must be carefully validated to avoid overcorrection. Open preprocessing and sensitivity analysis are consequently important elements of responsible integration pipelines, by which harmonization can in no way diminish but can and will enrich biological understanding (Yu et al., 2024).

Interpretability is one of the requirements that data integration and analysis methods in reproductive omics must have. To enable scientific knowledge and ethical control, integrated models should allow inferred results to be traced back to the biological processes and molecular layers that contribute to each prediction. The predictive performance and interpretability of AI prove especially useful in the reproductive sphere, where AI-informed decision-making, has great welfare and long-term consequences. Interpretability can also promote interdisciplinary communication between the stakeholders of molecular biology, veterinary science, data science, and management practice (Mihajlović, 2025).

Integration and analysis techniques are also designed with scalability and robustness. Reproductive datasets are becoming very large and complicated, and it is necessary to have analytical frameworks that can support the increasing dimensionality without compromising the stability or interpretation. AI models are becoming more and more hierarchically scalable, allowing new data sources to be gradually added and predictive representations to be refined continuously. Models should be validated across different breeds, production environments, and management systems to ensure robust performance beyond the original training population (Padmanaban, 2024).

In general, data integration and analysis methods permit reproductive omics to be used as a coherent system of analysis instead of an assortment of detached measurements. AI has the potential to transform livestock reproduction through generating predictive and prescriptive applications of its unifying molecular, physiological, and environmental information within structured and interpretable representations. The efficiency of these methods defines the analytical performance, as well as the plausibility, ethical appropriateness, and practical implications of AI-based reproductive decision-making (see Fig. 11.2). This research approach is the direct basis of the strategies of predictive modeling discussed in the following sections and the general introduction of AI in livestock reproduction. Data integration strategies for multimodal reproductive omics in livestock are presented in Table 11.4.

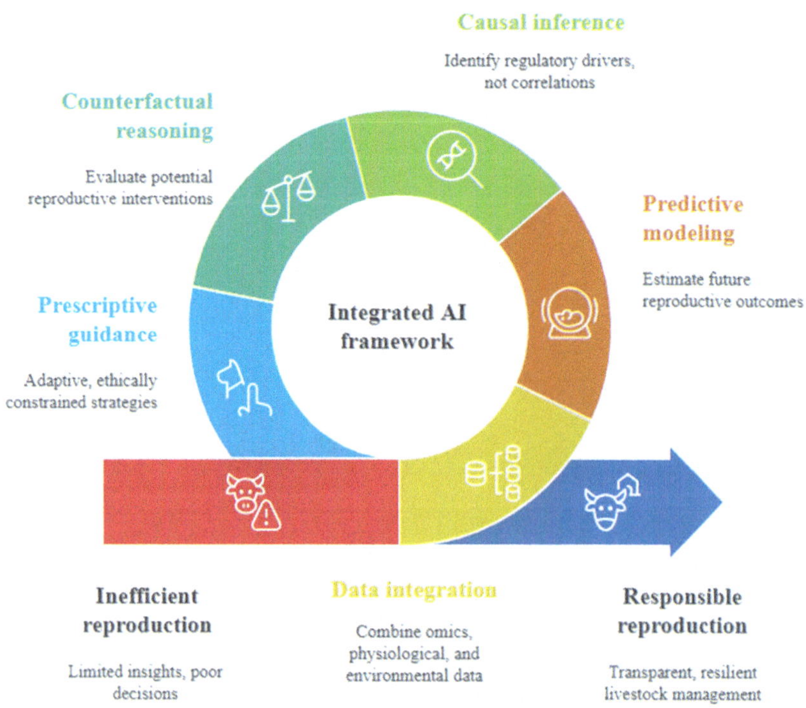

FIGURE 11.2 Conceptual framework of AI-driven predictive, causal, and prescriptive modeling in livestock reproduction. This figure depicts an integrated AI framework in which reproductive omics, physiological, and environmental data are unified to support progressive analytical layers. Data integration underpins predictive modeling of future reproductive outcomes, causal inference to identify regulatory drivers rather than simple correlations, and counterfactual reasoning to evaluate potential reproductive interventions. These components collectively enable prescriptive guidance that promotes transparent, adaptive, and ethically constrained decision-making, facilitating a transition from inefficient reproductive management toward responsible and resilient livestock reproduction.

TABLE 11.4 Data integration strategies for multimodal reproductive omics in livestock.

Integration strategy	Conceptual approach	Types of data integrated	Strengths and analytical advantages	Challenges and considerations for AI modeling
Early (feature-level) integration	• Concatenation of features into a unified input space • Joint representation before modeling	• Genomics • Epigenomics • Transcriptomics • Proteomics	• Captures cross-omics interactions directly • Simplifies downstream modeling pipeline	• High dimensionality increases overfitting risk • Requires extensive normalization and scaling
Intermediate (representation-level) integration	• Modality-specific feature extraction followed by fusion • Shared latent space learning	• All molecular omics • Physiological and sensor data	• Preserves modality-specific structure • Reduces noise and redundancy	• Requires careful architecture design • Latent space interpretability can be challenging
Late (decision-level) integration	• Independent modeling per modality • Aggregation of predictions	• Any omics layer • Environmental and management data	• Robust to missing modalities • Modular and scalable	• Limited capture of cross-modal interaction • Potential loss of biological coherence
Hierarchical integration	• Models reflecting biological regulatory hierarchy • Anchoring dynamic layers to stable priors	• Genomics as baseline • Regulatory and physiological layers	• Aligns with biological organization • Enhances interpretability	• Requires strong biological assumptions • Increased model complexity
Temporal integration	• Alignment of asynchronous time-series data • Trajectory-based modeling	• Longitudinal omics • Endocrine and physiometrics	• Captures reproductive dynamics and phase transitions • Distinguishes transient versus persistent signals	• Sparse sampling limits resolution • Temporal misalignment risk
Graph-based integration	• Representation of regulatory and interaction networks • Nodes and edges encode biological relationships	• Multi-omics networks • Protein-gene-metabolite interactions	• Biologically intuitive modeling • Supports causal inference	• Graph construction uncertainty • Computational scalability
Probabilistic integration	• Explicit modeling of uncertainty and missingness • Latent variable inference	• Sparse and incomplete omics datasets	• Reduces bias from missing data • Provides uncertainty-aware outputs	• Increased computational demand • Requires careful prior specification
Context-aware integration	• Joint modeling of biological and environmental data • Conditional inference	• Omics data • Nutrition, climate, and management factors	• Improves real-world applicability • Captures environment-dependent regulation	• Complex interaction modeling • Potential confounding
Population-individual hybrid integration	• Multilevel modeling across individuals and populations • Shared and specific representations	• Individual reproductive profiles • Population-level omics	• Balances generalization and personalization • Improves transferability	• Requires careful variance partitioning • Increased model complexity
Federated and distributed integration	• Decentralized model training across institutions • Privacy-preserving aggregation	• Reproductive omics across farms or regions	• Enables collaboration without data sharing • Reduces governance barriers	• Communication overhead • Data heterogeneity challenges

11.5 Machine learning for predictive modeling in livestock reproduction

Machine learning plays a key analytical role in the context of AI-enabled livestock reproductive omics, which involves reducing the high-dimensionality of data by predicting future reproductive outcomes. Nonlinear interactions between genetic predisposition, regulatory modulation, physiological condition, and environmental exposure form reproductive biology patterns that are difficult to quantify using traditional parametric or rule-based models. Machine learning copes with this complexity by directly learning data representations, which allows for the inference of the integrated and dynamic nature of reproductive systems (van Rumpt and Rabaglino, 2024).

The fundamental issue of predictive modeling in livestock reproduction is how to model reproductive state in a biologically and computationally feasible way. Reproductive omics data have several layers of regulation that vary in their stability, temporal sensitivity, and distance to phenotype. Genomic data offers a more stable background structure, whereas epigenomic, transcriptomic, proteomic, and metabolomic data vary in accordance with developmental and environmental signals. Machine learning models combine these layers by learning nonlinear mappings that encode coordinated regulatory states within latent representations, enabling reproductive status to be predicted as a system property that is continuous and changes over time, rather than a static classification (Wang et al., 2025).

A characteristic of reproductive processes is temporal dependency, and an important factor to consider in predictive modeling is that phenomena related to fertility occur over time-scales that are overlapping, such as cyclical hormonal cycles, development, and more adaptive changes that occur over a longer time. Observations are typically taken at random times, which is realistic for livestock systems rather than under idealized experimental conditions. Machine learning models that can predict temporal structure can be used to predict the future by learning predictive reproductive data through embedding observations in inferred processes. Such a time-based view increases the power of prediction by considering both phase, progression, and momentum of reproductive systems, as opposed to using only instantaneous measures (Rachakonda et al., 2024).

Reproductive omics predictive modeling should also respond to significant uncertainty due to biological uncertainty, stochastic effects, and unreliable observability. Many determinants of fertility cannot be measured directly or controlled experimentally, which makes the predictive structures remain elusive to reduction. Machine learning methods are being developed to increasingly use probabilistic reasoning as the measure of uncertainty, in addition to point estimates, so that predictions become expressed as a distribution, not as a deterministic value. This explicit expression of uncertainty is especially needed in situations involving reproductive decision-making, where predictions of overconfidence can result in inappropriate interventions, which have welfare or long-term impacts (Wu et al., 2025).

The other significant challenge related to predictive modeling in livestock reproduction is generalization with respect to different populations and production environments. Reproductive data are usually gathered in a particular genetic background, management regimes, and environmental conditions, restricting the relevance of models outside of their original environment. Predictive models are able to adjust to new circumstances without being constrained by biologically motivated commonalities through machine learning methods that explicitly represent shared biological structure and context-specific variation. These methods minimize the chances of poor performance in the domain shift and increase the prognostic strength of predictions made in heterogeneous livestock systems (Xue et al., 2024).

The concept of interpretability plays a primary role in the use of machine learning for reproductive modeling. The predictive accuracy is not enough in situations where decisions impact animal welfare, mating patterns, and future population dynamics. Stakeholders should be in a position to figure out the factors that model predictions are based on and to determine whether the factors would be consistent with the reproductive biology that is known. This can be achieved through machine learning models that are constructed to include interpretable structures or that are otherwise equipped to provide post hoc explanations, so that users can trace model predictions to the underlying regulatory and physiological characteristics and trust and use responsibly (Kanchepu, 2025).

More innovative applications to reproductive management are founded on predictive modeling and further prescriptive and adaptive applications. Machine learning models can be used in assessing alternative strategies before their implementation by predicting reproductive outcomes under different conditions. This is the ability that helps in proactive management by detecting any possible risks and opportunities at an earlier stage of the reproductive process. Nevertheless, predictive models must be used in a prescriptive way with stringent restrictions to ensure that the objectives of optimization incorporate ethical issues, biological boundaries, and long-term sustainability, and not simply the performance indicators (Hynniewta et al., 2025).

Another aspect of predictive modeling in reproductive omics is that of stratification and personalization. There is a high level of heterogeneity in reproductive response in the livestock populations due to variations in genetic background, developmental history, physiological state, and environmental exposure. This heterogeneity can be modeled in machine learning through learning individualized or subgroup-specific representations, which can represent different reproductive trajectories. This form of personalization would help achieve a better congruence between reproductive plans and biology, thereby minimizing inefficiencies and avoiding unintended side effects of generalized advice (Rao Suura, 2025).

Machine learning also has a role in the continuous improvement of reproductive knowledge through iterative learning. Predictive models can be updated with new reproductive data as it becomes available to capture new trends and increase accuracy and applicability. This adaptive ability is used to transform predictive modeling from a static analytical approach into a dynamic learning system that evolves along with the reproductive populations it constitutes. To maintain stability and prevent drift between such updates, validation and monitoring are strictly enforced, especially in cases in which the distributions of data vary with time (Adimoelja et al., 2024).

The application of machine learning models to conduct predictive reproductive analytics also raises significant concerns about the issue of validation and accountability. Predictive performance should be tested in a wide variety of situations and over a wide range of time scales to guarantee reliability, and models should be monitored to degrade as biological or management conditions change. There is a need to have transparency in the reporting of model assumptions, limitations, and uncertainty so that they can be responsibly integrated into reproductive decision-making processes. In the absence of these protective measures, predictive models will be vulnerable to misuse or overuse, which will reduce their scientific and practical usefulness (Doultani et al., 2024).

In general, predictive modeling using machine learning turns livestock reproductive omics into a future-driven analytic model that can predict reproductive behavior and act proactively. Predictive modeling creates a pivotal connection between reproductive biology and decision intelligence by combining heterogeneous regulatory layers, temporal dynamics, quantifying uncertainty, and being interpretable. This bridge will allow reproductive omics to move beyond retrospective analysis and transition to an anticipatory and adaptive mode of operation, yet remain compatible with biological realism, ethical accountability, and sustainability in the long term. Machine learning approaches for predictive modeling of reproductive outcomes in livestock are presented in Table 11.5.

11.6 Challenges in AI adoption and integration in livestock reproductive omics

Although the analytical capability and conceptual potential of AI in livestock reproductive omics are growing, its application and implementation are still limited by complex interactions between biological, technical, methodological, and governance-based issues. The obstacles are not caused by one limitation, but by the inherent nature of reproductive systems and socio-technical conditions in which reproductive decisions are made. Their solution involves a systemic view of the interaction of AI with aspects of biological complexity, data constraints, institutional behavior, and ethical obligations (Ghavi Hossein-Zadeh, 2025).

The nature and availability of reproductive data pose one of the most basic challenges. Reproductive phenotypes are, by their nature, difficult to quantify, time-based, and can be hindered by constraints in sampling frequency and scale due to economic, logistical, and welfare considerations. Consequently, the reproductive omics data often have a large proportion of relatively small samples relative to their dimensionality, which elevates the chances of overfitting and inaccurate inference. High representational capacity AI models are vulnerable to such imbalances, and they might learn patterns that encourage noise or sampling artifacts instead of true biological regulatory mechanisms. This problem is also exacerbated by an uneven spread of data on reproductive stages, tissues, and environmental conditions that may bias model learning toward the more common states (Doultani et al., 2024).

Biological heterogeneity is another factor that creates a further aspect of complexity and makes integrating AI a complicated issue. The reproductive systems of different individuals, sexes, breeds, and production settings differ significantly and indicate genetic diversity as well as regulation determined by the context. This heterogeneity makes more than mere noise, and is rather to be regarded as a feature, essential to reproductive biology, that gives it adaptability and resilience. AI models that do not account for this organized variation face the risk of making overly generalized predictions that are general rather than situation-specific. On the contrary, models that are overfit on small groups of data cannot be applied to other groups and do not have generalizability (Doultani et al., 2024; Moustakli et al., 2025).

The other great challenge is temporal complexity. The cycles of reproduction occur on an overlapping cyclical, developmental, and adaptive scale, but data collection in animal production systems is usually sparse, intermittent, and not in phase with the biology. AI models should thus be able to deduce latent reproductive states based on missing temporal data, which adds more uncertainty and makes them more difficult to validate. Unless the temporal structure is modeled carefully, prediction can confound a temporal fluctuation with a significant regulatory change or phase-varying difference with abnormality. This problem highlights the importance of considering temporal context in reproductive inference, and it is challenging to observe it in real-life conditions (Yao et al., 2025).

TABLE 11.5 Machine learning approaches for predictive modeling of reproductive outcomes in livestock.

Machine learning approach	Core modeling principle	Reproductive data types utilized	Predictive capabilities and outputs	Strengths and limitations in reproductive omics
Linear and regularized models	• Linear relationship modeling • Penalization for feature selection	• Genomics • Epigenomics • Transcriptomics	• Estimation of reproductive risk scores • Baseline fertility prediction	• Strengths: interpretability and stability • Limitations: limited nonlinear capacity
Tree-based ensemble methods	• Decision-tree aggregation • Nonlinear interaction capture	• Multi-omics features • Physiological indicators	• Classification of reproductive states • Outcome probability estimation	• Strengths: handles heterogeneous data • Limitations: reduced transparency
Support vector machines	• Margin-based classification and regression • Kernel-based transformation	• Transcriptomics • Proteomics	• Separation of fertile versus subfertile states • Boundary detection	• Strengths: effective in high-dimensional space • Limitations: kernel selection sensitivity
Artificial neural networks	• Multilayer nonlinear representation learning • Universal function approximation	• Integrated multi-omics • Sensor-derived physiometrics	• Complex reproductive outcome prediction • Latent state inference	• Strengths: expressive modeling capacity • Limitations: high data demand
Deep learning models	• Hierarchical feature learning • End-to-end optimization	• Genomics and sequence data • High-resolution omics	• Prediction of developmental competence • Trajectory forecasting	• Strengths: captures complex dependencies • Limitations: interpretability challenges
Temporal models	• Time-series and sequential learning • State transition modeling	• Longitudinal endocrine data • Time-resolved omics	• Cycle phase prediction • Early detection of reproductive disruption	• Strengths: models dynamics explicitly • Limitations: sparse sampling constraints
Probabilistic and Bayesian models	• Explicit uncertainty representation • Prior-informed inference	• Sparse and incomplete reproductive datasets	• Probabilistic outcome forecasting • Risk and confidence estimation	• Strengths: uncertainty awareness • Limitations: computational complexity
Graph-based learning	• Network and relational modeling • Dependency-aware inference	• Regulatory and interaction networks	• Identification of regulatory drivers • Causal pathway inference	• Strengths: biological interpretability • Limitations: graph construction challenges
Transfer and meta-learning	• Knowledge transfer across tasks or domains • Rapid adaptation	• Multipopulation reproductive datasets	• Generalization to new breeds or environments	• Strengths: improves robustness • Limitations: domain mismatch risk

Hybrid mechanistic-data-driven models	• Integration of biological priors with ML • Constraint-based learning	• Omics and physiological models	• Biologically plausible reproductive forecasting	• Strengths: interpretability and realism • Limitations: model complexity
Explainable AI frameworks	• Post hoc and intrinsic interpretability methods • Feature attribution	• Any reproductive omics modality	• Identification of key regulatory contributors	• Strengths: trust and transparency • Limitations: explanation variability
Reinforcement and adaptive learning	• Policy optimization via feedback • Sequential decision learning	• Integrated reproductive and management data	• Adaptive reproductive intervention modeling	• Strengths: supports prescriptive analytics • Limitations: reward definition complexity

There are also variations in the technical aspect and data quality that make AI adoption even more difficult. The production of reproductive omics data is on a variety of different experimental platforms, procedures, and analytical pipelines, dealing with batch effects and biases that serve to mask the biological signal. The AI models are especially skilled at taking advantage of regular data trends such as those produced by technical objects and not biology. Such artifacts, when left uncorrected, can be accidently learned as predictive features, resulting in shortcut learning and poor generalization. To reduce this risk, it is necessary to have stringent harmonization, validation, and transparency in preprocessing, all of which are costly and time-intensive processes (Adams and Bittremieux, 2025).

Interpretability is one of the ongoing issues when it comes to the implementation of AI in the field of reproductive omics. Most sophisticated machine learning models are based on complicated internal representations that are hard to agree with known biological insights. In reproductive uses, where the choices can have an impact on the welfare of animals, breeding patterns, and the future population dynamics, the inability to interpret data destroys trust and restricts adoption. The stakeholders need not only the right predictions, but also knowledge of the biological explanation that refers to the predictions. The tension between predictive and explanatory clarity is one of the key issues of AI integration (Van Hilten et al., 2024).

The other challenge exists because of the difference between correlation and causation when making inferences of reproduction. Reproductive omics data sets are mainly observational in nature, as the management practices and environmental exposures involved cannot be completely controlled. The AI models that are trained on this data can find powerful associations that are not associated with causal mechanisms. Noncausal signals have the potential to lead to ineffective or harmful interventions, especially in reproductive systems where regulatory systems are closely related. Causal reasoning is still both methodologically and computationally expensive to incorporate into AI models, but it is crucial to responsible AI usage (Ali, 2025).

Technical issues blend with government and moral issues. The possibilities of reproductive choices based on AI determine the genetic variation, the structure of the population, and the consequences of generations, not only in single animals. Models that maximize narrow goals can easily encourage welfare- or long-term resilience-destroying practices unless more general constraints are explicitly factored in. Moreover, accountability concerns emerge when the algorithms make reproductive choices, and the indicators of accountability are not clearly defined, nor is there a clear mechanism of human control and responsibility (Faria, 2025).

There are other barriers to integration with data governance. The information of reproductive omics may be part of proprietary breeding programs or regulated, which restricts accessibility and collaboration. The division of data between institutions and jurisdictions limits the creation of generalizable models, as well as the ability to independently verify them. AI solutions based on centralized data aggregation can thus be limited practically and legally, and decentralized ones can bring technical complexity of their own (Zelinschi et al., 2025).

The effects of AI in livestock reproductive omics are also restricted by translation problems. The implementation of advanced analytical tools into some of the current management processes necessitates congruence with operational limits, user proficiency, and regulatory provisions. Computationally expensive models, which are hard to interpret or which are not well-calibrated to the variability of the real world, can be kept within the confines of research environments, notwithstanding their analytical potential. It takes not only unbelievable refinement and knowledge to achieve good translations, but also education, infrastructure building, and testing and assessment in field application (Rosati, 2025).

Lastly, the idea of sustainability becomes an increasingly influential aspect of AI usage. The environmental and economic costs involved in the computational requirements of training and deploying complex models have to be considered in comparison with the expected benefits. The AI solutions need to show responsible resource consumption and not just an analytical value in the livestock systems that already face pressure to become more sustainable. Such a challenge makes efficiency and scalability, as well as alignment with the larger sustainability aims, even more significant (Pavunraj et al., 2025).

Taken together, these difficulties indicate that the example of AI application in livestock reproductive omics is not an exclusively technical task but a multidimensional process necessitating coordination of biological knowledge, data management, approach to methodology, ethical regulation, and practice. These are the challenges that must be overcome to come up with the AI that would improve reproductive knowledge and decision-making without undermining the standards of biological integrity, welfare, and societal confidence. Major challenges in AI-driven livestock reproductive omics and their implications are shown in Table 11.6.

11.7 Ethical considerations and public perception

The ethical aspects cannot be separated from the introduction of AI in livestock reproductive omics because reproductive decision-making occupies a special place at the boundary of biology, technology, and society. Reproduction has a direct

TABLE 11.6 Major challenges in AI-driven livestock reproductive omics and their implications.

Challenge category	Nature of the challenge	Underlying causes	Implications for reproductive inference	Broader scientific and practical consequences
Data scarcity and imbalance	• Limited sample sizes relative to data dimensionality • Uneven representation of reproductive stages	• High cost and logistical difficulty of sampling • Welfare and management constraints	• Increased risk of overfitting • Reduced predictive reliability for rare reproductive states	• Limits generalizability across populations • Slows translation into applied systems
Biological heterogeneity	• High variability across individuals, breeds, and environments • Context-dependent reproductive regulation	• Genetic diversity • Environmental and management variation	• Difficulty distinguishing signal from adaptive variation • Context-sensitive model performance	• Challenges the standardization of AI solutions • Necessitates personalized or stratified modeling
Temporal complexity	• Reproductive processes unfold across multiple timescales • Irregular and sparse longitudinal data	• Cyclical and developmental biology • Practical sampling constraints	• Misalignment of observations with the biological phase • Reduced accuracy of dynamic predictions	• Complicates validation and benchmarking • Limits real-time deployment
High dimensionality and noise	• Thousands of features per omics layer • Technical and biological variability	• Multi-omics technologies • Experimental and platform differences	• Difficulty isolating biologically meaningful patterns • Risk of spurious associations	• Increases computational burden • Demands advanced regularization and validation
Technical variability and batch effects	• Systematic differences between datasets • Nonbiological signal contamination	• Platform heterogeneity • Protocol and pipeline variation	• Shortcut learning by AI models • Reduced reproducibility	• Undermines cross-study integration • Complicates collaborative research
Missing and incomplete data	• Partial observation of omics layers • Nonrandom missingness	• Cost constraints • Sampling feasibility	• Biased inference if unmodeled • Increased uncertainty in predictions	• Limits the robustness of AI deployment • Requires probabilistic modeling frameworks
Interpretability limitations	• Complex internal representations of AI models • Limited biological transparency	• Use of deep and nonlinear architectures	• Reduced trust in model outputs • Difficulty validating biological plausibility	• Slows adoption by stakeholders • Raises ethical concerns
Correlation-causation gap	• AI models capture associations rather than mechanisms • Confounded observational data	• Limited experimental control • Complex regulatory feedback	• Risk of ineffective or harmful interventions • Misguided decision-making	• Necessitates causal modeling integration • Slows regulatory acceptance
Model generalization and transferability	• Performance degradation across contexts • Sensitivity to domain shift	• Context-specific training data	• Unreliable predictions in new environments	• Limits the scalability of AI solutions • Requires continuous retraining

Continued

TABLE 11.6 Major challenges in AI-driven livestock reproductive omics and their implications.—cont'd

Challenge category	Nature of the challenge	Underlying causes	Implications for reproductive inference	Broader scientific and practical consequences
Validation and benchmarking	• Lack of standardized evaluation protocols • Context-dependent performance metrics	• Heterogeneous data and objectives	• Unclear reliability of predictions • Difficulty comparing models	• Slows methodological convergence • Reduces scientific transparency
Computational and infrastructure demands	• High resource requirements for training and deployment • Scalability constraints	• Large multi-omics datasets • Complex model architectures	• Barriers to real-time or on-farm application	• Increases environmental and economic costs
Data governance and accessibility	• Fragmented data ownership • Legal and ethical constraints	• Proprietary breeding systems • Regulatory requirements	• Limited data diversity for model training	• Restricts collaboration • Slows innovation
Ethical and welfare concerns	• Risk of over-optimization • Long-term and intergenerational effects	• Productivity-focused objectives • Incomplete biological understanding	• Potential compromise of animal welfare	• Influences public trust and acceptance
Translational and adoption barriers	• Gap between research models and practical use • User expertise limitations	• Complex model interfaces • Misalignment with workflows	• Limited real-world impact	• Requires education and system redesign
Sustainability considerations	• Environmental and resource costs of AI • Long-term system resilience	• Computational intensity • Intensification pressures	• Misalignment with sustainability goals	• Necessitates efficiency-focused innovation

effect on animal welfare, their population structure, genetic diversity, and long-term adaptability, and thus has moral implications that cannot be reduced to technical optimization. The scope and ramifications of reproductive decision-making are magnified by AI because it allows making judgments that are more accurate and serves as a starting point for intervention, placing ethical considerations as a bottom line and not a side issue (Kılıç et al., 2024).

Another ethical concern is the possibility of a dilemma between productivity goals and the biological integrity of animals. Unless limited by welfare concerns, AI systems developed to maximize reproductive efficiency can give preference to short-term performance indicators. The data provided by reproductive omics can be associated with impressive regulatory plasticity and adaptive capacity, yet the interpretation of this plasticity as inexhaustible malleability gives reason to justify the intervention that is beyond the physiological tolerance threshold or the long-term health deterioration. Ethical application of AI thus must acknowledge that the adaptability of biology does not mean that it is morally permissible and that welfare needs to be considered a fundamental goal to be enshrined within systems of analysis (Faria, 2025).

AI utilization is also causing significant questions to be posed on transparency and accountability in the context of reproductive omics. The use of advanced machine learning models can make use of complex internal representations that are not readily interpretable and that are difficult to explain using established norms of scientific explanation and decision justification. In reproduction, where the products of the algorithm can affect the mating process, reproductive technologies, or population control, the lack of transparency compromises trust and restricts the ethical right to act. Stakeholders need to have the capability of knowing how conclusions are made, examining assumptions underlying the conclusions, and questioning recommendations when necessary. Ethical practice thus requires that AI systems be in favor of interpretability, traceability, and human supervision throughout the decision-making process (David et al., 2025).

Equity and fairness are other areas of ethical consideration that determine scientific integrity and public perception. The distribution of reproductive omics datasets among breeds, production systems, and geographic regions is commonly skewed and is related to patterns of investment and access in the past. This imbalance can be internalized by AI models that were trained on such data, and that generate a prediction or recommendation that is unevenly distributed across the population. This brings about the issue of equitable sharing of the technological advantage and the probable strengthening of the disparity. Ethical incorporation of AI necessitates a conscious focus on embodying the representation of balance, validation in various settings, and means of detecting and alleviating bias in reproductive inference (Sucharita and Shaw, 2025).

There is also the complex ethical environment that is complicated by data governance and ownership. The reproductive omics information is often incorporated into restrictive breeding software or may be regulated, which makes it unclear as to whether consent, benefit sharing, and access are being met. Such data, when trained as AI models, can create value that surpasses the contributors of the data, and leaves one wondering who owns, gets, or is responsible for downstream use. Open governance systems that explain data rights and responsibilities and accountability are a key to establishing cooperation and ensuring trust within the stakeholder community (Carlson et al., 2025).

Ethical considerations of AI in reproductive omics have been considered at different time scales, where effects relate not only to the short-term but also the long-term and intergenerational impacts. The effects of reproductive choices informed by AI can have long-term changes on developmental patterns and epigenetic control, as well as population structure, that impact future generations. There is scientific uncertainty about these long-term effects, and this has required a precautionary response in terms of monitoring, reversibility, and restraint. The ethical frameworks should thus be able to embrace the aspect of time responsibility, where there is a responsibility toward the population in the future as well as the current goals (Pradhan, 2024).

The social attitude is very important in determining the acceptability and uptake of AI in animal reproduction. Cultural values, historical experience, and trust in the institutions shape societal attitudes toward technological intervention on biological processes. Reproductive technologies are especially likely to trigger increased apprehension as they are associated with the process of creating and manipulating life itself. When combined with reproductive omics, AI can be perceived as a control domain, enhancing control over biological processes, which will result in resistance if it is considered opaque, coercive, and incongruous with social values. This means that ethical integration necessitates intentional involvement, open communication, and positioning AI as a decision aid tool, not as a decision maker (Doultani et al., 2024).

Ethically, communication strategies are relevant since they influence perceptions and anticipations of AI possibilities and constraints. Exaggeration of predictive certainty or effectiveness of intervention poses the risk of undermining trust when such outcomes are not consistent with predictions. Ethical communication should focus on uncertainty, context dependence, and supporting AI in wider decision-making contexts. Transparent communication also leads to informed consent and confidence among the population by creating realistic expectations (Abokhoza et al., 2024).

There is a need to establish accountability mechanisms to govern AI ethically in reproductive omics. Under situations where algorithmic suggestions affect reproductive choices, there should be clear lines of authority to avoid the emergence of accountability spreading throughout the technical frameworks. Human oversight is essential, not only to place the outputs of the algorithms in the context of biological and ethical considerations, but also to take action when the recommendations conflict with welfare or sustainability goals. Ethical practice demands institutional frameworks that attribute responsibility, provide auditing, and offer redress in case of any harm (Adeniyi and Monehin, 2024).

The ethical issues overlap with sustainability, as AI-based reproductive optimization imposes its effects on resource consumption, environmental impacts, and the resilience of the system in the long term. The ethical analysis should not focus only on individual reproductive performance but also on the overall cumulative impacts on livestock herds and production ecosystems. AI systems that foster intensification without considering ecological limitations may hinder sustainability objectives, public trust, and increase related environmental stress. Ethical integration involves matching reproductive analytics with larger sustainability goals (Manzoor et al., 2025).

In general, ethical concerns and social perception are integral to the AI adoption of livestock reproductive omics. Ethics should not be a post hoc consideration or a regulatory process but should be integrated into model design, validation, deployment, and communication. By foregrounding fairness, accountability, sustainability, and welfare with the use of AI, the latter becomes a responsible, socially acceptable, and legitimate aspect of reproductive science and management. Ethical issues are best handled through proactive approaches to protect both animals and stakeholders, which can also enhance the scientific validity and sustainability of AI-based reproductive systems. Ethical dimensions and governance considerations in AI-enabled livestock reproductive omics are presented in Table 11.7.

11.8 Future perspectives

The future of AI implementation in livestock reproductive omics is determined by the integration of the increasing biological knowledge, growing data ecosystem, and more complex computational paradigms. Due to the continued movement toward defining reproductive systems in finer molecular, cellular, and physiological detail, AI will play an increasingly leading role in making sense of this complexity into understandable, practical representations. It is hoped that future developments will not only result in more efficient analysis but also in a fundamental change in the way reproductive biology is conceptualized, investigated, and implemented in livestock systems (Giordano, 2024).

The further extension of reproductive data resolution and scope is one of the most important ways forward. Further developments in omics technologies will allow for a more detailed description of reproductive regulation in tissues, developmental stages, and physiological conditions. The data that will be produced by these advances will be richer and more complex, with capacities of spatial heterogeneity and temporal dynamics that can no longer be analyzed using traditional methods. AI will be needed to reconcile these streams of data into combined forms, which maintain biological form but allow making inferences at the system level. With a further increase in the resolution, the analysis focus will probably become more concerned with models that directly encode hierarchical structure, time dynamics, and regulatory dependence instead of viewing reproductive data as fungible characteristics (Sengupta et al., 2024; Wadood and Zhang, 2024).

It is believed that the evolution of the representation learning paradigms' maturation will be central in this development. Generalizable representations of reproductive regulation in a variety of biological settings can be learned by AI systems that do not rely on highly specific models and large labeled datasets. These representations would facilitate cross-breed, cross-environment, and cross-species transfer of reproductive insight and would provide a wider application of reproductive omics knowledge. This change in the foundational representations is an indication of a transition of task-specific modeling to more unified and flexible analysis frameworks that can support various inferential goals (Li et al., 2024).

The aspect of temporal modeling will also gain a higher role as reproductive omics is increasingly focused on dynamic processes. The future generations of AI systems will probably be able to extract a continuous, changing process of reproductive trajectories that combine molecular, physiological, and environmental information across time. This type of model would promote more accurate characterizations of developmental transitions, adaptive responses, and recovery dynamics, and allow reproductive management strategies to become more aligned with biological timing. With an increase in the temporal resolution, the difference between short-term fluctuations and long-term reprogramming will be better identified, leading to a higher interpretability and predictive reliability (Hijazin et al., 2024).

Causal inferences are also likely to become one of the hallmarks of the reproductive omics analytics of the next generation. Although most uses of AI have been in pattern recognition and prediction, future systems will have more inference of regulatory drivers and mechanistic pathways. This change is an awareness that effective intervention can only

TABLE 11.7 Ethical dimensions and governance considerations in AI-enabled livestock reproductive omics.

Ethical or governance dimension	Core ethical concern	Manifestation in reproductive omics	Governance and oversight requirements	Implications for responsible AI deployment
Animal welfare and biological integrity	• Protection of physical and physiological well-being • Respect for biological limits	• Risk of reproductive over-optimization • Potential stress from intensified reproductive intervention	• Welfare-centered regulatory frameworks • Inclusion of physiological and health constraints in models	• Ensures reproductive efficiency does not compromise well-being • Supports ethically sustainable breeding systems
Reproductive autonomy and naturalness	• Ethical limits of technological intervention • Preservation of biological processes	• Increased algorithmic influence over reproductive timing and selection • Perceived manipulation of life processes	• Ethical review of intervention scope • Clear delineation between decision support and decision automation	• Maintains societal legitimacy • Prevents erosion of ethical boundaries
Transparency and explainability	• Right to understand AI-driven decisions • Accountability for outcomes	• Use of complex, opaque predictive models • Difficulty linking outputs to biological rationale	• Explainability standards • Documentation and auditability requirements	• Builds trust among stakeholders • Enables informed and contestable decision-making
Fairness and bias mitigation	• Equitable treatment across breeds and systems • Avoidance of systemic bias	• Uneven representation of populations in data • Differential model performance across contexts	• Bias detection and mitigation protocols • Validation across diverse populations	• Promotes inclusive technological benefit • Reduces reinforcement of existing disparities
Data ownership and consent	• Rights over biological and omics data • Control of data use	• Use of proprietary reproductive datasets • Secondary use of data for model training	• Clear data governance agreements • Consent and access frameworks	• Encourages collaboration while protecting contributors • Clarifies responsibility and benefit sharing
Privacy and confidentiality	• Protection of sensitive biological information • Prevention of misuse	• Exposure of genetic and reproductive profiles • Cross-institutional data integration	• Secure data handling protocols • Privacy-preserving analytical methods	• Reduces risk of data misuse • Enhances stakeholder confidence
Accountability and responsibility	• Assignment of responsibility for AI-informed decisions • Liability for adverse outcomes	• Algorithmic recommendations influencing breeding decisions • Diffusion of responsibility across systems	• Human-in-the-loop requirements • Clear accountability hierarchies	• Prevents abdication of ethical responsibility • Ensures human oversight
Long-term and intergenerational impact	• Responsibility toward future populations • Preservation of genetic diversity	• AI-guided selection shaping population structure • Potential cumulative epigenetic effects	• Longitudinal monitoring frameworks • Sustainability-oriented governance	• Supports resilient and adaptable livestock populations

Continued

TABLE 11.7 Ethical dimensions and governance considerations in AI-enabled livestock reproductive omics.—cont'd

Ethical or governance dimension	Core ethical concern	Manifestation in reproductive omics	Governance and oversight requirements	Implications for responsible AI deployment
Sustainability and environmental ethics	• Alignment with ecological responsibility • Resource stewardship	• Intensification of reproductive systems • Increased computational resource use	• Integration of sustainability metrics • Evaluation of environmental trade-offs	• Aligns AI innovation with sustainability goals
Regulatory compliance and harmonization	• Adherence to legal and ethical standards • Cross-jurisdiction consistency	• Variation in regulatory acceptance of AI tools • Unclear classification of AI-driven decisions	• Harmonized regulatory guidelines • Adaptive policy frameworks	• Facilitates safe and lawful deployment
Public trust and social acceptance	• Societal perception of AI in reproduction • Ethical legitimacy	• Concerns over transparency and control • Cultural sensitivity to reproductive intervention	• Stakeholder engagement and communication strategies	• Enhances adoption and long-term viability
Scientific integrity and reproducibility	• Reliability and credibility of AI-driven knowledge	• Difficulty reproducing complex models • Context-dependent results	• Standards for validation and reporting • Open methodological practices	• Strengthens scientific confidence • Supports cumulative knowledge building

be achieved through knowing not only what is concomitant with reproductive outcomes but also the cause of such outcomes. AI systems with causal reasoning and predictive models will assist in assessing hypothetical scenarios by providing an opportunity to evaluate possible interventions prior to their implementation and decreasing the use of trial-and-error solutions (Oh, 2024).

Digital reproductive twins are just the idea that will tend to become a key structuring paradigm in livestock reproductive omics. These constructs would combine both a priori genomic constructions and dynamically decreed molecular, physiological, and environmental data to generate an individualized or population-level account of reproductive state. The digital reproductive twins would be able to facilitate constant tracking, modeling, and prediction of reproductive tracks, which would support adaptive administrative interventions that can occur in advance based on arising threats or opportunities. The vision will require the development of improvements in model scalability, real-time data assimilation, and uncertainty quantification, as well as sound validation with biological and practical results (Youssef et al., 2024).

The AI application in reproductive choice will also increasingly focus on prescriptive and adaptive control as opposed to prediction. It is anticipated that future systems will also consider alternative strategies to reproductive approaches as part of learned models of biological constraint and then optimize between productivity and welfare, resilience, and sustainability. This prescriptive ability will necessitate a keen formulation of goals and limitations, where optimization is in line with the moral and biological limits. AI will therefore not be a free decision-maker; it will be a formalized advisory system that assists in making informed human judgments (Kaulwar, 2025).

The role of ethical governance in the development of the AI-enabled reproductive omics will become even more important. The more analytical powers are developed, the more power can be abused or overused. The innovation of the future should therefore be such that it provides ethical considerations as part and parcel of model design, validation, and deployment processes. This entails direct expression of welfare limitations, openness concerning uncertainty and constraints, as well as accountability and control mechanisms. Ethical governing systems will have to evolve in tandem with advances in technology, ensuring that reproductive innovation aligns with societal norms and long-term sustainability (Mishra, 2024).

Social acceptance and trust are likely to be the key determinants in the use of AI in livestock reproduction. Transparency and communication will be crucial in gaining understanding and acceptance as analytical systems become increasingly complex and influential. Moving forward, intellectual work should focus on defining what AI can achieve and what it cannot, as a decision support system, rather than a replacement for biological knowledge or moral reasoning. Developing trust will take time, and ongoing communication between researchers, producers, policymakers, and the general population, based on transparency and a willingness to address concerns (Kılıç et al., 2024).

The future development will also be determined by data governance and collaborative infrastructure. The potential of AI in reproductive omics is conditional upon the availability of different, high-quality data on genetic backgrounds, environment, and management systems. The next-generation models need to balance transparency and concerns about data ownership and privacy, as well as regulatory adherence. The use of distributed and federated methods of analytics can allow collaboration without centralizing the aggregation of data, allowing wider participation to occur, and reducing the barriers of governance. Defining common standards of data representation, data model evaluation, and reporting will also increase reproducibility and cumulative knowledge building (Hynniewta et al., 2025; Zelinschi et al., 2025).

The concern for sustainability will continue to have an impact on the creation of technology and the appraisal of AI implementation in reproduction. Analytical performance will be joined with important criteria such as computational efficiency, resource use, and environmental impact. It will require AI systems to be able to fulfill not just predictive or prescriptive value in the future, but also alignment with larger sustainability goals in livestock production. This involves considering the long-term resilience of the population, ecological performance, and sustainable use of resources (Pandey, 2025).

The interdisciplinary cooperation will be necessary to achieve these perspectives in the future. Reproductive omics based on AI is at the intersection of biology, data science, ethics, and management, in which interdisciplinary knowledge is essential. The further evolution will be based on collaborative systems that would allow communication and integration between these fields, so that the models could be based on biological reality and yet be computationally competent and ethically knowledgeable (Biswas et al., 2023).

To operate effectively, AI-driven livestock reproductive omics will be in its mature form, a continuously learning system that also evolves together with the biological populations it describes. These systems will combine new data, improve representations, and adapt to evolving situations over time, which will help them to be resilient in the face of uncertainty. Such a vision can be contrasted with the inelastic analytical paradigms and can be seen as a general change on the way to more adaptive and systems-focused reproductive science (Fortes and Pegolo, 2023).

TABLE 11.8 Future directions, emerging technologies, and conceptual shifts in AI-enabled livestock reproductive omics.

Future domain	Emerging technologies and methodologies	Conceptual shift in reproductive omics	Anticipated scientific and analytical impact	Implications for livestock reproductive systems
High-resolution reproductive profiling	• Single-cell and spatial omics platforms • High-frequency physiological sensing	• From bulk averaging to cellular and spatial heterogeneity • From static snapshots to continuous profiling	• Improved resolution of regulatory dynamics • Identification of rare but critical reproductive states	• Enables precision reproductive assessment • Supports individualized reproductive management
Integrated multiscale modeling	• Hierarchical AI architectures • Multilevel systems modeling	• From single-layer analysis to systems-level integration • From isolated traits to emergent reproductive states	• Enhanced biological realism • Improved cross-scale interpretability	• Strengthens robustness of reproductive inference • Reduces unintended intervention effects
Foundation and representation learning	• Self-supervised and transfer learning models • Cross-species representation learning	• From task-specific models to reusable biological representations • From data scarcity to knowledge reuse	• Improved generalization across breeds and environments • Reduced reliance on labeled data	• Expands applicability of AI tools • Accelerates innovation in underrepresented systems
Temporal and trajectory-based inference	• Continuous-time modeling frameworks • Longitudinal AI analytics	• From event-based prediction to trajectory forecasting • From outcome focus to process understanding	• Improved prediction of reproductive transitions • Clearer distinction between transient and persistent states	• Enables proactive reproductive management • Enhances timing and precision of interventions
Causal and counterfactual modeling	• Causal discovery algorithms • Counterfactual simulation frameworks	• From correlation-driven inference to mechanism-aware modeling • From prediction to intervention reasoning	• Reduced risk of spurious conclusions • Improved intervention reliability	• Supports safer reproductive decision-making • Enhances ethical accountability
Digital reproductive twins	• Dynamic biological simulation platforms • Real-time data assimilation	• From population averages to individualized reproductive models • From retrospective analysis to continuous learning	• Enables scenario testing and adaptive management • Improves monitoring of reproductive health	• Supports precision breeding and management • Enhances system resilience
Prescriptive and adaptive decision systems	• Reinforcement and adaptive learning frameworks • Multiobjective optimization	• From descriptive analytics to guided decision support • From reactive to proactive reproductive strategies	• Balances productivity, welfare, and sustainability objectives	• Improves long-term reproductive efficiency • Reduces management uncertainty
Ethics-embedded AI design	• Constraint-based and value-aware modeling • Welfare-integrated optimization	• From post hoc ethics to ethics-by-design • From performance-only to responsibility-aware systems	• Aligns AI outputs with ethical and biological limits	• Strengthens public trust • Ensures socially legitimate deployment
Privacy-preserving and federated analytics	• Federated learning • Secure multiparty computation	• From centralized data ownership to collaborative intelligence • From data sharing to model sharing	• Enables large-scale collaboration • Reduces governance barriers	• Expands data diversity • Accelerates collective progress

Standardization and reproducibility infrastructure	• Harmonized data standards • Transparent model reporting	• From fragmented methodologies to cumulative science • From opaque models to auditable systems	• Improves reproducibility and comparability	• Strengthens scientific credibility • Facilitates regulatory acceptance
Human-AI collaborative frameworks	• Decision-support interfaces • Explainable AI integration	• From automation to augmentation of expertise • From algorithmic authority to *shared judgment*	• Improves interpretability and adoption	• Empowers stakeholders • Maintains human accountability
Sustainability-oriented AI development	• Energy-efficient model architectures • Resource-aware computation	• From performance maximization to sustainability optimization • From short-term gains to long-term resilience	• Reduces the environmental footprint of AI	• Aligns reproductive innovation with sustainability goals

Altogether, the future of AI integration in livestock reproductive omics can be described as growing biological, methodological, and ethical resolution. AI can make reproductive omics an active intelligence layer that can be used to predict, interpret, and take responsible action in complex biological and production systems. To achieve this potential, there will be a need for a long-lasting dedication to scientific rigor, transparency, governance, and engagement with society. With a systemic integration of technological innovation and biological knowledge, along with ethical care, AI can be useful in the sustainable, resilient, and socially acceptable reproduction of livestock in a more complex and unpredictable world. Future directions, emerging technologies, and conceptual shifts in AI-enabled livestock reproductive omics are shown in Table 11.8.

References

Abokhoza, R., Al-Tkhayneh, K.M., Al Qaruty, R., Hadi, S.A., Ellala, Z.K., 2024. The ethics of applying artificial intelligence (AI) for communication governance. In: International Conference on Intelligent Computing, Communication, Networking and Services, ICCNS 2024. Institute of Electrical and Electronics Engineers Inc., United Arab Emirates, pp. 236–240. https://doi.org/10.1109/ICCNS62192.2024.10776312. http://ieeexplore.ieee.org/xpl/mostRecentIssue.jsp?punumber=10774579.

Adams, C., Bittremieux, W., 2025. A 2025 Perspective on the Role of Machine Learning in Clinical Proteomics. ChemRxiv, Belgium. https://doi.org/10.26434/chemrxiv-2025-b9lpx.

Adeniyi, O.M., Monehin, V.B., 2024. The use of artificial intelligence (AI) in assisted reproductive technology (ART): examining the legal and ethical implications. ABUAD Law J. 12 (1), 135–152. https://doi.org/10.53982/alj.2024.1201.08-j.

Adimoelja, A., Mahmudy, W.F., Kurnianingtyas, D., 2024. A review of machine learning methods to build predictive models for male reproductive health. IAES Int. J. Artif. Intell. 13 (4), 3739–3749. https://doi.org/10.11591/ijai.v13.i4.pp3739-3749.

Ali, F., 2025. AutoMRAI: a multi-omics causal inference platform using structural equation modelling. In: medRxiv. Cold Spring Harbor Laboratory. https://doi.org/10.1101/2025.11.17.25340455.

Anandaram, H., Mittal, K., Gupta, P., Kulshrestha, S., Choudhury, A., Syed, M., Narad, P., Modi, D., Gupta, D., Sengupta, A., 2024. Proteomics and Metabolomics in Reproductive Health. Informa UK Limited, pp. 60–75. https://doi.org/10.1201/9781003487548-4.

Anjum, S., Khurshid, Y., Du Plessis, S.S., Omolaoye, T.S., 2025. Unmasking the epigenome: insights into testicular cell dynamics and reproductive function. Int. J. Mol. Sci. 26 (15). https://doi.org/10.3390/ijms26157305.

Biswas, A., Kumari, A., Gaikwad, D.S., Pandey, D.K., 2023. Revolutionizing biological science: the synergy of genomics in health, bioinformatics, agriculture, and artificial intelligence. OMICS A J. Integr. Biol. 27 (12), 550–569. https://doi.org/10.1089/omi.2023.0197.

Burton, A., Torres-Padilla, M.E., 2025. Epigenome dynamics in early mammalian embryogenesis. Nat. Rev. Genet. 26 (9), 587–603. https://doi.org/10.1038/s41576-025-00831-4.

Carlson, C.J., Granados, M., Phelan, A., Ramakrishnan, N., Poisot, T., 2025. The LISTEN principles for genetic sequence data governance and database engineering. Nat. Genet. 57 (9), 2099–2105. https://doi.org/10.1038/s41588-025-02270-7.

David, R., Shankar, H., Kura, P., Kowtarapu, K., Uma Maheswari, S.K.S., 2025. Advancement in explainable AI: bringing transparency and interpretability to machine learning models for use in high-stakes decisions. In: International Conference on Emerging Smart Computing and Informatics, ESCI 2025. Institute of Electrical and Electronics Engineers Inc., United States https://doi.org/10.1109/ESCI63694.2025.10988079. http://ieeexplore.ieee.org/xpl/mostRecentIssue.jsp?punumber=10987191.

Doultani, S., Sharma, P., Makwana, P., P Patil, S., S Layek, S., George, L.B., Highland, H.N., Hadiya, K.K., 2024. AI in reproductive biology: transforming fertility assessment, ART, and research. Annu. Res. Rev. Biol. 39 (9), 147–158. https://doi.org/10.9734/arrb/2024/v39i92129.

Emsen, E., Korkmaz, M.K., Odevci, B.B., 2025. Artificial Intelligence Assisted Selection Strategies in Sheep: Linking Reproductive Traits with Behavioral Indicators. https://doi.org/10.20944/preprints202506.0575.v1.

Faria, C., 2025. Creating life, creating strife? Assisted reproductive technologies, extinction, and wild animal welfare. J. Appl. Anim. Ethics Res. 7 (1), 68–90. https://doi.org/10.1163/25889567-bja10060.

Fatima, S., Sur, S., Tiwari, J., Rani, S., Malik, S., 2025. Endogenous regulation of behavior and reproductive physiology in a resident passerine songbird. Biol. Reprod. 113 (1), 154–168. https://doi.org/10.1093/biolre/ioaf073.

Fernandez, A.C.G., Pelnekar, S., Robinson, J.F., Shaw, G.M., Padula, A.M., Woodruff, T.J., Giudice, L.C., 2025. Climate change and reproductive health. Endocr. Rev. 46 (6), 908–921. https://doi.org/10.1210/endrev/bnaf026.

Fortes, M.R.S., Pegolo, S., 2023. Editorial: application of omics Technologies to improve robustness and resilience in livestock species. Front. Vet. Sci. 10. https://doi.org/10.3389/fvets.2023.1224630.

Garcia, R.G., Naas, I.de A., Burbarelli, M.F.de C., Caldara, F.R., Komiyama, C.M., Valentim, J.K., Sgavioli, S., 2025. Aplicações de inteligência artificial para otimização do bem-estar e sustentabilidade em sistemas de produção animal. Editora Cientifica Digital 8–34. https://doi.org/10.37885/250619568.

Ghavi Hossein-Zadeh, N., 2013. Effects of main reproductive and health problems on the performance of dairy cows: a review. Span. J. Agric. Res. 11 (3), 718–735. https://doi.org/10.5424/sjar/2013113-4140.

Giordano, J.O., 2024. Data-driven technologies and management practices for improving the sustainability of reproductive management. Reprod. Fertil. Dev. 37 (1). https://doi.org/10.1071/rd24148.

Hassani, R., Ghavi Hossein-Zadeh, N., 2025. Metabolic disorders are genetically correlated with the production and reproductive performance of dairy cows: a meta-analysis. J. Dairy Res. 92, 174–184. https://doi.org/10.1017/S0022029925101052.

Hegde, A., Nguyen, T., Cheng, J., 2025. Machine learning methods for gene regulatory network inference. Briefings Bioinf. 26 (5). https://doi.org/10.1093/bib/bbaf470.

Hijazin, M., Shi, P., Wang, J., Ding, J., 2024. DTPSP: A Deep Learning Framework for Optimized Time Point Selection in Time-Series Single-Cell Studies. bioRxiv, Canada. https://doi.org/10.1101/2024.12.18.629276.

Hossein-Zadeh, N.G., 2025. Artificial intelligence in veterinary and animal science: applications, challenges, and future prospects. Comput. Electron. Agric. 235, 110395. https://doi.org/10.1016/j.compag.2025.110395.

Hussein, A., Prasad, M., Braytee, A., 2024. Explainable AI Methods for Multi-Omics Analysis: A Survey. arXiv, Australia. https://doi.org/10.48550/arXiv.2410.11910.

Hynniewta, B.C., Navas, F., Marbaniang, K., 2025. Artificial intelligence in assisted reproduction: the future for reproductive medicine. Int. J. Reprod. Contracept. Obstet. Gynecol. 14 (9), 3181–3188. https://doi.org/10.18203/2320-1770.ijrcog20252767.

Kan, Y., Zhang, Z., Wang, Y., Qi, Y., Chang, H., Wang, W., Zhang, Z., Liu, Q., Shi, X., 2025. Integration of single cell multiomics data by deep transfer hypergraph neural network. Brief. Funct. Genom. 24. https://doi.org/10.1093/bfgp/elaf009.

Kanchepu, N., 2025. Unveiling the Black Box: The Crucial Role of Interpretability in Machine Learning Models. Bentham Science Publishers Ltd, pp. 22–44. https://doi.org/10.2174/9789815305548125010004.

Kaulwar, P.K., 2025. The Future of Intelligent Advisory: Agentic Systems and Beyond. Deep Science Publishing, pp. 180–195. https://doi.org/10.70593/978-93-49910-38-6_12.

Khan, R., Azhar, M., Umair, M., 2024. Decoding the genes orchestrating egg and sperm fusion reactions and their roles in fertility. Biomedicines 12 (12), 2850. https://doi.org/10.3390/biomedicines12122850.

Kılıç, K.D., Gökhan, A., Çavuşoğlu, T., 2024. The Transformative Role of Artificial Intelligence in Advancing Bovine Reproductive Biology. IGI Global, pp. 64–83. https://doi.org/10.4018/979-8-3693-3629-8.ch004.

Lac, L., Leung, C.K., Hu, P., 2024. Computational frameworks integrating deep learning and statistical models in mining multimodal omics data. J. Biomed. Inf. 152, 104629. https://doi.org/10.1016/j.jbi.2024.104629.

Li, Q., Sorscher, B., Sompolinsky, H., 2024. Representations and generalization in artificial and brain neural networks. Proc. Natl. Acad. Sci. U. S. A. 121 (27). https://doi.org/10.1073/pnas.2311805121.

Lingasamy, P., Modhukur, V., Mändar, R., Salumets, A., 2024. Exploring immunome and microbiome interplay in reproductive health: current knowledge, challenges, and novel diagnostic tools. Semin. Reprod. Med. 41 (05), 172–189. https://doi.org/10.1055/s-0043-1778017.

Liu, L., Chen, A., Li, Y., Mulder, J., Heyn, H., Xu, X., 2024. Spatiotemporal omics for biology and medicine. Cell 187 (17), 4488–4519. https://doi.org/10.1016/j.cell.2024.07.040.

Lu, Y., Li, M., Gao, Z., Ma, H., Chong, Y., Hong, J., Wu, J., Wu, D., Xi, D., Deng, W., 2024. Innovative insights into single-cell technologies and multi-omics integration in livestock and poultry. Int. J. Mol. Sci. 25 (23), 12940. https://doi.org/10.3390/ijms252312940.

Mansour, H.A.E.H., 2025. Integration of assisted reproductive technologies and artificial intelligence for optimizing fertility and genetic selection in livestock production. Discov. Appl. Sci. 7 (10). https://doi.org/10.1007/s42452-025-07380-9.

Manzoor, M., Khalid, S.M., Dhand, S., Singh, S., Kapoor, M., 2025. Dimensions of Environmental Sustainability and the Use of AI. Informa UK Limited, pp. 251–268. https://doi.org/10.1201/9781003476238-16.

Mihajlović, K., 2025. Data Integration of Longitudinal Single-Cell with Multi-Omics Data to Enable Precision Medicine for Complex Diseases. https://doi.org/10.5821/dissertation-2117-432840.

Mishra, A., 2024. Scalable AI Governance and Ethics. Springer Science and Business Media LLC, pp. 147–165. https://doi.org/10.1007/979-8-8688-0158-7_9.

More, A., Chimurkar, V., Mahajan, S., Dakre, S., Anjankar, N., More, D., 2025. The integration of artificial intelligence in in vitro fertilization: a comprehensive narrative review. J. Pharm. BioAllied Sci. 17 (Suppl. 1), S155. https://doi.org/10.4103/jpbs.jpbs_437_25.

Moustakli, E., Grigoriadis, T., Stavros, S., Potiris, A., Zikopoulos, A., Gerede, A., Tsimpoukis, I., Papageorgiou, C., Louis, K., Domali, E., 2025. Artificial intelligence in assessing reproductive aging: role of mitochondria, oxidative stress, and telomere biology. Diagnostics 15 (16), 2075. https://doi.org/10.3390/diagnostics15162075.

Oh, T.R., 2024. Integrating predictive modeling and causal inference for advancing medical science. Child. Kidney Dis. 28 (3), 93–98. https://doi.org/10.3339/ckd.24.018.

Padmanaban, H., 2024. Machine learning algorithms scaling on large-scale data infrastructure. J. Artif. Intell. Gen. Sci. 3 (1), 1–26. https://doi.org/10.60087/jaigs.vol03.issue01.p26.

Pandey, A., 2025. The Transformative Potential of Artificial Intelligence in Building Sustainable Agricultural Systems. Advances in Computational Intelligence and Robotics Book Series, pp. 161–182. https://doi.org/10.4018/979-8-3373-0680-3.ch006.

Pavunraj, D., Mathankumar, A., Anbumaheshwari, K., Daisy Merina, R., 2025. Green Artificial Intelligence (AI) and Machine Learning (ML). IGI Global, pp. 249–264. https://doi.org/10.4018/979-8-3373-0766-4.ch010.

Peña, F.J., Martín-Cano, F.E., Becerro-Rey, L., Ortega-Ferrusola, C., Gaitskell-Phillips, G., da Silva-Álvarez, E., Gil, M.C., 2024. Proteomics is advancing the understanding of stallion sperm biology. Proteomics 24 (14). https://doi.org/10.1002/pmic.202300522.

Pradhan, T., 2024. Navigating the promise and perils: artificial intelligence in assisted reproductive techniques. Birat Journal of Health Sciences 9 (1), 1–2. https://doi.org/10.62065/bjhs551.

Rachakonda, S.A., Pudipedi, S., Shiny Angel, T.S., 2024. Predictive modelling of menstrual cycle length using machine learning and time-series analysis. Int. J. Sci. Res. Eng. Manag 08 (008), 1–15. https://doi.org/10.55041/ijsrem37150.

Rao Suura, S., 2025. Machine Learning Models for Predictive Genomics: From Variant Interpretation to Early Risk Stratification. Deep Science Publishing, pp. 37–52. https://doi.org/10.70593/978-93-49307-76-6_3.

Rosati, A., 2025. Invited paper Artificial intelligence and the future of animal husbandry: rethinking knowledge, practice, and governance. Arch. Latinoam. Prod. Anim. 33 (Suppl. 1), 567–572. https://doi.org/10.53588/alpa.330509.

Rumpt, M.van, Belen Rabaglino, M., 2024. The role of machine learning in decoding the molecular complexity of bovine pregnancy: a review. Reprod. Fertil. Dev. 37 (1). https://doi.org/10.1071/rd24141.

Sengupta, P., Dutta, S., Liew, F., Samrot, A., Dasgupta, S., Rajput, M., Slama, P., Kolesarova, A., Roychoudhury, S., 2024. Reproductomics: exploring the applications and advancements of computational tools. Physiol. Res. 73 (5), 687–702. https://doi.org/10.33549/physiolres.935389.

Sucharita, S., Shaw, P.K., 2025. Bias detection and mitigation in AI: a framework for fair and inclusive machine learning models. Int. J. Sci. Res. Eng. Manag. 09 (07), 1–9. https://doi.org/10.55041/ijsrem51300.

Tedeschi, L.O., Guarnido Lopez, P., Menendez III, H.M., Seo, S., 2025. Advancing precision livestock farming: integrating artificial intelligence and emerging technologies for sustainable livestock management. Anim. Biosci. https://doi.org/10.5713/ab.25.0289.

Tsonis, O., Khlifa, N., 2025. Editorial: artificial intelligence in assisted reproductive treatments. Front. Reprod. Health 7. https://doi.org/10.3389/frph.2025.1704386.

Van Hilten, A., Katz, S., Saccenti, E., Niessen, W.J., Roshchupkin, G.V., 2024. Designing interpretable deep learning applications for functional genomics: a quantitative analysis. Briefings Bioinf. 25 (5). https://doi.org/10.1093/bib/bbae449.

Wadood, A.A., Zhang, X., 2024. The omics revolution in understanding chicken reproduction: a comprehensive review. Curr. Issues Mol. Biol. 46 (6), 6248–6266. https://doi.org/10.3390/cimb46060373.

Wadood, A.A., Bordbar, F., Zhang, X., 2025. Integrating omics approaches in livestock biotechnology: innovations in production and reproductive efficiency. Front. Anim. Sci. 6. https://doi.org/10.3389/fanim.2025.1551244.

Wang, X., Si, J., Wang, Y., Fang, L., Zhang, Z., Zhang, Y., 2025. Integrating gene expression data via weighted multiple kernel ridge regression improved accuracy of genomic prediction. Genet. Sel. Evol. 57 (1). https://doi.org/10.1186/s12711-025-00997-9.

Werner, T., Fahrner, M., Schilling, O., 2024. Advancements in mass spectrometry-based proteomics: a new era in pathology research and diagnostics. Die Pathologie 45 (S1), 56–62. https://doi.org/10.1007/s00292-024-01390-x.

Wu, Y., Xie, L., 2024. AI-driven multi-omics integration for multi-scale predictive modeling of causal genotype-environment-phenotype relationships. arXiv. https://doi.org/10.48550/arXiv.2407.06405.

Wu, Y.C., Chia-Yu Su, E., Hou, J.H., Lin, C.J., Lin, K.B., Chen, C.H., 2025. Artificial intelligence and assisted reproductive technology: a comprehensive systematic review. Taiwan. J. Obstet. Gynecol. 64 (1), 11–26. https://doi.org/10.1016/j.tjog.2024.10.001.

Xue, W., Zhang, Z., Du, H., Pfeiffer, C., Mészáros, G., Ding, X., 2024. Predictive ability of multi-population genomic prediction methods of phenotypes for reproduction traits in Chinese and Austrian pigs. Genet. Sel. Evol. 56 (1). https://doi.org/10.1186/s12711-024-00915-5.

Yao, K., McGahan, I.P., Zhu, J., Storm, D.J., Walsh, D.P., 2025. Animal trajectory imputation and uncertainty quantification via deep learning. Environmetrics 36 (6). https://doi.org/10.1002/env.70027.

Yin, H., Li, Q., Yang, S., Zhang, C., Zhai, Y., Hou, H., Qu, Y., 2024. The relationship between ethics and innovation: specifically regarding the application and ethical considerations of artificial intelligence in animal models. Holist. Integr. Oncol. 3 (1). https://doi.org/10.1007/s44178-024-00134-4.

Youssef, A., Vodorezova, K., Aarts, Y., Agbeti, W.E.K., Palstra, A.P., Foekema, E., Aguilar, L., da Silva Torres, R., Grübel, J., 2024. IUMENTA: A Generic Framework for Animal Digital Twins within the Open Digital Twin Platform. https://doi.org/10.48550/arxiv.2411.10466.

Yu, Y., Mai, Y., Zheng, Y., Shi, L., 2024. Assessing and mitigating batch effects in large-scale omics studies. Genome Biol. 25 (1). https://doi.org/10.1186/s13059-024-03401-9.

Zelinschi, D.O., Ursache, A., Mihoci, E.C., Anton, E., Năvolan, D., Nemescu, D., 2025. Generative Artificial Intelligence Applications in Reproductive Health: Opportunities and Challenges. https://doi.org/10.20944/preprints202507.2446.v1.

Chapter 12

Artificial intelligence integration in livestock phenomics

The implementation of artificial intelligence (AI) in livestock phenomics represents a paradigm shift from sparse and manually maintained records of traits to continuous and high-dimensional, context-sensitive characterization of animal form, functions, behavior, physiology, product properties, and adaptive resilience. In this context, phenomics involves the organized acquisition and integration of a variety of sensor, imaging, acoustic, kinematic, metabolic, behavioral, environmental, and management streams of interaction, which, in combination, can define the dynamic phenotypic state pathways across the life cycle. Phenotypic expression is fluid, multiscale, and environmentally contingent, in contrast to all highly sequence-anchored genomic layers, and represents nonlinear interactions between genotype, microbiome, nutrition, housing microclimate, health perturbation, social structure, and managerial interventions. AI offers the computational framework that can extract latent structure in this heterogeneity, and convert unstructured or weakly annotated torrents of data into representations that contain biologically meaningful information and can be used to make predictive, causal, and prescriptive insights at individual, group, and system levels.

The main methodological changes that have supported this transformation include multimodal representation learning to balance the differences in temporal resolution, spatial granularity, and noise properties; self-supervised goals to utilize redundancy and time continuity to alleviate sparse labeled data; and architectures that consider spatiotemporal topology, such as geometric encoders of three-dimensional conformation streams, and sequence models of irregular physiological signals. Fusion networks extend beyond simple feature concatenation by learning hierarchical abstractions, whereby encoders that operate at modality-specific scales are fed common latent manifolds, both of which maintain overall phenotypic signals as well as fine-grained oddities. Causal discovery and counterfactual simulation have the ambition of separating mechanistic contributors of performance, welfare, efficiency, and environmental footprint from confounded correlations, and uncertainty quantification helps avoid the tendency toward overconfidence in high-stakes decision pipelines by regulating the timing of health interventions, nutritional changes, and selective focus.

Adaptive AI-enabled digital twins are on the rise that incorporate a combination of static genomic priors and real-time phenotypic signals to predict any latent physiological reserves, cumulative stress, propensity toward disease preclinical development, and forecasted production paths in alternative management conditions. Their iterative policy refinement based on reinforcement and adaptive control paradigms, which is limited by welfare, sustainability, and biosecurity guardrails, has offered promising outcomes in precision livestock management, but will require rigorous interpretability, fairness, and robustness audits to ensure that the entrenchment of systemic biases or optimization of welfare-driven policy compromises is not entertained. Edge inference and federated learning mitigate latency, privacy, and bandwidth issues by moving analytics closer to data creation, as well as generalizing at the population level without random mixing of raw data.

The ethics and governance aspects are interwoven during this integration. Data stewardship must have the incentive to encourage innovation and be fair in terms of access, attribution, and protection of operations that carry sensitive knowledge. The idea of algorithmic fairness implies equal performance between breeds, production systems, geographic and climatic situations, and the level of technology adoption, to ensure that phenomics data are not used to support productivity differences. A clear description of model rationality, conformity to welfare-consistent optimization restrictions, biosecurity-aware sharing schemes, environmental responsibility accounting of computational resources, as well as inclusive participatory oversight systems, all influence socially legitimate implementation. The following chapter sections are placed in the context of an interdisciplinary space in which AI does not present itself as an analytic overlay but as an integrated analytical infrastructure that allows a holistic, responsible, and future-ready livestock phenomics.

Artificial Intelligence in Livestock Omics. https://doi.org/10.1016/B978-0-443-45663-3.00009-9

12.1 Introduction to AI in livestock phenomics

Livestock phenomics is an emerging scientific paradigm for the comprehensive and integrative observation of the full spectrum of the manifestation of biology in animals in space, time, and environmental context. In contrast to traditional methods of phenotyping based on sparse and manually documented trait measurements, phenomics focuses on the continuous, high-dimensional description of morphology, physiology, behavior, productivity, welfare status, and adaptive behavior. These manifested phenotypes result from intricate and heterogeneous interactions among genetic predisposition, epigenetic control, metabolic condition, microbial ecosystems, environmental exposures, and management interventions. In this regard, the inherent phenomics is multidimensional, time-structured, and context-dependent and thus requires analytic methods that can resolve heterogeneity, scale, and uncertainty. AI has thus come to be a vital part of phenomics, not only as a computational process, but also as an infrastructure that converts the raw complexity of the observable processes into biologically consistent, interpretable, and actionable knowledge (Upadhyay et al., 2024; Klingström et al., 2024).

AI in livestock phenomics represents a conceptual change in how traits are measured episodically to the estimation of phenotypic states continuously. Phenotypes are not being taken as a fixed endpoint but as a pathway that is being followed through developmental stages, production cycles, and environmental changes. This redefinition of time is what brings phenomics into contact with systems biology, in which phenotypic variation is seen as a product of control mechanisms acting at several biological scales. In AI, the methodological ability to learn the representations of unstructured and semistructured data streams, including sensor telemetry, imaging, acoustic data, and environmental sensors, and combine these modal data into functional latent representations that not only capture global phenotypic causations but also maintain detailed deviations. By means of this integration, AI provides a chance to extract latent phenotypic constructs that best estimate underlying biological conditions compared with any one observable measure (Fig. 12.1) (Ghavi Hossein-Zadeh, 2025; Zhu et al., 2025).

The main distinguishing feature of livestock phenomics is the use of high-throughput and automated data retrieval technologies that produce continuous flows of heterogeneous data. These data streams are quite different in sampling rate, spatial content, noise content, and semantic content, which puts traditional statistical assumptions of independence, stationarity, and homogeneity to the test. AI solves these problems by providing representation learning as well as sequence modeling and probabilistic inference frameworks, which explicitly support temporal dependence, missingness, and context sensitivity. Methods of self-supervised learning utilize redundancy and temporal coherence in phenomics data to avoid relying on much manual labeling, and multimodal architectures balance the differences between data types by acquiring hierarchical abstractions that adopt both modality-specific structure and cross-modal interrelation. Through this, AI converts phenomics data, which consists of fragmented observational data, to structures that can be used in downstream predictive analysis, interpretation, and decision support (Zhu et al., 2025; Klingström et al., 2024).

The development of AI-based phenomics has significant implications for livestock enhancement, control, and sustainability. The relationship between genotype and actual performance occurs through phenotypic expression, which is the main point of interaction between genetic potential and environmental and managerial conditions. Increasing the phenotypic resolution and the time scale, AI-based phenomics boosts the validity and biological insights of genetic assessment, allowing for making complex breeding values more precise. Furthermore, phenomics describes adaptive reactions to stressors, disease challenges, and environmental variation, and thus helps to establish selection strategies that emphasize resilience, robustness, and welfare in addition to productivity. The AI helps to realize this integration through nonlinear interaction modeling, discovering phenotypic latent axes associated with adaptive capacity, and measuring uncertainty in a manner that can be used to make risk-aware decisions (Džermeikaitė et al., 2025).

In addition to genetic enhancement, AI-enhanced phenomics facilitates the move toward anticipatory management of livestock. Monitoring of phenotypes continuously allows deviations in expected courses of action to be identified at an early stage, and therefore action can be taken before an individual starts losing performance or compromising their welfare. This capability is supported by AI baselines where individualized baselines are learned, the important biological signals are identified among noise, and phenotypic conditions of the future are predicted by altering the circumstances. This predictive ability is critical in the management of more complex production systems that are experiencing increased environmental unpredictability, resource instability, and social demands. Notably, AI-enhanced phenomics is not just a way to automate the process of decision-making but also supports human expertise through interpretable decision support,

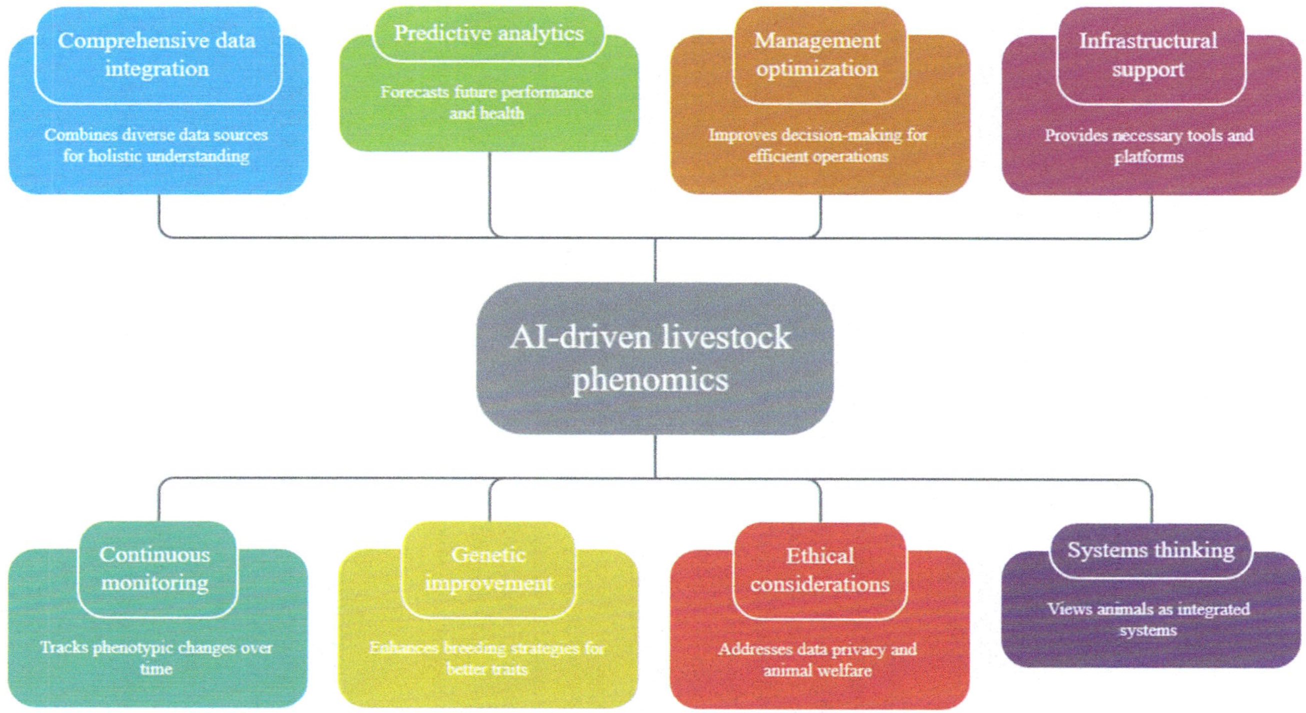

FIGURE 12.1 An overview of AI-driven livestock phenomics. This figure illustrates how AI serves as a central integrative framework that connects diverse phenotypic data streams with analytical and decision-support functions in livestock systems. Key components include comprehensive data integration, predictive analytics, continuous monitoring, and genetic improvement, all of which collectively enable more informed and adaptive management and breeding strategies. The framework also emphasizes infrastructural support, ethical considerations, and systems-level thinking, highlighting the role of AI in promoting responsible, holistic, and sustainable livestock phenomics.

and it presents interpretable, uncertainty-sensitive information to inform management decisions while maintaining liability and ethical control (Tedeschi et al., 2025; Sharma et al., 2025).

Livestock phenomics encompasses a broad spectrum of species, production systems, and geographic conditions, which include terrestrial and aquatic animals having varied physiological and behavioral repertoires. Such diversity requires the use of analytical frameworks that are sensitive to species-specificity and system-specificity and which are equally generalizable. AI facilitates this kind of balance using the transfer learning concept, domain adaptation, and hierarchical modeling methods that can inform inference on data-scarce conditions without making unsuitable assumptions. This ability is essential in determining the fact that innovation driven by phenomics will not contribute to existing inequities in production systems, but will instead be used in inclusive livestock improvement that is globally relevant (Upadhyay et al., 2024).

The concept of using AI in livestock phenomics cannot be discussed without considering ethical aspects. The lack of control and continuous monitoring, as well as automated inference, raises questions about data ownership, privacy, transparency, and what is right and appropriate when it comes to technological intervention in biological systems. AI should, then, be integrated into governance models that place the welfare of animals at the center, consider the rights of the stakeholders, and ensure that long-term sustainability or genetic diversity are not undermined by the optimization goals. The latter are not optional additions to explainability, bias auditing, and uncertainty quantification, but underlying requirements to be considered a responsible deployment of phenomics. This can be achieved by ensuring that model behavior is made open and also making uncertainty explicit so that AI systems can support informed human judgment instead of ousting it, increasing trust and legitimacy in phenomics-based decision-making (Rosati, 2025; Zangana et al., 2025).

Instructively, AI-based phenomics relies on the presence of effective data pipelines, interoperable standards, and scalable computing infrastructures, which can support the amount, speed, and types of phenotypic information. They are edge computing, federated learning, and privacy-preserving analytics that have become critical enablers to enable phenomics-based intelligence to be deployed near data sources and collaborative model development across organizational boundaries. Such infrastructures offer real-time inference, minimize latency, and overcome governance limitations that are related to centralizing the data aggregation. AI is therefore not only a layer of analysis but is an integrative platform that connects technologies of sensing, biological interpretation, and operational decision-making (Paul et al., 2026).

Conceptually, livestock phenomics conflicts with reductionist perspectives on the phenotype as a set of individual traits. Rather, it preempts the organism as a coherent system to which the expressed states are indicative of coordinated regulation in many of the biological domains. AI makes possible the translation of this view in the form of learning distributed representations to represent emergent phenotypic patterns and the implementation of causal reasoning structures that transcend correlation to a point of mechanistic understanding. These frameworks are necessary in making the distinction between the phenotypic variation caused by inherent biological differences and the variation caused by external constraints, hence enabling an intervention that fixes the underlying cause instead of the surface symptoms (Palmquist et al., 2024).

12.2 Key concepts in livestock phenomics

The interrelated conceptual foundations of livestock phenomics form the basis of a new understanding of the phenomenon of phenotypic expression, which is observed, interpreted, and applied in animal science. In its most basic sense, phenomics identifies a multidimensional and dynamic phenotype as a manifestation of biological function; the continuous interaction of intrinsic regulation of the biology and extrinsic environmental conditions. This view goes beyond conventional trait-based phenotyping in that it highlights the combined time-varying nature of phenotypic expression and its context-sensitivity. These are some of the main ideas that one must understand to appreciate the scientific and analytical needs of phenomics, especially in combination with AI (Upadhyay et al., 2024).

Reconceptualizing the phenotype as a process, rather than an endpoint, is one of the key aspects of livestock phenomics. Phenotypic states are dynamic and not fixed; they are further influenced by the development stage, physiological condition, exposure, and interaction with the environment, and the interventions put in place during management. This process perspective identifies phenomics with systems biology and criticizes the conceptions of analysis that focus on measurements at a single time or single characteristics. The phenomics pattern of change, coordination, and adaptation over time demands, as a form of meaningful biological insight, a process that is longitudinal and requires temporal modeling. In operationalizing this idea, AI is essential because it allows the designation of latent phenotypic states and transitions between temporally structured data (Yurochka, 2024; Sun and Ye, 2025).

The other concept underpinning this is phenotypic integration, which is the coordinated development of several biological dimensions within the organism. The phenotypes of livestock include morphology, physiology, behavior, productivity, health condition, and adaptation responses, which are controlled by intricate interactions of biological processes. These dimensions are not considered to be independent but interdependent, as in phenomics, covariance structures, and emergent patterns represent underlying regulatory coherence. This holistic perspective recognizes the fact that alterations in a single area of phenotypic expression tend to spread throughout others, which supports the importance of analytical methods that can be used to model multivariate relationships. The use of AI aids in the phenotypic integration, whereby cross-domain relationships are not collapsed by learning joint representations (Maiorano et al., 2024; Zhu et al., 2025).

The other important concept that forms the basis of livestock phenomics is phenotypic plasticity. Plasticity is the ability of a given organism to alter its phenotype with respect to environmental or physiological input without changing the genetic code. Phenotypic plasticity in livestock systems enables them to adapt to changing nutrition, climate, health-related issues, and management systems (Ehsaninia et al., 2019, 2020). Phenomics aims to describe the time-dependent, reversible, and long-term effects of phenotypic responses, in addition to their magnitude. This emphasis highlights the difference between adaptive plasticity and maladaptive responses, thereby distinguishing adaptive plasticity from maladaptive responses, which amplifies the resistance and resilience to changes in performance or welfare. AI enables the phenotypic plasticity to be studied and modeled by the response surfaces that are not linear, and by finding the patterns in the context that do not seem to be obvious in traditional analysis (Sun and Ye, 2025).

Intimately connected to plasticity is the concept of phenotypic resilience, which is the capacity of an organism to sustain or regain functional performance quickly in response to perturbation. Resilience is dynamic in nature and cannot be determined reliably by mere measurements that are fixed. Phenomics methods have been used to define resilience through tracking variations of anticipated phenotypic patterns and evaluating recovery processes in the long term. This idea is moving toward stability, robustness, and adaptability, which corresponds to changes of priorities in livestock production systems. The AI can provide resilience evaluation through the learning of individualized baselines, seeking minor deviations, and predicting the paths of recovery in different conditions (Laghouaouta et al., 2024).

Another characteristic of livestock phenomics is called phenotypic heterogeneity. There is a significant amount of variation in the expression of phenotypes by individuals, even in genetically similar populations, despite the standardized environments they may be exposed to. This heterogeneity is a manifestation of complex patterns between genetic background, developmental history, exposure to microenvironment, and random biological mechanisms (Ehsaninia et al., 2016). Phenomics accepts this variability as information and not noise, which aims to describe systematic laws of change that can be used in biological knowledge and management choices. AI supports this objective by enabling stratification, clustering, and individual modeling, which both reproduce shared population-level structure and significant individual differences (Upadhyay et al., 2024).

The fundamental concept of phenomics is the notion of scale that affects both spatial and temporal and biological dimensions. The expression of phenotypic activities exists at levels ranging from molecular activities down to the entire animal behavior and performance at the system level. Equally, temporal scales are covered by fast physiological reactions, day-by-day behavior cycles, and development over a long period or aging. Phenomics requires analytical frameworks that are able to combine information between these scales without favoring one level over the other. AI allows multiscale data integration because it learns hierarchical representations between fine-grained signals and higher-order phenotypic states to provide a coherent interpretation of the signal between levels of organization (Wu and Xie, 2024).

Another important concept that sets the difference between phenomics apart from traditional phenotyping is context dependency. Phenotypic expression is naturally dependent on environmental, managerial, and social circumstances, and the same biological states can be expressed differently in varying circumstances. Phenomics explicitly uses contextual data to prevent false interpretation of the phenotypic cues and to increase the external validity in systems. The use of AI models that conditionally infer with respect to contextual variables is better suited at separating inherent biological variability and extrinsic effects of the setting, rendering such models more likely to generalize and thus more accurate in prediction as well as more biologically realistic (Weine et al., 2025).

Such a conceptual consideration of phenomics as observability and measurement is significant. There are not only all the relevant phenotypic states that are directly measurable, but most must be inferred by proxy signals that are noisy, incomplete, or indirect. Phenomics recognizes such partial observability and focuses on the inference rather than direct measurement. AI offers tools in latent variable modeling, probabilistic inference, and quantification of uncertainties, enabling the estimation of the hidden phenotypic states explicitly defining the level of confidence and ambiguity. This strategy is opposed to deterministic interpretations that may be subject to overconfidence in imperfect observations (Ilan, 2025).

Phenotypic alignment, that is, the alignment of observed phenotypic expression with the underlying biological functionality, is another concept required. Misalignment may occur in cases where measurements are taken of superficial characteristics, which are not important in reflecting functional state, or when the model used to analyze them gives more emphasis to those signals that are easy to measure rather than biologically meaningful. Phenomics aims to enhance the achievement of alignment, both through a combination of multiple data sources and by testing inferred states of phenotypes based on established biological standards. AI also helps to approximate functional biology by determining the latent structures, which in turn are more accurate than raw measurements (Ramsden, 2023).

The problem between the phenotype and the genotype is not a new topic in animal science, and phenomics has brought a new conceptual clarity to this relationship. Phenomics describes the phenotype as an emergent phenomenon rather than a direct product of genotype, and it is based on the view that the phenotype is produced by regulatory processes, environmental modulation, and time dynamics. This school of thought highlights the significance of intermediate levels of phenotypic states and pathways as objects of analysis and intervention. The AI can be used to model the underlying layers, and it helps to understand a more detailed way of how genetic potential is brought into realization in a given circumstance (Sun and Ye, 2025).

Lastly, the epistemology of livestock phenomics is based on systems-level epistemology, which emphasizes integration, adaptability, and awareness of uncertainty. The generation of knowledge in phenomics is cyclic and probabilistic, and it takes into account the complexity of biological systems and the constraints of observation. The use of AI is consistent with this epistemology as it promotes the idea of constant learning, adaptive modeling, and the representation of uncertainty clearly. This correspondence supports phenomics as a prospective science that is capable of not only describing phenotypic variation but also guiding prediction, intervention, and making ethical decisions (Clarke, 2023). Conceptual dimensions and defining characteristics of livestock phenomics are presented in Table 12.1. Furthermore, major phenotypic data modalities in livestock phenomics and their biological significance are shown in Table 12.2.

12.3 AI technologies and their applications in livestock phenomics

The modern livestock phenomics is based on AI technologies, which allow the conversion of multidimensional, complex phenotypic data into interpretable forms of biological state and activity. Phenomics data are, by their nature, heterogeneous and contain signals that vary in terms of modality, time, space, and meaning. Conventional methods of analysis are unable to handle this diversity, especially when the phenotypic expression is dynamic and subject to other environmental and management conditions. AI overcomes these difficulties by offering elastic computational structures that can learn structure, pattern, and dependency directly through data, and thus enabling phenotypic inference at a degree of complexity that is comparable to biological reality (Zhu et al., 2025; Zhang et al., 2025).

Representation learning is one of the most basic applications of AI in livestock phenomics. Raw phenomics data streams are typically not directly biologically interpretable, and there is noise, redundancy, and nonresponse effects. AI models are trained to learn latent representations, which are compressed representations of these raw inputs where variation is preserved biologically. These representations serve as abstract phenotypic states that bring together information on many observable dimensions, which enables the phenotype to be a coherent system property and not a set of disjointed measurements. Representation learning then allows phenomics to go beyond the surface-level descriptions of organismal functionality into a more in-depth characterization of organismal functioning (Matentzoglu et al., 2025; Klingström et al., 2024).

The architectures of machine learning that work with high-dimensional data are focused on learning the latent phenotypic structure of complex observations. Phenomics data are often correlated, nonlinear, and interactive, and do not assume linearity and independence of relationships. To support these characteristics, AI models are trained to learn nonlinear functions and hierarchical representations of phenotypic control reflected in their structure. With these processes, AI technologies can unearth hidden phenotypic axes that can be related to functional capacity, adaptive behavior, or health conditions, despite the fact that these axes are not physically visible (Zhang et al., 2024).

The use of temporal modeling is another important field of AI use in livestock phenomics. Phenotypic expression is dynamic in nature and is governed by circadian rhythms, developmental courses, production cycles, and reactions to environmental disturbance. AI structures that can simulate sequences and paths enable the interpretation of phenotypic data based on time and not as discrete points. Patterns of continuity, transition, and recovery are found in these models, which make it possible to infer phenotypic states that are determined by the time context. Temporal modeling aids the early warning of deviations from the predicted phenotypic lineages and forms a basis for forecasting future states when faced with uncertainty (Sebia et al., 2024; Eckhardt et al., 2025).

Multimodal phenomics data can also be integrated into single analytical frameworks using AI. Phenotypic modalities represent different facets of biological functioning, but in most instances, they differ in magnitude, prevalence, and stability. Multimodal integration methods based on AI learn their interaction and co-varying nature, yielding joint

TABLE 12.1 Conceptual dimensions and defining characteristics of livestock phenomics.

Conceptual dimension	Definition and scope	Biological and analytical characteristics	Implications for phenomics data structure	Role of AI
Dynamic phenotype representation	• Phenotype viewed as a process evolving over time • Emphasis on trajectories rather than static traits	• Continuous biological regulation • Phase-dependent expression and transition states	• Longitudinal and time-series data • Irregular sampling and temporal dependency	• Temporal modeling and trajectory inference • Detection of state transitions
Phenotypic integration	• Coordinated expression across biological domains • System-level phenotype interpretation	• Interdependent regulation of behavior, physiology, and performance • Emergent phenotypic states	• Multivariate and multimodal datasets • High covariance structure	• Joint representation learning • Multimodal data fusion
Phenotypic plasticity	• Capacity for phenotype modification without genetic change • Context-responsive expression	• Environment- and management-sensitive regulation • Reversible and adaptive responses	• Context-conditioned observations • Nonlinear response patterns	• Context-aware modeling • Nonlinear pattern learning
Phenotypic resilience	• Ability to maintain or recover function under perturbation • Stability of phenotypic trajectories	• Homeostatic and adaptive regulatory mechanisms • Recovery dynamics	• Baseline deviation and recovery signals • Time-resolved response patterns	• Change-point detection • Resilience and recovery modeling
Phenotypic heterogeneity	• Structured variation among individuals • Diversity in phenotypic expression	• Individual-specific regulatory responses • Population-level diversity	• High interindividual variance • Clusterable phenotypic patterns	• Stratification and personalization • Individualized baseline modeling
Multiscale organization	• Phenotypic expression across biological scales • Integration from molecular to organismal levels	• Cross-scale regulatory coupling • Hierarchical biological organization	• Data spanning multiple resolutions • Nested data structures	• Hierarchical modeling • Cross-scale integration
Context dependency	• Dependence of phenotype on external conditions • Conditional expression	• Interaction between biology and environment • Management- and climate-sensitive responses	• Contextual metadata integration • Conditional variability	• Conditional inference models • Context-aware learning
Partial observability	• Incomplete and indirect measurement of phenotype • Latent biological states	• Hidden regulatory processes • Proxy-based observation	• Missing and noisy data • Indirect phenotypic indicators	• Latent variable modeling • Probabilistic inference
Uncertainty and variability	• Inherent biological and measurement uncertainty • Nondeterministic expression	• Stochastic regulation • Measurement noise	• Probabilistic data representation • Variance-rich datasets	• Uncertainty quantification • Bayesian and probabilistic models
Systems-level epistemology	• Holistic understanding of phenotype • Integration over reductionism	• Emergent properties • Networked biological processes	• High-dimensional integrated datasets • Nonlinear dependencies	• Systems modeling • Integrative learning frameworks

TABLE 12.2 Major phenotypic data modalities in livestock phenomics and their biological significance.

Phenotypic data modality	Biological scope and representation	Key phenotypic information captured	Temporal and spatial characteristics	Relevance for AI-driven phenomics analysis
Behavioral phenotyping	• Expression of movement, posture, and interaction patterns • Observable functional behavior	• Activity rhythms • Social interaction dynamics • Adaptive behavioral responses	• High temporal resolution • Spatially explicit patterns	• Enables pattern recognition of behavioral states • Supports early detection of deviation from baseline
Physiological phenotyping	• Internal functional state of the organism • Homeostatic regulation	• Cardiometabolic balance • Thermoregulation • Stress and recovery responses	• Continuous or high-frequency time series • Often context-dependent	• Supports inference of latent physiological states • Enables temporal trajectory modeling
Morphological phenotyping	• Structural and anatomical characteristics • Body conformation	• Growth patterns • Structural symmetry and proportionality	• Lower temporal frequency • High spatial specificity	• Enables structural feature extraction • Supports longitudinal growth modeling
Productivity-related phenotyping	• Expression of production performance • Functional output traits	• Yield dynamics • Efficiency-related indicators	• Medium to long temporal scale • Aggregated over production cycles	• Supports predictive modeling of performance trajectories
Health and welfare phenotyping	• Indicators of physical well-being and functional integrity • Stress and disease-related expression	• Subclinical variation • Recovery and resilience signals	• Variable temporal resolution • Event-driven patterns	• Enables early risk detection • Supports welfare-centered inference
Reproductive phenotyping	• Expression of reproductive function and timing • Physiological readiness	• Cyclicity and reproductive status • Developmental progression	• Phase-dependent temporal structure • Moderately high resolution	• Supports state classification and transition modeling
Metabolic phenotyping	• Biochemical and energetic state of the organism • Resource utilization	• Energy balance indicators • Metabolic adaptation patterns	• Medium temporal resolution • Highly context-sensitive	• Enables inference of energetic constraints • Supports predictive metabolic modeling
Environmental response phenotyping	• Phenotypic reaction to external conditions • Adaptive capacity	• Stress adaptation • Plasticity-related changes	• Coupled to environmental dynamics • Variable spatial scale	• Enables context-aware phenotypic modeling
Sensor-derived phenotyping	• Integrated signals from automated sensing systems • Proxy-based measurement	• Composite indicators of behavior and physiology	• Continuous high-frequency data streams	• Supports real-time inference • Enables scalable phenomics monitoring
Social and interaction phenotyping	• Expression of group-level behavior and interaction • Social dynamics	• Hierarchy and affiliation patterns • Social stress indicators	• Spatially explicit • Group-level temporal structure	• Enables network-based modeling • Supports collective behavior inference

representations that represent a coordinated phenotypic expression. This combination brings in robustness through missing or noisy data in one of the modalities, as well as ways to make better sense of cross-domain dependencies underlying complex phenotypes (Shamim et al., 2025).

Another significant application of AI that is of significance in phenomics is the ability to infer latent and partially observable phenotypic states. Numerous biologically important conditions are impossible to measure directly, but rather can be estimated using indirect effects. Latent variable inference is part of AI models used to approximate these states of hidden uncertainty. This is a probabilistic view, which is in line with the epistemological principles of phenomics, and which accepts that the knowledge of phenotypes is always incomplete and subject to change with new information surfacing. Through uncertainty quantification, AI systems account for taking risks and prevent overconfidence in noisy measurements (Liu et al., 2025).

Pattern recognition and classification of phenotypic states are other ways in which AI can be applied to phenomics. Although phenomics focuses on variation as continuous, discrete categorization is applicable in decision-making, monitoring, and communication. AI technologies can make it possible to have flexible classification schemes that take into consideration within-class heterogeneity and boundary uncertainty due to the graded character of biological states. These categories are not fixed terms but moving categories subject to change with the evolution of phenotypic paths, and support the process-based perspective of phenotype (Bansal, 2025).

Predictive modeling is one of the significant uses of AI in livestock phenomics. AI systems can anticipate phenotypic paths by learning the connections between the present phenotypic states and specific contextual variables and future results. This kind of prediction facilitates proactive management since it helps in the identification of emerging risks or opportunities before they become noticeable. Notably, predictive modeling in phenomics has to deal with a large degree of biological variability and environmental uncertainty. AI models can overcome this difficulty by generating probabilistic predictions that express not only the anticipated results but also the uncertainties and thus aid in decision-making (Dorea and Menezes, 2024; Zhu et al., 2025).

In addition to prediction, AI can be used in prescriptive and adaptive phenomics. Prescriptive modeling applies the expected actions along with the predicted phenotypic response when making decisions that can suit multiple objectives. Adaptive learning models dynamically update models as new phenotype information arrives to enable systems to improve inference and modify their recommendations. These functions make phenomics a proactive decision-support system; no longer a descriptive science, but one that continually adapts in response to its biological system (Paul et al., 2026).

Personalization in livestock phenomics is achieved through AI as well. The phenotypic manifestation is highly differentiated among individuals due to genetic and developmental backgrounds and environmental exposure. This heterogeneity is described by AI technologies, which learn individualized deviations and baselines and make phenotypic inference specific rather than average across groups. Personalization will improve accuracy and minimize the possibility of inappropriate generalization, especially where phenotypic responses to interventions are very heterogeneous (Ferreira and Dórea, 2025).

Another important aspect of AI technologies that should be applied to phenomics is its scalability. Livestock phenomics data may be massive, continuous, and located in more than one place and system. AI systems are being engineered to be able to work in these environments, and in this manner, facilitate incremental learning, decentralized computation, and real-time inference. This kind of scalability enables such insights about phenomics to be generated effectively and implemented in operational environments without compromising the analytical rigor (Lee et al., 2025).

Interpretability is also one of the primary issues when it comes to the use of AI in livestock phenomics. Models that are complex and not transparent can be highly predictive, but they do not provide much information on underlying biological processes. Interpretability is needed in phenomics, where the decisions can have an impact on animal welfare and long-term sustainability. The AI technologies can be increasingly equipped with mechanisms that help to map the learned representations onto the phenotypic features and contextual factors, and make the model output comprehensible and evaluable in a biological context for the users (Zhu et al., 2025; Rosati, 2025).

Ethical issues intersect with AI technologies in phenomics. Constant surveillance, automatic decision-making, and proactive advice raise concerns about data control, openness, and reasonable boundaries of technological intrusion. The AI systems should thus be designed in such a way that there are ethical limitations on their purposes and appraisal measures. With awareness of uncertainty, explainability, and human control, AI technologies can facilitate the responsible use of phenomics to support the expectations and welfare priorities of society (Stevens et al., 2025). AI technologies applied in livestock phenomics are presented in Table 12.3.

TABLE 12.3 AI technologies applied in livestock phenomics.

AI technology	Core analytical function	Phenomics data types addressed	Key analytical capabilities	Implications for phenomics insight
Supervised machine learning	• Prediction based on labeled data • Regression and classification	• Structured phenotypic measurements • Aggregated sensor outputs	• Phenotypic state classification • Outcome prediction with confidence estimates	• Enables targeted inference • Supports decision-oriented analytics
Unsupervised learning	• Pattern discovery without labels • Structure and cluster identification	• High-dimensional phenomics data • Multimodal observations	• Identification of latent phenotypic states • Discovery of phenotypic subpopulations	• Reveals emergent biological structure
Semisupervised learning	• Learning from limited labeled data • Leveraging unlabeled observations	• Sparse or partially annotated phenomics datasets	• Improved generalization under data scarcity	• Enhances scalability of phenomics analysis
Deep learning	• Hierarchical feature learning • Nonlinear representation extraction	• Imaging, acoustic, and high-frequency sensor data	• Complex pattern recognition • Automated feature extraction	• Captures subtle phenotypic variation
Temporal and sequence models	• Modeling time-dependent data • Trajectory inference	• Longitudinal phenotypic time series	• Detection of phenotypic transitions • Forecasting future states	• Supports dynamic phenotypic interpretation
Multimodal learning	• Integration of heterogeneous data sources • Joint representation learning	• Behavioral, physiological, and environmental data	• Cross-domain correlation capture • Robust inference under missing data	• Enables holistic phenotypic modeling
Probabilistic and Bayesian models	• Uncertainty-aware inference • Latent variable estimation	• Noisy and incomplete phenomics datasets	• Quantification of uncertainty • Robust inference under ambiguity	• Supports risk-aware decision-making
Graph-based learning	• Modeling relational and network structure • Dependency-aware inference	• Social interaction data • Phenotypic correlation networks	• Identification of interaction patterns • Network-level phenotypic insight	• Enhances biological interpretability
Representation learning	• Extraction of latent phenotypic embeddings • Dimensionality reduction	• High-dimensional continuous phenomics data	• Compression of complex signals • Preservation of biological structure	• Enables scalable phenomics analysis
Transfer and meta-learning	• Knowledge reuse across tasks or populations • Rapid adaptation	• Multipopulation phenomics datasets	• Improved performance in new contexts	• Enhances generalizability
Explainable AI frameworks	• Model transparency and interpretability • Feature attribution	• Any phenomics modality	• Linking predictions to phenotypic signals	• Builds trust and accountability
Reinforcement and adaptive learning	• Sequential decision optimization • Policy learning	• Phenomics data linked to management actions	• Adaptive strategy evaluation	• Supports prescriptive phenomics

12.4 Data acquisition and management in livestock phenomics

The basic infrastructure on which livestock phenomics and its connection with AI are based is data acquisition and management. Phenomics is data-intensive in nature as it involves a continuous, high-resolution recording of various phenotypic signals that span behavioral, physiological, morphological, and performance-related spectrums. Phenomics is also in contrast to the traditional methods of phenotyping, which in many cases require sparse and hand-curated measures, while phenomics relies on automated and scalable sensing systems that can be utilized over long time horizons and variable production conditions. The efficiency of AI in phenomics is thus inextricably linked to the quality, format, management, and intelligibility of the data on which the models of analysis are based (Billah et al., 2025; Zhu et al., 2025; Ghavi Hossein-Zadeh, 2026).

The peculiarity of phenomics data acquisition is the focus on continuity and temporal depth. Phenotypic expression is dynamic, context-dependent, and meaningful interpretation may not be possible without observation through entire biological cycles and not just over individual time intervals. Real-time data allows the phenotypic paths to be recreated, adaptive responses can be documented, and the patterns of stability or deviation can be found, which would otherwise be hidden when the phenotypic pattern is measured sporadically. This richness of time brings both opportunity and complexity, with an increasing rate of data volume and sampling of data with varying modalities. The solution to this is therefore to have effective data acquisition strategies that are able to balance the resolution and feasibility, to have the collected data be informative enough yet manageable within the operational limitations (Pommier et al., 2025).

The other key characteristic of phenomics data acquisition is heterogeneity. Livestock phenomics combines the signals measured by a variety of different sensing modalities, each of which has its own measurement principles, noise structures, and spatial and temporal resolutions. The perspectives on phenotypic state are unique to behavioral observations, physiological measures, environmental conditions, and productivity indicators. To cope with this heterogeneity, coordination of the streams of data is needed to achieve time correlation, semantic consistency, and compatibility with the analytical steps happening downstream. This diversity is an advantage of AI, provided that the information is curated and organized in such a manner that it maintains the biological meaning and contextual integrity (Pommier et al., 2025; Klingström et al., 2024).

The quality of the data is a significant factor in the reliability and analytical validity of phenomics. Phenomics data can be affected by measurement error, sensor drift, environmental interference, and missingness, and any of these can affect inference unless unchecked. Unlike controlled experimental environments, livestock production environments contain variability that can only be completely removed through instrumentation. Quality assurance procedures that will detect and reduce artifacts and retain real biological variation should then be included in data management frameworks. AI models are able to learn how to distinguish between signal and noise; however, they will be effective only in the case of systematic data validation and clear reporting of the data provenance and restrictions (Wu et al., 2025).

The idea of completeness of data is of particular relevance to phenomics, where complete observability is the exception and not the rule. Not every dimension of phenotype is either measurable on a single occasion or continuously, and the lapses in streams of data can be due to technical, logistical, or biological reasons. The data management strategies should take note of this incompleteness and directly reflect it instead of trying to cover up this incompleteness by deterministic imputation. Uncertainty-resultant probabilistic data processing and uncertainty-aware models are more consistent with the epistemological principles of phenomics, where AI uses the latent phenotypic state to communicate confidence and ambiguity. These methods will minimize the risk of overinterpretation and contribute to more vigorous decision-making (Cámara, 2025; Pommier et al., 2025).

Interoperability is a significant concern in managing phenomics data, especially with the evidence of data being produced on a variety of platforms, vendors, and institutional settings. Variations in the data formats, ontologies, and metadata standards may hinder integration and reuse, which constrains the scalability of phenomics programs. To build interoperable data structures, data schema harmonization, the consistent annotation of phenotypic variables, and the utilization of shared vocabularies with biological significance are needed. Such standardization is advantageous to AI models, which decreases the complexity of preprocessing, and it improves transferability between systems and populations (Matentzoglu et al., 2025).

Metadata is crucial to ensure phenomics data usability and understandability. Relevant information about the identity of animals, their physiological condition, environmental factors, animal management, and parameters of data collection is necessary to make meaningful analysis. Phenomics observations will be decontextualized clues that can be interpreted biologically only with rich metadata. The management practice of data must, therefore, not consider metadata as an afterthought from phenomics datasets but as an integral part of them. It is on this contextual information that AI systems draw to separate intrinsic phenotypic variation and extrinsic factors, enhancing predictive accuracy and biological plausibility (Yang et al., 2025).

The characteristic of data management in livestock phenomics is scalability. The ongoing data collection on large populations produces large amounts of data that need to be stored, processed, and accessed in an effective way. An architecturally centralized system might not be able to serve such a scale, especially where real-time inference is needed. Phenomics-based intelligence can be deployed closer to data sources and still coordinate across systems, and this is made possible through distributed and hierarchical approaches to data management. Such architectures facilitate progressive data assimilation, dynamic model revision, and prompt feedback, which is required in operational phenomics applications (Toth et al., 2025).

Data governance is a key aspect of phenomics data management, and it involves ownership, access, privacy, and ethical usage. Phenomics data can be an indicator of sensitive biological information and practices, which has become an issue of misuse, unauthorized access, and unfair distribution of benefits. Governance structures should be able to reconcile the necessity of data sharing to support science, concerning stakeholder rights and compliance with regulations. AI enhances the value of the phenomics information, and it is necessary to have transparent governance frameworks to ensure trust and encourage cooperation. Clear policies on access to data, its use in models, and responsibilities are some of the pillars of sustainable phenomics initiatives (Taddese et al., 2025).

Phenomics information has a lifecycle that goes beyond its initial acquisition and analysis to include long-term storage, curation, and re-use. The older the phenomics datasets, the more valuable they are as a resource in their retrospective analysis, model refinement, and hypothesis generation. Longitudinal consistency requires the adoption of data management approaches that allow the reconstruction of phenotypic trajectory over time, as well as the changing analytical models. This need explains why version control, documentation, and traceability are essential in phenomics data systems. AI models working with historical data must be interpretable with respect to the conditions of the data on which they are trained, which makes it necessary to strictly maintain data stewardship (Meijer et al., 2024).

The addition of phenomics data to other fields of biological data makes data management even more complex. Phenomics interfaces with genomic, epigenomic, and environmental data, the management of which must be coordinated in both the omics and nonomics spaces. This integration requires the identification of identifiers, time references, context variables, inconsistencies, and conflict resolution mechanisms. AI uses these combined datasets to gain a consistent representation of biological systems, and cross-domain data management is a vital aspect that allows phenomics-based insight (Brachmann and Jacobs, 2025).

Increasingly, phenomics data management addresses the consideration of security and resilience. Ongoing data feeds and distributed systems create data integrity vulnerabilities, system reliability vulnerabilities, and cyber threats. The data management structures must be strong enough to include mechanisms that prevent loss, corruption, and unauthorized manipulation of data. These precautions are needed not only to prevent sensitive information from being leaked but also to ensure the accuracy of AI models, which rely on the integrity of the data (Pommier et al., 2025).

Regardless of automation, human expertise will still be needed in the data acquisition and management of phenomics. The choice of what data to collect, how to annotate the data, and what quality signals to interpret requires domain knowledge, which cannot be entirely automated. Phenomics initiatives are consequently based on the synergistic cooperation of animal scientists, data engineers, and computation experts. AI supplements such experience by being able to scale and be complex, yet AI does not eliminate the requirement to make informed decisions regarding data stewardship (Pommier et al., 2025; Taddese et al., 2025). Phenomics data acquisition systems and management considerations are shown in Table 12.4.

12.5 Machine learning techniques for genetic improvement

Machine learning has moved to the forefront of methodological pillars in the quest to achieve genetic improvement in the context of livestock phenomics, essentially transforming complex phenotypic information into actionable selection decisions. Genetic enhancement is based on the proper prediction of heritable variation and the dependable forecast of future performance in varying environmental and managerial conditions. Conventional genetic assessment systems have offered strong support structures to these purposes, but have been limited by the extent, complexity, and dynamism of phenomics information. Machine learning can solve these issues by providing adaptable analytic frameworks that can incorporate high-dimensional phenotypic observations, nonlinear relationship modeling, and intricate interactions that can be used in genetic expression (Panigrahi et al., 2025; Zhu et al., 2025).

One of the best contributions of machine learning to genetic improvement is its ability to draw informative patterns out of highly detailed phenomics data. Phenomics is a continuous and multifactorial measure of animal performance,

TABLE 12.4 Phenomics data acquisition systems and management considerations.

Data acquisition or management dimension	Definition and scope	Key characteristics and requirements	Challenges in phenomics contexts	Role of AI
Continuous phenotypic monitoring	• Ongoing capture of phenotypic signals across time • Emphasis on temporal completeness	• High-frequency data streams • Longitudinal consistency	• Data volume accumulation • Temporal misalignment	• Temporal modeling and data compression • Trajectory-based inference
Multimodal data acquisition	• Collection of heterogeneous phenotypic signals • Integration across sensing modalities	• Diverse data formats and resolutions • Complementary biological perspectives	• Modality synchronization • Semantic integration	• Multimodal representation learning • Cross-domain data fusion
Data quality assurance	• Validation of phenomics data integrity • Noise and artifact mitigation	• Sensor calibration and consistency • Error detection	• Environmental interference • Sensor drift	• Automated anomaly detection • Noise-robust learning
Temporal alignment and synchronization	• Harmonization of time-stamped data streams • Alignment across modalities	• Accurate timestamping • Consistent temporal reference	• Irregular sampling frequencies • Missing intervals	• Time-series alignment algorithms • Temporal interpolation with uncertainty
Metadata and context annotation	• Documentation of biological and environmental context • Data provenance	• Rich contextual descriptors • Standardized annotation	• Incomplete or inconsistent metadata	• Context-aware inference • Conditional modeling
Handling missing and incomplete data	• Management of partial observability • Explicit representation of gaps	• Nonrandom missingness • Variable data coverage	• Bias in inference • Reduced model robustness	• Probabilistic modeling • Latent state inference
Scalability and data volume management	• Storage and processing of large datasets • Long-term data accumulation	• Efficient data pipelines • Incremental ingestion	• Computational burden • Latency constraints	• Automated feature extraction • Scalable learning architectures
Interoperability and standardization	• Compatibility across platforms and systems • Data reuse	• Common data formats and ontologies • Harmonized identifiers	• Fragmented standards • Integration overhead	• Automated data harmonization • Representation alignment
Data governance and ownership	• Control over phenomics data access and use • Stakeholder rights	• Transparent access policies • Consent frameworks	• Conflicting interests • Restricted data sharing	• Federated learning • Privacy-preserving analytics
Data security and privacy protection	• Safeguarding sensitive biological and operational data	• Secure storage and transmission • Access control	• Cybersecurity risks • Unauthorized access	• Secure model training • Encrypted computation
Long-term data curation and reusability	• Preservation of phenomics datasets over time • Support for retrospective analysis	• Version control and documentation • Traceability	• Data degradation • Loss of context	• Continuous model updating • Knowledge transfer across datasets
Integration with other biological data domains	• Linking phenomics with genomics and environment	• Consistent identifiers • Cross-domain alignment	• Semantic mismatch • Temporal inconsistency	• Cross-domain representation learning • Integrated inference frameworks

behavior, and physiologic condition, producing phenotypic signals that are more closely aligned with biological mechanisms compared with traditional summary characteristics. Latent representations learned from these data reduce noise and redundancy, and are also biologically meaningful variations, which are learned by machine learning models. Such representations provide a better input of phenotypes that are subject to genetic analysis, and genetic potential is more accurately inferred and displayed (Spangler, 2024; Crossa et al., 2025).

The methods of machine learning allow reconceptualization of the relationships between phenotype and genotype, beyond a linear association. Regulatory networks, developmental processes, and context-dependent modulation usually mediate genetic effects on phenotypic expression. Such complexities are learned in machine learning models, which use nonlinear relationships between genetic backgrounds, phenomics expression, and performance outcomes. This is especially useful in complex traits with multifactorial effects of loci with small effects and interactions between environmental and genetic factors. Machine learning can improve the biological realism of genetic improvement models by accommodating such structure (Hegde et al., 2025).

The assimilation of longitudinal phenomics information in genetic assessment is among the greatest developments of machine learning made possible. Genetic merit is conventionally estimated as the phenotypes accumulated over time or at particular phases. Phenomics, on the contrary, offers uninterrupted phenotypic courses that incorporate persistent genetic impacts as well as brief environmental impacts. Such trajectories are explicitly described using machine learning methods, and it is possible to separate genetic components of these trajectories, removing short-term variations and noise. This modeling over time enhances better genetic inferences and acts as an assistant in making selection decisions that are based on lifelong performance as opposed to a single observation (Guo and Li, 2023).

Intermediate phenotypes can also be used in genetic improvement with the help of machine learning. Numerous phenomics characteristics are upstream biological processes that come before and determine the end performance results. Machine learning allows the genetic potential to be assessed earlier and more accurately by introducing such intermediate phenotypic states into genetic assessment. This methodology eliminates the use of end-of-life or end-of-outcome assessments and enables a faster selection process. Notably, intermediate phenotypes tend to be more heritable or have more genetic signal than aggregated performance characteristics, which increases the selection efficiency further (Džermeikaitė et al., 2025).

The ability of machine learning to simulate interactions is especially applicable to genetic enhancement of heterogeneous production settings. The interactions between the genotype and the environment are complex, and they determine genetic performance, which can differ depending on the management systems and climatic conditions. Opportunities through machine learning methods learn conditional relationships that modify genetic predictions to particular settings. It is a context-sensitive modeling that encourages the generation of breeding approaches that prioritize durability and flexibility, so that the genetic acquisition is attained in a greater variety of production settings and not designed to optimize under limited circumstances (Fernandes et al., 2024).

Genetic improvement is also achieved by the use of machine learning to deal with missing, noisy, and imbalanced data. Phenomics datasets are frequently incomplete due to the practical limitations in the data collection, and the lack of data can be associated with a life aspect or management factor. The conventional genetic appraisal procedures might not perform well in this scenario, and machine learning designs have probabilistic inference and regularization frameworks that counteract bias and instability. Machine learning can be used to increase the strength of genetic prediction in imperfect datasets by expressing and taking advantage of the common structure among data, as well as by expressing uncertainty (Chafai et al., 2023).

The other significant aspect of machine learning in genetic enhancement is multipurpose and multiobjective selection. Breeding of livestock is known to focus more on balancing productivity, with traits of health, welfare, and sustainability, most of which are phenomically complex and interconnected. Machine learning models are capable of storing correlations and trade-offs among the various phenotypic dimensions, and allow for a combined assessment of genetic merit on a variety of goals. This broad view minimizes the likelihood of unintended consequences of the narrowly motivated selection on a limited list of traits (Panigrahi et al., 2025).

Genetic improvement also has machine learning methods that allow genetic assessments to be dynamically updated as new phenomics data becomes available. Conventional methods of evaluation are based on infrequent corrections, but machine learning models may be created to learn continuously, updating representations and predictions with new data. This flexibility capability makes genetic enhancement consistent with the dynamic aspect of phenomics, in that the selection decisions will be based on the most up-to-date data and minimize delays between observation and action (Wang et al., 2025).

The question of interpretability is an important factor in the use of machine learning in genetic enhancement. Although complex models can have high predictive performance, their applications in the breeding context require the capacity to connect the prediction to the biological knowledge and selective goals. Explainable machine learning has been advanced to help interpretation by connecting latent representations and predictive outputs with the phenotypic features and genetic inputs. This openness helps to justify, build trust, and make decisions, ensuring that machine learning and genetic understanding should complement rather than replace one another (Atzmueller et al., 2024).

The application of machine learning in genetic improvement cannot be discussed without considering ethical and sustainability issues. The use of AI-based phenomics to make selection choices affects the population structure, genetic diversity, and adaptability over the long term. The models of machine learning should thus be constructed to have clear limits to avoid over-strengthening or over-weakening of resilience. The integration of welfare-related phenomics parameters and sustainability-oriented goals into genetic assessment systems is a means of making genetic advancements and mainstreaming to meet the expectations and moral standards of society (Brito et al., 2025).

Genetic improvement is also enhanced by machine learning, which allows breeding to be stratified and personalized. Instead of using the same selection criteria on whole populations, machine learning models can identify subpopulations that share similar genetic and phenotypic similarities. This stratification facilitates the specific selection schemes to consider the biological diversity and production conditions to enhance the overall system operation without damaging genetic variation. These methods are especially useful in complex livestock systems, where universal strategies could be less efficient (Panigrahi et al., 2025).

Machine learning applications to genetic improvement must be validated and governed with care. The models should be tested on a variety of populations and conditions to be reliable and applicable. Continuous testing should also be used to indicate any drift in model performance as the genetic and environmental landscapes change. Responsible deployment requires open reporting of assumptions, limitations, and uncertainty, especially in situations where selection choices are long-term and irreversible (Huang, 2024).

Conceptually, machine learning transforms genetic improvement into a stationary estimation problem, as a dynamic and data-driven learning process. Genetic merit is no longer derived merely through a combination of historical records, and now it develops through an unbroken process of genotype interacting with phenotype and the environment. Phenomics offers the rich data of observation on which this transformation can take place, and machine learning offers the analytical power needed to find meaning within the complexity. By combining them, it is possible to gather a more delicate, receptive, and biologically based treatment of livestock breeding (Panigrahi et al., 2025). Machine learning paradigms supporting genetic improvement through phenomics are presented in Table 12.5.

12.6 Phenotypic data analysis and interpretation

The analytical essence of livestock phenomics lies in phenotypic data analysis and interpretation, which provides the transition between high-volume data acquisition and significant biological and managerial understanding. Although phenomics produces enormous amounts of nonuniform observational data, it is the possibility to convert such observations into coherent forms that determines its scientific and practical worth, representing the results of biological condition, activity, and variation. The core of this change is played by AI, although the analysis of the phenotypic data is still rooted in biological reasoning, systems thinking, and attention to uncertainty. Phenotypic analysis is thus a matter of both computing dexterity and conceptual cogency in terms of what phenotypic signals are and what sort of insights and decision-making they must provide (Olatinwo et al., 2025).

One of the most frequent problems with analyzing phenotypic data is that phenomics observations have a high dimension. Phenomics simultaneously measures many aspects of phenotype, most of which are correlated and some of which are redundant or only indirectly linked to the underlying biological processes (Matentzoglu et al., 2025). The way in which this complexity is analyzed thus requires methods of discerning structure within, and discriminating between useful patterns and noise without reducing the richness of biological reality. This is solved by AI with dimensionality reduction and representation learning, which reduces the size of high-dimensional data to lower-dimensional latent spaces and preserves important biological variation. These latent representations are the basis of interpretation, in that they systematize phenotypic data into consistent patterns, which are indicative of an entire biological activity and not a single measurement (Crossa et al., 2025).

In phenomics, the interpretation of phenotypic data focuses on relations and trends and does not focus on absolute values. Phenotypic expression is relative, and it is influenced by the development stage, physiological situation, and environmental factors. Analytical frameworks should then consider variation at the baseline and deviations, trends, and coordination of phenotypic dimensions. The AI models assist this relational view because they learn both individualized base and population-wide normativity, allowing the interpretation of phenotypic signals contextually. This method minimizes the chances of misclassification or overinterpretation on the basis of crude measurements only and synchrony of analysis with biological concepts of homeostasis and adaptation (Stager et al., 2025).

TABLE 12.5 Machine learning paradigms supporting genetic improvement through phenomics.

Machine learning paradigm	Primary role in genetic improvement	Phenomics information utilized	Key analytical strengths	Implications for breeding and selection
Linear and regularized models	• Baseline estimation of genetic merit • Feature selection under high dimensionality	• Aggregated phenomics traits • Derived phenotypic summaries	• High interpretability • Stability under limited data	• Supports transparent selection decisions • Provides reference benchmarks
Tree-based ensemble methods	• Modeling nonlinear genotype-phenotype relationships • Capturing feature interactions	• Multidimensional phenomics indicators • Integrated behavioral and physiological traits	• Robustness to heterogeneous data • Implicit interaction modeling	• Improves prediction of complex genetic effects • Enhances selection accuracy
Artificial neural networks	• Learning nonlinear mappings between phenotype and genetic potential	• High-dimensional continuous phenomics data	• Flexible function approximation • Capacity to capture complex dependencies	• Enables refined estimation of genetic potential • Supports complex trait selection
Deep learning architectures	• Hierarchical representation of phenotypic expression • Feature learning from raw phenomics signals	• Imaging-based phenotypes • High-frequency sensor data	• Automated feature extraction • High representational capacity	• Captures subtle genetic signals embedded in phenomics
Temporal and sequence models	• Disentangling genetic effects from temporal variation • Modeling lifetime performance	• Longitudinal phenomics trajectories	• Explicit temporal dependency modeling • Improved signal separation	• Enables selection based on stability and resilience
Multitrait learning frameworks	• Joint evaluation of correlated traits • Balanced genetic improvement	• Multiple phenomics dimensions across domains	• Captures covariance and trade-offs • Improves multiobjective inference	• Supports holistic breeding goals
Latent variable and representation learning models	• Extraction of intermediate phenotypes • Dimensionality reduction	• High-dimensional phenomics feature spaces	• Noise reduction • Discovery of biologically meaningful latent traits	• Enables early and precise genetic evaluation
Transfer and meta-learning approaches	• Knowledge transfer across populations or environments • Rapid model adaptation	• Cross-population phenomics datasets	• Improved generalization • Reduced data dependency	• Enhances applicability of breeding values across systems
Probabilistic and Bayesian models	• Uncertainty-aware genetic inference • Integration of prior knowledge	• Noisy and incomplete phenomics data	• Explicit uncertainty quantification • Robust inference under data gaps	• Supports risk-aware selection decisions
Hybrid statistical-machine learning models	• Integration of biological priors with data-driven learning	• Phenomics data aligned with genetic structure	• Balance of interpretability and flexibility	• Improves the biological plausibility of selection models
Personalized and individual-level models	• Individual-specific genetic potential estimation • Stratified selection	• Individual phenomics profiles	• Captures interindividual heterogeneity	• Enables precision breeding strategies
Adaptive and incremental learning systems	• Continuous updating of genetic evaluations • Responsiveness to new data	• Streaming phenomics observations	• Dynamic model refinement	• Keeps breeding decisions aligned with current phenotypic expression

The key element in the interpretation of phenotypic data is temporal structure. Most phenotypic signals only have meaning in the context of a time series, and they indicate a process, for example, adaptation, recovery, or decline. Phenomics data analysis at the static level is prone to concealing these dynamics and confounding their temporary changes with enduring biological states. Phenotypic trajectories can be explicitly modeled using AI methods on sequential data, and this can be interpreted in a way that represents the present condition as well as the past. The temporal analysis also allows for realizing the leading indicators and lagging responses, which improves the perception of the cause-and-effect relationships in phenotypic systems (Li et al., 2025).

Integration of different phenotypic domains is another important issue in the interpretation of phenotypic data. Phenomics acknowledges that a phenotype is a product of integrated regulation in the realms of behavior, physiology, morphology, and performance aspects. The methodologies of analysis should therefore take into consideration the whole of the data at phenotypic levels and not individually. This integration is made by AI through learning joint representations that learn covariance and interaction among phenotypic variables. These representations can only be interpreted through close mapping onto biological processes so that the integrated phenotypic states are related to meaningful functional constructs and not abstract statistical artifacts (Kocakaya and Kul, 2025).

Phenotypic data and data analysis are characterized by uncertainty. There is ambiguity in measurement error, biological variability, and incomplete observability, which cannot be completely removed without analytical sophistication. Phenotypic interpretation should be carried out responsibly, with explicit consideration and representation of uncertainty. Models of AI are becoming more probabilistic to measure the confidence in assessed phenotypic states and predictions. The perception of phenotypic information in the light of uncertainty can be used to aid uncertainty-aware decision-making and avoid overconfidence in model results, which is especially relevant in cases where relevant interventions may have serious biological or ethical impacts (Marques-Garcia et al., 2025).

In particular, the difference between correlation and causation is especially relevant to the interpretation of phenotypic data. Phenomics data are generally observational in nature, involving natural variation but not experimental control. Consequently, correlations between phenotypic variables could indicate common regulation or environmental confounding or indirect relations as opposed to cause-and-effect relationships. AI is good at identifying multifaceted correlations and needs to be supplemented by causal logic to prevent misinterpretation. The application of phenotypic data requires careful interpretation with consideration of biological plausibility, timeline, and contextual factors that can provide a causal hypothesis (Olaker et al., 2025).

There is also the fact that interpretation is a matter of scale, both in terms of biological organization and in terms of time. It is possible that phenotypic signals visible at one scale can have an alternative meaning when viewed at a different scale. Due to the adaptive nature of short-term physiological variation and the structural changes of long-term trends, an example can be given. Analytical systems need to permit the analysis of phenotypic data on various scales to aid in the interpretation of transient variation as compared to enduring phenotypic changes. From the viewpoint of supporting multiscale analysis, AI allows for the learning of hierarchical representations, where more fine-grained observations are connected with higher-level phenotypic constructs (Redekop, 2025).

Phenotypic heterogeneity is both difficult to analyze and interpret. Differences between individuals are a manifestation of both biological diversity and adaptive mechanisms, but also lead to the distortion of macrolevel trends. The analysis of phenotypic data should thus be characterized by a compromise between aggregation and stratification, where common structures should be found, and at the same time, the meaningful individual differences should be maintained. AI facilitates such a balance by facilitating clustering, personalization, and mixed effects model strategies that can model common and individual phenotypic traits. The heterogeneity should be interpreted sensitively with respect to the biological context so that variation is interpreted as informative and not as an error (Sharma and Goel, 2025).

Commentation of phenotypic data is also influenced by context, such as environmental influence, management style, and social structure. The same phenotypic signal can have varied implications in various contexts, and this dependency cannot be ignored; otherwise, faulty conclusions can be drawn. The interpretability of the AI models that condition inference upon contextual variables can be improved by placing the phenotypic states in the context of their operational environment. This context-sensitive interpretation is necessary in the translation of phenomics understanding into action-oriented knowledge that applies to different production systems (Baek et al., 2025).

Domain knowledge, as far as phenotypic data interpretation is concerned, is indispensable. Although AI is capable of detecting patterns and associations in the data, biological knowledge is needed to determine their feasibility and relevance. Interpretation is a process of repeated communication between computational output and biological thought, where theories are used to inform hypotheses and knowledge used by experts to validate. Such an interaction guarantees that phenomics analysis does not build up into solitary predictive artifacts, but instead, it can contribute to cumulative scientific knowledge (Melandri et al., 2025).

There is also an interpretation in terms of the perception of phenotypic insights. Complicated analytical products need to be reported in a format that is comprehensible and practical to various stakeholders, such as scientists, managers, and policymakers. Explainability mechanisms are becoming a common feature of AI models, where predictions and latent representations are connected to phenotypic characteristics that are observed. This type of transparency makes interpretation easier, promotes trust, and allows for making informed judgments about the implications of the findings of phenomics (Hossain et al., 2025).

The ethical considerations would play a significant role in the interpretation of data in phenotypes, especially those analytical findings that would be used in making decisions that impact animals, genetic diversity, or the sustainability of the system over the long term. To apply an ethical approach in interpreting phenotypic data, it is necessary to understand that the focus of analysis and optimization goals influences the results. AI models should be tested in terms of both accuracy and adherence to ethical priorities. Welfare and sustainability-driven interpretation will aid in the realization of welfare and sustainability-driven decisions, being socially responsible and biologically sound phenomics-driven decisions (Chakilam et al., 2025).

Lastly, interpretation of phenotypic data is a dynamic and continuous process of phenomics, and not a single analysis. Due to the acquisition of new information and the updating of models, interpretations can be revised, with more sophisticated knowledge and varying circumstances. AI facilitates this dynamic interpretation by facilitating incessant learning and updating of models. Nonetheless, this flexibility should be supported by strict documentation and verification to ensure that emerging interpretations are consistent and valid. Analytical strategies for phenotypic data analysis and interpretation are shown in Table 12.6.

12.7 Challenges in AI integration in livestock phenomics

The AI in livestock phenomics introduces a complex of interrelated problems related to the complexity of biological systems, the characteristics of phenomics data, and the socio-technical environments of deployment of analytical tools. Since AI is able to process and interpret high-dimensional phenotypic data with unparalleled capacity, its success in application is limited by scientific, technical, ethical, and operational constraints. The knowledge of these issues is crucial to the creation of AI-enabled phenomics as a sound, responsible, and sustainable science instead of a set of disconnected technological applications (Wilson and Anwar, 2024).

One of the most basic issues is that phenotypic expression is complex and variable in nature. Phenotypes in livestock are the resultant behavior of dynamic biological systems that are determined by genetic backgrounds, developmental cycles, exposure to the environment, and management conditions. This complexity creates phenomics information that is very heterogeneous, situational, and usually nonstationary. It is possible that AI models, which are optimized to be more predictive, even with variability of that level, will not generalize, due to the possible risk of training context-specific shortcuts that can fail in a new environment. However, it is not just a matter of creating correct models, but also ensuring that learned representations are reflecting biological processes, not surface correlations (Upadhyay et al., 2024).

The main issue of AI in phenomics is the data-related challenges. While the term phenomics is commonly used to describe data-rich environments, the quality of available, well-filtered, and representative datasets is still unequal. Phenomics streams of data can be incomplete, noisy, or biased against a specific population or management systems. These biases may become transmitted by AI models and cause unequal performance and increase existing inequality in the contexts of production. To solve this problem, it is necessary to pay special attention to data provenance, sampling methods, and representativeness, along with analytical methods that clearly reflect uncertainty and imbalance (Upadhyay et al., 2024; Ghavi Hossein-Zadeh, 2025).

The other significant problem is related to the time character of phenomics data. The expression of phenotype is manifested at various time scales, ranging from short-term physiological variations to long-term developmental and aging processes. To capture such dynamics, longitudinal data is needed with enough resolution and continuity, but in practice, irregular sampling and time series discontinuities are common. AI models are thus faced with time misalignment, data loss, and the length of variables. Poor treatment of the temporal structure may cause misunderstanding of the transient variation as a permanent change or may blur the initial signs of the phenotypic change (Aliferis and Simon, 2024).

Further complexity is introduced by the intermodal integration of phenotypes. Phenomics is based on the integration of various types of data of varying scale, reliability, and biological content. These modalities can be made to align not only technically but also conceptually, with respect to the relationship between various phenotypic dimensions. It is possible that AI models can combine heterogeneous inputs with success; however, the latent representations generated by these

TABLE 12.6 Analytical strategies for phenotypic data analysis and interpretation.

Analytical strategy	Conceptual focus	Type of phenomics patterns addressed	Interpretative value	Contribution of AI
Dimensionality reduction and feature extraction	• Simplification of high-dimensional phenomics data • Identification of dominant variation axes	• Correlated and redundant phenotypic signals • Latent structure across modalities	• Enhances interpretability by revealing core phenotypic constructs • Reduces analytical noise	• Automated representation learning • Preservation of biologically meaningful variation
Trajectory and longitudinal analysis	• Interpretation of phenotype as a dynamic process • Emphasis on temporal evolution	• Time-dependent phenotypic changes • Developmental and adaptive trends	• Distinguishes transient fluctuations from persistent states • Enables understanding of phenotypic progression	• Sequence and temporal modeling • Forecasting of phenotypic trajectories
Baseline and deviation modeling	• Contextual interpretation relative to expected states • Individualized reference frameworks	• Deviations from normative phenotypic patterns • Early indicators of change	• Supports early detection and risk assessment • Improves sensitivity to subtle variation	• Personalized baseline learning • Anomaly and change-point detection
Multimodal integration analysis	• Holistic interpretation across phenotypic domains • Cross-modal dependency capture	• Coordinated behavioral, physiological, and performance signals	• Reveals emergent phenotypic states • Reduces misinterpretation of isolated signals	• Joint representation learning • Cross-domain data fusion
Uncertainty-aware interpretation	• Explicit consideration of ambiguity and variability • Probabilistic reasoning	• Noisy, sparse, or incomplete phenomics observations	• Prevents overconfidence in inference • Supports risk-aware decision-making	• Probabilistic and Bayesian modeling • Confidence estimation
Stratification and phenotypic clustering	• Identification of structured heterogeneity • Subpopulation discovery	• Interindividual variation patterns • Distinct phenotypic profiles	• Enhances understanding of biological diversity • Supports targeted interpretation	• Unsupervised and semisupervised learning
Context-conditioned interpretation	• Phenotype interpretation relative to environment and management	• Environment- and management-sensitive variation	• Improves biological plausibility • Enhances generalizability	• Context-aware and conditional modeling
Cross-scale interpretation	• Linking signals across biological and temporal scales	• Fine-grained signals and higher-order phenotypic states	• Supports systems-level understanding • Distinguishes local from systemic effects	• Hierarchical and multiscale modeling
Correlation and dependency analysis	• Exploration of relationships among phenotypic variables	• Covariation and co-expression patterns	• Identifies coordinated regulation • Informs hypothesis generation	• Nonlinear dependency learning
Causal-oriented interpretation	• Differentiation between association and mechanism	• Directed phenotypic relationships	• Reduces risk of spurious inference • Supports intervention reasoning	• Causal inference frameworks integrated with AI
Comparative and longitudinal benchmarking	• Interpretation across time and populations	• Phenotypic shifts and stability patterns	• Enables assessment of consistency and drift	• Adaptive learning and model updating
Explainability and model transparency	• Linking analytical outputs to observable signals	• Model-derived phenotypic representations	• Builds trust and interpretability • Supports expert validation	• Explainable AI methods and attribution analysis

models are biologically difficult to interpret. This obscurity makes validation harder, and it destroys faith in model outputs, especially when ethical or economic outcomes are involved in making decisions informed by phenomics (Baek et al., 2025).

One of the most long-standing issues in the use of AI in the context of livestock phenomics is interpretability. Numerous AI designs with the ability to deal with high-dimensional phenomics data are constructed based on complex architectures that are not easily understandable in terms of transparency. Although impressive predictive accuracy can be attained by such models, the inner logic can conflict with biological knowledge. This disconnect between performance and interpretation is dangerous as it relates to trust, accountability, and responsible usage. Interpretability is not an option in phenomics, where knowledge can affect the welfare of animals, their breeding, and their sustainability over time as a system (Melandri et al., 2025).

In phenomics, the issue of separating correlation and causation is especially sharp. Phenomics datasets are mostly observational, the variation recorded in nature, and not the experimental manipulation. AI is effective in revealing statistical associations but fails to determine by default the difference between a causal association and a spurious relationship. The AI-based phenomics analysis can cause interventions that cannot achieve desired results and produce unintended effects without proper causal reasoning. This difficulty promotes the necessity to implement the principles of causal inference and biological knowledge in AI frameworks instead of using pattern recognition (Mohanan, 2025).

Another group of challenges is scalability and computational requirements. Phenomics data streams may be big, continuous, and distributed, which require large storage, processing, and energy requirements. State-of-the-art AI models, especially representation learning models and temporal modeling models, may be computationally expensive to train and implement. These needs can reduce accessibility, especially within resource-constrained environments, and also bring up the issue of environmental sustainability. It is also a continuing challenge to the field to balance analytical sophistication, efficiency, and inclusivity (Trirat et al., 2024).

This is further complicated by data governance and ownership as problems facing AI integration in livestock phenomics. The phenomics data are sometimes a product of two or more stakeholders, such as producers, researchers, and technology providers, each of whom has unique interests and responsibilities. The questions of who owns phenomics data, sharing, and the benefits of the derived insights are not just about legal considerations but also about ethical considerations. AI magnifies the value of data, increasing the necessity to have open systems of governance that safeguard the rights of stakeholders and allow partnerships and innovation (Upadhyay et al., 2024; Klingström et al., 2024).

Data governance is associated with privacy and security issues. Regular phenomics surveillance provides comprehensive biological and operational portraits, which may have many misuses if they are not secured properly. The AI systems that combine and analyze such data should be built to ensure high security and privacy-protective systems. The lack of consideration of such issues can jeopardize the willingness to trust and accept the phenomics technologies, no matter how scientifically worthy they are (Arigbabu et al., 2024).

Human factors are a somewhat underrated problem in AI-based phenomics. Successful integration involves more than technical ability, but also congruence with human knowledge, work processes, and decision-making. The excessive use of automated results can destroy the use of critical thinking, and the lack of integration can reduce the utility. To ensure that AI systems are used as tools of decision support and not decision-makers, the interfaces, training, and governance structures should be designed with care. The difficulty is indicative of more general concerns regarding the role of automation in biological systems, particularly their complex forms (Boy, 2024).

The issues of ethics are extensively found in every sphere of AI implementation in livestock phenomics. The AI models have optimization goals that affect phenotypic results and selection forces, which may have an impact on animal well-being, genetic diversity, and evolutionary adaptability. In the absence of formal ethical limitations, AI-based phenomics can be focused on the short-term efficiency of the solution at the cost of resilience and sustainability. To overcome this problem, it is essential to consider the ethical implications of models and implement them during their design, evaluation, and deployment instead of viewing them as peripheral considerations (Neethirajan, 2024).

A second important issue is that of validation and reproducibility. Often, phenomics AI models are trained to work on a particular dataset under strict conditions, which makes them questionable regarding their applicability to different populations, settings, and times. The absence of standard benchmarks and validation procedures makes comparing and evaluating the performance of a model quite difficult. The complexity of the AI models and the dynamic character of phenomics data further put the issue of reproducibility under pressure. Opening up the validation practices and reporting standards is the way to gain trust in AI-enabled phenomics (Yurkovich et al., 2024).

The changing character of biological and production systems poses the problem of model drift. Phenotypic patterns may vary as genetic composition, management practices change, and environmental conditions, leading to poor model performance. The AI systems should not be left alone; there is a need to monitor and update them continuously. This makes the implementation and governance of it harder because this requirement necessitates continuous monitoring and resource allocation, and not the imposition of implementation at once (Motsinger-Reif et al., 2024).

The interdisciplinary coordination is another issue. Livestock phenomics is at the intersection of the fields of animal science, data science, engineering, and ethics. The successful AI implementation needs cooperation between these areas, and language, priorities, and approach to methods may cause the inability to communicate effectively. To overcome such divides, institutional support is necessary, common structures, and respect toward various types of expertise (Upadhyay et al., 2024).

Lastly, there is an expanded conceptual issue that is based on expectations and scope. AI is commonly described as a revolutionary tool that can be used to solve complicated biological and managerial challenges. As a matter of fact, AI is a device, the power of which will be evaluated by the quality of data, the suitability of models, and the intelligibility of purposes. Claiming too much about the abilities of AI can be a way to disappoint and misuse it, whereas underestimating its limitations can cause unintended outcomes. There is thus a need to adopt a moderate and pragmatic approach toward the application of AI in livestock phenomics in a responsible manner. Major challenges in AI integration within livestock phenomics are listed in Table 12.7.

12.8 Ethics in AI integration in livestock phenomics

The integration of AI in livestock phenomics has ethical implications, as it alters the ways of perceiving, interpreting, and responding to animal biology. Phenomics provides unprecedented insights into the lives of animals by repeatedly recording minute details of their behavior, physiology, performance, and adaptive responses. The power of analysis of these data is multiplied by AI and turns the observation into a prediction and guidance. This increased ability is ethically burdened, with a set of phenomics-informed decisions based on AI affecting the welfare of animals, genetic outcomes, animal husbandry, and perceptions of livestock production in society. The ethical reflection is thus not an extrinsic limit to technological advancement, but a part of sound innovation in phenomics (Neethirajan, 2024; Upadhyay et al., 2024).

One of the key ethical concerns related to AI-powered livestock phenomics is animal welfare concerns. Ongoing observing and predicting modeling give the chance to enhance welfare by detecting stress, suffering, or failing health sooner than conventional methods. Simultaneously, the same technologies can be employed to step up production systems, and might result in greater physiological load or efficiency over well-being. The ethical implementation of AI in phenomics must be explicit about the fact that welfare is not only an implication of productivity but a goal on its own. Such an attitude requires welfare-related phenotypic indicators to be considered as primary but not secondary constraints, and the AI system is to be assessed not only based on performance results but also on animal experience (Neethirajan, 2024; Ghavi Hossein-Zadeh, 2025).

The issue of proportionality must be central in ethical phenomics. The breadth and persistence of phenotypic surveillance pose the question of the extent to which observation should be warranted and the purpose of the observation. Although animals do not have privacy in the human sense, ethical accountability can still be given to the amount of monitoring placed on living beings. AI-based phenomics should then be informed by the fact that the data collection and analysis should be commensurate with justifiable objectives regarding health, welfare, sustainability, and knowledge development. Oversurveillance, which is only marginally beneficial but has no obvious benefit, is a way of instrumentalizing animals as sources of data, instead of valuing them as biological beings that have intrinsic worth (Rubel et al., 2025).

The other ethical aspect is the interpretive power of AI. The AI is producing more and more suggestions or recommendations that are applicable in the management and breeding choices. There are ethical implications in applying the outputs of an algorithm as objective or final without proper investigation. Phenomics data are probabilistic in nature and context-dependent, and AI models are subject to assumptions in their design and training data. Ethical practice demands that AI outputs are interpreted as decision-support systems and not as decision-makers, and that human judgment is still at the center stage. Such a balance keeps the accountability intact and does not put the ethical responsibility on algorithms (Hamadani et al., 2024).

The phenomena of transparency and explainability are directly associated with ethical responsibility in AI-enabled phenomics. The decisions made concerning animal welfare, genetic selection, or the sustainability of a system should be capable of justification to a variety of stakeholders, such as scientists, producers, regulators, and society in general. Black-box models make good predictions, but not having interpretability, as they do not explain the reasons behind their choices, defies this. It is the ethics of AI, however, that focuses on the explainability to ensure that the stakeholders can know the way the phenotypic signals are interpreted, and the use of conclusions is made. Openness fosters trust and enhances control, as well as allows identification and resolution of ethical issues before they arise (Hassija et al., 2024).

TABLE 12.7 Major challenges in AI integration within livestock phenomics.

Challenge category	Nature of the challenge	Underlying causes and characteristics	Implications for phenomics analysis	AI-oriented mitigation perspectives
Biological complexity and variability	• Phenotype as an emergent and dynamic property • High inter- and intra-individual variability	• Nonlinear genotype-environment interactions • Context-dependent regulation	• Reduced generalizability of models • Risk of oversimplification	• Representation learning capturing latent structure • Context-aware modeling
Data quality and noise	• Presence of measurement errors and artifacts • Variable data reliability	• Sensor drift and environmental interference • Heterogeneous acquisition conditions	• Biased inference • Reduced model robustness	• Noise-robust learning algorithms • Automated anomaly detection
Incomplete and missing data	• Partial observability of phenotypic states • Irregular data streams	• Practical limitations in continuous monitoring • Nonrandom missingness	• Increased uncertainty in interpretation • Potential inference bias	• Probabilistic modeling • Latent state inference
Temporal misalignment	• Inconsistent temporal resolution across modalities • Fragmented longitudinal data	• Asynchronous data acquisition systems • Variable sampling frequencies	• Distorted phenotypic trajectories • Misinterpretation of dynamics	• Temporal alignment algorithms • Sequence modeling
Multimodal data integration	• Difficulty combining heterogeneous phenotypic signals	• Divergent data formats and scales • Semantic inconsistency	• Loss of biological meaning • Increased analytical complexity	• Multimodal representation learning • Data fusion frameworks
Model interpretability	• Limited transparency of complex AI models • Black-box behavior	• High model complexity • Abstract latent representations	• Reduced trust and accountability • Difficulty in biological validation	• Explainable AI frameworks • Feature attribution methods
Correlation-causation confounding	• Inference dominated by associations • Limited causal insight	• Observational nature of phenomics data • Confounding variables	• Ineffective or misleading interventions	• Integration of causal reasoning • Temporal and contextual constraints
Scalability and computational demands	• High storage and processing requirements • Computational intensity	• Continuous high-frequency data streams • Complex model architectures	• Restricted deployment • Increased operational cost	• Efficient architectures • Distributed and incremental learning
Data standardization and interoperability	• Fragmented data formats and ontologies	• Lack of shared phenomics standards • Platform heterogeneity	• Limited data reuse • Integration barriers	• Automated harmonization • Representation alignment
Generalizability across systems	• Limited transferability of models	• Population-and environment-specific patterns	• Reduced applicability of phenomics insights	• Transfer learning • Domain adaptation
Model drift over time	• Degradation of model performance	• Changing biology and environments • Evolution of management practices	• Outdated inference and recommendations	• Continuous model updating • Monitoring frameworks

Human-AI interaction	• Overreliance or underutilization of AI outputs	• Misalignment with expert workflows	• Suboptimal decision-making • Loss of contextual judgment	• Decision-support system design • Human-in-the-loop learning
Ethical and welfare considerations	• Potential prioritization of efficiency over welfare	• Narrow optimization objectives	• Risk of welfare compromise • Reduced societal trust	• Ethics-by-design modeling • Welfare-aware objectives
Data governance and ownership	• Unclear control over phenomics data use	• Multiple stakeholders • Competing interests	• Restricted collaboration • Inequitable benefit distribution	• Federated learning • Transparent governance frameworks
Validation and reproducibility	• Difficulty reproducing AI-based results	• Context-specific datasets • Complex model pipelines	• Reduced scientific credibility	• Standardized validation protocols • Transparent reporting

Another huge ethical issue in livestock phenomics is prejudice and equity. There can be historical biases in phenomics data, such as bias in representing breeds, breed production systems, or breed management. Introducing AI on such data will reproduce or enhance these biases and result in unfair performance or unwanted exclusion of groups of people. The systematic detection and reduction in bias, alongside a close interest in the way phenomics insights are applied to different contexts, would be one way to approach ethical integration. Fairness in phenomics goes beyond technical accuracy to include fair distribution of benefits as well as nonharm to systems and populations (Upadhyay et al., 2024).

Ethical implications can also be applied to the genetic implications of AI-driven phenomics. Phenomics analysis influences the selection that determines the genetic makeup of future populations. Although AI-enhanced phenomics may enhance precision and efficiency, it has a higher likelihood of reducing genetic diversity in the case of too narrow a concentration of selection based on idealized phenotypic records. To have ethical stewardship of genetic resources, there should be a balance between short-term benefits and long-term flexibility and resiliency. This balance will have to be manifested in the weighting and implementation of the phenomics implications into breeding programs, where AI will not exacerbate inadvertent homogenization of various genes, as well as diminish the evolutionary capacity (Yurkovich et al., 2024; Kocakaya and Kul, 2025).

Another AI integration aspect in phenomics that is ethically relevant is data ownership and governance. Phenomics data have a tendency to be created by various stakeholders and have scientific and economic importance. Two ethical issues are raised regarding who owns these data, how they can be exploited, and how the gains made out of them are shared. AI enhances the value of phenomics data, raising the issue of consent, access, and equity. Ethical data governance models should propose that the stakeholders have control over how they use data, that there is transparency in data sharing, and that the benefits of data sharing are not concentrated in the hands of a few players (Taddese et al., 2025).

The issues of privacy and security are related to data governance ethics. Phenomics monitoring produces ample biological and operational data continuously, which may be abused in case of insufficient protection. Although animals are not directly legal objects of privacy, the information that is stored in the phenomics systems can disclose sensitive data about production practices or management decisions. The ethical implementation of AI needs to have strong measures to ensure the integrity of data and their confidentiality, as well as informing people how the data will be stored, analyzed, and shared. The inability to discuss these issues would lead to the loss of confidence and credibility in phenomics technologies (Balagurunathan and Sethuraman, 2024).

Ethical consequences of automation are also of importance. Phenomics made possible by AI can be automated in things that could previously be done by human beings in terms of observation and analysis. Though automation may enhance uniformity and scalability, it may also lead to the dismissal of the human factor or the lackadaisical treatment of individual animals. The ethical integration involves prudent development of the human–AI interaction, such that human responsibility is augmented and not substituted by automation. The importance of the human touch and meaningful human engagement in the decision-making process can assist in keeping the ethical sense and contextual sensitivity (Upadhyay et al., 2024; Mohamed, 2025).

Societal perception and legitimacy are more comprehensive ethical issues in livestock phenomics. The perception of livestock production in the population is defined by the issues of animal welfare, sustainability, and technological intervention into biological processes. The AI-driven phenomics can either be viewed as a welfare and transparency enhancement tool or used as a means of heightened animal control. The process of integrating ethics, hence, incorporates not only the technical design but also communication and interaction with society. Honesty regarding goals, constraints, and protection is critical in establishing trust and ensuring that phenomics aligns with social values (Gupta, 2024).

Environmental responsibility is also included in ethics in phenomics. Without specifically considering sustainability, the AI-based phenomics insights can be applied to maximize productivity in a manner that consumes more resources or triggers greater harm to the environment. Ethical integration demands that the phenomics analysis aligns with the wider ecological objectives, understanding that livestock systems are integrated into the environmental systems. Including sustainability aspects in phenomics goals contributes to the idea of making AI-powered innovation a long-term contribution to the ecological balance, as opposed to optimization in the short term (Rahman et al., 2025).

Another ethical aspect that applies to AI-enabled phenomics is intergenerational responsibility. Informed phenomics decisions about the current animals and their future populations can be made based on the phenomics analysis, as it decides the genetic selection and management of future populations. Ethical assessment should thus take into account the long-term implications, such as possible cumulative impact on health, strengths, and flexibility. The AI systems should be created to embody long-term thinking as opposed to focusing on short-term gains and justify stewardship as opposed to exploitation (Yurkovich et al., 2024).

Namely, the ethics of uncertainty management are worth special consideration. Phenomics inference is by nature uncertain, and AI models reflect probabilities, not certainties. To act ethically, uncertainty should be honestly communicated, and the person should not be overconfident in the output of the models. Decisions formed in uncertain situations should be cautious, especially where their possible effects on well-being or sustainability are huge. Explicitly characterized AI systems allow making ethical decisions, since they allow users to consider risks and alternatives carefully (Chakraborti et al., 2025).

Lastly, to ensure the ethical integration of AI usage in livestock phenomics, there must be institutional and cultural commitment. Ethics cannot be turned into technical features or checklists; it should be integrated into organizational practices, culture of research, and structures of governance. Animal scientists, ethicists, data scientists, and stakeholders must engage in interdisciplinary communication so that they can pinpoint ethical dilemmas and establish common standards. This kind of dialog will ensure that the ethical reflection does not fall behind the technological capability. Ethical principles and governance considerations in AI-enabled livestock phenomics are presented in Table 12.8.

12.9 Future directions and perspectives on AI in livestock phenomics

The future of AI in livestock phenomics is predetermined by the combination of the progressive development of sensing technologies, the increased capacity of the computational part, and a change in theoretical concepts of conceptualizing phenotype as a dynamic and integrated biological system. As the science of phenomics evolves, AI will assume a more active role as not only an important analysis instrument but also as a key organizing system that shapes the structure of the phenotypic information creation, understanding, and use. The future will also be marked by a more extensive level of integration on the biological level, more focus on the temporal and contextual reasoning, and more direct correspondence to ethical, sustainability, and governance priorities (Upadhyay et al., 2024).

The further development of phenotypic resolution is one of the most important future directions of livestock phenomics. The development of sensing and data gathering will allow animal biology to be characterized to finer and finer scales in terms of behavioral, physiological, and morphological levels. The increasing complexity will require AI to handle it, and the representations of learning will be able to reconcile high-resolution signals with biologically meaningful constructs. As phenomics data develops to a deeper form, AI models will not only shift to recognizing coarse phenotypic patterns, but to recognizing subtle phenotypic regulatory processes and the earliest signs of phenotypic change. This advancement will facilitate the previous, more accurate, and more deliberate inference; it will advance not only scientific knowledge but also practical decision-making (Panda et al., 2024).

Temporal reasoning will become more and more central in the application of phenomics in the future. Instead of working with time as an auxiliary dimension, future AI systems will simulate the phenotypic expression as a continuous process that is being expressed on a multitemporal scale. Such systems will bring together short-term variations, medium-term adaptation, and long-term developmental patterns into cohesive projections of phenotypic trajectory. Temporal integration of this kind will facilitate the ability to predict phenotypic transitions and facilitate proactive management strategies based on the emerging situation, as opposed to responding to the established ones. This change is indicative of a larger shift in the direction of static evaluation of livestock biology toward dynamic knowledge (Ahmad et al., 2024).

The other valuable view of the future of AI-enabled phenomics is an increased focus on multiscale integration. Phenotypic expression is the expression of the interaction between molecular regulation, cellular activity, tissue structure, whole-animal behavior, and environmental context. To a greater extent, AI will be challenged with filling the gaps between the fine-grained biological signals and higher-order phenotypic states that are of management and breeding relevance. This will be aided by the hierarchical and systems-based architecture of AI, which will allow phenomics to operate as a coordinating platform that binds different fields of livestock science together (Garcia-Oliveira et al., 2026).

Representation learning and foundation models will also influence the future of phenomics. Generalized representations of phenotype learned in AI systems trained on large and diverse phenomics datasets will result in the ability to apply them in new settings with limited extra data. This ability will bolster transferability between species, production systems, and environments, and curb the issues that exist today in terms of data scarcity and context specificity. Such representations will facilitate cumulative learning, whereby the phenomics knowledge can be constructed in an incremental fashion as opposed to being rediscovered in separate datasets each time (Sharma and Goel, 2025).

Another area of AI critical frontier in livestock phenomics is causal reasoning. Future systems will not only stop at correlation-based inference, but will be based on frameworks that will explicitly represent causal relationships between phenotypic variables, environmental factors, and management interventions. This change will enhance the dependability of the phenomics knowledge and minimize the risk of unintended consequences in case AI-informed judgments are used. With causal structure, the AI systems will facilitate counterfactual reasoning and intervention outcomes, which will empower phenomics as a decision-support discipline over a purely descriptive discipline (Matthay et al., 2025).

TABLE 12.8 Ethical principles and governance considerations in AI-enabled livestock phenomics.

Ethical dimension	Core ethical concern	Key considerations and tensions	Implications for phenomics research and practice	Role of AI design and governance
Animal welfare	• Protection of physical and psychological well-being • Avoidance of harm	• Risk of prioritizing efficiency over welfare • Intensification enabled by automation	• Potential welfare trade-offs if poorly guided • Need for welfare-centered phenotypic indicators	• Incorporation of welfare-aware objectives • Continuous monitoring for early distress detection
Proportionality of monitoring	• Ethical limits to continuous observation	• Balance between insight generation and over-surveillance • Justification of data collection intensity	• Risk of instrumentalizing animals as data sources	• Purpose-driven data acquisition • Minimization of unnecessary monitoring
Transparency and explainability	• Accountability of AI-driven decisions	• Black-box models versus interpretability • Need for justifiable recommendations	• Reduced trust if decisions cannot be explained	• Explainable AI frameworks • Traceable model outputs
Human oversight and responsibility	• Preservation of human ethical judgment	• Overreliance on automated outputs • Diffusion of responsibility	• Risk of uncritical adoption of AI recommendations	• Human-in-the-loop system design • Decision-support rather than decision replacement
Fairness and bias	• Equitable treatment across populations and systems	• Dataset bias and uneven representation • Amplification of historical inequalities	• Unequal benefits and performance disparities	• Bias detection and mitigation strategies • Inclusive model training
Genetic stewardship	• Long-term integrity of genetic resources	• Narrow selection driven by optimized phenotypes • Loss of diversity and resilience	• Reduced adaptability of future populations	• Multiobjective optimization • Diversity-preserving constraints
Data ownership and consent	• Rights over phenomics data generation and use	• Multiple stakeholders with competing interests • Benefit-sharing concerns	• Conflicts over access and control	• Transparent data governance frameworks • Federated and consent-aware learning
Privacy and data security	• Protection of sensitive biological and operational data	• Risk of misuse or unauthorized access • Cybersecurity vulnerabilities	• Loss of trust and reluctance to adopt phenomics	• Secure data architectures • Privacy-preserving AI techniques
Sustainability and environmental ethics	• Alignment with long-term ecological goals	• Efficiency gains versus environmental burden	• Potential externalization of environmental costs	• Sustainability-informed phenomics objectives • Systems-level evaluation
Uncertainty and risk communication	• Ethical handling of probabilistic inference	• Overconfidence in AI predictions • Ambiguity in phenomics interpretation	• Misguided decisions under false certainty	• Explicit uncertainty quantification • Risk-aware decision support
Societal trust and legitimacy	• Public perception of AI use in animal systems	• Concerns about technological overreach • Moral acceptability	• Resistance to adoption if legitimacy erodes	• Transparent communication • Alignment with societal values

Intergenerational responsibility	• Long-term consequences of phenomics decisions	• Short-term gains versus future resilience	• Cumulative negative effects over time	• Long-horizon evaluation metrics • Stewardship-oriented AI design
Governance and accountability	• Clear assignment of responsibility	• Ambiguity in decision authority • Regulatory gaps	• Ethical and legal uncertainty	• Institutional oversight mechanisms • Auditable AI systems
Ethics-by-design integration	• Embedding ethics into system architecture	• Ethics treated as an external constraint rather than a core principle	• Reactive rather than proactive ethical management	• Value-aligned objective functions • Integrated ethical constraints

The concept of personalization and modeling at the individual level will gain more and more significance in the application of phenomics in the future. With the increase in data sources and analytical capabilities, AI systems will have the opportunity to define the individual phenotypic profiles with a higher degree of accuracy. Such personalization will assist in the implementation of management and breeding practices that consider biological diversity as opposed to using only the average populations. Individual-level phenomics is consistent with larger trends of precision livestock management and a consideration that variability is an inherent theme of biological systems, and not a nuisance to work out (Suchkov et al., 2025).

The future directions of AI in the context of phenomics also focus on the combination with other data areas, especially genomics and environmental data. Phenomics is an ideal place between genetic potential and actual performance, thus it is a natural point of contact between various domains (Ghavi Hossein-Zadeh, 2024). AI will be used more and more to enable the collective modeling of genotype, phenotype, and environment to amplify the knowledge of the interaction of genetic predisposition with context to produce phenotypic results. Such an integrative approach will be able to supplement genetic enhancement along with adaptive management, which will help achieve resilience of livestock systems in the environment of change (Sharma and Goel, 2025; Ghavi Hossein-Zadeh, 2026).

From a methodological perspective, future AI systems in phenomics will be more focused on uncertainty quantification and robustness. Since phenomics insights are being increasingly employed to make consequential decisions, the capacity to reflect confidence and establish inference limits becomes necessary. Explicitly quantified uncertainty AI models will allow implementing risk-aware decision-making and building trust among stakeholders. Resistance to data noise, missingness, and domain shift will be given a high priority, as it is what real-world livestock systems imply (Upadhyay et al., 2024).

Further development of the AI-powered phenomics will also be defined by the progress in computational hardware and strategies of deployment. Phenomics-based intelligence will be made more applicable to sources of data by edge computing, distributed learning, and real-time inference, which will decrease latency and improve responsiveness. Such advancements will help in maintaining constant monitoring and adaptive decision support with the consideration of scales and governance issues of central data aggregation. AI will, in turn, be more integrated into the livestock systems of operations, and it will be a part of the management instead of a layer of analysis (Ahmad et al., 2024).

The question of ethics and governance will remain a definitive factor in the direction of AI in livestock phenomics. With greater analytical capability, there is a greater possibility of abuse or backfiring. The next stage of evolution will focus more on ethics-by-design models, making welfare, sustainability, and fairness limitations directly linked to the AI goals and assessment measures. Ethical structures will be changed to deal with the ownership of data, transparency, accountability, and involvement of stakeholders, and will ensure that the introduction of phenomics-centered innovation is in line with the values and the regulatory demands of society (Neethirajan, 2024; Ghavi Hossein-Zadeh, 2025).

In the mounting of the AI potential in phenomics, education and cross-disciplinary efforts will be important in accomplishing the future. To make the most of the integration, it must be done professionally with knowledge in animal science, data science, engineering, and ethics. The future outlooks are on the significance of training courses and cross-functional organizations that cut across disciplinary lines and create mutual understanding in the workplace. This form of integration will allow for being more critical about the use of AI technologies and will allow phenomics to be based on biological reality, but will also allow it to take advantage of computational innovation (Muhammad et al., 2025).

The change in the definition and evaluation of success will also become part of the future of phenomics. Future models will not just emphasize predictive accuracy or immediate improvements on performance over the short-term, but they will also be based on long-term outcomes in terms of resilience, welfare, and sustainability. AI systems will not be evaluated based on what they are predicting, but on how their usage is influencing the biological systems in the long run. Such a wider evaluative approach strengthens the importance of phenomics as a stewardship and not an exploitation instrument (Adeyeye and Akanbi, 2024).

In theory, the directions of AI-enabled phenomics in the future are a shift toward comprehensive and adaptive knowledge of livestock biology. Phenotype will become more and more of an emergent and evolving property, which cannot be entirely defined reductionistically. The process of altering the perspective on complexity and appreciating it is facilitated by AI because it helps create models that learn without trying to simplify it too early. This view identifies phenomics with the modern understanding of biology as being a network of interacting processes that exist on scales and contexts (Garcia-Oliveira et al., 2026). Future directions and conceptual shifts in AI-driven livestock phenomics are presented in Table 12.9.

TABLE 12.9 Future directions and conceptual shifts in AI-driven livestock phenomics.

Future direction or theme	Conceptual focus	Emerging capabilities and innovations	Anticipated impact on phenomics understanding	Implications for livestock systems
High-resolution phenotypic characterization	• Deeper and more granular observation of phenotype	• Fine-scale sensing • Continuous high-density data streams	• Enhanced detection of subtle phenotypic variation • Earlier identification of phenotypic transitions	• Improved precision in management and breeding
Temporal intelligence and trajectory-centered modeling	• Phenotype as a continuous process • Emphasis on dynamics over static states	• Advanced sequence modeling • Multiscale temporal inference	• More accurate representation of biological progression	• Proactive and anticipatory decision-making
Foundation and pretrained phenomics models	• Generalizable phenotypic representations	• Large-scale representation learning • Cross-context adaptability	• Improved transferability of phenomics insight	• Reduced data dependency across systems
Multiscale biological integration	• Linking molecular, organismal, and environmental scales	• Hierarchical AI architectures • Systems-level modeling	• Holistic understanding of phenotype emergence	• Coherent integration of multidomain data
Causal and mechanistic inference	• Moving beyond correlation-driven analysis	• Causal modeling frameworks • Counterfactual reasoning	• Increased reliability of phenomics conclusions	• Safer and more effective interventions
Personalized and individual-level phenomics	• Recognition of biological individuality	• Individualized baseline learning • Adaptive phenotypic profiling	• Improved interpretation of heterogeneity	• Precision-oriented livestock management
Real-time and edge AI deployment	• Localized and immediate phenomics inference	• Edge computing • Low-latency analytics	• Faster response to phenotypic change	• Operational scalability and resilience
Uncertainty-aware and robust AI systems	• Explicit handling of ambiguity and variability	• Probabilistic modeling • Robust optimization	• Reduced risk of overconfident inference	• Risk-aware decision support
Ethics-by-design AI frameworks	• Integration of ethical principles into AI systems	• Welfare-aware objectives • Fairness constraints	• Alignment of phenomics with societal values	• Improved trust and legitimacy
Sustainability-oriented phenomics	• Long-term ecological and system-level outcomes	• Integrated sustainability metrics	• Broader evaluation of phenotypic success	• Balanced productivity and environmental impact

Continued

TABLE 12.9 Future directions and conceptual shifts in AI-driven livestock phenomics.—cont'd

Future direction or theme	Conceptual focus	Emerging capabilities and innovations	Anticipated impact on phenomics understanding	Implications for livestock systems
Federated and privacy-preserving learning	• Distributed analysis without centralized data sharing	• Federated learning • Secure computation	• Expanded collaboration with data protection	• Enhanced stakeholder participation
Adaptive and continual learning systems	• Ongoing model updating over time	• Incremental learning frameworks	• Alignment with evolving biological systems	• Long-term relevance of phenomics insights
Human-AI co-decision frameworks	• Collaborative intelligence rather than automation	• Interactive decision-support systems	• Preservation of expert judgment	• Responsible technology adoption
Standardization and interoperability evolution	• Harmonized phenomics data ecosystems	• Shared ontologies • Cross-platform compatibility	• Improved data integration and reuse	• Accelerated scientific progress
Interdisciplinary integration and education	• Bridging biological and computational expertise	• Cross-domain training paradigms	• Improved interpretability and innovation	• Sustainable phenomics workforce development

References

Adeyeye, O.J., Akanbi, I., 2024. Artificial intelligence for systems engineering complexity: a review on the use of AI and machine learning algorithms. Comput. Sci. IT Res. J. 5 (4), 787–808. https://doi.org/10.51594/csitrj.v5i4.1026.

Ahmad, A., Liew, A.X.W., Venturini, F., Kalogeras, A., Candiani, A., Di Benedetto, G., Ajibola, S., Cartujo, P., Romero, P., Lykoudi, A., De Grandis, M.M., Xouris, C., Lo Bianco, R., Doddy, I., Elegbede, I., D'Urso Labate, G.F., García del Moral, L.F., Martos, V., 2024. AI can empower agriculture for global food security: challenges and prospects in developing nations. Front. Artif. Intell. 7. https://doi.org/10.3389/frai.2024.1328530.

Aliferis, C., Simon, G., 2024. Overfitting, underfitting and general model overconfidence and under-performance pitfalls and best practices in machine learning and AI. In: Artificial Intelligence and Machine Learning in Health Care and Medical Sciences: Best Practices and Pitfalls, pp. 477–524.

Arigbabu, A.T., Olaniyi, O.O., Adigwe, C.S., Adebiyi, O.O., Ajayi, S.A., 2024. Data governance in AI—enabled healthcare systems: a case of the project nightingale. Asian J. Res. Comput. Sci. 17 (5), 85–107. https://doi.org/10.9734/ajrcos/2024/v17i5441.

Atzmueller, M., Fürnkranz, J., Kliegr, T., Schmid, U., 2024. Explainable and interpretable machine learning and data mining. Data Min. Knowl. Discov. 38 (5), 2571–2595. https://doi.org/10.1007/s10618-024-01041-y.

Baek, S., Song, K., Lee, I., 2025. Single-cell foundation models: bringing artificial intelligence into cell biology. Exp. Mol. Med. 57 (10), 2169–2181. https://doi.org/10.1038/s12276-025-01547-5.

Balagurunathan, Y., Sethuraman, R.R., 2024. An analysis of ethics-based foundation and regulatory issues for genomic data privacy. J. Inst. Eng. Ser. B 105 (4), 1097–1107. https://doi.org/10.1007/s40031-024-01058-3.

Bansal, P., 2025. AI pattern recognition and its features. Int. J. Eng. Adv. Technol. 14 (3), 18–25. https://doi.org/10.35940/ijeat.c4562.14030225.

Billah, M., Bermann, M., Hollifield, M.K., Tsuruta, S., Chen, C.Y., Psota, E., Holl, J., Misztal, I., Lourenco, D., 2025. Review: genomic selection in the era of phenotyping based on digital images. Animal 19, 101486. https://doi.org/10.1016/j.animal.2025.101486.

Boy, G.A., 2024. Human Systems Integration of Human-AI Teaming, 2024 IEEE 4th International Conference on Human-Machine Systems (ICHMS), Toronto, ON, Canada, pp. 1-4. https://doi.org/10.1109/ICHMS59971.2024.10555843.

Brachmann, M.K., Jacobs, C.T., 2025. Navigating through the noise: a roadmap for combining interdisciplinary high dimensional data in biological systems. Evol. Biol. https://doi.org/10.1007/s11692-025-09655-w.

Brito, L.F., Schinckel, A.P., Rojas de Oliveira, H., 2025. Genomics and phenomics: who will be the dairy cows of the future? JDS Commun. 6, S23. https://doi.org/10.3168/jdsc.2025-0872.

Cámara, J., 2025. Exploring the interaction of design variability and stochastic operational uncertainties in software-intensive systems through the lens of modeling. Software Syst. Model 24 (3), 629–650. https://doi.org/10.1007/s10270-024-01226-5.

Chafai, N., Hayah, I., Houaga, I., Badaoui, B., 2023. A review of machine learning models applied to genomic prediction in animal breeding. Front. Genet. 14. https://doi.org/10.3389/fgene.2023.1150596.

Chakilam, C., Kannan, S., Recharla, M., Suura, S.R., Nuka, S.T., 2025. The impact of big data and cloud computing on genetic testing and reproductive health management. Am. J. Psychiatr. Rehabil. 28 (1), 62–72.

Chakraborti, T., Banerji, C.R.S., Marandon, A., Hellon, V., Mitra, R., Lehmann, B., Bräuninger, L., McGough, S., Turkay, C., Frangi, A.F., Bianconi, G., Li, W., Rackham, O., Parashar, D., Harbron, C., MacArthur, B., 2025. Personalized uncertainty quantification in artificial intelligence. Nat. Mach. Intell. 7 (4), 522–530. https://doi.org/10.1038/s42256-025-01024-8.

Clarke, J., 2023. Leveraging AI/Ml to address critical challenges in livestock research. J. Anim. Sci. 101 (Suppl. 3), 142. https://doi.org/10.1093/jas/skad281.173.

Crossa, J., Montesinos-Lopez, O.A., Costa-Neto, G., Vitale, P., Martini, J.W.R., Runcie, D., Fritsche-Neto, R., Montesinos-Lopez, A., Pérez-Rodríguez, P., Gerard, G., Dreisigacker, S., Crespo-Herrera, L., Pierre, C.S., Lillemo, M., Cuevas, J., Bentley, A., Ortiz, R., 2025. Machine learning algorithms translate big data into predictive breeding accuracy. Trends Plant Sci. 30 (2), 167–184. https://doi.org/10.1016/j.tplants.2024.09.011.

Dorea, J., Menezes, G.L., 2024. Artificial intelligence and machine learning to improve livestock farming. J. Anim. Sci. 102 (Suppl. 3), 296. https://doi.org/10.1093/jas/skae234.338.

Džermeikaitė, K., Šidlauskaitė, M., Antanaitis, R., Anskienė, L., 2025. Enhancing genomic selection in dairy cattle through artificial intelligence: integrating advanced phenotyping and predictive models to advance health, climate resilience, and sustainability. Dairy 6 (5), 50. https://doi.org/10.3390/dairy6050050.

Eckhardt, R., Arablouei, R., Ingham, A., McCosker, K., Bernhardt, H., 2025. Livestock behaviour forecasting via generative artificial intelligence. Smart Agric. Technol. 11, 100987. https://doi.org/10.1016/j.atech.2025.100987.

Fernandes, I.K., Vieira, C.C., Dias, K.O.G., Fernandes, S.B., 2024. Using machine learning to integrate genetic and environmental data to model genotype-by-environment interactions. https://doi.org/10.1101/2024.02.08.579534.

Ehsaninia, J., Ghavi Hossein-Zadeh, N., Shadparvar, A.A., 2016. Homogeneity and heterogeneity of variance components for milk and protein yield at different cluster sizes in Iranian Holsteins. Livest. Sci. 188, 174–181. https://doi.org/10.1016/j.livsci.2016.04.020.

Ehsaninia, J., Ghavi Hossein-Zadeh, N., Shadparvar, A.A., 2019. Estimation of genetic variation for macro- and micro-environmental sensitivities of milk yield and composition in Holstein cows using double hierarchical generalized linear models. J. Dairy Res. 86 (2), 145–153. https://doi.org/10.1017/S0022029919000293.

Ehsaninia, J., Ghavi Hossein-Zadeh, N., Shadparvar, A.A., 2020. Estimation of genetic parameters for micro-environmental sensitivities of production traits in Holstein cows using two-step method. Anim. Prod. Sci. 60, 752–757. https://doi.org/10.1071/AN18687.

Ferreira, R.E.P., Dórea, J.R.R., 2025. International symposium on Ruminant physiology: leveraging computer vision, large language models, and multimodal machine learning for optimal decision making in dairy farming. J. Dairy Sci. 108 (7), 7493–7510. https://doi.org/10.3168/jds.2024-25650.

Garcia-Oliveira, A.L., Dwivedi, S.L., Chander, S., Nelimor, C., Abd El Moneim, D., Ortiz, R.O., 2026. Breeding smarter: artificial intelligence and machine learning tools in modern breeding—a review. Agronomy 16 (1). https://doi.org/10.3390/agronomy16010137.

Ghavi Hossein-Zadeh, N., 2024. An overview of recent technological developments in bovine genomics. Vet. Anim. Sci. 25, 100382. https://doi.org/10.1016/j.vas.2024.100382.

Ghavi Hossein-Zadeh, N., 2025. Artificial intelligence in veterinary and animal science: applications, challenges, and future prospects. Comput. Electron. Agric. 235, 110395. https://doi.org/10.1016/j.compag.2025.110395.

Ghavi Hossein-Zadeh, N., 2026. Advancing climate-resilient livestock systems: Next-generation emission mitigation strategies and integrated technological innovations. Vet. Anim. Sci. 31, 100588. https://doi.org/10.1016/j.vas.2026.100588.

Guo, T., Li, X., 2023. Machine learning for predicting phenotype from genotype and environment. Curr. Opin. Biotechnol. 79, 102853. https://doi.org/10.1016/j.copbio.2022.102853.

Gupta, S., 2024. AI Applications in Animal Behavior Analysis and Welfare Agriculture 4.0: Smart Farming with IoT and Artificial Intelligence. CRC Press, pp. 224–244. Available from: https://www.routledge.com/Agriculture-40-Smart-Farming-with-IoT-and-Artificial-Intelligence/Gupta-Hasan-Singh-Kumar-Ansari-Nisar/p/book/9781032943367.

Hamadani, H., Hamadani, A., Shabir, S., 2024. Artificial intelligence in animal farms for management and breeding. In: A Biologists Guide to Artificial Intelligence. Academic Press, pp. 167–182.

Hassija, V., Chamola, V., Mahapatra, A., Singal, A., Goel, D., Huang, K., Scardapane, S., Spinelli, I., Mahmud, M., Hussain, A., 2024. Interpreting black-box models: a review on explainable artificial intelligence. Cogn. Comput. 16 (1), 45–74. https://doi.org/10.1007/s12559-023-10179-8.

Hegde, A., Nguyen, T., Cheng, J., 2025. Machine learning methods for gene regulatory network inference. https://doi.org/10.48550/arXiv.2504.12610.

Hossain, Md I., Zamzmi, G., Mouton, P.R., Salekin, Md S., Sun, Y., Goldgof, D., 2025. Explainable AI for medical data: current methods, limitations, and future directions. ACM Comput. Surv. 57 (6), 1–46. https://doi.org/10.1145/3637487.

Huang, X., 2024. Developing machine learning applications for population genetic inference: ensuring precise Terminology and robust implementation. EcoEvoRxiv. https://doi.org/10.0410/cata/fdca2e1272bd4f87a8fdabaf13a0a2fe.

Ilan, Y., 2025. The constrained disorder principle overcomes the challenges of methods for assessing uncertainty in biological systems. J. Personalized Med. 15 (1). https://doi.org/10.3390/jpm15010010.

Klingström, T., Zonabend König, E., Zwane, A.A., 2024. Beyond the hype: using AI, big data, wearable devices, and the internet of things for high-throughput livestock phenotyping. Brief. Funct. Genom. https://doi.org/10.1093/bfgp/elae032.

Kocakaya, A., Kul, B.Ç., 2025. Phenomics evaluation in cattle breeding. Slovenian Vet. Res. 62 (2), 73–82. https://doi.org/10.26873/SVR-2047-2024.

Laghouaouta, H., Fraile, L.J., Pena, R.N., 2024. Selection for resilience in livestock production systems. Int. J. Mol. Sci. 25 (23), 13109. https://doi.org/10.3390/ijms252313109.

Lee, S.H., Lim, K.S., Lee, H., Heo, J., Kim, J., Kim, S.H., Kim, S.H., Park, J.E., Lim, D., Oh, J.D., Kim, B.M., Yoo, S.W., Shin, D., Kim, J.M., 2025. The vision of big data recirculation for smart livestock farming in South Korea. J. Anim. Sci. Technol. 67 (2), 303–313. https://doi.org/10.5187/jast.2025.e21.

Li, S., Agudelo Garcia, P.A., Aliferis, C., Becich, M.J., Calyeca, J., Cosgrove, B.D., Elisseeff, J., Farzad, N., Fertig, E.J., Glass, C., Gu, L., Hu, Q., Ji, Z., Königshoff, M., LeBrasseur, N.K., Li, D., Ma, A., Ma, Q., Menon, V., Mitchell, J.T., Mora, A.L., Nagaraj, S., Nelson, A.C., Niedernhofer, L.J., Rojas, M., Taha, H.B., Wang, J., Wang, S., Wu, P.H., Xie, J., Xu, M., Yu, M., Zhang, X., Zhao, Y., Adams, P.D., Aguayo-Mazzucato, C., Baker, D. J., Benz, C., Bernlohr, D.A., Bueno, M., Chen, J., Childs, B.G., Chuang, J.H., Chung, D., Dileepan, M., Ding, L., Dong, M., Duncan, F., Enninful, A., Flynn, W.F., Franco, A.C., Furman, D., Garovic, V., Halene, S., Herman, A.B., Hertzel, A.V., Iwasaki, K., Jeon, H., Kang, J.W., Karmakar, S., Kirkland, J.L., Korstanje, R., Kummerfeld, E., Lee, J.H., Liu, Y., Lu, Y., Lugo-Martinez, J., Martini, H., Melov, S., Musi, N., Passos, J.F., Peters, S.T., Rahman, I., Ramasamy, R., Rindone, A.N., Robbins, P.D., Robson, P., Rodriguez-Lopez, J., Rosas, L., Rosenthal, N., Schafer, M.J., Schilling, B., Schmidt, E.L., Schneider, K., Sengupta, K., Shu, J., So, P.T.C., Sun, L., Tchkonia, T., Teneche, M.G., Vanegas, N., Wang, J., Xie, J., Yin, S., Zhang, K., Zhu, Q., Fan, R., 2025. Advancing biological understanding of cellular senescence with computational multiomics. Nat. Genet. 57 (10), 2381–2394. https://doi.org/10.1038/s41588-025-02314-y.

Liu, Y., Xiong, X., Liao, Y., Qin, M., Huang, Z., Zhu, S., Yin, L., Fu, Y., Zhang, H., Xu, J., Yin, D., Huang, X., Quan, Y., Li, X., Jiang, T., Yang, W., Yuan, X., Frantz, L., Li, X., … Zhao, S., 2025. Bridging the genotype-phenotype gap with generative artificial intelligence (Version 1). https://doi.org/10.48550/ARXIV.2511.13141.

Maiorano, A.M., Ablondi, M., Qiao, Y., Steibel, J.P., Bernal Rubio, Y.L., 2024. Editorial: increasing sustainability in livestock production systems through high-throughput phenotyping approaches. Front. Genet. 15. https://doi.org/10.3389/fgene.2024.1403133.

Marques-Garcia, F., Nieto-Librero, A., Gonzalez-García, N., Tejedor-Ganduxé, X., Martinez-Bravo, C., 2025. Data science applied to the assessment of biological variation estimates. Adv. Lab. Med. Avances en Medicina de Laboratorio 6 (2), 154–159. https://doi.org/10.1515/almed-2025-0042.

Matentzoglu, N., Bello, S.M., Stefancsik, R., Alghamdi, S.M., Anagnostopoulos, A.V., Balhoff, J.P., Balk, M.A., Bradford, Y.M., Bridges, Y., Callahan, T.J., Caufield, H., Cuzick, A., Carmody, L.C., Caron, A.R., de Souza, V., Engel, S.R., Fey, P., Fisher, M., Gehrke, S., Grove, C., Hansen, P., Harris, N.L., Harris, M.A., Harris, L., Ibrahim, A., Jacobsen, J.O.B., Köhler, S., McMurry, J.A., Munoz-Fuentes, V., Munoz-Torres, M. C., Parkinson, H., Pendlington, Z.M., Pilgrim, C., Robb, S.M.C., Robinson, P.N., Seager, J., Segerdell, E., Smedley, D., Sollis, E., Toro, S., Vasilevsky, N., Wood, V., Haendel, M.A., Mungall, C.J., McLaughlin, J.A., Osumi-Sutherland, D., 2025. The unified phenotype ontology: a framework for cross-species integrative phenomics. Genetics 229 (3). https://doi.org/10.1093/genetics/iyaf027.

Matthay, E.C., Neill, D.B., Titus, A.R., Desai, S., Troxel, A.B., Cerdá, M., Díaz, I., Michele, S., Thorpe, L.E., 2025. Integrating artificial intelligence into causal research in epidemiology. Curr. Epidemiol. Rep. 12 (1). https://doi.org/10.1007/s40471-025-00359-5.

Meijer, P., Aggoune, Y., Ambrose, M., Beaubien, A., Harvey, J., Howard, N., Inala, N., Johnson, E., Kelsey, A., Kinsey, M., Liang, J., Mariz, P., Pister, S., Subramanian, S., Tereshchenko, V., Vetto, A., 2024. Reproducibility Needs Reshape Scientific Data Governance. https://doi.org/10.48550/arXiv.2410.12800.

Melandri, G., R-Radohery, G., Beaumont, C., de Cripan, S.M., Muller, C., Piras, L., Alcina Pereira, M., Salvador, A.F., Domingo-Almenara, X., Bolger, M., Colombié, S., Prigent, S., Arechederra, B.G., Canela, N.C., Pétriacq, P., 2025. Artificial intelligence: the human response to approach the complexity of big data in biology. GigaScience 14. https://doi.org/10.1093/gigascience/giaf057.

Mohamed, N., 2025. Augmented intelligence: a comprehensive review of the flexibility between human and artificial intelligence. J. Inst. Eng. Ser. B 106 (6), 2031–2058. https://doi.org/10.1007/s40031-025-01213-4.

Mohanan, P.V., 2025. Artificial Intelligence and Biological Sciences. CRC Press. https://doi.org/10.1201/9781003492726.

Motsinger-Reif, A.A., Reif, D.M., Akhtari, F.S., House, J.S., Campbell, C.R., Messier, K.P., Fargo, D.C., Bowen, T.A., Nadadur, S.S., Schmitt, C.P., Pettibone, K.G., Balshaw, D.M., Lawler, C.P., Newton, S.A., Collman, G.W., Miller, A.K., Merrick, B.A., Cui, Y., Anchang, B., Harmon, Q.E., McAllister, K.A., Woychik, R., 2024. Gene-environment interactions within a precision environmental health framework. Cell Genom. 4 (7). https://doi.org/10.1016/j.xgen.2024.100591.

Muhammad, A., Zheng, X.Y., Gan, H.L., Guo, Y.X., Xie, J.H., Chen, Y.J., Chen, J.J., 2025. AI-enhanced morphological phenotyping in humanized mouse models: a transformative approach to infectious disease research. Biophysica 5 (4). https://doi.org/10.3390/biophysica5040043.

Neethirajan, S., 2024. Artificial intelligence and sensor innovations: enhancing livestock welfare with a human-centric approach. Hum. Centr. Intell. Syst. 4 (1), 77–92. https://doi.org/10.1007/s44230-023-00050-2.

Olaker, V.R., Fry, S., Terebuh, P., Davis, P.B., Tisch, D.J., Xu, R., Miller, M.G., Dorney, I., Palchuk, M.B., Kaelber, D.C., 2025. With big data comes big responsibility: strategies for utilizing aggregated, standardized, de-identified electronic health record data for research. Clin. Transl. Sci. 18 (1). https://doi.org/10.1111/cts.70093.

Olatinwo, D.D., Myburgh, H.C., De Freitas, A., Abu-Mahfouz, A.M., 2025. A review of smart crop technologies for resource constrained environments: leveraging multimodal data fusion, edge-to-cloud computing, and IoT virtualization. J. Sens. Actuator Netw. 14 (5). https://doi.org/10.3390/jsan14050099.

Palmquist, K.H., Ko, C.S., Shyer, A.E., Rodrigues, A.R., 2024. Biological theories of morphogenesis based on holistic biophysical thinking. Biol. Theory 20 (2), 105–118. https://doi.org/10.1007/s13752-024-00477-1.

Panda, R.A., Ramanan, S.V., Kawale, S.R., Srinivas, D., Junnarkar, A.A., Behera, S.R., 2024. Smart sensors and IoT in studying animal morphology and developmental biology. Bull. Pure Appl. Sci. Zool. 43.

Panigrahi, M., Rajawat, D., Nayak, S.S., Bose, A., Bharia, N., Singh, S., Sharma, A., Dutt, T., 2025. Advancements in animal breeding: from mendelian genetics to machine learning. Int. J. Mol. Sci. 26 (23), 11352. https://doi.org/10.3390/ijms262311352.

Paul, K., Schweng, S., Kaul, H.-P., Gansberger, F., Holzinger, A., 2026. AI-powered Pheno-Farm server: making adaptive farming decisions. Agric. Syst. 231, 104563. https://doi.org/10.1016/j.agsy.2025.104563.

Pommier, C., Alic, I., Cabrera-Bosquet, L., Draye, X., Neveu, P., Reif, J.C., Robbins, K.R., Krajewski, P., Tardieu, F., 2025. Reassessing data management in increasingly complex phenotypic datasets. Trends Plant Sci. https://doi.org/10.1016/j.tplants.2025.09.001.

Rahman, M.A., Emon, M.M.H., Khan, T., Hamid, A.B.A., Yaakub, N.I., 2025. Eco-intelligence: AI's contribution to a sustainable world. In: AI and Green Technology Applications in Society. IGI Global Scientific Publishing, pp. 31–62.

Ramsden, J.J., 2023. Phenomics. Springer Nature (Netherlands), pp. 347–351. https://doi.org/10.1007/978-3-030-45607-8_25.

Redekop, E., 2025. Large-Scale Representation Learning and Generative Modeling for Multimodal Healthcare Data. Doctoral dissertation. University of California, Los Angeles.

Rosati, A., 2025. Invited paper artificial intelligence and the future of animal husbandry: rethinking knowledge, practice, and governance. Arch. Latinoam. Prod. Anim. 33 (Suppl. 1), 567–572. https://doi.org/10.53588/alpa.330509.

Rubel, A., Kaehrle, M., Streiffer, R., Platforms, D., 2025. Privacy, and the ethics of wildlife information sharing. Philos. Technol. 38 (1). https://doi.org/10.1007/s13347-025-00841-4.

Sebia, H., Guyet, T., Audureau, E., 2024. SWoTTeD: an extension of tensor decomposition to temporal phenotyping. Mach. Learn. 113 (9), 5939–5980. https://doi.org/10.1007/s10994-024-06545-8.

Shamim, M., Faridi, A.R., Masood, F., 2025. Analysis of emerging trends and challenges in multimodal data. In: 2025 12th International Conference on Computing for Sustainable Global Development (INDIACom), pp. 1–7. https://doi.org/10.23919/indiacom66777.2025.11115680.

Sharma, J., Goel, P., 2025. The use of AI for phenotype-genotype mapping. In: Artificial Intelligence (AI) in Cell and Genetic Engineering. Humana, New York, pp. 369–410.

Sharma, S., Maurya, H.K., Sharma, K., Kumar, V., 2025. Machine learning applications in livestock management. Adv. Comput. Intell. Robot. Book Ser. 221–238. https://doi.org/10.4018/979-8-3373-0842-5.ch009.

Spangler, M.L., 2024. Machine learning and AI to improve genetic prediction in beef cattle: potential uses and misuses. J. Anim. Sci. 102 (Suppl. 3), 296–297. https://doi.org/10.1093/jas/skae234.339.

Stager, M., Velotta, J.P., Cheviron, Z.A., Senner, N.R., 2025. Assessing the evidence for treating developmental plasticity and phenotypic flexibility as different phenomena. Funct. Ecol. 39 (11), 2949–2961. https://doi.org/10.1111/1365-2435.14590.

Stevens, I., Pacia, D., Hansen, J.N., Parker, J., Payne-Foster, P., Ideker, T., Bélisle-Pipon, J.C., Ravitsky, V., 2025. Towards the ethical development of functional genomics research. Trends Biotechnol. https://doi.org/10.1016/j.tibtech.2025.10.015.

Suchkov, S., Lustig, R.H., Escobedo-Monge, M., Blokh, S., Andronova, N., Goryacheva, O., Moyseyak, M.B., Vlasov, T., Polyakova, V., Antonova, E., Tuykavin, A., 2025. Personalized and precision medicine as a health-care model of the next step generation through translational applications of individualized nutrition- and food design-driven resources. Glob. Transl. Med. 4 (3), 60. https://doi.org/10.36922/gtm025080017.

Sun, P.S., Ye, K., 2025. Phenotypic plasticity: mechanisms, evolutionary implications, and research methodologies. Sci. Sinica Vitae 55 (12), 2475–2482. https://doi.org/10.1360/ssv-2025-0179.

Taddese, A.A., Addis, A.C., Tam, B.T., 2025. Data stewardship and curation practices in AI-based genomics and automated microscopy image analysis for high-throughput screening studies: promoting robust and ethical AI applications. Hum. Genom. 19 (1). https://doi.org/10.1186/s40246-025-00716-x.

Tedeschi, L.O., Guarnido Lopez, P., Menendez III, H.M., Seo, S., 2025. Advancing precision livestock farming: integrating artificial intelligence and emerging technologies for sustainable livestock management. Anim. Biosci. https://doi.org/10.5713/ab.25.0289.

Toth, G., Szabo, S., Haidegger, T., Alexy, M., 2025. Edge or cloud architecture: the applicability of new data processing methods in large-scale poultry farming. Technologies 13 (1), 17. https://doi.org/10.3390/technologies13010017.

Trirat, P., Shin, Y., Kang, J., Nam, Y., Na, J., Bae, M., Kim, J., Kim, B., Lee, J.G., 2024. Universal time-series representation learning: a survey. https://doi.org/10.48550/arXiv.2401.03717.

Upadhyay, V.R., Ramesh, V., Kumar, H., Somagond, Y.M., Priyadarsini, S., Kuniyal, A., Prakash, V., Sahoo, A., 2024. Phenomics in livestock research: bottlenecks and promises of digital phenotyping and other quantification techniques on a global scale. OMICS A J. Integr. Biol. 28 (8), 380–393. https://doi.org/10.1089/omi.2024.0109.

Wang, J., Zhang, J., Hao, W., Zong, W., Liang, M., Zhao, F., Zhang, L., Wang, L., Gao, H., Wang, L., 2025. An interpretable integrated machine learning framework for genomic selection. Smart Agric. Technol. 12, 101138. https://doi.org/10.1016/j.atech.2025.101138.

Weine, E., Smith, S.P., Knowlton, R.K., Harpak, A., 2025. Trade-offs in modeling context dependency in complex trait genetics. eLife. https://doi.org/10.7554/eLife.99210.3.

Wilson, A., Anwar, M.R., 2024. The future of adaptive machine learning algorithms in high-dimensional data processing. Int. Trans. Artif. Intell. 3 (1), 97–107. https://doi.org/10.33050/italic.v3i1.656.

Wu, Y., Xie, L., 2024. AI-driven multi-omics integration for multi-scale predictive modeling of causal genotype-environment-phenotype relationships. https://doi.org/10.48550/arXiv.2407.06405.

Wu, X.L., Cole, J.B., Legarra, A., Parker Gaddis, K.L., Dürr, J.W., 2025. Handling errors in the response: considerations for leveraging unsupervised or incomplete data for genetic evaluations. JDS Commun. 6 (5), 675–680. https://doi.org/10.3168/jdsc.2024-0668.

Yang, W., Fu, R., Amin, M.B., Kang, B., 2025. The impact of modern AI in metadata management. Hum. Centr. Intell. Syst. 5 (3), 323–350. https://doi.org/10.1007/s44230-025-00106-5.

Yurkovich, J.T., Evans, S.J., Rappaport, N., Boore, J.L., Lovejoy, J.C., Price, N.D., Hood, L.E., 2024. The transition from genomics to phenomics in personalized population health. Nat. Rev. Genet. 25 (4), 286–302. https://doi.org/10.1038/s41576-023-00674-x.

Yurochka, S.S., 2024. Concept of digital assessment of phenotype and behavior of cattle. Mach. Equip. Rural Area 9, 24–28. https://doi.org/10.33267/2072-9642-2024-9-24-28.

Zangana, H.M., Luckyardi, S., Mustafa, F.M., Li, S., 2025. Ethical and secure AI applications in agricultural research. In: Advances in Computational Intelligence and Robotics Book Series, pp. 43–76. https://doi.org/10.4018/979-8-3373-4252-8.ch002.

Zhang, S., Zhou, Y., Geng, P., Lu, Q., 2024. Functional neural networks for high-dimensional genetic data analysis. IEEE ACM Trans. Comput. Biol. Bioinf 21 (3), 383–393. https://doi.org/10.1109/tcbb.2024.3364614.

Zhang, S., Lai, C., Zhao, J., Wang, J., 2025. Big data and AI-powered modeling: a pathway to sustainable precision animal nutrition. Adv. Sci. 12 (41). https://doi.org/10.1002/advs.202507564.

Zhu, L., Yang, X., Xian, Y., Jiang, W., Pu, X., Wang, S., Cheng, J., Niu, L., Zhao, Y., Chen, L., Zhou, X., Wang, Y., Gan, M., Zhu, L., Shen, L., 2025. Empowering precision livestock farming: artificial intelligence applications in animal genomic breeding and multi-dimensional phenotypic measurement. Smart Agric. Technol. 12, 101655. https://doi.org/10.1016/j.atech.2025.101655.

Chapter 13

Systems biology and multi-omics integration

In livestock science, systems biology and multi-omics integration represent a conceptual and methodological shift toward systems-level understanding, going beyond isolated molecular inventories or compartmentalized trait associations to dynamic models that explain how physiological states, expressed phenotypes, microbial ecosystem dynamics, immune signaling, transcriptional regulation, protein interaction cascades, metabolic fluxes, genomic variation, physiological states, and microbial ecosystem dynamics interact in environmental, nutritional, managerial, and climatic contexts. The livestock organism is presented in this chapter as an adaptive, multiscale networked system whose emergent properties result from layered feedback, modular robustness, redundancy, and context-dependent plasticity rather than as a linear conduit from genotype to phenotype. Therefore, building coherent mechanistic narratives and predictive abstractions that maintain biological causality, temporal ordering, spatial compartmentalization, and cross-scale energy and resource constraints while remaining computationally tractable and generalizable across various production ecologies is more important for integration than simply combining multiple omics layers.

High-throughput assays, dense phenotyping, and ecological metadata are becoming more widely available, which simultaneously increases opportunities and complexity. This calls for principled approaches to balance heterogeneous resolution, variable noise structure, batch and management artifacts, missingness mechanisms, and unequal lineage representation. To balance empirical statistical strength with interpretability and intervention relevance, systems biology provides organizing frameworks based on networks, dynamical systems, constraint-based modeling, probabilistic graphical formalisms, and hybrid mechanistic machine learning. In this context, multi-omics integration turns into a hypothesis-refinement process where latent structure is iteratively extracted, tested against orthogonal evidence streams, and rigorously evaluated using sensitivity analysis, counterfactual simulation, and assessment of causal transportability across breeds, environments, and management paradigms. To more accurately attribute variation and better target selective, nutritional, health, and environmental interventions, emphasis is placed on separating hierarchical biological signals that distinguish stable genomic scaffolding from plastic regulatory responses and transient phenotypic excursions.

Artificial intelligence (AI) serves both as a catalyst and an enabling framework, providing generative models that investigate plausible unobserved states, self-supervised representations that preserve cross-layer dependencies, and causal discovery algorithms that prioritize mechanistic edges while reducing confounding. However, epistemic rigor, biological plausibility, welfare alignment, sustainability accounting, and equitable accessibility must co-evolve with computational advancements; methodological sophistication alone is insufficient. Data stewardship, algorithmic fairness, lineage, and environmental diversity preservation, transparency of inference provenance, and adaptive oversight capable of assessing novel composite biomarkers or intervention policies emerging from integrated frameworks are just a few of the ethical and governance aspects that permeate the integrative agenda. In order to enable resilient, resource-efficient, welfare-conscious, and climate-adaptive livestock production, the chapter guides the reader toward a transdisciplinary synthesis where systems thinking, multi-omics depth, high-resolution phenomics, and responsible AI converge. It lays the groundwork for in-depth investigation of conceptual architectures, analytical pipelines, illustrative integrative paradigms, validation philosophies, and translational pathways that together characterize the developing field of systems-scale livestock omics.

13.1 Introduction to systems biology

By focusing on integration, interaction, and dynamic regulation across various organizational levels rather than isolated components, systems biology has become a fundamental paradigm for comprehending biological complexity. Systems biology reframes the animal as an adaptive, multiscale system in the context of livestock science, where observable phenotypes are shaped by a combination of genomic information, regulatory mechanisms, metabolic processes,

Artificial Intelligence in Livestock Omics. https://doi.org/10.1016/B978-0-443-45663-3.00006-3

physiological functions, behavior, and environmental interactions. This viewpoint acknowledges that biological outcomes result from nonlinear networks of feedback, redundancy, and context-dependent modulation, departing from conventional linear interpretations of genotype-phenotype relationships. Systems biology offers a conceptual and analytical framework that can address these multifaceted issues in a coherent and predictive way as livestock production systems confront growing demands for efficiency, resilience, welfare alignment, and environmental sustainability (Pathak and Kim, 2024; Kant et al., 2025).

Fundamentally, systems biology aims to explain how interactions between components, rather than the characteristics of individual elements alone, give rise to biological function. Genes in livestock organisms function within regulatory networks that react dynamically to environmental stressors, microbial interactions, nutritional status, developmental stage, and health issues. Molecular interactions spread upward through cellular pathways, tissues, organs, and whole-animal physiology in these hierarchically structured networks, while molecular states are influenced downward by organism-level constraints such as energy balance, thermoregulation, and management techniques. Thus, network connectivity, where a biological component's functional meaning is determined by its position and connectivity within a network, and dynamics, where temporal trajectories and state transitions are frequently more informative than static measurements, are emphasized in systems biology (Fig. 13.1) (Pathak and Kim, 2024; Kitano, 2007).

In the past, reductionist methods that identified genes, pathways, and characteristics linked to production, reproduction, and health have been very beneficial to livestock research. However, it also became apparent that additive models and single-layer analyses were inadequate to capture the emergent nature of these phenotypes as the complexity of biologically and economically significant traits became apparent. Numerous interacting loci and regulatory processes affect traits like feed efficiency, disease resilience, fertility, stress tolerance, and product quality. The environment and management have an impact on these processes. Instead of rejecting reductionism, systems biology incorporates it into a broader integrative framework that enables the contextualization of mechanistic insights from targeted experiments within whole-system behavior. This synthesis makes it possible to reconcile population-level predictability with in-depth molecular knowledge (Choudhary et al., 2024; Wadood et al., 2025).

FIGURE 13.1 An overview of the systems biology cycle in livestock science. This figure illustrates the iterative systems biology workflow in livestock research, beginning with the integration of diverse biological data and progressing through interaction analysis, system modeling, and application of predictive insights. The cycle emphasizes the use of models to inform decision-making and guide interventions within livestock systems. Continuous evaluation of outcomes feeds back into data integration, highlighting the adaptive and self-refining nature of systems biology approaches.

Systems biology's explicit consideration of various biological scales is one of its distinguishing characteristics. Although a landscape of possible regulatory states is established by molecular variation at the genomic and epigenomic levels, actual phenotypic expression is dependent on transcriptional regulation, protein activity, metabolic fluxes, cellular interactions, and tissue-level coordination. From quick signaling events to long-term developmental programming and adaptation, these processes take place over a variety of temporal scales. Through computational and mathematical models that uphold biophysical constraints, maintain causal ordering, and permit information to flow across levels, systems biology integrates these scales. Such multiscale modeling is crucial for connecting genetic selection choices to physiological performance, welfare outcomes, and environmental impact in livestock systems (Parvatikar et al., 2023; Puniya et al., 2024).

Systems biology has greatly benefited from the development of high-throughput measurement technologies. Large amounts of data, capturing various aspects of biological state, are now produced by advances in sequencing, mass spectrometry, imaging, and sensor technologies. Although these data provide unprecedented resolution, they also present serious analytical difficulties due to heterogeneity, dimensionality, noise, and missingness. The organizing principles required to turn these disparate data streams into cohesive models are provided by systems biology. Systems biology stresses hypothesis-driven synthesis, in which patterns seen across several layers are assessed for consistency, plausibility, and robustness, as opposed to viewing data integration as a straightforward aggregation task. This method makes it easier to distinguish between spurious correlations that could result from shared confounders or technical artifacts and causal mechanisms (Lavandosque et al., 2024; Kitano, 2007).

Systems biology is closely related to the objectives of prescriptive and predictive modeling in livestock applications. Prescriptive models seek interventions that can move the system toward desired states, whereas predictive models predict phenotypic outcomes, such as growth trajectories, health risks, or reproductive performance, under particular conditions. Systems biology facilitates the achievement of these goals by integrating prediction into mechanistic understanding, which enhances interpretability and extrapolation beyond observed data. For instance, knowing how regulatory networks reorganize in response to nutritional stress enables more accurate forecasting of performance under unfamiliar diets or environmental circumstances. In livestock systems, where experimental manipulation is expensive, ethically challenging, and frequently impractical at scale, this mechanistic foundation is especially crucial (Lavandosque et al., 2024; Pathak and Kim, 2024).

Analytical capabilities have been further enhanced by the fusion of systems biology and AI. Systems biology provides structure, limitations, and interpretive context, while machine learning techniques are excellent at identifying patterns in high-dimensional data. These methods work together to create hybrid models that strike a balance between biological plausibility and flexibility. While mechanistic priors direct model training toward solutions consistent with known physiology, representation learning can reveal latent structures spanning several biological layers. This collaboration improves the capacity to model complex traits across a range of genetic backgrounds and production environments in livestock science, while upholding the transparency necessary for ethical decision-making (Ghavi Hossein-Zadeh, 2024; Lawrence et al., 2024).

To address variability and adaptability, which are inherent characteristics of biological systems, systems biology is also essential. Due to environmental exposure, management techniques, and genetic diversity, livestock populations exhibit significant heterogeneity. Systems biology views variability as informative, showing how various system configurations achieve similar functional outcomes or react differently to perturbations, rather than as noise that should be minimized. By identifying network motifs and regulatory architectures that provide robustness, this viewpoint aids in the development of resilient livestock systems. These insights are becoming increasingly crucial as animal agriculture faces unpredictable pressures from resource limitations, emerging diseases, and climate variability (Pathak and Kim, 2024; Singh et al., 2024).

Systems biology and ethical and sustainable issues are closely related in livestock contexts. Animal welfare, genetic diversity, and environmental impact can all be significantly impacted by decisions made using integrated biological models. Systems biology promotes balanced decision frameworks rather than single-trait optimization by evaluating trade-offs across several objectives. Systems approaches encourage more responsible innovation by explicitly modeling the relationships between productivity, health, welfare, and environmental factors. This all-encompassing approach harmonizes scientific progress with societal demands for sustainable food production and humane treatment of animals (Vlaicu et al., 2024). Core principles and conceptual foundations of systems biology in livestock research are presented in Table 13.1.

13.2 Overview of multi-omics integration

A key operational pillar of systems biology is multi-omics integration, which offers a methodological way to make a biologically meaningful connection between molecular variation and higher-order biological function. To fully capture the

TABLE 13.1 Core principles and conceptual foundations of systems biology in livestock research.

Core principle	Conceptual definition	Biological interpretation in livestock systems	Analytical implications	Relevance to multi-omics integration
Holism	• Biological function emerges from the whole system rather than isolated parts	• Traits reflect coordinated molecular, physiological, and environmental interactions • System behavior cannot be inferred from single components	• Requires integrated modeling rather than single-factor analysis	• Necessitates simultaneous consideration of multiple omics layers
Emergence	• System-level properties arise from interactions among components	• Productivity, resilience, and health are emergent outcomes • No single gene or pathway fully explains complex traits	• Focus on interaction patterns rather than individual effects	• Justifies integrating genomics, regulation, and metabolism
Biological hierarchy	• Organization across nested levels of biological scale	• Molecular regulation shapes cellular function • Cellular processes aggregate into tissues and whole-animal phenotypes	• Encourages hierarchical and multi-scale modeling	• Supports structured integration from genome to phenotype
Nonlinearity	• System responses are not proportional to inputs	• Small molecular changes may cause large phenotypic shifts • Threshold and saturation effects are common	• Linear models are insufficient for complex traits	• Motivates nonlinear AI and network-based integration
Network organization	• Biological systems are structured as interacting networks	• Regulatory, metabolic, and signaling networks govern traits	• Analysis centers on connectivity and modularity	• Enables network-based multi-omics integration
Feedback regulation	• Outputs influence upstream system components	• Homeostasis is maintained through feedback loops • Dysregulation propagates system-wide effects	• Requires dynamic and recursive modeling frameworks	• Integrates omics layers with bidirectional influence
Redundancy and robustness	• Multiple components can perform similar functions	• Alternative pathways maintain function under perturbation	• Identifies robustness rather than single points of failure	• Highlights the value of multi-omics redundancy
Context dependence	• Biological meaning depends on internal and external context	• Gene effects vary with environment, development, and management	• Models must incorporate contextual variables	• Encourages environment-aware omics integration
Dynamic regulation	• Biological states change continuously over time	• Development, adaptation, and stress responses are temporal	• Static analysis is insufficient	• Requires longitudinal multi-omics integration
Causal structure	• Distinction between correlation and mechanism	• Molecular associations do not imply functional control	• Promotes causal and mechanistic modeling	• Guides integration toward biologically meaningful inference
Multiscale interaction	• Cross-talk between biological levels	• Organism-level demands constrain molecular regulation	• Requires bidirectional modeling across scales	• Integrates top-down and bottom-up omics signals

Variability as information	• Biological variation reflects adaptive potential	• Individual heterogeneity reveals system flexibility	• Variability is analyzed, not eliminated	• Multi-omics captures diverse regulatory strategies
System adaptability	• Ability to reconfigure under perturbation	• Livestock adapt to nutrition, climate, and stress	• Models must capture plasticity	• Integrates regulatory and metabolic flexibility
Predictive orientation	• Understanding enables anticipation of system behavior	• Future performance inferred from system state	• Shifts focus from description to prediction	• Multi-omics strengthens predictive modeling
Integration over reductionism	• Emphasis on synthesis rather than isolation	• Traits cannot be decomposed into single causal units	• Encourages composite analytical frameworks	• Foundational justification for multi-omics systems biology

complexity of biological regulation in livestock systems, multi-omics integration aims to integrate data from genomics, epigenomics, transcriptomics, proteomics, metabolomics, microbiomics, and related molecular layers. Each omics layer represents a different but related facet of the biological state, and none of them is adequate on its own to provide a thorough explanation of a phenotype. Understanding how genetic potential is realized, modulated, and constrained under particular environmental and management conditions thus requires integration across these layers (Mudadla et al., 2024; Morabito et al., 2025).

Multi-omics integration is conceptually based on the understanding that biological systems are dynamically regulated and hierarchically organized. The fundamental blueprint of biological potential is established by genomic variation, but the dynamic expression of this potential is determined by gene expression, protein activity, and metabolic flux. The microbiome adds another level of biological interaction that affects metabolism, immunity, and development, while epigenomic mechanisms mediate the interaction between the genome and the environment, influencing regulatory accessibility and responsiveness. Instead of treating these layers as separate information streams, multi-omics integration views them as interacting parts of a single system, where causality, feedback, and temporal coordination are key components (Kant et al., 2025).

Because many traits of interest in livestock biology are complex, polygenic, and context-dependent, multi-omics integration is especially important. Numerous molecular processes work in concert to produce resilience, performance, health, and reproduction. These processes have different effects depending on the stage of development and the environment. Single-omics analyses frequently reveal relationships that are hard to understand mechanistically or that do not apply to other populations. By offering a more comprehensive depiction of biological state and facilitating the distinction between primary regulatory drivers and downstream effects, multi-omics integration overcomes these constraints. Both explanatory power and predictive reliability are improved by this systems-level perspective (Khanzadeh et al., 2021; Choudhary et al., 2024; NASIR et al., 2024).

It is possible to view multi-omics integration as a computational and conceptual challenge. From a conceptual standpoint, integration necessitates a thorough comprehension of the relationships between various molecular layers and phenotype. This involves acknowledging feedback loops, regulatory hierarchies, and causal ordering. Reconciling data that vary in scale, dimensionality, noise structure, and temporal resolution is a computational aspect of integration. Metabolomic and microbiomic data show quick reactions to both internal and external stimuli, transcriptomic and proteomic data are dynamic and high-dimensional, and genomic data are comparatively static and sparse. While allowing for significant cross-layer inference, effective integration must maintain the distinct information content of each layer (Chetty and Blekhman, 2024).

Heterogeneity in data quality and structure is one of the main obstacles to multi-omics integration. Different omics platforms produce data with varying levels of measurement error, technical biases, and distributions. Direct alignment is further complicated by the fact that omics data are frequently gathered under various experimental designs or sampling schedules. In order to overcome these difficulties, systems biology approaches prioritize representation, abstraction, and normalization over raw data fusion. Integration focuses on relationships and patterns that are resilient to technical variation by mapping various omics signals into shared latent spaces or network structures. This abstraction reduces sensitivity to platform-specific artifacts and makes biological interpretation easier (Kant et al., 2025).

Another crucial component of multi-omics integration is temporal alignment. The temporal relationships between omics layers are crucial for comprehending regulatory dynamics because molecular processes operate on various timescales. For instance, transcriptional changes may occur before changes in the abundance of proteins, which then affect the metabolic state. Integrative frameworks that disregard temporal ordering run the risk of misinterpreting fleeting responses as stable characteristics or confusing cause and effect. In order to infer regulatory cascades and adaptive responses, systems biology places a strong emphasis on dynamic modeling techniques that explicitly represent temporal dependencies. Such temporal integration is especially crucial for comprehending development, stress adaptation, and reactions to dietary or environmental changes in livestock contexts (Moscardó García et al., 2025; Tjandrawinata et al., 2025).

The identification of emergent properties that are not visible within individual data layers is another benefit of multi-omics integration. Emergent properties frequently appear at higher levels of organization and are the result of interactions between components. For example, rather than the activity of a single gene or metabolite, metabolic efficiency or disease resilience may be the result of coordinated regulation across several molecular pathways. Systems biology uncovers these emergent patterns by combining omics layers, offering an understanding of intricate trait architecture. This ability is essential for improving livestock, since selection choices must take system-level results into consideration rather than discrete molecular effects (De Filippis et al., 2025).

The interpretability of AI models used in livestock science is also improved by the incorporation of multi-omics data. Although machine learning is very good at identifying patterns, when it is trained on single-layer data, its results can be challenging to understand biologically. Contextual richness from multi-omics integration limits model inference and makes interpretation easier. Predictions are more likely to represent true biological mechanisms than erroneous correlations when they are based on coordinated signals across molecular layers. For decision-making applications where accountability and transparency are necessary, this integrative context is especially crucial (Yetgin, 2025).

Methodologically speaking, multi-omics integration refers to a variety of approaches that vary in the degree of coupling between data layers. While some methods work at the model level, connecting disparate analyses through common parameters or constraints, others integrate data at the feature level, combining molecular measurements into unified representations. Integration techniques that maintain biological structure, like network-based representations or hierarchical models, are preferred in systems biology. These methods take into account the fact that molecular layers are connected by particular regulatory relationships rather than being interchangeable. Such biologically informed integration enhances robustness and generalizability across populations and environments in livestock systems (Kant et al., 2025; Morabito et al., 2025).

To address biological variability and heterogeneity, multi-omics integration plays a crucial role. The diversity of livestock populations and the substantial variation in molecular responses to similar conditions among individuals is known as diversity. Multiple pathways to functional equivalence can be revealed by integrating omics analysis, which enables the identification of alternative regulatory strategies that lead to similar phenotypic outcomes. This understanding recognizes that different biological configurations can be used to achieve resilience and performance, which supports more flexible and inclusive breeding and management strategies. Instead of focusing on eliminating problems, systems biology reframes biological variability as an adaptive resource (Choudhary et al., 2024; NASIR et al., 2024).

The integration of multi-omics data is also driven by practical and ethical considerations. Animal welfare, resource use, and environmental impact are all factors that affect livestock production and research. By increasing knowledge of underlying mechanisms, multi-omics integration reduces the need for general or invasive approaches and allows for more targeted and precise interventions. Finding regulatory bottlenecks or early signs of dysfunction, for example, can help implement preventive measures that reduce stress or damage. Thus, systems-level understanding balances scientific progress with ethical obligation (Choudhary et al., 2024).

In livestock systems, where data may be generated across large populations and diverse production environments, the scalability of multi-omics integration remains a major challenge. By focusing on modularity and abstraction, systems biology approaches address scalability by enabling models to be expanded or modified in response to new data. Since biological systems and production conditions change over time, this adaptability is crucial for long-term use. Real-world livestock contexts are better served by integrated models that can accept updated measurements or new omics layers without requiring a full reconstruction (Wadood et al., 2025).

Multi-omics integration is a step toward a comprehensive and predictive understanding of living systems within the larger framework of systems biology. Integrated analysis asks how molecular networks coordinate to produce functional outcomes and how these networks react to perturbation, as opposed to identifying the genes or pathways linked to a trait. This change has significant ramifications for sustainability, health management, and breeding in livestock science. It makes it possible to transition from reactive to anticipatory and adaptable tactics based on mechanistic knowledge (Kant et al., 2025). Omics layers contributing to systems-level biological understanding in livestock are shown in Table 13.2.

13.3 Key technologies in systems biology

Systems biology's key technologies provide the operational framework for observing, integrating, and modeling biological systems as dynamic, coordinated wholes. These technologies facilitate the shift in livestock science from fragmented measurements to a comprehensive system-level understanding, making it possible to capture biological complexity in ways that are both biologically meaningful and analytically rigorous. Systems biology technologies are characterized by their ability to facilitate integration across molecular layers, biological scales, and temporal dimensions, rather than just their ability to produce large datasets. Together, they make it possible to reconstruct biological organization, regulation, and adaptation in a way that accurately depicts how living systems function (Kant et al., 2025; Kitano, 2007).

Systems biology relies heavily on high-throughput molecular profiling technologies. Advances in molecular measurement have made it possible to characterize genetic variation, regulatory states, and downstream molecular products in great detail across a wide range of conditions and large populations. These technologies enable the observation of biological systems at a resolution that shows both context-specific and conserved patterns of regulation. Systems biology places a strong emphasis on measuring the coordinated behavior of molecular components under various conditions and over time, in addition to identifying them. By connecting molecular variation to phenotypic expression through organized

TABLE 13.2 Omics layers contributing to systems-level biological understanding in livestock.

Omics layer	Primary biological scope	Type of information captured	Systems-level interpretation	Role in multi-omics integration
Genomics	• Inherited biological blueprint	• DNA sequence variation • Structural genomic features	• Defines biological potential rather than realized function • Sets constraints on downstream regulation	• Serves as the foundational reference layer • Anchors all other omics data
Epigenomics	• Regulatory modulation of the genome	• Chromatin accessibility • DNA and histone modifications	• Encodes biological memory and plasticity • Mediates environment-genome interaction	• Links static genomics to dynamic expression layers
Transcriptomics	• Gene expression activity	• RNA abundance and isoform diversity	• Reflects regulatory decisions under specific conditions • Captures rapid system responses	• Acts as an intermediate layer between regulation and function
Proteomics	• Functional molecular machinery	• Protein abundance and modification states	• Represents executed biological programs • Integrates regulation and post-transcriptional control	• Bridges gene expression with biochemical activity
Metabolomics	• Cellular and systemic metabolic state	• Small-molecule metabolites and flux indicators	• Reflects real-time physiological condition • Sensitive to environmental and nutritional changes	• Provides functional readout of system performance
Lipidomics	• Structural and signaling lipid dynamics	• Lipid species and composition	• Captures membrane integrity and signaling processes	• Refines metabolic interpretation within systems models
Microbiomics	• Host-associated microbial ecosystems	• Microbial composition and functional potential	• Extends the biological system beyond the host genome • Influences metabolism and immunity	• Enables host-microbiome systems integration
Immunomics	• Immune system regulation	• Immune gene and protein activity	• Represents system defense and tolerance mechanisms	• Integrates health and resilience into systems models
Phenomics	• Observable organism-level traits	• Behavioral, physiological, and performance data	• Reflects emergent system outcomes • Integrates all upstream layers	• Serves as the ultimate validation layer
Physiological omics	• Organ and system function	• Hormonal and physiological indicators	• Links molecular states to organismal regulation	• Supports cross-scale integration
Interactomics	• Molecular interaction networks	• Protein-protein and regulatory interactions	• Reveals system architecture and connectivity	• Enables network-based integration
Fluxomics	• Dynamic biochemical flows	• Reaction rates and pathway utilization	• Captures system dynamics and efficiency	• Strengthens causal and dynamic modeling
Exposomics	• Environmental and management influences	• External chemical and physical exposures	• Contextualizes biological responses	• Anchors systems biology in real-world conditions

and comprehensible frameworks, this ability facilitates the inference of regulatory architectures and network dynamics that underlie complex traits in livestock (Pathak and Kim, 2024; Cao et al., 2025).

Because proteomic and metabolomic technologies capture functional states that are closer to the phenotype than genomic or transcriptomic information alone, they play a particularly significant role in systems biology. The combined effects of enzymatic activity, post-translational modification, gene regulation, and environmental influence are reflected in proteins and metabolites. These technologies are used by systems biology to close the gap between actual biological function and molecular potential. It is possible to deduce metabolic fluxes, regulatory bottlenecks, and adaptive responses that influence physiological performance by integrating proteomic and metabolomic data into integrated models. Such functional insight is crucial for a mechanistically grounded understanding of robustness, efficiency, and health in livestock systems (Mehrotra and Mendes, 2006; Sanches et al., 2024).

By capturing the mechanisms through which environmental and developmental signals modulate gene activity, technologies for characterizing epigenetic and regulatory landscapes further advance systems biology. These technologies shed light on the molecular encoding of biological memory, plasticity, and context sensitivity. Systems biology explains how similar genetic backgrounds can result in different phenotypes under various circumstances by integrating regulatory information with other omics layers. This ability is essential for comprehending developmental programming, environmental adaptation, and long-term reactions to management techniques in livestock contexts (Kant et al., 2025).

By integrating host-associated microbial communities into integrated biological models, microbiome profiling technologies have broadened the conceptual boundaries of systems biology. These technologies acknowledge that complex interactions between the host and its microbial partners, rather than just host biology, are responsible for livestock phenotypes. The microbiome is viewed by systems biology frameworks as an essential part of the biological system that affects development, immunity, and metabolism. The integration of microbial data with host molecular and phenotypic information, especially in settings where diet and management have a significant impact on microbial ecosystems, enables more thorough models of biological function and adaptation (Ye et al., 2025; Venegas et al., 2026).

Beyond molecular profiling, phenotyping technologies are essential to systems biology because they establish a connection between molecular regulation and outcomes at the organism level. Detailed observations of behavior, physiology, and performance are produced over time by automated and continuous phenotyping technologies. These measurements are crucial for verifying that deduced molecular mechanisms match observable biological function and validating systems-level models. Phenotyping technologies are valued in systems biology for their ability to capture the dynamics, transitions, and variability that characterize living systems, rather than just their volume of data. These technologies enable longitudinal analysis across environmental changes and developmental stages in livestock science, offering a rich data source for dynamic modeling (Pathak and Kim, 2024; Klingström et al., 2024).

Systems biology gains a spatial dimension from imaging technologies that is crucial to comprehending biological organization. The distribution of molecular and cellular processes within tissues and organs is revealed by spatial information, which shapes function through structural limitations and compartmentalization. Systems biology creates models that represent the spatial heterogeneity of biological systems by combining molecular and phenotypic measurements with imaging data. Spatially informed models improve knowledge of tissue development, growth, and pathological processes in livestock contexts, allowing for more precise interpretation of system behavior across scales (Hui et al., 2025).

Advanced data management and computational infrastructure are required due to the size and diversity of systems biology data. Scalable storage systems, effective data pipelines, and high-performance computing platforms are essential for managing the complexity of multi-layer biological data. Large-scale integration, iterative model improvement, and simulation-based analysis are made possible by these technologies. Robust computational infrastructure ensures that analytical capabilities can scale with data availability and support ongoing systems biology efforts in livestock systems, where data may be continuously generated across geographically dispersed operations (Mardinoglu et al., 2025).

Systems biology's analytical foundation is made up of computational and mathematical modeling tools. With the use of these technologies, empirical observations are converted into formal representations that can be examined, tested, and simulated. In order to explain how biological behavior results from interrelated processes, systems biology models place a strong emphasis on interaction, feedback, and dynamics. Such models facilitate the prediction of phenotypic outcomes under various genetic and environmental conditions in livestock science, allowing for the investigation of scenarios that would be difficult or unethical to test experimentally. These models are valuable because they can combine different kinds of data into cohesive representations that aid in comprehension and decision-making (Domínguez-Romero et al., 2024; Kitano, 2007).

Because network modeling technologies offer mathematically tractable and intuitive representations of biological interactions, they have a particularly significant impact on systems biology. The metabolic, signaling, and regulatory connections that determine the structure and operation of a system are captured by networks. Network models are used in systems biology to identify important regulators, modular structure, and emergent characteristics like redundancy and robustness. Network analysis explains how systems compensate for disruption and why some perturbations have disproportionate effects in livestock applications. Designing interventions that improve resilience while preventing unforeseen consequences can benefit from these insights (Pathak and Kim, 2024; Qiao et al., 2025).

Systems biology is further distinguished from static analytical methods by dynamic modeling technologies. Because biological systems are time-dependent by nature, it is impossible to fully comprehend their behavior without a clear depiction of temporal processes. Analysis of stability, oscillation, adaptation, and transition is made possible by dynamic models, which record changes in system state over time. Dynamic modeling is used in livestock science to better understand aging, stress reactions, recovery, and development. These models are especially useful for analyzing longitudinal data and forecasting how systems will react to persistent or recurring problems (Banga and Villaverde, 2025).

Because AI and machine learning technologies allow for flexible analysis of high-dimensional and nonlinear data, they are becoming an increasingly useful addition to traditional systems biology approaches. These tools are useful for exploratory analysis and representation learning because they are good at spotting patterns that might be challenging to define beforehand. AI techniques work best in systems biology when combined with mechanistic knowledge to create hybrid models that combine biological constraints with data-driven discovery. Such hybrid approaches allow for scalable analysis across environments and populations in livestock systems, while preserving biological relevance and interpretability (Melandri et al., 2025).

Another important area of technology in systems biology is simulation and virtual experimentation. Integrated models can be used to simulate system behavior under fictitious circumstances once they are created. Without conducting direct experiments, this capability enables the exploration of parameter space, the evaluation of intervention strategies, and critical thresholds. Virtual experimentation lowers risk and increases efficiency in livestock contexts by facilitating the evaluation of management or breeding strategies prior to their implementation (Tiukova et al., 2024).

To facilitate integration across datasets, platforms, and studies, standardization and interoperability technologies are crucial. To enable the combination and comparison of various data sources, systems biology relies on common data formats, ontologies, and metadata frameworks. Interoperability fosters cooperation and the development of cumulative knowledge in livestock science, where data is frequently produced by several stakeholders using various technologies. The promise of systems-level integration would be compromised by fragmentation and incompatibility in the absence of such standards (Doniparthi et al., 2024).

In systems biology, visualization technologies are essential because they enable human experts to understand complex models and integrated data. Effective visualization facilitates comprehension, validation, and communication of insights, and systems-level representations can be abstract and multidimensional. Visualization technologies help stakeholders understand system behavior, trade-offs, and uncertainty in livestock applications by bridging the gap between computational analysis and practical decision-making. Thus, visualization promotes transparency, confidence, and the responsible use of systems biology knowledge (Selli et al., 2024). Key technologies enabling systems biology and multi-omics analysis in livestock are listed in Table 13.3.

13.4 Data integration methods

The analytical foundation of systems biology and multi-omics research is composed of data integration techniques, which enable the conversion of heterogeneous biological data into logical, comprehensible, and predictive depictions of living systems. Since phenotypes result from the coordinated interaction of molecular, cellular, physiological, and environmental processes, data integration in livestock biology is not just a technical exercise but a conceptual requirement. Meaningful insight only emerges when these viewpoints are integrated in ways that respect biological organization, temporal dynamics, and causal structure. Each data modality captures only a partial and context-dependent view of the biological state. To maintain biological meaning while facilitating scalable and reliable analysis, systems biology approaches to data integration place more emphasis on principled synthesis rather than simple aggregation (Pathak and Kim, 2024; Morabito et al., 2025).

The understanding that biological data differ significantly in scale, resolution, and informational content serves as a fundamental guide for data integration in systems biology. While transcriptomic, proteomic, and metabolomic data are dynamic and represent changing regulatory and functional states, genomic data are essentially static and define inherited potential. This hierarchy is further expanded by phenotypic and environmental data, which capture outcomes at the organism level and outside factors that impact biological expression. In order to align data across dimensions without collapsing their unique biological roles, effective integration techniques must balance these disparities. Abstraction techniques that preserve the directional flow of biological regulation while mapping disparate measurements into common representational frameworks are necessary for this alignment (Brachmann and Jacobs, 2025; Kant et al., 2025).

TABLE 13.3 Key technologies enabling systems biology and multi-omics analysis in livestock.

Technology domain	Primary functional role	Type of biological information generated	Systems biology contribution	Integration with AI and multi-omics
High-throughput sequencing technologies	• Comprehensive molecular profiling	• Genome-wide sequence variation • Regulatory and expression landscapes	• Enables observation of system components at scale • Supports reconstruction of regulatory architecture	• Provides foundational data for multi-omics modeling • Fuels AI-driven pattern discovery
Epigenomic profiling platforms	• Characterization of regulatory modulation	• Chromatin accessibility and modification states	• Reveals plasticity and environment-responsive regulation	• Bridges static genomic data with dynamic system behavior
Proteomic technologies	• Quantification of functional molecular machinery	• Protein abundance and modification patterns	• Captures executed biological processes • Reflects post-transcriptional regulation	• Enhances functional interpretation of omics integration
Metabolomic and lipidomic technologies	• Measurement of biochemical system outputs	• Metabolite and lipid composition	• Provides a real-time snapshot of physiological state	• Anchors molecular models in functional outcomes
Microbiome profiling technologies	• Characterization of host-associated microbial systems	• Microbial composition and functional capacity	• Extends system boundaries beyond the host genome	• Enables host-microbiome systems integration
Advanced phenotyping technologies	• Continuous observation of organism-level traits	• Behavioral, physiological, and performance data	• Links molecular regulation to emergent phenotypes	• Validates and contextualizes systems models
Imaging technologies	• Spatial mapping of biological structure and function	• Tissue and cellular organization data	• Adds spatial dimension to system understanding	• Supports spatially informed integration frameworks
Physiological monitoring systems	• Real-time assessment of system regulation	• Hormonal and physiological signals	• Captures dynamic regulatory responses	• Enables temporal integration and adaptive modeling
Computational infrastructure	• Data storage and high-performance processing	• Scalable and distributed computational resources	• Supports large-scale integration and simulation	• Enables AI training and systems-level computation
Mathematical modeling frameworks	• Formal representation of biological systems	• Mechanistic and statistical models	• Translates data into predictive system behavior	• Integrates AI with biological constraints
Network modeling technologies	• Representation of biological interactions	• Regulatory, metabolic, and signaling networks	• Reveals connectivity, modularity, and robustness	• Facilitates network-based multi-omics integration
Dynamic and temporal modeling tools	• Analysis of time-dependent system behavior	• Longitudinal state transitions	• Captures adaptation and feedback processes	• Supports trajectory-based AI modeling
AI and machine learning	• Pattern discovery and representation learning	• Latent biological structures	• Enables scalable systems inference	• Integrates heterogeneous omics layers efficiently
Simulation and virtual experimentation platforms	• Exploration of hypothetical system scenarios	• Predicted system responses	• Enables safe evaluation of interventions	• Reduces reliance on physical experimentation
Data standardization and interoperability tools	• Harmonization of diverse datasets	• Shared formats and ontologies	• Enables cross-study and cross-platform integration	• Facilitates collaborative AI-enabled systems biology
Visualization and decision-support technologies	• Interpretation and communication of complexity	• System-level representations	• Enhances transparency and expert understanding	• Bridges AI outputs with human decision-making

Feature-level integration, which combines measurements from various data layers into cohesive analytical representations, is a significant class of data integration techniques. This method allows models to simultaneously learn associations across layers by treating integrated features as joint descriptors of biological state. The direct application of numerous statistical and machine learning techniques makes feature-level integration appealing. This method must be used carefully in systems biology, though, since careless feature concatenation can distort biological hierarchy and increase dimensionality. Therefore, in order to guarantee that integrated representations reflect biologically meaningful variation rather than technical artifacts, systems-oriented feature integration depends on normalization, scaling, and dimensionality reduction (Kant et al., 2025; Morabito et al., 2025).

Before integration takes place, each data modality is first converted into a latent representation that captures its fundamental structure. This is another well-known class of integration techniques. These latent representations could be abstract constructs that summarize complex data, such as pathway activity profiles or regulatory states. Systems biology approaches manage dimensionality, reduce noise, and enhance interpretability by integrating at the level of representations rather than raw features. Representation-level integration is especially useful for managing longitudinal and population-scale data in livestock systems, because it allows for consistent comparison across individuals and conditions, while accounting for missing or unevenly sampled data (Yang et al., 2024).

A unique approach known as "model-level integration" involves creating distinct analytical models for various data layers and connecting them via shared parameters, constraints, or outputs. This method allows for coordinated inference across layers while maintaining the integrity of each data modality. Since model-level integration reflects the modular structure of biological systems and enables the explicit representation of causal relationships, it is closely aligned with the principles of systems biology. Model-level integration facilitates the integration of phenotypic performance models with molecular regulation models in livestock research, allowing for the investigation of how molecular perturbations lead to organism-level outcomes (Agmon, 2024; Lavandosque et al., 2024).

Because network-based integration techniques naturally depict interactions between biological components, they are essential to systems biology. By mapping various data types onto common interaction structures, networks offer an adaptable framework for integrating them. It is possible to depict genes, proteins, metabolites, and phenotypic traits as nodes with edges that represent statistical, functional, or regulatory relationships. The identification of important regulators, modules, and pathways that mediate system behavior is made possible by network integration techniques. By demonstrating how various molecular processes converge on functional outcomes and how alternative pathways can compensate for perturbations, network-based integration in livestock systems aids in the understanding of complex traits (Al Musawi et al., 2025; Kant et al., 2025).

By explicitly modeling how relationships between data layers change over time, temporal integration techniques address the dynamic nature of biological systems. Temporal integration acknowledges the sequential nature of biological regulation and the dependence of causal interpretation on temporal ordering. Time-based methods enable the inference of feedback loops, adaptive responses, and regulatory cascades. Understanding development, stress reactions, and long-term adaptation to environmental or management changes in livestock contexts requires temporal integration. Instead of treating time as a static covariate, systems biology approaches place a strong emphasis on temporal coherence, ensuring that integrated models respect biological timing (Velten and Stegle, 2023; Khazaal and Vafaee, 2025).

By organizing data integration across biological organization levels, hierarchical integration techniques mirror the multi-scale structure of biological systems. Organism-level phenotypes are shaped by processes at the cellular and tissue levels, which are informed by molecular data. By explicitly encoding these relationships, hierarchical models maintain the causal structure while enabling information to move across scales. Hierarchical integration facilitates the use of intermediate regulatory and physiological layers in livestock systems to connect genetic variation to phenotypic performance. This method improves predictive robustness and interpretability, especially when extrapolating beyond observed data (Vasseur et al., 2022; Pathak and Kim, 2024).

Because probabilistic integration techniques explicitly capture the uncertainty and variability present in biological data, they are becoming increasingly significant in systems biology. Deterministic modeling is insufficient to remove the ambiguity caused by measurement error, biological heterogeneity, and incomplete observability. Uncertainty can spread across data layers using probabilistic integration frameworks, resulting in integrated conclusions with confidence estimates. Uncertainty-aware integration helps livestock applications make risk-aware decisions and avoids misinterpreting noisy or incomplete data. This viewpoint is consistent with the practical and ethical realities of biological systems, where choices are frequently made in the face of uncertainty (Balsa-Canto et al., 2025).

Known biological concepts, such as stoichiometric balance, regulatory logic, or physical limitations, are incorporated into the integration process by constraint-based integration techniques. These techniques reduce the solution space and direct models toward biologically plausible configurations by imposing constraints derived from prior knowledge.

Constraint-based integration is used by systems biology to balance established knowledge of biological processes with data-driven inference. Such limitations are useful in livestock research because they allow metabolic, physiological, and performance data to be integrated in ways that represent homeostatic regulation, energy balance, and resource allocation (Ahmed Sidi et al., 2025).

Because machine learning-based integration techniques can handle heterogeneous and high-dimensional data, they are becoming increasingly popular. These techniques do not require explicit interaction specification to learn intricate relationships across data layers. Machine learning works best in systems biology when it is integrated into frameworks that take biological structure and interpretability into account. Hybrid methods allow for scalable integration while preserving alignment with biological knowledge by combining mechanistic constraints with data-driven learning. These methods facilitate population-scale analysis and adaptation to various production environments in livestock systems (Procopio et al., 2023; Noordijk et al., 2024).

A particular class of machine learning integration techniques known as "multi-view learning" views each data modality as a separate but connected perspective of the underlying biological system. These techniques seek to learn shared representations that preserve modality-specific information while capturing common structure across views. Because multi-view integration acknowledges the complementary nature of various data layers, it is closely aligned with systems biology principles. Multi-view learning supports comprehensive interpretation of system behavior in livestock applications by facilitating the integration of molecular, phenotypic, and environmental data (Sun et al., 2024).

Practical issues with data imbalance and sparsity must also be addressed by integration techniques. In livestock systems, sampling intensity may differ between populations and over time, and not all data layers are available for every individual. As a result, systems biology integration techniques place a strong emphasis on adaptability and resilience, allowing inference even in situations where certain modalities are absent or poorly represented. Methods that make use of shared structure among individuals and data layers lessen the need for full datasets and enable scalable use in practical settings (Henao et al., 2023).

In systems biology, evaluation and validation are essential elements of data integration. It is necessary to evaluate integrated models for biological plausibility, robustness, and generalizability, in addition to predictive accuracy. Integrated inferences are frequently validated by contrasting them with independent data, established biological relationships, or observed system behavior under perturbation. Validation is especially crucial in livestock research, since integrated models can influence choices that have long-term effects on sustainability, genetic diversity, and animal welfare. Therefore, systems biology places a strong emphasis on clear and understandable integration techniques that facilitate thorough assessment (Kitano, 2007; Morabito et al., 2025).

Data integration strategies are also influenced by ethical considerations. Which facets of biology are highlighted and how trade-offs are portrayed depend on integration decisions. Techniques that give priority to particular data layers or goals may subtly influence results in ways that impact sustainability or welfare. Integrative frameworks that balance several goals and explicitly state underlying assumptions are encouraged by systems biology. In livestock systems, this openness promotes ethical oversight and well-informed decision-making (Ducrot et al., 2024). Data integration strategies and methodological paradigms in systems biology are presented in Table 13.4.

13.5 Applications of integrated systems biology in livestock multi-omics

A strong and cohesive framework for using multi-omics data toward thorough comprehension and management of livestock biological systems is provided by integrated systems biology. Systems biology enables applications that go beyond descriptive analysis to include prediction, interpretation, and well-informed intervention by combining data from molecular, cellular, physiological, and phenotypic layers. These applications are especially important in livestock science because they deal with intricate characteristics and processes that cannot be sufficiently managed or explained by single-layer analyses. Thus, integrated systems biology aligns biological knowledge with the objectives of productivity, resilience, welfare, and sustainability by acting as a link between molecular insight and useful results (Gutierrez Reyes et al., 2024; Pathak and Kim, 2024).

Clarifying the biological architecture of complex traits is one of the most fundamental uses of integrated systems biology in livestock multi-omics. Performance efficiency, health resilience, stress tolerance, and successful reproduction are examples of traits that result from coordinated activity across several molecular layers, rather than from discrete genetic effects. Systems biology reconstructs the networks and pathways that control these characteristics by integrating genomic, regulatory, functional, and phenotypic data. This integrative approach makes it possible to identify regulatory structures that explain why some molecular state combinations consistently result in positive outcomes, while others put animals at risk of dysfunction. Integrated approaches go beyond cataloging associations to reveal mechanistic coherence within biological complexity by concentrating on system-level organization (NASIR et al., 2024; Venegas et al., 2026).

TABLE 13.4 Data integration strategies and methodological paradigms in systems biology.

Integration strategy	Conceptual basis	Primary data handling approach	Strengths in systems biology	Limitations and considerations
Feature-level integration	• Direct combination of measurements across data layers	• Concatenation of multi-omics features • Unified feature matrices	• Enables joint modeling across omics layers • Compatible with many statistical and AI methods	• High dimensionality and redundancy • Risk of obscuring biological hierarchy
Representation-level integration	• Abstraction of each modality into latent representations	• Dimensionality reduction before integration • Shared latent spaces	• Reduces noise and complexity • Enhances interpretability and robustness	• Latent features may be difficult to map to biology • Requires careful validation
Model-level integration	• Modular treatment of biological layers	• Separate models linked via shared parameters or outputs	• Preserves the biological structure of each omics layer • Supports causal reasoning	• Increased model complexity • Higher computational demands
Network-based integration	• Biology represented as interacting networks	• Mapping omics data onto interaction graphs	• Captures connectivity and modularity • Reveals emergent system properties	• Network inference sensitive to data quality • Interpretation depends on prior knowledge
Hierarchical integration	• Explicit multi-scale biological organization	• Nested models across molecular to phenotypic levels	• Reflects biological causality and scale • Improves mechanistic interpretability	• Model specification can be complex • Requires extensive biological insight
Temporal integration	• Emphasis on system dynamics over time	• Longitudinal alignment and trajectory modeling	• Enables inference of regulatory cascades • Distinguishes cause from consequence	• Requires dense time-series data • Sensitive to sampling irregularity
Probabilistic integration	• Explicit modeling of uncertainty	• Bayesian and stochastic frameworks	• Propagates uncertainty across layers • Supports risk-aware inference	• Computationally intensive • Interpretation may be less intuitive
Constraint-based integration	• Use of known biological laws and limits	• Application of stoichiometric or regulatory constraints	• Ensures biological plausibility • Reduces solution space	• Limited by the completeness of prior knowledge
Multi-view learning	• Each omics layer as a complementary system view	• Joint learning of shared and modality-specific features	• Preserves complementary information • Improves integration across heterogeneous data	• Model complexity increases with views
Machine learning-driven integration	• Data-driven pattern discovery	• Deep learning and nonlinear modeling	• Handles high-dimensional and complex relationships	• Risk of black-box inference • Requires interpretability safeguards
Hybrid mechanistic-AI integration	• Synergy between biological theory and data-driven learning	• AI models constrained by mechanistic rules	• Balances flexibility and interpretability • Enhances generalizability	• Development and validation are resource-intensive

Context-aware integration	• Biological meaning depends on the environment and state	• Conditioning models on context variables	• Improves relevance and robustness • Reduces spurious associations	• Requires comprehensive contextual data
Cross-population integration	• Learning across diverse biological systems	• Transfer learning and domain adaptation	• Enhances generalizability • Supports scalable deployment	• Risk of negative transfer
Incremental and adaptive integration	• Systems evolve over time	• Continual model updating	• Maintains long-term relevance	• Governance and version control challenges
Federated integration	• Distributed data governance	• Decentralized learning without data sharing	• Preserves data ownership and privacy	• Increased coordination complexity

Enhancing the biological interpretability of multi-omics data is another important application. High-throughput technologies produce detailed molecular profiles, but because of the dimensionality and interdependence of signals, their interpretation is frequently still difficult. By placing molecular variation within functional networks and phenotypic outcomes, integrated systems biology tackles this problem. Systems-level analysis makes the biological significance of multi-omics patterns clear by connecting molecular changes to physiological processes and observable traits. This interpretive ability reduces the need for proxy indicators or simplistic trait definitions in livestock systems and promotes more significant biological inference (Brachmann and Jacobs, 2025; Li et al., 2024).

Predictive modeling in livestock science is also greatly advanced by integrated systems biology. The goal of predictive applications is to foresee biological outcomes, such as health risks, performance trajectories, or adaptability under different circumstances. By combining complementary data from several biological layers and capturing both potential and actual system states, multi-omics integration improves prediction accuracy. By incorporating causal structure and mechanistic constraints, systems biology frameworks enhance prediction even further by enabling models to extrapolate beyond observed data. In livestock contexts, where robust prediction under uncertainty is crucial and experimental testing of all scenarios is impractical, this capability is especially valuable (Lu et al., 2024; Qiao et al., 2025).

Integrated systems biology is particularly useful for applications in biological robustness and resilience. Robustness is the stability of function under changing circumstances, whereas resilience is the capacity of biological systems to preserve or restore function in the face of disruption. Rather than coming from individual components, these characteristics are the result of network architecture, redundancy, and adaptive regulation. Integrated multi-omics analysis demonstrates how alternative pathways compensate for disruption and how regulatory networks reorganize under stress. A systems-level understanding of resilience in livestock systems enables management approaches that prioritize long-term stability and welfare over just short-term optimization (Proverbio et al., 2024; Bullinger et al., 2024).

Applications in decision support and precision management are also supported by integrated systems biology. Systems-level models provide a complex picture of an animal's condition that can guide management decisions by integrating molecular indicators with phenotypic and environmental data. These models reduce unnecessary intervention while facilitating a prompt response when necessary by enabling the distinction between transient variation and significant changes in biological condition. Multi-omics data integration improves the sensitivity and specificity, allowing management choices that are more in line with underlying biology. This strategy respects biological limitations and welfare concerns while increasing efficiency (Pathak and Kim, 2024; Vlaicu et al., 2024).

The improvement of breeding and selection techniques is another crucial application area. Conventional methods of selection frequently depend on aggregated phenotypic measurements that mask underlying biological processes. Selection decisions can be influenced by system-level characteristics, such as metabolic efficiency, regulatory stability, and adaptive capacity, thanks to integrated systems biology. Systems-based methods enable earlier and more accurate evaluation of genetic potential by capturing intermediate biological states that connect genotype to phenotype through the integration of multi-omics data. Breeding strategies that strike a balance between productivity, resilience, and long-term sustainability are supported by this integration (Montesinos-López et al., 2025; Ghavi Hossein-Zadeh, 2026).

A key component of integrated systems biology in livestock multi-omics is health-related applications. The intricate relationships between immune regulation, metabolism, microbiome composition, and environmental exposure control disease susceptibility and recovery. These layers are combined by systems biology to describe health as a dynamic state, as opposed to a binary condition. This viewpoint makes it possible to identify early departures from homeostasis and gain a better understanding of how biological systems react to stress. Preventative and adaptive health management strategies that reduce dependency on reactive interventions and improve overall system stability are supported by integrated analysis (Choudhary et al., 2024; Pathak and Kim, 2024).

The importance of systems-level integration is further demonstrated by applications in life-course analysis and developmental biology. Coordinated regulation across molecular layers takes place over long periods of time during developmental processes. Reconstructing developmental trajectories and identifying crucial regulatory transitions that influence resilience and performance throughout life are made possible by integrated systems biology. Understanding developmental programming in livestock contexts enables tactics that maximize early-life conditions in ways that produce long-lasting advantages throughout the production lifespan. These applications highlight the significance of longitudinal multi-omics integration and the temporal aspect of systems biology (NASIR et al., 2024; Kant et al., 2025).

By making it possible to assess trade-offs between biological and environmental aspects, integrated systems biology also supports applications that are focused on sustainability. Livestock systems must balance biological efficiency with resource consumption and environmental impact because they operate within ecological constraints. Investigating how molecular and physiological regulation affects waste production, efficiency, and resource allocation is made possible by systems-level integration. Integrated approaches facilitate decision-making that balances environmental stewardship with biological performance by modeling these relationships. The wider role of systems biology in promoting ethical and sustainable livestock production is reflected in this application (Herzon et al., 2024; Halpern et al., 2025).

Improving adaptability to environmental variability is another application area. Livestock systems must have biological systems that can adapt to changing climatic and management conditions. Integrated systems biology shows how regulatory networks facilitate adaptation and how multi-omics profiles change in response to environmental change. This knowledge facilitates the identification of biological configurations linked to robustness and flexibility, guiding breeding and management approaches meant to improve adaptability. These applications are especially pertinent in light of changing production challenges and global change (Shashank et al., 2024).

Additionally, knowledge transfer and generalization across populations and contexts are made easier by integrated systems biology. Systems-based models are more likely to generalize beyond particular datasets because they focus on underlying system organization rather than surface-level associations. By capturing conserved regulatory principles that function across genetic backgrounds and environments, multi-omics integration supports the wider applicability of findings. This generalizability increases the return on research investments and facilitates the application of knowledge across various production systems in livestock science (Wang et al., 2025).

Applications for decision support are a useful extension of integrated systems biology. Complex data is synthesized into structured representations by systems-level models, which can inform strategic decisions. By offering biologically based insights into system behavior and possible outcomes, these applications supplement human judgment rather than replace it. Systems biology helps make decisions based on a thorough understanding of the biological state and trajectory by combining multi-omics data with phenotypic and environmental information. This position is in line with the growing focus on the ethical and open application of AI in livestock systems (Ghavi Hossein-Zadeh, 2025; Tedeschi et al., 2025).

Applications of integrated systems biology are intricately linked to ethical issues. Long-term system sustainability, genetic diversity, and animal welfare are all impacted by decisions made using systems-level analysis. By modeling trade-offs and interactions instead of optimizing discrete goals, integrated approaches allow for the explicit consideration of these dimensions. This ability ensures that technological advancement is in line with societal values by promoting ethical reflection and oversight. Therefore, systems biology advances both science and the ethical management of livestock systems (Pathak and Kim, 2024; Richter et al., 2024).

In general, the development of livestock science as an anticipatory and predictive field is aided by integrated systems biology. Systems-level applications allow for the investigation of possible futures and the assessment of alternative approaches before implementation, as opposed to responding to observed results. By offering comprehensive and mechanistically informed representations of biological systems, multi-omics integration enhances this capacity for anticipation. In order to successfully navigate the intricate and unpredictable challenges that livestock production will face in the upcoming decades, this forward-looking orientation is crucial. Major application domains of integrated systems biology in livestock multi-omics are outlined in Table 13.5.

13.6 Challenges in systems biology and multi-omics integration

The intrinsic complexity of biological systems, the technical constraints of data generation, and the conceptual and operational requirements of integration across scales and domains present a number of significant obstacles to systems biology and multi-omics integration, despite its transformative potential. These difficulties are made more challenging in livestock science by the variety of production settings, moral restrictions, and the need for models that are both practically useful and biologically sound. In order to ensure that systems biology develops as a solid and responsible framework rather than continuing to be an aspirational ideal limited by unresolved limitations, these challenges must be addressed (Kant et al., 2025; Kitano, 2007).

The intrinsic complexity of biological systems is one of the most fundamental problems in systems biology and multi-omics integration. Multilayered regulatory architectures with feedback loops, redundancy, nonlinearity, and context-dependent behavior are characteristics of livestock organisms. Although these characteristics increase robustness and adaptability, modeling and interpretation are more difficult. Simple cause-and-effect relationships are frequently absent from biological processes, and depending on the state of the system, minor perturbations may have delayed or disproportionate effects. Although systems biology aims to embrace this complexity, models that are either oversimplified or unmanageable can result from the sheer dimensionality and interconnectedness of biological data overwhelming analytical frameworks (Kant et al., 2025; Morabito et al., 2025).

One major and ongoing issue is data heterogeneity. The scale, resolution, noise properties, and temporal dynamics of multi-omics datasets vary greatly. While transcriptomic, proteomic, and metabolomic data are dynamic and susceptible to changing circumstances, genomic data are comparatively stable and sparse. Additional layers of contextual dependence

TABLE 13.5 Major application domains of integrated systems biology in livestock multi-omics.

Application domain	Systems-level objective	Integrated biological layers involved	Analytical focus	Value added by systems biology
Complex trait architecture	• Understand emergent biological traits	• Genomics • Epigenomics • Transcriptomics • Proteomics	• Network reconstruction • Interaction mapping	• Reveals regulatory coherence beyond single-gene effects
Predictive phenotype modeling	• Anticipate future biological outcomes	• Multi-omics • Phenomics • Environmental data	• Trajectory and state prediction	• Improves generalizability and robustness
Resilience and robustness analysis	• Identify system stability mechanisms	• Regulatory • Metabolic • Microbiomics	• Stress-response network analysis	• Shifts focus from optimization to adaptability
Precision management support	• Inform targeted biological interventions	• Molecular indicators • Physiological signals	• Context-aware decision support	• Reduces unnecessary or harmful interventions
Breeding and selection strategies	• Select for system-level performance	• Genomics • Regulatory omics • Phenomics	• Integration of intermediate biological states	• Balances productivity with long-term sustainability
Health state characterization	• Define health as a dynamic system state	• Immunomics • Metabolomics • Microbiomics	• Homeostasis and deviation detection	• Enables early and preventative health management
Developmental systems analysis	• Understand life-course biological regulation	• Epigenomics • Transcriptomics • Physiology	• Trajectory reconstruction	• Identifies critical regulatory transitions
Environmental adaptation modeling	• Assess biological response to external change	• Omics • Environmental data	• Plasticity and response mapping	• Enhances adaptability to climate and management variation
Sustainability-oriented evaluation	• Balance biological and ecological outcomes	• Metabolomics • Phenomics • Environmental indicators	• Resource-use efficiency modeling	• Aligns productivity with environmental stewardship
Host-microbiome systems analysis	• Integrate microbial contributions to phenotype	• Microbiomics • Metabolomics • Immunomics	• Cross-system interaction modeling	• Expands biological system boundaries
Decision-support systems	• Translate complex data into actionable insight	• Integrated multi-omics • Phenotypic data	• Scenario evaluation and trade-off analysis	• Enhances transparency and accountability
Longitudinal monitoring and surveillance	• Track system evolution over time	• Time-series omics • Phenomics	• Change-point and trend detection	• Supports proactive system management
Knowledge generalization and transfer	• Enable cross-population learning	• Multi-population omics	• Conserved pattern identification	• Improves scalability and applicability
Ethical and welfare-informed analysis	• Incorporate welfare into system objectives	• Behavioral phenomics • Physiological indicators	• Trade-off and constraint evaluation	• Supports responsible innovation
Virtual experimentation and simulation	• Explore hypothetical biological scenarios	• Integrated systems models	• Predictive simulation	• Reduces reliance on physical experimentation

and variability are added by phenotypic and environmental data. Reconciling incompatible data structures and measurement paradigms is necessary to integrate such disparate data types into a cohesive framework. Integration runs the risk of confusing biologically significant variation with technical artifacts unless careful alignment and abstraction are used, which compromises interpretability and dependability (Velten and Stegle, 2023; Baião et al., 2025).

Reproducibility and data quality present additional difficulties. Batch effects, measurement error, and platform-specific biases can affect high-throughput omics technologies. When data is produced across various labs, time periods, or production systems, these problems are exacerbated. Environmental variability makes data consistency even more challenging in livestock research, where controlled experimental conditions are frequently impractical. Instead of capturing the true biological signal, systems biology models based on such data may unintentionally capture noise or confounding factors. Therefore, a significant barrier to dependable integration and downstream application is ensuring data quality and reproducibility across omics layers (Yu et al., 2024; Morabito et al., 2025).

Another major issue is uneven and incomplete data coverage. It is rarely feasible to gather all omics layers for every individual at every time point in real-world livestock systems. Missing data is frequently not random and can result from logistical limitations, financial concerns, or technical malfunctions. If not handled properly, these gaps can skew inference and make integration more difficult. As a result, systems biology approaches have to deal with partial observability by creating models that can propagate uncertainty and infer latent states without exaggerating confidence. This requirement raises the bar for both computational power and methodological sophistication (Flores et al., 2023).

Additional challenges include dynamic consistency and temporal alignment. Respecting the temporal ordering of molecular and phenotypic events is essential for meaningful integration in biological regulation, which occurs over several timescales. Alignment is difficult, however, because omics data is frequently gathered at various frequencies and times. Static integration techniques that disregard temporal structure run the risk of obscuring dynamic processes and misinterpreting causal relationships. In systems biology, creating integration frameworks that reliably capture temporal dependencies while maintaining scalability and interpretability remains a challenge (Snead and Clark, 2022; Morabito et al., 2025).

Finding a balance between mechanistic understanding and data-driven discovery is another significant challenge. While statistical and machine learning techniques are excellent at finding patterns in high-dimensional data, they may produce models that are not biologically interpretable. On the other hand, the size and diversity of contemporary omics data may be too much for mechanistic models based on biological theory. Systems biology aims to balance these methods, but it is difficult to achieve this balance. Overly constrained mechanistic models may overlook emergent patterns in the data, while an over-reliance on purely data-driven approaches runs the risk of producing models that perform well in particular datasets but are unable to generalize or offer useful insights (Procopio et al., 2023; Balsa-Canto et al., 2025).

In livestock applications, where decisions based on systems biology may have long-term effects on animal welfare, genetic diversity, and sustainability, model interpretability and transparency are especially important issues. Even experts may find it challenging to understand complex integrated models, which raises questions about trust and accountability. Stakeholders may be reluctant to rely on systems-level insights or may abuse them without understanding the underlying assumptions and limitations if they do not have a clear understanding of how predictions are generated. Therefore, responsible application requires the development of integration techniques that strike a balance between complexity and interpretability (White, 2025).

Another significant barrier is scalability and computational load. Large datasets and intricate models used in multi-omics integration require a significant amount of processing power. Scalability becomes a practical limitation in livestock systems, where data may be generated continuously across large populations and dispersed locations. Accessibility may be restricted by high computational costs, leading to differences between well-resourced research environments and more general industry contexts. In order to address scalability, careful prioritization of analytical goals and effective model design are just as important as technological innovation (Tulpan, 2024).

The problem of generalizability across environments and populations remains ongoing. Livestock systems differ significantly in terms of genetics, environmental factors, and management techniques. If integrated models created in one context capture context-specific patterns rather than fundamental biological principles, they may not function consistently in another. The goal of systems biology is to identify emergent properties and conserved regulatory architectures that can be applied to a variety of situations, but doing so requires meticulous validation and iterative improvement. The practical impact of systems-level insights may be limited if generalizability is neglected (Pathak and Kim, 2024; Kazanski et al., 2025).

Technical challenges in systems biology and multi-omics integration are closely linked to ethical and governance issues. Concerns regarding data ownership, consent, and fair benefit distribution are raised by integrated models, which concentrate biological knowledge and decision-making authority. Concerns regarding animal welfare, the impact on the environment, and public trust are all intertwined in livestock contexts. Systems biology methods that prioritize particular goals or data layers may subtly influence results in ways that are significant from an ethical standpoint. Therefore, a crucial challenge that goes beyond technical considerations is ensuring that integration methods are transparent, inclusive, and aligned with ethical principles (Liao et al., 2024; Kant et al., 2025).

Validation and benchmarking of integrated systems models present another difficulty. Systems biology models, in contrast to single-layer analyses, frequently seek to capture dynamic processes and emergent behavior that are challenging to directly validate. Experimental validation may be expensive or impractical, and traditional performance metrics might not be adequate to evaluate the model's credibility. Establishing confidence in integrated models requires developing suitable validation frameworks that integrate robustness assessment, biological plausibility, and predictive performance. Rigorous validation is especially crucial in livestock science, where interventions based on model outputs may have irreversible effects (Tatka et al., 2023; Heerden et al., 2024).

The problem of model drift arises because biological systems are dynamic. The relationships captured by integrated models may change over time due to changes in environmental conditions, management strategies, and genetic composition. Static models run the risk of becoming outdated, which could result in poor performance and false conclusions. In order to enable models to update in response to new data while maintaining accumulated knowledge, systems biology approaches must include mechanisms for continuous learning and adaptation. Maintaining this equilibrium between stability and flexibility is a difficult task with implications for governance and technology (Pisupati and Niv, 2022; Davis et al., 2024).

Another major challenge is interdisciplinary coordination. At the nexus of biology, mathematics, computer science, and ethics are systems biology and multi-omics integration. Collaboration between specialists with various viewpoints, approaches, and priorities is necessary for successful integration. Communication can be hindered and progress slowed by differences in terminology, presumptions, and evaluation standards. Transforming systems-level insights into useful applications in livestock science requires bridging these disciplinary divides. It is still difficult to create training programs and institutional frameworks that facilitate interdisciplinary cooperation (Pathak and Kim, 2024; Kant et al., 2025).

In general, there is a more general conceptual issue concerning scope and expectations. Multi-omics integration and systems biology are frequently offered as all-encompassing answers to challenging biological problems. These methods are limited in practice by uncertainty, model assumptions, and data quality. Underestimating their limitations could result in an unsuitable reliance on incomplete models, while overestimating their capabilities runs the risk of disappointment and misuse. Therefore, a realistic and thoughtful approach is crucial, acknowledging that systems biology is a developing framework that enhances rather than replaces other biological research methods. Key challenges and limitations in systems biology and multi-omics integration are presented in Table 13.6.

13.7 Future perspectives and directions

The field of systems biology and multi-omics integration has evolved from an exploratory and primarily descriptive endeavor to a predictive, adaptive, and ethically grounded scientific framework, as evidenced by future perspectives and directions. The convergence of increasing data accessibility, developing computational techniques, and rising public expectations for sustainability, animal welfare, and responsible innovation is driving this evolution in livestock science. Technological advancements alone will not determine systems biology's future; rather, it will depend on how well these advancements are incorporated into conceptual models that respect biological complexity and provide useful insights (Kant et al., 2025; Yamanouchi et al., 2026).

The expansion of biological integration across domains and scales is a key future path. Future frameworks will more thoroughly integrate molecular regulation with cellular organization, tissue function, whole-animal physiology, behavior, and environmental interaction, even though current systems biology approaches already incorporate multiple omics layers. Given that biological processes spread in both directions across organizational levels, this multiscale integration will place an increasing emphasis on continuity rather than compartmentalization. Such integration will facilitate a deeper understanding of how genetic and molecular variation affects phenotypic performance, resilience, and adaptation in livestock systems (Karim et al., 2024; Pathak and Kim, 2024).

Future systems biology frameworks will increasingly emphasize temporal intelligence. Despite the inherent dynamic nature of biological systems, many contemporary integration techniques still mainly rely on static or snapshot-based representations. Models that treat time as a primary organizing dimension and capture trajectories, transitions, and long-term adaptations rather than isolated states will be given priority in future directions. Reconstructing developmental pathways, stress responses, recovery dynamics, and aging processes will be made possible by longitudinal multi-omics integration. This temporal emphasis will facilitate anticipatory rather than reactive decision-making in livestock

TABLE 13.6 Key challenges and limitations in systems biology and multi-omics integration.

Challenge category	Nature of the challenge	Underlying causes	Implications for systems-level analysis	Conceptual and methodological considerations
Biological complexity	• Emergent and nonlinear system behavior	• Feedback loops • Redundancy and pleiotropy	• Difficulty isolating causal mechanisms • Model oversimplification risk	• Embrace network and dynamic modeling rather than linear inference
Data heterogeneity	• Incompatible data structures across omics	• Diverse measurement platforms • Variable resolution and scale	• Integration artifacts • Loss of biological meaning	• Abstraction and representation-based integration
Data quality and noise	• Measurement error and technical bias	• Batch effects • Environmental variability	• Reduced reliability and reproducibility	• Rigorous normalization and quality control
Incomplete data coverage	• Missing omics layers or time points	• Cost and logistical constraints	• Biased inference and uncertainty	• Probabilistic modeling and latent state inference
Temporal misalignment	• Inconsistent sampling across modalities	• Asynchronous data collection	• Misinterpretation of causal relationships	• Time-aware integration and dynamic models
Scalability constraints	• High computational and storage demands	• Large datasets and complex models	• Limited accessibility and deployment	• Efficient algorithms and distributed computing
Model interpretability	• Opacity of complex integrated models	• High-dimensional latent representations	• Reduced trust and accountability	• Explainable and transparent modeling frameworks
Correlation-causation confounding	• Reliance on associative patterns	• Observational data dominance	• Ineffective or harmful interventions	• Integration of causal inference principles
Generalizability across contexts	• Context-specific model performance	• Population and environment variability	• Limited transferability of insights	• Domain adaptation and cross-context validation
Validation and benchmarking	• Difficulty evaluating system-level models	• Lack of ground truth for emergent properties	• Reduced scientific credibility	• Multi-criteria and longitudinal validation
Model drift over time	• Degradation of model relevance	• Biological and environmental change	• Outdated predictions	• Continual and adaptive learning strategies
Ethical and welfare risks	• Potential misalignment with welfare priorities	• Narrow optimization objectives	• Societal and ethical concerns	• Ethics-by-design integration
Data governance and ownership	• Unclear rights over integrated datasets	• Multiple stakeholders	• Barriers to collaboration	• Transparent governance and federated learning
Interdisciplinary barriers	• Communication gaps across domains	• Differing conceptual frameworks	• Fragmented integration efforts	• Cross-disciplinary training and collaboration
Overinterpretation of models	• Inflated expectations of predictive power	• Black-box inference	• Misuse of systems-level outputs	• Explicit uncertainty communication

contexts, allowing interventions to be timed to biological windows of opportunity rather than being applied after dysfunction has already manifested (Banga and Villaverde, 2025; Khazaal and Vafaee, 2025).

The incorporation of causal reasoning into systems biology and multi-omics analysis is another important future path. Although correlation-based models have proven useful in identifying patterns, their shortcomings in directing intervention are becoming more widely acknowledged. To facilitate reasoning about how perturbations spread through biological networks and how different interventions might result in different outcomes, future systems biology frameworks will more explicitly incorporate causal structure. This move toward causality will improve the safety and reliability of systems-level forecasts, especially in livestock systems where experimental manipulation is limited, and intervention effects may be long-lasting (Hockings et al., 2023; Lobentanzer et al., 2024).

AI developments will continue to transform systems biology, but they will prioritize synergy over substitution in the future. Increasingly, mechanistic and constraint-based models will be combined with machine learning and deep learning techniques to create hybrid frameworks that strike a balance between interpretability and flexibility. These methods will allow for the scalable analysis of intricate multi-omics data while maintaining conformity to established biological principles. Such hybrid systems will enable generalizable insights across populations and environments in livestock science, resolving present issues with robustness and model transferability (Noordijk et al., 2024).

Another significant future path is the creation of shared representations and foundation models. It will be feasible to train generalized models that capture basic patterns of biological regulation across contexts as multi-omics datasets increase in size and diversity. With comparatively little additional data, these models can then be tailored to particular populations, characteristics, or environments. Foundation models have the potential to improve cross-system learning, reduce data requirements, and expedite the application of systems biology insights in various production environments in livestock systems (Li et al., 2023).

Future frameworks for systems biology will also prioritize robustness and uncertainty quantification. Integrated models must explicitly reflect the inherent noise and incompleteness of biological data, rather than hiding it behind deterministic predictions. Future methods will spread uncertainty across data scales and layers, allowing users to evaluate prediction confidence and make risk-aware choices. Such transparency will be crucial for responsible application and stakeholder trust in livestock contexts, where decisions frequently involve trade-offs under uncertainty (Balsa-Canto et al., 2025).

Future systems biology will increasingly be characterized by ethical integration. Ethical issues must be directly incorporated into analytical frameworks, rather than being addressed after the fact, as integrated models become more influential in breeding, management, and policy decisions. Future developments will include ethics-by-design strategies that integrate equity, sustainability, and welfare constraints into model goals and assessment standards. As a result, systems biology will develop as both a scientific framework and a tool for governance, facilitating the alignment of societal values with technological capability (Saranya et al., 2024; Paleri and Hens, 2025).

Another crucial future path is modeling that is focused on sustainability. Future systems biology techniques will increasingly incorporate environmental data and ecosystem-level considerations into biological models, as livestock systems function within ecological boundaries. Evaluation of trade-offs between productivity, resource consumption, and environmental impact will be made possible by this integration, assisting in decision-making that strikes a balance between several goals over extended periods of time. Systems biology can support more comprehensive evaluations of sustainability in livestock production by connecting molecular and physiological regulation to environmental outcomes (Pathak and Kim, 2024; Kumar et al., 2025).

The development of data infrastructure and interoperability will also influence systems biology in the future. Scalable and standardized data ecosystems will be necessary for integration and reuse as multi-omics and phenotypic data continue to accumulate. Interoperable data standards, shared ontologies, and federated analysis frameworks that facilitate cooperation without jeopardizing data ownership or privacy are examples of future directions. Such infrastructure will respect stakeholder interests while promoting cross-institutional learning and accelerating scientific advancement in livestock systems (Liao et al., 2024; Morabito et al., 2025).

To achieve future goals in systems biology, education and capacity building will be essential. Systems biology is an interdisciplinary field that necessitates knowledge of biology, data science, mathematics, and ethics. Thus, training initiatives and cooperative settings that promote systems thinking and interdisciplinary fluency are future directions. Building such capacity will be crucial in livestock science to ensure that a deeper biological understanding, rather than just technical considerations, drives innovation and translates systems-level insights into practice (Didier et al., 2025).

The co-evolution of real-world systems and systems biology models is another new viewpoint. Future frameworks will prioritize continuous learning, where models are updated as new data becomes available and as biological systems change, as opposed to treating models as static representations. Because livestock systems are dynamic, this adaptive modeling

TABLE 13.7 Emerging directions and conceptual shifts shaping the future of systems biology in livestock.

Emerging direction	Conceptual shift	New analytical capabilities	Impact on biological understanding	Implications for livestock systems
From reductionism to holistic modeling	• *Focus on system behavior rather than isolated components*	• *Integrated multi-layer modeling*	• *Captures emergent biological properties*	• *More realistic interpretation of complex traits*
Temporal-centric systems biology	• Phenotype viewed as a dynamic trajectory	• Longitudinal and sequence-based modeling	• Improved understanding of development and adaptation	• Enables proactive management strategies
Causal and mechanistic inference	• Shift from association to causation	• Counterfactual and causal modeling frameworks	• Increased reliability of system-level predictions	• Safer and more effective interventions
Hybrid AI-mechanistic models	• Integration of data-driven and theory-driven approaches	• AI constrained by biological rules	• Balances flexibility with interpretability	• Scalable yet biologically grounded models
Foundation models and transfer learning	• Learning shared biological representations	• Pretrained systems-level models	• Generalizable insight across contexts	• Reduced data requirements
Uncertainty-aware systems biology	• Explicit representation of ambiguity	• Probabilistic and Bayesian frameworks	• Prevents overconfident inference	• Risk-aware decision-making
Ethics-by-design modeling	• Ethical principles embedded in system objectives	• Welfare- and sustainability-aware optimization	• Aligns biological insight with societal values	• Increased trust and legitimacy
Sustainability-integrated systems biology	• Inclusion of ecological constraints	• Multi-objective systems evaluation	• Holistic assessment of biological impact	• Balanced productivity and environmental stewardship
Host-microbiome co-system modeling	• Expansion of system boundaries	• Integrated host-microbial network models	• Improved understanding of system complexity	• Enhanced resilience and efficiency
Adaptive and continual learning systems	• Models evolve with biological systems	• Incremental updating frameworks	• Maintains long-term relevance	• Sustained applicability in changing environments
Federated and privacy-preserving integration	• Distributed collaboration without data centralization	• Secure decentralized learning	• Broader data participation	• Enhanced stakeholder engagement
Multi-scale digital twins	• Virtual representation of biological systems	• Simulation of hypothetical scenarios	• Deeper mechanistic insight	• Reduced reliance on physical experimentation
Human-AI co-decision frameworks	• Collaboration rather than automation	• Interactive decision-support systems	• Preserves expert judgment	• Responsible technology adoption
Standardization and interoperability evolution	• Harmonized data ecosystems	• Shared ontologies and formats	• Improved integration and reuse	• Accelerated innovation
Resilience-focused biological evaluation	• Emphasis on stability and adaptability	• Network robustness analysis	• Identification of adaptive architectures	• Improved long-term system sustainability

approach allows for long-term relevance. Future systems biology frameworks must specifically address the governance, validation, and version control issues that this approach raises (Pessoa, 2024).

Additionally, human-system interaction will become a more significant area of study. Effective interfaces are required to ensure that insights are comprehensible and useful as systems biology models become more sophisticated and significant. In the future, decision-support tools that combine human expertise with systems-level understanding will be developed, allowing for collaborative reasoning as opposed to automated decision-making. Through thorough biological understanding, these tools will help maintain the role of expert judgment in livestock contexts while increasing its effectiveness (Martires et al., 2024).

Future directions in systems biology can be better understood in the context of global change and uncertainty. Livestock systems face previously unheard-of difficulties due to climate variability, new diseases, and changing social norms. Systems biology is well-positioned to handle these issues because of its focus on integration, adaptability, and prediction. In order to identify biological configurations and management techniques that function well in unpredictable and changing environments, future frameworks will place a greater emphasis on resilience and adaptability (Shashank et al., 2024; Ghavi Hossein-Zadeh, 2026).

In general, the capacity of systems biology and multi-omics integration to produce significant results while upholding both ethical and scientific integrity will determine their future. This calls for constant consideration of model limitations, open communication of uncertainty, and inclusive stakeholder engagement. Systems biology will continue to develop as a social and scientific field, influencing the creation, interpretation, and use of biological knowledge. Emerging directions and conceptual shifts shaping the future of systems biology in livestock are shown in Table 13.7.

References

Agmon, E., 2024. Prelude to a compositional systems biology. In: Other Quantitative Biology. https://doi.org/10.48550/arxiv.2408.00942.

Ahmed Sidi, M.L., Bocquillon, R., Cabret, F., Mohamed Babou, H., Dhib, C., Néron, E., Soukhal, A., Nanne, M.F., 2025. Constraint programming approaches for finding conserved metabolic and genomic patterns. Comput. Oper. Res. 183. https://doi.org/10.1016/j.cor.2025.107166.

Al Musawi, A.F., Roy, S., Ghosh, P., 2025. A review of link prediction applications in network biology. In: Institute of Electrical and Electronics Engineers Inc., Iraq IEEE Access, vol 13, pp. 54997–55016. https://doi.org/10.1109/ACCESS.2025.3553732.

Baião, A.R., Cai, Z., Poulos, R.C., Robinson, P.J., Reddel, R.R., Zhong, Q., Vinga, S., Gonçalves, E., 2025. A technical review of multi-omics data integration methods: from classical statistical to deep generative approaches. Briefings Bioinf. 26 (4). https://doi.org/10.1093/bib/bbaf355.

Balsa-Canto, E., Campo-Manzanares, N., Moimenta, A.R., Roudaut, G., Troitiño-Jordedo, D., 2025. Quantifying and managing uncertainty in systems biology: mechanistic and data-driven models. Curr. Opin. Syst. Biol. 42, 100557. https://doi.org/10.1016/j.coisb.2025.100557.

Banga, J.R., Villaverde, A.F., 2025. Mechanistic dynamic modelling of biological systems: the road ahead. Curr. Opin. Syst. Biol. 42. https://doi.org/10.1016/j.coisb.2025.100553.

Brachmann, M.K., Jacobs, C.T., 2025. Navigating through the noise: a roadmap for combining interdisciplinary high dimensional data in biological systems. Evol. Biol. https://doi.org/10.1007/s11692-025-09655-w.

Bullinger, E., Findeisen, R., Kalamatianos, D., Wellstead, P., 2024. System and control theory furthers the understanding of biological signal transduction. In: Lecture Notes in Control and Information Sciences. Springer Berlin Heidelberg, pp. 123–135. https://doi.org/10.1007/978-3-540-71988-5_6.

Cao, J., Zhou, X., Liu, J., Liu, D., Liu, G., 2025. Single-cell multi-omics sequencing technologies and their applications in livestock and poultry research. J. Agric. Food Chem. 73 (51), 32508–32520. https://doi.org/10.1021/acs.jafc.5c13040.

Chetty, A., Blekhman, R., 2024. Multi-omic approaches for host-microbiome data integration. Gut Microbes 16 (1). https://doi.org/10.1080/19490976.2023.2297860.

Choudhary, R.K., Sunil Kumar, B.V., Mukhopadhyay, C.S., Kashyap, N., Sharma, V., Singh, N., Tazerji, S.S., Kalantari, R., Hajipour, P., Malik, Y.S., Rabbani, I., 2024. Animal wellness: the power of multiomics and integrative strategies. Vet. Med. Int. 2024 (1). https://doi.org/10.1155/2024/4125118.

Davis, S.E., Embı, P.J., Matheny, M.E., 2024. Sustainable deployment of clinical prediction tools—a 360° approach to model maintenance. J. Am. Med. Inform. Assoc. 31 (5), 1195–1198. https://doi.org/10.1093/jamia/ocae036.

De Filippis, G.M., Amalfitano, D., Russo, C., Tommasino, C., Rinaldi, A.M., 2025. A systematic mapping study of semantic technologies in multi-omics data integration. J. Biomed. Inf. 165. https://doi.org/10.1016/j.jbi.2025.104809.

Didier, J., Croce, S., Bayoumi, S., Valceschini, E., Escoffier, H., Gonzalez, E., Kishk, A., Badkas, A., De Landtsheer, S., Sauter, T., 2025. Challenges and opportunities in systems biology education. Endocr. Relat. Cancer 32 (6). https://doi.org/10.1530/erc-25-0024.

Domínguez-Romero, E., Mazurenko, S., Scheringer, M., Martins Dos Santos, V.A.P., Evelo, C.T., Anton, M., Hancock, J.M., Zupanič, A., Suarez-Diez, M., 2024. Making PBPK models more reproducible in practice. Briefings Bioinf. 25 (6). https://doi.org/10.1093/bib/bbae569.

Doniparthi, G., Mühlhaus, T., Deßloch, S., 2024. Integrating FAIR experimental metadata for multi-omics data analysis. Datenbank-Spektrum 24 (2), 107–115. https://doi.org/10.1007/s13222-024-00473-6.

Ducrot, C., Barrio, M.B., Boissy, A., Charrier, F., Even, S., Mormède, P., Petit, S., Pinard-van der laan, M.-H., Schelcher, F., Casabianca, F., Ducos, A., Foucras, G., Guatteo, R., Peyraud, J.-L., Vayssier-Taussat, M., Veysset, P., Friggens, N.C., Fernandez, X., 2024. Animal board invited review: improving animal health and welfare in the transition of livestock farming systems: towards social acceptability and sustainability. Animal 18 (3), 101100. https://doi.org/10.1016/j.animal.2024.101100.

Flores, J.E., Claborne, D.M., Weller, Z.D., Webb-Robertson, B.J.M., Waters, K.M., Bramer, L.M., 2023. Missing data in multi-omics integration: recent advances through artificial intelligence. Front. Artif. Intell. 6. https://doi.org/10.3389/frai.2023.1098308.

Ghavi Hossein-Zadeh, N., 2024. An overview of recent technological developments in bovine genomics. Vet. Anim. Sci. 25, 100382. https://doi.org/10.1016/j.vas.2024.100382.

Ghavi Hossein-Zadeh, N., 2025. Artificial intelligence in veterinary and animal science: applications, challenges, and future prospects. Comput. Electron. Agric. 235, 110395. https://doi.org/10.1016/j.compag.2025.110395.

Ghavi Hossein-Zadeh, N., 2026. Advancing climate-resilient livestock systems: next-generation emission mitigation strategies and integrated technological innovations. Vet. Anim. Sci., 100588 https://doi.org/10.1016/J.VAS.2026.100588.

Gutierrez Reyes, C.D., Alejo-Jacuinde, G., Perez Sanchez, B., Chavez Reyes, J., Onigbinde, S., Mogut, D., Hernández-Jasso, I., Calderón-Vallejo, D., Quintanar, J.L., Mechref, Y., 2024. Multi omics applications in biological systems. Curr. Issues Mol. Biol. 46 (6), 5777–5793. https://doi.org/10.3390/cimb46060345.

Halpern, C., Schreefel, L., Amon, B., McAllister, T.A., Zanten, H.van, 2025. 11 Assessing the role of livestock in a circular bioeconomy. J. Anim. Sci. 103 (Suppl. 3), 64. https://doi.org/10.1093/jas/skaf300.080.

Heerden, N.v., Grobbelaar, S.S., Meyer, I., Vermeulen, E., 2024. Emerging smallholder cattle farming from a One Health perspective: a system dynamics model of the highveld region of South Africa. Environ. Sci. Pol. 152. https://doi.org/10.1016/j.envsci.2023.103646.

Henao, J.D., Lauber, M., Azevedo, M., Grekova, A., Theis, F., List, M., Ogris, C., Schubert, B., 2023. Multi-omics regulatory network inference in the presence of missing data. Briefings Bioinf. 24 (5). https://doi.org/10.1093/bib/bbad309.

Herzon, I., Mazac, R., Erkkola, M., Garnett, T., Hansson, H., Jonell, M., Kaljonen, M., Kortetmäki, T., Lamminen, M., Lonkila, A., Niva, M., Pajari, A. M., Tribaldos, T., Toivonen, M., Tuomisto, H.L., Koppelmäki, K., Röös, E., 2024. Both downsizing and improvements to livestock systems are needed to stay within planetary boundaries. Nat. Food 5 (8), 642–645. https://doi.org/10.1038/s43016-024-01030-w.

Hockings, N., Iwersen, M., Hancock, A., Oczak, M., 2023. Causal structure search and modeling of precision dairy farm data for automated prediction of ketosis risk, and the effect of potential interventions. Front. Anim. Sci. 4. https://doi.org/10.3389/fanim.2023.1180610.

Hui, T., Zhou, J., Yao, M., Xie, Y., Zeng, H., 2025. Advances in spatial omics technologies. Small Methods 9 (5). https://doi.org/10.1002/smtd.202401171.

Kant, S., Deepika, Roy, S., 2025. Integrative multi-omics and artificial intelligence: a new paradigm for systems biology. OMICS A J. Integr. Biol. 29 (12), 576–587. https://doi.org/10.1177/15578100251392371.

Karim, A.S., Brown, D.M., Archuleta, C.M., Grannan, S., Aristilde, L., Goyal, Y., Leonard, J.N., Mangan, N.M., Prindle, A., Rocklin, G.J., Tyo, K.J., Zoloth, L., Jewett, M.C., Calkins, S., Kamat, N.P., Tullman-Ercek, D., Lucks, J.B., 2024. Deconstructing synthetic biology across scales: a conceptual approach for training synthetic biologists. Nat. Commun. 15 (1). https://doi.org/10.1038/s41467-024-49626-x.

Kazanski, C.E., Balehegn, M., Jones, K., Bartlett, H., Calle, A., Garcia, E., Hawkins, H.J., Mayberry, D., McDonald-Madden, E., Odadi, W.O., Zionts, J., Clark, M., Garnett, T., Herrero, M., VanZanten, H., Ritten, J., Mallmann, G., Harrison, M.T., Bossio, D., Gennet, S., 2025. Context is key to understand and improve livestock production systems. Global Food Secur. 45. https://doi.org/10.1016/j.gfs.2025.100840.

Khanzadeh, H., Ghavi Hossein-Zadeh, N., Ghovvati, Sh., 2021. Statistical power and heritability in whole-genome association studies for quantitative traits. Meta Gene 28, 100869. https://doi.org/10.1016/j.mgene.2021.100869.

Khazaal, A., Vafaee, F., 2025. From static to dynamic: exploring temporal networks in systems biology. arXiv. https://doi.org/10.48550/arXiv.2505.10741.

Kitano, H., 2007. Scientific Challenges in Systems Biology Introduction to Systems Biology. Humana Press, Japan, pp. 3–13. https://doi.org/10.1007/978-1-59745-531-2_1.

Klingström, T., Zonabend König, E., Zwane, A.A., 2024. Beyond the hype: using AI, big data, wearable devices, and the internet of things for high-throughput livestock phenotyping. Brief. Funct. Genom. 24. https://doi.org/10.1093/bfgp/elae032.

Kumar, A., Singh, H., Singh, L., Kaur, J., Kumar, P., 2025. Mathematical biology and ecology for sustainability: network and systems biology approaches to ecological and environmental systems. Int. J. Multidiscip. Res. 7 (5). https://doi.org/10.36948/ijfmr.2025.v07i05.59341.

Lavandosque, L.L., Salazar, I.A.S., Monteiro, L.de F.R., Winck, F., 2024. Systems biology: from the fundamentals of complexity to applications in plant biology. Concilium 24 (11), 551–564. https://doi.org/10.53660/clm-3569-24j22.

Lawrence, E., El-Shazly, A., Seal, S., Joshi, C.K., Liò, P., Singh, S., Bender, A., Sormanni, P., Greenig, M., 2024. Understanding biology in the age of artificial intelligence. arXiv. https://doi.org/10.48550/arXiv.2403.04106.

Li, J., Xu, M., Xiang, L., Chen, D., Zhuang, W., Yin, X., Li, Z., 2023. Large Language models and foundation models in smart agriculture: basics, opportunities, and challenges. arXiv. https://doi.org/10.48550/arXiv.2308.06668.

Li, W., Ballard, J., Zhao, Y., Long, Q., 2024. Knowledge-guided learning methods for integrative analysis of multi-omics data. Comput. Struct. Biotechnol. J. 23, 1945–1950. https://doi.org/10.1016/j.csbj.2024.04.053.

Liao, X., Ederveen, T.A., Niehues, A., de Visser, C., Huang, J., Badmus, F., Doornbos, C., Orlova, Y., Kulkarni, P., van der Velde, K.J., Swertz, M.A., Brandt, M., van Gool, A.J., 't Hoen, P.A.C., 2024. FAIR Data Cube, a FAIR data infrastructure for integrated multi-omics data analysis. J. Biomed. Semant. 15 (1). https://doi.org/10.1186/s13326-024-00321-2.

Lobentanzer, S., Rodriguez-Mier, P., Bauer, S., Saez-Rodriguez, J., 2024. Molecular causality in the advent of foundation models. arXiv. https://doi.org/10.48550/arXiv.2401.09558.

Lu, Y., Li, M., Gao, Z., Ma, H., Chong, Y., Hong, J., Wu, J., Wu, D., Xi, D., Deng, W., 2024. Innovative insights into single-cell technologies and multi-omics integration in livestock and poultry. Int. J. Mol. Sci. 25 (23), 12940. https://doi.org/10.3390/ijms252312940.

Mardinoglu, A., Turkez, H., Shong, M., Srinivasulu, V.P., Nielsen, J., Palsson, B.O., Hood, L., Uhlen, M., 2025. Longitudinal big biological data in the AI era. Mol. Syst. Biol. 21 (9), 1147–1165. https://doi.org/10.1038/s44320-025-00134-0.

Martires, P.Z.D., Derkinderen, V., De Raedt, L., Krantz, M., 2024. Automated reasoning in systems biology: a necessity for precision medicine. arXiv. https://doi.org/10.48550/arXiv.2410.13487.

Mehrotra, B., Mendes, P., 2006. Bioinformatics approaches to integrate metabolomics and other systems biology data. Biotechnol. Agric. For. 57, 105–115. https://doi.org/10.1007/3-540-29782-0_8.

Melandri, G., R-Radohery, G., Beaumont, C., de Cripan, S.M., Muller, C., Piras, L., Alcina Pereira, M., Salvador, A.F., Domingo-Almenara, X., Bolger, M., Colombié, S., Prigent, S., Arechederra, B.G., Canela, N.C., Pétriacq, P., 2025. Artificial intelligence: the human response to approach the complexity of big data in biology. GigaScience 14. https://doi.org/10.1093/gigascience/giaf057.

Montesinos-López, O.A., Montesinos-López, A., Mosqueda-González, B.A., Delgado-Enciso, I., Chavira-Flores, M., Crossa, J., Dreisigacker, S., Sun, J., Ortiz, R., 2025. Genomic prediction powered by multi-omics data. Front. Genet. 16. https://doi.org/10.3389/fgene.2025.1636438.

Morabito, A., De Simone, G., Pastorelli, R., Brunelli, L., Ferrario, M., 2025. Algorithms and tools for data-driven omics integration to achieve multilayer biological insights: a narrative review. J. Transl. Med. 23 (1). https://doi.org/10.1186/s12967-025-06446-x.

Moscardó García, M., Aalto, A., Montanari, A.N., Skupin, A., Gonçalves, J., 2025. Multi-omic network inference from time-series data. npj Syst. Biol. Appl. 11 (1). https://doi.org/10.1038/s41540-025-00591-1.

Mudadla, T., Sharma, G., Mishra, A., Gola, S., 2024. Multifaceted landscape of omics data. Bio. Algorithm Med. Syst. 20 (1), 22–36. https://doi.org/10.5604/01.3001.0054.8093.

Nasir, T., Rehman, A., Tariq, M., Kubra, K.T., Ashraf, N., Jamal Nasir, F., 2024. Role of multi-omics in functional genomic, transcriptomic, proteomic and metabolic mechanisms for milk production, growth, fertility and health in livestock animals. Zeugma Biol. Sci. 6, 38–46. https://doi.org/10.55549/zbs.1680746.

Noordijk, B., Garcia Gomez, M.L., ten Tusscher, K.H.W.J., de Ridder, D., van Dijk, A.D.J., Smith, R.W., 2024. The rise of scientific machine learning: a perspective on combining mechanistic modelling with machine learning for systems biology. Front. Syst. Biol. 4. https://doi.org/10.3389/fsysb.2024.1407994.

Paleri, V.A., Hens, K., 2025. From consultors to collaborators – an SOP for advancing ethics engagement in science. Synth. Syst. Biotechnol. 10 (4), 1180–1189. https://doi.org/10.1016/j.synbio.2025.06.006.

Parvatikar, P.P., Bagali, S., Kanthe, P.S., Patil, A.V., Das, K.K., 2023. The Perspective of Physiome Modelling in Systems Biology: New Horizon Systems Biology, Bioinformatics and Livestock Science. Bentham Science Publishers, India, pp. 51–75. https://doi.org/10.2174/9789815165616123010008.

Pathak, R.K., Kim, J.M., 2024. Veterinary systems biology for bridging the phenotype-genotype gap via computational modeling for disease epidemiology and animal welfare. Briefings Bioinf. 25 (2). https://doi.org/10.1093/bib/bbae025.

Pessoa, L., 2024. Beyond networks, towards adaptive systems. arXiv. https://doi.org/10.48550/arXiv.2411.03621.

Pisupati, S., Niv, Y., 2022. The challenges of lifelong learning in biological and artificial systems. Trends Cognit. Sci. 26 (12), 1051–1053. https://doi.org/10.1016/j.tics.2022.09.022.

Procopio, A., Cesarelli, G., Donisi, L., Merola, A., Amato, F., Cosentino, C., 2023. Combined mechanistic modeling and machine-learning approaches in systems biology – a systematic literature review. Comput. Methods Progr. Biomed. 240, 107681. https://doi.org/10.1016/j.cmpb.2023.107681.

Proverbio, D., Katz, R., Giordano, G., 2024. Bridging robustness and resilience for dynamical systems in nature. IFAC-PapersOnLine 58 (17), 43–48. https://doi.org/10.1016/j.ifacol.2024.10.111.

Puniya, B.L., Verma, M., Damiani, C., Bakr, S., Dräger, A., 2024. Perspectives on computational modeling of biological systems and the significance of the SysMod community. Bioinform. Adv. 4 (1). https://doi.org/10.1093/bioadv/vbae090.

Qiao, L., Khalilimeybodi, A., Linden-Santangeli, N.J., Rangamani, P., 2025. The evolution of systems biology and systems medicine: from mechanistic models to uncertainty quantification. Annu. Rev. Biomed. Eng. 27 (1), 425–447. https://doi.org/10.1146/annurev-bioeng-102723-065309.

Richter, S., Scherer, L., Hegwood, M., Bartlett, H., Bossert, L.N., Frehner, A., Schader, C., 2024. Conceptual framework for considering animal welfare in sustainability assessments of foods. Sustain. Prod. Consum. 52, 179–209. https://doi.org/10.1016/j.spc.2024.10.004.

Sanches, P.H.G., de Melo, N.C., Porcari, A.M., de Carvalho, L.M., 2024. Integrating molecular perspectives: strategies for comprehensive multi-omics integrative data analysis and machine learning applications in transcriptomics, proteomics, and metabolomics. Biology 13 (11). https://doi.org/10.3390/biology13110848.

Saranya, S., Thamanna, L., Chellapandi, P., 2024. Unveiling the potential of systems biology in biotechnology and biomedical research. Syst. Microbiol. Biomanuf. 4 (4), 1217–1238. https://doi.org/10.1007/s43393-024-00286-4.

Selli, A., Miller, S.P., Ventura, R.V., 2024. The use of interactive visualizations for tracking haplotypic inheritance in livestock. Ruminants 4 (1), 90–111. https://doi.org/10.3390/ruminants4010006.

Shashank, C.G., Sejian, V., Silpa, M.V., Devaraj, C., Madhusoodan, A.P., Rebez, E.B., Kalaignazhal, G., Sahoo, A., Dunshea, F.R., 2024. Climate resilience in farm animals: transcriptomics-based alterations in differentially expressed genes and stress pathways. BioTech 13 (4). https://doi.org/10.3390/biotech13040049.

Singh, A., Verma, A., Dutta, G., Gowane, G.R., Ludri, A., Alex, R., 2024. Functional transcriptome analysis revealed major changes in pathways affecting systems biology of Tharparkar cattle under seasonal heat stress. 3 Biotech 14 (7). https://doi.org/10.1007/s13205-024-04018-2.

Snead, A.A., Clark, R.D., 2022. The biological hierarchy, time, and temporal 'omics in evolutionary biology: a perspective. Integr. Comp. Biol. 62 (6), 1872–1886. https://doi.org/10.1093/icb/icac138.

Sun, S., Wang, B., Tian, Y., 2024. Decoupled representation for multi-view learning. Pattern Recogn. 151, 110377. https://doi.org/10.1016/j.patcog.2024.110377.

Tatka, L.T., Smith, L.P., Hellerstein, J.L., Sauro, H.M., 2023. Adapting modeling and simulation credibility standards to computational systems biology. J. Transl. Med. 21 (1). https://doi.org/10.1186/s12967-023-04290-5.

Tedeschi, L.O., Guarnido Lopez, P., Menendez III, H.M., Seo, S., 2025. Advancing precision livestock farming: integrating artificial intelligence and emerging technologies for sustainable livestock management. Anim. Biosci. https://doi.org/10.5713/ab.25.0289.

Tiukova, I.A., Brunnsåker, D., Bjurström, E.Y., Gower, A.H., Kronström, F., Reder, G.K., Reiserer, R.S., Korovin, K., Soldatova, L.B., Wikswo, J.P., King, R.D., 2024. Genesis: towards the automation of systems biology research. arXiv. https://doi.org/10.48550/arXiv.2408.10689.

Tjandrawinata, R.R., Rebeca, C., Nurcahyanti, A., 2025. Spatiotemporal omics: integrating multi-omics data for translational research and drug development. Asian J. Eng. Soc. Health 4 (3), 686–705. https://doi.org/10.46799/ajesh.v4i3.561.

Tulpan, D., 2024. 458 Advanced computing and information technology to address challenges in livestock production. J. Anim. Sci. 102 (Suppl. 3), 295–296. https://doi.org/10.1093/jas/skae234.337.

Vasseur, F., Westgeest, A.J., Vile, D., Violle, C., 2022. Solving the grand challenge of phenotypic integration: allometry across scales. Genetica 150 (3–4), 161–169. https://doi.org/10.1007/s10709-022-00158-6.

Velten, B., Stegle, O., 2023. Principles and challenges of modeling temporal and spatial omics data. Nat. Methods 20 (10), 1462–1474. https://doi.org/10.1038/s41592-023-01992-y.

Venegas, L., Ramayo-Caldas, Y., Bohannan, B.J.M., Derome, N., Yáñez, J.M., 2026. From genomes to hologenomes: integrating host and microbiome data for complex trait prediction in livestock and aquaculture. Front. Anim. Sci. 6. https://doi.org/10.3389/fanim.2025.1678538.

Vlaicu, P.A., Gras, M.A., Untea, A.E., Lefter, N.A., Rotar, M.C., 2024. Advancing livestock technology: intelligent systemization for enhanced productivity, welfare, and sustainability. AgriEngineering 6 (2), 1479–1496. https://doi.org/10.3390/agriengineering6020084.

Wadood, A.A., Bordbar, F., Zhang, X., 2025. Integrating omics approaches in livestock biotechnology: innovations in production and reproductive efficiency. Front. Anim. Sci. 6. https://doi.org/10.3389/fanim.2025.1551244.

Wang, H., Bai, F., Lu, H., Zhou, Y.J., 2025. Understanding methanol metabolism through systems biology: advances and future perspectives. Curr. Opin. Biotechnol. 96, 103370. https://doi.org/10.1016/j.copbio.2025.103370.

White, R.R., 2025. 205 Artificial intelligence and the internet of things: responsible and sustainable intensification of livestock production. J. Anim. Sci. 103 (Suppl. 3), 260–261. https://doi.org/10.1093/jas/skaf300.302.

Yamanouchi, H.M., Notomi, Y., Hisamoto, S., 2026. Toward understanding the principles of animal behaviors: a systems ethology approach. Artif. Life Robot. https://doi.org/10.1007/s10015-025-01112-3.

Yang, X., Mann, K., Wu, H., Ding, J., 2024. scCross: a deep generative model for unifying single-cell multi-omics with seamless integration, cross-modal generation, and in silico exploration. Genome Biol. 25 (1). https://doi.org/10.1186/s13059-024-03338-z.

Ye, X., Sahana, G., Lund, M.S., Cai, Z., 2025. The crosstalk between host and rumen microbiome in cattle: insights from multi-omics approaches and genome-wide association studies. World J. Microbiol. Biotechnol. 41 (8). https://doi.org/10.1007/s11274-025-04504-6.

Yetgin, A., 2025. Revolutionizing multi-omics analysis with artificial intelligence and data processing. Quant. Biol. 13 (3). https://doi.org/10.1002/qub2.70002.

Yu, Y., Mai, Y., Zheng, Y., Shi, L., 2024. Assessing and mitigating batch effects in large-scale omics studies. Genome Biol. 25 (1). https://doi.org/10.1186/s13059-024-03401-9.

Chapter 14

Artificial intelligence in precision breeding of livestock

In this chapter, artificial intelligence (AI) is positioned as the integrative intelligence layer that is changing precision breeding from periodic genomic evaluation to an ongoing, context-aware optimization of genetic advancement, resilience, welfare, and sustainability. Growing data dimensionality from whole genome sequences, dense imputed variant panels, structural and epigenetic markers, transcriptomic and proteomic endophenotypes, microbiome-mediated functional signatures, longitudinal sensor-derived behavioral and physiological phenotypes, advanced imaging, environmental telemetry, and life cycle sustainability metrics challenges traditional selection architectures based on fixed trait definitions, static indices, and linear additive assumptions. AI enables the discovery of latent biological axes and dynamic genotype-by-environment contingencies that elude traditional linear or strictly parametric formulations by providing representational depth and algorithmic flexibility to reconcile heterogeneity in scale, noise structure, missingness, temporal granularity, and cross-population diversity. To protect genetic diversity, ethical acceptability, and socio-environmental accountability, the chapter presents precision breeding as an iterative decision ecosystem where data ingestion, multimodal fusion, feature abstraction, causal disentanglement, predictive inference, uncertainty quantification, and decision optimization co-evolve under governance.

The shift from static genomic prediction to adaptive modeling that incorporates environmental perturbations and streaming phenotypes, recalibrating expected breeding values in almost real-time, and separating stable heritable signal from plastic regulatory or transient stress responses are some of the central themes. To create transferable embeddings that preserve evolutionary constraint and functional modularity while allowing for localized fine-tuning, foundation, and self-supervised models trained across multibreed and cross-species corpora will be investigated. To maintain interpretability and causal plausibility, hybrid mechanistic statistical frameworks that incorporate quantitative genetic priors, biological ontologies, and physiological constraints are positioned alongside graph-based and sequence-aware architectures, attention mechanisms, and probabilistic deep ensembles. A rigorous foundation for striking a balance between short-term gain and long-term robustness, inbreeding control, and adaptive capacity under climate volatility is provided by the integration of causal discovery and intervention modeling, which introduces counterfactual reasoning for expected responses to selection emphasis shifts, genomic edits, microbiome modulation, or management adaptations.

The chapter delves deeper into closed-loop breeding pipelines, where AI-mediated decision agents negotiate multiple objective trade spaces that include economic viability, welfare indicators, health resilience, product quality, production efficiency, and environmental footprint. Principled uncertainty decomposition is emphasized to distinguish between stochastic variability and epistemic gaps, directing strategic genotyping, targeted phenotyping, or exploratory crosses that optimize information gain per resource used. To mitigate bias amplification and proprietary data leakage, advanced predictive performance across regions and species with uneven data infrastructure can be democratized through the use of transfer learning, privacy-preserving federated learning, and synthetic data generation. The story is infused with ethical, legal, and societal issues, including model provenance transparency, fair trait prioritization, algorithmic fairness across breeds and production contexts, and compliance with frameworks for sustainability and animal integrity. The chapter creates a coherent conceptual framework by outlining methodological underpinnings, infrastructural facilitators, governance principles, and translational pathways. This framework is then used in later sections to elaborate on technical details and practical examples of AI-powered precision breeding.

14.1 Introduction to precision breeding of livestock

A conceptual and practical development in the science of animal improvement, precision breeding of livestock reflects the convergence of advances in AI-driven analytics, digitalization of production systems, and the expansion of biological knowledge. Fundamentally, precision breeding reframes genetic improvement as an ongoing, data-driven process that

Artificial Intelligence in Livestock Omics. https://doi.org/10.1016/B978-0-443-45663-3.00004-X

explicitly takes into account biological complexity, environmental variability, ethical responsibility, and long-term sustainability, rather than as a static or periodic optimization exercise. Through integrative modeling, real-time feedback, and multiobjective decision frameworks that balance genetic advancement with societal and ecological expectations, it expands the scope of quantitative genetics and genomic selection while building upon their foundations (Fig. 14.1) (Tedeschi et al., 2025; Zhu et al., 2025).

Livestock breeding has historically undergone several methodological stages, starting with phenotype-based selection based on observable performance and ending with statistically formalized breeding value estimation (Panigrahi et al., 2025). Mathematical rigor was introduced by the development of quantitative genetic theory, which supported systematic selection strategies and made it possible to divide phenotypic variance into heritable and environmental components. By capturing actual genetic relationships between individuals, the subsequent integration of molecular markers and dense genomic data further improved prediction accuracy. Although these developments improved breeding efficiency, they mainly relied on linear, additive, and equilibrium-oriented assumptions that treated environments as background noise rather than dynamic, interacting forces and abstracted biological mechanisms. The increasing realization that such simplifications are inadequate for negotiating the complexity, volatility, and ethical scrutiny that define modern livestock systems gives rise to precision breeding (Ullah and Waheed, 2025; Ghavi Hossein-Zadeh, 2024).

The current state of livestock production is characterized by unprecedented data richness and heterogeneity. These days, fragmented data ecosystems include whole-genome sequence variation, functional regulatory information, transcriptomic and proteomic profiles, microbiome composition, metabolomic signatures, high-resolution phenomics, behavioral telemetry, environmental sensing, and management records. The goal of precision breeding is to bring these disparate streams together into cohesive representations that allow for scalable inference while maintaining biological meaning. To convert unprocessed data into useful knowledge, AI offers the representational flexibility and computational power needed to balance variations in scale, noise structure, missingness, and temporal resolution. AI systems can extract latent biological dimensions that represent physiological states, regulatory modules, or adaptive strategies, all of which are essential for well-informed selection but are not directly observable, through sophisticated representation learning techniques (Baião et al., 2025; Wadood et al., 2025; Zhu et al., 2025).

The explicit recognition of livestock as complex, adaptable biological systems embedded within equally complex production and ecological contexts is a defining characteristic of precision breeding. Regulatory networks, developmental trajectories, microbiome interactions, management strategies, and environmental factors all mediate genetic effects. Therefore, context-aware modeling, which views genetic merit as contingent upon particular environmental and management circumstances rather than as a universal characteristic of a single genotype, is emphasized in precision breeding.

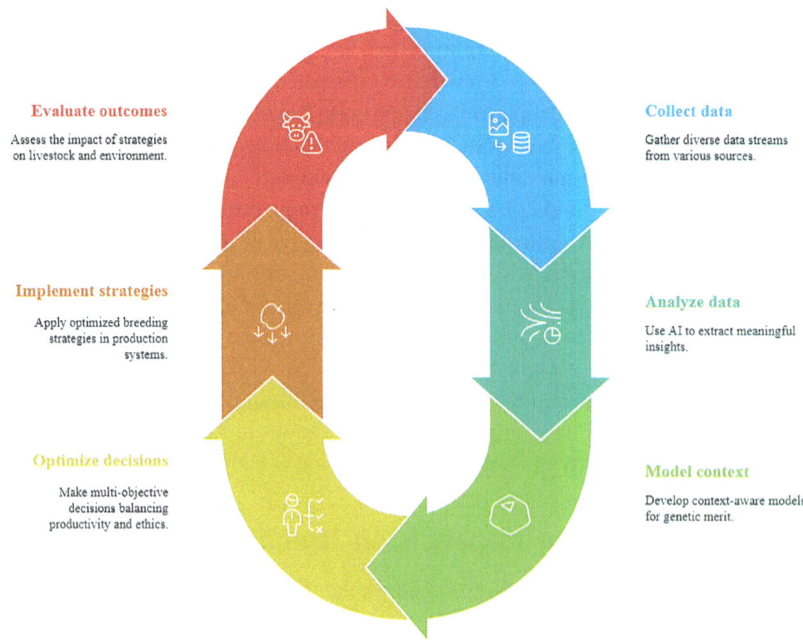

FIGURE 14.1 An overview of the precision breeding of livestock cycle. This figure illustrates the iterative cycle underlying precision livestock breeding, beginning with the collection of diverse biological and management data and followed by AI-driven data analysis and context-aware modeling of genetic merit. The cycle emphasizes multiobjective decision optimization that balances productivity, environmental considerations, and ethical constraints, leading to the implementation of optimized breeding strategies. Continuous evaluation of outcomes feeds back into subsequent data collection and model refinement, highlighting the adaptive and self-improving nature of AI-enabled precision breeding systems.

This change casts doubt on conventional ideas of breeding values as static rankings and substitutes probabilistic performance distributions that change in response to new information and circumstances (Ghavi Hossein-Zadeh, 2026a; Venegas et al., 2026).

AI's contribution to this change goes beyond a marginal increase in prediction accuracy. Continuous learning, causal reasoning, and decision optimization under uncertainty are all made possible by AI, which functions as an integrative intelligence layer. To update predictive models in almost real-time and identify early indicators of genotype-by-environment interactions or emerging stressors, machine learning architectures can integrate streaming data from sensors and laboratories. By breaking down uncertainty into elements related to biological variability, measurement error, and knowledge gaps, probabilistic modeling frameworks facilitate focused data collection and risk-aware decision-making. By further differentiating between correlation and mechanism, causal inference techniques enable breeders to assess the likely outcomes of management interventions or selection choices before their implementation (Garcia-Oliveira et al., 2026; Ghavi Hossein-Zadeh, 2025).

Multiobjective breeding is a fundamental aspect of precision breeding. Although efficiency and productivity are still important objectives, they are increasingly weighed against economic stability, animal welfare, environmental impact, and health resilience. By creating multidimensional optimization landscapes where different selection strategies can be transparently compared, AI makes it easier to explicitly navigate these trade-offs. Precision breeding frameworks enable dynamic adjustment of priorities in response to changing market conditions, regulatory constraints, climate variability, or societal values, as opposed to depending on fixed economic indices with static weights. In light of global issues such as climate change, antibiotic resistance, and the growing ethical concern for animal welfare, this flexibility is especially crucial (Ducrot et al., 2024; Kistanova et al., 2024).

Ethical considerations are not peripheral to precision breeding but integral to its conceptual foundation. The ability to quantify and forecast ever subtler facets of animal biology raises concerns regarding accountability, openness, and appropriate intervention boundaries. Therefore, principles that protect animal welfare, maintain genetic diversity, guarantee equity across breeds and production systems, and respect data ownership and sovereignty must guide precision breeding. In order for stakeholders to comprehend not only which decisions are suggested but also why, AI systems used in breeding decisions must be auditable and interpretable. Unintended consequences that could result from limited efficiency-driven incentives can be avoided by directly incorporating ethical constraints into model objectives and optimization procedures (Nallappan and Kanathasan, 2024; Ghavi Hossein-Zadeh, 2025).

The temporal orientation of precision breeding is another essential component. Decisions made today influence the ability of future populations to adapt, and genetic improvement occurs over generations. By simulating the cumulative effects of selection on genetic diversity, inbreeding trajectories, and resilience to future uncertainties, precision breeding frameworks take long-term viewpoints into account. The use of simulation-based methods and digital representations of breeding populations enables the exploration of alternative evolutionary pathways, which highlights possible dangers related to excessively aggressive selection or inadequate conservation of diversity. This forward-thinking approach harmonizes precision breeding with sustainable resource stewardship and intergenerational responsibility (Lozada-Soto et al., 2024; Panwar et al., 2025).

Integration with production systems further distinguishes precision breeding from earlier paradigms. Genetic decisions are now co-optimized within unified decision architectures, rather than being separated from management practices. Breeding programs and farms exchange information in both directions, allowing management interventions to be customized to genetic predispositions and genetic selection to take into consideration practical management constraints. This integration lowers the welfare risks related to genotype-environment mismatches, while promoting more effective expression of genetic potential. By coordinating decision-support interfaces, optimization algorithms, and predictive models across organizational and temporal scales, AI coordinates these interactions (Neethirajan et al., 2024; Zhu et al., 2025).

The concept of uncertainty in decision-making is also redefined by precision breeding. Conventional breeding methods frequently place a strong emphasis on rankings and point estimates, implicitly assuming predictability and stability. Precision breeding uses uncertainty as a guide for strategic exploration, recognizing it as an intrinsic characteristic of biological systems. Adaptive phenotyping techniques and resource allocation are informed by uncertainty-aware models, which pinpoint areas where more data could most significantly boost confidence. Precision breeding facilitates more robust decisions that continue to be effective in a variety of realistic future scenarios by explicitly representing uncertainty (Deglurkar et al., 2024). Conceptual pillars and defining characteristics of precision livestock breeding are described in Table 14.1.

TABLE 14.1 Conceptual pillars and defining characteristics of precision livestock breeding.

Conceptual pillar	Core definition	Key characteristics	Implications for breeding practice	Contribution of AI
Systems-level perspective	• Livestock viewed as complex adaptive biological systems	• Integration of molecular, physiological, and phenotypic layers • Explicit consideration of environmental context	• Moves breeding beyond single-trait optimization • Supports holistic evaluation of biological outcomes	• Enables multilayer data integration • Captures nonlinear system interactions
Precision and individualization	• Tailoring breeding decisions to fine-scale biological differences	• Recognition of individual and subgroup variability • Context-dependent genetic merit	• Replaces uniform selection strategies with targeted approaches	• Learns individualized predictive representations
Multiobjective optimization	• Simultaneous consideration of multiple breeding goals	• Productivity, health, welfare, adaptability, and sustainability	• Makes trade-offs explicit and transparent	• Supports dynamic weighting and scenario evaluation
Data-driven decision-making	• Reliance on integrated biological and phenotypic data	• Continuous incorporation of new information • Evidence-based inference	• Reduces reliance on heuristic or static indices	• Extracts latent patterns from high-dimensional data
Dynamic and temporal orientation	• Breeding is viewed as an evolving, long-term process	• Emphasis on trajectories rather than static endpoints • Anticipation of future conditions	• Enables proactive rather than reactive breeding	• Supports longitudinal modeling and continual learning
Context awareness	• Genetic effects interpreted within specific environments	• Explicit modeling of genotype-environment interaction	• Aligns selection with real production conditions	• Learns context-dependent relationships
Predictive and anticipatory capacity	• Focus on forecasting future biological outcomes	• Probabilistic performance distributions	• Improves planning under uncertainty	• Generates forward-looking predictions

Uncertainty management	• Explicit representation of biological and data uncertainty	• Risk-aware inference and decision support	• Prevents overconfident or fragile decisions	• Quantifies and propagates uncertainty
Integration of multi-omics information	• Use of multiple biological data layers	• Genomic, regulatory, functional, and phenotypic data	• Improves biological interpretability	• Harmonizes heterogeneous data modalities
Ethics-by-design	• Ethical considerations embedded in system objectives	• Welfare, diversity, and sustainability constraints	• Aligns breeding outcomes with societal values	• Formalizes ethical priorities in optimization
Preservation of genetic diversity	• Maintenance of long-term adaptive potential	• Avoidance of excessive homogenization	• Supports intergenerational resilience	• Models diversity trajectories and risks
Human-AI co-decision frameworks	• Collaboration between computational models and expert judgment	• AI as decision support, not replacement	• Preserves accountability and expertise	• Enhances human reasoning with comprehensive analysis
Scalability and adaptability	• Applicability across populations and systems	• Capacity to evolve with new data	• Enables broad and sustained implementation	• Supports scalable and adaptive learning systems
Integration with production systems	• Alignment of breeding with management practices	• Bidirectional feedback between genetics and environment	• Improves expression of genetic potential	• Connects predictive models with real-world data
Long-term sustainability orientation	• Emphasis on durable and responsible improvement	• Balancing short-term gain with future viability	• Redefines success beyond immediate productivity	• Supports multigenerational scenario modeling

14.2 Role of AI in livestock breeding

Livestock breeding has evolved from a primarily statistical and retrospective field to a dynamic, predictive, and adaptable decision science, thanks in large part to AI. Its influence on the representation of biological data, the management of uncertainty, and the formulation of breeding decisions within intricate and dynamic systems extends far beyond computational efficiency or small increases in accuracy. AI serves as an integrative intelligence layer in precision breeding, synthesizing various data streams, learning nonlinear and context-dependent relationships, and facilitating ongoing optimization across various biological and societal goals (Zhu et al., 2025).

AI fundamentally changes how livestock breeding models the genetic and phenotypic data. While useful for many applications, traditional breeding frameworks are fundamentally limited in their capacity to capture nonlinear interactions, higher-order dependencies, and emergent biological behavior due to their strong reliance on linear models and additive assumptions. By learning adaptable representations directly from data, AI techniques circumvent these restrictions and enable the modeling of genetic effects as conditional and context-sensitive rather than fixed and universal. This ability is especially crucial in livestock systems, since the environment, management, developmental stage, and interactions between biological pathways all have a significant impact on gene expression and phenotypic results (Shokor et al., 2025; Li et al., 2024).

The ability of AI to integrate heterogeneous and high-dimensional data is one of its most important contributions to livestock breeding. Information from genomics, epigenomics, transcriptomics, phenomics, environmental sensors, and management records, all of which have different scales, noise structures, and levels of completeness, is increasingly produced by modern breeding programs. By extracting latent biological features that summarize intricate patterns across data layers, AI offers the representational tools required to reconcile these disparities. A more comprehensive understanding of breeding value that goes beyond single-trait or single-layer analysis is made possible by AI systems using representation learning to find shared structures that connect molecular variation to physiological performance and long-term outcomes (Ballard et al., 2024; Zhu et al., 2025).

The temporal aspect of breeding inference is also significantly altered by AI. Conventional breeding assessments are frequently carried out at specific intervals, yielding static estimates that are periodically updated in response to new information. AI-driven breeding systems, on the contrary, facilitate ongoing learning and update predictions in response to streaming data from production settings and phenotyping platforms. Breeding programs can identify new trends, changes in genotype-by-environment interactions, or early warning signs of unexpected consequences thanks to this temporal adaptability. Instead of relying on assumptions from the past that may no longer be true, AI supports breeding decisions that remain relevant under changing conditions by maintaining models that evolve alongside biological systems (Sangjan et al., 2025).

AI's capacity to explicitly model and manage uncertainty plays a crucial role in livestock breeding. Due to the inherent variability of biological systems, breeding decisions must be made in the face of stochastic outcomes and incomplete information. Probabilistic reasoning, which separates signal from noise, measures prediction confidence, and spreads uncertainty via decision pipelines, is incorporated into AI frameworks. This uncertainty-aware approach supports risk-sensitive breeding strategies, allowing breeders to assess not only anticipated gains but also their robustness across likely future scenarios. Avoiding overconfident selection decisions that could jeopardize genetic diversity or long-term resilience requires this kind of reasoning (Valone, 2024; Sun et al., 2026).

By facilitating the investigation of intricate optimization landscapes involving numerous, frequently conflicting goals, AI further improves livestock breeding. Productivity, health, welfare, adaptability, and sustainability must all be taken into account simultaneously for precision breeding; none of these factors can be reduced to a single scalar metric without losing significance. These goals can be represented collectively thanks to AI-based optimization techniques, which highlight trade-offs and synergies that are hidden by conventional index-based methods. AI increases the flexibility and responsiveness of breeding programs by supporting dynamic reweighting of objectives in response to changing constraints, market demands, or ethical considerations, rather than enforcing fixed priorities (Zhu et al., 2025; Boktar, 2025).

In contemporary livestock breeding, AI's interpretive function is becoming increasingly crucial. Although early AI applications were frequently criticized for being opaque, more recent advancements emphasize explainability and a biological foundation. Interpretable AI frameworks explain why particular individuals or genetic configurations are preferred by connecting predictions to underlying biological characteristics, regulatory pathways, or system-level patterns. This openness fosters stakeholder trust, scientific understanding, and regulatory compliance, all of which are essential for the responsible application of AI in breeding contexts. AI bridges the gap between biological reasoning and computational inference by linking prediction and explanation (Džermeikaitė et al., 2025).

AI is essential for simulating genotype-by-environment interactions, which have long been a problem in livestock breeding. It can be challenging to find genotypes that perform consistently or adaptively because environmental heterogeneity across production systems can significantly alter genetic expression. AI systems are excellent at identifying intricate patterns of interaction and learning how genetic effects differ across environmental gradients and management regimes. This ability makes it possible to identify breeding strategies that are context-specific and match realistic production conditions with genetic selection. AI promotes targeted breeding strategies that optimize performance and welfare within specified ecological and managerial contexts, as opposed to searching for universally optimal genotypes (Spoelstra et al., 2026).

AI's incorporation into livestock breeding also modifies institutional structures and decision-making processes. A growing number of breeding decisions come from collaborative human-AI systems, where expert judgment is used to interpret and improve predictions, scenarios, and recommendations generated by computational models. Through thorough data synthesis and scenario exploration, this co-decision paradigm enhances human expertise while maintaining its role. Therefore, AI serves as a cognitive partner that increases the breadth and depth of decision-making ability, rather than taking the place of breeders (Rosati, 2025).

Another area where AI is essential is ethical governance. Narrow optimization goals run the risk of taking precedence over welfare, diversity, or social acceptance as breeding decisions become more data-driven and automated. To ensure that welfare indicators, sustainability metrics, and equity considerations impact selection outcomes, AI systems can be built to directly integrate ethical constraints into their objective functions. AI assists in coordinating breeding practices with broader societal values and legal requirements by integrating ethical priorities into computational frameworks (Ghose et al., 2024; Ghavi Hossein-Zadeh, 2025).

Through simulation and foresight analysis, AI also helps with long-term strategic planning of breeding programs. AI-powered models are able to forecast the effects of selection tactics over several generations, taking into consideration feedback effects on genetic diversity, inbreeding, and adaptability. By proactively identifying risks and opportunities, these simulations help breeders modify their strategies before unfavorable trends become ingrained. In the context of climate change and changing production challenges, where future conditions may differ significantly from historical baselines, this anticipatory capability is especially valuable. Roles and functions of AI across the livestock breeding pipeline are shown in Table 14.2.

14.3 Techniques in machine learning and data analysis

The methodological basis for AI-enabled precision breeding of livestock comprises machine learning and data analysis techniques. These methods offer the formal processes that convert complicated biological data into knowledge that is predictive, comprehensible, and useful. Machine learning functions as an integrative analytical paradigm that works at the intersection of biology, statistics, and decision science in the context of livestock breeding, rather than as a stand-alone computational exercise. Its value is found in its capacity to manage uncertainty, accommodate nonlinearity and interaction, learn structured representations from heterogeneous data, and continuously adapt as new information becomes available (Panigrahi et al., 2025; Zhu et al., 2025).

Machine learning's departure from strict parametric assumptions is one of its distinguishing features in livestock breeding. Conventional data analysis techniques frequently depend on predetermined functional forms that enforce linearity, additivity, or stationarity, assumptions that are rarely true for biological systems. By enabling models to directly infer relationships from data, machine learning techniques loosen these limitations and make it possible to capture intricate dependencies between genetic, molecular, phenotypic, and environmental variables. Because trait expression results from complex regulatory processes, and the effects of genetic variation are frequently context-dependent, this flexibility is especially crucial for breeding applications (Reinoso-Peláez et al., 2022; Panigrahi et al., 2025).

A key class of techniques in breeding-oriented machine learning is supervised learning. These methods discover relationships between input data and desired results, such as resilience, fitness-related characteristics, or expected performance. Supervised learning in precision breeding uses high-dimensional inputs, including environmental descriptors, functional omics signals, phenotypic trajectories, and genomic variation. The ability of supervised learning to combine these inputs into cohesive predictive models that can extrapolate beyond observed data is what gives it its power. However, managing bias, variance, and overfitting carefully is necessary to achieve robust generalization, especially in situations where data dimensionality greatly exceeds sample size (Lourenço et al., 2024; Ghavi Hossein-Zadeh, 2025).

For supervised models to function well on complicated biological data, representation learning is essential. Representation learning techniques convert data into latent spaces that capture crucial biological structure, rather than depending on raw input features. These latent representations may represent physiological states, regulatory modules, or

TABLE 14.2 Roles and functions of AI across the livestock breeding pipeline.

Breeding pipeline stage	Primary breeding objective	Role of AI	Key functions performed by AI	Added value over conventional approaches
Data acquisition and harmonization	• Capture comprehensive biological and phenotypic information	• Acts as a data integration engine	• Harmonizes heterogeneous data sources • Handles missing, noisy, and unbalanced data	• Enables use of large, complex datasets • Reduces information loss during preprocessing
Genetic and genomic evaluation	• Estimate genetic merit and biological potential	• Learns complex genotype-phenotype relationships	• Models nonlinear and interaction effects • Incorporates functional and regulatory data	• Improves prediction beyond linear additive models
Multi-omics integration	• Contextualize genetic variation biologically	• Extracts latent system-level representations	• Integrates genomics with regulatory and functional layers • Identifies biologically coherent patterns	• Enhances biological interpretability of breeding values
Phenotypic prediction	• Forecast future performance and robustness	• Functions as a predictive inference system	• Generates probabilistic performance distributions • Accounts for uncertainty and variability	• Supports anticipatory and risk-aware selection
Genotype-by-environment modeling	• Align genetics with production environments	• Learns context-dependent genetic effects	• Captures nonlinear environmental modulation • Adapts predictions to local conditions	• Enables environment-specific breeding strategies
Selection decision support	• Identify optimal candidates for breeding	• Acts as a decision-support system	• Ranks individuals under multiobjective criteria • Explores trade-offs among goals	• Makes selection logic explicit and transparent
Multiobjective optimization	• Balance productivity, welfare, and sustainability	• Optimizes across competing objectives	• Evaluates alternative weighting scenarios • Identifies Pareto-efficient solutions	• Moves beyond single-index optimization
Risk and uncertainty management	• Minimize unintended long-term consequences	• Quantifies and propagates uncertainty	• Assesses confidence in predictions • Highlights fragile decisions	• Reduces overconfident or risky selection
Genetic diversity monitoring	• Preserve long-term adaptive capacity	• Tracks population-level genetic structure	• Models inbreeding and diversity trajectories • Detects homogenization risks	• Supports sustainable and resilient breeding
Long-term breeding strategy simulation	• Anticipate multigenerational outcomes	• Enables virtual experimentation	• Simulates alternative selection scenarios • Projects long-term impacts	• Reduces reliance on irreversible real-world trials

Integration with management systems	• Ensure effective expression of genetic potential	• Aligns breeding with production practices	• Links genetic predictions to management data • Supports co-optimization	• Improves genotype-environment matching
Model updating and continual learning	• Maintain relevance under changing conditions	• Learns continuously from new data	• Updates predictions as environments evolve • Detects shifts in system behavior	• Prevents model obsolescence
Explainability and transparency	• Enable trust and accountability	• Provides interpretable outputs	• Links predictions to biological drivers • Supports expert validation	• Facilitates ethical and regulatory oversight
Ethical constraint integration	• Align breeding with societal values	• Encodes ethical priorities in models	• Incorporates welfare and sustainability limits • Evaluates ethical trade-offs	• Embeds responsibility directly into decision-making
Human-AI collaboration	• Support informed expert judgment	• Acts as a cognitive partner	• Augments breeder expertise • Enables scenario exploration	• Enhances decision quality without replacing humans

adaptive strategies that arise from the data but are not labeled explicitly. Representation learning enables models to reason across molecular layers and phenotypic scales in livestock breeding by harmonizing disparate data modalities. This abstraction facilitates transferability across populations and environments, reduces noise, and improves interpretability (Pathak and Kim, 2024).

By exposing inherent structure in data without depending on predetermined results, unsupervised learning approaches support precision breeding. These techniques are especially useful for latent biological state detection, population stratification, and exploratory analysis. Unsupervised learning can reveal patterns of phenotypic clustering, regulatory coordination, or genetic similarity in breeding contexts that guide subsequent modeling and decision-making. Unsupervised methods facilitate the creation of hypotheses and direct focused data collection tactics by spotting natural groupings and emerging patterns (Soloshenkov et al., 2024; Ghavi Hossein-Zadeh, 2025).

Given the realities of livestock data availability, semisupervised learning occupies a crucial middle ground between supervised and unsupervised approaches. Only a small portion of individuals in many breeding programs have complete phenotypic labels, leaving vast amounts of molecular or sensor data unlabeled. By taking advantage of the shared structure throughout the entire dataset, semisupervised techniques improve model performance and robustness by utilizing both labeled and unlabeled data. In situations where thorough phenotyping is impractical, this capability is especially important for scaling precision breeding techniques across large and diverse populations (Qian et al., 2024).

Ensemble learning, which integrates several models to increase predictive accuracy and stability, is another crucial class of methods. Because biological systems are intricate and multifaceted, it is unlikely that a single model architecture will fully capture all pertinent patterns. By combining predictions from various models, ensemble approaches reduce sensitivity to noise and model-specific biases. Ensemble learning improves resilience to changes in data distribution brought about by environmental or management changes, and it supports more accurate estimation of breeding merit in precision breeding (Gu et al., 2024).

Techniques for feature selection and dimensionality reduction are crucial parts of machine learning pipelines in livestock breeding. Redundant or poorly informative features are frequently present in high-dimensional omics and phenomics data, which can obscure important patterns and increase computational load. While feature selection techniques give priority to inputs that contribute most to predictive performance or interpretability, dimensionality reduction techniques find compact representations that preserve biologically relevant variation. By identifying important factors influencing system behavior, these methods enhance model efficiency and promote biological understanding in precision breeding (Andersen and Reading, 2024).

As breeding programs integrate longitudinal phenotypic and environmental data, temporal data analysis techniques become increasingly crucial. Static models are unable to capture the trajectories, trends, and transitions that time-series analysis techniques do. Temporal approaches make it possible to model growth curves, developmental patterns, stress reactions, and adaptation over time in livestock breeding. Machine learning models advance from snapshot predictions to dynamic forecasting by incorporating temporal information, enabling earlier and better-informed breeding decisions (Fu et al., 2024; Panigrahi et al., 2025).

Another strong analytical framework for precision breeding is provided by graph-based and network-oriented methods. Genes, proteins, metabolites, and phenotypes are all connected by regulatory and functional interactions, making biological systems inherently networked. Inference that respects biological connectivity is made possible by graph-based machine learning techniques that work directly on these relational structures. Network-based analysis facilitates the identification of important regulatory hubs, modular organization, and emergent properties that affect robustness and trait expression in breeding contexts. These methods enhance the interpretability of complex models and are closely aligned with the principles of systems biology (Subirana-Granés et al., 2025; Amelkina and Comizzoli, 2026).

To manage uncertainty and variability in breeding data, probabilistic machine learning techniques are essential. Probabilistic models explicitly represent confidence and risk by generating distributions over potential outcomes instead of single-point predictions. This probabilistic viewpoint facilitates decision-making under uncertainty in precision breeding, enabling breeders to assess trade-offs between robustness and expected gain. Furthermore, probabilistic methods make it easier to integrate existing biological knowledge, allowing models to integrate empirical data with well-established knowledge of physiological and genetic processes (Shoyombo et al., 2024).

One new and growing aspect of machine learning in livestock breeding is the application of causal inference techniques. Although many machine learning models are very good at making predictions, they are not naturally able to discern between causation and correlation. By simulating the directional relationships between variables and assessing the results of hypothetical interventions, causal inference techniques overcome this limitation. Causal analysis helps determine how changes in management or selection are likely to affect results in precision breeding, thereby lowering the possibility of unforeseen consequences. Breeding decisions are more reliable and ethically sound when causal reasoning is incorporated into machine learning frameworks (Clarke, 2023).

Techniques for model interpretability and explainability are crucial for the ethical use of machine learning in livestock breeding. Stakeholders must be able to comprehend the reasoning behind model outputs because breeding decisions have long-term, system-wide effects. By connecting predictions to underlying characteristics or representations, explainability techniques shed light on the biological forces behind selection. Interpretability in precision breeding ensures that AI-driven recommendations can be examined and improved by human experts, supporting scientific validation, regulatory compliance, and stakeholder trust (Dorsch and Moll, 2024).

All machine learning applications in livestock breeding are based on data preprocessing and normalization techniques. Before modeling, biological data must be carefully prepared because it is frequently noisy, heterogeneous, and incomplete. To ensure that models learn significant biological patterns rather than technical artifacts, preprocessing techniques address problems such as missing values, batch effects, scaling differences, and outlier detection. Given the integration of several data sources gathered under various circumstances, thorough data preprocessing is especially crucial in precision breeding (Ullah and Waheed, 2025; Zhu et al., 2025).

Techniques for model evaluation and validation are essential for evaluating the dependability and generalizability of machine learning models. Evaluation in breeding contexts must take into account biological plausibility, stability over time, and robustness across environments, in addition to predictive accuracy. To evaluate transferability, validation strategies frequently involve testing models on separate populations or under various management circumstances. Thorough testing prevents overfitting and ensures that machine learning models are useful outside of the datasets on which they were trained (Crossa et al., 2025).

Practical factors such as scalability and computational efficiency influence the selection and application of machine learning methods in precision breeding. Large populations and constant data streams may be present in livestock breeding programs, necessitating models that can grow without incurring unaffordable computational costs. The application of advanced machine learning techniques in practical breeding contexts is made possible by developments in distributed computing, parallel processing, and algorithmic efficiency. Methodological decisions are influenced by the continuous challenge of striking a balance between computational viability and model complexity (Roy et al., 2022).

Data analysis for precision breeding also heavily relies on human-machine interaction techniques. To integrate computational insight with experiential knowledge, machine learning outputs must be effectively communicated to breeders and decision-makers. This interaction is made easier by visualization, interactive interfaces, and decision-support systems, which allow users to examine scenarios, evaluate uncertainty, and adjust priorities. These methods support collaborative decision-making frameworks by ensuring that machine learning supplements human judgment rather than replacing it (Liu and Peng, 2025). Machine learning and data analysis techniques applied in precision livestock breeding are listed in Table 14.3.

14.4 Ethical considerations and challenges

Given the significant biological, societal, and intergenerational effects of breeding choices, ethical issues and challenges play a crucial role in the development and implementation of AI-enabled precision breeding of livestock. Because precision breeding operates at the nexus of scientific advancement, animal life, food systems, environmental stewardship, and public trust, ethical reflection is a fundamental necessity rather than an optional supplement. AI increases the power, scope, and speed of breeding decisions, but it also increases ethical responsibility, necessitating careful consideration of the objectives' definition, the use of data, the justification of decisions, and the management of long-term effects (Ghavi Hossein-Zadeh, 2025; Tedeschi et al., 2025).

Animal welfare is one of the most important ethical issues in precision breeding. AI makes it possible to evaluate and forecast biological characteristics, such as those pertaining to efficiency, productivity, and physiological function, with ever-greater precision. These abilities have the potential to enhance health and lessen suffering, but they also run the risk of increasing selection for characteristics that might unintentionally jeopardize welfare. Narrow attention to production-related traits can result in unintended physiological burdens, behavioral constraints, or decreased resilience, as historical breeding practices have shown. By optimizing biological systems toward extremes that surpass animals' capacity for adaptation, precision breeding could increase these risks if it is only motivated by efficiency-driven goals. Therefore, to ensure that welfare is treated as a primary outcome rather than a secondary constraint, ethical precision breeding necessitates the explicit integration of welfare considerations into model objectives, evaluation criteria, and decision-making processes (Ghavi Hossein-Zadeh, 2025; Panwar et al., 2025; Zhu et al., 2025).

TABLE 14.3 Machine learning and data analysis techniques applied in precision livestock breeding.

Technique category	Conceptual purpose	Type of data utilized	Analytical strengths	Limitations and considerations
Supervised learning	• Learn mappings between biological inputs and breeding outcomes	• Genomic data • Multi-omics profiles • Phenotypic and environmental data	• High predictive accuracy • Direct relevance to breeding objectives	• Sensitive to biased training data • Risk of overfitting under selection
Unsupervised learning	• Discover intrinsic structure without predefined outcomes	• Genomic variation • Molecular profiles • Phenomic data	• Reveals latent biological states • Supports population structure analysis	• Does not directly optimize breeding targets
Semi-supervised learning	• Leverage limited labeled data with abundant unlabeled data	• Sparse phenotypes • Large-scale omics datasets	• Improves scalability and robustness • Reduces phenotyping burden	• Performance depends on shared data structure
Representation learning	• Extract compact, informative biological features	• High-dimensional omics • Sensor-derived phenotypes	• Reduces noise and redundancy • Enhances interpretability	• Latent features may be biologically abstract
Ensemble learning	• Combine multiple models to improve stability	• Integrated multisource data	• Reduces model-specific bias • Enhances robustness	• Increased computational complexity
Dimensionality reduction	• Simplify high-dimensional biological data	• Omics and phenomics data	• Improves efficiency and visualization • Highlights dominant variation axes	• Potential loss of subtle biological signals
Feature selection methods	• Identify biologically informative variables	• Genomic markers • Molecular features	• Enhances interpretability • Reduces computational load	• Risk of excluding context-dependent features
Temporal and longitudinal modeling	• Capture biological dynamics over time	• Time-series phenotypes • Environmental data	• Reflects development and adaptation • Enables early prediction	• Requires dense and consistent sampling
Graph-based and network learning	• Model biological interactions explicitly	• Regulatory networks • Interaction data	• Captures system connectivity and modularity	• Network inference sensitive to data quality
Probabilistic modeling	• Quantify uncertainty in predictions	• Multi-omics and phenotypic data	• Supports risk-aware decision-making • Represents outcome distributions	• Computationally demanding
Causal inference methods	• Distinguish causation from correlation	• Observational multi-omics data	• Enables intervention-aware reasoning • Reduces unintended consequences	• Strong assumptions may be required
Multitask learning	• Learn shared structure across related traits	• Multitrait phenotypes	• Improves efficiency and consistency	• Negative transfer risk if tasks differ

Transfer learning	• Apply knowledge across populations or systems	• Cross-breed or cross-environment data	• Enhances generalizability • Reduces data requirements	• Context mismatch may degrade performance
Active learning	• Optimize data collection strategies	• Candidate individuals for phenotyping	• Reduces the cost of data acquisition	• Requires adaptive experimental design
Explainable AI techniques	• Enhance transparency and trust	• Model outputs and latent features	• Supports biological validation • Facilitates ethical oversight	• Trade-off between accuracy and simplicity
Anomaly and outlier detection	• Identify unusual biological patterns	• Phenotypic and omics data	• Detects data errors or novel phenotypes	• May flag biologically meaningful extremes
Optimization algorithms	• Identify optimal breeding strategies	• Predicted breeding values and constraints	• Supports multiobjective decision-making	• Solutions depend on objective formulation

The ethical dilemma of defining what constitutes improvement in livestock breeding is closely tied to welfare. AI systems optimize toward human-specified targets, but these targets invariably reflect value judgments about desirable biological states. Choosing which qualities to emphasize, how to balance conflicting goals, and how to measure success are normative rather than purely technical decisions. As a result, precision breeding needs to consider the ethical implications of value encoding in algorithms. The resulting breeding trajectories may be at odds with societal expectations regarding animal dignity, natural behavior, or ecological responsibility if economic efficiency is prioritized over all other considerations. To ensure that AI-driven breeding is consistent with more general ethical commitments, transparent articulation of breeding goals and inclusive discussion of value trade-offs are crucial (Foris et al., 2025; Ryan and Blok, 2026).

Another significant ethical issue is genetic diversity and long-term population resilience. AI makes extremely accurate selection possible, which could speed up genetic gain but also raise the possibility of genetic homogenization. Genetic diversity loss can reduce future breeding options, increase susceptibility to new diseases, and compromise adaptability. Since present breeding choices influence the biological options accessible to future populations and societies, this raises ethical concerns about intergenerational responsibility. Therefore, precision breeding frameworks must incorporate mechanisms that maintain diversity and prevent irreversible genetic resource narrowing, balancing short-term gains against long-term resilience. By simulating diversity trajectories and assessing long-term effects, AI can support this ethical imperative, but only if these factors are given specific priority (Ukpaka, 2024; Zsolnai et al., 2025).

Precision breeding raises serious ethical issues related to data governance, ownership, and consent when using large amounts of biological and phenotypic data. Breeders, farmers, businesses, and occasionally public institutions, all of whom have legitimate interests and expectations, create livestock data across intricate networks. These data are aggregated and analyzed by AI systems at scales that may mask individual contributions and make value attribution more difficult. The distribution of risks and rewards, who controls the data, and who gains from insights gleaned from it all raise ethical questions. Therefore, transparent and equitable data governance frameworks that uphold stakeholder rights, encourage fair benefit sharing, and protect against the exploitation or exclusion of participants with fewer resources must go hand in hand with precision breeding (Akinyemi et al., 2025; Ghavi Hossein-Zadeh, 2025).

Even in livestock systems, privacy and confidentiality serve as additional ethical dimensions. While animal data are closely linked to human livelihoods, business strategies, and competitive advantage, they do not possess privacy rights. If not properly regulated, AI-driven breeding platforms that centralize and analyze data across farms or regions could potentially expose sensitive information. Strong safeguards are necessary to protect data confidentiality and ensure that data-sharing initiatives are voluntary, informed, and reversible to foster ethical precision breeding. Analytical approaches that are federated and privacy-preserving offer technical means to tackle these concerns. However, their ethical efficacy relies on the establishment of transparent governance and trust among stakeholders (Coghlan and Cardilini, 2025; Neethirajan, 2025).

Another crucial ethical issue is algorithmic explainability and transparency. Precision breeding AI systems frequently rely on intricate, hard-to-understand models, which raises questions about trust and accountability. It is unethical to rely on vague recommendations that cannot be examined or supported because breeding decisions can have permanent effects on animal populations and production systems. Therefore, explainability is an ethical requirement rather than just a technical preference. Stakeholders must be able to comprehend the assumptions underlying models, how and why specific breeding decisions are advised, and what uncertainties still exist. Informed consent, ethical oversight, and bias or error correction before harm occurs are all made possible by transparent AI systems (Chaudhary, 2024; Nannini et al., 2024).

The ethical landscape of AI-enabled precision breeding is further complicated by concerns about bias and fairness. AI systems gain knowledge from past data that may represent current disparities between breeds, production systems, or geographical areas. These biases have the potential to be reinforced by limited definitions of genetic merit or to disadvantage specific populations if left unchecked. To ensure that models perform equally across a range of genetic backgrounds and environmental contexts, ethical precision breeding necessitates the proactive identification and mitigation of bias. Access to AI-enabled breeding technologies and the distribution of their advantages are examples of fairness considerations that go beyond technical performance, especially in environments with limited resources (Verma et al., 2024; Panwar et al., 2025).

Ethical concerns regarding human agency and responsibility are raised by the growing automation of breeding decisions. There is a risk that decision-makers will become less accountable as AI systems take on more responsibility for assessing genetic merit and suggesting selection strategies. To maintain human accountability for choices and results, ethical precision breeding must maintain distinct lines of accountability. Instead of making decisions on its own, AI should be used as a tool to support human judgment while still being subject to review, criticism, and modification (Neethirajan, 2025).

Precision breeding and environmental ethics are closely related. Breeding choices affect resource use, emissions, and ecosystem interactions over extended periods of time, and livestock systems contribute to environmental stresses. AI-enabled precision breeding presents opportunities to choose for characteristics that increase productivity and lessen environmental impact, but it also runs the risk of prioritizing productivity gains over ecological constraints. Since livestock systems are integrated into larger ecological systems, ethical precision breeding requires incorporating environmental sustainability into breeding goals and assessment frameworks. By simulating environmental trade-offs, AI can facilitate this integration; however, ethical dedication is necessary to ensure that these factors have a significant impact on decisions (Neethirajan, 2023; Panwar et al., 2025).

In precision breeding, the ethical dilemma of uncertainty warrants special consideration. Because of biological variability, incomplete data, and uncertain future conditions, AI models will always operate under uncertainty. Predictions presented without proper context can lead to dangerous decisions and false confidence. In order for stakeholders to comprehend the boundaries of model knowledge and to properly assess risks, ethical practice demands open communication of uncertainty. By recognizing the tentative nature of forecasts and promoting caution in situations with dire or irreversible consequences, uncertainty-aware modeling facilitates ethical decision-making (Okengwu et al., 2023).

Another crucial ethical factor is public opinion and societal legitimacy. Public scrutiny of livestock breeding practices is growing, and the application of AI may raise questions about technological overreach, the loss of naturalness, or the commercialization of animal life. Therefore, rather than assuming implicit acceptance, ethical precision breeding must actively engage with societal values, promoting communication and transparency. Establishing trust requires proving that AI-driven breeding promotes welfare, sustainability, and ethical responsibility in addition to economic interests. If public concerns are not addressed, the social license required for ongoing innovation may be compromised (Ryan and Blok, 2026).

To overcome these obstacles, interdisciplinary ethical governance is crucial. Precision breeding raises ethical issues in biology, data science, economics, philosophy, and law, requiring interdisciplinary cooperation. Governance frameworks must develop in tandem with technological advancements, offering flexible yet ethically sound guidance. A promising approach to balancing innovation and responsibility is ethics-by-design, which incorporates ethical considerations directly into system architecture and goals. Ethical dimensions and governance challenges in AI-enabled precision breeding are shown in Table 14.4.

14.5 Future trends and innovations in AI for precision livestock breeding

Future developments in AI for precision livestock breeding indicate a significant change in the way biological improvement is conceived, implemented, and regulated. AI is set to transform from a potent analytical tool into an adaptive, integrative, and ethically embedded infrastructure that continuously shapes genetic trajectories as breeding systems advance further into the era of digital biology. These future advancements are characterized by conceptual innovations that align biological complexity, technological capability, and societal responsibility within unified decision-making ecosystems rather than just faster algorithms or larger datasets (Rosati, 2025; Tedeschi et al., 2025).

The development of fully integrated, multiscale breeding intelligence systems is one of the most important upcoming developments in AI-enabled precision breeding. Future frameworks will integrate these elements into cohesive architectures that model the livestock organism as a dynamic system embedded within production and environmental contexts, rather than treating genomic prediction, phenotypic evaluation, and management optimization as distinct analytical tasks. AI will increasingly function simultaneously across molecular, physiological, behavioral, and ecological scales, enabling system-level knowledge rather than discrete metrics to guide breeding decisions. Selection strategies that maximize individual traits as well as the stability, adaptability, and resilience of the entire biological system over time will be supported by this integration (Ullah and Waheed, 2025; Zhu et al., 2025).

This change will be greatly aided by developments in representation learning. Future AI systems will create representations of livestock genotypes and phenotypes that are more abstract and biologically significant, capturing latent structures that correlate to physiological states, regulatory modules, and adaptive tactics. Breeding decisions will be based on comprehensive biological profiles rather than specific performance indicators, thanks to these representations, which will go beyond conventional trait definitions. AI will speed up innovation in various breeding contexts by reducing reliance on large amounts of local data and facilitating knowledge transferability by learning shared representations across populations, environments, and production systems (Dorea and Menezes, 2024; Zhu et al., 2025).

The incorporation of causal reasoning into AI-driven breeding frameworks will be another significant innovation. Future systems will place a greater emphasis on comprehending the cause-and-effect relationships that control biological outcomes, even though current machine learning techniques are excellent at prediction. Breeders will be able to assess the

TABLE 14.4 Ethical dimensions and governance challenges in AI-enabled precision breeding.

Ethical dimension	Core ethical concern	Manifestation in precision breeding	Governance challenges	Role of AI in mitigation
Animal welfare	• Avoidance of harm and suffering	• Risk of selecting for extreme productivity traits • Potential mismatch between genotype and welfare needs	• Defining enforceable welfare thresholds • Monitoring long-term welfare outcomes	• Integrates welfare indicators into optimization • Enables early detection of welfare risks
Genetic diversity	• Preservation of long-term adaptive potential	• Accelerated homogenization through precise selection	• Balancing short-term gain with long-term resilience	• Models diversity trajectories and inbreeding risk
Transparency and explainability	• Accountability of algorithmic decisions	• Opaque model recommendations	• Ensuring interpretability for stakeholders	• Provides explainable outputs and decision rationales
Fairness and bias	• Equitable treatment across populations	• Unequal model performance across breeds or systems	• Identifying and correcting historical bias	• Detects and mitigates systematic bias
Data ownership and sovereignty	• Control over biological and phenotypic data	• Centralized data aggregation	• Clarifying rights and benefit sharing	• Supports federated and decentralized analysis
Privacy and confidentiality	• Protection of sensitive information	• Exposure of farm-level or business data	• Implementing secure data governance	• Enables privacy-preserving learning
Human responsibility and agency	• Avoidance of automation bias	• Overreliance on algorithmic recommendations	• Maintaining human accountability	• Functions as decision support, not replacement
Ethical objective encoding	• Translation of values into models	• Implicit prioritization of efficiency	• Making value trade-offs explicit	• Formalizes ethical constraints in optimization
Intergenerational responsibility	• Long-term consequences of selection	• Irreversible genetic changes	• Assessing multigenerational impact	• Simulates long-term outcomes
Environmental sustainability	• Minimization of ecological harm	• Selection without ecological context	• Integrating environmental metrics	• Models environmental trade-offs
Uncertainty communication	• Avoidance of false confidence	• Presentation of point estimates only	• Ensuring risk-aware decision-making	• Quantifies and visualizes uncertainty
Regulatory oversight	• Compliance with evolving standards	• Rapid technological advancement	• Keeping regulation aligned with innovation	• Supports auditability and traceability
Stakeholder inclusion	• Participatory governance	• Exclusion of small-scale or resource-limited actors	• Ensuring inclusive decision-making	• Facilitates transparent and collaborative platforms
Public trust and social license	• Societal acceptance of AI-driven breeding	• Perception of unnatural manipulation	• Addressing public concerns	• Enables transparent communication
Ethics-by-design implementation	• Embedding ethics into system architecture	• Ethics treated as a post hoc constraint	• Integrating ethical principles early	• Encodes ethics within model objectives

probable outcomes of selection choices under fictitious future circumstances thanks to causal modeling, which will promote safer and more dependable interventions. This move toward causal intelligence will lessen the possibility of unforeseen outcomes that occur when correlations are confused with mechanisms, especially in complex systems where selection pressures can interact with environmental change and spread across generations (Kosaraju, 2022; Kitov, 2025).

Another significant future trend is the creation of systems for continuous and adaptive learning. Over extended periods of time, genetic architectures, environmental stresses, and management techniques change as livestock breeding takes place. Static models that were developed using historical data run the risk of becoming outdated and producing recommendations that are either inaccurate or subpar. Continuous learning mechanisms will be incorporated into future AI systems, enabling models to retain accumulated knowledge while updating gradually as new data becomes available. To remain relevant and resilient over time, these adaptive systems will allow breeding programs to react dynamically to new challenges, such as climate variability, emerging diseases, or changing market demands (García-Vázquez, 2024; Hamadani et al., 2024).

AI will play a bigger part in precision breeding thanks to simulation-based innovation. With the use of sophisticated AI-driven simulation frameworks, breeding strategies can be virtually explored over several generations, assessing the long-term effects on genetic diversity, resilience, welfare, and sustainability. These simulations will serve as virtual testing grounds where different selection scenarios can be evaluated without making a permanent commitment to the real world. AI-driven simulations will support risk mitigation and strategic foresight by fusing biological realism with computational experimentation, giving breeders more confidence to navigate uncertainty (Ghavi Hossein-Zadeh, 2026b; Pérez-Enciso, Zingaretti and de los Campos, 2025).

The idea of digital twins is likely to have a bigger impact on precision breeding in the future. Digital twins are virtual versions of biological systems that mimic and develop in tandem with their physical counterparts. Digital twins of animals, populations, or entire breeding programs could incorporate phenotypic trajectories, multi-omics data, and environmental inputs to mimic responses to management and selection interventions in livestock breeding. To continuously align model predictions with observed biological outcomes, AI will play a key role in maintaining and updating these digital twins. As a result, breeding will become a predictive and anticipatory discipline due to these innovations (Arulmozhi et al., 2024; Ghavi Hossein-Zadeh, 2025).

One of the key innovations in AI-enabled precision breeding in the future will be ethics-by-design. Future systems will incorporate sustainability, diversity, and welfare goals into their optimization frameworks, rather than viewing ethical issues as external limitations. AI will be used to codify ethical priorities, converting them into quantifiable metrics and limitations that direct choices. Through this integration, ethical trade-offs will be openly examined, assisting stakeholders in making well-informed decisions. Precision breeding can develop in a way that balances technological power with social values by explicitly incorporating ethical considerations into computational architectures (Šekrst et al., 2024).

Another crucial future path is the incorporation of environmental intelligence. Future AI frameworks will increasingly integrate environmental data and ecosystem-level considerations into breeding models because livestock breeding is not isolated from ecological systems (Ghavi Hossein-Zadeh, 2022, 2023). AI will make it possible to evaluate how genetic selection affects emissions, resource consumption, and environmental resilience over time. By balancing productivity with ecological responsibility, this integration will support breeding practices that support sustainable food systems. AI-driven precision breeding will match biological advancement with more general sustainability objectives by connecting molecular and physiological regulation to environmental outcomes (Vlaicu et al., 2024; Džermeikaitė et al., 2025).

The future of AI in precision breeding will also be shaped by innovation in data ecosystems and interoperability. The capacity to integrate data across platforms, institutions, and geographical areas will become increasingly crucial as data volumes and diversity continue to grow. Federated and decentralized learning architectures that allow for cooperative analysis without centralized data ownership will be the main focus of future developments. These methods will respect data sovereignty and privacy while enabling breeding programs to benefit from collective knowledge. By learning common representations and models across dispersed data sources, AI will enable this cooperation and promote large-scale innovation (Ghavi Hossein-Zadeh, 2025; Ogunbajo et al., 2025).

The co-evolution of humans and AI will be a major theme in precision breeding in the future. AI will increasingly serve as a cognitive partner that enhances decision-making ability, rather than taking the place of human expertise. Breeders will be able to engage more deeply with complex trade-offs thanks to interactive AI systems that facilitate scenario exploration, uncertainty assessment, and ethical reflection. To bridge the gap between computational complexity and practical judgment, visualization and decision-support interfaces will be developed to present systems-level insights in understandable forms. AI-driven breeding decisions will be more legitimate and effective thanks to this cooperative paradigm (Hemmer et al., 2024).

TABLE 14.5 Future trends and technological innovations shaping AI-driven precision breeding.

Emerging trend or innovation	Core conceptual shift	Enabling AI technologies	Anticipated impact on precision breeding	Key considerations and challenges
Systems-level breeding intelligence	• From trait-centric to system-centric selection	• Deep representation learning • Multimodal data fusion	• Captures emergent biological behavior • Improves robustness and resilience	• Increased model complexity • Higher data integration demands
Causal AI in breeding decisions	• From correlation-based to mechanism-aware inference	• Causal modeling frameworks • Counterfactual reasoning	• Reduces unintended consequences • Enables safer intervention planning	• Requires strong assumptions • Limited experimental validation
Continual and adaptive learning systems	• From static to evolving predictive models	• Online learning • Incremental model updating	• Maintains relevance under changing conditions • Adapts to new environments and data	• Risk of model drift • Governance of updates
Digital twins of animals and populations	• Virtual representation of biological systems	• Simulation-based AI • Systems modeling	• Enables virtual experimentation • Supports long-term strategy evaluation	• High computational cost • Model validation complexity
Foundation models for livestock biology	• Learning universal biological representations	• Large-scale pretrained AI models	• Improves transferability across populations • Reduces local data requirements	• Risk of hidden bias • Resource-intensive training
Multiobjective optimization frameworks	• Explicit navigation of trade-offs	• Evolutionary algorithms • Pareto optimization	• Balances productivity, welfare, and sustainability	• Objective weighting subjectivity
Ethics-by-design AI architectures	• Ethics embedded directly in models	• Constraint-aware optimization • Value-aligned learning	• Aligns breeding outcomes with societal values	• Formalizing ethical concepts
Environmental and climate-aware breeding models	• Integration of ecological context	• Context-aware AI • Environmental data fusion	• Enhances climate resilience • Supports sustainable production	• Uncertainty in long-term climate projections

Federated and privacy-preserving learning	• Distributed collaboration without data centralization	• Federated learning • Secure computation	• Enhances data sharing and trust	• Coordination complexity
Human-AI co-decision platforms	• From automation to collaboration	• Explainable AI • Interactive decision-support tools	• Preserves human accountability • *Improves decision quality*	• Training and usability *requirements*
Uncertainty-aware breeding strategies	• From point estimates to risk-sensitive decisions	• Probabilistic AI • Bayesian modeling	• Reduces overconfident selection • Enhances long-term stability	• Communication of uncertainty
Integration of real-time phenotyping	• Continuous monitoring of biological state	• Streaming data analytics • Edge AI	• Enables rapid feedback loops	• Data volume management
Cross-system knowledge transfer	• Learning across breeds and environments	• Transfer and meta-learning	• Accelerates innovation in under-resourced systems	• Risk of negative transfer
AI-guided diversity preservation	• Diversity treated as a system objective	• Population genetics modeling • Optimization under constraints	• Protects long-term adaptability	• Balancing gain and diversity
Regulatory and audit-ready AI systems	• Built-in transparency and traceability	• Model interpretability tools • Logging and version control	• Facilitates ethical oversight and regulation	• Increased system complexity

TABLE 14.6 Integration of biological, environmental, and ethical objectives in precision breeding decision frameworks.

Objective dimension	Core focus	Key biological or system indicators	Decision trade-offs addressed	Role of AI in integration
Biological performance	• Enhancement of genetic and physiological function	• Productivity-related traits • Functional robustness indicators	• Maximizing performance without compromising stability	• Learns nonlinear genotype-phenotype relationships • Predicts system-level outcomes
Animal health	• Maintenance of physiological integrity	• Immune and metabolic indicators • Disease susceptibility signals	• Balancing production gain with health resilience	• Integrates multi-omics health predictors • Detects early deviation from homeostasis
Animal welfare	• Minimization of stress and suffering	• Behavioral and physiological welfare proxies	• Avoiding selection-induced welfare burdens	• Embeds welfare constraints into optimization • Monitors welfare-related risk
Genetic diversity	• Preservation of adaptive potential	• Population structure metrics • Inbreeding trajectories	• Short-term gain versus long-term resilience	• Models diversity dynamics • Constrains selection to maintain variability
Developmental stability	• Consistency across life stages	• Growth and maturation trajectories	• Rapid gain versus developmental robustness	• Captures longitudinal biological patterns
Environmental sustainability	• Reduction of ecological footprint	• Resource-use efficiency indicators • Emission-related proxies	• Productivity versus environmental impact	• Integrates environmental context into predictions
Climate adaptability	• Resilience to environmental variability	• Stress-response and adaptation markers	• Specialization versus adaptability	• Learns context-dependent genetic effects
Production system compatibility	• Alignment with management conditions	• Phenotype-environment matching indicators	• Genetic potential versus management capacity	• Supports environment-specific optimization
Economic viability	• Long-term financial sustainability	• Cost-benefit and risk indicators	• Immediate profit versus long-term stability	• Evaluates probabilistic economic outcomes
Ethical acceptability	• Alignment with societal values	• Welfare, sustainability, and fairness criteria	• Technical efficiency versus ethical responsibility	• Formalizes ethical priorities in decision models
Transparency and accountability	• Justifiability of decisions	• Model interpretability indicators	• Model complexity versus explainability	• Provides interpretable and auditable outputs
Risk and uncertainty management	• Robustness under uncertainty	• Variance and confidence measures	• Aggressive optimization versus safety margins	• Quantifies and propagates uncertainty
Intergenerational responsibility	• Long-term population health	• Multigenerational performance projections	• Present gain versus future flexibility	• Simulates long-term selection outcomes
Stakeholder equity	• Fair access to breeding benefits	• Distributional impact indicators	• Centralized advantage versus inclusivity	• Enables decentralized and federated learning
Regulatory and governance alignment	• Compliance and oversight	• Traceability and audit indicators	• Innovation speed versus governance control	• Supports traceable and compliant AI systems

The customization of breeding techniques is another new trend. More precise customization of selection choices to particular production contexts, environmental conditions, and management objectives will be possible with future AI systems. AI will enable context-specific optimization that takes into account local opportunities and constraints, rather than applying consistent breeding objectives across various systems. The impact and relevance of breeding programs will be increased by this personalization, especially in environments with varying resources and heterogeneity (Rosa et al., 2023).

Future precision breeding will see an increase in the use of risk-sensitive and uncertainty-aware AI. Errors have more serious repercussions as breeding choices have a greater impact on long-term system trajectories. Breeders will be able to assess not only expected results but also worst-case scenarios and resilience under stress, thanks to future AI systems that explicitly represent uncertainty. This risk-sensitive viewpoint will encourage more cautious and sound decision-making, thereby lowering susceptibility to unanticipated difficulties (Borgohain et al., 2026).

To fully utilize upcoming AI advancements in precision breeding, education, and capacity building will be crucial. The need for interdisciplinary knowledge spanning biology, data science, ethics, and systems thinking will increase as systems become more complex. Future developments will include the creation of cooperative platforms and training courses that enable scientists, breeders, and other stakeholders to interact with AI-driven systems in an efficient manner. To ensure that precision breeding remains rooted in biological knowledge and ethical responsibility, developing this human capacity is just as crucial as technological innovation itself (Ryan et al., 2023; Borgohain et al., 2026).

The future of AI in precision livestock breeding will also be shaped by innovations in governance and regulations. Regulatory frameworks will need to change as AI-driven decision-making gains traction to address concerns about fairness, accountability, and transparency. Future models of governance will likely place a strong emphasis on flexible regulation that strikes a balance between oversight and innovation, allowing for responsible experimentation while defending the interests of the general public. By offering auditability, traceability, and scenario analysis that guide policy development, AI can assist these governance initiatives (Rosati, 2025; White, 2025).

In general, a change in perspective from optimization to stewardship will define the future of AI in precision livestock breeding. Future breeding frameworks will prioritize long-term system health, adaptability, and ethical alignment over maximizing short-term gains. By enabling a comprehensive assessment of breeding results across biological, environmental, and societal dimensions, AI will facilitate this shift. AI-driven precision breeding can support robust and sustainable livestock systems that satisfy the demands of the current and future generations by fusing ethics, foresight, and systems-level knowledge. Future trends and technological innovations shaping AI-driven precision breeding are presented in Table 14.5. To synthesize the biological, technological, and ethical dimensions discussed throughout this chapter, an integrative framework for AI-enabled precision breeding is presented in Table 14.6.

References

Akinyemi, B.E., Jessiman, L., Turner, S.P., Siegford, J.M., 2025. Beyond the farm: stakeholder perspectives on precision livestock farming in the swine industry. Front. Anim. Sci. 6. https://doi.org/10.3389/fanim.2025.1710969.

Amelkina, O., Comizzoli, P., 2026. Making sense of expanding transcriptomic data: network-based approaches for studying reproduction in domestic and wild animal species. Front. Vet. Sci. 12. https://doi.org/10.3389/fvets.2025.1728981.

Andersen, L.K., Reading, B.J., 2024. A supervised machine learning workflow for the reduction of highly dimensional biological data. Artif. Intell. Life Sci. 5. https://doi.org/10.1016/j.ailsci.2023.100090.

Arulmozhi, E., Deb, N.C., Tamrakar, N., Kang, D.Y., Kang, M.Y., Kook, J., Basak, J.K., Kim, H.T., 2024. From reality to virtuality: revolutionizing livestock farming through digital twins. Agriculture (Switzerland) 14 (12). https://doi.org/10.3390/agriculture14122231.

Baião, A.R., Cai, Z., Poulos, R.C., Robinson, P.J., Reddel, R.R., Zhong, Q., Vinga, S., Gonçalves, E., 2025. A technical review of multi-omics data integration methods: from classical statistical to deep generative approaches. Briefings Bioinf. 26 (4). https://doi.org/10.1093/bib/bbaf355.

Ballard, J., Wang, Z., Li, W., Shen, L., Long, Q., 2024. Deep learning-based approaches for multi-omics data integration and analysis. BioData Min. 17 (1). https://doi.org/10.1186/s13040-024-00391-z.

Boktar, M., 2025. Revolutionizing animal breeding: quantifying the impact of genomic selection, CRISPR, and artificial intelligence on animal breeding. Anim. Rep. 1 (2), 100–117. https://doi.org/10.64636/ar.22.

Borgohain, T., Suma, R., Muttappagol, M., Saikia, B., Keithellakpam, A., Laskar, A., Hiremath, S.S., Basu, U., Velmurugan, N., Palakolanu, S.R., Chikkaputtaiah, C., 2026. Precision breeding in a changing climate: unlocking resilience through omics and gene editing. Funct. Integr. Genomics 26 (1). https://doi.org/10.1007/s10142-025-01796-7.

Chaudhary, G., 2024. Unveiling the black box: bringing algorithmic transparency to AI. Masaryk Univ. J. Law Technol. 18 (1), 93–122. https://doi.org/10.5817/mujlt2024-1-4.

Clarke, J., 2023. Leveraging AI/Ml to address critical challenges in livestock research. J. Anim. Sci. 101 (Suppl. 3), 142. https://doi.org/10.1093/jas/skad281.173.

Coghlan, S., Cardilini, A., 2025. Animal privacy and information technology. Philos. Technol. 38 (2). https://doi.org/10.1007/s13347-025-00900-w.

Crossa, J., Montesinos-Lopez, O.A., Costa-Neto, G., Vitale, P., Martini, J.W.R., Runcie, D., Fritsche-Neto, R., Montesinos-Lopez, A., Pérez-Rodríguez, P., Gerard, G., Dreisigacker, S., Crespo-Herrera, L., Pierre, C.S., Lillemo, M., Cuevas, J., Bentley, A., Ortiz, R., 2025. Machine learning algorithms translate big data into predictive breeding accuracy. Trends Plant Sci. 30 (2), 167–184. https://doi.org/10.1016/j.tplants.2024.09.011.

Deglurkar, S., Shen, H., Muthali, A., Pavone, M., Margineantu, D., Karkus, P., Ivanovic, B., Tomlin, C.J., 2024. System-Level Analysis of Module Uncertainty Quantification in the Autonomy Pipeline. https://doi.org/10.48550/arXiv.2410.12019.

Dorea, J., Menezes, G.L., 2024. Artificial intelligence and machine learning to improve livestock farming. J. Anim. Sci. 102 (Suppl. 3), 296. https://doi.org/10.1093/jas/skae234.338.

Dorsch, J., Moll, M., 2024. Explainable and Human-Grounded AI for Decision Support Systems: The Theory of Epistemic Quasi-Partnerships. https://doi.org/10.48550/arXiv.2409.14839.

Ducrot, C., Barrio, M.B., Boissy, A., Charrier, F., Even, S., Mormède, P., Petit, S., Pinard-van der laan, M.-H., Schelcher, F., Casabianca, F., Ducos, A., Foucras, G., Guatteo, R., Peyraud, J.-L., Vayssier-Taussat, M., Veysset, P., Friggens, N.C., Fernandez, X., 2024. Animal board invited review: improving animal health and welfare in the transition of livestock farming systems: towards social acceptability and sustainability. Animal 18 (3), 101100. https://doi.org/10.1016/j.animal.2024.101100.

Džermeikaitė, K., Šidlauskaitė, M., Antanaitis, R., Anskienė, L., 2025. Enhancing genomic selection in dairy cattle through artificial intelligence: integrating advanced phenotyping and predictive models to advance health, climate resilience, and sustainability. Dairy 6 (5), 50. https://doi.org/10.3390/dairy6050050.

Foris, B., Sheng, K., Dürnberger, C., Oczak, M., Rault, J.L., 2025. AI for one welfare: the role of animal welfare scientists in developing valid and ethical AI-based welfare assessment tools. Front. Vet. Sci. 12. https://doi.org/10.3389/fvets.2025.1645901.

Fu, L., Wang, Y.G., Wu, J., 2024. Recent advances in longitudinal data analysis. Handb. Stat. 50, 173–221. https://doi.org/10.1016/bs.host.2023.10.007.

Garcia-Oliveira, A.L., Dwivedi, S.L., Chander, S., Nelimor, C., Abd El Moneim, D., Ortiz, R.O., 2026. Breeding smarter: artificial intelligence and machine learning tools in modern breeding—a review. Agronomy 16 (1). https://doi.org/10.3390/agronomy16010137.

García-Vázquez, F.A., 2024. Artificial intelligence and porcine breeding. Anim. Reprod. Sci. 269, 107538. https://doi.org/10.1016/j.anireprosci.2024.107538.

Ghavi Hossein-Zadeh, N., 2022. Estimates of the genetic contribution to methane emission in dairy cows: a meta-analysis. Sci. Rep. 12 (1), 12352. https://doi.org/10.1038/s41598-022-16778-z.

Ghavi Hossein-Zadeh, N., 2023. A meta-analysis of the genetic contribution to greenhouse gas emission in sheep. J. Anim. Breed. Genet. 140 (1), 49–59. https://doi.org/10.1111/jbg.12744.

Ghavi Hossein-Zadeh, N., 2024. An overview of recent technological developments in bovine genomics. Vet. Anim. Sci. 25, 100382. https://doi.org/10.1016/j.vas.2024.100382.

Ghavi Hossein-Zadeh, N., 2025. Artificial intelligence in veterinary and animal science: applications, challenges, and future prospects. Comput. Electron. Agric. 235, 110395. https://doi.org/10.1016/j.compag.2025.110395.

Ghavi Hossein-Zadeh, N., 2026a. Advancing climate-resilient livestock systems: next-generation emission mitigation strategies and integrated technological innovations. Vet. Anim. Sci., 100588 https://doi.org/10.1016/J.VAS.2026.100588.

Ghavi Hossein-Zadeh, N., 2026b. Generative artificial intelligence in animal genomics for smart agriculture: Applications, challenges, and future prospects. Vet. Anim. Sci. 33, 100702. https://doi.org/10.1016/J.VAS.2026.100702.

Ghose, S., Tse, Y.F., Rasaee, Ķ. Sebo, J., Singer, P., 2024. The Case for Animal-Friendly AI. https://doi.org/10.48550/arXiv.2403.01199.

Gu, L.L., Yang, R.Q., Wang, Z.Y., Jiang, D., Fang, M., 2024. Ensemble learning for integrative prediction of genetic values with genomic variants. BMC Bioinf. 25 (1). https://doi.org/10.1186/s12859-024-05720-x.

Hamadani, H., Hamadani, A., Shabir, S., 2024. Artificial Intelligence in Animal Farms for Management and Breeding. Elsevier BV, pp. 167–182. https://doi.org/10.1016/b978-0-443-24001-0.00011-7.

Hemmer, P., Schemmer, M., Kühl, N., Vössing, M., Satzger, G., 2024. Complementarity in human-AI collaboration: concept, sources, and evidence. https://doi.org/10.48550/arXiv.2404.00029.

Kistanova, E., Yotov, S., Zaimova, D., 2024. Intelligent animal husbandry: present and future. Animals 14 (11), 1645. https://doi.org/10.3390/ani14111645.

Kitov, I., 2025. Combining causal analysis and machine learning to predict the effects of interventions. Am. J. Eng. Technol. 07 (04), 134–140. https://doi.org/10.37547/tajet/volume07issue04-18.

Kosaraju, D., 2022. Causal inference and AI: elevating decision-making in uncertain environments. Galore Int. J. Appl. Sci. Hum. 6 (3), 179–184. https://doi.org/10.52403/gijash.20220720.

Li, M., Hall, T., MacHugh, D.E., Chen, L., Garrick, D., Wang, L., Zhao, F., 2024. KPRR: a novel machine learning approach for effectively capturing nonadditive effects in genomic prediction. Briefings Bioinf. 26 (1). https://doi.org/10.1093/bib/bbae683.

Liu, Z., Peng, Y., 2025. Human-machine interactive collaborative decision-making based on large model technology: application scenarios and future developments. In: 2025 2nd International Conference on Artificial Intelligence and Digital Technology (ICAIDT), pp. 106–110. https://doi.org/10.1109/icaidt66272.2025.00027.

Lourenço, V., Ogutu, J., Rodrigues, R.P., Posekany, A., Piepho, H.P., 2024. Genomic prediction using machine learning: a comparison of the performance of regularized regression, ensemble, instance-based and deep learning methods on synthetic and empirical data. BMC Genom. 25 (1). https://doi.org/10.1186/s12864-023-09933-x.

Lozada-Soto, E.A., Maltecca, C., Jiang, J., Cole, J.B., VanRaden, P.M., Tiezzi, F., 2024. Effects of germplasm exchange strategies on genetic gain, homozygosity, and genetic diversity in dairy stud populations: a simulation study. J. Dairy Sci. 107 (12), 11149–11163. https://doi.org/10.3168/jds.2024-24992.

Nallappan, D., Kanathasan, J.S., 2024. Navigating ethics and animal welfare. In: Advances in Environmental Engineering and Green Technologies. IGI Global, pp. 67–91. https://doi.org/10.4018/979-8-3693-4115-5.ch004.

Nannini, L., Huyskes, D., Panai, E., Pistilli, G., Tartaro, A., 2024. Nullius in explanans: an ethical risk assessment for explainable AI. Ethics Inf. Technol. 27 (1). https://doi.org/10.1007/s10676-024-09800-7.

Neethirajan, S., 2023. The significance and ethics of digital livestock farming. AgriEngineering 5 (1), 488–505. https://doi.org/10.3390/agriengineering5010032.

Neethirajan, S., 2025. Agency in livestock farming—a perspective on human–animal–computer interactions. Hum. Centric Intell. Syst. 5 (3), 291–307. https://doi.org/10.1007/s44230-025-00108-3.

Neethirajan, S., Scott, S., Mancini, C., Boivin, X., Strand, E., 2024. Human-computer interactions with farm animals—enhancing welfare through precision livestock farming and artificial intelligence. Front. Vet. Sci. 11. https://doi.org/10.3389/fvets.2024.1490851.

Ogunbajo, A., Taiwo, I., Quddus Abidola, A., Fisayo Adediran, O., Agbo-Adediran, I., 2025. Privacy preserving AI models for decentralized data management in federated information systems. GSC Adv. Res. Rev. 22 (2), 104–112. https://doi.org/10.30574/gscarr.2025.22.2.0043.

Okengwu, U.A., Onyejegbu, L.N., Oghenekaro, L.U., Musa, M.O., Ugbari, A.O., 2023. Environmental and ethical negative implications of AI in agriculture and proposed mitigation measures. Sci. Afr. 22 (1), 141–150. https://doi.org/10.4314/sa.v22i1.13.

Panigrahi, M., Rajawat, D., Nayak, S.S., Bose, A., Bharia, N., Singh, S., Sharma, A., Dutt, T., 2025. Advancements in animal breeding: from Mendelian genetics to machine learning. Int. J. Mol. Sci. 26 (23), 11352. https://doi.org/10.3390/ijms262311352.

Panwar, D., Sai Krishna Reddy, B., Sai Harini, A., Mohapatra, R., Giri, D., Karthickraja, A., Kumar, M., 2025. Emerging technologies in precision breeding for sustainable agriculture: a review. J. Adv. Biol. Biotechnol. 28 (4), 666–680. https://doi.org/10.9734/jabb/2025/v28i42226.

Pathak, R.K., Kim, J.M., 2024. Veterinary systems biology for bridging the phenotype-genotype gap via computational modeling for disease epidemiology and animal welfare. Briefings Bioinf. 25 (2). https://doi.org/10.1093/bib/bbae025.

Pérez-Enciso, M., Zingaretti, L.M., de los Campos, G., 2025. Generative AI for predictive breeding: hopes and caveats. Theor. Appl. Genet. 138 (7). https://doi.org/10.1007/s00122-025-04942-8.

Qian, W., Wan, L., Shu, W., 2024. Semi-supervised feature selection based on discernibility matrix and mutual information. Appl. Intell. 54 (13–14), 7278–7295. https://doi.org/10.1007/s10489-024-05481-3.

Reinoso-Peláez, E.L., Gianola, D., González-Recio, O., 2022. Genome-enabled prediction methods based on machine learning. Methods Mol. Biol. 2467, 189–218. https://doi.org/10.1007/978-1-0716-2205-6_7.

Rosa, G.J.M., Lourenco, D., Rowan, T.N., Brito, L.F.F., Gondro, C., Huang, J., Valle de Souza, S., 2023. Integrating enviromics, genomics, and machine learning for precision breeding of resilient beef cattle. J. Anim. Sci. 101 (Suppl. 3), 49. https://doi.org/10.1093/jas/skad281.060.

Rosati, A., 2025. Invited paper artificial intelligence and the future of animal husbandry: rethinking knowledge, practice, and governance. Arch. Latinoam. Prod. Anim. 33 (Suppl. 1), 567–572. https://doi.org/10.53588/alpa.330509.

Roy, R., Baral, M.M., Pal, S.K., Kumar, S., Mukherjee, S., Jana, B., 2022. Discussing the present, past, and future of machine learning techniques in livestock farming: a systematic literature review. In: 2022 International Conference on Machine Learning, Big Data, Cloud and Parallel Computing, COM-IT-CON 2022. Institute of Electrical and Electronics Engineers Inc., pp. 179–183. https://doi.org/10.1109/COM-IT-CON54601.2022.9850749

Ryan, M., Blok, V., 2026. Artificial intelligence and animal farming: a scenario of drivers, barriers, and impacts in 2032. Agric. Hum. Val. 43 (1). https://doi.org/10.1007/s10460-025-10806-0.

Ryan, M., Isakhanyan, G., Tekinerdogan, B., 2023. An interdisciplinary approach to artificial intelligence in agriculture. NJAS Impact Agric. Life Sci. 95 (1). https://doi.org/10.1080/27685241.2023.2168568.

Sangjan, W., Kick, D.R., Washburn, J.D., 2025. Improving plant breeding through AI-supported data integration. Theor. Appl. Genet. 138 (6). https://doi.org/10.1007/s00122-025-04910-2.

Šekrst, K., McHugh, J., Cefalù, J.R., 2024. AI Ethics by Design: Implementing Customizable Guardrails for Responsible AI Development. https://doi.org/10.48550/arXiv.2411.14442.

Shokor, F., Croiseau, P., Gangloff, H., Saintilan, R., Tribout, T., Mary-Huard, T., Cuyabano, B.C.D., 2025. Deep learning and genomic best linear unbiased prediction integration: an approach to identify potential nonlinear genetic relationships between traits. J. Dairy Sci. 108 (6), 6174–6189. https://doi.org/10.3168/jds.2024-26057.

Shoyombo, A.J., Popoola, M.A., Kuusu, D.J., Yisah, L.J., Adebayo, O.M., Yakubu, H., Ndiomu, E.P., Moses, A.A., 2024. Computational approaches in livestock breeding: a review. In: International Conference on Science, Engineering and Business for Driving Sustainable Development Goals, SEB4SDG 2024. Institute of Electrical and Electronics Engineers Inc. https://doi.org/10.1109/SEB4SDG60871.2024.10630145

Soloshenkov, A.D., Soloshenkova, E.A., Semina, M.T., Spasskaya, N.N., Voronkova, V.N., Stolpovsky, Y.A., 2024. Artificial intelligence and classical methods in animal genetics and breeding. Russ. J. Genet. 60 (7), 843–856. https://doi.org/10.1134/s1022795424700297.

Spoelstra, M., Duguma, G., Ebro, A., van der Lee, J., ten Napel, J., 2026. Assessing genotype-by-environment interaction in dairy cattle in case of a wide range of production environments, unstructured crossbreeding and no pedigree information. Animal, 101760. https://doi.org/10.1016/j.animal.2026.101760.

Subirana-Granés, M., Hoffman, J., Zhang, H., Akirtava, C., Nandi, S., Fotso, K., Pividori, M., 2025. Genetic studies through the lens of gene networks. Annu. Rev. Biomed. Data Sci. 8 (1), 125–147. https://doi.org/10.1146/annurev-biodatasci-103123-095355.

Sun, J., Zhang, X., You, X., Montesinos-López, O.A., Montesinos-López, A., Crossa, J., Sorrells, M.E., 2026. Bayesian neural networks for genomic prediction: uncertainty quantification and SNP interpretation with SHAP and GWAS. Theor. Appl. Genet. 139 (1). https://doi.org/10.1007/s00122-025-05127-z.

Tedeschi, L.O., Guarnido Lopez, P., Menendez III, H.M., Seo, S., 2025. Advancing precision livestock farming: integrating artificial intelligence and emerging technologies for sustainable livestock management. Anim. Biosci. https://doi.org/10.5713/ab.25.0289.

Ukpaka, P., 2024. The ethical implications of using artificial intelligence to manipulate or enhance natural ecosystems and biodiversity. Bioética UNAM (1). https://doi.org/10.22201/pub.bioeticaunam.2024.1.6.

Ullah, K., Waheed, A., 2025. Review: genetic improvement in livestock: a journey from conventional breeding to genomic precision. Vet. Biomed. Clin. J. 7 (1), 30–44. https://doi.org/10.21776/ub.vetbioclinj.2025.007.01.3.

Valone, T.J., 2024. Probabilistic inference and Bayesian-like estimation in animals: empirical evidence. Ecol. Evol. 14 (7). https://doi.org/10.1002/ece3.11495.

Venegas, L., Ramayo-Caldas, Y., Bohannan, B.J.M., Derome, N., Yáñez, J.M., 2026. From genomes to hologenomes: integrating host and microbiome data for complex trait prediction in livestock and aquaculture. Front. Anim. Sci. 6. https://doi.org/10.3389/fanim.2025.1678538.

Verma, S., Paliwal, N., Yadav, K., Vashist, P.C., 2024. Ethical considerations of bias and fairness in AI models. In: 2nd International Conference on Disruptive Technologies, ICDT 2024. Institute of Electrical and Electronics Engineers Inc., pp. 818–823. https://doi.org/10.1109/ICDT61202.2024.10489577

Vlaicu, P.A., Gras, M.A., Untea, A.E., Lefter, N.A., Rotar, M.C., 2024. Advancing livestock technology: intelligent systemization for enhanced productivity, welfare, and sustainability. AgriEngineering 6 (2), 1479–1496. https://doi.org/10.3390/agriengineering6020084.

Wadood, A.A., Bordbar, F., Zhang, X., 2025. Integrating omics approaches in livestock biotechnology: innovations in production and reproductive efficiency. Front. Anim. Sci. 6. https://doi.org/10.3389/fanim.2025.1551244.

White, R.R., 2025. 205 Artificial intelligence and the internet of things: responsible and sustainable intensification of livestock production. J. Anim. Sci. 103 (Suppl. 3), 260–261. https://doi.org/10.1093/jas/skaf300.302.

Zhu, L., Yang, X., Xian, Y., Jiang, W., Pu, X., Wang, S., Cheng, J., Niu, L., Zhao, Y., Chen, L., Zhou, X., Wang, Y., Gan, M., Zhu, L., Shen, L., 2025. Empowering precision livestock farming: artificial intelligence applications in animal genomic breeding and multi-dimensional phenotypic measurement. Smart Agric. Technol. 12, 101655. https://doi.org/10.1016/j.atech.2025.101655.

Zsolnai, A., Nagy, A., Szalai, G., Curik, I., Anton, I., Hudák, P., Varga, L., 2025. A novel conservation genomic strategy: selection for the probability of offspring heterozygosity. Animals 15 (15), 2217. https://doi.org/10.3390/ani15152217.

Glossary

Accuracy The degree to which model predictions correspond to true or observed values, reflecting the overall correctness of an analytical or predictive system.

Active learning A machine learning paradigm in which the algorithm selectively queries the most informative data points for labeling, thereby improving learning efficiency with minimal data.

Adaptive systems Computational or biological systems capable of dynamically adjusting their behavior or structure in response to environmental changes or new data inputs.

Algorithm A finite sequence of well-defined computational steps designed to perform a specific task such as data processing, optimization, or prediction.

Allele One of two or more alternative forms of a gene found at a specific locus, contributing to genetic variation within a population.

Annotation The process of identifying, labeling, and describing functional elements within genomic or other biological datasets, including genes, regulatory regions, and variants.

Anomaly detection The identification of rare or unusual patterns in data that deviate significantly from expected behavior, often used in quality control or disease detection.

Approximation The process of estimating a function or value when an exact solution is difficult or impossible to obtain.

Artificial intelligence (AI) A broad field of computational science focused on developing systems capable of performing tasks that typically require human cognitive functions such as learning, reasoning, and decision-making.

Artificial neural network (ANN) A computational architecture composed of interconnected nodes that mimic biological neurons, used for pattern recognition and predictive modeling.

Association mapping A statistical approach used to identify relationships between genetic variants and phenotypic traits across populations.

Autoencoder A type of neural network designed to learn compressed representations of data by encoding inputs into a latent space and reconstructing them.

Backpropagation An optimization algorithm used to train neural networks by propagating prediction errors backward through the network to update weights.

Batch effect Systematic non-biological variation introduced during data generation or processing, often requiring correction in omics analyses.

Batch normalization A technique used in deep learning to normalize inputs within a layer, improving training speed and stability.

Bayesian inference A statistical framework that updates prior beliefs with observed data to produce posterior probability distributions.

Bias (statistical) A systematic deviation of estimated values from the true parameter, often arising from flawed data or model assumptions.

Big data Extremely large and complex datasets that require advanced computational tools for storage, processing, and analysis.

Bioinformatics An interdisciplinary field combining biology, computer science, and statistics to analyze and interpret biological data.

Biomarker A measurable biological indicator used to assess physiological states, disease conditions, or responses to interventions.

Bootstrap A resampling technique used to estimate the variability and reliability of statistical estimates.

Breeding value An estimate of an individual's genetic potential to pass desirable traits to its offspring.

Causal inference A set of methods used to determine cause-and-effect relationships rather than mere associations between variables.

Classifier A predictive model that assigns input data into predefined categories or classes.

Clustering An unsupervised learning method that groups data points based on similarity without prior labeling.

Computational biology The use of computational techniques to model and analyze biological systems.

Confounding variable A variable that influences both the independent and dependent variables, potentially biasing results.

Convolutional neural network (CNN) A deep learning architecture particularly effective for analyzing spatial or image-based data.

Correlation A statistical measure indicating the strength and direction of association between two variables.

Covariate An independent variable that may influence the outcome of a model or experiment.

Cross-entropy A loss function commonly used in classification tasks to measure the difference between predicted and true distributions.

Cross-species data integration (Xeno-data integration) The integration of biological data across different species to identify conserved mechanisms.

Cross-validation A model evaluation technique that partitions data into training and testing subsets to assess generalizability.

Cross-validation (X-validation) A statistical method for assessing model performance using multiple data partitions.

Data augmentation Techniques used to artificially expand training datasets by applying transformations to existing data.

Data drift Changes in the statistical properties of input data over time, which can degrade model performance.

Data fusion The integration of multiple data sources into a unified representation.

Data integration The process of combining heterogeneous datasets into a unified analytical framework to enable comprehensive analysis.

Data normalization The process of scaling data to a standard range or distribution.

Data pipeline A structured sequence of processes for collecting, transforming, and analyzing data.

Data wrangling The process of cleaning, transforming, and preparing data for analysis.

Decision tree A hierarchical model that makes decisions based on sequential feature-based splits.

Deep learning A subset of machine learning involving neural networks with multiple layers capable of learning complex patterns.

Digital twin A virtual representation of a physical or biological system used for simulation and prediction.

Dimensionality reduction Techniques used to reduce the number of variables in a dataset while preserving essential information.
Dropout A regularization technique that randomly disables neurons during training to prevent overfitting.
Edge computing A decentralized computing approach where data processing occurs near the data source rather than in centralized systems.
Elastic net A regression technique that combines L1 and L2 regularization.
Embedding A method of representing complex data as vectors in a lower-dimensional space for machine learning tasks.
Enrichment analysis A method for identifying biological features that are overrepresented in a dataset.
Ensemble learning A strategy that combines multiple models to improve predictive performance.
Environmental covariates External environmental factors that influence biological traits.
Epigenetics The study of heritable changes in gene expression that do not involve alterations in DNA sequence.
Epigenomics The genome-wide analysis of epigenetic modifications.
Error rate The proportion of incorrect predictions made by a model.
Explainable AI (XAI) AI systems designed to provide transparent and interpretable decision-making processes.
False discovery rate The expected proportion of false positives among significant results.
Feature engineering The process of creating and transforming variables to improve model performance.
Feature selection The identification of the most relevant variables for inclusion in a model.
Federated learning A decentralized learning approach where models are trained across multiple datasets without sharing raw data.
Feed efficiency A measure of how effectively animals convert feed into production outputs.
Fine mapping The process of identifying specific causal variants within genomic regions.
Fitness trait A biological characteristic that influences reproductive success.
Forecasting The prediction of future outcomes based on historical data.
Functional genomics The study of gene functions and interactions.
Gene expression The process by which genetic information is transcribed and translated into functional products.
Gene network A representation of interactions among genes.
Genetic correlation The degree to which two traits share a common genetic basis.
Genetic gain The improvement in population performance achieved through selection.
Genome The complete set of genetic material in an organism.
Genome editing Techniques used to modify DNA sequences.
Genomic selection A method that uses genome-wide markers to predict breeding values.
Genotype The genetic constitution of an organism.
Gradient descent An optimization algorithm used to minimize a loss function.
Graph neural network (GNN) A neural network designed to model relational data structured as graphs.
Haplotype A group of alleles inherited together from a single parent.
Heritability The proportion of phenotypic variation attributable to genetic factors within a population.
Heterogeneity The presence of diversity or variability within data.
Heuristic A practical approach to problem-solving that may not guarantee optimal solutions.
Hidden layer Intermediate layers in a neural network where feature transformations occur.
High-throughput sequencing Technologies that enable rapid sequencing of large volumes of DNA.
Homogeneity The degree of similarity within a dataset.
Hybrid vigor The enhanced performance observed in crossbred individuals.
Hyperparameter A parameter set before model training that influences learning behavior.
Hypothesis testing A statistical method used to evaluate assumptions about data.
Imputation A statistical technique used to estimate and replace missing values in a dataset based on observed patterns and relationships.
Inbreeding The mating of genetically related individuals, often resulting in increased homozygosity and potential expression of deleterious traits.
Index selection A breeding strategy that combines multiple trait values into a single index to guide selection decisions based on overall genetic merit.
Inference The process of concluding underlying patterns, relationships, or parameters from observed data using statistical or computational models.
Information gain A metric used in decision tree learning to quantify the reduction in uncertainty achieved by splitting data based on a specific feature.
Input feature An individual measurable variable or attribute used as an input in a computational or statistical model.
Instance-based learning A learning paradigm in which predictions are made based on similarity to stored examples rather than explicit generalization.
Integration (data integration) The systematic combination of heterogeneous datasets, often from multiple omics layers or data modalities, into a unified analytical framework to enable comprehensive biological interpretation.
Intelligent system A system capable of perceiving its environment, processing information, and making decisions to achieve defined objectives.

Interoperability The ability of different computational systems, databases, or analytical platforms to exchange, interpret, and utilize data in a seamless and standardized manner.

Interpretability The extent to which the internal mechanisms and outputs of a computational model can be understood and explained by humans.

Iteration A repeated computational process in which model parameters or outputs are progressively refined through successive cycles.

Jaccard index A similarity coefficient used to measure the overlap between two sets, commonly applied in clustering and classification tasks.

Jackknife method A resampling technique used to estimate the bias and variance of statistical estimators by systematically leaving out observations.

Jensen's inequality A mathematical principle stating that the value of a convex function of an expectation is greater than or equal to the expectation of the function.

Job scheduling The process of allocating computational tasks to available resources in an efficient and organized manner.

Joint distribution A probability distribution that describes the likelihood of multiple random variables occurring together.

Joint modeling The simultaneous statistical analysis of multiple related traits or datasets to capture shared variation and interdependencies.

Joint optimization The process of simultaneously optimizing multiple objective functions or parameters within a model.

JSON (JavaScript Object Notation) A lightweight data-interchange format commonly used for storing and transmitting structured data.

Junction analysis The study of genomic breakpoints or splice junctions, particularly relevant in transcriptomics and structural variation analysis.

Just-in-time learning A machine learning approach in which models are updated dynamically at the time of prediction using newly available data.

K-means clustering An unsupervised learning algorithm that partitions data into clusters based on minimizing within-cluster variance.

K-nearest neighbors (KNN) A non-parametric algorithm that classifies or predicts values based on the closest data points in feature space.

Kappa statistic A measure of agreement between categorical classifications, adjusted for chance agreement.

Kernel method A class of algorithms that transform data into higher-dimensional spaces to enable linear separation of non-linear patterns.

Kernel trick A computational technique that enables implicit mapping of data into higher-dimensional spaces without explicit transformation.

Key performance indicator (KPI) A measurable value used to evaluate the success of a system, process, or model in achieving objectives.

Kinetic modeling The mathematical representation of dynamic biological processes, often describing rates of biochemical reactions.

Knowledge graph A structured representation of entities and their relationships, enabling advanced reasoning and data integration.

Knowledge representation The process of encoding information in a structured format that can be utilized by computational systems for reasoning.

Kriging A geostatistical method used for spatial interpolation and prediction based on observed data points.

Label The known output or target variable used in supervised learning.

LASSO (Least Absolute Shrinkage and Selection Operator) A regression technique that performs variable selection by imposing L1 regularization.

Latent space A lower-dimensional representation of data learned by models such as autoencoders.

Latent variable An unobserved variable inferred from observed data, representing underlying structures or processes.

Layer (neural network) A collection of interconnected computational units within a neural network that perform transformations on input data.

Learning curve A graphical representation of model performance as a function of training data size or iterations.

Learning rate A hyperparameter that determines the magnitude of updates to model parameters during training.

Likelihood function A function that quantifies the probability of observing data given a set of model parameters.

Linear model A statistical model that assumes a linear relationship between input variables and output responses.

Logistic regression A classification model used to estimate the probability of binary or categorical outcomes.

Longitudinal data Data collected repeatedly over time from the same subjects, enabling temporal analysis.

Loss function A mathematical function that quantifies the discrepancy between predicted and actual values.

Machine learning (ML) A field of AI focused on developing algorithms that learn patterns from data to make predictions or decisions.

Markov model A stochastic model that describes transitions between states based on probabilistic rules.

Meta-learning A learning paradigm in which models improve their ability to learn new tasks based on prior experience.

Metabolomics The comprehensive study of metabolites within a biological system, reflecting physiological and biochemical states.

Microbiome The collection of microorganisms and their genetic material within a specific environment or host.

Model generalization The ability of a model to perform well on unseen data.

Model training The process of optimizing model parameters using a dataset to minimize prediction error.

Model validation The evaluation of model performance using independent data to assess generalizability.

Multi-omics The integrative analysis of multiple omics datasets, such as genomics, transcriptomics, and proteomics.

Multicollinearity A condition in which predictor variables are highly correlated, potentially affecting model stability.

Multivariate analysis Statistical analysis involving multiple variables simultaneously.

Mutation A change in the DNA sequence that can introduce genetic variation.

Natural selection An evolutionary process in which traits that enhance survival and reproduction become more prevalent.

Negative selection The removal of deleterious alleles from a population.

Network topology The structural arrangement of nodes and connections within a network.
Neural network A computational model composed of interconnected nodes that process information through weighted connections.
Neuroevolution The application of evolutionary algorithms to optimize neural network structures and parameters.
Node An individual computational unit within a network or graph.
Noise Random variability in data that does not represent meaningful information.
Nominal data Categorical data without inherent ordering.
Nonlinear model A model that captures complex relationships that are not adequately described by linear functions.
Normalization The process of scaling data to a standard range or distribution to improve model performance.
Null hypothesis A default assumption that there is no effect or relationship between variables.
Objective function A function that defines the goal of an optimization process.
Observation An individual data point in a dataset.
Omics A collective term for high-throughput biological disciplines studying large-scale molecular datasets.
Online learning A learning paradigm in which models are updated continuously as new data becomes available.
Ontology A structured framework for organizing knowledge into categories and relationships.
Operationalization The process of defining abstract concepts in measurable terms.
Optimization The process of finding parameter values that minimize or maximize a specified objective function.
Outlier A data point that deviates significantly from the overall distribution.
Output variable The predicted or dependent variable in a model.
Overfitting A modeling error in which a model captures noise rather than underlying patterns, reducing generalizability.
Overparameterization A condition in which a model contains more parameters than necessary, potentially leading to overfitting.
Parameter A variable within a model that is adjusted during training to optimize performance.
Phenomics The large-scale study of phenotypes using high-throughput measurement technologies.
Phenotype The observable characteristics of an organism resulting from genetic and environmental influences.
Phenotypic plasticity The ability of an organism to alter its phenotype in response to environmental conditions.
Policy learning A framework in which models learn optimal decision-making strategies based on feedback.
Polynomial regression A regression technique that models non-linear relationships using polynomial functions.
Precision agriculture A data-driven approach to managing agricultural systems for improved efficiency and sustainability.
Precision breeding The application of data-driven approaches to optimize genetic improvement in livestock populations.
Predictive modeling The use of statistical or machine learning techniques to forecast outcomes based on input data.
Principal component analysis (PCA) A dimensionality reduction technique that transforms data into orthogonal components capturing maximum variance.
Probability distribution A mathematical function describing the likelihood of different outcomes.
Proteomics The comprehensive study of proteins, including their structure, function, and interactions.
Quadratic model A mathematical model that includes squared terms to capture non-linear relationships between variables.
Quality control A set of procedures implemented to ensure the accuracy, reliability, and consistency of data before analysis.
Quantification The process of measuring and expressing biological or computational variables in numerical form.
Quantile regression A statistical technique that models relationships between variables at different points of the outcome distribution, rather than focusing solely on the mean.
Quantitative trait A measurable phenotype influenced by multiple genes and often environmental factors, typically exhibiting continuous variation within a population.
Quantitative trait loci (QTL) Genomic regions associated with variation in quantitative traits, identified through statistical linkage or association analyses.
Quasi-experimental design A research design that approximates experimental conditions without full randomization, often used in observational studies.
Query A structured request for retrieving or manipulating data within a database or information system.
Queue processing A computational mechanism for managing tasks in an ordered sequence, typically following a first-in, first-out structure.
Quick sort An efficient sorting algorithm based on partitioning data into smaller subsets.
Random forest An ensemble learning method that constructs multiple decision trees and aggregates their predictions to improve accuracy and robustness.
Recursive algorithm An algorithm that solves a problem by repeatedly applying itself to smaller instances of the same problem.
Regression A statistical approach used to model and analyze relationships between dependent and independent variables.
Regularization A set of techniques used to prevent overfitting by adding constraints or penalties to model complexity.
Reinforcement learning A machine learning paradigm in which agents learn optimal behaviors through interactions with an environment based on rewards and penalties.
Representation learning A machine learning approach focused on automatically discovering meaningful features from raw data.
Reproducibility The capacity to obtain consistent results using the same data, methods, and analytical procedures.
Residual The difference between observed values and model-predicted values, often used to assess model performance.
Response variable The outcome variable that a model aims to predict or explain.

Risk assessment The systematic evaluation of potential adverse outcomes associated with a decision or system.
Robustness The ability of a model or system to maintain performance under varying conditions or in the presence of noise.
Sampling The process of selecting a subset of data from a larger population for analysis.
Selection index A mathematical formula combining multiple trait values into a single score for guiding breeding decisions.
Sensitivity analysis An assessment of how variation in model inputs affects outputs.
Signal processing The analysis and transformation of signals to extract meaningful information.
Simulation The use of computational models to replicate and study real-world processes.
Sparse model A model that uses a limited number of features, often enhancing interpretability.
Standardization The transformation of data to have a consistent scale, typically with zero mean and unit variance.
Stochastic model A model that incorporates randomness and probabilistic behavior in its structure.
Supervised learning A machine learning approach in which models are trained using labeled datasets to predict known outcomes.
Sustainability The capacity of a system to maintain functionality and productivity over time without depleting resources.
Systems biology An interdisciplinary approach that studies complex biological systems through the integration of multiple data types and modeling techniques.
Target variable The dependent variable that a model seeks to predict.
Tensor A multi-dimensional array used in numerical computations, particularly in deep learning.
Threshold A predefined value used to make decisions, such as classification boundaries.
Time-series analysis The analysis of data points collected over time to identify patterns and trends.
Topology The structural arrangement of elements within a network or system.
Training data The dataset used to train a machine learning model by adjusting its parameters.
Trait A measurable characteristic of an organism, influenced by genetic and environmental factors.
Transcriptomics The study of RNA transcripts produced by the genome under specific conditions.
Transfer learning A technique in which knowledge gained from one task is applied to improve performance on a related task.
Transformer model A deep learning architecture based on attention mechanisms, widely used for sequential and structured data.
Tuning (hyperparameter tuning) The process of optimizing model hyperparameters to improve performance.
Ubiquitous computing The integration of computational capabilities into everyday environments and objects.
Uncertainty The degree of confidence or variability associated with predictions or estimates.
Underfitting A modeling issue where a model is too simple to capture underlying patterns in data.
Uniform distribution A probability distribution in which all outcomes are equally likely.
Unit variance A condition in which data have been scaled to have a variance of one.
Unsupervised learning A machine learning approach that identifies patterns in data without labeled outcomes.
Update rule A formula that defines how model parameters are adjusted during training.
Upsampling A technique used to increase the representation of underrepresented classes in a dataset.
User interface The component of a system that enables interaction between users and computational models.
Utility function A mathematical representation of preferences used in decision-making processes.
Validation The process of evaluating a model's performance using independent data.
Value function A function that estimates expected outcomes in decision-making frameworks.
Variable A measurable attribute or characteristic that can take different values.
Variance A statistical measure of the dispersion of data points around the mean.
Variational inference A technique for approximating complex probability distributions in Bayesian analysis.
Vector An ordered array of numerical values representing data points or features.
Velocity (optimization context) A parameter used in optimization algorithms to accelerate convergence.
Virtualization The creation of virtual versions of physical systems or resources.
Visualization The graphical representation of data to facilitate understanding and interpretation.
Volatility The degree of variation or fluctuation in data over time.
Wasserstein distance A metric used to measure differences between probability distributions.
Weak learner A simple model that performs slightly better than random guessing, often used in ensemble methods.
Web-based analytics Data analysis performed through web-based platforms and tools.
Weight A parameter in a model that determines the importance of input features.
Weighted average An average in which different values contribute unequally based on assigned weights.
Whole-genome sequencing The determination of the complete DNA sequence of an organism's genome.
Windowing The segmentation of data into smaller subsets for analysis.
Workflow A structured sequence of analytical steps used in data processing and modeling.
Workload balancing The distribution of computational tasks across resources to optimize efficiency.
X-coordinate The horizontal axis value in a coordinate system.
X-factor An unmeasured or unknown variable that influences outcomes.
X-inactivation A biological process in which one of the X chromosomes in females is transcriptionally silenced.
X-optimization Optimization processes involving unknown or latent variables.

Xenobiology A field of synthetic biology focused on creating novel biological systems.
XGBoost (Extreme Gradient Boosting) An optimized gradient boosting algorithm widely used for structured data analysis.
XML (eXtensible Markup Language) A flexible markup language used for structuring and storing data.
Y-axis The vertical axis in graphical representations.
Y-chromosome A sex chromosome associated with male-specific traits in many species.
Y-intercept The value of the dependent variable when the independent variable is zero.
Y-variable The dependent variable in regression or predictive models.
Year effect Variation in data attributable to differences across years.
Yield prediction The estimation of production outputs using statistical or computational models.
Yield stability The consistency of production traits across varying environmental conditions.
Yield trait A production-related characteristic such as milk yield, growth rate, or egg production.
Yoked design An experimental design in which subjects are paired to control for variability.
Yule process A stochastic process used to model evolutionary branching patterns.
Z-score A standardized value representing the number of standard deviations a data point is from the mean.
Z-statistic A value used in hypothesis testing under the normal distribution.
Z-transformation The process of standardizing data using the mean and standard deviation.
Zero-inflation A condition in which a dataset contains an excessive number of zero values.
Zero-shot learning A machine learning approach that enables prediction for unseen classes without direct training examples.
Zipf's law A statistical distribution describing the frequency of events in ranked data.
Zonal analysis The evaluation of spatial data within defined regions or zones.
Zonal breeding strategy A breeding approach tailored to specific environmental or geographic regions.
Zoonosis A disease that can be transmitted between animals and humans, often relevant in livestock systems.
Zygote The initial cell formed by the fusion of gametes during fertilization.

Index

'*Note:* Page numbers followed by "f" indicate figures and "t" indicate tables.'

Printed and bound by CPI Group (UK) Ltd, Croydon, CR0 4YY
14/09/2026
02216388-0011